Brill's Companion to War in the Ancient Iranian Empires

Brill's Companions in Classical Studies

Warfare in the Ancient Mediterranean World

VOLUME 9

The titles published in this series are listed at *brill.com/wamw*

Brill's Companion to War in the Ancient Iranian Empires

Edited by

John Hyland
Khodadad Rezakhani

BRILL

LEIDEN | BOSTON

Cover illustration: Battle between Khusrow Parviz and the rebellious Wahram Chobin, from the *Book of Kings* (*Shahnameh*) by Ferdowsi. Chester Beatty Library, Per. 270.189.

The Library of Congress Cataloging-in-Publication Data is available online at https://catalog.loc.gov
LC record available at https://lccn.loc.gov/2024033601

Typeface for the Latin, Greek, and Cyrillic scripts: "Brill". See and download: brill.com/brill-typeface.

ISSN 2452-1493
ISBN 978-90-04-70821-1 (hardback)
ISBN 978-90-04-71077-1 (e-book)
DOI 10.1163/9789004710771

This book is printed on acid-free paper and produced in a sustainable manner.

Contents

Preface

The genesis and scope of this book owe a great debt to Lee Brice, who first proposed the idea of an overarching companion volume on ancient Iranian warfare. The initial charge was equally exciting and intimidating, but the formation of a collaborative partnership between an Achaemenid and a Sasanian scholar made it possible to realise a project that would contribute to the many sub-fields of ancient Iranian studies, while generating conversation across disparate chronological and departmental boundaries. Khodadad Rezakhani has helped to shape the volume and added his wisdom and insights through many stimulating editorial conversations; in addition, I have to thank him for the kind invitation to participate in a Sasanian Zoom reading group, which greatly enriched my knowledge of Iranian Late Antiquity and helped to enlighten the darkest days of the pandemic. We found common ground not only in thematic comparisons between different periods and dynasties, but in a mutual commitment to the study of war and empire in the broadest sense, encompassing military operations and organisations but also extending to the realms of imperial worldviews, ideology, and diplomacy.

We have also been fortunate in the assembly of numerous contributors who shared this vision and added a wide variety of perspectives and methodological approaches. Of course, we had no idea in 2019 of what lay in store when we asked them to submit their chapters in the following summer. The chaos of the pandemic naturally disrupted our timeline, and prevented the completion of chapters by some of the original authors, whose insights are sadly missed. A silver lining was the opportunity to bring in additional scholars whose papers enhanced the collection. We are deeply grateful to all of the authors for their shared wisdom and perseverance through revisions, edits, and delays, and hope that they will share our satisfaction in seeing the completed volume in print.

Lee Brice has been an exemplary editor and mentor, offering generous encouragement, support, and advice throughout the project. We are very grateful to the publishing staff at Brill, and to the anonymous referees for thoughtful comments on the finished chapters. My colleagues in History and Classics in Christopher Newport University have provided moral support and encouragement as I worked to balance the edited volume with a parallel monograph project, teaching, and university service. I could not have completed the editorial work, to say nothing of my own writing, without the support of Trible Library's exceptional staff, especially our university librarian Mary Sellen for her enthusiastic support and expansion of our ancient history collection, and Jesse

Spencer, our interlibrary loan librarian, who has worked tirelessly to acquire the outside resources needed for research in ancient Iranian studies. Above all, my heartfelt thanks go out to Sharon, Liam, and Patrick, who have given their unconditional love and support throughout all the hard-working ups and downs of this hectic half-decade.

John Hyland
Newport News, VA

I would like to first and foremost thank my colleague John Hyland for his collegiality, patience, and mentorship during this process. It was only with his diligence, organisation, and vast intellectual capabilities that this volume now sees the light of the day. On my own contributions to the volume, various friends and colleagues, only a few of whom are mentioned here, have helped me in thinking through questions, answering queries, and ploughing through my own chapter. These include my intellectual and life partner, Sara Mashayekh, who always patiently listens to my ramblings, as well as Michael Alram, Touraj Daryaee, Shervin Farridnejad, Simcha Gross, Rob Haug, Robert Hoyland, Ab de Jong, Michael Morony, Ekaterina Nechaeava, Yusef Saadat, Petra Sijpesteijn Alison Vacca, Peter Webb, and Arash Zeini. I want to dedicate this book to the memory of my dearly missed late father, the first person who encouraged me to follow my passion, Javanshir Bonyab (1949–2022).

Khodadad Rezakhani
Leiden, Netherlands

Figures and Maps

Figures

Maps

Notes on Names and Abbreviations

The rendering of ancient Iranian names and terminology in English and other European languages is complicated by the frequency of transmission through non-Iranian sources, as well as variations in transliteration between ancient scripts and the modern Roman and Farsi alphabets. With this in mind, we have sought where possible to promote clarity for readers through consistency in spelling conventions. The chapters on the Achaemenid period use the royal names bestowed in Greek historiography and Latinised in Anglophone parlance, such as Cyrus, Darius, and Xerxes, while usually referring to other individuals in the Hellenised forms that appear in the sources, rather than the attested or reconstructed Old Persian versions. Variations in transliteration of personal names grow more complex in the Arsacid and Sasanian periods, and especially in the cases of royal names, the editors have aimed at the selection of common spellings most consistently used in current scholarship.

The abbreviation conventions for citations of Greek and Roman sources follow the *Oxford Classical Dictionary*, 4th edition; others used for important Iranian, Mesopotamian, Armenian, and early Islamic sources are summarised below. Abbreviations for most of the journals cited in chapter bibliographies follow the standard format employed by the *Oxford Classical Dictionary*, 4th edition; additional periodicals and collections of relevance to the study of the Ancient Near East and Iran, referred to in a number of chapters, are also listed below.

ABC	A.K. Grayson (1975) *Assyrian and Babylonian Chronicles*. Locust Valley.
AD	Babylonian Astronomical Diaries. H. Hunger and A. Sachs (1988–1996) *Astronomical Diaries and Related Texts from Babylonia*, 3 vols. Vienna.
AmH	Ariaramnes, Hamadan inscription
A^{3}P	Artaxerxes III, Persepolis inscription
AO	*Der Alte Orient*
AS	Artaxerxes II, Susa inscription
BCHP	R. Van der Spek, I. Finkel, R. Pirngruber, and K. Stevens (forthcoming) *Babylonian Chronographic Texts of the Hellenistic Period.*
BM	British Museum
BAI	*Bulletin of the Asia Institute*
Chron. Pasch.	*Chronicon Paschale*. L. Dindorf (1832) *Chronicon Paschale*. Bonn; M. Whitby and M. Whitby (1989) *Chronicon Paschale*. Liverpool.
CT 56/57	Babylonian tablet published in T.G. Pinches (1982) *Cuneiform Texts from Babylonian Tablets in the British Museum, Part LVII: Neo-Babylonian and Achaemenid Economic Texts*. London.

DB	Darius I, Bisotun (Bisitun, Behistun) inscription (Pers. = Old Persian version; Bab. = Babylonian (Akkadian) version; El. = Elamite version). (R. Schmitt (1991) *The Bisitun Inscriptions of Darius the Great. Old Persian Text.* London; E. von Voigtlander (1978) *The Bisitun Inscriptions of Darius the Great. Babylonian Text.* London.)
Dīn.	Abu Ḥanifa Aḥmad al-Dīnawarī. (V. Guirgass and I. Kratchkovsky (eds) (1888) *Kitāb al-Akhbār al-Tiwāl.* Leiden; A. ʿAmir (1960) *al-Akhbār al-ṭiwāl.* Cairo.)
DN	Darius I, Naqš-e Rostam inscription. (R. Schmitt (2000) *The Old Persian Inscriptions of Naqsh-i Rustam and Persepolis.* London.)
DP	Darius I, Persepolis inscription
DS	Darius I, Susa inscription
DZ	Darius I, Suez inscription
EKI	F.W. König (1965) *Die elamischen Königsinschriften.* Graz.
Epic Histories	S. Malhasjanc (1968) Փավստոս Բուզանդ Պատմություն Հայոց. (Faustus Buzand History of Armenia). Yerevan; N.G. Garsoïan (1989) *The Epic Histories Attributed to P'awstos Buzand, Buzandaran Patmut'iwnk'.* Cambridge, MA.
Fort.	Persepolis Fortification texts from draft editions by G. Cameron, R.T. Hallock, C. Jones, and M.W. Stolper, collated by W. Henkelman for forthcoming publication by the Persepolis Fortification Archive Project.
GC 2	Babylonian tablet published in R. Dougherty (1933) *Archives from Erech: Neo-Babylonian and Persian Periods, Goucher College Cuneiform Inscriptions*, vol. 2. New Haven.
IStratonikeia	M.Ç. Şahin (1981) *Die Inschriften von Stratonikeia. Teil 2.1: Lagina, Stratonikeia und Umgebung.* Bonn.
JAEI	*Journal of Ancient Egyptian Interconnections*
JANEH	*Journal of Ancient Near Eastern History*
JAR	*Journal of Anthropological Research*
JCS	*Journal of Cuneiform Studies*
JPS	*Journal of Persianate Studies*
JSOT	*Journal for the Study of the Old Testament*
JESHO	*Journal of the Economic and Social History of the Orient*
Jos. Styl.	Joshua the Stylite. F.R. Trombley and J.W. Watt (2000) *The Chronicle of Pseudo-Joshua the Stylite.* Liverpool.
Khuz. Chron.	Khuzistan Chronicle. Rezakhani, K. and Amiri Bavandpour, S. (eds) (2016) (*Rūdādnāme-Ye Soryāni Mosūm Be Rūydādnāme-Ye Khūzestān* (*The Syriac Chronicle Known as the Khuzistan Chronicle*). Tehran.)
MDP	*Mémoires de la Délégation en Perse* series: *Mémoires de la Délégation en Perse*, vols. 1–13 (1900–12); *Mémoires de la Mission archéologique de*

	Susiane vol. 14 (1913); *Mémoires de la Mission archéologiques de Perse – Mission de Susiane*, vols. 16–28 (1921–39); *Mémoires de la Mission archéologiques en Iran – Mission de Susiane*, vol. 29–38 (1943–65); *Mémoires de la Délégation archéologiques en Iran – Mission de Susiane*, vols. 39–52 (1966–92).
NPi	Narseh, Paikuli inscription. (H. Humbach and Skjærvø, P.O. (1983) *The Sassanian Inscription of Paikuli. Restored text and translation.* Wiesbaden.)
OGIS	W. Dittenberger (1903–1905) *Orientis Graeci Inscriptiones Selectae.* Leipzig.
PAT	D.R. Hillers and E. Cussini (1996) *Palmyrene Aramaic Texts*. Baltimore.
PF	Persepolis Fortification tablet published in R.T. Hallock (1969) *Persepolis Fortification Tablets*. Chicago.
PFa	Persepolis Fortification tablet published in R.T. Hallock (1978) 'Selected Fortification Texts', *Cahiers de la Délégation Archéologique Française en Iran* 8: 109–36.
PF-NN	Persepolis Fortification tablet from draft editions by R.T. Hallock, collated by W. Henkelman for forthcoming publication by the Persepolis Fortification Archive Project.
RINAP	Royal Inscriptions of the Neo-Assyrian Period
SAA	State Archives of Assyria
Seert Chron.	*Chronicle of Seert.* (A. Scher (1909) 'Histoire Nestorienne (Chronique de Séert), seconde partie, fasc. 1'. *Patrologia Orientalis* 7: 95–203.)
ŠKZ (=*RGDS*)	Shapur I, Ka'ba-ye Zardusht inscription (= '*Res Gestae Divi Saporis*'). (M. Back (1978) *Die sassanidischen Staatsinschriften. Studien zur Orthographie und Phonologie des Mittelpersischen der Inschriften zusammen mit einem etymologischen Index des mittelpersischen Wortgutes und einem Textcorpus der behandelten Inschriften* (Acta Iranica 18), 284–371. Leiden; P. Huyse (1999) *Die dreisprachige Inschrift Šābuhrs I. an der Kaba-i Zardušt* (*ŠKZ*). London.)
Ṭab.	Muhammad b. Jarir al-Ṭabarī. (M.J. de Goeje (1879) *Ta'rīkh al-rusul wa'l-mulūk*. Leiden; M.J. de Goeje, P. de Jong, and E. Prym (1890) *History of Prophets and Kings*. Vol. IV. Leiden; C.E. Bosworth (1999) *The History of al-Ṭabarī, Vol. V: The Sāsānids, the Byzantines, the Lakhmids, and Yemen*. Albany.)
XP	Xerxes, Persepolis inscription (R. Schmitt (2000) *The Old Persian Inscriptions of Naqsh-i Rustam and Persepolis*. London.)
ZAR	*Zeitschrift für altorientalische und biblische Rechtsgeschichte*
ZA	*Zeitschrift für Assyriologie und vorderasiatische Archäologie*

Notes on Contributors

Javier Álvarez-Mon
holds degrees in ancient Near Eastern art and archaeology from the École du Louvre (Paris) and the University of California at Berkeley, respectively. He is presently Professor in Near Eastern Archaeology and Art at Macquarie University (Australia). Javier has pursued specialised studies in the art and archaeology of the civilisations of the ancient Near East. Within this field, he has become a leading specialist in the art of Elam (ca. 4200–525 BCE).

Daniel Beckman
holds a Ph.D. in Near Eastern Studies from the department of Near Eastern Languages and Cultures at the University of California, Los Angeles. His research focuses on the history of Pre-Islamic Iran, specifically on the interactions between the Greek and Iranian worlds in the Achaemenid period.

Henning Börm
is professor of Ancient History at the University of Rostock, Germany. He received his doctorate at the University of Kiel in 2006 with a thesis on Procopius of Caesarea and the Persians. Between 2003 and 2019 he taught at universities in Kiel, Münster, Konstanz, Berlin, Tübingen, and Innsbruck. In 2017, Börm completed his habilitation with a study on 'stasis' in Hellenistic Greece. From 2020 to 2022 he was a professor at Bochum University.

Omar Coloru
is a Senior Lecturer in Greek history at the University of Bari 'Aldo Moro'. He is a specialist of the Hellenistic period and his research interests focus on the relations between the Iranian and Greco-Roman worlds. Since 2018, he has collaborated with the Italian Archaeological Mission in Pakistan, where he studies the Indo-Greek site of Barikot, the ancient Bazira conquered by Alexander the Great.

Touraj Daryaee
is the Maseeh Chair in Persian Studies and Culture and Director of the Samuel M. Jordan Center for Persian Studies and Culture at the University of California, Irvine. He is the author of numerous books and articles on the history of ancient and early medieval Iran, specifically the Sasanian Empire, as well as Middle Persian literature and the history of Zoroastrianism.

Jenn Finn
is associate professor and chair of Classical Studies at Loyola University Chicago. She has published widely on all aspects of Mediterranean history, including cross-cultural contacts, Roman reception of the Classical Greek world, and ancient military history. Her second book, *Contested Pasts: A Determinist History of Alexander the Great in the Roman Empire*, was published in 2022 with the University of Michigan Press.

Leonardo Gregoratti
was educated at the Universities of Udine (Italy) and Trier (Germany). He has conducted research in Udine, Trier, Kiel and Bergen. From 2013 until 2018 he collaborated with the Department of Classics and Ancient History of Durham University as IAS Fellow. Recently he has been a visiting researcher at the University of Freiburg (Germany). His research interests include the Roman Near East, Palmyra, long-distance trade, and the Parthian kingdom.

Robert Haug
is Associate Professor of Islamic World History, pre-1500 in the Department of History at the University of Cincinnati, where he teaches courses on the history of Southwest and Central Asia and World History. His research focuses on the transformation of Iran and Central Asia during the early Islamic period. He is the author of *The Eastern Frontier: Limits of Empire in Late Antique and Early Medieval Central Asia* (I.B. Tauris, 2019).

Waldemar Heckel
is Professor Emeritus of Greek and Roman Studies at the University of Calgary and a research fellow of the Centre for Military and Strategic Studies. He taught at the University of Calgary from 1977 until his retirement in 2013. His publications include *Alexander's Marshals: A Study of the Makedonian Aristocracy and the Politics of Military Leadership* (Routledge, 2016) and *Who's Who in the Age of Alexander the Great and His Successors: From Chaironeia to Ipsos* (Casemate, 2021).

John Hyland
is Professor of History at Christopher Newport University in Newport News, Virginia. He holds a Ph.D. from the Committee on the Ancient Mediterranean World at the University of Chicago (2005), and specialises in the history of Achaemenid Persia and its Aegean borderlands. He is the author of *Persian Interventions: The Achaemenid Empire, Athens, and Sparta 450–386 BCE* (Johns

Hopkins University Press, 2018) and *Persia's Greek Campaigns: Kingship, War, and Spectacle on the Achaemenid Frontier* (Oxford, forthcoming).

Katarzyna Maksymiuk
is Associate Professor of History at the Faculty of Humanities, the Siedlce University, Poland. She is the author of *Geography of Roman-Iranian Wars: Military Operations of Rome and Sasanian Iran* (Siedlce UPH, 2015) and co-editor of *Crowns, Hats, Turbans and Helmets: The Headgear in Iranian History, vol. 1: Pre-Islamic Period* (Siedlce UPH, 2017). Her research deals with the history of Iran, with particular attention to the problems of military conflicts.

Craig Morley
is an Academic Developer at the University of Salford, having previously taught at the University of Chester and the University of Manchester. He graduated with a doctorate from the University of Liverpool in 2015. His research interests include Ammianus Marcellinus, Roman-Sasanian competition and cooperation, the relationship between the Eastern and Western Roman Empires after 395 CE and cross-frontier travel.

Jake Nabel
is the Tombros Early Career Professor of Classical Studies and an Assistant Professor of Classics and Ancient Mediterranean Studies at Pennsylvania State University. He is a historian of ancient Rome, pre-Islamic Iran, and the points of contact between the two. Jake's research interests include Roman-Parthian relations, ancient political thought, early imperial Latin poetry, and late antique Armenia. He is currently writing a book on a group of Arsacid princes who lived at the court of the Roman emperor in the first century CE.

Nikolaus Overtoom
is Associate Professor in Ancient History at Washington State University. He has published several articles on Alexander the Great's image, ancient imperialism and propaganda, and ancient international relations. His current scholarship focuses on the Parthians' rivalry with their Greek, Persian, and Roman neighbours in the Hellenistic period. Oxford University Press published his monograph, *Reign of Arrows: The Rise of the Parthian Empire in the Hellenistic Middle East*, in 2020.

Reinhard Pirngruber
obtained his Ph.D. from the Vrije Universiteit Amsterdam in 2013 with a thesis on commodity prices in Late Achaemenid and Hellenistic Babylonia. Until 2022, he was a post-doctoral researcher in various projects dedicated to

the socio-economic and cultural history of first millennium BCE Babylonia at the University of Vienna and the Austrian Academy of Sciences.

Khodadad Rezakhani
is a global historian of the first millennium CE, concentrating on Central and West Asia. Currently the leader of the Gerda Henkel Stiftung Funded project, *Memory of Ruins: Ctesiphon and Baghdad*, and a Lecturer at Leiden University, Netherlands, he is the author of *ReOrienting the Sasanians: East Iran in Late Antiquity* (Edinburgh University Press, 2017) and the forthcoming *Creating the Silk Road: Travel, Trade and Myth-Making* (I.B. Tauris/Bloomsbury).

Jeffrey Rop
is Associate Professor of History at the University of Minnesota, Duluth. His publications include *Greek Military Service in the Ancient Near East, 401–330 BCE* (Cambridge University Press, 2019) and articles and book chapters on the military and political history of ancient Greece and Achaemenid Persia.

Eduard Rung
is Professor in the Department of Ancient History and the Institute of Oriental Studies and Antiquity at the Russian State University for the Humanities in Moscow. He is the author of more than 140 publications, including two monographs, articles and book chapters. His research is focused on the history of the Achaemenid Persian Empire, Greco-Persian diplomatic relations and political and cultural interactions, and Greek perceptions of the East.

Patryk Skupniewicz
is an art historian (MA) and historian (Ph.D.), collaborating with Siedlce University, Poland. He is the author of several studies on Sasanian armaments and art as well as the aspects of the remnants of Hellenistic art in Asia. He studies the transfer of visual structures and military technology between the cultures of the Eurasian landmass.

Yasmina Wicks
(Ph.D. 2017, University of Sydney) specialises in the ancient Near East with a focus on the mortuary practices of southwest Iran from the Chalcolithic to the Iron Age. She has published several papers on the archaeology of Elam and two monographs: *Profiling Death: Neo-Elamite Mortuary Practices, Afterlife Beliefs, and Entanglements with Ancestors* (Brill, 2019) and *Bronze 'Bathtub' Coffins in the Context of 8th–6th Century B.C.E. Babylonian, Assyrian and Elamite Funerary Practices* (Archaeopress, 2015).

MAP 1 The Achaemenid Empire

PRODUCED BY K. REZAKHANI

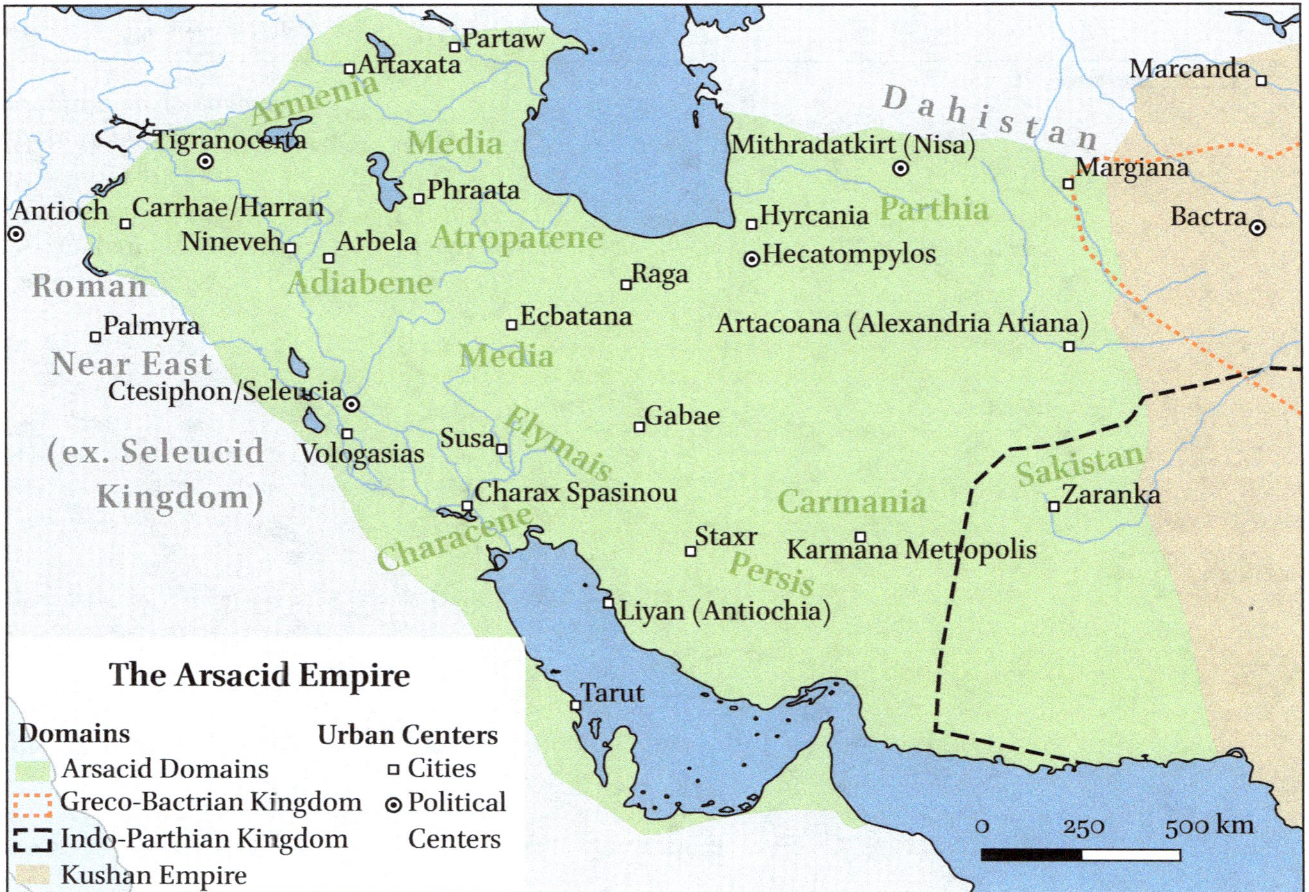

MAP 2 The Arsacid Empire
PRODUCED BY K. REZAKHANI

MAP 3 The Sasanian Empire
PRODUCED BY K. REZAKHANI

CHAPTER 1

War and Peace in Ancient Iran: Ideologies, Structures, and Practice

John Hyland and Khodadad Rezakhani

> When you think, how many lands were those, which Darius the King held? – Look at the figures which support the throne. Then will you recognise, then will it be clear to you – the Persian man's spear has gone far. Then will it be clear to you – the Persian man, far from Persia, has beaten back the foe.
>
> DARIUS I, Naqš-e Rostam a, §4

The martial boast that adorns the final resting place of the Achaemenid dynasty's most powerful ruler draws an explicit connection between his unprecedented earthly authority and Persian military action. Monumental reliefs carved seven centuries later on the same cliffs celebrate the outcomes of battlefield success for the early Sasanian monarchs – Ardashir I rides over the corpse of Ardawan IV, the final leader of the Arsacid dynasty, to accept divine investiture as king of Eranshahr, and Shapur I glories over humbled emperors of Rome. Naqš-e Rostam's celebrations of conquest, separated by three quarters of a millennium, encapsulate the continuous significance of war in the history of ancient Iran.

The ancient Iranian empires dominated large portions of western and central Asia for more than a millennium, and rank with those of China and Rome among the most extensive and influential polities of world history. Warfare played a critical role in their development, contributing to the formation of successive imperial states, their economic development and administrative practices, and their periodic dissolution and reformation under new ruling families. Bronze Age and Iron Age Elamite dynasties projected power across the Zagros and greater Mesopotamia, and despite military misfortunes at the hands of aggressive Neo-Assyrian rivals, established organisational precedents that laid the foundations for the rise of Cyrus and the Persians. The

 | DOI:10.1163/9789004710771_002

Achaemenid kings conquered most of the lands between the Indus and the Mediterranean, projecting power into Egypt and briefly southeastern Europe. After the fall of the Achaemenids to Alexander of Macedon, the successor dynasty established by his general Seleucus incorporated Iranian subjects and military precedents as it governed former Achaemenid territories from Syria to the Hindu Kush. The Seleucids gave way in time to the Arsacids, and the Arsacids to the Sasanian dynasty, both of whom projected power across an imperial expanse that stretched from the Oxus valley to the Caucasus and upper Euphrates. At the zenith of late Sasanian power in the early seventh century CE, Khosrow II reached further west to conquer Anatolia, the Levant, and Egypt, unconsciously reviving the territorial reach of his distant Achaemenid predecessors, before the mantle of imperial power passed on to the rulers of the early Islamic caliphate.

Each of the ancient Iranian empires in turn coalesced in rapid bursts of conquest and longer periods of consolidation, succeeding not only in battle and siege, but also in the organisation of human and material resources, logistical infrastructure, intelligence-gathering, communication, and projections of diplomatic power. Their military structures facilitated wide-ranging political authority and extensive duration, and when ruling dynasties succumbed to external challengers or internal turbulence in the wake of battlefield defeats, surviving elements of their military systems contributed to long-term continuities across periods of political transition. The study of war in ancient Iran, therefore, is not only vital for contextualising the military history of its Greek and Roman neighbours, but for gaining a deeper understanding of the Iranian empires in their own right as central actors in the history of ancient Eurasia.

1 Historiography and Challenges

Despite the importance of warfare in ancient Iran, its historical study poses unique difficulties in comparison with the military histories of its Roman, Chinese, or Mesopotamian peers. These arise from two fundamental problems of evidence. First, the surviving documentary corpus for the ancient Iranian empires contains comparatively little military material, in contrast with the documentation available for Roman imperial legions and *limes*, or the copious references to logistics, intelligence-gathering, and mobilisations of human and animal resources in the Neo-Assyrian palace correspondence. Second, ancient Iranian conflicts are only rarely narrated by internal written sources, and therefore require reconstruction from Assyrian, Greek, Roman, Armenian, Georgian, Arabic, Chinese, and other external and secondary historiographical narratives.

In both situations, the study of Iranian wars and military institutions requires complex processes of contextualisation, with reference to surviving fragments of documentary material, numismatic and glyptic evidence, and archaeological findings, often requiring interaction between multiple academic specialists across disciplinary boundaries.

In the early stages of modern historiography on ancient Iranian empires and warfare, European scholars applied rationalising critiques to certain aspects of the ancient narrative evidence, such as the outrageous numbers bestowed on Iranian armies in order to amplify the heroism of those Greek and Roman armies that withstood or conquered them.[1] Early Iranologists also engaged with Greco-Roman descriptions of ancient Iranian armed forces through philological analyses of military titles and contingent names attested in Classical authors.[2] But historical approaches that emerged in an era of modern European imperialism were naturally slow to detach themselves from assumptions about the inferior military capabilities of ancient Western Asian states, rooted in colonial prejudices that seemed to find confirmation in selected Greco-Roman battle descriptions. Another significant and related problem was the influence of Aryan racial theories, which romanticised early phases of Iranian expansion and conquest but postulated the dilution of military effectiveness, and eventual stagnation and decline, blamed on the crossing of bloodlines with non-Indo-European subject peoples.[3] In the context of the World Wars and Cold War, retellings of ancient conflicts between the Iranian empires and their Mediterranean neighbours easily slipped into stereotypical associations between the invaders of Greece and the enemies of the present day. Documentary evidence for the internal workings of the Iranian empires was also notoriously slow to emerge and did not offer obvious correctives to the battle-centred narratives of ancient authors. Specialised archives of cuneiform tablets, or glyptic and numismatic collections, remained most familiar to scholars in their own disciplines with training in their linguistic or methodological intricacies, and interpretation involved significant difficulties. The result was a prolonged lack of engagement between historians trained in the Classical tradition and the internal Iranian documentation.[4]

Inside Iran, the earliest studies of antiquity that emerged in the nineteenth century drew on classical Persian traditions, embodied in Ferdowsi's epic

1 See Delbrück (1887), 137–45; Meyer (1901), 325–26, 375–76; Beloch (1916) 67–73.

2 See for instance Marquart (1896, 19 n. 84), reconstructing **baivarapati* as an Old Persian title for the myriarchs named by Herodotus.

3 See Wiesehöfer (1992) 71–76.

4 See Manning (2021) 22–45.

Shahnameh, which celebrated the fictional Pishdadi and Kayani kingdoms while remaining unaware of most historical details preceding the Sasanian dynasty. A growing consciousness of an earlier imperial history, stoked by archaeological excavations of Achaemenid palaces and resentment of colonial encroachments on Iran's ancient riches, led to the writing of nationalist histories that reframed Western academic narratives to celebrate the glories and selected military exploits of the Achaemenid, Arsacid, and Sasanian kings. Aryan racial theories and anti-Arab sentiments played a notable role in framing popular understandings of the Sasanian Empire's final collapse in the face of the Islamic conquests. A foundational work of historiography was Hasan Pirnia's *Iran-e Bastan*, first published in 1927 and reissued in many subsequent editions, which drew on European studies such as the first edition of the *Cambridge Ancient History* for edifying accounts of ancient Iranian wars with Greece and Rome, and stressed imperial grandeur even in moments of defeat. Such studies fed into the nationalist ideologies promoted by the regimes of Reza Shah and Mohammad Reza Shah Pahlavi, which sought to enhance their dynastic image and status on the international stage through evocations of Iran's ancient world historical importance.[5] The Pahlavi-era interest in ancient Iran's military history culminated in the infamous 1971 celebrations for the 2500th anniversary of the Iranian Empire (*Shahanshahi*), which included a grand review of thousands of soldiers in historic costumes against the backdrop of the Persepolis Apadana, for the entertainment of assembled world dignitaries. The spectacle generated a spate of publications on ancient military dress and material equipment that drew on iconographic and archaeological evidence as well as Greco-Roman historiography.[6] Despite the inherent methodological weakness of some of this work, such as a frequent reliance on material from Xenophon's laudatory but fictional *Cyropaedia*, it began to lay foundations for more critical study of ancient Iranian military institutions; the potential for further progress was exemplified by Alireza Shapur Shahbazi's essay for an early fascicle of the *Encyclopaedia Iranica*, 'Army. i. Pre-Islamic Iran', which carefully synthesised Avestan military references together with Greco-Roman evidence for military organisation, armaments, and tactics in

5 Pirnia (1927). For the evolution of modern Iranian scholarship on ancient Iran, see Abdi (2001); Grigor (2021); Mousavi (2021). For the broader contexts of nationalist historiography, see Ansari (2010); Zia-Ebrahimi (2016) pays particular attention to Aryanism and racial theories in nationalist identity formation.

6 The parade program (*anon.* (1971)) contains an order of march and extensive watercolor illustrations of the costumes. See Grigor (2005, 26–27) for discussion of the military pageantry; for more detail and context on the Persepolis celebrations, see Steele (2020).

the Achaemenid, Arsacid, and Sasanian periods.[7] Unfortunately, the revolution of 1979 and its tumultuous aftermath disrupted the pace of ancient historical studies inside Iran, but archaeological work of significance for Iran's ancient military history resumed in the twenty-first century, with collaborative efforts between Iranian and European scholars resulting in great progress on topics such as the Sasanian frontier fortifications in Gorgan.[8]

In Western academia, the last two decades of the twentieth century saw the evolution of Elamite, Achaemenid, Arsacid, and Sasanian studies into full-fledged disciplines, dedicated to the study of ancient Iranian empires in their own contexts and efforts at more critical approaches to the Greco-Roman sources. Their studies have profited from a steadily increasing engagement with Assyriological scholarship and the vast well of documentary evidence from Mesopotamia, including comparative material on imperial military structures and practices from the Assyrian and Babylonian Empires, and direct evidence on the Mesopotamian heartland as a crucial region within each of the ancient Iranian empires.[9] Much of the new historiography focused on aspects of social, economic, and cultural history, in light of the emerging riches of the documentary evidence, but was slow to reengage with military topics.[10] This reticence is understandable in the context of a post-Vietnam decline in the wider academic study of military history, and the inherent problems in Greco-Roman tropes of Iranian empires' failures against Western armies as indicators of decadence and decline. By the end of the twentieth century, though, scholars were beginning to notice that the study of war and armies in ancient Iran was lagging behind the growing sophistication of approaches to military topics in the history of other ancient empires (Rome in particular). In 1999, Pierre Briant lamented the lack of a critical monograph on Achaemenid military structures and practices, but the gap would not be rectified for two more decades, either for the Achaemenids or their Arsacid and Sasanian successors.[11] The principal

7 Shahbazi (1986).

8 See Sauer, Rekavandi, Wilkinson, and Nokandeh (2013).

9 See Gombert (2018); Manning (2021).

10 For Elamite studies, see Álvarez-Mon and Garrison (2011); Álvarez-Mon, Basello, and Wicks (2018). On the evolution of modern Achaemenid historiography, see McCaskie (2012); Degen and Manning (2021); Root (2021). For an introduction to Arsacid historiography, see Hauser (2012) 1001–1003, 1019–20. The impressive website by Chris Hopkins, www.parthia.com, presented an almost complete bibliography of works on the Parthians, but is sadly no longer updated. For a summary of recent Sasanian studies, see Daryaee and Rezakhani (2016, particularly xiii–xvi).

11 Briant (1999) 107–108. In contrast, Bar-Kokhva (1976) had surveyed Seleucid military organisation in depth using Classical and Biblical evidence.

treatments of ancient Iranian warfare outside the Pahlavi-era Iranian literature remained illustrated surveys for popular Western audiences, highlighting typologies of arms and military dress with colourful illustrations while retelling traditional narratives of Iranian conflicts with Greece and Rome.[12] As late as 2013, a scholarly synthesis such as the *Oxford Handbook of Ancient Iran* could introduce scholars and students to the Iranian empires without devoting a single chapter to conflict or military institutions.[13]

The twenty-first century, though, has witnessed a broad disciplinary revival of historiographical interest in the role of warfare and armed forces, within the larger context of a surge of attention to the comparative dynamics of historical empires. A growing number of ancient Iranian specialists have now turned their attention to military topics. Journals such as *Historia i Świat* have devoted considerable space to ancient Iranian military topics, and numerous publications have appeared in edited volumes, both thematic collections on Greek and Roman warfare making space for chapters on Iranian perspectives,[14] and larger compendia on the history of the Iranian empires themselves, most recently Wiley's *Companion to the Achaemenid Persian Empire*.[15] The first major scholarly books on the Sasanian and Achaemenid armed forces have come out within the last half-decade,[16] and notable monographs have attempted to reconstruct and contextualise major Iranian campaigns and conflicts.[17] Our objective in this collection is to build on this progress and contribute to the foundations for future work on the military dynamics of ancient Iran, by assembling a body of scholarship in a single volume that spans the breadth of the imperial states and dynasties from the Elamite through the Sasanian periods.

2 Themes and Questions

The serious study of war as a component of ancient Iranian imperialism requires attention not only to the minutiae of battles and soldiers' material trappings, but to 'big picture' questions regarding the causation, objectives,

12 See Sekunda and Chew (1992); Head (1992); Nicolle (1998); Farrokh (2007).

13 Potts (2013).

14 See Wiesehöfer (2007); Tuplin (2010); Laforse (2013); McDonough (2013); Lee (2013); Waters (2016); Lee (2016a); Lee (2016b); Dusinberre (2016); Wiesehöfer (2016); J. Rop (2018); Hyland (2018a); Wiesehöfer (forthcoming).

15 Konijnendijk (2021); Hassan (2021); Tuplin (2021); Tuplin and Jacobs (2021).

16 See Farrokh, Karamian, and Maksymiuk (2018) on the Sasanian military and Manning (2021) on Achaemenid armed forces and warfare.

17 See for instance Greatrex (1998); Howard-Johnston (2021); Hyland (forthcoming).

and impacts of military activities at the level of the state. This volume places particular emphasis on four overarching themes within the study of ancient imperial warfare, each of which is essential to advancing our understanding of the Iranian empires' worldviews, structures, and external relations. A fundamental starting point is the question of imperial ideology, what war meant at certain periods for the self-perception and public imagery of rulers and elites, and what they sought to gain through military conflict. If ideology helps to bring the objectives of imperial wars into clearer focus, the second topic entails the pursuit of those objectives through military means, including the planning and leadership of military campaigns, logistical preparations, and battlefield methods that contributed to performative success or failure. A third aspect of imperial warfare involves its relationship with imperial diplomacy, that is, the spectrum of communicative activities employed to complement or complete the military activities of the state, and the ensuing state-building as the result of peace processes. Last but not least comes the question of how empires organised their human and material resources in response to their military needs and ambitions, and how expectations of military service and the periodic experience of warfare interacted with the social and economic currents of life under empire. Despite variations in the quality and quantity of surviving historical evidence for ancient Iranian warfare, there is plentiful material for engagement and expansion on older approaches within each of these topics.

2.1 *The Purpose and Meaning of War: Imperial Ideologies*

There is no doubt that war was central to Iranian dynastic ideologies, regardless of evolutions in their linguistic and iconographic features over different dynastic eras. Representations of warrior kingship and the divinely favoured glory of imperial arms helped to perpetuate and legitimise rule over disparate imperial spaces and populations, as expansion gave way to consolidation. The internal evidence for Iranian imperial ideology is abundant, including not only royal inscriptions from the Cyrus Cylinder and Darius' Bisotun monument to Shapur's Ka'ba-ye Zardusht and Narseh's Paikuli inscriptions, but also artistic depictions of royal heroism and military success, ranging from miniature scenes of combat on administrative seals, to images of royal archers in Achaemenid and Arsacid coin series, to rock carvings of Elamite warriors and the triumphal reliefs of Achaemenid and Sasanian rulers on the cliff faces of Bisotun, Firuzabad, Naqš-e Rostam, and Taq-e Bostan.[18] Relevant material

18 For Assyrian precedents, see Reade (1979); for Elamite reliefs and royal imagery, see Álvarez-Mon (2019). For imperial ideology in Achaemenid reliefs, which avoid explicit

can also be found in post-Sasanian religious and literary compendia such as the *Denkard*, to say nothing of the various external depictions of the Iranian empires and their worldviews.

At their heart, the ideological aspects of Iranian imperial warfare began with idealised depictions of kings as heroic warriors whose efforts brought about peace and world order, and engaged in performative courtship of elite and subject loyalty through displays of royal justice and protection of the empire against external (and internal) threats. This association between military power and the fundamental legitimacy of imperial rule intersected with imagined mental maps that circumscribed the imperial world and the hierarchical interactions between its rulers and the peoples and polities outside their realm.[19]

Throughout each of the empires of ancient Iran, religious ideologies informed relationships between kings, subjects, and enemies, justifications (or pretexts) for military action, and understandings of the reasons for victory or defeat through divine patronage or disapproval. Religious elements in inscriptions, reliefs, coins, and textual evidence are therefore of special importance for the study of Iranian warfare and its socio-political contexts. Within this broad spectrum, though, there are points of uncertainty and contention over the details, chronology, and impact of critical religious concepts and structures. Debate continues to swirl around the extent to which Mazdaean concepts impacted Achaemenid imperial worldviews and policy, and attributions of religious motives to a military action such as Xerxes' invasion of Greece remain fraught with methodological peril.[20] Less information survives about Arsacid religious ideology and its military significance, although this gap can be remedied in part through the sources available for the Indo-Parthian and

---

battle imagery while celebrating a universal dominance achieved through military success, the critical starting point is Root (1979); essential studies of battle scenes in Achaemenid seals are Wu (2014) and Tuplin (2020). For royal archer imagery, see Root (1989) 43–50; Nimchuk (2002); Lerner (2017); Skupniewicz (2021). For Sasanian reliefs and royal power display, see Canepa (2010); see Herrmann, Mackenzie, and Howell (1989) for a critical edition of Shapur's reliefs at Naqš-e Rostam.

19 For Achaemenid royal ideology and worldviews, see Brosius (2005); Rollinger (2013); Tuplin (2017); Rollinger and Degen (2021). For the ideology of Arsacid kingship, see Olbrycht (2016b). For Sasanian conceptions of kingship, see for instance Choksy (1988); Daryaee (2003); on the military dimensions, see Whitby (1994).

20 For an introduction to the current state of the field regarding Achaemenid religion and Mazdaism, see Henkelman and Redard (2017). On the problems with characterisation of Xerxes' campaign as a religious mission akin to a 'jihad' or 'crusade', as in Rahe (2015, 164), see Macgregor Morris (2021) 29–30.

Kushan empires, as well as Armenian historical works.[21] For the Sasanians, Zoroastrianism is often assumed to be an ever present religious ideology, closely connected with state institutions, but the situation is complicated by challenges to the orthodoxy and interaction with subjects' belief systems.[22]

Sasanian kings naturally cloaked themselves in sacral imagery and attributed military victories to divine favour (as in ŠKZ §17), although it becomes more difficult to extract explicitly religious motivations as opposed to economic, political, or strategic strands from accounts of their activities on the western or eastern frontiers.[23] Battles, requiring preliminary prayer and ritual, were easily framed in religious contexts, but wars involving attempts at religious conversion, such as Yazdgird II's efforts to compel the Armenian nobles back to Zoroastrianism, are isolated events requiring careful contextual attention.[24] Although Heraclius' use of Christian 'holy war' propaganda is well-known, religious ideology played into the wars of 602–630 on both sides, and not necessarily on confessional dividing lines between Christians and Zoroastrians. Khosrow II's intentions in taking the True Cross as hostage are subject to debate; perhaps he intended to transfer its magical powers from the Orthodox Church to the Sasanian-sanctioned Church of the East, but there may have been personal motives at stake in pleasing his wife Shirin and the treasurer Yazdin, and its ideological usage remains open to question.[25]

Further questions on the relationship between ideology and imperial warfare concern its evolution beyond periods of initial dynastic expansion, and its continuity or disruption and transformation in periods of turbulence and transition from one dynasty to another. For the Sasanians, for example, an alleged ideology of reclaiming the territorial legacy of the Achaemenids might have been a creation of Roman historiography rather than reality.[26] One of the most interesting evolutionary aspects of Iranian imperial ideology involves the gradual movement from concepts of world rule in situations of unrivalled empire, most notable in the early Achaemenid period, to the acceptance of external states such as Rome and the Turkish khaganate as peers on an equal standing with the Sasanian kings and deserving of diplomatic respect, while

21 For a sweeping study of this and other matters relating to Arsacid ideology, see Shayegan (2011).

22 See Rezakhani (2015).

23 For later claims about the priestly origins of Ardashir's family, see Daryaee (2003); on Sasanian kingship and divinity, see further Choksy (1988); Overlaet (2013).

24 For religious rituals relating to battle, see Azarnouche and Petitjean (2022) 364–67, 371; Daryaee (this volume). For the Armenian case, see Honarchiansaky (2018).

25 See Stoyanov (2011); Howard-Johnston (2021) 87–102.

26 See Shahbazi (2001); Shayegan (2011).

posing periodic threats to the empire's interests in contentious and fluid border regions.[27]

When it comes to the causes of specific wars, the absence of an internal narrative historiography throughout much of ancient Iranian history complicates reconstruction. Outside of the inscriptions of Darius, Shapur, and Narseh, we tend to lack direct contemporary accounts that could provide Iranian perspectives on the decision-making processes of kings and counsellors. An episode such as Xerxes' decision to campaign against the trans-Aegean Greeks, as described in Herodotus' *Histories*, remains visible only through a Greek literary construction of questionable relationship to underlying historical realities. Among the shifting polities of eastern Iran, from the Greco-Bactrians to the Kushans and Hephthalites, coinage provides crucial iconographic glimpses into military ideology despite the lack of written chronicles of conquest. Finally, traditions of royal performance and military ideals find their richest expression in the later Iranian literary tradition, above all the *Shahnameh*, which preserve key Sasanian ideological concepts despite the natural historiographic challenges resulting from extended preservation over time. At all periods, the study of imperial ideology and worldviews provides an essential starting point for any attempts to contextualise state objectives and strategic decisions in Iranian terms.

2.2 *The Practice of War: Campaigns and Battles*

It is often rather easier to comprehend what war meant to the Iranian empires and their rulers, and what they expected to gain from it, than to explain the strategies and methods of combat that affected its success or failure. But to understand the effectiveness of particular dynasties and rulers in building empires and maintaining them against rebellions and external enemies, or the causes of imperial collapse and transition to new regimes, it is essential to grapple with sources for the waging of military campaigns, from the movements and logistical requirements of armies and fleets to battlefield tactics and the employment of particular weapons systems.

As a rule, we have more information at the operational than at the tactical level. We have a few internal descriptions of royal campaigns, most famously in the multilingual inscriptions of Darius I and Shapur I, although they are concise and formulaic in description, and do not offer explicit insight into the 'generalship' of their authors – for example, why Darius might have chosen to move against rebels in Media while sending a subordinate to tackle

27 See Canepa (2010); Howard-Johnston (2016).

opponents in Fars, or how he addressed problems of numbers or topography in selecting the sites of the decisive battles.[28] For explanation of operational decisions – whether to raid the enemy's countryside or besiege fortified positions, seek out battle or avoid it, defend a frontier or retire to the interior – historians usually have to depend on the non-Iranian narrative historiography, and must take care to contextualise its assertions through judicious appeal to the evidence for Iranian worldviews and political and social contexts.[29] In the late Sasanian period, this situation improves somewhat with awareness of Iranian military manuals such as the epitome of the *Artēštārestān* in a passage of the Denkard; this suggests the development of a technical literature on military operations produced in other ancient societies such as Sunzi's *Art of War*, Kautilya's *Arthashastra*, or Maurice's *Strategikon*, the last of which also offers an interesting representation of 'typical' Sasanian tactics from a Roman perspective (11.1). Study of the limited evidence for Iranian strategic literature is beginning to broaden our understanding of the theoretical underpinnings of military practice.[30]

There is also great potential for the application of internal documentation as well as judicious comparative study to questions of supply, essential for understanding the capabilities and limitations of armed forces at all historical periods. Campaign logistics provide a linkage between the institutional structures of the Iranian empires and their armies' victorious or unsuccessful performance in the field; their reconstruction offers invaluable opportunities to contextualise narrative accounts of military movement through documentary and archaeological evidence for the supporting imperial infrastructure, survey of the geography of the conflict zone, and quantitative study of the needs of human and animal consumption (assisted by comparative evidence from better-attested periods of conflict history).[31]

---

28 For the Bisotun inscription and Darius' strategy in the succession wars of 522–521, see Briant (2002) 114–21; Young (2004); Tuplin (2005).

29 On Achaemenid strategy, see among others Cawkwell (2005); Ruzicka (2012); Lee (2016a); Lee (2016b); Tuplin (2014). For aspects of Arsacid strategy, see Olbrycht (2012); Overtoom (2017). For Sasanian strategy, see Syvänne and Maksymiuk (2018); Howard-Johnston (2016); Potts (2018); Howard-Johnston (2021).

30 See Azarnouche and Petitjean (2022).

31 As of yet there are no comprehensive studies of ancient Iranian logistics, in contrast with the attention paid to this aspect of Greek and Roman military systems. For aspects of the topic in the Achaemenid context, see Tuplin (1997); O'Connor (2015); Henkelman (2017) 55–63, 82–97; Hyland (2018a) 85–89; Briant (2019); Potts (2021). Archaeological study of Sasanian fortification systems and military bases is greatly expanding the potential for study of late antique Iranian campaign logistics and their supporting infrastructure; see Lawrence and Wilkinson (2017); Sauer, Nokandeh, Pitskhelauri, and Rekavandi (2017). For

Battles and tactics are more challenging, not least because of the general difficulties of reconstructing sequential accounts of large-group combat behaviour and individual soldiers' experiences through the 'fog of battle', even in situations for which the documentation is far more extensive.[32] The Greek, Roman, and Arabic historical narratives are at their most dramatic, and by extension, their most dangerous, in recounting tales of how the heroic protagonists overcame menacing Iranian armies. Such problematic external battle accounts lie at the core of many broader popular misconceptions regarding the ancient Iranian empires (and their respective downfalls). In other cases, the historical narratives are so brief, or hyper-focused on the deeds of individuals, as to provide little information for any meaningful tactical reconstruction. Even in cases of Iranian victories, the limited number of battle descriptions in surviving sources leaves room for debate over the extent to which individual episodes are illustrative of systemic long-term practice.[33]

But in spite or perhaps because of these problems, scrutiny of battlefield dynamics remains important for our understanding of key moments in the construction, maintenance, and collapse of ancient Iranian imperial power. Effective re-evaluations of battle practice must begin with rigorous attention to the relevant historiographic contexts and agendas, while carefully weighing each of the sources' claims regarding generalship and battlefield movement in reference to plausible army sizes and the physical environments in which each encounter occurred. When available, in cases of siege or fighting in close proximity to urban sites, archaeological reports on identifiable military structures, materiel, and human remains are invaluable for such discussions.[34] Above all, both siege and battle studies require the integration of all available data on military equipment and weaponry, including physical artefacts, iconographic depictions, and where applicable, ballistic testing of reconstructed missile weapons. Greek accounts of Achaemenid military defeats, for example, place heavy emphasis on materially deterministic explanations for failure – shorter spears, inferiority of bow to spear, lack of adequate armour, and so on; but

document-based studies of military supply during the late Sasanian conquest of Egypt, see Sänger (2008) and Sänger (2011).

32 See Whatley (1964); Whitby (2007).

33 A classic case is the Arsacid triumph over Crassus at Carrhae; see Overtoom (2017), and Overtoom's and Gregoratti's chapters in this volume.

34 See for example Cahill (2010) for Cyrus' sack of Sardis and Lillington-Martin (2007) for the Sasanian-Roman battle at Dara. Identifiable traces of open battles are few and far between; we can only lament, for example, at an unverified report by a late nineteenth-century Prussian officer of amateur excavations in an alleged Persian mass grave at Marathon (Eschenburg (1886) 10).

modern narratives too often repeat these claims without adequate attention to the complexities of the material remnants and images of Iranian military equipment.[35] The assembly and study of the diverse surviving evidence for arms and armour entails its own methodological challenges and debates, but as specialist work in this area intensifies, it will offer increasingly important material with which to test the assertions of the textual narratives and reassess the combat performance of Iranian armies.[36]

2.3 *The Pursuit of Objectives beyond the Battlefield: Diplomacy in War and Peace*

Despite the prominence of campaigns and battles in the historiography of ancient wars, we must not forget that these were hardly the only means of achieving objectives in conflict with empires' rivals. Diplomacy functioned not merely with respect to the limitation or conclusion of conflicts, but as a vital complement to military operations, and much of the time before, during, and after campaigns was occupied with ceasefires, talks, negotiations of alliance, and the construction of short-term or long-term peace treaties.[37] Previous scholarship has shed light on selected aspects of this imperial diplomacy as described in the Greco-Roman authors, such as the Achaemenid treaties with Greek city-states and Arsacid and Sasanian dealings with Roman legates.[38] Formal or informal negotiators, representing imperial sovereigns, presented the power and glory of each side as awesome and irresistible, and provided idealised depictions of empires and their military capabilities.[39] But their ideological presentations required a complex and ever-changing balance with such pragmatic realities as the immediate availability of military resources and manpower, the state of frontier defences, or the presence of internal threats or conflicts on other frontiers that might weaken negotiators' hands.

35 On problematic tropes in Greek accounts of Achaemenid combat techniques, see Gazzano (2018); Manning (2021) 278–92. For critical approaches to individual battle reconstructions, see for instance Tuplin (2010); Konijnendijk (2012).

36 See Manning (2021) 223–59 for a region-by-region survey of material evidence for Achaemenid military equipment. For Achaemenid combat iconography see Summerer (2007); Ma (2008); Wu (2014); Tuplin (2020). For Sasanian military imagery and problems of technical reconstruction, see Skupniewicz (2015); Skupniewicz (2019).

37 See Nissinen (2008).

38 For detailed discussions of Achaemenid-Spartan diplomacy, see Lewis (1977); Hyland (2018b). For diplomacy in the contexts of Roman-Iranian wars, see Dodgeon and Lieu (1992); Greatrex (1998); Campbell (2002); Greatrex and Lieu (2005); Dignas and Winter (2007); Schlude and Rubin (2017); Schlude (2020).

39 See Wiesehöfer (2007).

Many questions arise when considering the issue of diplomacy, alliance, and constructions of peace during the wars of ancient Iran. Among these are the identity and position of those who acted as emissaries,[40] the languages of negotiation, the terms and conditions that governed formal and informal communications between the Iranian empires and their opponents, and the authority of emissaries, relative to their political sponsors, to enforce the terms of treaties. As in the conduct of military campaigns, our major sources for these questions are narrative in character, including Greek and Roman histories for early periods as well as Armenian and Arab accounts of relevance for the Sasanians, especially in their later centuries. This often poses questions regarding the authenticity of the negotiating techniques and purposes ascribed to the Iranian diplomats. Due to their military focus, many such narratives allude to important topics of negotiation only by accident or in passing, or in the worst case, may exclude such important events as peace treaties, a problem all too familiar to any who have delved into the voluminous modern literature on the so-called Peace of Kallias which may have concluded the fifth-century BCE hostilities between the Achaemenids and Athens, but is notoriously absent from the relevant section of Thucydides.[41] On more fortunate occasions, though, certain narrative sources preserve invaluable evidence such as direct quotations of correspondence or treaty terms, providing reliable material for diplomatic practice that may be further contextualised against the larger imperial worldview.[42] Such records of diplomatic negotiation, which might encompass not only strictly strategic questions but other matters of state from trade to religious confrontation, can often tell us more than battle narratives about the broader issues at stake in imperial wars and the causes of their ultimate resolution.

2.4 *Armed Forces, Society, and the Imperial State*

No study of imperial warfare can be complete without attention to the formation of military institutions, practices of internal and external recruitment, and their interactions with diverse social and economic structures across the vast trans-regional spaces of the imperial state. Only by studying these social and organisational foundations of the armed forces can one appreciate the

40 See Blockley (1980).

41 The bibliography on this topic is too immense to summarise here; for introduction and further references, see Hyland (2018b, 16–18, 176–77), as well as Beckman's chapter in this volume.

42 See the correspondence between Heraclius and Kavad II in *Chron. Pasch.* 727–736 (Whitby and Whitby (1989) 182–88).

military capabilities and limitations of imperial powers. Within this larger topic, particular attention is due to the following problems and questions:

a.) Can we speak of 'The Army' as a unitary institution under the successive dynasties of imperial Iran? To what extent, and for what purposes, did the rulers of Iranian empires rely on standing forces, including 'royal guards' and fortress garrisons in regional centres, and how were these bodies of 'professional' soldiers recruited, maintained, and compensated for their service? The most prominent evidence for standing armies occurs under the Achaemenids, not only in Greek references to units such as Xerxes' 'Immortals', but in the documentary and archaeological evidence for garrison communities, and the growing body of evidence on royal spearbearers and related service personnel in the Persepolis Fortification archives – although in the latter case, the appearance of many of these personnel in 'non-military' contexts complicates their interpretation.[43] The scale and significance of standing forces at later periods has been the subject of greater debate, for example in the case of the Arsacid kingdom, for which the written sources emphasise levies by aristocrats while perhaps unjustly passing over the relationship between heavy cavalry forces, frontier garrisons, and the crown.[44] In the Sasanian period, while there may have been a long-term tradition of standing forces, reconstruction of the empire's military organisation is complicated by its extended lifespan and the occurrence of significant reforms in its later centuries.[45]

b.) How did the maintenance of standing armies intersect with the short-term employment of additional soldiers through diverse methods of recruitment? To what extent did extensions of imperial territory, frontier defence, and the enforcement of internal order depend on units of professional 'mercenaries', or contingents of 'allies' provided by (semi)autonomous rulers or tribal entities, or on soldiers obligated to serve in fulfilment of *corvée* or land-grant obligations or dependency on provincial aristocrats? The documentary evidence, of critical importance to questions of internal recruitment, suffers from problems of uneven survival, but a particularly rich corpus of Babylonian cuneiform records from the Achaemenid and Hellenistic periods allows detailed

43 Key studies include Henkelman (2002) and Henkelman (2017) 74–77; see also Tuplin (1987) for Achaemenid garrisons. Charles (2011) provides a useful survey of the Greek evidence for Achaemenid elite troops; see also Charles (2021).

44 See Hauser (2006); Winkelmann (2012); Olbrycht (2016).

45 See Isaac (1998) 437–66; Gyselen (2001); Howard-Johnston (2012); McDonough (2013); Farrokh, Kharamian, and Maksymiuk (2018).

exploration of the relationship between tax obligations and military service in Mesopotamia, a critical region for all of the Iranian empires.[46] Research on hired and allied soldiers is often obliged to push back against pejorative characterisations in the textual sources, such as the notorious emphasis of Greek authors on Achaemenid recruitment of Greek mercenaries as an alleged response to perceptions of internal military deficiency; recent and forthcoming studies are developing greater nuance in dealing with categories of hired soldiers across different regions, chronological periods, and political contexts.[47]

c.) How did the recruitment, command, and compensation of permanent and temporary military forces intersect with the political dynamics of the royal court and provincial elite networks? Was the empire's armed manpower tightly controlled by the king and his delegates, or did the assembly of soldiers constitute a tool for fractious nobles seeking semi-autonomous authority in resistance to the crown? The broader context is a sweeping structural debate about the characteristics of ancient Iranian imperial states, expressed in terms of the strength or weakness of the monarchy vis-à-vis its nobles and generals, or alternatively, the centralised or decentralised nature of their administrative and provincial systems. Such questions have taken on particular prominence in studies of Arsacid and Sasanian military, political, and social institutions. There is substantial evidence for the military importance of semi-autonomous Parthian aristocratic families in both empires;[48] but important studies have also stressed the necessity not to undervalue the monarchy's role in contexts such as the mobilisation of military resources for the frontier fortification systems of Gorgan and the Caucasus, or the royal reforms and expanded military hierarchy articulated during the reigns of Kavad and Khosrow I.[49]

46 For recent attention to the complexities of Achaemenid Babylonian military structures and recruitment, see Gombert (2018); Manning (2021) 155–221. For the tradition of 'bow lands' imposing the obligation of military service (or commutation through payment) on lessees, see Stolper (1985). A recent surge in scholarship on the Neo-Babylonian and Achaemenid economies has shed new light on the complex relationship between evolving forms of taxation and compulsory military service; see in particular Jursa (2011); Jursa and Schmidl (2017). For the economic impacts of military recruitment and warfare in Babylonian documentation, see Pirngruber (2017).

47 See Rop (2019); Tuplin (2021).

48 For the traditional thesis of Arsacid decentralisation and aristocratic dominance, see Wolski (1981); for counter-argument, see Hauser (2005) and Hauser (2006). For decentralisation under the (late) Sasanians, see Pourshariati (2008); Daryaee (2010).

49 See Howard-Johnston (2014) 148–57; for the Gorgan Wall and its implications for Sasanian military policy, see Sauer, Rekavandi, Wilkinson, and Nokandeh (2013).

d.) A final question is that of continuity or disruption in military organisation and hierarchy across periods of dynastic transition – from the Elamite kingdoms to the early Achaemenids, from the Achaemenids to Alexander and his successors, from the Seleucids to the Arsacids, the Arsacids to the Sasanians, and from the Sasanians to the Islamic caliphates. The problem is especially urgent in cases of external conquest such as Alexander's in the fourth century BCE; to what extent did his exploitation of Achaemenid infrastructure and recruitment of Iranian soldiers interact with Macedonian traditions of military organisation?[50] In the case of the Islamic conquest in the seventh century CE, on the other hand, many of the Arab forces involved in the defeat of the Sasanian armies had previously fought for Khosrow II in the Roman wars, and a simplistic nationalist paradigm of conquest by foreign Arab invaders cannot stand up to critical scrutiny.[51]

3 Approaches and Concluding Reflections

Each of the chapters that follow engages with at least one, and sometimes more than one, of the themes and questions outlined above, which resurface as unifying strands throughout the study of ancient Iranian warfare in each separate dynasty and chronological period. For the purpose of accessibility and sequential coherence, the volume is structured chronologically, beginning with the Elamite and Achaemenid periods, then moving on to the Hellenistic and Arsacid eras, and concluding with the Sasanian Empire. Two chapters deal with topics that straddle these dynastic boundaries, but all of the essays are meant both to provide a synthesis of the sources and state of the field on their respective topics and to facilitate comparison with similar dynamics in earlier and later periods.

Javier Álvarez-Mon and Yasmina Wicks open with a sweeping study of early warfare in the Elamite regions of southwestern Iran, from prehistory through the middle of the first millennium BCE. Tracing the initial appearance and evolution of the bow as a formidable battlefield weapon and symbol of royal power, they connect the cultivation of archery skills with the effectiveness in siege warfare that allowed the early Elamite kingdom to dominate its Mesopotamian neighbours in the late third and early second millennia. These victories put Elam's Sukkalmah rulers in a position of imperial authority over coalitions of vassal polities, including Hammurabi's Babylon, which

50 See Olbrycht (2011); Olbrycht (2015).

51 See Rezakhani (2019) 241–46.

they drew on for further projections of military power. Several centuries later, the Shutrukid dynasty's military victories over Babylon are also characterised by references to Elamite archery as the decisive factor. In the final phase of Elamite power, the so-called Neo-Elamite kingdom of the first millennium, their Assyrian enemies continued to respect and fear the prowess of Elamite bowmen. Álvarez-Mon and Wicks astutely connect the longevity of the archery tradition with Elam's administrative organisation and production of the necessary material, and close with a note of continuity between the role of the bow in Elam and the place of archery in Achaemenid warfare.

Eduard Rung opens the discussion of the Achaemenid Empire by surveying the textual evidence for warfare's ideological significance, with special attention to the Cyrus Cylinder and the inscriptions of Darius I and Xerxes. After tracing the warlike elements in the royal image and expressions of Persian supremacy over non-Persian subjects, he focuses on the tensions between the celebration of conquest and the promotion of 'peace' as a stated goal of Achaemenid imperialism. Rewards to favoured subjects and displays of exemplary punishment were employed in support of this constructed vision of imperial order.

Jenn Finn explores questions of Achaemenid grand strategy and its evolution over time, stressing unpredictability and adaptation in response to a variety of contexts. Strategic goals might entail political performance as well as more pragmatic territorial and economic objectives; the Achaemenid rulers had to balance their military priorities between numerous regions, notably Egypt as well as the Aegean, and achieved considerable success for long periods despite setbacks such as Xerxes' Greek expedition.

John Hyland examines the organisation, recruitment methods, and logistical practices of the Achaemenid armed forces at the empire's centre, drawing on the Persepolis Fortification Archive as well as Classical sources. Achaemenid military structures, which built on both Elamite and Mesopotamian precedents, encompassed both a standing palace army associated with the mobile royal court and varied means of shorter-term recruitment for other categories of military personnel. The royal army should be associated with the 'Spear-Bearers' documented in various administrative contexts in the Persepolis tablets, but other categories of non-elite soldiers are visible there as well, in contingents of troops, fortress guards, and transport personnel attested in contexts of long-distance travel. The storehouses associated with the famous imperial road network facilitated rapid movement by minimising the necessity for long-distance food transport, but naval transport, advanced stockpiling arrangements with local allies, and exploitation of dietary supplements such as meat on the hoof were all essential elements in the supply capabilities of large campaign armies beyond the imperial frontiers.

Jeffrey Rop turns a revisionist lens on the famous battles between Achaemenid armies and Greek opponents, from the period of Darius' and Xerxes' trans-Aegean expeditions to the defensive struggle against Alexander's Macedonian invasion. Whereas traditional histories have drawn on Classical historians' simplistic dichotomy between static Persian bowmen and armoured Greek spearmen, Rop uses Achaemenid visual evidence to remind us of the spear's equal significance in Persian infantry combat. He argues for a model of 'cross-training' in both hand-to-hand fighting and long-distance skirmishing that gave Achaemenid armies versatility in their responses to different tactical situations, and rejects the notion of an innate battlefield inferiority, closing with a provocative reflection on the possibility of Persian tactical influence on Macedonian battle methods.

Persian-Greek encounters, of course, encompassed diplomatic engagement as well as battlefield confrontations, and Daniel Beckman explores strategies of communication and negotiation with Greek polities as 'war by other means'. Noting the lack of internal evidence for diplomatic practice, he draws on Near Eastern precedents such as the Neo-Assyrian *adê*-treaties to contextualise Achaemenid interaction with smaller states conceptualised as vassals rather than equals. Applying this framework to case studies of 'earth and water' demands, the Peloponnesian War alliance with Sparta, and the peace which Artaxerxes II imposed on the Greek city-states, Beckman shows that Achaemenid diplomacy aimed at the achievement of stability and suppression of disorder in tandem with the king's frontier military efforts.

Reinhard Pirngruber brings the discussion of war and empire from the Aegean frontiers back to Babylonia, a core region for the Achaemenids and their successors. Temple and entrepreneurial archives from the Neo-Babylonian and early Achaemenid periods, along with the astronomical diaries that continued into the Arsacid era, facilitate the long-term study of social and institutional processes related to war. Pirngruber draws attention to the social impacts of revolts and external invasions that punctuated periods of imperial order, the regional settlement of deportee populations generated by more distant conflicts, and the intimate relationship between land usage, taxation, and military service.

Waldemar Heckel's chapter studies the military role of Iranian satraps in Alexander's post-conquest regime. Through a series of prosopographic case studies in the Alexander sources, Heckel examines the circumstances that affected prominent nobles' retention or loss of the conqueror's favour after the fall of Darius III. Tracing the evolving contribution of Iranian satraps and soldiers to Alexander's armed forces, he argues that Iranian elite resistance in Alexander's final years was not as common as often postulated. Above all, he rejects views of Alexander as uninterested in imperial governance, and shows that he recognised the importance of Iranian military participation in a viable state.

Omar Coloru surveys the military organisation of the Iranian 'Upper Satrapies' during the Seleucid Empire, whose kings pursued regional stability through delegations of overlapping administrative and military responsibility to an interconnected network of satraps, *stratēgoi*, and vassal rulers, sometimes coordinated under the overarching jurisdiction of a crown prince or trusted general; the emergence of the Arsacid threat in the northeast seems to have prompted a period of reorganisation and streamlining of defences in the late third and early second centuries. Although the Seleucid armed forces drew on personnel from the rural settlements known as the *katoikiai*, these also included local civilian communities and should not be viewed exclusively as military colonies. The state of the evidence often imposes difficulties in assessing Seleucid soldiers' ethnic origins, but catalogues of multi-ethnic armies in Polybius and Livy offer some clear examples of Iranian regional contingents, particularly among the cavalry.

Nikolaus Overtoom introduces Arsacid Parthia's strategic struggle with the Roman Republic, which replaced the Seleucids as its predominant western rival in the first century BCE. Viewing the opening stages of conflict from an Iranian rather than a Roman perspective, Overtoom untangles the Arsacid political contexts and internal conflicts that preceded the famous overthrow of the triumvir Crassus at Carrhae. He explicates the events that led Mithridates IV to seek external allies to counter his brother Orodes II's support by the aristocratic Suren family, and shows how victory in civil war and the need to consolidate newly won control over Mesopotamia set up Orodes' aggressive stance towards Armenia and the Romans beyond the Euphrates. The Arsacid rulers, while undermined by the vicissitudes of fratricidal conflict, thus emerge as active drivers of imperial confrontation rather than simple foils for Roman ambition.

In a chapter covering the last two and half centuries of Arsacid rule, Leonardo Gregoratti studies the instances of war between the Arsacid Empire and Imperial Rome. Building on the fact that the Battle of Carrhae (53 BCE) has often been essentialised as the sole representative of such conflicts, Gregoratti provides a detailed study of Arsacid wars against Augustus and Nero, often focused in the Caucasus, examining changes in tactics and Roman adaptations to Parthian battle styles. He then gives thorough attention to the wars between the Arsacids and Trajan in the early second century, and closes the discussion of Parthian and Roman fighting methods with the dwindling Parthian defences against Caracalla in the early third century.

Stepping away from the strategic and tactical aspects of war, Jake Nabel's chapter takes on a theoretical approach to the issue of diplomacy and tests Ernest Satow's definition against the history of Arsacid-Roman relations. The chapter provides an overview of the two empires' engagements and diplomatic relations in two sections, studying first state-level engagements and then non-state level relations. This analysis allows us to see that our modern, post-Westphalian understanding of state-level diplomatic relations is misleading and cannot be simply applied to ancient entities.

Katarzyna Maksymiuk focuses on the establishment and expansion of the Sasanian empire, particularly its western half. Starting with a study of the sources, the chapter provides a narrative of Ardashir's rise to power and early conquests. The campaigns of Shapur I, in particular in confrontation with Rome, are studied thoroughly, and the conflict between Shapur II and Julian provides a last phase of the first two centuries of Sasanian imperial expansion. The second part of the article uses literary and material evidence to study the structure of the Sasanian army, including its command structure. In the process, we learn about the centrality of the army in the Sasanian state structure and its evolution from its Arsacid origins.

Patryk Skupniewicz's chapter, written in collaboration with Maksymiuk, is a detailed study of cavalry, often viewed as the most important part of various Iranian armies, from the Achaemenid to Sasanian periods. Starting with a discussion on the importance of art historical sources for such study, the authors focus on details of cavalry armour and weapons. Each section provides a study of related sources and their implications for each part of the cavalry's function. While mostly focused on visual sources, the authors compare the evidence with similar evidence from the Roman Empire, thus allowing for a study of cross-cultural influences. A catalogue of iconographic variations in the depiction of cavalry concludes the essay.

In a welcome shift to the east, Robert Haug details the background of the war of the Sasanian king Peroz with the Hephthalites and his disastrous campaign and death in 484. By surveying the available, mostly numismatic sources, Haug studies the background of Sasanian relations with the 'Iranian Huns' and the complications of Sasanian relations with the east. The rise and fall of the Kidarites and the division of their state between the Alkhans, Nezaks, and the Hephthalites provides us with the necessary details for understanding the rise of the Hephthalites. In turn, the Hephthalite rise and consolidation of power after the death of Peroz provides a useful picture of war and politics in East Iran.

Craig Morley returns to the issue of diplomacy and state relations in the fourth and fifth centuries CE. While providing a survey of treaties between the Sasanian and Roman empires, Morley focuses on the detailed provisions of each treaty how they provided for peace. Considering the century-long 'peace' between the two empires in the fifth century, the study of this progression is essential to our understanding of the history of both. The second half of the paper is concerned with aspects of diplomatic practice, focusing on the mechanics of treaty-making, the nature of negotiations, and the ritualisation of treaty-making.

Henning Börm's chapter focuses on the wars of the sixth century, between Kavad I and his son Khosrow I on one side and Roman emperors from Anastasius to Justin II on the other. The chapter narrates the strategic history in detail, studying the motivations behind each phase. Ending almost a century of peace, wars dominated Sasanian-Roman relations throughout the sixth century. The author, however, posits the futility of these wars and their lack of gain for either empire, suggesting that campaigns such as those of Khosrow I were meant only as means of increasing the glory of the king.

Khodadad Rezakhani's chapter turns to the wars of the seventh century and the prelude to the rise of Islam. The chapter begins with the sixth-century economic and military reforms of Khosrow I, examined as foundations for the wars of Khosrow II in the following century. It further contextualises Khosrow II's conquest of the Levant and Egypt between 602 and 619 with attention to the place of the Arabs in the Sasanians' frontier activities. Arguing for the development of a Sasanian grand strategy that focused on expansion to the west and incorporation of the eastern Mediterranean, it stresses the under-appreciated continuities between the final Sasanian conquests and the wars of early Islamic expansion, and offers a critical reading of Heraclius' counter-attacks and their limitations.

Touraj Daryaee's epilogue closes the book with a summative overview of the evolution of ancient Iranian armed forces and military systems, from the early Indo-Europeans through the Sasanian period, and offers final reflections on the significance of ritual in Iranian warfare.

While the chapters vary in their treatment of military practice and the significance of warfare across different periods, they interact through respective attention to the common themes and questions traced above. Rung addresses the centrality of war in the imperial ideology of the Achaemenid period, and Finn and Rezakhani touch on the interplay between ideological and pragmatic goals in the formation of Achaemenid and Sasanian strategic priorities. Álvarez-Mon and Wicks, Overtoom, Maksymiuk, Börm, Haug, and Rezakhani explore operational goals and decision-making in Iranian powers' campaigns against major state rivals, east and west. Several essays address different aspects of military armament and battlefield tactics, including the exploration of Elamite archery by Álvarez-Mon and Wicks, Rop's study of Achaemenid infantry performance, Gregoratti's study of later Arsacid cavalry tactics and their discrepancies with those employed at Carrhae, and Skupniewicz's and Maksymiuk's survey of iconographic evidence for cavalry armament in the Achaemenid and Sasanian eras. Beckman, Nabel, and Morley investigate the complex relationships between state objectives, warfare, and diplomatic practice (by both formal embassies and non-state actors) at different periods. Institutional questions regarding the intersections between imperial structures, administration, economies, and practices of military recruitment run through Hyland's study of the Achaemenid military system, Pirngruber's discussion of Achaemenid through Arsacid Babylonia, Heckel's chapter on Iranian elites in Alexander's empire, Coloru's study of the Seleucid military system, and Maksymiuk's discussion of Sasanian military structures.

The editors do not make a claim to comprehensive treatment of every topic, which would hardly be feasible in a single study given the chronological scope of the volume. The usual challenges of assembling a large collection of contributions by busy scholars have been amplified by the challenges of academic production during the COVID-19 pandemic, with the unfortunate result that certain topics which would have been desirable additions had to remain unexplored in the current volume. Among the most important advances in the recent study of Sasanian imperial dynamics is the chronological fixing of the frontier fortification systems of the Gorgan and Caucasus regions in the fifth century, and full analysis and interpretation of their implications will provide historians with ample material for years to come. It was regrettably impossible to include a detailed study of this topic, and future scholarship might benefit not only from further study of frontier defences in this period but from broader comparison with military structures and fortress systems at earlier periods, incorporating attention to the Arsacid, Hellenistic, and Achaemenid eras. Another topic that would benefit from expanded study is the naval practice of Iranian empires, with greater comparative attention to the structures and practices of the Achaemenids and the late Sasanian and early Islamic periods. It would finally be beneficial to devote more explicit attention to the impact of warfare on the civilian populations of the imperial borderlands across a long spectrum of Iranian imperial history, addressing matters from economic impacts to the social effects of military deportation in selected periods.

These desiderata aside, the assembly of studies on so many aspects of warfare in the ancient Iranian empires provides an exciting opportunity to benefit the respective subfields and the broader historical study of ancient Iran. We hope that this volume can provide a foundation for further research and an impetus for dialogue between scholars of war and imperialism across chronological and disciplinary boundaries.

Bibliography

Abdi, K. (2001) 'Nationalism, Politics, and the Emergence of Archaeology in Iran'. *AJA* 105: 51–76.

Álvarez-Mon, J. (2019) *The Monumental Reliefs of the Elamite Highlands: A Complete Inventory and Analysis (from the Seventeenth to the Sixth Century* BC). Philadelphia.

Álvarez-Mon, J., Basello, G., and Wicks, Y. (eds) (2018) *The Elamite World*. London.

Álvarez-Mon, J. and Garrison, M. (eds) (2011) *Elam and Persia*. Winona Lake.

anon. (1971) *Râhnamâ-ye rezhe-ye Takht-e Jamshid (Parseh) = Parade a Persépolis = Parade at Persepolis*. Tehran.

Ansari, A. (2012) *The Politics of Nationalism in Modern Iran*. Cambridge.

Armstrong, J. (2016) *Circum Mare: Themes in Ancient Warfare*. Leiden.

Azarnouche, S. and Petitjean, M. (2022) 'Sasanian Warriors in Context: Historical and Religious Commentary on a Middle Persian Chapter on *Artēštārān* (*Dēnkard* VIII.26)'. *Revue Internationale d'Histoire Militaire Ancienne* 11: 331–84.

Balakhvantsev, A.S. and Makkaveev, N.A. (eds) (2019) *Эпоха империй. Восточный Иран от Ахеменидов до Сасанидов: история, археология, культура. Материалы международной научной конференции, посвященной памяти Бориса Анатольевича Литвинского*, Moskva.

Bar-Kokhva, B. (1976) *The Seleucid Army: Organization and Tactics in the Great Campaigns*. Cambridge.

Beloch, K.J. (1916) *Griechische Geschichte. Band II: Bis auf die sophistische Bewegung und den Peloponnesischen Krieg*. Abt. II. Strassburg.

Bernheimer, T. and Silverstein, A. (eds) (2012) *Late Antiquity: Eastern Perspectives*. Exeter.

Binder, C., Börm, H., and Luther, A. (eds) (2016) *Diwan: Studies in the History and Culture of the Ancient Near East and the Eastern Mediterranean. Festschrift für Josef Wiesehöfer zum 65. Geburtstag*. Duisburg.

Blockley, R.C. (1980) 'Doctors as Diplomats in the Sixth Century AD'. *Florilegium* 2: 89–100.

Briant, P. (1999) 'The Achaemenid Empire', in Raaflaub and Rosenstein (1999) 105–28.

Briant, P. (2002) *From Cyrus to Alexander: A History of the Persian Empire*. Winona Lake.

Briant, P. (2019) 'L'approvisionnement de l'armée Macédonienne: Alexandre le Grande et l'organisation logistique de l'Empire Achéménide', in Gondet and Haerinck (2019) 55–70.

Brosius, M. (2005) 'Pax Persica: königliche Ideologie und Kriegführung im Achämenidenreich', in Messner, Schmitt, and Sommer (2005) 135–61.

Buraselis, K. and Meidani, K. (eds) (2010) *Marathon: Battle and Deme*. Athens.

Cahill, N. (2010a) *The Lydians and their World*. Istanbul.

Cahill, N. (2010b) 'The Persian Destruction of Sardis', in Cahill (2010a) 339–61.

Campbell, B. (2002a) 'War and Diplomacy: Rome and Parthia, 31 BC–AD 235', in Campbell (2002b) 225–52.

Campbell, B. (ed) (2002b) *War and Society in the Roman World*. New York.

Campbell, B. and Tritle, L. (eds) (2013) *Oxford Handbook of Warfare in the Classical World*. Oxford.

Canepa, M. (2010) 'Technologies of Memory in Early Sasanian Iran: Achaemenid Sites and Sasanian Identity'. *AJA* 114: 563–96.

Cawkwell, G. (2005) *The Greek Wars: The Failure of Persia*. Oxford.

Charles, M. (2011) 'Immortals and Apple-Bearers: Towards a Better Understanding of Achaemenid Infantry Units'. *CQ* 61: 114–33.

Charles, M. (2021) 'The Royal Elite of the Achaemenid Army', in Heckel, Naiden, Garvin, and Vanderspoel (2021) 173–84.

Choksy, J.K. (1988) 'Sacral Kingship in Sasanian Iran'. *BAI* 2: 35–52.

Clark, J. and Turner, B. (eds) (2018) *Brill's Companion to Military Defeat in Ancient Mediterranean Society*. Leiden.

Cohen, R. and Westbrook, R. (eds) (2008) *Isaiah's Vision of Peace in Biblical and Modern International Relations*. Berlin.

Collins, J. and Manning, J. (eds) (2016) *Revolt and Resistance in the Ancient Classical World and the Near East: In the Crucible of Empire*. Leiden.

Compareti, M. and Daryaee, T. (eds) (2020) *Studi sulla Persia sasanide e suoi rapporti con le civiltà attigue*. Bologna.

Dabrowa, E. (1994) (ed) *The Roman and Byzantine Army in the East*. Krakow.

Daryaee, T. (2003) 'The Ideal King in the Sasanian World: Ardaxšīr ī Pābagān or Xusrō Anōšagruwān?' *Nāme-ye Irān-e Bāstān* 3: 33–45.

Daryaee, T. (2010) 'Barbarized Armies and Barracks Kings of Late Antique Iran', in Macuch, Weber, and Durkin-Meisterernst (2010) 43–52.

Daryaee, T. and Rezakhani, K. (2016) *From Oxus to Euphrates: The World of Late Antique Iran*. Irvine and Los Angeles.

Degen, J. and Manning, S. (2021) 'Western Europe', in Jacobs and Rollinger (2021) 1561–608.

Delbrück, H. (1887) *Die Perserkriege und die Burgunderkriege*. Berlin.

Delemen, I., Casabonne, O., Karagöz, Ş., and Tekin, O. (eds) (2007) *The Achaemenid Impact on Local Populations and Cultures in Anatolia (6th–4th Centuries B.C.)*. Istanbul.

Di Cosmo, N. and Maas, M. (eds) (2018) *Empires and Exchanges in Eurasian Late Antiquity: Rome, China, Iran, and the Steppe, ca. 250–750*. Cambridge.
Dignas, B. and Winter, E. (2007) *Rome and Persia in Late Antiquity: Neighbours and Rivals*. Cambridge.
Dodgeon, M. and Lieu, S. (1992) *The Roman Eastern Frontier and the Persian Wars A.D. 226–363: A Documentary History*. London.
Dunsch, B. and Ruffing, K. (eds) (2013) *Herodots Quellen – die Quellen Herodots*. Wiesbaden.
Dusinberre, E. (2016) 'Resistance, Revolt, and Revolution in Achaemenid Persia: Response', in Collins and Manning (2016) 122–37.
Dusinberre, E., Garrison, M., and Henkelman, W.F.M. (eds) (2020) *The Art of Empire in Achaemenid Persia: Studies in Honour of Margaret Cool Root*. Leiden.
Ellerbrock, U. and Winkelmann, S. (eds) (2012) *Die Parther: die vergessene Großmacht*. Darmstadt.
Eschenburg, A. (1886) *Topographische, archaeologische und militärische Betrachtungen auf dem Schlachtfelde von Marathon*. Berlin.
Fagan, G. and Trundle, M. (eds) (2010) *New Perspectives on Ancient Warfare*. Leiden.
Farrokh, K. (2007) *Shadows in the Desert: Ancient Persia at War*. London.
Farrokh, K., Karamian, G., and Maksymiuk, K. (2018) *A Synopsis of Sasanian Military Organization and Combat Units*. Siedlce and Tehran.
Frame, G. (ed) (2004) *From the Upper Sea to the Lower Sea: Studies on the History of Assyria and Babylonia in honour of A.K. Grayson*. Leiden.
Gaspa, S., Greco, A., Bonacossi, D., Ponchia, S., and Rollinger, R. (eds) (2014) *From Source to History: Studies on Ancient Near Eastern Worlds and Beyond (Festschrift for Giovanni B. Lanfranchi)*. Münster.
Gazzano, F. (2018) 'Uno stereotipo dell'armata persiana nella tradizione classica'. *Historika* 8: 91–128.
Gombert, B. (2018) 'L'Armée en Babylonie du VIe au IVe Siècle av. N.È.', PhD Dissertation, Université Paris 1 Panthéon-Sorbonne.
Gondet, S. and Haerinck, E. (eds) (2019) *L'Orient est son Jardin. Hommages à Rémy Boucharlat*. Leuven.
Greatrex, G. (1998) *Rome and Persia at War, 502–532*. Leeds.
Greatrex, G. and Lieu, S. (2005) *The Roman Eastern Frontier and the Persian Wars A.D. 363–628*. London.
Grigor, T. (2005) 'Preserving the Antique Modern: Persepolis '71'. *Future Anterior* 2: 22–29.
Grigor, T. (2021) *The Persian Revival: The Imperialism of the Copy in Iranian and Parsi Architecture*. University Park, Pennsylvania.
Gyselen, R. (2001) *The Four Generals of the Sasanian Empire: Some Sigillographic Evidence*. Rome.

Hassan, C. (2021) 'Structure of the Army and Logistics', in Jacobs and Rollinger (2021) 1151–59.

Hauser, S. (2005) 'Die ewigen Nomaden? Bemerkungen zu Herkunft, Militär, Staatsaufbau und nomadischen Traditionen der Arsakiden', in Meissner, Schmitt, and Sommer (2005) 163–208.

Hauser, S. (2006) 'Was There No Paid Standing Army? A Fresh Look on Military and Political Institutions in the Arsacid Empire', in Mode and Tubach (2006) 295–319.

Hauser, S. (2012) 'The Arsacid (Parthian) Empire', in Potts (2012) 1001–20.

Head, D. (1992) *The Achaemenid Persian Army*. London.

Heckel, W., Müller, S., and Wrightson, G. (eds) (2015) *The Many Faces of War in the Ancient World*. Newcastle.

Heckel, W., Naiden, F., Garvin, E., and Vanderspoel, J. (eds) (2021) *A Companion to Greek Warfare*. Hoboken, NJ.

Henkelman, W.F.M. (2002) 'Exit der Posaunenbläser: On Lance-Guards and Lance-Bearers in the Persepolis Fortification Archive'. *ARTA* 2002.007: 1–35.

Henkelman, W.F.M. (2017b) 'Imperial Signature and Imperial Paradigm: Achaemenid Administrative Structure and System across and beyond the Iranian Plateau', in Jacobs, Henkelman, and Stolper (2017) 45–256.

Henkelman, W.F.M. and Redard, C. (eds) (2017) *Persian Religion in the Achaemenid Period*. Wiesbaden.

Herrmann, G., Mackenzie, D.N., and Howell, R. (1989) *The Sasanian Rock Reliefs at Naqsh-i Rustam: Naqsh-i Rustam 6, The Triumph of Shapur I*. Berlin.

Honarchiansaky, A. (2018) 'Connected Histories in Late Antiquity: A Study of a Peace between the Roman and Sasanian Empires and Its Diverging Impacts on Christianity in Greater Armenia and Iran', PhD Dissertation, University of California, Los Angeles.

Howard-Johnston, J. (1999) 'Heraclius' Persian Campaigns and the Revival of the East Roman Empire, 622–630'. *War in History* 6: 1–44.

Howard-Johnston, J. (2012) 'The Late Sasanian Army', in Bernheimer and Silverstein (2012) 87–127.

Howard-Johnston, J. (2014) 'The Sasanian State: The Evidence of Coinage and Military Construction'. *JAH* 2: 144–81.

Howard-Johnston, J. (2016) 'The Grand Strategy of the Sasanian Empire', in Binder, Börm, and Luther (2016) 591–613.

Howard-Johnston, J. (2021) *The Last Great War of Antiquity*. Oxford.

Howe, T. and Brice, L. (2016) *Brill's Companion to Insurgency and Terrorism in the Ancient Mediterranean*. Leiden.

Hyland, J. (2018a) 'Achaemenid Soldiers, Alexander's Conquest, and the Experience of Defeat', in Clark and Turner (2018) 74–95.

Hyland, J. (2018b) *Persian Interventions: The Achaemenid Empire, Athens, and Sparta, 450–386* BCE. Baltimore.

Hyland, J. (forthcoming) *Persia's Greek Campaigns. Kingship, War, and Spectacle on the Achaemenid Frontier.* Oxford.

Isaac, B. (1998) *The Near East under Roman Rule.* Leiden.

Jacobs, B. and Rollinger, R. (eds) (2021) *A Companion to the Achaemenid Persian Empire.* Hoboken, NJ.

Jacobs, B., Henkelman, W.F.M., and Stolper, M.W. (eds) (2017) *Verwaltung im Achämenidenreich.* Wiesbaden.

Jursa, M. (2011) 'Taxation and Service Obligations in Babylonia from Nebuchadnezzar to Darius and the Evidence for Darius' Tax Reform', in Rollinger, Truschnegg, and Bichler (2011) 431–48.

Jursa, M. and Schmidl, M. (2017) 'Babylonia as a Source of Imperial Revenue from Cyrus to Xerxes', in Jacobs, Henkelman, and Stolper (2017) 715–35.

Klinkott, H., Luther, A., and Wiesehöfer, J. (eds) (2021) *Beiträge zur Geschichte und Kultur des alten Iran und benachbarter Gebiete. Festschrift für Rüdiger Schmitt.* Stuttgart.

Konijnendijk, R. (2012) '"Neither the Less Valorous nor the Weaker:" Persian Military Might and the Battle of Plataia'. *Historia* 61: 1–17.

Konijnendijk, R. (2021) 'Legitimisation of War', in Jacobs and Rollinger (2021) 1139–50.

Laforse, B. (2013) 'Fighting the Other Part I: Greeks and Achaemenid Persians', in Campbell and Tritle (2013) 569–87.

Larsen, M. (ed) (1979) *Power and Propaganda: A Symposium on Ancient Empires.* Copenhagen.

Lawrence, D. and Wilkinson, T. (2017) 'The Northern and Western Borderlands of the Sasanian Empire: Contextualizing the Roman/Byzantine and Sasanian Frontier', in Sauer (2017) 99–125.

Lee, A.D. (2013) 'Roman Warfare with Sasanian Persia', in Campbell and Tritle (2013) 708–25.

Lee, G., Whittaker, H., and Wrightson, G. (eds) (2015) *Ancient Warfare: Introducing Current Research.* Newcastle.

Lee, J.W.I. (2016a) 'Cyrus the Younger and Artaxerxes II, 401 BC: an Achaemenid Civil War Reconsidered', in Collins and Manning (2016) 103–21.

Lee, J.W.I. (2016b) 'Tissaphernes and the Achaemenid Defense of Western Anatolia, 412–395 B.C.', in Armstrong (2016) 262–82.

Lerner, J. (2017) 'Mithradates I and the Parthian Archer', in Schlude and Rubin (2017) 1–24.

Lewin, A. and Pellegrini, P. (eds) (2007) *The Late Roman Army in the Near East from Diocletian to the Arab Conquest.* London.

Lewis, D.M. (1977) *Sparta and Persia.* Leiden.

Lillington-Martin, C. (2007) 'Archaeological and Ancient Literary Evidence for a Battle near Dara Gap, Turkey, AD 530: Topography, Texts, and Trenches', in Lewin and Pellegrini (2007) 299–311.

Ma, J. (2008) 'Mysians on the Çan Sarcophagus: Ethnicity and Domination in Achaimenid Military Art'. *Historia* 57: 243–52.

Macuch, M., Weber, D., and Durkin-Meisterernst, D. (eds) (2010) *Ancient and Middle Iranian Studies: Proceedings of the 6th European Conference of Iranian Studies*. Wiesbaden.

Manning, S. (2021) *Armed Force in the Teispid-Achaemenid Empire: Past Approaches, Future Prospects*. Stuttgart.

Marquart, J. (1896) *Untersuchungen zur Geschichte von Eran*. Vol. 1. Göttingen.

McCaskie, T. (2012) '"As on a Darkling Plain": Practitioners, Publics, Propagandists, and Ancient Historiography'. *Comparative Studies in Society and History* 54: 145–73.

McDonough, S. (2013) 'Fighting the Other Part III: Military and Society in Sasanian Iran', in Campbell and Tritle (2013) 601–20.

Meissner, B., Schmitt, O., and Sommer, M. (eds) (2005) *Krieg, Gesellschaft, Institutionen: Beiträge zu einer vergleichenden Kriegsgeschichte*. Berlin.

Meyer, E. (1901) *Geschichte des Altertums. Band III: Das Perserreich und die Griechen. Erste Hälfte: Bis zu den Friedensschlüssen von 448 und 446 v. Chr.* Stuttgart.

Millard, A., Bienkowski, P., Mee, C., and Stater, E. (eds) (2005) *Writing and Ancient Near Eastern Society: Essays in Honor of Alan Millard*. London.

Mode, M. and Tubach, J. (2006) *Arms and Armour as Indicators of Cultural Transfer: The Steppes and the Ancient World from Hellenistic Times to the Early Middle Ages*. Wiesbaden.

Mousavi, A. (2021) 'Iran', in Jacobs and Rollinger (2021) 1637–47.

Nicolle, D. (1998) *Sassanian Armies: The Iranian Empire Early 3rd to Mid 7th Centuries AD*. London.

Nimchuk, C. (2002) 'The "Archers" of Darius: Coinage or Tokens of Royal Esteem'? *Ars Orientalis* 32: 55–79.

Nissinen, M. (2008) 'From Holy War to Holy Peace', in Cohen and Westbrook (2008) 181–97.

O'Connor, S. (2015) 'The Problem of the 400 Wagons: The Provisioning of the Ten Thousand on the March to Cunaxa', in Lee, Whittaker, and Wrightson (2015) 124–46.

Olbrycht, M.J. (2011) 'First Iranian Military Units in the Army of Alexander the Great'. *Anabasis: Studia Classica et Orientalia* 2: 67–84.

Olbrycht, M.J. (2012) 'The Political-Military Strategy of Artabanos/Ardawān II in AD 34–37'. *Anabasis* 3: 215–37.

Olbrycht, M.J. (2015) 'The Epigonoi – The Iranian Phalanx of Alexander the Great', in Heckel, Müller, and Wrightson (2015) 196–212.

Olbrycht, M.J. (2016a) 'Manpower Resources and Army Organisation in the Arsakid Empire'. *AncSoc* 46: 291–338.

Olbrycht, M.J. (2016b) 'The Sacral Kingship of the Early Arsacids I: Fire Cult and Kingly Glory'. *Anabasis* 7: 91–106.

Overlaet, B. (2013) 'And Man Created God? Kings, Priests and Gods on Sasanian Investiture Reliefs'. *Ir. Ant.* 48: 313–54.

Overtoom, N. (2017) 'The Parthians' Unique Mode of Warfare: A Tradition of Parthian Militarism and the Battle of Carrhae'. *Anabasis* 8: 95–122.

Pirngruber, R. (2017) *The Economy of Late Achaemenid and Seleucid Babylonia*. Cambridge.

Pirnia, H. (1927) *Iran-e Bastan*. Tehran.

Potts, D.T. (ed) (2012) *A Companion to the Archaeology of the Ancient Near East*. Oxford.

Potts, D.T. (ed) (2013) *The Oxford Handbook of Ancient Iran*. Oxford.

Potts, D.T. (2018) 'Sasanian Iran and Its Northeastern Frontier: Offense, Defense, and Diplomatic Entente', in Di Cosmo and Maas (2018) 287–301.

Potts, D.T. (2021) 'Camels and Their Rations in the Persepolis Fortification Archive: An Enigma and Its Variations'. *Egitto e Vicino Oriente* 44: 231–47.

Pourshariati, P. (2008) *Decline and Fall of the Sasanian Empire: The Sasanian-Parthian Confederacy and the Arab Conquest of Iran*. London.

Raaflaub, K. (ed) (2007) *War and Peace in the Ancient World*. Oxford.

Raaflaub, K. and Rosenstein, N. (eds) (1999) *War and Society in the Ancient and Medieval Worlds*. Cambridge, MA.

Rance, P. (ed) (forthcoming) *Brill's Companion to Military Culture in Late Antiquity*. Leiden.

Reade, J. (1979) 'Ideology and Propaganda in Assyrian Art', in Larsen (1979) 329–43.

Rezakhani, K. (2015) 'Mazdakism, Manichaeism and Zoroastrianism: In Search of Orthodoxy and Heterodoxy in Late Antique Iran', *Iranian Studies* 48: 55–70.

Rezakhani, K. (2019) 'The End of Sasanian Rule: the Center and Periphery of Ērānšahr in the Seventh Century', in Compareti and Daryaee (2020) 229–55.

Rollinger, R. (2013) 'Dareios und Xerxes an den Rändern der Welt und die Inszenierung von Weltherrschaft. Altorientalisches bei Herodot', in Dunsch and Ruffing (2013) 95–116.

Rollinger, R. and Degen, J. (2021) 'Conceptualizing Universal Rulership: Considerations on the Persian Achaemenid Worldview and the Saka at the "End of the World"', in Klinkott, Luther, and Wiesehöfer (2021) 187–224.

Rollinger, R., Truschnegg, B., and Bichler, R. (eds) (2011) *Herodot und das persische Weltreich*. Wiesbaden.

Root, M.C. (1979) *King and Kingship in Achaemenid Art: Essays on the Creation of an Iconography of Empire*. Leiden.

Root, M.C. (1989) 'The Persian Archer at Persepolis: Aspects of Chronology, Style, and Symbolism'. *Rev. Ét. Anc.* 91: 33–50.

Root, M.C. (2021) 'The USA', in Jacobs and Rollinger (2021) 1609–28.

Rop, J. (2018). 'The Assassination of Tissaphernes: Royal Responses to Military Defeat in the Achaemenid Empire', in Clark and Turner (2018) 51–73.

Rop, J. (2019) *Greek Military Service in the Ancient Near East, 401–330 BCE*. Cambridge.

Ruzicka, S. (2012) *Trouble in the West: Egypt and the Persian Empire, 525–332 BCE*. Oxford.

Sabin, P., Van Wees, H., and Whitby, M. (eds) (2007) *The Cambridge History of Greek and Roman Warfare. Volume 1: Greece, the Hellenistic World and the Rise of Rome*. Cambridge.

Sänger, P. (2008) 'Saralaneozan und die Verwaltung Ägyptens unter den Sassaniden'. *ZPE* 164: 191–201.

Sänger, P. (2011) 'The Administration of Sasanian Egypt: New Masters and Byzantine Continuity'. *GRBS* 51: 653–65.

Sauer, E. (ed) (2017) *Sasanian Persia: Between Rome and the Steppes of Eurasia*. Edinburgh.

Sauer, E., Rekavandi, H., Wilkinson, T., and Nokandeh, J. (eds) (2013) *Persia's Imperial Power in Late Antiquity: Sasanian Frontier Walls, Forts and Landscapes of Northern Iran*. Oxford.

Sauer, E., Nokandeh, J., Pitskhelauri, K., and Rekavandi, H. (2017) 'Innovation and Stagnation: Military Infrastructure and the Shifting Balance of Power Between Rome and Persia', in Sauer (2017) 241–67.

Schlude, J.M. and Rubin, B.B. (eds) (2017) *Arsacids, Romans, and Local Elites: Cross-Cultural Interactions of the Parthian Empire*. Oxford; Philadelphia.

Schlude, J.M. (2020) *Rome, Parthia, and the Politics of Peace: The Origins of War in the Ancient Middle East*. London.

Sekunda, N., and Chew, S. (1992) *The Persian Army 560–330 BC*. London.

Shahbazi, A.S. (1986) 'Army i: Pre-Islamic Iran'. *EncIr*. (http://www.iranicaonline.org/articles/army-i).

Shahbazi, A.S. (2001) 'Early Sasanians' Claim to Achaemenid Heritage'. *Name-ye Iran-e Bastan* 1: 61–73.

Shayegan, M.R. (2011) *Arsacids and Sasanians: Political Ideology in Post-Hellenistic and Late Antique Persia*. Cambridge.

Skupniewicz, P. (2015) 'The Iconographic Function of Armor in Sasanian Art'. *Rivista Degli Studi Orientali* 88: 235–65.

Skupniewicz, P. (2019) 'On the Shortfalls of Iconography in the Reconstruction of Swordsmanship: The Sasanian Case', in Balakhvantsev and Makkaveev (2019) 195–212.

Skupniewicz, P. (2021) 'The Bow as an Insignia of Power in the Art of Ancient Iran'. *Historia i Świat* 10: 153–70.

Steele, R. (2020). *The Shah's Imperial Celebrations of 1971: Nationalism, Culture, and Politics in Late Pahlavi Iran*. London.

Stolper, M.W. (1985) *Entrepreneurs and Empire: The Murašû Archive, the Murašû Firm, and Persian Rule in Babylonia*. Leiden.

Stoyanov, Y. (2011) *Defenders and Enemies of the True Cross*. Vienna.

Summerer, L. (2007) 'From Tatarlı to Munich: The Recovery of a Painted Wooden Tomb Chamber in Phrygia', in Delemen, Casabonne, Karagöz, and Tekin (2007) 131–58.

Syvänne, I. and Maksymiuk, K. (2018) *The Military History of the Third Century Iran*. Siedlce.

Tuplin, C. (1987) 'Xenophon and the Garrisons of the Persian Empire'. *AMI* 20: 167–245.

Tuplin, C. (1997) 'Achaemenid Arithmetic: Numerical Problems in Persian History'. *Topoi. Orient-Occident* suppl. 1: 365–421.

Tuplin, C. (2005) 'Darius' Accession in (the) Media', in Millard, Bienkowski, Mee, and Stater (2005) 217–44.

Tuplin, C. (2010a) 'All the King's Horse: In Search of Achaemenid Persian Cavalry', in Fagan and Trundle (2010) 101–82.

Tuplin, C. (2010b) 'Marathon: In Search of a Persian Dimension', in Buraselis and Meidani (2010) 251–74.

Tuplin, C. (2014) 'From Arshama to Alexander: Reflections on Persian Responses to Attack', in Gaspa, Greco, Bonacossi, Ponchia, and Rollinger (2014) 668–96.

Tuplin, C. (2017) 'War and Peace in Achaemenid Imperial Ideology'. *Electrum* 24: 31–54.

Tuplin, C. (2020) 'Sigillography and Soldiers: Cataloguing Military Activity on Achaemenid Period Seals', in Dusinberre, Garrison, and Henkelman (2020) 329–459.

Tuplin, C. (2021) 'Mercenaries', in Jacobs and Rollinger (2021) 1183–95.

Tuplin, C. and Jacobs, B. (2021) 'Military Organisation and Equipment', in Jacobs and Rollinger (2021) 1161–81.

Waters, M. (2016) 'Xerxes and the Oath-Breakers: Empire and Rebellion on the Northwestern Front', in Collins and Manning (2016) 93–102.

Whatley, N. (1964) 'On the Possibility of Reconstructing Marathon and Other Ancient Battles'. *JHS* 84: 119–39.

Whitby, M. and Whitby, M. (1989) *Chronicon Paschale* (*Translated Texts for Historians* 7). Liverpool.

Whitby, M. (1994) 'The Persian King at War', in Dabrowa (1994) 227–63.

Whitby, M. (2007) 'Reconstructing Ancient Warfare', in Sabin, Van Wees, and Whitby (2007) 54–82.

Wiesehöfer, J. (1992) '"Den es sind welthistorische Siege ...": Nineteenth- and Twentieth-century German Views of the Persian Wars'. *Culture and History* 11: 61–83.

Wiesehöfer, J. (2007) 'From Achaemenid Imperial Order to Sasanian Diplomacy: War, Peace, and Reconciliation in Pre-Islamic Iran', in Raaflaub (2007) 121–40.

Wiesehöfer, J. (2016) 'Fourth-century Revolts against Persia: The Test Case of Sidon', in Howe and Brice (2016) 93–112.

Wiesehöfer, J. (forthcoming) 'Military Conflict in Late Antiquity: The Iranian World', in Rance (forthcoming).

Winkelmann, S. (2012) 'Das Militärwesen – die Armee der Parther', in Ellerbrock and Winkelmann (2012) 185–96.

Wolski, J. (1981) 'L'aristocratie et l'organisation de l'armée parthe'. *Klio* 63: 105–22.

Wu, X. (2014) '"O Young Man ... Make Known of What Kind You Are": Warfare, History, and Elite Ideology of the Achaemenid Persian Empire'. *IrAnt* 49: 209–99.

Young, T.C. (2004) 'Darius I, Commander-in-Chief: Bisitun I:35 to III:92 as Military History', in Frame (2004) 277–88.

Zia-Ebrahimi, R. (2016) *The Emergence of Iranian Nationalism: Race and the Politics of Dislocation*. New York.

CHAPTER 2

Like a Raining Cloud: Archery as the Pillar of Elamite Warfare

Javier Álvarez-Mon and Yasmina Wicks

His bow was nocked to let the arrows fly
But we ourselves, we did not know how to grasp a quiver
Elam overwhelmed our sacred localities.

Babylonian Temple Rituals, reign of Nabu-kudurri-usur I (Nebuchadnezzar I), ca. 1100 BCE

Elam, an ancient polity with a territorial configuration of lowlands and highlands roughly equating to today's southwest Iran, was a major, yet often overlooked, political player of the ancient Near East. In recent years an increasing body of academic publications have suggested that the legacy of Elam was more considerable and long-lasting than previously estimated. A recognition of this fact must now be regarded as essential for any scholar interested in Near Eastern military history, and in the genesis and development of Achaemenid Persia and later Iranian empires.[1] Elam boasted a remarkable longevity that hinged on the strategic possibilities and resources afforded by its geographical composition, political dynamism, astute diplomacy, and, as this chapter will demonstrate, a military renowned for its archery finesse, particularly with the composite bow, and a large-scale state production of archery paraphernalia. This investigation, which draws together archaeological, iconographic and textual evidence for archery in Elam, commences in the late Neolithic when lowland settlement sites on the Susiana plain close to Mesopotamia began to exhibit signs of the persistent threat of warfare, and travels into the historical era up to the rise of the Persian Empire, whose rulers perpetuated the Elamite cultural ideology surrounding archery. Though historical sources are lacking

1 For general recent syntheses of Elamite history, archaeology and art, see Potts (2016); Álvarez-Mon, Basello and Wicks (2018); and Álvarez-Mon (2020).

 | DOI:10.1163/9789004710771_003

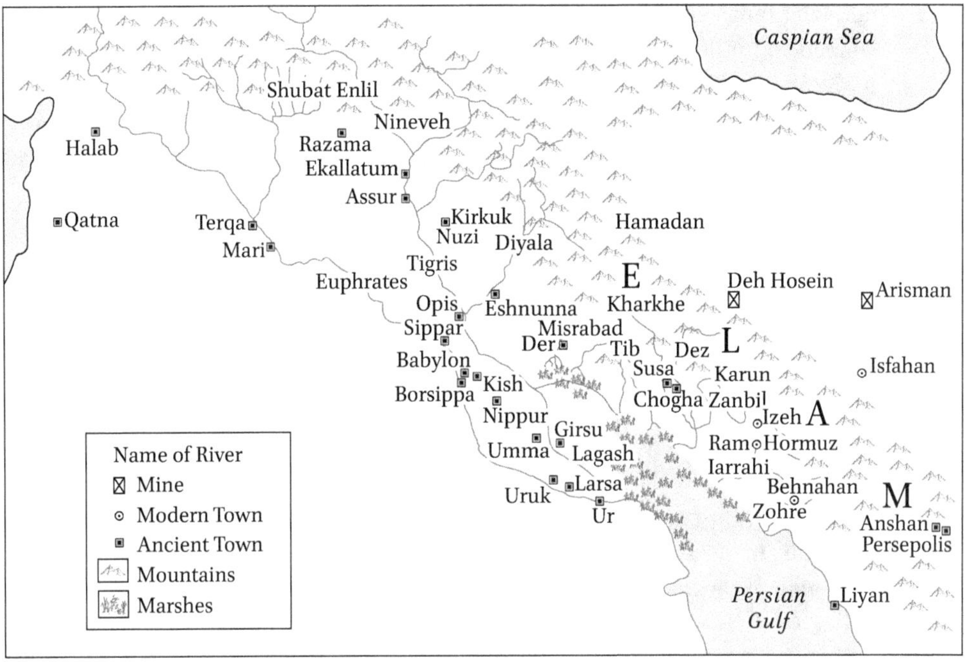

FIGURE 2.1 General map with sites mentioned in text
J. ÁLVAREZ-MON © 2021

for Elam's political and military activities on its east and south borders, it will become clear that a strong tradition of archery played a key role in historic Elamite victories that changed the political trajectories of its western neighbours [Fig. 2.1].

1 Warfare in Elamite Prehistory (ca. 4200–2700 BCE): the Critical Invention of the Composite Bow

The origins of Elam's lowland culture can be traced back to the foundation of Susa on the Susiana plain in ca. 4200 BCE. Here the threat of warfare and the early military potential of the local populations were manifested in the form of defensive city walls and an array of weaponry: short-range weapons including limestone maces[2] and polished stone or primitive copper axes,[3]

2 Morgan (1912) 13, figs. 36–39, 18, figs. 87–93; Le Brun (1978) fig. 39.
3 Morgan (1912) 11, figs. 27–28; Tallon (1987a) 312, fig. 49.

and long-range weapons including javelin/spearheads,[4] arrowheads, and ovoid-shaped sundried clay sling missiles,[5] which were widely deployed as weapons in the ancient Near East and probably appear in a rare early depiction of warfare at a fortified city in a glyptic image (ca. 3800–3100 BCE) from the nearby site of Chogha Mish [Fig. 2.4e].[6] Much of the archaeological evidence for archery – bows, quivers, and protective hand- and arm-gear – was made in organic materials and has perished through time, leaving us only with arrowheads, which are defined as projectile points of less than 7/8 cm in length,[7] made of flint, metal, and occasionally bone.[8] Arrowheads cut from variously coloured flint, and occasionally obsidian, were widespread in the earliest layers of Susa. They varied in form and in finish, with some retouched or carefully denticulated [Fig 2.2]. Finely worked lanceolate-shaped forms were most abundant, and of these the early director of the French archaeological mission at Susa, Jacques de Morgan, wrote, 'flint was never fashioned with more perfection'.[9] In contrast to hand-to-hand combat weapons, which were sometimes deposited with the dead, arrowheads are conspicuously absent from burial assemblages of the two main excavated late Neolithic cemeteries in the Acropole mound at Susa and at Chogha Sofla on the Behbehan plain,[10] suggesting that they played no role yet in funerary ideology. Further into the highlands at Bakun A in Fars, which exhibits cultural links to sites in Susiana, and at other contemporary settlements in the area, which would become part of the Elamite political and cultural sphere by the end of the third millennium BCE, projectile points were absent from the flint industry.[11] Nevertheless, use of projectile points made of organic materials, such as wood and bone, cannot be excluded.

---

4 E.g., copper/copper alloy spearheads in Le Brun (1971) fig. 67: 1; Tallon (1987a) 139–40; (1987b) no. 197.

5 Note, however, that weapons (except perhaps mace-heads) were also deployed in hunting and other activities, so martial use is rarely assured (Hamblin (2006) 15).

6 Delougaz and Kantor (1996) 146–47; Pl. 151C. Sling missiles are common finds at Neolithic and Chalcolithic sites in Iran (Potts (2020) 19–20), including Elamite areas: Susa, where missiles mixed with architectural decoration may result from enemy offensives (Steve and Gasche (1971) 31, 143, 167–68, Pls. 28.25, 59.2; 60:1–2; additional missiles in Le Brun (1971) 187; (1978) 84). Chogha Mish (Alizadeh (2008) 33, 42; Delougaz and Kantor (1996) 108, n. 42, 147, 253, Pl. 65L); Tall-i Ghazir (Alizadeh (2014) 23, 50); and Tell-e Bakun in the highlands (Alizadeh (2006) 79, fig. 63K–N).

7 Gernez (2018) 69.

8 Mecquenem (1936) 2; (1937) 2; (1938) 6.

9 Morgan (1912) 17, figs. 60–73.

10 See Moghaddam (2020). At Tal-e Ghazir on the neighbouring Ram Hormuz plain, chipped stone objects were rarely collected by archaeologists (Alizadeh (2014) 50).

11 Crowfoot in Langsdorff and McCown (1942) 76, fn. 19.

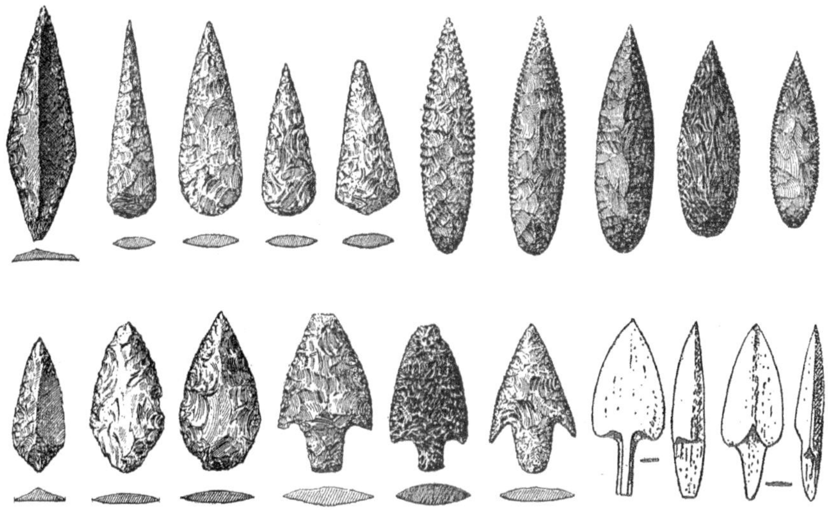

FIGURE 2.2 Flint arrowheads from Susa
AFTER MORGAN 1912: 17, FIGS. 60–73

FIGURE 2.3 Archers from Susiana (ca. 4200–3800 BCE) [a] painted bowl from Susa; [b] painted pottery sherd from Djowi
PHOTOGRAPHS © J. ÁLVAREZ-MON

The late fifth/early fourth millennium BCE artistic record offers only very limited evidence for the bow, and the context of its use is unclear. One schematic human figure preserved on a painted pottery sherd is shown frontally with W-shaped arms holding a bow in one hand [Fig. 2.3b]. Another is depicted inside a painted bowl, this time in profile, holding (aiming?) a bow and arrow

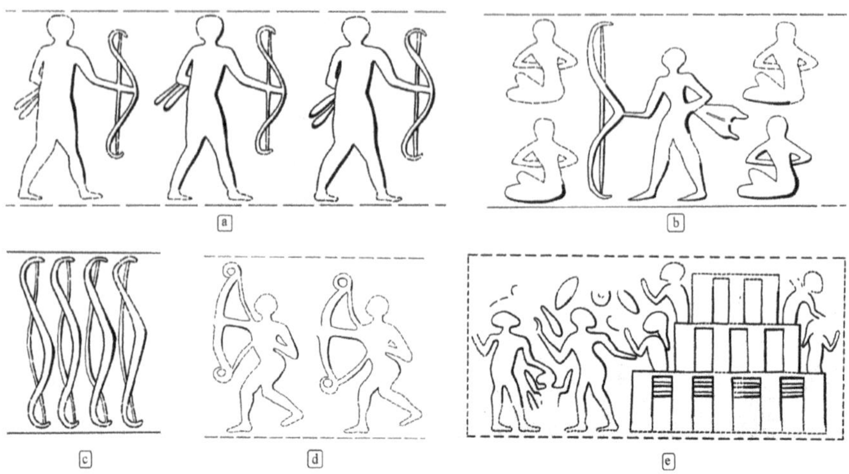

FIGURE 2.4 Composite bows in Chogha Mish sealings (ca. 3800–3100 BCE) [a] line of bow-bearers, [b] oversize bow; [c] four bows; [d] two archers; [e] depiction of warfare at a fortified city
LINE DRAWINGS AFTER DELOUGAZ AND KANTOR 1996: PLS. 150–151

[Fig. 2.3a].[12] In both images the bows appear to be of the single-stave type (the all-wood or self-bow), which was the earliest and simplest to manufacture of the two main bow categories attested in the ancient Near East, the other being the composite reflexed bow.

An evolution from the simple single-stave type, the composite reflexed bow represents a *tour de force* of engineering, requiring access to specific materials, skilled craftsmen and knowledge derived from an established tradition of archery. Essential material components of the composite bow are a non-resinous wooden core fixed with an adhesive to a sinew backing to resist tension, a layer of horn facing the belly to resist compression, and bone tips.[13] Production of a composite bow was time-consuming: each layer of material had to thoroughly cure and bond together before the next could be added, a process that could last between one and five years.[14] An extremely effective weapon in skilled hands, the composite bow boasts up to twice the power of a simple wooden bow of the same draw weight, allowing far greater penetration and range, and use of heavier arrows.[15] Classical and Islamic sources,

12 Álvarez-Mon (2020) Pl. 14a,f.

13 Based on composite bows preserved in the tomb of Tutankhamun in Egypt (McLeod (1970) 31).

14 Álvarez-Mon forthcoming.

15 Klopsteg (1947) 90; Moorey (1986) 208–210; Miller, McEwen, and Bergman (1986) 182–87; Lorenz and Schrakamp (2011) 137; Genz (2013) 98.

supported by modern experiments, indicate that it could be fired quite accurately up to a distance of about 60 m, with an effective range extending as far as perhaps 200 m, but factors such as the structure and size of the bow, arrowhead type, skill of the archer, terrain, and climate all introduce variation into the equation.[16] The composite bow's smaller size and weight, and high capacity to withstand tensile and compressive forces without damage to the limb made it more transportable, more reliable, and able to be braced at the ready for long periods without losing strength.

Composite reflexed bows are notoriously difficult to study. They are seldom preserved in the archaeological record,[17] and while Mesopotamian texts mention bows (Sumerian GIŠ.BAN, Akkadian *qaštum*) from the mid-third millennium, they do not clearly discriminate between self-bows and composite bows.[18] This leaves only iconography, which can demonstrate the presence of composite bows in Susiana by the late fourth millennium BCE[19] through the combined evidence of a double-concave (or B-shape) profile when drawn [Fig. 2.5b][20] and arms curving away from the string when braced but not drawn [Figs. 2.4a–c, 5a].[21] Glyptic evidence from Chogha Mish and Susa dated to ca. 3800–3100 BCE displays bows alone in groups of four [Fig. 2.4c], bows held by striding individuals in a strung but disengaged position [Fig. 2.4a], and bows put to use in hunting [Fig. 2.5b]. Bows appearing alone have a reflexed body and bowstring secured to sharply curved terminals.[22] Under drawback tension the terminals are sometimes shown in an exaggerated circular form [Fig. 2.4d]. Where humans are present to give an idea of scale, the bows reach from the top of the head to about the knee [Fig. 2.4a, d],[23] though size is sometimes deliberately overstated [Fig. 2.4b].

16 McLeod (1965) 8; Lorenz and Schrakamp (2011) 144.

17 Álvarez-Mon (2023).

18 Civil (2003) 51; Schrakamp with refs. (2010) 154.

19 Delougaz and Kantor (1996) 146; Potts (2016) 64; and adopting a more precise methodology, Randall (2016) 55.

20 With the caveat that there is, as noted by Collon (1983, 53), 'a sufficient degree of recurvature, i.e., more than that which can be imparted by steam bending', since a single-piece wooden bow can also be given a curved or reflexed form by exposing it to steam/heat (curing) and soaking it in water, a practice perhaps attested later in texts from Mari (Wilke (1991)).

21 When unbraced, the composite reflexed bow bends in the opposite direction. Recognition of the composite bow variously informed by comments in McLeod (1958) 396; Yadin (1963) 8; Zutterman (2003) 121.

22 Amiet (1972) nos. 684, 687, 688, 689; Delougaz and Kantor (1996) Pls. 33H, 150A.

23 Amiet (1972) nos. 600–602, 606; Le Brun (1971) fig. 44, no. 2.

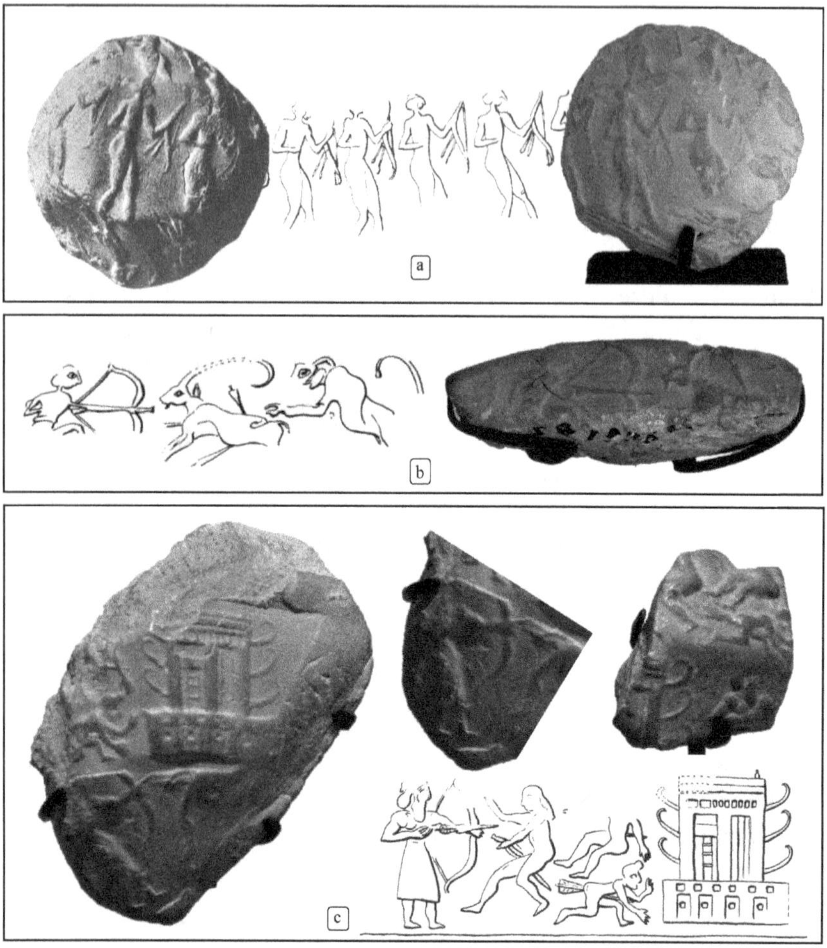

FIGURE 2.5 Composite bows in Susa sealing (ca. 3800–3100 BCE) [a] row of archers; [b] hunter with drawn bow; [c] 'priest king'
PHOTOGRAPHS © J. ÁLVAREZ-MON; LINE DRAWINGS AFTER P. AMIET

From the depictions of groups of disengaged composite bows at Chogha Mish, P. Delougaz and H. Kantor inferred the presence of an armoury.[24] The military application of a composite bow in a glyptic image from Susa dating to ca. 3800–3200 BCE illuminates a revolution in the history of weaponry and warfare. The image shows a 'priest-king' standing outside a monumental horned building aiming his drawn bow towards three naked figures who have been

24 Delougaz and Kantor (1996) 146; Álvarez-Mon (2020) 43, Pl. 18.

hit by his arrows [Fig. 2.5c].[25] This first image of warfare,[26] accompanied by an outburst of glyptic scenes of manufacture, processing, management, and storage of goods at Susa, must be contextualised within the rise of urbanism and social complexity, increased control of wealth by elites and religious institutions, and Susiana's participation in the Uruk-centred economic network.[27] It cannot go unnoticed that a similar 'priest-king' archer is shown hunting on a relief-carved boulder from Uruk (bow type undiscernible);[28] however, the direction of transfer of the iconography is unclear.[29] While a rich body of imagery indicates the manufacture and use of the composite bow by the late fourth millennium in Susiana, bows were still rarely depicted well into the third millennium in Mesopotamia, and there are no obvious examples of the composite type.[30] This inevitably raises the question of whether the composite bow had initially diffused from the highlands or foothill fringes of Susiana, or was even invented in Susiana thanks to its population's access to quality bow-making materials.

Following the collapse of the Uruk organization around 3100 BCE, changes in material culture and a new 'Proto-Elamite' written language attest to the rise of an entity that was asserting itself from Susa through the Iranian highlands and plateau. Now a wider array of weapons are attested: a biconical clay sling bullet type alongside the typical ovoid form,[31] copper/copper alloy projectile points for spears[32] and arrows with elongated triangular blades[33] or bifid points, which became popular through the third millennium [Fig. 2.6], flat axes,[34] elongated triangular spearheads, and new forms such as javelin heads, daggers, and knives with curved blades.[35] The diversity of weapons in copper and copper alloys must have been closely linked to exploitation of metal sources on the western central Iranian plateau at Arisman, which reached a high point of copper production on an industrial scale at around 3000 BCE [location in Fig. 2.1].[36]

25 Amiet (1972) no. 695; Álvarez-Mon (2020) 44, Pl. 19.

26 Another warfare-related scene from Chogha Mish depicts a large-scale ruler (?) returning by boat from a military expedition with two prisoners (Álvarez-Mon (2020) 41, Pl. 17b).

27 For which see Potts (2016) 64–66; Álvarez-Mon (2020) 34–58.

28 Image in Collon (2008) fig. 3.

29 Álvarez-Mon (2020) 45.

30 Collon (2008, 95), however, perceives them in Mesopotamian art already in the Uruk period.

31 Le Brun (1971) 195.

32 Le Brun (1971) fig. 67.2–3.

33 Tallon (1987a) 148; (1987b) nos. 222–24.

34 Tallon (1987b) nos. 426–33, 442.

35 Le Brun (1971) fig. 67.1, 2, 3, 6; Tallon (1987a) 320, fig. 51.

36 Helwing (2011) 529–31.

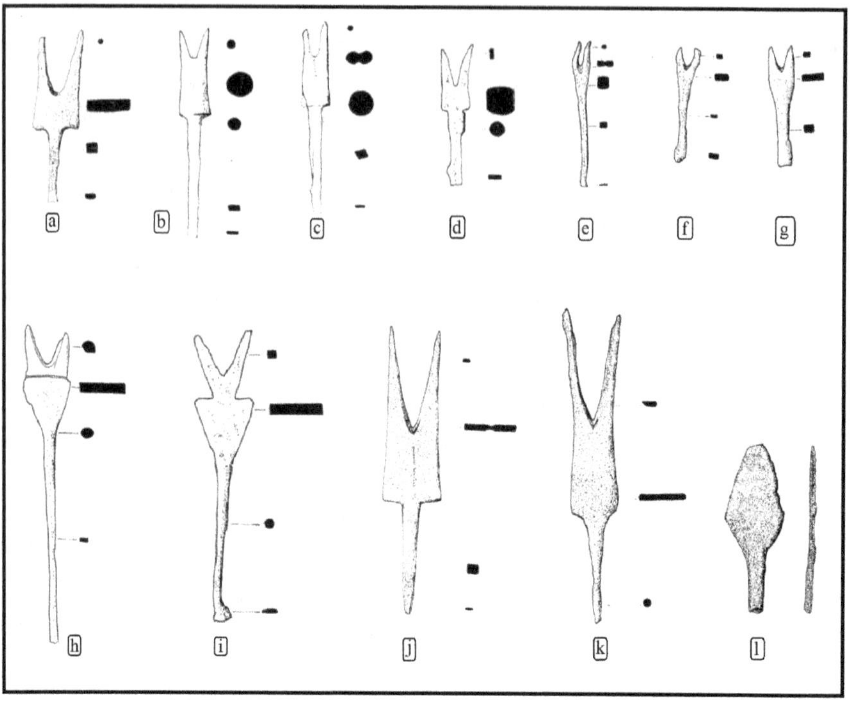

FIGURE 2.6 Third millennium BCE bifid copper/copper alloy arrowheads
AFTER TALLON 1987B [A] NO. 347; [B] NO. 360; [C] NO. 359; [D] NO. 357; [E] NO. 319; [F] NO. 308; [G] NO. 313; [H] NO. 336; [I] NO. 337; [J] NO. 352; [K] NO. 353; [L] NO. 218

2 Wars between East and West (ca. 2700–2004 BCE)

A literary-mythical composition known as the Sumerian King List (SKL) highlights Elam's involvement from around 2675 BCE in a millennia-long historically documented pattern of antagonism, exchanges, and alliances between polities of Mesopotamia and Susiana, and various other Zagros entities associated with Elam.[37] Experience with the bow, particularly its more developed composite version, and the technology and access to materials to make it, surely aided Elam in both its defensive and offensive military activities.

37 See Schrakamp (2010) for a comprehensive analysis of third millennium Mesopotamian warfare. Fragments of SKL copies from Susa published by V. Scheil (1939) 16–29. Discussion of the relationship between Susa and the highland polities can be found in Steinkeller (2018).

Iconography suggests that onager- or oxen-pulled carts or chariots must have been used for transport of people and weapons, shock attacks, and presumably high-platform projectile throwing.[38] Four wooden chariot wheels preserved in a grave at Susa (Donjon B280)[39] are close in construction to ED II (ca. 2700 BCE) chariots in burials at Kish,[40] suggesting shared technological know-how and elite warrior ideology at the time of the SKL events. The buried individual may have enjoyed a high military status, going by the vehicle and weapons in the grave: a lead hatchet, copper knife, cleaver, two axes, and two arsenical copper arrowheads of an unusual bifid form with a triangular base with V-shaped extension and very long tang found only at Susa [Fig. 2.6i].[41] Flint arrowheads had not been altogether discarded in favour of metal ones, going by a remarkable collection of flint arrowheads in a similarly dated burial from the same mound.[42]

Some scholars believe that the composite bow had accompanied the appearance of carts in Mesopotamian warfare. A siege warfare scene engraved on an Early Dynastic III plaque from the palace of Mari [Fig. 2.7],[43] the earliest depiction of its kind known from Mesopotamia,[44] shows a pair of helmeted soldiers, one holding a spear and a large shield, the other standing behind him about to engage a composite bow with the double-curved stave and reflexed tips and bowstring familiar from much earlier depictions in Susiana. With its long firing range, this bow type, which here appears ready to shoot a flaming arrow,[45] had potential to play a particularly key role in the execution of sieges, which depend heavily on long-range weapons on both the offensive and defensive sides.[46] As the plaque demonstrates, archery was most effective in warfare when units of spearmen and close-quarters infantry protected the archers and followed through after the arrow showers.[47]

38 Moorey (1986); Littauer and Crouwel (2002).

39 Made from three pieces of wood held together by two transversal boards and copper nails around the rims (Mecquenem (1943) 122, figs. 89.1–4, Pl. X; Tallon (1987a) 297–301, figs. 42–43; (1987b) 335, nos. 1290–1303).

40 Watelin and Langdon (1934) 30, fig. 3.

41 Mecquenem (1943) fig. 73.3; Tallon (1987a) 152, type A 2.

42 Mecquenem (1939) 9, Pl. 29, burial at −9.6 to −11 m.

43 Parrot (1971) 269, Pl. XIV, fig. 4.

44 Yadin (1963) 91–92.

45 Miller, McEwan and Bergman (1986) fig. 3, 191 perceive a fire arrow.

46 Genz (2020) 25, after Korfmannn (1986) 144.

47 Yadin (1963) 91–92.

FIGURE 2.7 Plaque from Mari
PHOTOGRAPH AFTER *VOUÉS À ISHTAR, SYRIE, JANVIER 1934, ANDRÉ PARROT DÉCOUVRE MARI*. EXPOSITION AU MUSÉE DE L'INSTITUT DU MONDE ARABE, 23 JANUARY–4 MAY 2014. PARIS 2014, PG. 125, III.6

Bows are still conspicuously absent, however, from the military equipment of the heavily armed Sumerian infantry on the Standard of Ur (ca. 2550 BCE) and the Stele of the Vultures from Lagash (ca. 2525 BCE). The latter's fragmentary inscription, which commemorates a (second) victory of Eanatum of Lagash (ca. 2450 BCE) over Umma during a border dispute, presents survival of an arrow wound as a motif of heroic character: 'towards Eanatum; a man shot an arrow; He was penetrated by the arrow; but he broke it off(?) ....'[48] The arrow is presumed to have been fired from the side of Umma, even if the list of adversaries 'obliterated by Lagash' was extensive, reaching as far afield as Susa, Elam, Mari, and Subartu. Even the later Sumerian armies of Urukagina of Lagash (ca. 2350 BCE) and Lugalzageshi of Umma (ca. 2358–2334 BCE) did not privilege the bow, being instead composed primarily of civilian conscripts (farmers, fishermen, etc.) allocated lances or single or double-edged shaft-hole axes and shields. Bows are only known to have been used by a small contingent of elite (professional?) 'arrow people' at Lagash who were directly answerable, it seems, to the palace.[49]

Immediately striking from the archaeological evidence is that Elam produced a far greater quantity of metal arrowheads than its adversaries who, insofar as the scant evidence suggests, must have used arrowheads made of organic materials such as bone and wood, and flint forms that are difficult to identify as arrowheads (e.g., transverse arrowheads).[50] In a database of metal weaponry from forty-eight sites in and around the Near East collated by G. Gernez, close to 35% of the arrowheads dating to ca. 2600–1800 BCE come from Susa, a site that also showed a comparatively low ratio of spearheads to arrowheads.[51] From this a higher reliance on archery at Susa can be inferred. Copper arrowheads are advantageous in that, despite requiring more expensive materials, they are quicker to produce than flint arrowheads and can be made with a very thin tang, facilitating easier fitting to the shaft.[52] A range of mainly bifid copper/copper alloy arrowheads appear around 3000 BCE and remained in use across the third millennium at Susa, often simultaneously [Fig. 2.6]. The bifid types, which are concentrated in the middle of the millennium, include a common short, flat, tanged version that disappeared in the early Akkadian period,[53] and two other distinct types: the abovementioned type unique to Susa,

48 Translation: composite text, 155–157': https://cdli.ucla.edu/search/search_results.php?CompositeNumber=Q001056 (accessed 6/4/2021).

49 Schrakamp (2010) 9, 11.

50 Collon (2008) 94; Genz (2020) 27–29.

51 Gernez (2007) 415–16.

52 Gernez (2007) 403; Gernez (2018) 69.

53 Tallon (1987a) 151–54, nos. 300–328, bifid type A 1.

characterised by a triangular base with V-shaped extension and long tang of circular or quadrangular section [Fig. 2.6i];[54] the other with a rectangular base extending into two points and short quadrangular-section tang [Fig. 2.6a–d, j–k].[55] Except for a single tin-bronze example of the rectangular-base type, all bifid arrowheads subject to chemical composition analyses were made of arsenical copper.[56] A rarer arrowhead form, which may also have originated in Iran, has a lozenge-shaped head, in one case with a slightly rhomboidal shape, and a thick, flat tang that tapers towards the end [Fig. 2.6l].[57] As R. Miller, E. McEwen and C. Bergman highlight, different styles of arrowhead were probably more a matter of taste than function, whereas variations in weight were linked to suitability for different bow types and to different tactics. A single military archer could carry arrowheads of different weights to enable powerful close shots with a heavier arrow and longer-distance shots with a lighter arrow to harass the enemy.[58]

When southwestern Iran came under the geopolitical control of Sargon of Akkad (2335–2279 BCE), a confederacy of Awanite/Elamite powers held the western part of the Iranian plateau as far as Anshan and Marhashi.[59] The possibility that archery played a critical role in Akkadian military success has been raised based on an increased presence of the bow in inventories of military equipment, and the appearance of a composite bow with a straight grip and double-curved limb in Naram-Sin's left hand in his famous victory stele [Fig. 2.8].[60] However, neither iconographic nor textual evidence point to a large contingent of archers in the Sargonic armies. Rather, the bulk of the non-professional army used lances, and sometimes axes, and were perhaps supported by an elite group of soldiers (*nisqū*) with archery skills under the command of a 'general of the archers'.[61] Inventory texts from Susa dated to the classic Sargonic period (late Naram-Sin to Sharkalisharri) refer to battle

54 Tallon (1987a) 151–54, nos. 329–337, bifid type A 2.
55 Tallon (1987a) 151–54, nos. 338–354, bifid type A 3.
56 Tallon (1987b) 37–40.
57 Tallon (1987b) no. 218–21, type A 1 a.
58 Miller, McEwen, and Bergman (1986) 189. A text from Mari records a single order of bronze arrows of three different weights for use in a siege: 50 × 40 g (heavy); 50 × 24 g, 100 × 16 g; 200 × 8 g (ultralight) (Dalley and Postgate (1984) 63; ARM 18 54).
59 Potts (2016) 92–101. Awan likely described roughly the same geographical area and even the same political organism as Elam at this time (Steinkeller (2018) 177).
60 Discussion in Schrakamp (2010) 3–4 n. 46–47. Below Naram-Sin, a quiver-bearing soldier holds a similar bow. Collon (2008, 96, fig. 8) identifies more composite bows on a fragmentary victory stele of Manishtusu (2269–2255 BCE) from Tello.
61 Evidence for this remains inconclusive (Abrahami (2008) 7, 14, and n. 67; Schrakamp (2010) 14).

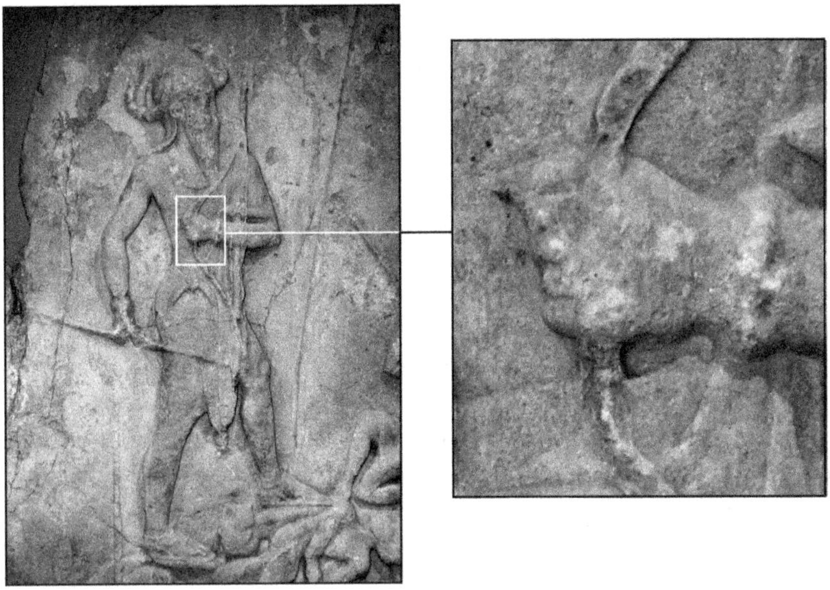

FIGURE 2.8 Depiction of Naram-Sin as archer in his victory stele; detail of left hand and bow
PHOTOGRAPHS © J. ÁLVAREZ-MON

equipment assigned to a group of Akkadian military officers (NU-bànda), including bows, arrows, and quivers,[62] as well as copper spears, axes, helmets in cowhide and bronze, helmets with silver bird decoration, and protective (textile) clothing,[63] which may have equipped an army of similar composition. However, it is not known whether Susa's socio-economic structures were comparable with those of the centres of southern Mesopotamia, and hence whether the makeup of the troops was the same.[64]

After the collapse of the Sargonic state rose the enigmatic leader Puzur-Inshushinak (ca. 2100 BCE), the 'first native ruler to unite most of Iran' and 'creator of the first Iranian empire'.[65] He sought to control the route, later referred to as the Great Khorasan road, connecting southern Babylonia to the Iranian Plateau and claims to have conquered 81 towns or regions over a large territory of Iran, incorporating Elam's two future capital cities, Susa and Anshan,

62 Quivers at this time held thirty arrows each (Abrahami (2008) 12).
63 Legrain (1913) nos. 85, 86; Scheil (1912). Recent translations and comments on the texts in Schrakamp (2010) 344–52; Sallaberger and Schrakamp (2015) 5, table 1.
64 Schrakamp (2010) 351.
65 Steinkeller (2018) 192.

into the Awanite kingdom. The extent of his conquests can be inferred from a text of Ur-Nammu (ca. 2112–2095 BCE), who confronted him and took over a number of towns in the Jebel Hamrin, Diyala and central Tigris, including Eshnunna (Tell Asmar) and Akkad, which Puzur-Inshushinak had held for at least eight years.[66] After his reign, Susiana returned to the Mesopotamian political orbit under Ur-Nammu, founder of the Ur III dynasty, whose successors sought alliances with the eastern rulers, sometimes via dynastic marriages.[67] But in 2004 BCE, the dynasty fell to the highland Elamite-Shimashki power that had been consolidating in the east. In both Mesopotamia and Elam this event marked a turning point: for the former, the end of the Sumerian renaissance; for the latter, the beginning of a long-lasting union between lowlands and highlands.

It is clear that the military use and symbolism of the bow and arrow had become increasingly important in the last quarter of the third millennium BCE,[68] and the persistence of the king as archer motif is demonstrated by a text in which Shulgi (ca. 2094–2046 BCE) boasts of acquiring the skill of archery at Dabrum and hunting wild boars with barbed arrowheads 'let free as a spitting snake',[69] a metaphor that brings to mind more vividly the popular bifid arrowheads of the third millennium [Fig. 2.6]. Various interactions between Elam and Mesopotamia are known to have resulted in the transfer of the 'Elamite (Anshanite) bow'. An 'Elamite Bow' appears in Ur III period administrative records documenting an arsenal at Puzrish-Dagan (modern Drehem) near Nippur,[70] and the Epic of Gilgamesh presents the hero of Uruk as an archer using the 'Anshan-style' bow, implying that the Elamite highland city of Anshan was a reputed centre of military expertise and craftsmanship during the Ur III period.[71] Considering that the composite bow was probably still rare in Mesopotamia, one could speculate that this 'Elamite (Anshanite) bow' was in fact a composite bow.

---

66 Potts (2016) 112–14, 116.

67 Shulgi (2094–2047 BCE) and Shu-Sin (ca. 2037–2029 BCE) each married a daughter to an Anshanite ruler, and Ibbi-Sin (ca. 2028–2004 BCE) married Tukin-hatti-migrisha, daughter of the governor of Zabshali (considered 'nuclear' Elamite territory; Potts (2016) 127, table 5.1).

68 Schrakamp (2010) 166.

69 Vermaak (1993) 13–15; for the location of Dabrum on the Tigris, northeast of Umma, see Powell (1980) 51.

70 Louvre Museum, TCL 02, 5488; trans. Schrakamp (2010) 163.

71 Fleming and Milstein (2010) 148, Col. VI. 240–242. The text dates to the early second millennium BCE, but has Ur III roots (Vanstiphout (2003) 1; Michalowski (2003)). In the early versions, the journey to the 'cedar' forest passes through Anshan and Aratta suggesting an acquaintance with these eastern locations (Tigay (2002) 12, n. 40; Fleming and Milstein (2010) 9).

3 The Fall of Ur and Elamite Siege Warfare (2004–1760s BCE)

The dramatic fall of the Ur III Empire in ca. 2004 BCE and subsequent Elamite occupation of Mesopotamia effected by Kindattu, the sixth ruler of the Shimashkian dynasty, ushered in one of Elam's most conspicuous apogees of power. From the Sumerian literary account of events, *The Lamentation over the Destruction of Sumer and Ur*, we can gather that in the last years of the Ur III period, Ibbi-Sin (ca. 2028–2004 BCE) declined to confront the Elamite military in open battle – instead placing his faith in the defensive capacity of the city walls of Ur. Thus Elam's success rested on skill in planning and implementing siege warfare in order to starve out the inhabitants and capture Ibbi-Sin.[72] An Elamite 'storm' gathered outside the city wall, ready 'to strike like a flood'. 'Large axes were sharpened in front of Ur, the spears, the arms of battle, were being launched, the large bows, javelins, and siege-shields gathered together to strike, the barbed arrows covered its outer side like a raining cloud, large stones, one after another, fell with great thuds'.[73] Employment of slingers in conjunction with archers became a typical strategy of Near Eastern warfare,[74] and here the latter use 'large', presumably composite, bows capable of powerful far-reaching shots, and barbed arrows, moulds for which are known from early second millennium BCE contexts at Susa [Fig. 2.9f, g]. Though certainly manufactured at Susa, just one has been documented in its archaeological record.[75] The seventy-plus late third/early second millennium copper/copper alloy arrowheads from Susa kept in the Louvre Museum are mainly typified by a squat or more elongated triangular blade and a flat tang [Fig. 2.10].[76] Chlorite and sandstone moulds (mostly bivalve) for these arrowheads are also attested at Susa [Fig. 2.9a–e].[77] An increasing reliance on tin-bronze rather than arsenical copper is demonstrated by composition analyses of a sample of thirty-three of these arrowheads: twenty-three contained more than 1% tin, and almost half are certainly deliberate copper-tin alloys with an average of 10% tin.[78] A recently discovered tin-copper mine at Deh-Hosein on the northeast border of Luristan may have played a critical role in the supply of tin for weapons and other objects at this time and later [location

72 In *The Lament for Sumer and Urim*, Ibbi-Sin is taken to Anshan in fetters and never returns (https://etcsl.orinst.ox.ac.uk/cgi-bin/etcsl.cgi?text=t.2.2.3#, accessed 19/03/2021).

73 After Michalowski (1989) lines 382–86.

74 Genz (2020) 26.

75 Tallon (1987a) 149, variant A 2 d.

76 Tallon (1987a) 148–49, variants A 2 a–c; (1987b) nos. 225–99. Almost 93% of the corpus of these arrowheads in the Louvre Museum is from Susa.

77 Tallon (1987a) 151; (1987b) 132–33, II, IV–XIII, XV–XVI.

78 Tallon (1987a) 149.

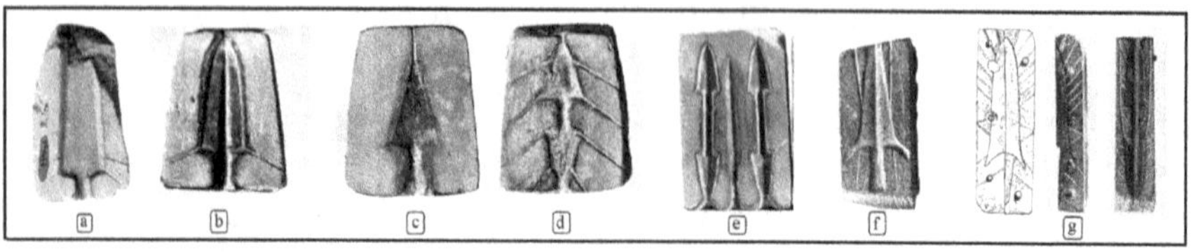

FIGURE 2.9 Chlorite and sandstone arrowhead moulds (early second millennium BCE) from Susa. [a–e] triangular blade, ranging from squat to elongated, and a flat tang; [f, g] barbed arrows
AFTER TALLON 1987B [A] PG. 345, II; [B] PG. 346, VI; [C] PG. 346, IIIA; [D] PG. 346, IIIB; [E] PG. 350, XVI; [F] PG. 348, XI; [G] PG. 347, VIII

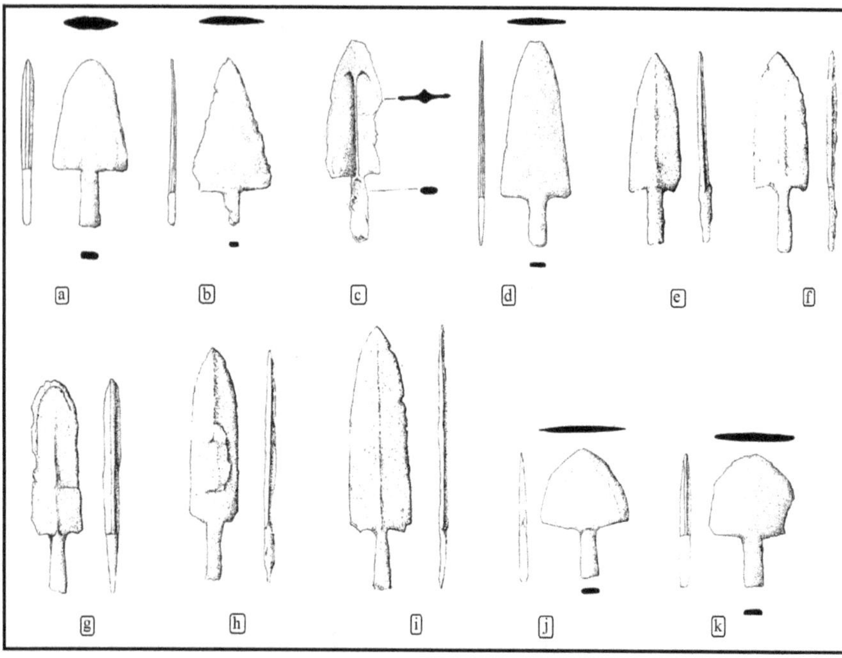

FIGURE 2.10 Arrowheads (second millennium BCE) from Susa
AFTER TALLON 1987B [A] NO. 255; [B] NO. 256; [C] NO. 290; [D] NO. 271; [E] NO. 292; [F] NO. 294, BOTTOM (LEFT TO RIGHT); [G] NO. 296; [H] NO. 297; [I] NO. 298; [J] NO. 230; [K] NO. 233

in Fig. 2.1].[79] The significant number of metal arrowheads at Susa contrasts with Mesopotamia, where they were still quite rare.[80] Some were collected from terracotta coffin burials, several of which included other weapons such

79 Weeks (2008) 336; Nezafati, Pernicka and Momenzadeh (2009).
80 Flint arrowheads were predominant until 1700–1600 BCE (Gernez (2018) 69).

as daggers and shaft-hole axes, perhaps marking a military office linked to socio-economic status.[81]

Siege warfare continued into the second millennium as the strategy *par excellence*. Prolonged sieges were time- and resource-demanding, and their success relied on the effective blockade and assault of the city's defensive walls, beginning with the construction of a packed-earth siege ramp rising towards the top of the city wall. This initial phase required 'headpad' troops to move baskets of earth. Freestanding timber siege towers were then erected in-situ atop the ramp for use as firing platforms, providing a vantage point from which to fire down upon defending soldiers on the city's battlements. Fire-proofing would have been a necessary step if fire arrows were used.[82] Since an important element of siege success was the ability to shower opponents with missiles,[83] archers would have played a leading role, positioned on and/or behind the towers, and Elam must have gained an edge from its access to metal sources to quickly produce large numbers of arrows. Undoubtedly a strong contingent of archers and archery equipment was key to the military strategy that enabled Elam to capture Ur, followed closely by conquests of Eridu, Girsu, Kish, Lagash, Uruk and Nippur, and to occupy part of lower Mesopotamia for several years.[84] The available sources lack information on Elam's recruitment of archers, but since use of the bow and in particular the powerful composite bow required several years of training and experience,[85] they may have been semi-professionals, or conscripts from the citizenry whose archery skills were part of their education or way of life, such as shepherds who needed to protect their flocks, as is attested in Mesopotamia.[86]

Snippets of documentation illustrate Elam's continued involvement in Mesopotamian affairs over the next century or so. Interactions alternated between marriage alliances with, and attacks on, Elam's highland capital of Anshan,[87] and around 1835 BCE a new ruling dynasty of Elamite origin was established

81 Donjon A14, A16, B70 (Mecquenem (1943) 77, 112); Ville Royale B VI (Ghirshman (1968) 7–8; Gasche (2000) 209–12, S. 179).

82 Fire arrows could be used to set alight buildings and fuel sources (Miller, McEwan, and Bergman (1986) 183, 190–91).

83 Genz (2020) 29.

84 Steinkeller (2018) 196 estimates 'some 20 years'.

85 E.g., Lorenz and Schrakamp (2011) 137.

86 Patterson (2018) 315–18.

87 Partnership with Isin, now a dominant power in southern Mesopotamia, is hinted at by the king of Anshan's marriage to Matumniatum, daughter of the king of Isin, Iddin-Dagan (ca. 1974–1954 BCE) (reconstructed after year names of kings of Lagash, http://cdli.ox.ac.uk/wiki/doku.php?id=year_names, accessed 26/10/2019). Gugunum of Larsa (ca. 1932–1906 BCE) claims to have destroyed Anshan in ca. 1927 BCE.

at Larsa by Kudur-mabuk (ca. 1850–1830 BCE).[88] Elam's continued military capacity is evidenced by a letter of 1785 BCE documenting the commitment of a remarkable 12,000 troops by the Elamite Sukkalmah, Siruktuh, to an alliance between Elam, Assyria, Eshnunna and perhaps the Turukkeans from the Zagros to fight the Gutians.[89] The Mari archives document an eventful period of Elamite relations with its western neighbours between 1771 and 1764 BCE, unveiling the authority of the Elamite Sukkalmah above any other ruler, including Zimri-Lim of Mari (ca. 1775–1762) and Hammurabi of Babylon (1792–1750 BCE).[90] He appears as orchestrator of diplomacy between Ashur, Babylon, Eshnunna, Larsa, Mari, Halab (Aleppo) and Qatna – seven kingdoms of rapidly shifting alliances who shared power in the wake of the death of Shamshi-Adad of Assyria. Under Siwe-palar-huppak (Akkadian Sheplarpak) – 'king of Anshan', 'Great King', Sukkalmah ('Grand Vizier'), and Sukkal of Elam – a viceroy named Kutu-zulush (Akkadian Kudushulush) was placed to manage the western lowland territories. This imperial pattern of political power has elicited comparisons with the Persian Empire and fostered recognition of the Elamite ruler as an 'emperor'.[91] In the context of Elamite diplomacy the bow enjoyed a preeminent symbolic role; for example, one letter reports that the Elamite emperor had agreed to present a bow as a votive gift to the god Addu (Adad) of Aleppo as a symbol of the establishment of his alliance with the western kingdom of Yahmad.[92]

The political strategy of the Elamite emperor consisted in exploiting regional divisions and engaging proxy rulers and armies of vassal states for its military campaigns. His success was facilitated by access to the abundant resources of the Iranian plateau and control of trade routes, including those supplying tin, enabling him to choke Mesopotamian tin-bronze production at will.[93] In 1765 BCE, the Elamite emperor commanded troops from Mari and Babylon to

88 Sulaiman and Dalley (2012) 155. Both Rim-Sin and his father, Shemti-Shilhak, bore Elamite names.

89 Laessoe (1965) 194–95; Eidem (1985) 90.

90 Abrahami (1997); Charpin (2012) 44–53.

91 Charpin and Durand (1991) 66.

92 Charpin (2013) 349–51.

93 A short supply of tin, over which Elam exerted its control, directly impacted military equipment production, as exemplified by a text of Shamshi-Adad reducing an order of 10,000 bronze arrowheads to 5,000 due to a bronze shortage (Durand (1998) 393–94, no. 663).

lay siege to Eshnunna, then the strongest kingdom in Mesopotamia.[94] Tactical information can be gleaned from a letter ordering Hammurabi to secure the 'troops of the headpad' to raise an assault ramp. Conquest of Eshnunna opened the door to the west and the Elamite expeditionary forces were divided into a northern army and a southern army. This capacity to act simultaneously on two fronts unveils both the extent of the emperor's ambitions and his ability to muster substantial troops.

The northern Elamite army, headed by a general named Kunnam and bolstered by newly acquired vassal troops from Eshnunna and Gutium, laid siege to and conquered Subat-Enlil, former capital of Shamshi-Adad and capital of the kingdom of Apum [location Fig. 2.1]. Another successful siege at Ramaza, the capital of the kingdom of Yussan on the plain to the east of the Sinjar Mountains,[95] was led by Atamrum, king of Allahad and Andarig, backed by troops from Elam and Eshnunna. Details of the siege have been differently translated and interpreted, but include the 'heaping up' of a ramp reaching the base of the wall of the lower city and surprise exit attacks by defenders who breached or tunnelled under the city walls and captured soldiers armed with bronze lances and shields.[96] Elam's swift conquest over the northern Mesopotamian regional powers that had emerged after the death of Shamshi-Adad brought the flourishing Old Assyrian-Anatolian trade network under Elamite control.[97] As the Elamite army 'devoured' all of Shubartum (i.e., northeastern Mesopotamia between the Tigris and the Habur triangle) and several of the kingdoms in the region became Elamite vassals, Mari's independence hung in the balance.[98]

Meanwhile, the southern army had plotted its way towards Babylon, laying siege to and capturing Upi (Opis), a strategic border fortress near the confluence of the Diyala and Tigris rivers that Hammurabi had annexed upon Eshnunna's defeat [location Fig. 2.1]. In 1764 BCE, about 40,000 troops were on the march towards Babylon, but their advance was blocked at the fortress of Hiritum on the Irnina channel linking the Tigris and Euphrates rivers. The Elamites laid siege to the garrison, raising a packed-earth assault ramp for two siege towers, but they lost the first tower to fire (suggesting involvement of

94 Elam's dominance over the ruler Ishme-Dagan, son of Shamshi-Adad, is also attested (Charpin (1986) 133, A.428: 21–28; Charpin and Ziegler (2003) 247, 3.7. Annex I; Heimpel (2003) Appendix 6).

95 Heimpel (2003) 32, n. 99.

96 Charpin and Ziegler (2003) 216; Heimpel (2003) 65.

97 Larsen (2015).

98 Charpin and Ziegler (2003) 217. For S(h)ubartu(m) see Bryce (2009) 663–64.

fire arrows) and abandoned the effort when one side of the ramp started to wash away, preventing the second assault tower from reaching the city wall.[99] The Sukkal of Elam called the assault to a halt, gathered the army and headed upstream along the Tigris to Mankisum but, for reasons unknown, turned around towards Eshnunna, looted it, and concluded the campaign. Elam then withdrew from direct involvement in the affairs of Mesopotamia, leaving the geo-political field open to Hammurabi's ambitions, and diplomatic relations between the two powers were swiftly re-established.[100] Afterwards, there are mere snapshots of individuals from Elamite backgrounds in the military corps of Sippar in Hammurabi's last years (ca. 1750 BCE) and a possible military clash between Elam and Babylon in ca. 1706–5 BCE at the time of Hammurabi's grandson, Abieshuh, involving the Sukkalmah Kutir Nahhunte I, second successor of Siwe-palar-huppak.[101] Elamite mercenaries are later recorded serving under the last Amorite ruler, Samsu-ditana (1625–1595 BCE), but there are no signs of direct involvement of the Elamite state.[102]

4 Chariot Revolution and Renewed Elamite Hegemony (ca. 1760s–1100 BCE)

Texts from the Mari archive clearly attest for the first time to the manufacture of the composite bow (*tilpânu*) as distinct from the straight bow (*qashtum*), the former distinguished by the inclusion of bone in its construction (wood, glue, and tendons are common to both).[103] Soon after, archery, and in particular the composite bow and metal arrowheads,[104] assumed a preeminent role in Late Bronze Age Near Eastern warfare. These changes can be seen in

99 Lacambre (1997).

100 Heimpel (2003) 108, 460–61, nos. 27 149, 27 150.

101 Scheil (1932) 75–76; van Koppen (2013) 380–81.

102 Lambert (2007) 25, 144; Beaulieu (2018) 118.

103 Abrahami (1997); Durand (1998) 387–93; Arkhipov (2012) 105–106; but see cautionary remarks by Wilke (1991). Schrakamp (2010, 165, fn. 1062) argues for earlier attestations of the composite bow in the 'artisan archives' at Isin (reigns of Ishbi-Erra 2017–1985 BCE and Shu-Ilishu 1984–1975 BCE) based on the use of glue in their production; however, the critical element of bone is missing.

104 Lorenz and Schrakamp (2011) 137; Genz (2013) 98. Note, however, that flint arrowheads still appear in mid/late second millennium Elamite contexts at Tell-i Ghazir (in a grave), Bard-e Kargar (in a sanctuary) (Alizadeh (2014) 18) and Chogha Zanbil (various locations/contexts) (Ghirshman 1966, Pls. LXVI, G.T.Z.58, LXXII, G.T.Z.373; 1968, Pl. LXXVII, G.T.Z.952, LXXXVII, G.T.Z.737).

conjunction with the rise of the light, two-wheeled, horse-drawn chariot with spoked wheels, which revolutionised warfare upon its introduction into Near Eastern armies around 1700 BCE.[105] Put to action as a military vehicle, the chariot offered a dynamic mobile firing platform for archers, and therefore armies competent with the composite bow were well-positioned to further enhance their formidable fire-power.[106] The powerful composite bow gave archers two options: to deliver a lightweight projectile over a long distance or a projectile of greater weight at short range with the capacity to pierce armour.[107] The invention of scale armour, which offered the highest level of protection for chariot-borne archers while still offering flexibility for movement of the arms, was closely associated with the light chariot and chariot tactics [Fig. 2.11a].[108]

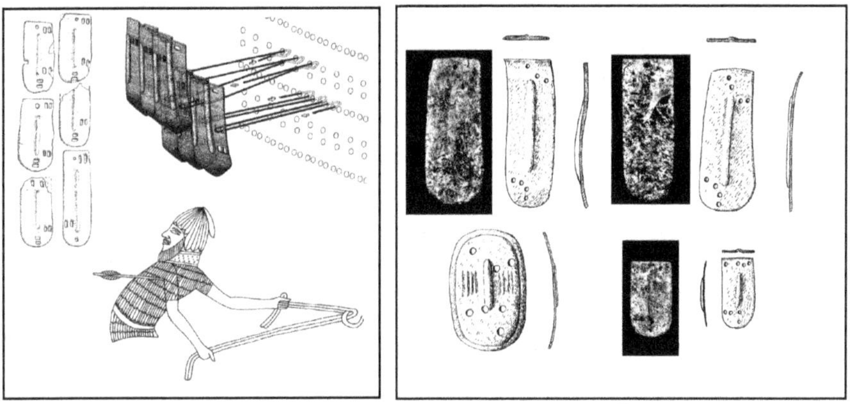

a b

FIGURE 2.11 Scale armour [a] armour reconstruction (line drawings after Lorenz and Schrakamp 2011: 129, figs. 1–3); [b] bronze scale armour from Chogha Zanbil (photographs after Ghirshman 1966, Pl. 55.4a,c, Pl. 93 GTZ213, 214)

105 Littauer and Crouwel (1979) 74–81; Moorey (1986) 197. Early second millennium copper/bronze tyre segments attest to wheeled vehicles at Susa (Mecquenem (1943) 89–90, fig. 74, Pl. x; (1922b) 137–38, fig. 14), but they probably belonged to much heavier wooden single-piece or tripartite disk wheels. Comparable tyres appear in 'royal' burials at Gonur Depe (ca. 2200–1700 BCE) in Turkmenistan (Sarianidi (2005) 240, fig. 99; Lamberg-Karlovsky (2013) 21, 33, fig. 9a), perhaps native territory of the Shimashki (Potts (2008)).

106 Moorey (1986) 209–10; Drews (1993) 119–26; Genz (2013). The bow was the primary weapon of the elite Hurrian *maryannu* chariot warriors (Dezsö (2002) 195), and Nuzi's chariot crews were equipped with bows, quivers, and arrows, as well as whips, helmets, body armour, and swords (Kendall (1975) 210–13).

107 McEwen (1978) 189.

108 Kendall (1975) 276–77; Dezsö (2002) 195–96.

Besides critical reliance on archers skilled in the use of the composite-bow, chariot warfare was expensive and time-consuming, necessitating substantial resources and technological know-how to build vehicles, breed and train teams of horses and drivers, and manufacture protective gear.[109] Therefore, even if Elam's political status within the Near East in the mid-second millennium BCE is still poorly understood, the maintenance of its military capacity is known from administrative texts from Kabnak (modern Haft Tepe) near Susa attesting to an on-site production and storage of military vehicles and military equipment in ca. 1400 BCE.[110] The texts list archery paraphernalia including bows, arrows and quivers with capacities of twenty and thirty arrows, and excavations at the site have yielded metal arrowheads in lanceolate and triangular forms with projecting midribs [Fig. 2.14b].[111] Scale armour plates (*kurzindu*) also appear in the texts,[112] and although none have been found at Haft Tepe itself, a small ca. fourteenth/thirteenth-century BCE sample came from neighbouring Chogha Zanbil [Fig. 2.11].[113] This production can be seen in context with external threats around 1400 BCE implied by a year name of the Elamite Kidinuid dynast, Tepti-Ahar, signalling that he had repelled an invasion from Babylonia.

Pahhir-Ishshan, son of the subsequent dynasty's founder, Igi-halki, married the eldest daughter of the Middle-Babylonian Kassite king Kurigalzu I (ca. 1400–1369 BCE) ushering in an era of diplomatic marriages between Elamite princes and Kassite princesses. Yet despite these unions – or perhaps because of them – animosity flared up between the Elamite and Kassite ruling houses, as documented by at least one episode of hostility during the time of Untash-Napirisha (1340–1300 BCE).[114] The leitmotif of Elamite interventionism in Mesopotamia truly returned to the fore under Kiddin-Hutran II (ca. 1240–1210 BCE). Believed to be the author of the so-called 'Berlin letter', and hence a descendant of Kurigalzu with a perceived claim over the Babylonian throne, he inaugurated a century of unremitting invasions into territories to Elam's west.[115] Assyrian ambitions in the south were not extraneous to Elamite-Babylonian dynamics. Indeed Kiddin-Hutran's first attested campaign may have been a response to Tukulti-Ninurta I's (1240–1205 BCE) overthrow of the Kassite king Kashtiliashu IV (1232–1225 BCE) and installation of a puppet ruler,

109 Deszö (2002); Genz (2013) 97, 101.
110 Álvarez-Mon and Wicks (2021).
111 Copper and tin-bronze arrowheads have been identified (Oudbashi et al. (2019), table 2).
112 Herrero and Glassner (1990) nos. 67, 68, 69.
113 Ghirshman (1966) Pl. 55.4a,c, Pl. 93 GTZ213, 214.
114 Paulus (2013) 439.
115 Potts (2016) 224, table 7.6 and Roaf (2017) 183, 189., sources 2, 3 and 6.

Enlil-nadin-shumi (1224 BCE).[116] Kiddin-Hutran removed Enlil-nadin-shumi and captured Der and Nippur, and in a second campaign destroyed Isin and Marad (west of Nippur) and overthrew another Assyrian appointee, Adad-shuma-iddina (1222–1217 BCE). His military strategy is not clarified by the source on these campaigns, the Babylonian Chronicle P;[117] but if the Berlin letter is correctly assigned to him, attacks on settlements rather than battlefield encounters can be inferred: 'I will demolish your cities, destroy your ramparts, I will fill in your moats, [I will uproot] your orchards, [I will fix (?)] locks in the mouths of your canals'.[118]

Elamite military hegemony over Mesopotamia was consolidated through conquests of a considerable number of cities by the first three rulers of the subsequent Shutrukid House, Shutruk-Nahhunte (ca. 1184–1155 BCE) and his sons Kutir-Nahhunte (ca. 1155–1150 BCE) and Shilhak-Inshushinak (ca. 1150–1120 BCE). The success of the Shutrukids' western campaigns, which encroached into Assyrian territory, suggests an unrivalled military capacity. Fragments of one or more inscribed stelae recovered from Susa seem to speak of Shutruk-Nahhunte's conquest of hundreds of towns,[119] and a short-lived period of Elamite control of Babylonia ensued in 1158 BCE when, according to a Neo-Assyrian copy of a text attributed to Nebuchadnezzar I, Shutruk-Nahhunte 'be[stowed royal authority] upon his eldest son' Kutir-Nahhunte (ca. 1155–1150 BCE).[120]

In the year he succeeded his father, Kutir-Nahhunte, who repeated Kiddin-Hutran II's earlier claims to the Babylonian throne, immediately launched his own successful Mesopotamian campaign(s).[121] Tactically, the only information we have is the swift movement of troops on chariots to launch attacks on cities including Nippur and Borsippa and the capture of Babylon and the Esagila: '[...] he sent like the deluge, all the cult centers of Akkad and their sanctuaries he burned [with fi]re'.[122] Kutir-Nahhunte deported Enlil-nadin-ahi and the Marduk statue of Esagila to Elam, toppling the three-centuries-old Kassite dynasty and ensuring his own infamy in Babylonian memory. A remarkable

116 Paulus (2013) 437; Potts (2016) 224.

117 Grayson (1975) 56.

118 After Roaf (2017) 185.

119 Scheil (1911) 17–20, no. XCI, fragments A–E; EKI 28.

120 Tablet K.2660, translation after Foster (1996) 295, line 2. Tadmor (1958, 138–39, obv. line 3) translates: '[handed] over [the reign] to his first-born Kudur-nanhundi'. The status of Babylon remains uncertain, but Tadmor restored a later line in the text (obv. line 14) as '[He installed a governor,] not of Babylonian descent, an enemy [of Marduk]'; Foster avoids restoration and suggests '[ ] not of Babylonian descent, hostile [to Marduk]'.

121 Foster (1996) 284.

122 Lambert (1994) 67; Foster (1996) 283–88.

fragmentary tablet of much later date preserves details of a ritual performed in the temple of Ishtar of Babylon, which commenced with a petition to Shamash describing an Elamite attack, probably at the end of the Kassite period.[123] It emphasises the Babylonians' helplessness against Elam's archers: 'The foe captured us. His bow was nocked to let the arrows fly. But we ourselves, we did not know how to grasp a quiver. Elam overwhelmed our sacred localities'.[124]

Elamite forces may have occupied parts of Mesopotamia for some time afterwards,[125] but mentions of campaigning against Babylon (Karintash), Akkad, Sippar, and Nippur by Kutir-Nahhunte's successor and brother, Shilhak-Inshushinak (1150–1120 BCE), imply that these areas presented ongoing challenges to Elamite control. A series of Shilhak-Inshushinak's inscriptions highlight a reign dedicated to a strategic expansion of the Elamite geo-political frontier, first consolidating the gains made by his father and brother in central Mesopotamia and to the north in the Zagros, then expanding into northern Mesopotamia all the way to Nuzi and Arrapha (Kirkuk) and into the heart of Assyria where he may have clashed with the Assyrian king Ashur-dan I (1179–1134 BCE) [location Fig. 2.1].[126] One inscription enumerates a series of perhaps 250 or more conquered settlements (up to half of the toponyms illegible),[127] structured into eight sections, which might reflect either a sequence of military campaigns or an administrative division of the conquered settlements into zones or vassal provinces.[128]

Elamite confrontation with Mesopotamia continued under Hutelutush-Inshushinak (ca. 1120–? BCE), bringing the first evidence for a Shutrukid clash in

---

123 George (2000) 272, 278.

124 The continued reputation of Elamite archers is also suggested by slightly earlier documents of ca. 1220 BCE (reign of Tukulti-Ninurta I) from Tell Chuera in North Syria listing rations for Elamite bowmen and their families (Jakob (2009) nos. 40, 46, and possibly 64, 69, 70–73, 77; Postgate (2013) 19–20, 87–89).

125 Brinkman (1968) 466; Beaulieu (2018) 154.

126 Various inscriptions on stele fragments from Susa probably refer to campaigns of this king (Scheil (1904) nos. 72 = EKI 51, 73 = EKI 54b, 74 = EKI 54a, 75 = EKI 68 [author less certain] and 76 = EKI 49; Scheil (1911) no. 92 = EKI 54). Additional unpublished inscriptions on stone door-sockets mention his campaigns in Akkad, Sippar, Nippur, the Kirkuk area, and the Lower Zab (Steve, Gasche and De Meyer (1980) 80–82; Steve, Vallat, and Gasche (2002) 466).

127 See Scheil (1911) 21–57, no. 92; EKI 54. Two stelae fragments published by Scheil may belong to this or similar stelae (Scheil (1904) nos. 73, 74). See also Potts (2016) 235–38; Table 7:11.

128 Labat (1975) 489. Middle Assyrian texts from Ashur-dan I's successor, Ninurta-tukul-Ashur (1133 BCE), attest to production of composite bows using sinew, glue, *kisanû* woods (poplar, ash or maple?) and ibex horns (Frahm (2002) 75–80).

the open battlefield. A tablet from Nineveh recounts a failed military foray into Susiana, probably by the Babylonian king Nebuchadnezzar I (1121–1100 BCE). The ruler waited with his troops 'at the head(?) of the Uqnu-river' seeking to confront the Elamite army, but plague overtook his warriors and horses, forcing his retreat upon the Elamites' advance.[129] An inscription on a kudurru of Nebuchadnezzar I (known as the Shitti-Marduk stele) from Sippar[130] and a text from Babylon[131] appear to describe the same event, but place it on the 'bank of the Ulai river' instead of the Uqnu.[132] The kudurru details the launch of a Babylonian surprise attack against Elam in the month of Tammuz (June–July), the height of summer, taking advantage of the legendary scorching heat of Khuzestan to take the Elamites off-guard. Together with the chariot commander Shitti-Marduk, the king charged from Der in the direction of the river Ulai on a journey said to have taken thirty double hours.[133] Hutelutush-Inshushinak's defensive strategy relied on the guarding of watering places, presumably with strategically positioned infantry, and avoiding any delays, Nebuchadnezzar pushed his troops forward 'by night (and) d[ay]' via waterless routes renouncing rest: 'I did not give (them) water to drink or allow them (time) to recover from their fatigue'.[134] The armies clashed on the riverbank: 'fire ignited between them. The sun's face was obscured by their dust'.[135] Hutelutush-Inshushinak is said to have retreated and vanished, and Nebuchadnezzar I claims to have 'seized Elam and looted its property' and retrieved a statue of Marduk from Susa. Most commentators therefore assume that he must have conquered the Elamite capital;[136] however, there is no evidence of a twelfth-century destruction of the site or any long-term Babylonian control.[137]

129 Foster (1996) 296, tablet K. 2660.
130 Nielsen (2018) 27, e.
131 Nielsen (2018) 29, f; Foster (1996) 386.
132 Identification of the location(s) of the Uqnu and/or Ulai rivers is fraught with difficulties, in no small part due to meandering river behaviour, fluvial processes, and intertidal coastal and marshland fluctuations of fresh and salt water. The Uqnu has been identified variously with the Karkhe (Foster (1996) 296); the lower Karun (Bagg (2000) 65; Cole and Gasche (2007) 7); the eastern branch of the Tigris in Khuzestan (Fuchs (1994) 459; Lipinski (2000) 434; Potts (2016) 253); and as another name for the Ulai, resulting in the united streams of the Karkhe, Dez and Karun rivers as they merged in the vicinity of Ahwaz (Cole and Gasche (2007) 30).
133 The distance between Der (Tell Aqar) and Susa in a straight line is about 250 km.
134 Bloch (2017) 511–12.
135 Translation after Cooley (2006); see also Beaulieu (2018) 160; Nielsen (2018) 54–55, 161.
136 Nielsen (2018) 64.
137 Miroschedji (1981) 36–37; Stolper (1984) 43.

FIGURE 2.12 Elamite royal family at Shekaft-e Salman II
PHOTOGRAPH AND LINE DRAWING © J. ÁLVAREZ-MON

We know little of the role of archery in enabling the Shutrukids to achieve their political ends, but the (composite) bow and arrow did go hand-in-hand with chariot warfare, and the language of archery deployed in a stele inscription of Shilhak-Inshushinak provides a quintessential metaphor for peace and prosperity: 'The quiver is laid down (held down), may the country harvest be brought in, may peacetime be, may abundance of blessings spread'.[138] The ideology of the worshipping ruler as archer also appears to have been integrated into Shutrukid propaganda going by the forearm-guard straps (or bracers) – probably accoutrements related to archery – depicted on the twelfth-century Elamite ruler in a relief (no. II) in the open-air sanctuary of Shekaft-e Salman in Izeh in the Elamite highlands [Fig. 2.12].[139] Some centuries later, a bronze plaque from the Acropole at Susa dating to around the ninth–eighth century BCE shows seven divine composite bow-carrying warriors with a similar series of straps wrapped around the left arm and wrist and a large thumb, probably also representing protective devices [Fig. 2.15].[140]

In the lowlands, small (composite?) bows with bent terminals are depicted in cultic and hunting contexts on a Middle-Elamite glyptic at Chogha Zanbil [Fig. 2.13],[141] where archaeological investigations have uncovered deposits of copper/bronze arrows in a range of different types, including lanceolate and pyramidal forms, together with other military equipment in the temple of Kiririsha East and in the so-called 'hypogeum palace' tomb complex

138 EKI 54, §15, §72.
139 Álvarez-Mon (2019) 37, Pl. 22.
140 Álvarez-Mon (2015).
141 Álvarez-Mon (2020) Pl. 124a, e, f.

FIGURE 2.13 Seals and sealings from Chogha Zanbil
PHOTOGRAPH © J. ÁLVAREZ-MON

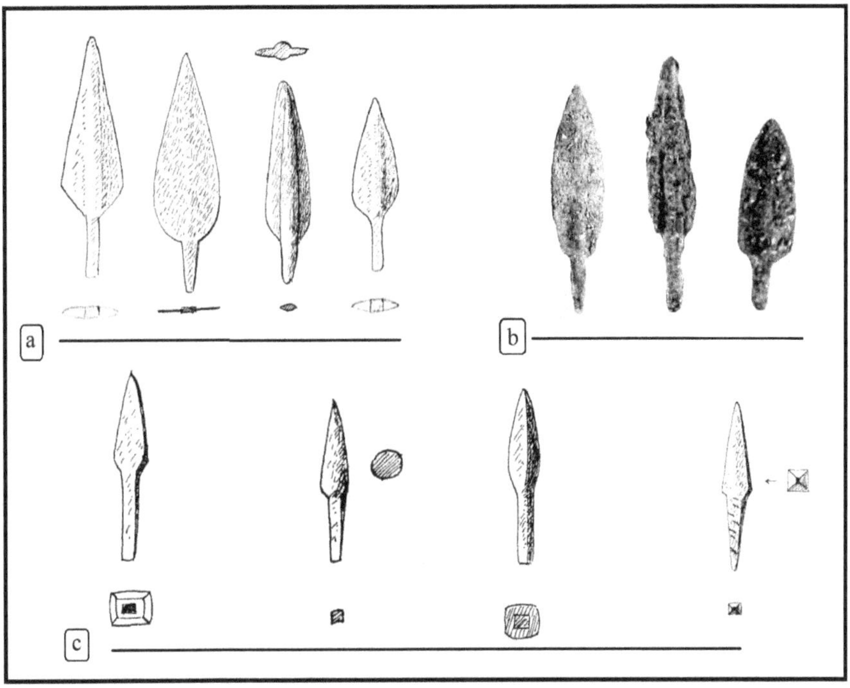

FIGURE 2.14 Bronze arrowheads [a] and [c] from Chogha Zanbil (after Ghirshman 1966, Pl. XCIII, G.T.Z.204–211; 1968, Pls. LXXXV, G.T.Z.736, XC, G.T.Z.838, G.T.Z.996); [b] Haft Tepe (after Negahban 1991, Pl. 30, No. 209)

[Fig. 2.14a, c].[142] A similar range of copper/bronze arrowheads were recovered at Susa amongst the Middle Elamite 'Inshushinak Temple Hoard' in the Acropole mound.[143]

142 Ghirshman (1966) Pl. XCIII, G.T.Z.204–211; (1968) Pls. LXXXV, G.T.Z.736, XC, G.T.Z.838, G.T.Z.996.

143 Mecquenem (1905) 81–83. A typical late third/early second millennium BCE triangular arrowhead also made its way into the deposit (Mecquenem (1905) fig. 192).

5 Contested Relations (ca. 1100–525 BCE)

Late second and early first millennium Elamite military and political history is shrouded by a dearth of written sources and identifiable archaeological remains, and when Elam re-emerges as a force around the mid-eighth century BCE, our view of it is shaped by biased and often contradictory Assyrian and Babylonian reports. Nevertheless, it is clear that Elamite rulers were actively engaged in developing strategies to protect, and possibly even expand, their political and economic interests. They maintained the military capacity to oppose Assyrian expansionistic ambitions in the south and in the Zagros, to harass and conquer Assyrian garrisons, to lay siege to pro-Assyrian towns, to neutralise and frustrate territorial gains, and to directly confront, and even defeat, Assyrian armies in pitched battles. Archers and archery come to the fore in external references to Elam as a critical aspect of the Elamite military, demonstrating that its military continued to lean on well-trained archers supported by cutting-edge archery equipment.

Already in 814 BCE Elamite military forces are recorded with Babylonian, Chaldean, Aramean, and Kassite allies executing an operation to rescue inhabitants of hundreds of cities of 'the land of Akkad' sheltering from the Assyrian king Shamshi-Adad V (824–811 BCE) in the island city Dur-Papsukkal on the Diyala river. Assyrian records report a defeat of the alliance, tallying 5,000 deaths and the capture of 2,000 troops, 200 cavalry and 100 chariots.[144] The next year, the Assyrians besieged and captured the strategic city of Der, situated in a contested buffer zone of intersecting Assyrian-Babylonian-Elamite interests that frequently hosted military conflicts. A peace treaty between Babylon and Adad-ninari III (811–783 BCE) attempted to address persistent tensions over this zone by reaching a mutual agreement on a suitable boundary line,[145] but any efficacy it may have had was short-lived. The documented presence of an Elamite ambassador in the Nimrud court in 784 BCE and the manufacture of the 'Elamite bow' (distinguished from the 'Assyrian bow') for Assyrian troops may be an outcome of this treaty.[146]

Frequent external references to the 'bow of Elam'/'Elamite bow' and deployment of sometimes staggering (though undoubtedly exaggerated) numbers of Elamite archers imply an organised, large-scale production of archery

144 Grayson (1996) 188, v 31–45. Also http://oracc.museum.upenn.edu/riao/pager/, accessed 27/05/2020.

145 Grayson (1975) 169; Synchronistic-History iv: 15–22, Chronicle 21.

146 Dalley and Postgate (1984) 256, no. 145. iii.12–13; Zadok (1994) 47.

FIGURE 2.15 Bronze plaque from Susa depicting Elamite divine warriors (ca. 9th–8th century BCE)
PHOTOGRAPHS AND LINE DRAWING © J. ÁLVAREZ-MON

equipment.[147] Locally, a new sculptural program in the highland Izeh valley depicting the Elamite ruler surrounded by elites participating in communal rituals highlights the centrality of the bow in Elamite political, military, and sociocultural life. The earliest of the series of relief carvings at the open-air sanctuary of Kul-e Farah (relief IV), carved on a cliff-face sometime around the ninth-eighth centuries BCE, illustrates no less than 141 individuals partaking in the ritual consumption of food. It incorporates two groups of archers: one group, shown in a register directly below the king, is composed of two archers holding small composite bows in the left hand and, in front of them, a weapon-bearer/chief archer with a bow, quiver, and sword [Fig. 2.16a]; the second group is composed of seven unevenly distributed archers holding small composite bows with bent terminals in the left hand. Besides the weapon-bearer's sword, the bows and arrows are privileged as the only weapons in the scene. Another weapon-bearer with archery equipment and a long-sword was carved around a century or so later on a nearby boulder (relief VI). He stands behind a large-scale ruler carried atop a platform, holding a bow with backwards-arched limb to his chest with his left hand, and carrying a quiver with decorative horizontal bands and projecting arrow feathers

147 Exaggerating the numbers of troops, cities conquered, booty, dead enemies, and prisoners, etc., was a literary device used by the Assyrian scribes to amplify the scope of the task and enhance the heroic nature of the Assyrian victory (see De Odorico (1995) and Frahm (2003), 146, with refs.).

[a] [b] [c]

FIGURE 2.16 Elamite weapon-bearers [a] Kul-e Farah IV (ca. 9th–8th century BCE); [b] Kul-e Farah VI (ca. 7th–6th century BCE); [c] Kul-e Farah I (ca. 7th–6th century BCE)
LINE DRAWINGS © J. ÁLVAREZ-MON

[Fig. 2.16b]. Dating to approximately the same time as the first relief is the bronze plaque from Susa showing seven fearsome divine warriors with protective arm-gear. All carry arrow-filled quivers and small bows with duck-head terminals [Fig. 2.15]. This is the earliest known local attestation of the composite 'duck-headed bow', which is considered an Elamite form. Initially the duck heads were purely decorative and the string was attached below them, but they assumed a functional aspect from the late eighth century when the string was placed around them.[148]

The long reign of the Elamite king Humban-nikash I (743–716 BCE) overlapped with those of the Assyrian rulers Tiglath-pileser III (745–727 BCE), Shalmaneser V (726–722 BCE) and Sargon II (722–705 BCE). By 731 BCE we know that he was actively protecting Elamite interests in the contested Transtigridian/Diyala region where Tiglath-pileser had been attempting to annex territories of Aramean tribes, and lending his support to secure a Chaldean leader, Nabu-Mukin-zeri, on the Babylonian throne.[149] Intelligence letters report movements of Humban-nikash and his army around the Assyrian garrison at Der and his conquest of Borsippa, just 17 km from Babylon.[150] He still presided over Elam in 722 BCE when Sargon II and the Chaldean 'king of the Sealand' Merodach-Baladan (Marduk-apla-iddina) II, respectively, assumed power in Assyria and Babylonia, and he thwarted an early attempt by Sargon to seize

148 Zutterman (2003) 128–29.
149 Tadmor and Yamada (2011) 18.
150 Luukko (2012) nos. 127, 133.

the Babylonian throne in 720 BCE, intercepting him outside Der. No details exist about what must have been a formidable pitched battle, fought alone by the Elamites on behalf of their Babylonian allies, who arrived too late to participate, but clues as to the composition of the armies the Elamites faced can be obtained from the ideal ratio of a provincial Assyrian army at this time, namely, one chariot to ten cavalry to 200 infantry, the latter composed of two archers to one spear/shield-bearer. Sargon's attested deployment elsewhere of 20,000 archers in this ratio highlights the potential numerical weight of archers involved.[151] There is no indication that Humban-nikash seized the garrison after the battle, but Sargon's claim to victory lacks credibility in view of Babylonian reports claiming an Elamite triumph,[152] Merodach-Baladan's continuation on the throne, and the expansion of Elamite presence in the area with the capture of the border fortresses of Bit-Ha'iri and Rasa.[153]

The subsequent Elamite ruler Shutruk-Nahhunte II (716–699 BCE) spent the initial years of his reign reinforcing the region of Yadburu, an Aramean tribal area in the buffer zone east of Babylonia. But in 710 BCE he faced a more calculated campaign by Sargon to remove Merodach-Baladan from the

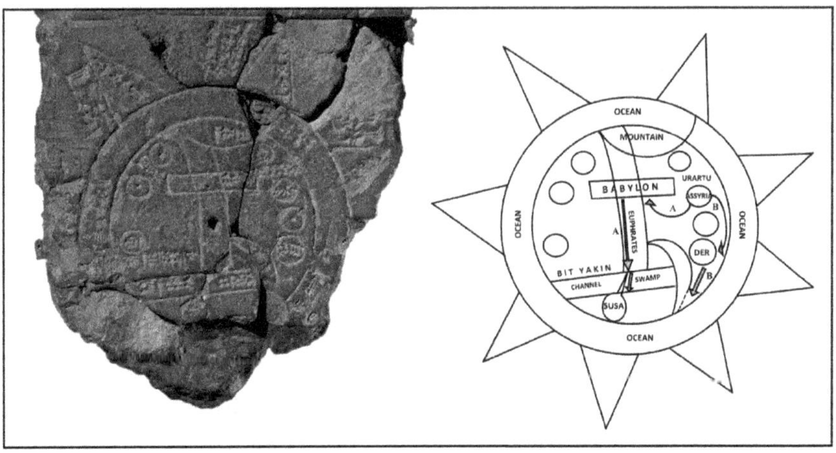

FIGURE 2.17 Babylonian 'map of the world' (ca. 7th/6th century BCE) showing how two main invasion routes into Elamite territory must have been envisaged [a] maritime route via Babylon; [b] in-land route via Der
PHOTOGRAPH COURTESY OF © THE TRUSTEES OF THE BRITISH MUSEUM; LINE DRAWING © J. ÁLVAREZ-MON

151 Fuchs (2011) 388.

152 Grayson (1975) 73, i 33–37; Glassner (2004) 195; Beaulieu (2018) 198.

153 Grayson (1965) 340–41; Grayson and Novotny (2012) 153; Sen. 18: iv.7'–17'.

Babylonian throne and temporarily lost Yadburu and its Elamite garrisons.[154] Shutruk-Nahhunte sheltered Merodach-Baladan when he afterwards escaped Sargon's two-year siege of his hometown Dur-Yakin in the Chaldean heartland. By 707 BCE Shutruk-Nahhunte was at the Elamite Zagros border garrison of Bit-Burnakka co-ordinating resources to support his involvement in a succession dispute in the wake of the death of king Dalta of Ellipi, the highland kingdom bordering on Elam and Media. While mobilising additional troops from regions such as Parsumash,[155] he sent 4,500 Elamite bowmen on ahead to assist Dalta's son Nibe. Though the course of events is unknown, the Assyrians seized the capital of Ellipi, Murubisu, and installed Dalta's brother Ashpabara. Another campaign of Shutruk-Nahhunte involving a successful siege of the Assyrian border town Malaku and a possible campaign against Der can be inferred from undated correspondence of the time of Sargon.[156]

Nervous letters from Assyrian fortress commanders written between 704 and 693 BCE reveal continued Elamite incursions in the Der area. One undated letter reported that 'the Palace Herald (field marshal) and entire army of Upper Elam' (counterpart to an army of Lower Elam?) were stationed east of Der at the fortress of Bit-Imbiya ready to cross the river.[157] In 703 BCE an immense army including Merodach-Baladan and his southern Babylonian-Aramean-Chaldean alliance gathered under the Elamite banner of Shutruk-Nahhunte to quell increasing Assyrian military intervention under Sennacherib (705–680 BCE). Substantial resources and coordination would have been required to plan, mobilise, equip, and deploy the 80,000 archers (and) spearmen, 12,200 horses and 850 wagons, divided in cohorts, probably in a regular organisation of 8,000 soldiers, overseen by ten commanders 'who did not know fear of death'. These commanders were answerable to a 'second-in-command'/'third man' named Tannanu and headed by a field-marshal (*turtan*) named Imbappa who took direct orders from the king.[158] The theatre of war lay within 35 km of Babylon, between Kutha and Kish, and it appears that three separate clashes took place on the plain, at Kutha and at Kish.[159] Sennacherib emerged the victor, placing a puppet king, Bel-ibni, on the throne of Babylon; however, three years later Bel-ibni rebelled, 'having listened to the Elamites and

154 Luckenbill (1927) no. 41; Fuchs and Parpola (2001) XXXV.

155 Fuchs and Parpola (2001) no. 111, 129. For the debates on the distinction between Parsumash and Parsuash, see Potts (2016) 264–66.

156 Fuchs and Parpola (2001) XXXV.

157 Dietrich (2003) nos. 120, 136.

158 Brinkman (1968) 43; Frahm (2003); Grayson and Novotny (2012), Sen. 1.8–9.

159 Grayson and Novotny (2012), Sen. 1.19–29.

Merodach-Baladan', and Sennacherib placed his own son, Ashur-nadin-shumi, on the southern throne.[160]

A new Elamite king named Hallutush-Inshushinak came to the throne in 700/699 BCE, after, according to the Babylonian Chronicle, he seized his brother Shutruk-Nahhunte and 'shut the door in his face', an intriguing Elamite idiom(?) understood to mean imprisonment.[161] He still reigned in 694 BCE when Elam was invaded by Sennacherib, who claimed to have sought the Chaldean asylum seekers living in its territories.[162] He sent a massive fleet to storm the Ulai shoreline, suggesting that his ultimate target was to penetrate deep into Elamite heartland using a traditional fluvial and commercial route (the Babylonian 'map of the world' gives a fascinating insight into how the two main invasion routes into Elamite territory must have been envisaged, see Fig. 2.17).[163] The Chaldeans and Elamites had gathered tens of thousands of archers, as well as wagons, horses and mules, and formed a battle line at the mouth of the Ulai river. Sennacherib claimed that his troops swarmed 'like locusts' out of the boats, defeated the defenders, and captured '30,500 archers (and) 20,200 shield bearers and added (them) to my royal [contingent]'.[164] Nonetheless, his strategy ended in disaster: taking advantage of the engagement of the Assyrian troops at the Ulai River, Hallutush-Inshushinak marched with his army into northern Babylonia, conquered Sippar and captured the Assyrian crown prince Ashur-nadin-shumi, who then vanished from history, and put Nergal-ushezib of the Gahal tribe, a relative of the Elamite royal family, on the throne, if only for a short period before he was removed again by Sennacherib.[165]

An anti-Assyrian Elamite policy seems to have prevailed into the reign of Humban-menana III (692–689 BCE), who mustered the armies of 'Elam and Akkad' and highland troops of Parsuash, Anzan, Pasheru and Ellipi, and lowland Babylonians, Aramean and Chaldeans to attack Assyria at Halule, located somewhere on the Tigris near Baghdad and the Diyala.[166] The alliance formed a battle line, with the river at their backs, blocking access to water sources. In an all-too-familiar scenario, the accounts of the outcome are conflicting:

160 Grayson and Novotny (2012) 98–105, Sen.15 v 2b–9.
161 Grayson and Novotny (2012) 26, ii 32–35.
162 Grayson and Novotny (2012) 12–13.
163 Ur III documents attest to direct maritime trade between Lagash and Susa (Sigrist (1986) 55–58). Shoreline geography discussed in Cole and Gasche (2007), Heyvaert, Verkinderen, and Walstra (2013); Gorris (2019).
164 Grayson and Novotny (2014) 83–84, Sen. 46 v 81b–106a.
165 Grayson (1975) Chronicle 1: ii.36–ii. 45.
166 Potts (2016) 266.

Sennacherib claims he enacted a heroic chariot charge forcing the enemy to retreat, killing 150,000 fighters and capturing chariots, wagons, and royal tents, whereas the Babylonian Chronicle records an Elamite victory – a more likely scenario given that the Babylonian king Mushezib-Marduk held his throne for the two subsequent years.[167]

In 675, a shift in Elamite-Assyrian relations was induced by the signing of a peace treaty between the Elamite king Urtak (675–664 BCE) and the Assyrian king Esarhaddon (681–669 BCE), who somewhat owed his throne to the Elamites' earlier removal of Ashur-nadin-shumi. Peace endured into the reign of his son Ashurbanipal (669–631 BCE) and was broken only in 664 BCE when Urtak allied with the Aramean Gambulu tribe and a governor of Nippur to attack Babylonia.[168] Within a year Urtak had died and the Elamite throne came into the hands of his in-law, Teumman (664–653 BCE), who sent the princes of the Houses of Urtak and Humban-haltash II (Urtak's brother and predecessor) into flight to Assyria with 'sixty of the royal family, countless bowmen, and free Elamites'. Urtak's sons may be the three Elamites represented in a relief in the North Palace at Nineveh laying five duck-headed bows at Ashurbanipal's feet.[169] The duck-headed bow was Ashurbanipal's own bow of choice and frequently appears in the hands of Elamite archers in late Neo-Assyrian palace reliefs.[170] The reliefs demonstrate two different methods of bracing these bows: one involves pulling the extremities of the limb using both hands and pinning down the centre with one knee while a second person pulls the looped end of the bowstring to the terminal; the second method, requiring just one person, is to bend the limb by kneeling on one end and pulling the other towards the body, using the free hand to secure the bowstring [Fig. 2.18].[171]

Taking up the cause of Urtak's sons, Ashurbanipal sent an army from Der with the eldest, Humban-nikash, to engage Teumman.[172] Adopting the conventional defensive strategy, Teumman's troops were positioned beside the Ulai river on a prominent hill barring access to the river. The Southwest and North Palaces at Nineveh illustrate the final episode of the battle: Elamite archers flee with horsemen toward the river while a vastly superior Assyrian

167 Grayson and Novotny (2012), Sen. 18 cols. v–vi 15; Sen. 22 cols. v–vi 35; Sen. 23 cols. v–vi 30; Sen. 34 44b–55a. The account of the battle of Halule is a remarkable piece of propaganda whose literary erudition and sophisticated intertextual borrowing from epics and myths is discussed by Pongratz-Leisten (2015) 307–308.

168 Frame (2007) 83.

169 Álvarez-Mon (2009) 136–38, n. 11.

170 Potts (2016) 338.

171 Barnett (1976) Plate XVI, Room C and North East Wall.

172 Waters (2002) 82.

FIGURE 2.18 Elamite bowman bracing a duck-headed bow
AFTER BARNETT 1976: PL. XVI, ROOM C AND NORTH EAST WALL

cavalry aided by Chaldean archers chases them down. The Elamite royal chariot, a typical low-platform, mule-drawn Elamite vehicle, tumbles over, and Teumman is decapitated. A caption next to Ituni, the Elamite *rab qašti* 'bow commander/chief', demonstrates a symbolic act of Elamite surrender: 'with his own hand he drew the iron dagger from his belt and cut his bow, the sign of his strength',[173] thus answering Ashurbanipal's plea to Ishtar to 'break the bow of Te'umman'.[174] Ashurbanipal later placed a bow over the severed head of Teumman in an offering ceremony and dedicated it to Ishtar of Arbela.[175] This 'breaking the bow' motif is by no means applied solely to Elamites, but it *is* applied to them with special emphasis,[176] and the defeat, humiliation and execution of Elamite archers[177] and their 'bow commanders', *rab qašti*,[178] who belonged exclusively to Elamite aristocracy and royalty, feature heavily in Ashurbanipal's propaganda.

Only a year after Ashurbanipal had restored the House of Urtak by placing its princes Humban-nikash II and Tammaritu on two Elamite thrones at Madaktu and Susa (creating a kind of co-regency?), the former defected, joining

173 Barnett (1976) Pl. XXIV; and Barnett, Bleibtreu, and Turner (1998) Pl. 295; Russell (1999) 182.
174 Borger (1996) 103, lines 44–54.
175 Weissert (1997) 350.
176 The motif also occurs in Esarhaddon's treaties (Parpola and Watanabe (1988) nos. 4:20–21, 6:453, 573).
177 For a prominent example, see the installation of Humban-nikash II in Madaktu exhibited in the Southwest Palace of Sennacherib, in Nimrud (Barnett et al. (1998) Pls. 286, 301–309 and 313; Room XXXIII).
178 Waters (2000) 49, 54, n. 79, 69–71.

an alliance with Ashurbanipal's brother, Shamash-shum-ukin (668–648 BCE) the king of Babylon, and the inhabitants of Akkad, Chaldea (now ruled by Merodach-Baladan's grandson), Arameans, and rulers of Gutium, Amurru (to the west, including Arabs), and Meluhha (Ethiopia).[179] In a pitched battle at Hiritu, the strategic fortress in the Irnina channel, Ashurbanipal triumphed over the alliance – Ashur 'smashed the [bo]ws of Elam, and strengthened your (Ashurbanipal's) bow' – and then laid siege to Babylon for around two years (650–648 BCE).[180] Afterwards, the Assyrian army marched into Elam seeking retribution, destroying dozens of Elamite cities in its path as the Elamite king retreated to the highlands. Ashurbanipal's capture of the Elamite city Hamanu is illustrated in both his palaces. In one, Elamite archers are using duck-headed composite bows to defend a riverside city surrounded by walls incorporating fifteen towers; in the other, the city is burning on a hilltop.

Internal power disputes and Assyria's punitive invasion in 647 BCE weakened Elamite military capacity but did not bring the Elamite kingdom to an end. And while the role of Elam in Babylonian military alliances and expansionism, the fall of Assyria, and the emergence of Persia as a military superpower are obscure, Nabopolassar's revolt against Assyria in 626 BCE, followed by his restoration of Elamite gods to Susa, and the rebellion of Der against Assyria in 623 BCE, suggest a continuation of the long-standing Elamite-Babylonian (Chaldean, Aramean) alliance.[181] The documentation in a "ration list" of a military squadron of 713 Elamites at Babylon in ca. 592 BCE under the command of a Babylonian officer endorses the view of a close association.[182]

Archery as the pillar of Elam's military success is memorialised in the oracle of Yahweh in the Bible (Jeremiah 49:35): 'Behold, I will break the bow of Elam, the mainstay of their might'.[183] Dated to as late as 586 BCE, this oracle is approximately contemporary with the Elamite royal archive (ca. 590–555 BCE) excavated in the Acropole mound at Susa.[184] Composed of 299 Elamite cuneiform documents belonging to a central administrative system managed from Susa, it preserves names of about 600 individuals and its geographic scope embraces much of Khuzestan and Fars, extending to Mesopotamia and the Persian Gulf shore. It records transfers of primarily (military?) clothing and weapons,

179 Beaulieu (2018) 214.
180 http://oracc.museum.upenn.edu/saao/corpus (retrieved 31/05/15); Potts (2016) 275.
181 Potts (2016) 283.
182 Moukarzel (2014) 144, n. 60.
183 Isaiah 22:6 also has Elam 'taking up the quiver' to proceed into battle.
184 Date based on the historical background covered in Jeremiah (ca. 627–586 BCE).

including a significant amount of archery equipment,[185] suggesting management of a central military arsenal.[186] At least sixty-seven tablets preserve mentions of bows, fifty list reed-arrows, twenty list arrowheads and twenty-eight list quivers. Bows are variously described as *rashmi*, *kakpin* and *parzimak*, which may refer to either the specific characteristics[187] or the types of bows.[188] There are also possible references to Egyptian and Assyrian bows,[189] and five bows perhaps made of bronze, calling to mind two votive bronze bows (ca. 65 cm long) excavated in undated find-contexts in the Acropole.[190] A large number of bows were notably transferred to/from Zari, situated in the historically problematic Elamite-Babylonian border region. The texts provide information on the material in which the arrowheads were manufactured. For example, one records bronze for production of arrowheads,[191] and another more precisely records that two pounds of iron had been forged into sixty-five arrowheads. Assuming they were equally sized, each arrowhead weighed around 15 grams.[192] Burials of the seventh and early sixth century BCE at Susa have also preserved limited evidence of iron and occasionally stone arrowheads.[193] 'Median' socketed trilobate bronze arrowheads of a type attested since 612 BCE and considered standard equipment of Achaemenid bowmen have also been excavated at Susa [Fig. 2.19].[194]

Also belonging to the seventh-sixth centuries BCE is the last relief (no. I) carved in the sanctuary of Kul-e Farah in Izeh, which carries a long inscription of Hanni, *kutur* of Aiapir and vassal of the Elamite king Shutur-Nahhunte (EKI 75–76). Hanni is shown presiding over an animal sacrifice with two high

185 Basello (2018) 232–33. It is impossible to estimate the total count of items, as the number 1 can also be read as 60 in the sexagesimal system and, furthermore, items may have been counted as packages (Gian Pietro Basello, pers. comm.).

186 Scheil (1907) 15, n. 1.

187 Scheil (1907) 37.

188 Hinz and Koch (1987) 162, *kakpin* (qa-am-ban); (1987) 405, 178, *parzimak* (bar-zí-ma-ak-k[a_4?]).

189 Relevant tablets, references and discussion in Álvarez-Mon (forthcoming).

190 Lampre (1900) 107, 151, fig. 170, size based on image scale of 2/15. A number of gods receive weapons in the tablets (Basello (2017)).

191 Scheil (1907) 29 [= MDP 9, no. 24: 11]; Hinz and Koch (1987) 459, GIŠ.GI.lg.

192 Scheil (1907) 88 [= MDP 9, no. 98:4]; Hinz and Koch (1987) 1058, sa-h.

193 Since none are shown or described, the forms of the arrowheads are unknown. It is possible that the early to mid-first millennium BCE forms at Susa were similar to those well-attested in contemporary graveyards in the Pusht-e Kuh region of Luristan (organised typologically by Overlaet (2003) 172–79, Iron Age I–II; Haerinck and Overlaet (2004) 41–42, Iron Age III).

194 Ghirshman 1954: 31–32, Pls. 43–44. Examples also found at Nineveh (Pickworth (2005)) and Pasargadae (Stronach (1978) 218–19, Pl. 165a–c).

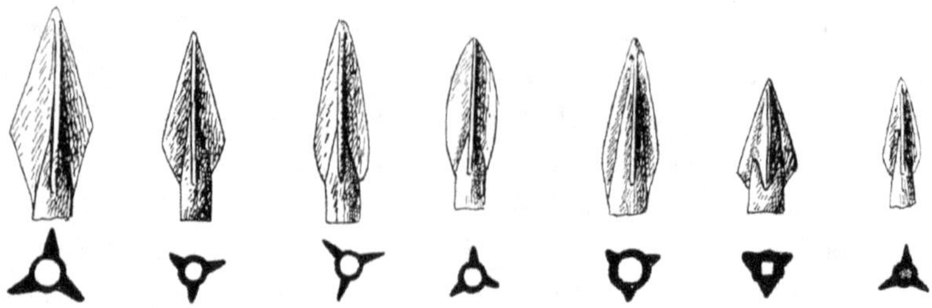

FIGURE 2.19 Sample of bronze trilobate arrowheads from Susa (ca. 6th century BCE or slightly later)
AFTER GHIRSHMAN 1954: 31–32, PLS. 43–44

officials behind him. One is identified as Shutruru, Hanni's *ragipal* and *gilira*.[195] He carries a quiver with small rectangular sections at the top indicating arrows and in his left hand holds a duck-headed bow braced in the later fashion with the string around the terminals [Fig. 2.16c]. In a similarly dated stone pedestal from Susa (Sb 5) showing the aftermath of the capture of a city-fortress, possibly celebrating a victory by an Elamite ruler over a town located in the highlands, one of the victors carries a quiver and holds a weapon in the right hand and a small duck-headed bow in the left hand.[196] The small size of this bow and the bows on the Kul-e Farah reliefs may indicate use of short bows in both the highlands and lowlands of Elam.

6 Legacy of Elamite Archery

This outline of military history unveils the bow as the quintessential weapon that enabled the political independence and longevity of the Elamite polity, whose state authorities commissioned industrial-scale manufacture of quality composite bows and metal arrows necessitating high-level technical skills and access to significant material resources. Throughout the millennia, bow, quiver, and arrow emerged as visual symbols of power which, wrapped in metaphorical language of war and peace, elevated archery to a prestigious and noble activity defining the Elamite elites. Depictions of the Achaemenid Persian ruler Darius I (522–486 BCE) as an archer on his Bisotun relief, tomb façade [Fig. 2.20a], and on coins (inaugurating the portrayal of the royal persona

195 EKI 75B, EKI 76:12; Henkelman (2008) 22.
196 Álvarez-Mon (2020), Pl. 191.

FIGURE 2.20 Representations of [a] Darius I depicted on his tomb at Naqš-e Rostam; [b] royal archer coins of Darius (Louvre Museum), types I, II, and IIIa; [c] bow-bearer of Darius I at Bisotun; [d] Elamite tributary carrying Elamite duck-headed bow from the Apadana staircase at Persepolis
PHOTOGRAPHS © J. ÁLVAREZ-MON

on coinage) [Fig. 2.20b],[197] and the depiction of his bow-bearer at Bisotun [Fig. 2.20c] reveal that Elam bequeathed its ideology of bowmanship upon its successors.[198] Darius and his bow-bearer carry none other than the esteemed Elamite duck-headed bow, which is shown being delivered by Elamites in the Persepolis reliefs [Fig. 2.20d]. Indeed, Darius' acceptance of bows exclusively from Elamite gift-bearers highlights his veneration of the Elamite tradition,[199] and a tradition of excellence in bowmanship would persevere in this area of the Zagros Mountains even beyond the Persian Empire.[200]

Strikingly, the Chinese philosopher and educator Confucius (551–479 BCE), a contemporary of Darius I, believed that virtue could be taught via the bow, and articulated an ethical apparatus defining personal qualities and exemplary behaviour based on analogies with the art of archery.[201] This ethos invites reflection on the teaching and practice of archery as instrumental in defining

197 Stronach (1989); Root (1989) 46.

198 Elam also provides a different lens through which to examine the symbolism of the bow and arrow in the Hebrew Bible, particularly in the hands of Yahweh (Bonfiglio (2012)).

199 Root (2010) 433 observes that no other gift-bearers are shown with bows.

200 Potts (2016) 370, 375.

201 Yu (2007); Behuniak (2010).

key aspects of Elamite political and individual identities, flowing through into Achaemenid Persia and later Iranian culture and religion. Indeed, Herodotus encapsulated the significance of this heritage in his description of Persian education: 'Their sons are carefully instructed from their fifth to their twentieth year in three things alone – to ride, to draw the bow, and to speak the truth'.[202]

Bibliography

Abrahami, P. (1997) 'L'armée à Mari', PhD Dissertation, Université de Paris I.

Abrahami, P. (2008) 'Le char de guerre en Syrie et Palestine au bronze récent', in Abrahami and Battini (2008) 57–70.

Abrahami, P. and Battini, L. (eds) (2008) *Les armées du Proche-Orient ancien: IIIe–Ier mill. av. J.-C. Actes du Colloque International Organisé à Lyon (les 1er et 2 décembre 2006), Maison de l'Orient et de la Méditerranée*. BAR International Series. London.

Alizadeh, A. (2006) *The Origins of State Organizations in Prehistoric Highland Fars, Southern Iran: Excavations at Tall-e Bakun*. Chicago.

Alizadeh, A. (2008) *Chogha Mish, Volume II: The Development of a Prehistoric Regional Center in Lowland Susiana, Southwestern Iran: Final Report on the Last Six Seasons of Excavations, 1972–1978*. Chicago.

Alizadeh, A. (2014) *Ancient Settlement Systems and Cultures in the Ram Hormuz Plain, Southwestern Iran*. Chicago.

Álvarez-Mon, J. (2009) 'Ashurbanipal's Feast: A View from Elam'. *IrAnt* 44: 131–80.

Álvarez-Mon, J. (2015) 'A Highland Elamite Archer from Kūl-e Farah IV, CI:4'. *IrAnt* 50: 251–78.

Álvarez-Mon, J. (2019) *The Monumental Reliefs of the Elamite Highlands: A Complete Inventory and Analysis (from the Seventeenth to the Sixth Century BC)*. Pennsylvania.

Álvarez-Mon, J. (2020) *The Art of Elam ca. 4200–525 BC*. London.

Álvarez-Mon, J. (2023) 'The Bow of Elam: Mainstay of their Might', in De Graef, Tavernier and Gorris (2023) 259–298.

Álvarez-Mon, J. and Garrison, M.B. (eds) (2011) *Elam and Persia*. Winona Lake.

Álvarez-Mon, J. and Wicks, Y. (2021) 'Elamite War Chariots and Military Equipment at Ancient Kabnak (ca. 1400 BCE)'. *JCS* 73: 121–34.

Álvarez-Mon, J., Wicks, Y., and Basello, G.P. (eds) (2018) *The Elamite World*. London.

Amiet, P. (1972) *Glyptique Susienne*. MDP 43. Paris.

Arkhipov, I. (2012) *Le vocabulaire de la métallurgie et la nomenclature des objets en métal dans les textes de Mari*. Archives Royales de Mari 32. Leuven.

202 Hdt. 1.136.

Bács, T.A. (ed) *A Tribute to Excellence: Studies offered in Honor of Erno Gaal, Ulrich Luft and Laszlo Torok*. Budapest.

Bagg, A.M. (2020) 'The Unconquerable Country: The Babylonian Marshes in the Neo-Assyrian Sources'. *Water History* 12: 57–73.

Barnett, R.D. (1976) *Sculptures from the North Palace of Ashurbanipal (668–627 B.C.)*. London.

Barnett, R.D., Bleibtreu, E., and Turner, G. (1998) *Sculptures from the Southwest Palace of Sennacherib at Nineveh*. London.

Baruchi-Unna, A., Forti, T.L., Ahituv, S., Eph'al, I., and Tigay, J.H. (eds) (2017) *'Now It Happened in Those Days': Studies in Biblical, Assyrian, and Other Ancient Near Eastern Historiography Presented to Mordechai Cogan on His 75th Birthday*. Vol. 1. Winona Lake.

Bartelmus, A. and Sternitzke, K. (eds) (2017) *Karduniaš: Babylonia Under the Kassites. The Proceedings of the Symposium Held in Munich (30 June to 2 July 2011)*. Berlin.

Basello, G.P. (2017) 'Of Gods and Men in the Persepolis Bronze Plaque', in Henkelman and Redard (2017) 347–83.

Basello, G.P. (2018) 'Administrative Topography in Comparison: Overlapping Jurisdiction between the Susa Acropole Tablets and the Persepolis Fortification Tablets', in Tavernier, Gorris, Abraham, and Boschloos (2018) 217–65.

Beaulieu, P.A. (2018) *A History of Babylon, 2200 BC–AD 75*. London.

Behuniak, Jr., J. (2010) 'Hitting the Mark: Archery and Ethics in Early Confucianism'. *Journal of Chinese Philosophy* 37: 588–604.

Bloch, Y. (2017) 'A Letter of Nebuchadnezzar I to the Babylonians: Literary and Historical Considerations', in Baruchi-Unna, Forti, Ahituv, Eph'al, and Tigay (2017) 493–524.

Blum, S.W.E., Efe, T., Kienlin, T.L., and Pernicka, E. (eds) (2020) *From Past to Present: Studies in Memory of Manfred O. Korfmann*. Studia Troica Monographien 11. Bonn.

Bonfiglio, R. (2012) 'Archer Imagery in Zechariah 9:11–17 in Light of Achaemenid Iconography'. *Journ. Bib. Lit.* 131: 507–27.

Borger, R. (1996) *Beiträge zum Inschriftenwerk Assurbanipals*. Wiesbaden.

Brinkman, J.A. (1968) *A Political History of Post-Kassite Babylonia, 1158–722 B.C.* Rome.

Bromiley, G.W. (ed) (1988) *The International Standard Bible Encyclopedia*. Vol. 4. Oxford.

Bryce, T. (2009) *The Routledge Handbook of the Peoples and Places of Ancient Western Asia*. London.

Carter, E. and Stolper, M.W. (1984) *Elam, Surveys of Political History and Archaeology*. Berkeley.

Charpin, D. (1986) 'Les Elamites à Subat-Enlil', in Meyer, Gasche, and Vallat (1986) 129–38.

Charpin, D. (2003; repr. 2012) *Hammurabi of Babylon*. New York.

Charpin, D. (2013) '"Ainsi parle l'empereur": À propos de la correspondance des sukkalmah'. *MDP* 58: 341–53.

Charpin, D. and Durand, J.-M. (1991) 'La Suzeraineté de l'empereur (Sukkalmah) d'Elam sur la Mésopotamie et le "Nationalisme" Amorrite'. *Actes 26ème Rencontre Assyriologique Internationale*: 59–66.

Charpin, D. and Ziegler, N. (2003) *Mari et le Proche-Orient a l'époque Amorrite: essai d'histoire politique*. Mémoires de N.A.B.U. 6. Florilegium marianum V. Paris.

Chavalas, M.W. (ed) (2006) *Ancient Near East: Historical Sources in Translation*. London.

Civil, M. (2003) 'Of Bows and Arrows'. *JCS* 55: 49–54.

Cole, S. and Gasche, H. (2007) 'Documentary and Other Archaeological and Environmental Evidence Bearing on the Identification and Location of the Rivers of Lower Khuzestan and the Position of the Head of the Persian Gulf ca. 1200 BC–200 AD'. *Akkadica* 128: 5–72.

Collon, D. (1983) 'Hunting and Shooting'. *Anat. St.* 33: 51–56.

Collon, D. (2008) 'Le développement de l'arc en Mésopotamie', in Abrahami and Battini (2008) 93–112.

Cooley, J.L. (2006) 'The Shitti-Marduk Stele', in Chavalas (2006) 160–64.

Dalley, S. and Postgate, J.N. (1984) *The Tablets from Fort Shalmaneser*, iii, Cuneiform Tablets from Nimrud. London.

Delougaz, P. and Kantor, H. (1996) *Chogha Mish, Volume 1: The First Five Seasons of Excavations, 1961–1971*. Chicago.

De Graef, K. and Tavernier, J. (eds) (2013) *Susa and Elam. Archaeological, Philological, Historical and Geographical Perspectives: Proceedings of the International Congress Held at Ghent University, December 14–17, 2009*. MDP 58.

De Graef, K., Tavernier, J., and Gorris, E. (eds) (2023) *Susa and Elam II. History, Language, Religion and Culture: Proceedings of the Second Susa and Elam Conference Held at Université Catholique de Louvain, July 6–9, 2015*. Leiden.

Denwood, P. (ed) (1978) *Arts of the Eurasian Steppelands*. London.

De Odorico, M. (1995) *The Use of Numbers and Quantifications in the Assyrian Royal Inscriptions* (SAAS 3). Helsinki.

Dezső, T. (2002) 'Scale Armour of the 2nd Millennium B.C.', in Bács (2002) 195–216.

Dietrich, M. (2003) *The Babylonian Correspondence of Sargon and Sennacherib*. State Archives of Assyria 17. Helsinki.

Drews, R. (1993) *The End of the Bronze Age: Changes in Warfare and the Catastrophe ca. 1200 B.C.* Princeton.

Durand, J.-M. (1998) *Documents épistolaires du palais de Mari, II*. Paris.

Edwards, I., Gadd, C., Hammon, N., and Sollberger, E. (eds) (1975) *The Cambridge Ancient History*, ii. 3rd ed. Cambridge.

Eidem, J. (1985) 'News from the Eastern Front: The Evidence from Tell Shemshāra'. *Iraq* 47: 83–107.

Fleming, D.E. and Milstein, S.J. (2010) *The Buried Foundation of the Gilgamesh Epic. The Akkadian Huwawa Narrative*. Leiden.

Foster, B.R. (1996) *Before the Muses: An Anthology of Akkadian Literature*. Bethesda.

Frame, G. (2007) *Babylonia 689–627 B.C. A Political History*. Publications de l'Institut historique et archéologique néerlandais de Stamboul 69. Leiden.

Frahm, E. (2002) 'Assur 2001: Die Schriftfunde'. *Mitteilungen der Deutschen Orient-Gesellschaft* 134: 47–86.

Frahm, E. (2003) 'New Sources for Sennacherib's "First Campaign"'. *ISIMU: Revista sobre Oriente Próximo y Egipto en la antigüedad* 6: 129–64.

Fuchs, A. (1994) *Die Inschriften Sargons II. aus Khorsabad*. Göttingen.

Fuchs, A. (2011) 'Das Osttigrisgebiet von Agum II. bis zu Darius I', in Miglus and Mühl (2011) 229–320.

Fuchs, A. and Parpola, S. (2001) *The Correspondence of Sargon II, part III: Letters from Babylonia and the Eastern Provinces*. State Archives of Assyria 15. Helsinki.

Gasche, H. (2000) 'Tell Ed-Der (Sippar-Amnanum) et Suse. Les tombes et les coutumes funéraires au 2e millénaire av. notre ère', PhD Dissertation, École Pratiques des Hautes Études.

Gasche, H., Tanret, M., Janssen, C., and Degraeve, A. (eds) (1994) *Cinquante-deux Réflexions sur le proche-orient ancien: offertes en hommage à Léon De Meyer*. Ghent.

Geller, M.J. (ed) (2014) *Melammu: The Ancient World in an Age of Globalization*. Berlin.

Genz, H. (2013) 'The Introduction of the Light, Horse-Drawn Chariot and the Role of Archery in the Near East at the Transition from the Middle to the Late Bronze Ages: Is There a Connection?', in Veldmeijer and Ikram (2013) 95–105.

Genz, H. (2020) 'Attacking and Defending Fortified Sites in the Early Bronze Age Levant: The Role of Archery', in Blum, Efe, Kienlin, and Pernicka (2020) 25–32.

Genz, H. and Mielke, D.P. (eds) (2011) *Insights into Hittite History and Archaeology*. Leuven.

George, A.R. (2000) 'Four Temple Rituals from Babylon: 2. Simanu', in Lambert, George, and Finkel (2000) 270–80.

Gernez, G. (2007) 'L'armement en métal au Proche et Moyen-Orient : des origines à 1750 av. J.-C', PhD Dissertation, Sciences de l'Homme et Société, Université Panthéon Sorbonne – Paris I.

Gernez, G. (2018) 'Metal Weapons', in Lebeau (2018) 39–76.

Ghirshman, R. (1954) *Village perse-achéménide*, MDP 36, Paris.

Ghirshman, R. (1966) *Tchogha Zanbil (Dur-Untash) Volume I: La Ziggurat*. MDP 39. Paris.

Ghirshman, R. (1968) 'Suse au tournant du IIIe au IIe millénaire avant notre ère'. *Arts asiatiques* 17: 3–44.

Glassner, J.-J. (2004) *Mesopotamian Chronicles*. Atlanta.

Gorris, E. (2019) 'The Harbour(s) of Nagitu: Mesopotamian Mooring Places, Elamite Garrisons and Aramean Settlements'. *IrAnt* 54: 59–82.

Grayson, A.K. (1965) 'Problematical Battles in Mesopotamian History', in Güterbock (1965) 337–42.

Grayson, A.K. (1975) *Assyrian and Babylonian Chronicles* (report. 2000). Winona Lake.

Grayson, A.K. (1996) *Assyrian Rulers of the Early First Millennium BC, II (858–745 BC)*. The Royal Inscriptions of Mesopotamia 3. Toronto.

Grayson, A.K. and Novotny, J. (2012) *The Royal Inscriptions of Sennacherib, King of Assyria (704–681 BC), Part 1*. Winona Lake.

Güterbock, H.G. (ed) (1965) *Studies in Honor of Benno Landsberger on His Seventy-Fifth Birthday, April 21, 1965*. Chicago.

Haerinck, E. and Overlaet, B. (2004) *The Iron Age III Graveyard at War Kabud Posht-e Kuh, Lorestān: Luristan Excavation Documents 5*. AI 42. Leuven.

Hamblin, W.J. (2006) *Warfare in the Ancient Near East to 1600 BC: Holy Warriors at the Dawn of History*. London.

Helwing, B. (2011) 'The Arisman Copper Production in a Wider Context', in Vatandoust, Parzinger, and Helwing (2011) 523–31.

Heimpel. W. (2003) *Letters to the King of Mari: A New Translation, with Historical Introduction, Notes, and Commentary*. Winona Lake.

Henkelman, W.F.M. (2008) *The Other Gods Who Are, Studies in Elamite Iranian Acculturation Based on the Persepolis Fortification Texts*. Leiden.

Henkelman, W.F.M. and Redard, C. (eds) (2017) *Persian Religion in the Achaemenid Period*. Wiesbaden.

Herrero, P. and Glassner, J.J. (1990) 'Haft-Tépé: choix de textes I'. *IrAnt* 25: 1–45.

Heyvaert, M.A., Verkinderen, P., and Walstra, J. (2013) 'Geoarchaeological Research in Lower Khuzestan: State of the Art', in De Graef and Tavernier (2013) 493–534.

Hinz, W. and Koch, H. (1987) *Elamisches Wörterbuch*. Vol. 1. Berlin.

Hinz, W. and Koch, H. (1987) *Elamisches Wörterbuch*. Vol. 2. Berlin.

Hoffmeier, J.K. (1988) 'Weapons of War', in Bromiley (1988) 1033–43.

Jakob, S. (2009) *Die mittelassyrischen Texte aus Tell Chuera in Nordost-Syrien*. Wiesbaden.

Kendall, T. (1975) 'Warfare and Military Matters in the Nuzi Tablets', PhD Dissertation, Brandeis University.

Klopsteg, P.E. (1947) *Turkish Archery and the Composite Bow: A Review of an Old Chapter in the Chronicles of Archery and a Modern Interpretation*. Illinois.

Korfmannn, M. (1986) 'Die Waffe Davids. Ein Beitrag zur Geschichte der Fernwaffen und zu den Anfängen organisierten kriegerischen Verhaltens'. *Saeculum* 37: 129–49.

Labat, R. (1975) 'Elam, c. 1600–1200 BC', in Edwards, Gadd, Hammond, and Sollberger (1975) 379–416.

Lacambre, D. (1997) 'La Bataille de Hiritum'. *Mari* 8: 431–54.

Laessoe, J. (1965) 'IM 62100: A Letter from Tell Shemshara', in Güterbock (1965) 189–95.

Lamberg-Karlovsky, C.C. (2013) 'The Oxus Civilization'. *Cuadernos de Prehistoria y Arquelogia, Universidad de Autonoma de Madrid* 39: 21–63.

Lambert, W.G. (1994) 'The Fall of the Cassite Dynasty to the Elamites: An Historical Epic', in Gasche, Tanret, Janssen, and Degraeve (1994) 67–72.

Lambert, W.G. (2007) *Babylonian Oracle Questions*. Winona Lake.

Lambert W.G., George, A.R., and Finkel, I.L. (eds) (2000) *Wisdom, Gods and Literature: Studies in Assyriology in Honour of W.G. Lambert*. Winona Lake.

Lampre, G. (1900) 'Travaux de l'hiver 1897–1898, tranchées nos. 7 et 7a'. *MDP* 1: 100–10.

Lansdorff, A. and McCown, D.E. (1942) *Tall-I Bakun A: Season 1932*. Chicago.

Larsen, M.T. (2015) *Ancient Kanesh: A Merchant Colony in Bronze Age Anatolia*. Cambridge.

Lebeau, M. (ed) (2018) *ARCANE Interregional 2: Artefacts*. Turnhout.

Le Brun, A. (1971) 'Recherches stratigraphiques à l'Acropole de Suse (1969–1971)'. *Cahiers de la Délégation archéologique francaise en Iran* 1: 163–216.

Le Brun, A. (1978) 'Le niveau 17B de l'Acropole de Suse (campagne de 1972)'. *Cahiers de la Délégation archéologique francaise en Iran* 9: 57–156.

Legrain, L. (1913) 'Tablettes de comptabilité, etc. de l'époque de la dynastie d'Agadé'. *MDP* 14: 62–126.

Lipinski, E. (2000) *The Arameans: Their Ancient History, Culture, Religion*. Leuven.

Littauer, M.A. and Crouwel, J.H. (1979) *Wheeled Vehicles and Ridden Animals in the Ancient Near East*. Leiden.

Littauer, M.A. and Crouwel, J.H. (1980; repr. 2002) 'Kamfwagen B. Archäologisch', in Raulwing (ed) (2002) 26–37.

Lorenz, J. and Schrakamp, I. (2011) 'Hittite Military and Warfare', in Genz and Mielke (2011) 125–52.

Luckenbill, D.D. (1927; repr. 1989) *Ancient Records of Assyria and Babylonia*. Vol. 1. London.

Luckenbill, D.D. (1927; repr. 1989) *Ancient Records of Assyria and Babylonia*. Vol. 2. London.

Luukko, M. (2012) *The Correspondence of Tiglath-pileser III and Sargon II from Calah/Nimrud*. State Archives of Assyria 19. Helsinki.

McEwen, E. (1978) 'Nomadic Archery: Some Observations on Composite Bow Design and Construction', in Denwood (1978) 188–202.

McLeod, W. (1958) 'An Unpublished Egyptian Composite Bow in the Brooklyn Museum'. *AJA* 62: 397–401.

McLeod, W. (1965) 'The Range of the Ancient Bow'. *Phoenix* 19: 1–14.

McLeod, W. (1970) *Composite Bows from the Tomb of Tutankhamun*. Oxford.

Mecquenem, R. de (1905) 'Offrandes de fondation du Temple de Chouchinak'. *MDP* 7: 71–130.

Mecquenem, R. de (1922) 'Fouilles de Suse: Campagnes 1914–1921–1922'. *Revue d'Assyriologie et d'Archéologie Orientale* 19: 109–40.

Mecquenem, R. de, 'Ministère de l'éducation nationale', *Roland de Mecquenem Archives de Suse 1912–1939* (1936), http://www.mom.fr/mecquenem/index/rapports, accessed 1 June, 2016.

Mecquenem, R. de, '1ere partie: travaux de fouilles', *Roland de Mecquenem Archives de Suse 1912–1939* (1937), http://www.mom.fr/mecquenem/index/rapports, accessed 1 June, 2016.

Mecquenem, R. de, 'Fouilles de archéologiques de l'Iran. Mission de Susiane. Campagne de Fouilles 1938. Rapport de Mission', *Roland de Mecquenem Archives de Suse 1912–1939* (1938), http://www.mom.fr/mecquenem/index/rapports, accessed 1 June, 2016.

Mecquenem, R. de, 'Fouilles de archéologiques de l'Iran. Mission de Susiane. Campagne de Fouilles 1939. Rapport de Mission', *Roland de Mecquenem Archives de Suse 1912–1939* (1939), http://www.mom.fr/mecquenem/index/rapports, accessed 1 June, 2016.

Mecquenem, R. de (1943) 'Fouilles de Suse 1933–1939'. *MDP* 29: 3–161.

Meyer, L. de, Gasche, H., and Vallat, F. (eds) (1986) *Fragmenta Historiae Elamicae, mélanges offerts à M.J. Steve*. Paris.

Michalowski, P. (1989) *The Lamentation over the Destruction of Sumer and Ur*. Winona Lake.

Miglus, P.A. and Mühl, S. (eds) (2011) *Between the Cultures: The Central Tigris Region from the 3rd to the 1st Millennium BC (Conference at Heidelberg January 22nd–24th, 2009)*. Heidelberger Studien zum alten Orient 14. Heidelberg.

Miller, R., McEwen, E., and Bergman, C. (1986) 'Experimental Approaches to Ancient Near Eastern Archery'. *World Archaeology* 18: 178–95.

Miroschedji, P. de (1981) 'Fouilles du chantier Ville Royale II à Suse (1975–1977). Les niveaux élamites'. *Cahiers de la Délégation archéologique francaise en Iran* 12: 9–136.

Moghaddam, A. (2020) *A Sanctuary at Tol-e Chega Sofla, the Third Season of Excavation*. Archaeological Report Monograph Series 2, No. 1. Tehran.

Moorey, P.R.S. (1986) 'The Emergence of the Light, Horse-Drawn Chariot in the Near East c. 2000–1500 B.C.'. *World Archaeology* 18: 196–215.

Morgan, J. de (1912) 'Observations sur les couches profondes de l'Acropole de Suse'. *MDP* 13: 1–25.

Moukarzel, K. (2014) 'Some Observations about "Foreigners" in Babylonia during the VI Century BCE', in Geller (2014) 129–55.

Nezafati, N., Pernicka, E., and Momenzadeh, M. (eds) (2009) 'Introduction of the Deh Hosein Ancient Tin-Copper Mine, Western Iran: Evidence from Geology, Archaeology, Geochemistry and Lead Isotope Data'. *TÜBA-AR* 12: 223–36.

Nielsen, J.P. (2018) *The Reign of Nebuchadnezzar I in Historical Memory*. London.

Oudbashi, O., Agha-Aligol, D., Mishmastnehi, M., and Barnoos, V. (2019) 'The Elamite Metalworkers: Multianalytical Study on Copper Objects and Ingots from Second Millennium BCE of Southwestern Iran'. *Archaeological and Anthropological Sciences* 11: 2059–72.

Overlaet, B. (2003) *The Early Iron Age in the Posht-e Kuh, Lorestān. Luristan Excavation Documents IV*. AI 40. Leuven.

Parpola, S. and Watanabe, K. (1988) *Neo-Assyrian Treaties and Loyalty Oaths*. State Archives of Assyria 2. Helsinki.

Parpola, S. and Whiting, R.M. (eds) (1997) *Assyria 1995*. Helsinki.

Parrot, A. (1971) 'Les fouilles de Mari, XIXe campagne (Printemps, 1971)'. *Syria* 48: 253–70.

Patterson, D. (2018) 'Elements of the Neo-Sumerian Military', PhD Dissertation, University of Pennsylvania.

Paulus, S. (2013) 'Beziehungen zweier Großmächte – Elam und Babylonien in der 2. Hälfte des 2. Jt. v. Chr. Ein Beitrag zur internen Chronologie', in De Graef and Tavernier (2013) 429–49.

Pickworth, D. (2005) 'Excavations at Nineveh: The Halzi Gate'. *Iraq* 67: 295–316.

Pongratz-Leisten, B. (2015) *Religion and Ideology in Assyria*. Berlin.

Postgate, J.N. (2013) *Bronze Age Bureaucracy, Writing and the Practice of Government in Assyria*. Cambridge.

Potts, D.T. (2008) 'Puzur-Inšušinak and the Oxus Civilization (BMAC): Reflections on Šimaški and the Geo-political Landscape of Iran and Central Asia in the Ur III Period'. *ZA* 98: 165–94.

Potts, D.T. (2016) *The Archaeology of Elam: Formation and Transformation of an Ancient Iranian State*. 3rd edn. New York.

Potts, D.T. (2020) 'Slings and Slingers in Ancient Iran', in Blum, Efe, Kienlin, and Pernicka, (2020) 17–24.

Powell, M.A. (1980) 'Karkar, Dabrum, and Tall Gidr: An Unresolved Geographical Problem'. *JNES* 39: 47–52.

Randall, K.C. (2016) 'Origins and Comparative Performance of the Composite Bow', PhD Dissertation. University of South Africa.

Raulwing, P. (ed) (2002), *Selected Writings on Chariots and other Early Vehicles, Riding and Harness*. Leiden.

Roaf, M. (2017) 'Kassite and Elamite Kings', in Bartelmus and Sternitzke (2017) 166–95.

Root, M.C. (1989) 'The Persian Archer at Persepolis: Aspects of Chronology, Style and Symbolism'. *REA* 91: 33–50.

Root, M.C. (2010) 'Elam in the Imperial Imagination: From Nineveh to Persepolis', in Álvarez-Mon and Garrison (2010) 419–74.

Russell, J.M. (1999) *The Writing on the Wall: Studies in the Architectural Context of Late Assyrian Palace Inscriptions*. Mesopotamian Civilizations 9. Winona Lake.

Sarianidi, V. (2005) *Gonur Depe, City of Kings and Gods*. Aşgabat.

Scheil, V. (1904) *Textes élamites-anzanites, deuxième série*. MDP 5. Paris.

Scheil, V. (1907) 'Traduction et commentaire du texte de la statue de Manistusu', in *Comptes rendus des séances de l'Académie des Inscriptions et Belles-Lettres*, 51e année 7: 413–15.

Scheil, V. (1911) *Textes élamites-anzanites, (IV) quatrième série*. MDP 11. Paris.

Scheil, V. (1912) 'L'armure de Narâm Sin, d'après une tablette de comptabilité trouvée à Suse', *Comptes rendus des séances de l'Académie des Inscriptions et Belles-Lettres*, 56e année 4: 296–301.

Scheil, V. (1932) *Actes Juridiques Susiens*. MDP 23. Paris.

Scheil, V. (1939) *Mélanges Épigraphiques*. MDP 28. Paris.

Steinkeller, P. (2018) 'The Birth of Elam in History', in Álvarez-Mon, Wicks, and Basello (2018) 177–202.

Sulaiman, M. and Dalley, S. (2012) 'Seven Naptanum-texts from the Reign of Rim-Sin I of Larsa'. *Iraq* 74: 153–65.

Sallaberger, W. and Schrakamp, I. (eds) (2015) *ARCANE III. Associated Regional Chronologies for the Ancient Near East and the Eastern Mediterranean: History and Philology*. Turnhout.

Sallaberger, W. and Schrakamp, I. (2015) 'Part I: Philological Data for a Historical Chronology of Mesopotamia in the 3rd Milennium', in Sallaberger and Schrakamp 1–135.

Schrakamp, I. (2010) 'Krieger und Waffen im frühen Mesopotamien: Organisation und Bewaffnung des Militärs im früdynastischer und sargonischer Zeit', PhD Dissertation, Philipps-Universität Marburg.

Sigrist (1986) 'Les courriers de Lagaš', in Meyer, Gasche, and Vallat (1986) 51–63.

Steve, M.-J. and Gasche, H. (1971) *l'Acropole de Suse*. MDP 46. Paris.

Steve, M.-J., Gasche, H., and Meyer, L. de (1980) 'La Susiane au deuxième millénaire: à propos d'une interpretation des fouilles de Suse'. *IrAnt* 15: 49–154.

Steve, M.-J., Vallat, H., and Gasche, H. (2002) 'Suse'. *Supplément au Dictionnaire de la Bible* 73: 360–511.

Stolper, M. (1984) 'Part I: Political History', in Carter and Stolper (1984) 3–102.

Stronach, D. (1978) *Pasargadae*. Oxford.

Stronach, D. (1989) 'Early Achaemenid Coinage, Perspectives from the Homeland'. *IrAnt* 24: 255–83.

Tadmor, H. (1958) 'Historical Implications of the Correct Rendering of Akkadian dâku'. *JNES* 17: 129–41.

Tadmor, H. and Yamada, S. (2011) *The Royal Inscriptions of Tiglath-pileser III (744–727 BC) and Shalmaneser V (726–722 BC), Kings of Assyria*. Royal Inscriptions of the Neo-Assyrian Period 1. Winona Lake.

Tallon, F. (1987a) *Métallurgie susienne: De la fondation de Suse au 18eme siècle avant J.-C.* Vol. 1. Paris.
Tallon, F. (1987b) *Métallurgie susienne: De la fondation de Suse au 18eme siècle avant J.-C.* Vol. 2. Paris.
Tavernier, J., Gorris, E., Abraham, K., and Boschloos, V. (eds) *Topography and Toponymy in the Ancient Near East.* Louvain-la-Neuve.
Tigay, J.H. (1982; repr. 2002) *The Evolution of the Gilgamesh Epic.* Pennsylvania.
Van Koppen, F. (2013) 'Abieshu, Elam and Ashurbanipal: New Evidence from Old Babylonian Sippar', in De Graef and Tavernier (2013) 377–97.
Vanstiphout, H. (2003) *Epics of Sumerian Kings: The Matter of Aratta.* Atlanta.
Vatandoust, A., Parzinger, H., and Helwing, B. (eds) (2011) *Early Mining and Metallurgy on the Western Central Iranian Plateau.* Mainz.
Veldmeijer, A.J. and Ikram, S. (eds) (2013) *Chasing Chariots: Proceedings of the First International Chariot Conference (Cairo 2012).* Leiden.
Vermaak, P.S. (1993) 'Shulgi as Sportsman in the Sumerian Self-laudatory Royal Hymns'. *Nikephoros* 6: 7–21.
Watelin, L. and Langdon, S. (1934) *Excavations at Kish Volume 4 (1925–30).* Paris.
Waters, M.W. (2000) *A Survey of Neo-Elamite History.* SAA 12. Helsinki.
Waters, M.W. (2002) 'A Letter from Ashurbanipal to the Elders of Elam (BM 132980)'. *JCS* 54: 79–86.
Waters, M.W. (2011) 'Parsumaš, Anšan, and Cyrus', in Álvarez-Mon and Garrison (2011) 285–96.
Weeks, L. (2008) 'The 2007 Early Iranian Metallurgy Workshop at The University of Nottingham'. *Iran* 46: 335–45.
Weissert, E. (1997) 'Royal Hunt and Royal Triumph in a Prism Fragment of Ashurbanipal (82-5-22,2)', in Parpola and Whiting (1997) 339–58.
Wilke, C. (1991) Zur Herstellung von Reflexbögen. *NABU* 1991, no. 17.
Yadin, Y. (1963) *The Art of Warfare in Biblical Lands in Light of Archaeological Discovery.* London.
Yu, J. (2007) *The Ethics of Confucius and Aristotle: Mirrors of Virtue.* London.
Zadok, R. (1994) 'Elamites and Other Peoples from Iran and the Persian Gulf Region in Early Mesopotamian Sources'. *Iran* 32: 31–51.
Zutterman, C. (2003) 'The Bow in the Ancient Near East: A Re-evaluation of Archery from the Late 2nd Millennium to the End of the Achaemenid Period'. *IrAnt* 38: 119–65.

CHAPTER 3

War and Ideology in Achaemenid Persia

Eduard Rung

This chapter focuses on the military aspect of Achaemenid ideology as reflected by Persian evidence, primarily the royal monumental inscriptions, which illustrate major themes in the propaganda circulated across the empire. Greek sources acquire special importance in reflecting Achaemenid Persia's ideology only when they confirm, specify or supplement the data obtained from Persian epigraphic and visual evidence, and priority should always be given to the latter. Close study of the Achaemenid texts, as well as Classical authors such as Herodotus, reveals an imperial ideology based, first of all, on the idea of subjecting all other peoples; and, secondly, on the idea of maintenance of peace and stability in the empire. One of the chief means available for the pursuit of these two objectives was war. The Persian Empire was a state created by conquest. It maintained its vitality by military efforts which protected it from disintegration and collapse for more than two centuries. The extant evidence from the territory of the Persian Empire, both textual and visual, provides information about the main features of the Achaemenid imperial ideology. According to these data, the Persian kings represented themselves as ideal rulers and perfect warriors and their empire as multi-ethnic and far-stretching, emphasised their generosity towards their loyal supporters and enmity towards their adversaries and rebels, and finally, referred to some military campaigns in which they demonstrated all their capacities. However, these sources are notable for their incompleteness in their descriptions of the events which were reported by the Greek and Roman authors. And if there are records of the events from Cyrus' reign (such as the conflict with Media, possibly, the subjugation of Lydia, and finally, the conquest of Babylon), no information is available from the Old Persian sources on the great battles of the Greco-Persian wars from the period of Darius I and Xerxes (the battles of Marathon, Thermopylai, Salamis, Plataia, etc). Moreover, all written information characterising the Achaemenid imperial ideology comes to us only from Cyrus the Great's Babylonian inscriptions and the multi-lingual inscriptions of Darius and Xerxes; no relevant data survived from Cambyses and from the Persian kings after Xerxes.

Most scholars have considered the Achaemenid ideology in the context of the relationship between the Persian imperial centre and the periphery. They

 | DOI:10.1163/9789004710771_004

often represent Persian relations with subject peoples in terms of cooperation aimed at the prosperity of the empire. This idea has been expressed by Richard Frye[1] and adopted by several scholars in more recent works.[2] However, representations of the Achaemenid Empire as based on the conciliation and cooperation of various ethnic groups may be considered as over-simplified. Although political practice demonstrates that representatives of non-Persian peoples were included in the power structures (especially at a local level), there was a clear distinction between Persia/Persians and other lands/peoples in the Achaemenid Empire. Luiz Alexandre Solano Rossi rightly asserts that 'the new world order, proposed by the empire, produced a scenario of great misery and violence' – that is, the *Pax Persica* and the stability of the new world could not be achieved without disastrous consequences for the conquered peoples.[3] Not all subjects experienced this violence equally, but for some, 'stability' could entail disastrous outcomes including punitive violence, deportations, and other harsh treatment of populations deemed rebels or opponents of order.

In his recent article, Christopher Tuplin considered Persian and Greek attitudes regarding the military factor in Achaemenid ideology, concluding that

> both war and peace do figure in the Persian worldview, and war perhaps more than is sometimes suggested ... Indeed, methodologically speaking, it is quite striking that the Persian and Greek data-sets, radically different in content and character, turn out not to be that far apart in their depiction of an ideological environment in which military values played a larger role than is sometimes acknowledged but were less fundamental than one might have expected.[4]

This chapter will follow up on Tuplin's discussion by reflecting on the Achaemenid sources and how their depictions of war and peace interact with and reinforce one another. The initial two sections of this chapter discuss the king's image and the empire's image in the Achaemenid ideology, and the next two sections consider more specific issues of Persian ideology and propaganda and their influence on expansionist policy (conquest, peace, stability), as well as means and methods of imperial control (punishment and reward).

---

1 Frye (1962) 86.
2 Briant (2002) 79; Brosius (2005); (2010), 33; Silverman (2020) 11–12.
3 Solano Rossi (2020) 11.
4 Tuplin (2017) 51.

1 The King's Image in Achaemenid Ideology

The king's image was the most important in the Achaemenid ideology because the Persian Empire was built on a monarchical idea that was meant to consolidate all the subject peoples. The most well-known, striking visual representation of this Achaemenid monarchy was the image of the Persian king on a throne platform supported by the representatives of all nations which were brought into the empire by the conquest (a figure of the Persian is notably present among the throne-bearers).[5] The king's authority in his empire surely depended not only on his sacral status, but also on his personal reputation as a military leader that was also spread across the empire by royal propaganda. Roel Konijnendijk rightly states, 'Needless to say, the best proof of a king's worth was conquest. In this sense royal ideology both legitimised and actively demanded war'.[6]

We have no knowledge of how Cyrus the Great, the founder of the Persian Empire, presented himself to his Persian subjects. Herodotus (3.89) reports that the Persians called Cyrus father because he was merciful and always worked for their well-being. Diodorus (9.24.1) paraphrases this as follows: 'Cyrus, we are told, was not only a courageous man in war, but he was also considerate and humane in his treatment of his subjects. And it was for this reason that the Persians called him Father'. The Cyrus Cylinder was written in the Babylonian dialect of Akkadian, and while it may reflect Persian propaganda, it was intended for the people of Babylonia and reflected traditional Mesopotamian themes and content.[7] As Amélie Kuhrt pointed out, it presented Cyrus not as a foreign and barbarian invader but as a restorer of what was right in the tradition of an earlier, venerated predecessor, Ashurbanipal.[8] The document represents Cyrus as a righteous king, dear to Marduk's heart (12), a shepherd in truth and righteousness (13), and claims that Marduk saw with pleasure Cyrus' good deeds and his righteous heart (13–14). It describes how the people of Mesopotamia greeted Cyrus after he entered into the city:

> All the people of Babylon, the entire land of Sumer and Akkad, nobles and governors, bowed down before (Cyrus) and kissed his feet, with shining faces they rejoiced at his kingship. Sweetly they hailed him as the lord through whose help they had come to life again from the perils of death, praising his name as the one who has saved them all from distress and disaster (18–19).

5 See Root (1979) 147–61; Brosius (2021) 87.
6 Konijnendijk (2021) 1144.
7 Rollinger (2021) 818.
8 Kuhrt (1983) 92.

The Cyrus Cylinder also stresses the military strength of the king when reporting that '[Cyrus'] vast troops whose number, like the water in a river, could not be counted, marched at his side, girt with their weapons' (16).[9] However, it does not record Cyrus' personal military capacities, in contrast with the inscriptions of his Mesopotamian predecessors and the texts of later Achaemenid kings. Cyrus presents himself in the Cylinder in several aspects: 1) a lover of the gods (22a, 25–28a, 35); 2) a peacemaker (24–25, 36); 3) a restorer of the population and its shrines (31–34); and 4) a successor of former Mesopotamian kings (39–40, 43).

The most detailed information on the ideological image of the Great Kings is contained in Darius I's inscriptions, which appraise his deeds in laudatory tones. In the texts on the Bisotun cliff and on the Naqš-e Rostam tomb, he says, 'All the peoples who came under my rule were my subjects (*manā bandakā*) (DB §7E), bore tribute to me (*manā bājim*), what was said to them by me either by night or by day, that was done' (DB §7F–H; DNa §3G–I).[10]

The king mentions further that he esteemed the best people and punished his enemies, and 'all the countries that had become subject to me showed respect toward my law (*manā dātā*)' (DNa §3J–K).[11] In the Naqš-e Rostam inscription, Darius also adds that he was a friend of the right, was not a friend of the wrong nor of a lie-follower, and protected the weak against harm by the strong (DNb §2–3). He then draws attention to his personal (military) qualities, which had to correspond to the image of Persia's ruler as perfect in all respects (DNb §7–9):[12]

> I am not hot-tempered. Whatever occurs to me in a quarrel, I firmly hold in my thinking. I am firmly in control of myself ... When you shall see or hear, what has been done by me both at court and in battle, that (is) my ability in addition to thinking and intelligence. Moreover this (is) my ability that my body is strong (and that), as a battle-fighter I am a good battle-fighter. At once my intelligence stands in its (proper) place,

9 See further Manning (2021) 127–28: 'The word for troops, *ummānu*, was just as ancient and generic ... Depending on context it could refer to soldiers, workers or people in general. Describing *ummānu* as *rapšātu* "great in size and or quantity" was also a standard poetic device meant to inspire awe ... The noun *kakku* "weapon" probably originally referred to some sort of mace. On the one hand, the *kakku* represented armed forces and destructive power ... On the other hand, *kakku* was the weapon which the forces of order and legitimacy wielded against the enemies'.

10 In citing Old Persian texts hereafter, references are to Schmitt (2009).

11 On the concept of *dāta-*, see Tuplin (2015). In the *Daiva* Inscription Xerxes says that law (*dāta-*) is set down by Ahuramazda (XPh. §6I–J).

12 Llewellyn-Jones (2015) 211–12.

> whether I see a rebel, (before me) or not. Both by intelligence and by command at that time I regard myself as superior to panic, when I see a rebel (before me) just as when I do not see (one). I am fervent in counter-attack with both hands as well as with both feet; as a horseman I am a good horseman; as a bowman I am a good bowman both on foot and on horseback; as a spearman I am a good spearman both on foot and on horseback.[13]

The bow was a traditional royal weapon (though the figure of the Persian king may be depicted in coins and seal images with various types of weapon including spear, dagger and battle-axe).[14] The portrayal of a king as a bowman characterised his military qualities and personal role in battle as well as the hunt. Some scholars believe that the idea of portraying a Persian king as a bowman was borrowed from the Mesopotamian predecessors, as the Assyrian reliefs often depict a king holding a bow.[15] Margaret Root suggests that Darius' adoption of the Assyrian bow as his own symbolises his appropriation of Assyrian military might.[16] Joan Westenholz assumes that this symbol refers to general might but not necessarily to specific military victories;[17] however, a triumphal meaning seems plausible in the case of Bisotun, despite the lack of explicit combat scenes in the relief.

The monument of Naram-Sin celebrates the king's victory over the Lullubi and depicts the king standing with a bow on his shoulder and an arrow or battle-axe in his left hand. The rock relief of Annubanini, the Lullubi king, shows him trampling a foe thrown underfoot and holding a bow in his right hand and a battle axe in his left.[18] The figure of Darius in the relief of the Bisotun cliff holds nothing but a bow. Persian kings depicted with bows may also be seen on the façades of the royal tombs in Naqš-e Rostam and Persepolis. The hieroglyphic inscription on the Statue of Darius found at Susa (DSab) reads, 'The goddess Neith has given him (sc. Darius) the bow she holds, to throw back all his enemies … so that he may be effective in repelling those who rebel against him'; the king himself is called 'he who crushes the Nine

13 Schmitt (2000) 40–41. Compare the Greek reports on the importance of archery and riding in Persian education (Hdt. 1.136; Xen. *Cyr.* 1.3; Xen. *An.* 1.9.3, 5; Pl. *Alc.* *1* 121e–122a; Strabo 15.13.8); for further analysis, see Lincoln (2012) 341; Panaino (2018).

14 For coins, see Tuplin (2014) 137, 146–47; for seal images, Tuplin (2020) fig. 7–9.

15 Westenholz (2000) 114–17. The bow was also important in Egyptian depictions of kingship; see Wilkinson (1991) 83.

16 Root (1979) 166–67.

17 Westenholz (2000) 123.

18 See Rollinger (2016) 13–14.

Bows', 'he who rushes into battle, shooting precisely, his arrow never missing his goal'. A Persian king's figure was depicted with various kinds of weapons, but the presentation of the king as a bowman was a distinctive feature of the royal coins known as darics, which led the Greeks to refer to them as 'archers'. Margaret Root categorises four types: 1) a royal figure carrying a bow in his left hand, a spear in his right hand; 2) a royal figure shooting a bow; 3) a royal figure carrying a bow in his left hand, a dagger in his right hand; 4) a royal figure carrying a bow in his left hand, two arrows in his right hand.[19] She arrives at the conclusion that 'there can be no doubt that the archers on Achaemenid coins are meant to represent the king, at least to the extent of symbolising the concept of kingship'.[20]

The Greeks perceived the bow's royal connotations. Aeschylus' *Persians* presents Darius as 'a lord of the bow' (556), as well as calling the Persians 'invincible archers' (26, 30, 926) and 'skilled in archery' (56). In Herodotus' *Histories* there are also several famous stories about the use of a bow by a Persian king, including Cambyses' madness and Darius' prayer for vengeance on the Athenians (3.21.3, 3.35–36, 5.105).

Darius' self-image as an archer, firstly, corresponds to the Near Eastern monarchical tradition that depicted the king, both visually and textually, in a heroic aspect (in Mesopotamia, kings had positioned themselves in that perspective since Sargon of Akkad).[21] Secondly, it correlates with Darius' own imperial ideology concerning the creation of his state by means of conquest.[22]

Darius presented himself as perfect ruler in all respects, and Xerxes echoed Darius' self-description as warrior as part of a strategy of emphasising his continuity with his father's royal image.[23] On a tablet of stone that was found in the vicinity of Persepolis in 1967, Xerxes reproduces, almost verbatim (XPl §2F–H), the characteristics that his father Darius gives of himself in the tomb inscription at Naqš-e Rostam:[24]

> Of such a kind (are) my intelligence and (my) command; when you shall see or hear what has been done by me, both at court and in battle, that (is) my ability in addition to thinking and intelligence. Moreover this (is) my ability, that my body is strong (and that) as a battle-fighter I am a good

19 Root (1979) 117. See also Nimchuk (2002) 63–64; Tuplin (2014) 137.

20 Root (1979) 117.

21 Westenholz (2000); Llewellyn-Jones (2015) 230.

22 See further Sancisi-Weerdenburg (1993); Kuhrt (2010).

23 Bridges (2014) 83.

24 Mayrhofer (1979).

> battle-fighter. At once intelligence stands in its (proper) place, whether I see a rebel (before me) or not. Both by intelligence and by command at that time I regard myself as superior to panic, when I see a rebel (before me) just as when I do not see (one). I am fervent in counter-attack with both hands as well as with both feet; as a horseman I am a good horseman; as a bowman I am a good bowman, both on foot and on horseback; as a spearman I am a good spearman, both on foot and on horseback.[25]

Remarkably, Herodotus (6.98) assigns comparable military capacities to Xerxes and his son Artaxerxes, erroneously believing that the names of these kings were translated from the Persian language as 'Warrior' and 'Great Warrior' respectively. When the historian describes Xerxes' army crossing over the Hellespont in 480 BCE, he states, 'Of all those tens of thousands of men, for goodliness and stature there was not one worthier than Xerxes himself to hold that command' (Hdt.7.187). But Herodotus (3.89; 6.98), contradicting the self-representation in the inscriptions, does not assign Darius any military qualities when reporting his Persian nickname as 'the merchant'[26] and translating his name from the Persian language as 'the Doer'.[27] Strabo (15.3.21) calls Darius 'the most handsome of men', also avoiding reference to his military capacities. However, it was Darius the Great himself who 'created' his own image of the perfect king by hailing himself as an excellent ruler, general and warrior in his monumental inscriptions. The Persian kings after Xerxes also receive characteristics from Classical authors. Cornelius Nepos (*De reg.* 1) reports that Artaxerxes I, nicknamed *Macrochir* ('long-armed'), was 'celebrated for a most noble and handsome person, which he rendered still more remarkable by extraordinary bravery in the field because no one of the Persians was more valorous in action than he'. Ammianus Marcellinus (30.8.4) describes Artaxerxes I as 'the most powerful king of the Persians' (*Persarum ... rex potentissimus*). However, there are no royal inscriptions after Xerxes which may be used to demonstrate the kings' self-representation. The visual evidence that survives from the later Achaemenid period demonstrates that these Persian kings followed the stereotypical royal imagery established by Darius; the typology of royal coins continued in use throughout the Achaemenid period, and the tombs of Xerxes, Artaxerxes I, and Darius II in Naqš-e Rostam and of Artaxerxes II and III in Persepolis were constructed with the tomb of Darius I as a model.

---

25 Schmitt (2000) 104.

26 See Descat (1994); Tuplin (1997) 379–81; Kurke (1999) 65–100.

27 Scott (2005, 349) believed that the name of Darius in the text of Herodotus should not correspond to ἐρξίης but rather to *ἐξίης, derived from the verb ἔχω and meaning 'possessor', thus bringing the explanation closer to the actual meaning of the name Darius.

2 The Image of Empire in Achaemenid Ideology

As the Persian Empire was established by military efforts, it is important to survey how Achaemenid ideology depicted the world that the kings had conquered and wished to control. Old Persian inscriptions make references to the earth (*būmī-*) as an important part of the concept of Achaemenid world rule.[28] Its sacred status is attested in the mention of Ahuramazda, 'who created this earth'.[29] The Persian kings made regular allusions to their vast territorial possessions. This may be seen in the Persian monarchs' assertions about themselves as 'kings in this great earth far and wide',[30] 'in this great earth',[31] 'in all the earth',[32] 'in this earth'.[33] A number of Old Persian texts call Darius 'king of countries containing all kinds of men' (*xšāyaϑiya dahyūnām vispazanānām*).[34] Another kind of definition – 'king of countries containing many kinds of men' (*xšāyaϑiya dahyūnām paruzanānām*) – comes to replace the previous term and may be read as early as Xerxes' and Darius II's inscriptions.[35] The Babylonian texts only use a designation of 'King of countries of the totality of all tongues' (LUGAL KUR.KUR.MEŠ *šá nap-ḫa-ar li-šá-nu*) for Darius I, Xerxes, and Artaxerxes I.[36] A number of Old Persian texts state that the kingdom (*xšaçam*) given to the king by Ahuramazda has good horses and good men (ASh, AMh, DPd, DSf, DZc), good horses and good chariots (DSs), or good charioteers, good horses, and good men (DSp). Many of the inscriptions of the Persian kings emphasise the idea of the Achaemenid Empire's territorial expanse.[37] This idea may be read as early as the Cyrus Cylinder, which describes the submission to Cyrus' rule in Babylon on the part of other rulers (28b–30a):

28 On the meaning of the word *būmī-*, see Schmitt (2014) 154–55. On the role of the concept of the earth in the Achaemenid royal ideology, see Ahn (1992) 258–71. Herrenschmidt (1976, 44) argues that in some inscriptions there were the contexts in which *būmī-* is used in the meaning of *xšaça-* and therefore did not signify 'the earth', but rather 'the political entity', i.e. 'kingdom'.

29 DEa §1B; DNa §1B; DSe §1B; DSf §1B; DSt §1B; DSab §1B; DZc §1C; XEa §1C; XPa §1B; XPb §1B; XPc §1B; XPd §1B; XPf §1B; XPh §1B; XVa §1C; A2Ha §1B; A2Hc §1C; A3Pa §1B.

30 DNa §2E; DEa §2E; DZc §2E; XEa §2E; XPa §2E; XPb §2E; XPc §2E; XPd §2E; XPf §2E; XPh §2E; XVa §2E; D2Ha §2E; A2Hc §3C.

31 DSab §3E; DZb §1E.

32 DSb §1E; DSf §5B; D2Ha §3D.

33 DSd §1E; DSf §4E; DSi §2C; DSj §1D; DSt §1B; DSy §1E; A2Ha §1E; A2Hc §2E; A2Sa §1E; A2Sc §2E; A2Sd §1E; A3Pa §2E.

34 DNa §2D; DSe §2D; DZc §2D.

35 XEa §2D; XPa §2D; XPb §2D; XPc §2D; XPd §2D; XPf §2D; XPh §2D; XVa §2D; DHa §2D.

36 DPa Bab.2; XPa Bab.7; XPb Bab.10; XPc Bab.6; XPd Bab.5; XPf. Bab. 0 a7; XPh Bab.0 a8; A[1]Pa Bab.4.

37 On the presentation of the Persian Empire's boundaries in Near Eastern and Classical sources, see Bichler and Rollinger (2017) 1–10; Rollinger (2021) 820.

> All the kings of the entire world, those who are seated on thrones, living in [distant] coun[tries] from the Upper to the Lower Sea as well as the kings of Amurru, dwelling in tents, (all these kings) brought their heavy tribute and kissed my feet in Babylon.[38]

Another of Darius I's inscriptions from Persepolis, written in Babylonian, states that this king ruled over the country from sea to sea, from desert to desert, and makes special mention of Persia and Media as central areas, different from all other territories of the empire (DPg):[39]

> §1 A great (god is) Auramazda, who is the greatest among all the gods, who created heaven and earth, created mankind, who gave all well-being to mankind who dwell therein, who made Darius king, and bestowed on Darius the king kingship over this wide earth, in which there are many lands: Persia, Media and the other lands of other tongues, of mountains and plains, from this side of the sea to that side of the sea, from this side of the desert to that side of the desert.
>
> §2 Darius the king speaks: With the protection of Auramazda, these (are) the lands, who did this, who gathered here: Persia, Media, the lands of other tongues, of mountains and plains, from this side of the sea to that side of the sea, from this side of the desert to that side of the desert, as I commanded them. All that I did, I did with the protection of Auramazda. May Auramazda, together with all the gods, protect me, me and all I love.[40]

A similar conception regarding water boundaries may be found in the report of the Greek historian Deinon of Colophon (Plut. *Alex.* 36), who claimed that Persian kings considered the Danube (Ister) River in the north and the Nile River in the south to be the borderlines of their empire: 'Deinon says that the Persian kings had water also brought from the Nile and the Ister and stored up among their treasures, as a sort of confirmation of the greatness of their empire and the universality of their sway'.[41] However, whereas the deserts and seas show the boundaries of the territorial possessions in the Achaemenid Babylonian inscriptions,[42] in Old Persian texts the sea (*drayah-*) sometimes delineates

38 Schaudig (2019) 23–24.
39 Kuhrt (2010) 95.
40 Kuhrt (2007) 483.
41 See Rung (2015).
42 Rollinger (2013) 95–116.

territories,[43] but is not mentioned as the empire's borderline. Instead, its margins are described in terms of the most distant countries/peoples (*dahyāva*), as shown by the following Old Persian inscription which describes the king's territorial possessions (DPh §2D–H): 'This is the kingdom (*xšaçam*) which I hold, from the Scythians who are beyond Sogdiana, thence to Kush; from India thence to Sardis'.[44]

An important piece of testimony, suggesting that the Achaemenids adopted an Assyrian geographical worldview, is the report of the Athenian orator Aeschines that Xerxes wrote in his letters that he was lord of all men from the rising of the sun unto its setting (3.132.5). A somewhat similar concept may be found in Herodotus' presentation of Xerxes' speech at the war-council before his expedition against Greece: 'No land that the sun beholds will border ours, but I will make all into one country, when I have passed over the whole of Europe' (7.8c.2). Aeschylus in his *Persians* (232) also gave the Chorus' answer to Atossa's question as to where Athens was situated: 'Far away towards the west, where the Lord Sun sets'. In the Achaemenid inscriptions, there are no references to holding supremacy over all the countries from the rising to the setting sun. But they are abundant in documents by the Assyrian rulers, sometimes in combination with other indicators of the empire's size.[45] Assyrian kings as well as Persian monarchs claimed that they ruled over the whole world, washed by waters, from one end to the other.[46] Darius' inscriptions may have referred to a more distant 'lower sea' than the Assyrian one, perhaps the Indian Ocean instead of the Persian Gulf, but the ideological grounds underlying the empire's boundaries were based on a similar perception of the Achaemenid Empire as stretching from one side of the *oikoumenē* to the other. This is confirmed by

43 This occurs in several expressions: '(those) who are beside the sea' (DB §6H); 'Ionians who are of the mainland and (those) who are by the sea, and countries which are across the sea' (DPe §2K–M); 'Scythians who are across the sea' (DNa §3T); 'Ionians, (those) who are by the sea and (those) who are across the sea' (DSe §4I–K); 'Ionians, those who dwell by the sea and those who dwell across the sea' (XPh §3Q–R). For details on these definitions as regards the Ionians and the Scythians, see Seager and Tuplin (1980) 148–149; Sancisi-Weerdenburg (2001b) 323–346; Klinkott (2001) 107–148; Rollinger (2021) 824–5. Heleen Sancisi-Weerdenburg (2001a, 11) believed that since *tayaiy drayahyā* in DB 6I is listed in between Egypt on the one hand and Sparda and Yauna on the other side, a place between Egypt and Lydia is the most likely one.

44 On the actual limits of the Achaemenid Empire, see Briant (2021) 55–56.

45 Tiglath-Pileser III: RINAP I. 35. ii.18–19. Sargon II: RINAP II. 73. 10. Sennacherib: RINAP 15. i.14; 16. i.15; 17. i.11; 22. 1.10; 23. i.9b; 24. i.9b; 27. ii.1; 46.2b; 136. i.10; 152.1; 155.1; 230.5b. Esarhaddon: RINAP IV. 1. ii. 25; 2. i. 11–12; 48. 44; 49. 1; 77.13–14; 78.12–13; 93.6. Ashurbanipal: RINAP VI. 3. i.39; 4. i. 31.

46 Rollinger (2012) 731; (2013) 95–116.

Xenophon's testimony (*Cyr.* 8.6.21) about the greatness of the Persian Empire from Cyrus the Great onwards:

> From that time on his empire was bounded on the east by the Indian Ocean, on the north by the Black Sea, on the west by Cyprus and Egypt, and on the south by Ethiopia. The extremes of his empire are uninhabitable, on the one side because of the heat, on another because of the cold, on another because of too much water, and on the fourth because of too little.

Darius I's inscription cited above, which speaks of the territorial boundaries of his supremacy (DPh §2D–H), gives a more dimensional orientation of the Achaemenid Empire, from north to south and east to west. A more 'complex' geographic organisation of the empire may be seen in the Persian royal 'catalogues' of countries and peoples contained in the inscriptions (DB §6; DNa §3; DNe; DPe §2; DSe §4; XPh §3; A^{3}Pb 1–2), as well as the iconography of the Apadana reliefs of Xerxes' royal palace.[47] In most of these texts the Achaemenids enumerated the countries and peoples starting from the empire's centre – Persia, Media, Elam – and moved on to the periphery in a varying order.[48] Herodotus lists the peoples twice in accordance with the Greek approach – from the west to the east – from the Ionians to the Lydians (Hdt. 3.90–94; 5.49), and only follows the Persian principle – starting with the Persians and the Medes – in the 'catalogue' of Xerxes' troops (7.61–81).[49] It is important to compare the 'catalogues' of the countries and peoples of the Achaemenid Empire with various Assyrian lists of subject territories, especially with the famous rock inscription of Sargon II in the vicinity of the village of Tang-e Var in Iran.[50] The latter outlines the empire's boundaries (*pāṭu*) by means of listing countries and cities, but the order of listing the lands is different to that employed by the Persians. Firstly, it enumerates only Assyria's 'external' possessions, and, secondly, they are listed southwards and northwards from the west (from Cyprus to Egypt and

47 For detailed analysis of the royal catalogues, see Vogelsang (1992) 96–119; on identification of the delegations in the Apadana reliefs, see Schmidt (1953) 85–90; Gropp (2009) 283–359.

48 On the Persian worldview and its reflection in the sources, see Dan (2013) 83–121. Rollinger (2021) 821 emphasises that in the mental map the 'Persian land', i. e. Parsa/Fars, constitutes the centre of the empire, followed by neighbouring lands such as Media, Babylonia, and Elam.

49 On Herodotus' representation of Persian political geography, see Dan (2013); Romney (2017).

50 See Frame (1999) 40–41.

Phrygia) and southwards from the east (from Media to Arabia).[51] According to Amélie Kuhrt, there was a difference in the treatment by the Persians of various peoples with the treatment of these by the Assyrians:

> Assyrian ideology propounded the idea that the defeated were incorporated into the Assyrian realm and 'Assyrianised' – 'as Assyrians I counted them', is the refrain of Assyrian royal inscriptions ... By contrast, in the Achaemenid world it was the persistent colourful diversity of the empire's subjects, their individual contributions, uniting to create the harmonious whole, which was the Persian Empire, that was celebrated in word and image.[52]

This approach may explain an absence of Assyria and the Assyrians, the conquerors, from the list of subjects in the Assyrian inscriptions, and the inclusion of Persia and Persians, subjects of the king like the others, in the Achaemenid inscriptions. But it is quite obvious that a special significance was attached to Persia in the royal 'catalogues' of countries and peoples. Maria Brosius, while considering the concept of *Pax Persica* in the Achaemenid Empire, erroneously believes that, firstly, no seeming distinction was made between core lands and those of the periphery of the empire, and, secondly, the idea of *Pax Persica* could not function with a strict separation between the Persian ruling class and the subject peoples.[53] However, Jason Silverman rightly notes that the Persians themselves did hold a special status, although several other groups (Medes, Elamites, eastern Iranians) also held important esteem and functions in administration: 'While the imperial elite was largely Persian, a more important criterion for social standing was royal favor'.[54] Finally, Pierre Briant says that the royal inscriptions serve as a testimony to the eminent and privileged place of Persia proper.[55] Indeed, the list in the Bisotun inscription starts with Persia and includes twenty-three countries 'which submitted to Darius' (DB §21E).[56] Persia is explicitly present in the catalogue in the hieroglyphic inscription on

51 On Assyrian understanding of their empire's size and political geography, see Rollinger and Ruffing (2014) 93–134.

52 Kuhrt (2010) 100. DeGrado (2019, 107–108) speaks of a program of 'Assyrianization', forced or voluntary, which required that rulers and subjects distinguish between 'Assyrian' and 'Other'.

53 Brosius (2010) 33.

54 Silverman (2020) 11–12; see also Kuhrt (2010) 95.

55 Briant (2021) 56.

56 It is possible that the initial inclusion of Persia in the Bisotun list might be connected with Darius' reconquest from the 'Liar King' Vahyazdāta (DB §40–43).

the statue of Darius from Susa (DSab). But in the rest of the inscriptions Persia is not mentioned among the subjected countries which brought their tribute to the king. In Darius' inscription from Persepolis the list of twenty-five countries starts with Elam (DPe §2H). In the Naqš-e Rostam inscription Media is listed first among the thirty countries which, 'by the favour of Ahuramazda', Darius seized outside of Persia (DNa §3L). The same may be seen in Xerxes' Daiva inscription (XPh §3L).

The ideology of the universal empire which was found in the Achaemenid royal inscriptions is reflected in the sculptural reliefs. They included 'gift-bearing delegations' at Persepolis along the stairways of the Apadana of Darius, and 'throne-bearers' on the façades of the royal tombs of Naqš-e Rostam and Persepolis, as well as the jambs of the tripylon of the 'One-Hundred-Column Hall' built by Xerxes and Artaxerxes I.[57]

Herodotus' testimony (1.134) is an illustration of the Greeks' perception and interpretation of one of the elements of the Persian imperial ideology, the relationships between the centre and the periphery, in which the historian emphasises the Persians' superior status in relation to other peoples of the empire:

> They [i.e., Persians] honour most of all those who live nearest them, next those who are next nearest, and so going ever onwards they assign honour by this rule: those who dwell farthest off they hold least honourable of all; for they think that they are themselves in all regards by far the best of all men, that the rest have only a proportionate claim to merit, until those who live farthest away have least merit of all.[58]

We may conclude that Darius and Xerxes often chose to treat Persia as an entity separate from the other countries under their rule. Herodotus (3.97) asserts that the Persian country was not recorded as tributary as the Persians lived free from all taxes.[59] Of course, there is no need to doubt that the Persians, like other peoples in the Achaemenid Empire, had some obligations before the king, but bringing the tribute (*bāji-*) was more of an ideological rather than only an administrative and fiscal proposition.[60] That is why a tribute-free status

57 Callieri (2023) 841.

58 Cf. Tuplin (2011) 42–43.

59 The Persepolis Fortification Archive suggests that elements of taxation such as labour obligations and payments associated with landholding and estate tenure were present in Achaemenid Fars. Kleber (2015) notes that Herodotus' report that the Persians were exempted from the payment of tribute may merely reflect the historian's idea of a tributary empire.

60 Tuplin (2008) 384.

should have marked the privileged position of Persia and the Persians among other provinces and peoples in the Achaemenid Empire from Darius onwards.

3 Conquest, Peace, and Stability in Achaemenid Ideology

The surviving inscriptions of Cyrus emphasise peaceful methods and outcomes in describing the foundation of the Persian Empire. Cyrus' inscription from Ur declares the establishment of supremacy over other countries ('The great gods have delivered all the lands into my hand'), and states that he has made the land to dwell in a peaceful habitation.[61] It was apparently inspired by the style of royal inscriptions found as early as in the Old Babylonian period and repeated throughout the Neo-Assyrian and Neo-Babylonian periods, in which the kings, among other merits, take credit for causing the people to dwell in peace.[62] Sargon II (721–705 BCE) used the phrasing 'I allowed the people to live in peace' to describe putting a foreign country under his control (RINAP II, 1, 208b; 2, 463b; 8, 6; 82, 11b; 117, ii 64); and Esarhaddon (680–669 BCE) claims to have gathered the scattered people of Babylon and resettled them in peaceful dwellings (RINAP IV 48. 42–43). The Nabonidus Chronicle (ABC 7) records that Cyrus was victorious at the battle of Opis on the river Tigris against the Babylonians (3.12–13) and carried off the plunder and slaughtered the people,[63] but then he captured Sippar without a fight (3.14) and entered Babylon without a fight (3.15–16), so that peace came to the city (3.19). According to the Cyrus Cylinder, Marduk made Cyrus enter his city Babylon without fighting or battle (17). Further, Cyrus speaks of his deeds in the first person:

> When I entered Babylon in peace, I took up my lordly abode in the royal palace amidst jubilation and rejoicing (22–23) … My vast troops walked around in Babylon in peace, and I did not permit anybody to frighten (the people of) the land of S[umer] and Akkad. Of Babylon and all its sacred places I took care in peace and sincerity (24–25) … Marduk, the great lord, rejoiced at [my good] deeds, and sent friendly blessings to me, Cyrus, the king who reveres him, to Cambyses, my son, the fruit of [my] loins, as well as to all my troops, and so we w[alked] in peace and

61 Gadd and Legrain (1928) 58, no. 194.

62 Waters (2019) 36.

63 Waters (2019) 27 supposes that interpretation of Cyrus' actions after his victory may take different turns based on how the verb *dâku* is translated in these contexts, whether Cyrus killed or just defeated the Babylonian army (or people) after his victory at Opis.

> happiness before him (27–28) … And (the statues of) the gods of the land of Sumer und Akkad, which Nabonidus – to the anger of the lord of the gods – had brought into Babylon, I had them dwell in peace in their beloved sanctuaries at the command of Marduk, the great lord (33–34) … [A]ll of the people of Babylon blessed my kingship, and I took care that all the countries live in peace (36).[64]

There is little doubt that such frequent use of the word 'peace' (*salīmu*) in the Cyrus Cylinder text intended to create an image of a king-peacemaker among the population of Mesopotamia and was, therefore, a manifestation of Cyrus' propaganda.[65] Aeschylus' *Persians* (768–769), like Cyrus' inscriptions, also presents him as a peacemaker: '… Cyrus, a most fortunate man, whose rule brought peace (εἰρήνην) to all his friends'. This phrasing indicates that Cyrus' propaganda of his peaceful policy toward conquered populations was possibly known to the Greeks.

Richard Frye introduced the term *pax Achaemenica* to define a new policy of conciliation and cooperation of the subject peoples that originated with Cyrus the Great.[66] However, scholars now recognise that Cyrus' treatment of subdued nations did not introduce new elements. Robartus van der Spek believes that non-interference with local government was a common characteristic of the empires of the ancient Near East. He reminds us that in the inscriptions of Sargon II as well as the Cyrus Cylinder, the king enters Babylon without violence, and just like his Assyrian predecessor, Cyrus presents himself as the one who removes the yoke from the Babylonians and restores a damaged city; but this idea contradicts the historical evidence reported by the Nabonidus Chronicle.[67]

The Achaemenid royal inscriptions use the word *šiyāti-* in the meaning of 'welfare', 'peace', 'happiness', 'prosperity'.[68] *Šiyāti-* was perceived, like all other things in the world – earth, heaven, mankind – as Ahuramazda's creation; however, it may reflect a concept of peace in the following passage:

> Saith Darius the King: If thus thou shall think, 'May I not feel fear of (any) other', protect this Persian people; if the Persian people shall be protected,

64 Schaudig (2019) 23–24.
65 See Medenieks (2017) 137.
66 Frye (1962) 86.
67 Van der Spek (2014) 255–56.
68 Kent (1950) 210; Schmitt (2014) 248.

> thereafter for the longest while happiness (*šiyātiš*) unbroken – this will by Ahura come down upon this royal house (DPe §3).[69]

The concept of peace is also suggested in another text. The Susa inscription underlines Darius' role in pacifying his empire:

> Provinces were in commotion; one man was smiting the other. The following I brought about by the favor of Ahuramazda, that the one does not smite the other at all, each one is in his place. My law – of that they feel fear, so that the stronger does not smite nor destroy the weak (DSe §5–6).[70]

All these principles could be applied by the Achaemenids to foreign relations for maintaining peace among the peoples beyond the borders of the Persian Empire.

Darius' references to territorial conquest are more explicit than the language of the Cyrus Cylinder. In his Naqš-e Rostam inscription, Darius precedes the list of countries under his rule by stating that he seized these countries outside of Persia (DNa §3C–E); in a Persepolis inscription, he says that he gained possession of them with this Persian army, and these countries felt fear of him (DPe §2C–E).[71] The political aspect of Darius I's imperial ideology is manifested in the inscription of the royal tomb in Naqš-e Rostam (DNa §4L–V), carved together with a relief depicting peoples bearing the royal throne, where Darius himself addresses his descendants:

> But if you shall think: 'How many are those countries which Darius the king held?', look at the sculptured figures which bear the throne platform. Then shall you perceive, then shall it become known to you: 'The spear of the Persian man has gone forth far away', then it shall become known to you: 'The Persian man has repulsed the enemy far away from Persia'.[72]

---

69 Kent (1950) 136. Lincoln (2012, 259) interprets this sentence as follows: 'If the Persian people/army (*kāra*) should be protected, he promised, happiness will be undestroyed for the longest time' (Lincoln always translates *šiyāti-* as happiness). However, Lincoln (2021, 20) also states that *šiyāti-* was an absolute bliss marked by peace, calm, and freedom from friction or strife.

70 Kent (1950) 142.

71 Briant (2021, 56) emphasises that the Persians are very frequently designated as conquerors in a number of inscriptions (DPe, DNa, DZe, DSab).

72 Schmitt (2000) 30. Cf. Rung (2008) 28–31.

References to the spear, as a rule, symbolise conquest of new lands, and in the Naqš-e Rostam inscription, associate conquest with Darius I's orders.[73] The figure of the Persian king is also depicted with a bow and a spear on the Type III royal coins minted at Sardis.[74] The numerous Persian and Median warriors on the Persepolis reliefs are depicted only with spears, while the bodyguards of the king from Susa are depicted with a bow and a spear.

The catalogue of 65 Achaemenid-period seal images depicting battle scenes, carefully assembled and investigated by Christopher Tuplin,[75] includes ten images, depicting battle scenes, in which one of the opponents fights with a bow, and the other with a spear (here a spear is usually found in the hands of a Persian warrior who defeats his opponent or dominates him).[76] On three seal images one of the combatants (the Persian king!) is depicted with both a bow and a spear in his hands, but it is the spear that acts as a weapon of battle.[77] The battle scenes on thirty-two seal images depict the Greeks and the Anatolians as the Persians' adversaries: both opponents fight with each other only with a spear.[78] The Persian King is also depicted on six seal images as striking his opponent with a spear.[79] Finally, even Aeschylus' *Persians* (754–755) cites the Queen's words that the King Darius has gained great wealth for his sons with a spear. Although it is very unusual to meet the mention of a spear in this context, since elsewhere in the play one can trace the opposition of the spear and the bow as respective weapons of Greeks and Persians (*Pers.* 239–240), it may be explained that the Greeks, like the Persians, considered a spear as the symbol of conquest.[80]

The inscription on the royal tomb in Naqš-e Rostam (DNa §4L–V) was an allusion to Darius' expansionist expeditions on multiple frontiers. The Persian royal inscriptions, which primarily focus on the maintenance of peace and stability in the empire, provide only a few details on specific foreign conquests: 1) the conquest of the *Sakā tigraxaudā*; 2) the (re)conquest of Egypt.

73 Degen (2019a) 70–73.
74 Nimchuk (2002) 64.
75 Wu (2014, 219–20) emphasises that the victors in these seal images are always 'Persian' or 'Iranian', occasionally the Persian king himself, while the defeated are depicted as Greeks, Egyptians, or inhabitants of Central Asia and the Eurasian steppe.
76 Tuplin (2020) fig. 1–4, 6, 13, 18, 29, 34–35.
77 Tuplin (2020) fig. 27–29.
78 Tuplin (2020) fig. 17–33, 37–51.
79 Tuplin (2020) fig. 16, 18, 24, 27–29.
80 On the Greek concept of spear-won land, see Murray (2020).

The conquest of the *Sakā tigraxaudā* (localised somewhere in Central Asia on the eastern fringes of the empire[81]) is described in the fifth column of the Bisotun inscription (DB §74):

> Afterwards with an army I went against Scythia; after that the Scythians who wear the pointed cap, these came against me, when I had come down to the sea. By means of a tree-trunk with the whole army I crossed it. Afterwards, I defeated those Scythians; another (part of them) they captured; that was led to me in fetters. And (the man) who was their chief, Skunkha by name, him they captured (and) led to me in fetters. There I made another (their) chief, as was my desire. After that the country became mine.[82]

The fifth column (in Old Persian only) was carved after the other four columns had been completed and, consequently, it depicts the events that took place in the third year after Darius became King (in 519/18 BCE). Bruce Lincoln draws attention to some peculiarities of this Persian campaign in Central Asia which make it different from all other campaigns described in the Bisotun inscription:

> Among these, his victory over one tribe of Scythians (the *Sāka tigraxauda*, 'Scyths who wear pointed hats') was prominent, and the text makes three things clear about this campaign: 1) These people were not previously part of the Persian empire, and thus could not be considered rebels; 2) Skunxa was not their 'King', but merely their 'Chief' (*maϑišta*) 3) He did not 'lie' in any fashion, nor was the Liar current among his people. At worst, the Scythians were 'gullible, vulnerable to deception', Old Persian *arīka*, and this was the argument Darius used to justify his aggression as a preemptive strike on 'the Lie' made by a coalition of the willing.[83]

This analysis makes clear that this Persian expedition to Central Asia was not a campaign against rebels (like all other campaigns recorded in the Bisotun inscription), but surely belonged to the expansionist policy of Darius which took place after the Great King had finally crushed all uprisings and completed the consolidation of empire.[84]

---

81 On the problem of localisation, see Rollinger and Degen (2021).

82 Schmitt (1991) 76.

83 Lincoln (2005) 178; (2012) 404.

84 Manning (2021, 151) notes that 'Darius seems to present his campaign against the Saka as the conquest of a new land'. However, the Saka are listed among the nine rebellious

The Bisotun inscription mentions Cambyses' expedition to Egypt: 'Cambyses went to Egypt' (DB §10N–O; in the Babylonian version of the inscription – 'Cambyses went to Egypt with his army': DB Bab. §10). However, this phrasing does not present this expedition as completely successful and resulting in the conquest of Egypt. Although in reality, Darius might have merely suppressed a revolt, mentioned in the Bisotun inscription (DB §21F), his Suez inscription ascribes the conquest to himself: 'I am a Persian; from Persia I seized Egypt' (DZs §3C).[85] The inscription on the statue of Darius from Susa makes a reference to capture of Egypt (DSab): 'This is the statue made of stone, which Darius the king ordered to be made in Egypt that he who sees it in the future will know that the Persian man held Egypt in possession/conquered Egypt'.[86] The reference to the 'Persian man' appears in this text as well as the Naqš-e Rostam inscription (DNa §4L–V); as Pierre Briant supposes, the 'Persian man' is generally taken to point implicitly to the king himself, but at the same time refers to all the Persian warriors who accompany him.[87] It seems probable that Darius intentionally wished to dissociate his achievements from Cambyses' deeds. One can also remember that in the Bisotun inscription Darius avoided reference to Cyrus or Cambyses as his predecessors when he mentioned how all the subject countries submitted to him (DB §6C; §7C), and in the Naqš-e Rostam (DNa §3C–E) and Persepolis (DPe §2C–E) inscriptions, he claimed to have seized or taken possession of these countries.

After the conquest, the main task for the king was to protect what had been gained. Darius asks Ahuramazda in one of his texts from Persepolis to protect the Persian country 'from a (hostile) army, from famine, from the Lie' (DPd §3).

The idea of keeping long-term peace in the empire became the keynote of Achaemenid ideology and propaganda, but the Bisotun inscription of Darius and the *Daiva* inscription of Xerxes emphasise a special role of war in maintaining that peace and stability. Quite similarly, both texts mention outbreaks of unrest that took place in different parts of the empire. According to the Bisotun inscription, the cause of the rebellions is declared to be the Lie (*draug̯a-*) that spread across all subject countries:

---

countries in DB §21G, and their leader Skunkha may be seen on the Bisotun relief along with other liar-kings.

85 The event in question is the defeat of the Egyptian revolt by Petubastis IV, c. 518: Kaper (2015); Wijnsma (2018). On the context of the Suez inscription, see Tuplin (1991).

86 Wasmuth (2019) 53.

87 Briant (1999) 110.

> When Cambyses had set out for Egypt, after that the people became disloyal (*arīka*), and Falsehood grew greatly in the land, in Persia as well as in Media and the other countries' (DB §10) … These (are) the countries which became rebellious (*hamiçiyā*). Falsehood made them rebellious, because these (men) lied to the people' (DB §54) … You, whosoever shall be king hereafter, be on your guard very much against Falsehood! The man who shall be a follower of Falsehood – punish him severely, if thus you shall think: "Let my country be consolidated!" (DB §55).[88]

The Bisotun inscription's phrasing is standard when it refers to defeat of the rebels by the King's army; it records in the King's name that 'by the favour of Auramazdā my army defeated that rebellious army utterly' (DB §25–30; 33; 38; 41–42; 45–46), or that Darius himself or some of his generals defeated the rebellious army utterly (DB §18–19; 35). The language of battle description is so brief here that Hilmar Klinkott concludes, 'The crucial battle fight is never really mentioned, and the description of the fight in particular is missing'.[89] Sean Manning also notes the inscription's lack of interest in the details of combat.[90] However, such details in the battle descriptions may appear lacking only if one contrasts them to those which are in the Assyrian royal inscriptions.[91] It seems to have been natural for Darius in his Bisotun inscription not to focus on the conquests or triumphal victories. The Achaemenid imperial ideology in this aspect is surely differentiated from the Assyrian one. The majority of the Assyrian royal annals usually emphasised the military expansion of the kings in order to spread fear among the unsubmissive people, pressuring them to submit to Assyrian domination.[92] However, Darius, though presenting himself as the king of countries, in the Bisotun inscription stressed the restoring of peace and stability by means of crushing the rebel armies. In this situation the reference to extreme cruelty in battle was unnecessary. Moreover, in two cases, the rebellions of Açina and Martiya in Elam, Darius states that he quelled the unrest even without resorting to military force (DB §17, 23).

88 Schmitt (1991) 51, 69. On the depiction of 'the Lie', see further Lincoln (2015) 139.

89 Klinkott (2020) 48.

90 Manning (2021, 144–45) intentionally cites the Babylonian version of the Bisotun inscription including more specific battle details than Old Persian one. He also stresses the rhetoric of the decisive battles which are followed by the capture of the enemy leader and the end of a revolt.

91 Klinkott (2020) 47. Manning (2021, 147) notes that the inscription's lack of interest in the details of combat was shared by the inscriptions of most Neo-Babylonian kings.

92 Parker (2011) 372.

Darius celebrated both conquests and victories over rebels by erecting monuments, following a practice started by Persia's Mesopotamian predecessors.[93] The earliest of such monuments – the stele of Naram-Sin – celebrates his victory over the Lullubi tribes in Iran. Scholarly literature shows its unquestionable influence on subsequent monuments up to the Achaemenid period.[94] The inscription and relief in Bisotun were the grandest monuments of the Achaemenid imperial ideology. As the text, located on a cliff at a height of 61.8 meters, appeared totally illegible from the ground,[95] a special ideological and propagandistic significance was acquired by the relief depicting a line of nine prisoners that stand before the king with their hands tied behind their backs, bound and led by a rope. The scantily preserved fragments of a stele from Babylon that were discovered by German archaeologists in front of the Ishtar gate during the excavation in 1899 and 1913/14 reproduce, to all appearances, a part of the same relief.[96] Stelae erected after Darius' other campaigns, following Neo-Assyrian precedents, not only celebrated victory but symbolised power and marked the territory that the conqueror had reached.[97] Examples include the Suez stelae (DZa–c), those erected at the Bosporus and the Tearos River in Thrace (Hdt. 4.87, 90),[98] and probably the one attested in a fragment found in Phanagoreia in 2016.[99]

Darius' monumental reliefs never illustrated acts of violence, brutal punishments, or cruelty which may be clearly seen on the Assyrian reliefs.[100] It is well-known that the Assyrian rulers liked to portray enemies in a very 'colorful' way and the depictions in the pictures are just as vivid: rampant battle scenes, the killing of enemies, flaying, cutting off limbs and collecting heads, the punishment of the leaders, deportation, looting, cutting down fruit trees, burning down cities.[101] The comparative restraint of Achaemenid monumental art

93 Feldman (2007); Head (2010); Rollinger (2016); Eppihimer (2019) 82–88.

94 Eppihimer (2019) 85.

95 Schmitt (1991) 18; Canepa (2015) 16.

96 Seidl (1999).

97 For Assyrian examples, see Radner (2010) 440; Frame (2006) 60.

98 See Bichler and Rollinger (2017) 8–9.

99 Some scholars see a fragment of the Old Persian inscription from Phanagoreia as a manifestation of the Persian kings' power that stretched as far as the Taman peninsula in the region of the Black Sea and the Azov Sea; see Tsetskhladze (2019); Avram (2018–2019). Contra: Rung and Gabelko (2019).

100 See further Manning (2021) 152–53: 'Persian court art reflected the *ahistoricism* of Achaemenid ideology. Sculptors did not depict specific military events. Instead, they depicted recurrent types of events, such as guards protecting the palace, obedient subjects bringing tribute, the lion devouring the bull, or the royal hero slaying monsters'.

101 See Jacobs (2009) 147–49; Klinkott (2020) 47, 55.

from emphasis on violence and cruelty may be due to the goal of creating a favourable image of the Persian kings as peace-keepers, and not only conquerors among their subject peoples. Besides, in fact, a close reading of the Bisotun inscription shows that Darius presented the world as already conquered.[102]

A final piece of textual evidence providing information on the military aspects of the Achaemenid ideology is the *Daiva* inscription of Xerxes (XPh). While Darius in the Naqš-e Rostam inscription (DNa §3C–E) lists the countries which he seized outside of Persia, Xerxes in the *Daiva* inscription (XPh §3E–F) speaks of countries which he ruled over outside of Persia, that is, he *inherited* them from his father. In the text of this inscription the king pledges loyalty to Ahuramazda and Arta and describes the measures taken against a country in which they were not worshipped, and worship took place instead at *daivadāna* (XPh §4):

> When I became king, there is among those countries which are inscribed above (one, which) was in turmoil. Afterwards Auramazdā brought me aid; by the favour of Auramazdā I defeated that country and put it in its proper place. And among those countries there were some where formerly the Daivas have been worshipped. Afterwards by the favour of Auramazdā I destroyed that place of the Daivas, and I gave orders: 'The Daivas shall not be worshipped any longer!' Wherever formerly the Daivas have been worshipped, there I worshipped Auramazdā at the proper time and in the proper ceremonial style.[103]

Amélie Kuhrt dates this inscription tentatively to the period after the unsuccessful attempt to conquer the Greeks in 480,[104] but its historical context cannot be accurately determined. Ernst Herzfeld and Roland Kent originally supposed that the events mentioned in this inscription lie in the period 486 to 480 BCE, but there is no way to prove it.[105] We can determine neither the date nor the countries involved; the expedition against certain rebels might have been mounted against Egypt, Babylon, Greece, or others. Xerxes' statements of loyalty to Ahuramazda did not in any way contradict the religious beliefs of his father Darius, who also asserted in all his inscriptions that he worshipped Ahuramazda and received unfailing support from the deity.[106] In historiography,

102 Manning (2021) 153.
103 Schmitt (2000) 93.
104 Kuhrt (2007) 305.
105 Herzfeld (1936) 63–65; Kent (1937) 305; see also Stoneman (2015) 106–107.
106 Lincoln (2007) 45; Lincoln (2008).

the evidence provided by the *Daiva* inscription of Xerxes is sometimes viewed to the opposite effect – as a sign that this king abandoned the religious policy that his father Darius had carried out.[107] But it is obvious that from Xerxes' perspective such a contradiction could not have existed: the action against the followers of *Daivas* who did not worship Ahuramazda was quite a natural means of pursuing his religious policy.[108] Heleen Sancisi-Weerdenburg has rightly drawn attention to the textual connection between the *Daiva* inscription of Xerxes and the last lines of the Bisotun inscription (DB §72, 75) that mention Darius' expeditions against the Elamites and Scythians, who, as the text states, were 'hostile' (*arīkā*) and did not worship Ahuramazda.[109] In short, Xerxes appears to have followed his father's policy by connecting military victory with divine order and the maintenance of peace in the empire.

4 Punishment and Reward in Achaemenid Ideology

Both punishment and reward were so significant in the Achaemenid ideology that they determined an actual royal policy towards the various peoples. Punishment was considered as the most effective way to cause fear among subjects and to prevent them from uprisings, while reward was thought of as stimulating their loyalty toward the King and his administration. The data on punishment of rebels provided by Achaemenid correlate with similar claims asserted in Greek authors. Old Persian inscriptions repeatedly emphasise punishment and reward on Darius' behalf:

> The man who was loyal, him I treated well, who was disloyal, him I punished severely (DB §8B–D) ... The man who strove for my (royal) house, him I treated well, who did harm, him I punished severely (DB §63K–N).[110]
>
> The man who cooperates, for him according to the cooperation, thus I care for him; who does harm, according to the harm done, thus I punish him ... What a man achieves or brings according to his powers, by that I become satisfied, and it is very much my desire; and I am pleased and give generously to loyal men (DNb §6E–G).[111]

107 Ahn (1992) 111–22; Shannon (2007) 84.

108 Bridges (2014) 92.

109 Sancisi-Weerdenburg (1980) 16–17, 35; (2002) 581, 589.

110 Schmitt (1991) 50, 71.

111 Schmitt (2000) 40; Xerxes replicates the statement (XPl §2C–E = Schmitt (2000) 103).

The Bisotun inscription mentions the punishment of a number of rebels, and particular stress is laid on the public character of execution, chosen, no doubt, in order to intimidate the people.[112] Darius gave an order to cut off the Median rebel Fravartish's nose and ears and tongue and to put out one eye, and to keep him bound at the palace entrance in Ecbatana so that all the people could see him. Afterwards the king commanded the impalement of Fravartish, and the hanging up of his foremost followers within the fortress (DB §32G–M). The Babylonian version of the Bisotun inscription specifies the beheading of Fravartish's followers and the display of their heads (DB Bab. §25). Darius also commanded his men to cut off both the nose and ears of the Sagartian rebel Tritantaichmes, and put out one eye; he was kept bound at the palace entrance and all the people saw him. Afterwards the king impaled him at Arbela (DB §33S–W). A less inventive execution awaited the Persian Vahyazdāta, the Babylonian Arakha, and their foremost followers: they were impaled (DB §42M–N; 50Q–R).

Impalement and cutting off the nose and ears were methods of execution used in Mesopotamia long before the Persians, especially in the Neo-Assyrian Empire.[113] Cutting off the nose and ears among the Persians is also mentioned on a number of occasions by Herodotus (3.69, 3.118.2, 3.154–59). Bruce Lincoln believes that cutting off the ears – perhaps along with other organs of sense and communication – was to mark those convicted of lying.[114]

The methods of punishing rebels mentioned in the Bisotun inscription are supplemented by ancient Greek and Roman authors, whose reports refer to a later period. Ctesias (FGrH 688 F15 52–53) says that Darius II Ochus executed the Persian nobles Arsites, Artyphios, and Pissouthnes by throwing them into burning ashes. Similar methods of putting political opponents to death are described by Valerius Maximus (9.ii.ext.6).

The punishment of larger rebel populations is also reflected in the Achaemenid monumental inscriptions. The Babylonian and Aramaic versions of Bisotun record not only the counting of casualties inflicted on rebel armies, but also the numbers of prisoners of war who were taken and/or executed.[115] However, the Bisotun inscription makes no mention of deportations of any population in the course of suppression of an uprising; and it is not clear if

112 On the practice of public execution in the Achaemenid Empire, see Jacobs (2009); McKeon (2020) 358 emphasises its visual emphasis as a deterrent against future insurrection. On the mercy of the Persian kings towards rebels, see Beckman (2020).

113 Jacobs (2009) 134–137; Radner (2015).

114 Lincoln (2012) 222; (2015) 143.

115 DB Bab. §19, 22–25, 28–29, 31, 34–35; DB Aram. §2–9. On the evaluation of this evidence, see Hyland (2014).

this method was ever employed against the rebels at the beginning of Darius' reign. It is well-attested, however, that the Achaemenids resorted to forcible displacement of peoples as a punishment for disobedience and disloyalty.[116] The use of deportations in the Achaemenid Empire mainly resembled practices in the Neo-Assyrian and Neo-Babylonian empires in which displacement of people was a method of imperial control.[117] Marc Mendoza names three types of Achaemenid deportations: 1) economic (displacement and movement of labour groups);[118] 2) punitive (punishment for disloyalty of rebellious people); 3) political (relocation of pro-Persian political elites into the Persian heartland).[119]

The only example of a punitive deportation (as prisoners of war) given by a Near Eastern source is the mention in the Babylonian Chronicle of a relocation of insurgent Sidon's citizens after the suppression of Tennes' revolt by Artaxerxes III:

> The fourteenth [year] of Umashu, who is called Artaxerxes (III): in the month Tishri the prisoners which the King took (from) Sidon (were brought) to Babylon and Susa. On the sixteen day the ... women, prisoners from Sidon, which the King sent to Babylon – on that day they entered the palace of the King.[120]

The most explicit examples of a punitive deportation by Darius' order were preserved only in Greek authors' works. The best-known examples of deportation are the following: 1) Darius I established Barkaian captives from northwestern Africa in a village in Bactria (Hdt. 4.204); 2) he also settled Paionians of Thrace in Asia Minor, though most of them returned home during the Ionian revolt (Hdt. 5. 14–15; 5.17; 5.98); 3) Milesians were deported to Ampe at the mouth of the Tigris (Hdt. 6. 20); 4) Eretrians were settled at Arderikka in Kissia of Elam (Hdt. 6.119).[121]

116 Van der Spek (2014, 258) finds examples of deportation in Cyrus the Great's policy, but the given examples (the existence of a community of Lydians in Nippur or Mazares' enslavement of the inhabitants of Priene) do not seem conclusive. On the deportation of the Egyptians by Cambyses' order, see Degen (2019b) 195–98.

117 On Assyrian practice of deportations, see Oded (1979).

118 See Rollinger (2018) 430–34.

119 Mendoza (2020) 143–52. On the Persian deportations in general, see Matarese 2021.

120 Grayson (1975) 114.

121 On the full list of Achaemenid deportations reflected in Classical sources, see Wiesehöfer (2021) 872.

Herodotus' description portrays Darius and Xerxes as merciless in punishment and generous in reward; on the whole, this conforms to what these monarchs say about themselves in their inscriptions, which declared that they were friends of the right and enemies of the wrong and lie-followers (DNb; XPl). Punishment was inflicted not only on those citizens who rose up against the king or were accused of giving their support to the rebels. Among those who were severely punished there were also private individuals who aroused the king's anger. Darius had all of the three sons of Oiobazos executed because the latter had asked him to release at least one of them from service before the expedition to Scythia (4.84). Xerxes had the eldest son of Pythios the Lydian cruelly executed because the father had asked the king to release him from the march against Hellas (7.38). But the kings could be merciful as well. As a reply to the exhortations of Intaphernes' wife, Darius agreed to pardon her and one of her family members after they had been found guilty of a conspiracy against the king (3.119). Xerxes, referring to his nobility of character, refused to take vengeance on Sperthias and Boulis, who had been sent by the Spartans to Susa to make atonement for the heralds who had been killed at Sparta (7.135), and then set free the Greek spies captured by the Persians (7.146). Herodotus speaks of the generosity of both kings on several occasions. Among those who were rewarded, there were citizens in general (7.116; 8.120) as well as the most zealous of his subjects who had rendered important services to the king; the rewards were money, cities and lands (3.140; 5.11, 94; 6.70; 8.136), houses and Persian wives (6.39), troops to command (9.109), and honorary titles (3.140; 8.85). Let us quote two examples in which the king's reward was given to citizens of two Greek communities:

> When Xerxes came to Akanthos, he declared the Akanthians his guests and friends, and gave them Median clothing, praising them for the zeal with which he saw them furthering his campaign, and for what he heard of the digging of the canal (Hdt. 7.116).
>
> It is known that when Xerxes came to Abdera in his return, he made a compact of friendship with its people and gave them a golden sword and a gilt tiara (Hdt. 8.120).

On several occasions, Herodotus cites the title of 'king's friend' (7.134, 150) and that of 'king's benefactor' (3.139–140; 8.85); they were received by the king's subjects, including the Greeks, for their services.[122] The historian draws

122 On these titles, see Wiesehöfer (1980). On *philia* in Persian-Greek diplomacy, see also Waters (2014).

attention to the fact that the *king's benefactors* were called *orosangai* in the Persian language (Hdt. 8.85), which may stem from the Old Persian word **var-usanka*, 'widely renowned, world-famous'.[123]

Nymphis of Herakleia (F 12) reports that *orosangai* had the highest precedence of *proedria* ('front seat') in royal audiences and were called royal guest-friends in their language. Nymphis' evidence explains the connection between the title of 'king's benefactor' with the 'front seat' privilege. In fact, Diodorus (17.14.2) speaks of such a privilege with respect to the Thebans: '(Thebans) alone of the Greeks were honoured as benefactors by the Persian kings, so that the ambassadors of the Thebans were seated on thrones set in front of the kings'. Plutarch (*Pel.* 30) also reports that the Thebans were regarded as the king's hereditary friends. We can also make an assumption that while 'friendship' marked one possible stage of relationship between a Persian king and his distant allies, the title of 'king's benefactor' could immediately raise the bearer's status in the Achaemenid Empire. The Thebans' case clearly demonstrates that both privileged statuses given to them as royal rewards were due to their military support for Xerxes in the Greco-Persian wars. So, it is quite evident that both punishment of rebels and rewards to the loyal subjects were closely related to the importance of war in Achaemenid ideology.

5 Conclusion

To sum up, the military aspect of the Achaemenid imperial ideology was of primary importance. Although the idea of conquest is clearly expressed in the monumental royal inscriptions and supported by visual evidence, nevertheless concrete details are not well reflected in Persian sources. The maintenance of peace and stability found even greater expression in Achaemenid ideology than the idea of conquest, as it was this very idea that determined the empire's vitality by preventing its disintegration. Punishment and reward were among the most important methods in the arsenal of imperial ideology, because they supported the overall theme of peace and stability from one side of the Achaemenid *oikoumenē* to the other. Of course, the royal propaganda addressed to the King's subjects misrepresented real events in support of this ideological programme. Cyrus presented himself as a peacemaker, Darius attributed his predecessors' conquests to himself, and Xerxes marked himself as a true follower of his father Darius.

123 Tavernier (2014) 319, n. 182.

There are some final issues relating to Achaemenid imperial ideology which can benefit from further investigation, but unfortunately, many of these issues cannot be resolved by use of Old Persian sources. The combination of military and non-military methods of expansion of the Persian Empire may be further clarified by investigation of the ideological concept of earth-and-water, though this concept is not reflected in the evidence of the Achaemenid Empire's main languages, and its significance is shown only in reports by Herodotus.[124] Lastly, although this subject is also not covered by Old Persian evidence and therefore may be considered only from a Greek perspective, we should not forget to pay more attention to the later period of Achaemenid ideology from Xerxes to Darius III.[125,126]

Bibliography

Ahn, G. (1992) *Religiöse Herrscherlegitimation im achämenidischen Iran*. Leiden.

Aro, S. and Whiting, R.M. (eds.) (2000) *The Heirs of Assyria: Proceedings of the Opening Symposium of the Assyrian and Babylonian Intellectual Heritage Project. Held in Tvärminne, Finland, October 8–11, 1998*. Helsinki.

Avram, A. (2018–2019) 'Les Perses en mer Noire à l'époque de Darius Ier: Nouveaux documents et nouvelles interpretations'. *Dacia* 67–69: 169–98.

Baker, H.D., Robson, E., and Zólyomim, G. (eds) (2010) *Your Praise is Sweet: A Memorial Volume for Jeremy Black from Students, Colleagues and Friends*. London.

Bakir, K. (ed) (2001) *Achaemenid Anatolia: Proceedings of the First International Symposium on Anatolia in the Achaemenid Period, Bandirma 15–18 August 1997*. Leiden.

Bakker, E.J., de Jong, I.J.F., and van Wees, H. (eds) (2002) *Brill's Companion to Herodotus*. Leiden.

Bernholz, P. and Vaubel, R. (eds) (2014) *Explaining Monetary and Financial Innovation*. Heidelberg.

Beckman, D. (2020) 'Law, Mercy, and Reconciliation in the Achaemenid Empire'. *JAH* 8.2: 127–51.

Bichler, R. and Rollinger, R. (2017) 'Universale Weltherrschaft und die Monumente an ihren Grenzen', in Rollinger (2017) 1–30.

124 On this concept, see Kuhrt (1988); Rung (2015); Waters (2014).

125 On Achaemenid ideology after Xerxes' death, see Hyland (2018) 30–34.

126 This study was performed with financial support of the Russian Science Foundation, project No. 20-18-00374, implemented on the basis of Lobachevsky University of Nizhni Novgorod.

Binder, C., Börm, H., and Luther, A. (eds) (2016) *Dīwān. Studies in the History and Culture of the Ancient Near East and the Eastern Mediterranean: Festschrift für Josef Wiesehöfer zum 65. Geburtstag*. Duisburg.

Boschung, D., Shapiro, A., and Waschenk, F. (eds) (2015) *Bodies in Transition: Dissolving the Boundaries of Embodied Knowledge*. Paderborn.

Briant, P. (1999) 'The Achaemenid Empire', in Raaflaub and Rosenstein (1999) 105–28.

Briant, P. (2002) *From Cyrus to Alexander: A History of the Persian Empire*. Winona Lake.

Briant, P. (2021) 'From the Mediterranean to the Indus Valley: Modalities and Limitations of the Achaemenid Imperial Space', in Pines, Biran, and Rüpke (2021) 49–78.

Briant, P. and Henkelman, W. (eds) (2008) *L'Archive des Fortifications de Persépolis: état des questions et perspectives de recherches*. Paris.

Bridges, E. (2014) *Imagining Xerxes: Ancient Perspectives on a Persian King*. London.

Brisch, N. (ed) (2008) *Religion and Power: Divine Kingship in the Ancient World and Beyond*. Chicago.

Brosius, M. (2005) '*Pax persica*: Königliche Ideologie und Kriegsfürung im Achämenidenreich', in Meißner, Schmitt, and Sommer (2005) 135–61.

Brosius, M. (2010) '*Pax Persica* and the People of the Black Sea Region: Extent and Limits of Achaemenid Imperial Ideology', in Rehm and Nieling (2010) 29–40.

Brosius, M. (2021) *A History of Ancient Persia: The Achaemenid Empire*. Hoboken, NJ.

Callieri, P. (2023) 'The Southeastern Regions of the Persian Empire on the Indo-Iranian Frontier: Arachosia, Drangiana, Gedrosia, Sattagydia, Gandhara, and India', in Radner, Moeller, and Potts (2023) 837–86.

Canepa, M.P. (2015) 'Inscriptions, Royal Spaces and Iranian Identity: Epigraphic Practices in Persia and the Ancient Iranian World', in Eastmond (2015) 10–35.

Cheng, J. and Feldman, M.H. (eds) (2007) *Ancient Near Eastern Art in Context Studies in Honor of Irene J. Winter by Her Students*. Leiden.

Dan, A. (2013) 'Achaemenid World Representations in Herodotus' Histories: Some Geographic Examples of Cultural Translation', in Geus, Irwin, and Poiss (2013) 83–121.

Degen, J. (2019a) 'Alexander III., Dareios I. und das speererworbene Land (Diod. 17, 17, 2)'. *JANEH* 6(1): 53–95.

Degen, J. (2019b) 'Deportationen zur Zeit der ersten persischen Eroberung Ägyptens. Kambyses' Sicherungspolitik im Lichte hellenischer und achaimenidischer Quellen', in Rollinger, and Stadler (2019) 183–217.

DeGrado, J. (2019) 'King of the Four Quarters: Diversity as a Rhetorical Strategy of the Neo-Assyrian Empire'. *Iraq* 81: 107–25.

Descat, R. (1994) 'Darius, le Roi "Kapelos"', in Sancisi-Weerdenburg, Kuhrt, and Root (1994) 161–66.

DeSouza, P. and France, J. (eds) (2008) *War and Peace in Ancient and Medieval History*. Cambridge.

Dusinberre, E., Garrison, M.B., and Henkelman, W.F.M. (eds) (2020) *The Art of Empire in Achaemenid Persia: Studies in Honour of Margaret Cool Root*. Leiden.

Dunsch, B. and Ruffing, K. (eds) (2013) *Herodots Quellen. Die Quellen Herodots*. Wiesbaden.

Eastmond, A. (ed) (2015) *Viewing Inscriptions in the Late Antique and Medieval World*. Cambridge.

Eppihimer, M. (2019) *Exemplars of Kingship: Art, Tradition, and the Legacy of the Akkadians*. Oxford.

Feldman, M.H. (2007) 'Darius I and the Heroes of Akkad: Affect and Agency in the Bisitun Relief', in Cheng and Feldman (2007) 265–93.

Fitzpatrick-McKinley, A. (ed) (2015) *Assessing Biblical and Classical Sources for the Reconstruction of Persian Influence, History and Culture*. Wiesbaden.

Fortin, M. (ed) (2006) *Tell' Acharneh 1998–2004. Rapports préliminaires sur les campagnes de fouilles et saison d'études/Preliminary Reports on Excavation Campaigns and Study Season* (Subartu, 18). Turnhout.

Frame, G. (1999) 'The Inscription of Sargon II at Tang-i Var'. *Orientalia* 68: 31–57.

Frame, G. (2006) 'The Tell Acharneh Stela of Sargon II of Assyria', in Fortin (2006) 49–68.

Frye, R.N. (1962) *The Heritage of Persia*. London.

Gadd, C.J. and Legrain, G. (1928) *Ur Excavations Text. I: Royal Inscriptions*. London.

Geus, K., Irwin, E., and Poiss, T. (eds) (2013) *Herodots Wege des Erzählens: Logos und Topos in den Historie*. Frankfurt.

Grayson, A.K. (1975). *Assyrian and Babylonian Chronicles*. Locust Valley.

Gropp, G. (2009) 'Die Darstellung der 23 Völker auf den Reliefs des Apadana von Persepolis'. *IrAnt* 44: 283–359.

Head, R. (2010) 'Assyria at Bisitun and the Universal Kingship of Darius I of Persia', in Baker, Robson, and Zólyomi (2010) 117–24.

Herrenschmidt, C. (1976) 'Désignation de l'empire et concepts politiques de Darius Ier d'après ses inscriptions en Vieux Perse'. *StIr* 5: 33–65.

Herzfeld, E. (1936) 'Xerxes' Verbot des Daiva-Cultes'. *AMI* 8: 56–77.

Hyland, J.O. (2014) 'The Casualty Figures in Darius' Bisitun Inscription'. *JANEH* 1(2): 1–27.

Hyland, J.O. (2018) *Persian Interventions: The Achaemenid Empire, Athens and Sparta, 450–386 BCE*. Baltimore.

Jacobs, B. (2009) 'Grausame Hinrichtungen – friedliche Bilder. Zum Verhältnis der politischen Realität zu den Darstellungsszenarien der achämenidischen Kunst', in Zimmermann (2009) 121–53.

Jacobs, B. and Rollinger, R. (eds) (2021) *A Companion to the Achaemenid Persian Empire*. Hoboken, NJ.

Kaper, O.E. (2015) 'Petubastis IV in the Dakhla Oasis: New Evidence about an Early Rebellion against Persian Rule and Its Suppression', in Silverman and Waerzeggers (2015) 125–50.

Kent, R.G. (1937) 'The Daiva-Inscription of Xerxes'. *Language* 13: 292–305.

Kent, R.G. (1950) *Old Persian: Grammar, Texts, Lexicon*. New Haven.

Kleber, K. (2015) 'Taxation in the Achaemenid Empire', in Oxford Handbooks online (https://www.oxfordhandbooks.com/view/10.1093/oxfordhb/9780199935390.001.0001/oxfordhb-9780199935390-e-34?rskey=oE1Sow&result=2).

Klinkott, H. (2001) 'Yauna – Die Griechen aus persischer Sicht?', in Klinkott (2001) 107–48.

Klinkott, H. (ed) (2001) *Anatolien in Lichte kultureller Wechselwirkungen. Akkulturationsphänomene in Kleinasien und seinen Nachbarregionen wärend des 2. und 1. Jahrtausends v. Chr*. Tübingen.

Klinkott, H. (2020) 'Victor without Victory? The Lack of Battle Descriptions in the Achaeamenid Empire', in Luggin and Fink (2020) 47–59.

Klinkott, H., Luther, A., and Wiesehöfer, J. (eds) (2021) *Beiträge zur Geschichte und Kultur des alten Iran und benachbarter Gebiete: Festschrift für Rüdiger Schmitt*. Stuttgart.

Konijnendijk, R. (2021) 'Legitimization of War', in Jacobs and Rollinger (2021) 1141–50.

Kozuh, M., Henkelman, W.F.M., Jones, C.E., and Woods, C. (eds) (2014) *Extraction & Control: Studies in Honor of Matthew W. Stolper*. Chicago.

Kuhrt, A. (1983) 'The Cyrus Cylinder and Achaemenid Imperial Policy'. *JSOT* 25: 83–97.

Kuhrt, A. (1988) 'Earth and Water', in Kuhrt and Sancisi-Weerdenburg (1988) 87–99.

Kuhrt, A. (2007) *The Persian Empire: A Corpus of Sources from the Achaemenid Period*. Oxford.

Kuhrt, A. (2010) 'Achaemenid Images of Royalty and Empire', in Lanfranchi and Rollinger (2010) 87–106.

Kuhrt, A. and Sancisi-Weerdenburg, H. (eds) (1988) *Achaemenid History 3: Method and Theory. Proceedings of the London 1985 Achaemenid History Workshop*. Leiden.

Kuhrt, A. and Sancisi-Weerdenburg, H. (eds) (1991) *Achaemenid History 6: Asia Minor and Egypt: Old Cultures in a New Empire*. Leiden.

Kurke, L. (1999) *Coins, Bodies, Games, and Gold: The Politics of Meaning in Archaic Greece*. Princeton.

Lanfranchi, G. and Rollinger, R. (eds) (2010) *Concepts of Kingship in Antiquity: Proceedings of the European Science foundation Exploratory Workshop held in Padova, November 28th–December 1st, 2007*. Padova.

Lanfranchi, G., Morandi Bonacossi, D., Pappi, C., and Ponchia, S. (eds) (2012) *Leggo! Studies Presented to Frederik Mario Fales on the Occasion of his 65th Birthday*. Wiesbaden.

Llewellyn-Jones, L. (2015) '"That My Body is Strong": The Physique and Appearance of Achaemenid Monarchy', in Boschung, Shapiro, and Waschenk (2015) 211–48.

Lincoln, B. (2005) 'Rebellion and Treatment of Rebels in the Achaemenid Empire'. *Archiv für Religionsgeschichte* 7: 167–79.

Lincoln, B. (2007) *Religion, Empire, and Torture: The Case of Achaemenid Persia*. Chicago.

Lincoln, B. (2008) 'The Role of Religion in Achaemenian Imperialism', in Brisch (2008) 221–42.

Lincoln, B. (2012) *'Happiness for Mankind': Achaemenid Religion and the Imperial Project*. Leuven.

Lincoln, B. (2015) 'Representing the Lie in Achaemenian Persia', in Swennen (2015) 135–44.

Lincoln, B. (2021) *Religion, Culture, and Politics in Pre-Islamic Iran: Collected Essays*. Leiden.

Luggin, J. and Fink, S. (eds) (2020) *Battle Descriptions as Literary Texts: A Comparative Approach*. Wiesbaden.

Manning, S. (2021) *Armed Force in the Teispid-Achaemenid Empire: Past Approaches, Future Prospects*. Wiesbaden.

Matarese, C. (2021) *Deportationen im Perserreich in teispidisch-achaimenidischer Zeit*. Wiesbaden.

Mayerhofer, M. (1979) 'Archaismus und Innovation. Eine Grammatische Unterscheidung in Keilschrift-Paralleltexten aus achämenidischer Zeit'. *AO* 47: 96–99.

McKeon, K.P.J. (2020) 'Barbarians at the Gate: Herodotus, Bisotun, and a Persian Punishment in Egypt'. *AJP* 141.3: 349–80.

Medenieks, S. (2017) 'Cyrus the Great: An Unconventional Peacemaker', in Moloney and Williams (2017) 133–47.

Meißner, B., Schmitt, O., and Sommer, M. (eds) (2005) *Krieg – Gesellschaft – Institutionen. Beiträge zu einer vergleichenden Kriegsgeschichte*, Berlin.

Mendoza, M. (2020) 'Achaemenid Deportations: Greek Perceptions, Persian Realities'. *Semitica* 62: 137–62.

Moloney, E.P. and Williams, M.S. (eds) (2017) *Peace and Reconciliation in the Classical World*. London.

Murray, S.R. (2020) '"Won by the Spear": The Importance of the Dory to the Ancient Greek Warrior'. *Hypothekai* 4: 74–88.

Nimchuk, C.L. (2002) 'The "Archers" of Darius: Coinage or Tokens of Royal Esteem?' *Ars Orientalis* 32: 55–79.

Oded, R. (1979) *Mass Deportations and Deportees in the Neo-Assyrian Empire*. Wiesbaden.

Panaino, A. (2018) 'Temper and Self-control in the Persian King's Ideal Portrait'. *DABIR* 6.

Parker, B. (2011) 'The Construction and Performance of Kingship in the Neo-Assyrian Empire'. *JAR* 67(3): 357–86.

Pines, Y., Biran, M., and Rüpke, J. (eds) (2021) *The Limits of Universal Rule: Eurasian Empires Compared*. Cambridge.

Radner, K. (2010) 'The Stele of Sargon II of Assyria at Kition: A Focus for an Emerging Cypriot Identity?', in Rollinger, Gufler, Lang, and Madreiter (2010) 429–49.

Radner, K. (2015) 'High Visibility Punishment and Deterrent: Impalement in Assyrian Warfare and Legal Practice'. *ZAR* 21: 103–28.

Radner, K., Moeller, N., and Potts, D.T. (eds) (2023) *The Oxford History of the Ancient Near East. Volume 5: The Age of Persia*. Oxford.

Raaflaub, K. (ed) (1993) *Anfänge politischen Denkens in der Antike. Die nahöstlichen Kulturen und die Griechen. [Kolloquium, München 10.–13. Juni 1990]*. München.

Raaflaub, K. and Rosenstein, N. (eds) (1999) *War and Society in the Ancient and Medieval Worlds: Asia, The Mediterranean, Europe and Mesoamerica*. Washington, DC.

Rehm, E. and Nieling, J. (eds) (2010) *Achaemenid Impact in the Black Sea Region: Conference Held in Sandbjerg, Denmark, 10–12 January 2008*. Aarhus.

Rollinger, R. (2012) 'From Sargon of Agade and the Assyrian Kings to Khusrau I and Beyond: On the Persistence of Ancient Near Eastern Traditions', in Lanfranchi, Morandi Bonacossi, Pappi, and Ponchia (2012) 725–43.

Rollinger, R. (2013) 'Dareios und Xerxes an den Rändern der Welt und die Inszenierung von Weltherrschaft. Altorientalisches bei Herodot', in Dunsch and Ruffing (2013) 95–116.

Rollinger, R. (2016) 'The Relief at Bisitun and Its Ancient Near Eastern Setting: Contextualizing the Visual Vocabulary of Darius' Triumph over Gaumata', in Binder, Börm, and Luther (2016) 5–51.

Rollinger, R. (2018) 'Between Deportation and Recruitment: Craftsmen and Specialists from the West in Ancient Near Eastern Empires (from Neo-Assyrian Times through Alexander III)', in Woytek (2018) 425–44.

Rollinger, R. (2021) 'Empire, Borders, and Ideology', in Jacobs and Rollinger (2021) 815–30.

Rollinger, R. (ed) (2017) *Die Sicht auf die Welt zwischen Ost und West (750 v. Chr.–550 n. Chr.). Looking at the World, from the East and the West (750 BCE–550 CE)*. Wiesbaden.

Rollinger, R. and Degen, J. (2021) 'Conceptualizing Universal Rulership: Considerations on the Persian Achaemenid Worldview and the Saka at the "End of the World"', in Klinkott, Luther, and Wiesehöfer (2021) 187–224.

Rollinger, R. and Ruffing, K. (2014) 'World View and Perception of Space. 1. The View from East to West: World View and Perception of Space in the Neo-Assyrian Empire', in Zenzen, Hölscher, and Trampedach (2014) 93–135.

Rollinger, R., Gufler, B., Lang, M., and Madreiter, I. (eds) (2010) *Interkulturalität in der Alten Welt: Vorderasien, Hellas, Ägypten und die vielfältigen Ebenen des Kontakts* (Philippika 34). Wiesbaden.

Rollinger, R., Truschnegg, B., and Bichler, R. (eds) (2011) *Herodot und das persische Weltreich – Herodotus and the Persian Empire*. Wiesbaden.

Rollinger, R. and Stadler, H. (eds) (2019) *7 Millionen Jahre Migrationsgeschichte. Annäherungen zwischen Archäologie, Geschichte und Philologie*. Innsbruck.

Romney, J. (2017) 'Herodotean Geography (4.36–45): A Persian *oikoumenē*?' *GRBS* 57: 862–81.

Root, M.C. (1979) *The King and Kingship in Achaemenid Art: Essays on the Creation of an Iconography of Empire*. Leiden.

Rung, E. (2008) 'War, Peace and Diplomacy in Graeco-Persian Relations', in DeSouza and France (2008) 28–50.

Rung, E. (2015) 'The Language of the Achaemenid Imperial Diplomacy towards the Greeks: The Meaning of Earth and Water'. *Klio* 97: 503–15.

Rung, E. and Gabelko, O. (2019) 'From Bosporus … to Bosporus. A New Interpretation and Historical Context of the Old Persian Inscription from Phanagoreia'. *IrAnt* 54: 83–125.

Salano Rossi, L.A. (2020) 'A pax Persica: o contexto imperial persa'. *Ribla* 81(1): 11–19.

Sancisi-Weerdenburg, H. (1980) 'Yauna en Persai: Grieken en Perzen in een ander perspectief', PhD Dissertation, University of Leiden.

Sancisi-Weerdenburg, H. (1993) 'Political Concepts in Old-Persian Royal Inscriptions', in Raaflaub (1993) 145–63.

Sancisi-Weerdenburg, H. (2001a) 'The Problem of the Yaunā', in Bakir (2001) 1–11.

Sancisi-Weerdenburg, H. (2001b) 'Yaunā by the Sea and across the Sea', in Malkin (2001) 323–46.

Sancisi-Weerdenburg, H. (2002) 'The Personality of Xerxes, King of Kings', in Bakker, de Jong, and van Wees (2002) 579–90.

Sancisi-Weerdenburg, H., Kuhrt, A. and Root, M.C. (eds) (1994). *Achaemenid History 8: Continuity and Change. Proceedings of the Last Achaemenid History Workshop, Ann Arbor, April 6–8, 1990*. Leiden.

Schaudig, H. (2019) 'The Text of the Cyrus Cylinder' in Shayegan (2019) 16–25.

Schmidt, E.F. (1953) *Persepolis. 1. Structures, Reliefs, Inscriptions*. Chicago.

Schmitt, R. (1991) *The Bisitun Inscriptions of Darius the Great: Old Persian Text*. London.

Schmitt, R. (2000) *The Old Persian Inscriptions of Naqsh-i Rustam and Persepolis*. London.

Schmitt, R. (2009) *Die altpersischen Inschriften der Achaimeniden. Editio minor mit deutscher Übersetzung*. Wiesbaden.

Schmitt, R. (2014) *Worterbuch der altpersischen Königsinschriften*. Wiesbaden.

Scott, L. (2005) *Historical Commentary on Herodotus Book 6*. Leiden.

Seager, R. and Tuplin, C.J. (1980) 'The Freedom of the Greeks of Asia: On the Origins of a Concept and the Creation of a Slogan'. *JHS* 100: 141–54.

Seidl, U. (1999) 'Ein Monument Darius' I. aus Babylon'. *ZA* 98: 101–14.

Shannon, A.R. (2007) 'The Achaemenid Kings and the Worship of Ahura Mazda: Proto-Zoroastrianism in the Persian Empire'. *Studia Antiqua* 5: 79–85.

Shayegan, M.R. (ed) (2019) *Cyrus the Great: Life and Lore*. Cambridge, MA.

Silverman, J.M. (2020) *Persian Royal-Judaean Elite Engagements in the Early Teispid and Achaemenid Empire: The King's Acolytes*. London.

Silverman, J.M. and Waerzeggers, C. (2015) *Political Memory in and After the Persian Empire*. Atlanta.

Stoneman, R. (2015) *Xerxes: A Persian Life*. New Haven.

Swennen, P. (ed) (2015) *Démons iraniens. Actes du colloque international organisé à l'Université de Liège les 5 et 6 février 2009 à l'occasion des 65 ans de Jean Kellens*. Liége.

Tavernier, J. (2014) 'Some Thoughts on the *ustarbaru*', in Kozuh, Henkelman, Jones, and Woods (2014) 296–322.

Tsetskhladze, G. (2019) 'An Achaemenid Inscription from Phanagoria: Extending the Boundaries of Empire'. *Ancient West & East* 18: 113–51.

Tuplin, C.J. (1991) 'Darius' Suez Canal and Persian Imperialism', in Kuhrt and Sancisi-Weerdenburg (1991) 237–83.

Tuplin, C.J. (1997) 'Achaemenid Arithmetic: Numerical Problems in Persian History'. *Topoi. Orient-Occident* 1: 365–421.

Tuplin, C.J. (2008) 'Taxation and Death: Certainties in the Fortification Archive?', in Briant and Henkelman (2008) 317–86.

Tuplin, C.J. (2011) 'Managing the World: Herodotus on Achaemenid Imperial Organization', in Rollinger, Truschnegg, and Bichler (2011) 39–63.

Tuplin, C.J. (2014) 'The Changing Pattern of Achaemenid Imperial Coinage', in Bernholz and Vaubel (2014) 127–68.

Tuplin, C.J. (2015) 'The Justice of Darius: Reflections on the Achaemenid Empire as a Rule-bound Environment', in Fitzpatrick-McKinley (2015) 73–126.

Tuplin, C.J. (2017) 'War and Peace in Achaemenid Imperial Ideology'. *Electrum* 24: 31–54.

Tuplin, C.J. (2020) 'Sigillography and Soldiers: Cataloguing Military Activity on Achaemenid Period Seals', in Dusinberre, Garrison, and Henkelman (2020) 329–459.

Van der Spek, R. (2014) 'Cyrus the Great, Exiles, and Foreign Gods: A Comparison of Assyrian and Persian Policies on Subject Nations', in Kozuh, Henkelman, Jones, and Woods (2014) 233–64.

Vogelsang, W.J. (1992) *The Rise and Organisation of the Achaemenid Empire: The Eastern Iranian Evidence*. Leiden.

Wasmuth, M. (2019) 'Transient or Eternal? Cross-regional Identity Display Reconsidered: The Missing Head of the Statue of Darius (NMI 4112)'. *JAEI* 24: 49–66.

Waters, M.W. (2014) 'Earth, Water, and Friendship with the King: Argos and Persia in the Mid-fifth Century', in Kozuh, Henkelman, Jones, and Woods (2014) 331–36.

Waters, M.W. (2019) 'Cyrus Rising: Reflections on Word Choice, Ancient and Modern', in Shayegan (2019) 26–45.

Westenholz, J. (2000) 'The King, the Emperor, and the Empire: Continuity and Discontinuity of Royal Representation in Text and Image', in Aro and Whiting (2000) 99–125.

Wiesehöfer, J. (1980) 'Die "Freunde" und "Wohltäter" des Grosskönigs'. *StIr* 9: 7–21.

Wiesehöfer, J. (2021) 'Deportations', in Jacobs and Rollinger (2021) 871–78.

Wijnsma, U.Z. (2018) 'The Worst Revolt of the Bisitun Crisis: A Chronological Reconstruction of the Egyptian Revolt under Petubastis IV'. *JNES* 77: 157–73.

Wilkinson, R.H. (1991) 'The Representation of the Bow in the Art of Egypt and the Ancient Near East'. *JANES* 20: 83–99.

Woytek, B. (ed) (2018) *Infrastructure and Distribution in Ancient Economies: Proceedings of a Conference Held at the Austrian Academy of Sciences, 28–31 October 2014*. Vienna.

Wu, X. (2014) '"O Young Man … Make Known of What Kind You Are": Warfare, History, and Elite Ideology of the Achaemenid Persian Empire'. *IrAnt* 49: 209–99.

Zenzen, N., Hölscher, T., and Trampedach, K. (eds) (2016) *Aneignung und Abgrenzung: Wechselnde Perspektiven auf die Antithese von 'Ost' und 'West' in der griechischen Antike*. Heidelberg.

Zimmermann, M. (ed) (2009) *Extreme Formen von Gewalt in Bild und Text des Altertums*. Munich.

CHAPTER 4

The Grand Strategy of Achaemenid Persia, 539–331 BCE

Jenn Finn

The Achaemenid Persian Empire was the largest territorial empire of its time, stretching from the Balkans to the Indus until its fall under Alexander the Great in 331 BCE. The Achaemenids were constantly embroiled in warfare, whether the quelling of internal revolts or projects of a more expansionist nature. Despite our predominantly Greek sources, cuneiform documents, material and numismatic evidence, and (albeit rare) corrective Classical texts have provided many scholars with the opportunity to successfully extract a Persian viewpoint on their imperial strategy. These materials return a picture of Persian grand strategy that reveals a lack of a consistent plan across geographical or chronological boundaries. Indeed, a hallmark of theoretical grand strategy is unpredictability, as an empire always acts according to its estimation of the best decisions for its future goals.[1] Furthermore, grand strategy is rarely ever preconceived, and requires a great deal of adaptability.[2] As we will see, the Persians successfully considered the dynamics of their imperial situation for more than two centuries, as their emphasis on performative aspects of strategy differed depending on imperial circumstances. But ultimately, dynastic instability, an inability to prevent and manage satrapal revolts, and the mismanagement of political and financial affairs in the eastern Mediterranean caused serious mishaps in the achievement of their desired imperial outcomes.

1 Early Achaemenid Kings

In the earliest stages of its formation, the primary strategy of the Persian kings was that of imperialism. Indeed, in the first chapter of his *From Cyrus to Alexander*, Briant dubbed the Teispid Cyrus and Cambyses 'the land-collectors'.[3] To achieve his supremacy over other influential empires in the Ancient Near

1 Murray (2011) 5.
2 Sinnreich (2011) 254–70. This aspect is also highlighted as crucial throughout Gaddis (2018).
3 On the Teispid branch of the Achaemenid dynasty, see Rollinger (2014).

 | DOI:10.1163/9789004710771_005

East and functionally create the empire, Cyrus campaigned against Croesus of Lydia, who maintained treaties with Sparta (Hdt. 1.69.3), Babylon, and Egypt (Hdt. 1.77.2).[4] In the 540s BCE, Cyrus conquered Croesus, taking the king prisoner and effectively ending the powerful Lydian Empire (Hdt. 1.86.1).[5] He divided Lydia up into two separate provinces, one based in Sardis and one in Daskyleion (Hdt. 3.120.1–2). Importantly, he maintained the Lydian mint that had been implemented under Croesus (which produced coinage often known as 'Croeseids') and established the Lydian capital of Sardis as the centre of his western satrapies.[6] However, he refused to treat the Ionian Greeks on the same terms as had Croesus, and they resolved on rebellion (Hdt. 1.141; Just. 1.7.11). Once the rebellion had been quashed by his Median commander Mazares, Cyrus determined to govern those states through the imposition of Persian garrisons[7] and the establishment of military overlords.[8] The Greeks in Asia Minor all quickly fell under Persian hegemony, and were forced to pay high taxes and provide contingents of troops to the Persian army.[9] Following the defeat of the Lydian kingdom, Cyrus himself had already set his sights on Bactria and the Sakai, although our sources are unclear about his activities following the defeat of Croesus (Hdt 1.153).[10] At some point, Cyrus subjugated the eastern Iranian and west central Asian domains of Drangiana, Areia, Margiana, Chorasmia, Sogdiana, Bactria, the Sakai, Sattagydia, Arachosia, Gedrosia, and Gandhara.[11] Here Cyrus seems to have established a cluster of garrison sites on the frontier zone (Arr. *Anab.* 4.2.2, 4.3.1; Curt. 7.6.16, 19; Just. 12.5.12; Strabo 11.11.4) as well as founding a new city, Cyropolis. He also installed his younger son, Bardiya, in Bactra, seemingly a recognition of the strategic importance of the East. The aims in this early period seem to have been to guarantee supply

4 The Greek sources often paint Croesus as 'the foolhardy aggressor', but see now Kuhrt (2021) 405.

5 Herodotus places the conquest of Lydia in the 540s BCE, a date that scholars have traditionally followed based on (tenuous) corroborating data from a Babylonian Chronicle (Nabonidus Chronicle II 16). However, Rollinger (2008, 56–57) provides a persuasive argument against the traditional reading of the lacuna in the Chronicle and notes that the conquest of Lydia could have occurred even after Cyrus' capture of Babylon in 539 BCE (see below).

6 Alram (2012) 63.

7 It is possible that this strategy is attested in ABC 7 III.15–18, although a lacuna at the end of line 16 makes it impossible to know where exactly Cyrus established garrisons.

8 Briant (2002) 38.

9 Dandamaev (1989) 30.

10 The chronology is confused in the sources; Ctesias (*BNJ* F9 §2–3) puts this campaign prior to the defeat of Lydia.

11 Callieri (2023, 838–40) points out that DB §6 (I 12–17) lists these countries as inherited by Darius, further noting that there are no known campaigns in this region during the reign of Cyrus' successor Cambyses.

of garrisons, the levy of troops, and the collection of money, although some places received special tax exemptions for their guarantee of submission and support (Arr. *Anab.* 3.27.4–5; Curt. 7.3.1–3).[12]

We cannot establish absolute chronologies for Cyrus' campaigns, but we know his conquests culminated with the subjugation of Babylon in 539 BCE (Cyrus Cylinder ls. 15–19; ABC 7 iii.18).[13] Although the Cyrus Cylinder alludes to a peaceful entry into the city, we know that Cyrus had defeated the Babylonians at Opis a mere two days before his march into the city (ABC 7 iii.12–24; Hdt. 1.190.1). He had become ruler of all the lands in the Neo-Babylonian king Nabonidus' purview, including Phoenicia, Syria and Palestine, Asia Minor, and districts of the Arabian Peninsula. Those who voluntarily recognised Cyrus' authority, such as the kings of Phoenicia, remained in office.[14] Following the capture of Babylon, Cyrus united Babylonia and Trans-Euphrates in a large satrapal administrative bloc which had its own governor, a man called Gūbaru in Akkadian texts and Gobryas in Greek texts.[15] To avert danger from the Massagetae in particular, Cyrus installed fortified frontier posts along the north-eastern limits of his empire. Recognising their potential danger, in 530 BCE he launched a campaign against them. His death is clouded in mystery.[16]

Although Cyrus did not attempt a campaign against the Egyptians, under his son Cambyses (r. 530–522 BCE) the territory was conquered after a long and difficult war (*BNJ* 688 F13.10). He was accompanied by subject nations, including the Greeks (Hdt. 3.1.1). The subjugation of Cyprus and Phoenicia also allowed Cambyses to build the first semblance of a Persian navy.[17] When confronted with the idea that the Persian position in the Mediterranean was acutely threatened by Egyptian sea power, Cambyses enlisted the aid of coastal naval forces such as that of Samos (Hdt. 3.44.2) and Mytilene (Hdt. 3.13.1); with these smaller and periodically levied contingents, he was able to subdue both Egypt

12 Stark (2021) 697–98.

13 A reconstruction of the timeline can be found in Kuhrt (2021) 406.

14 Dandamaev (1989) 60.

15 Briant (2002) 49. On Gūbaru and his jurisdiction, see Kleber (2021) 910–11.

16 The war and the death of Cyrus is reported in several Classical authors: Amm. Marc. 23.6.7ff.; Arr. *Anab.* 4.11.9; Diod. Sic. 2.44.2; Frontin. *Str.* 2.5.5; Joseph. *AJ* 11.2.1; Polyaenus, *Strat.* 8.28; Strabo 11.6.2, 11.8.6. There are three different – and seemingly irreconcilable – versions of Cyrus' death, which appear in Herodotus (1.214, who says that Cyrus died fighting the Massagetae, but acknowledges multiple versions), Ctesias (*BNJ* F9§7, who says that Cyrus died fighting the Derbicians), and Xenophon (*Cyr.* 8.7.2, who says that he died peacefully after seeing a god in a dream). On these three versions, see Beckman (2018); Bichler (2021).

17 Briant (2002) 53.

and Libya.[18] Herodotus connects Cambyses' conquering of Egypt together with the acquisition of 'the Sea' (Hdt. 3.34.4), setting the precedent for future Persian control over the territory. Briant puts Cambyses' motivation down to the necessity of subjugating the countries between the Euphrates and the Nile because of his father's annexation of Trans-Euphrates to Babylonia.[19] However, according to Ctesias the impetus for the expedition was Cambyses' desire to avenge his new consort Neitetis, whose father had been unjustly murdered by the Egyptian pharaoh Amasis. According to Herodotus (3.17.1), after conquering Egypt, Cambyses set his sights on the Carthaginians, the Ammonians, and the Ethiopians, territories which had all been goals of the most recent pharaohs. But in these endeavours, he was unsuccessful. Ultimately, the early phases of Persian empire formation under Cyrus and Cambyses (namely, the conquest and incorporation of previously independent kingdoms and federations) were not necessarily a direct result of their military initiative so much as a natural consequence of the gradual geographic expansion of their empire.[20] Nevertheless, these territorial additions recollected Neo-Assyrian imperial policy, bringing an end to the expansionist standstill that had dictated affairs during the Neo-Babylonian period.[21]

2 The Reign of Darius I

Despite the muddled nature of the story behind his accession to the throne,[22] the reign of Darius I (r. 522–486 BCE) saw the Persian territorial empire at its greatest extent. In his inscription at Naqš-e Rostam, Darius emphasised his 'seizure' of imperial lands, which were now responsible for paying tribute and abiding by his law and directives (DNa ls. 15–30; a similar list can be found at DB i ls. 6–17). Most conspicuously, Darius also established connections between centre and periphery in his palace at Persepolis, whose construction

18 Contra Wallinga (1987) 54–55, who argues that Cambyses' creation of the navy with which he conquered Egypt was 'a creation *ab ovo*, not the *ad hoc* assemblage of "national" squadrons of long standing, and thus a most radical measure'. See instead the balanced discussion of Cawkwell (2005) 255–73.

19 Briant (2002) 51.

20 Wiesehöfer (2009) 75.

21 Rollinger (2021) 368–77.

22 Briant (2002) 107–14; Young, Jr. (1988) 53–57; Olmstead (1948) 107–18; Dandamaev (1989) 83–113; Waters (2014) 58–76; Schwinghammer (2021). The primary sources for the incident in question are Hdt. 3.61–88 and Darius' inscription at Bisotun (DB), with another version of the magus' assassination from Jacoby (2021) F13, 16.

was begun circa 518 BCE and whose iconography focused on the universality of his empire.[23] Yet, in contrast with his predecessor Cyrus, almost immediately upon taking the throne Darius was faced with massive rebellions (in fact, the period is often referred to as 'the Great Rebellion'). First were Elam (DB §73) and Babylon (DB §18–20). The latter satrapy revolted under Darius I from 522 to 521 BCE, when a group of rebels, calling themselves after the Babylonian king Nebuchadnezzar, rejected Persian rule; to discourage further disobedience, Darius erected a stele on Babylon's processional way.[24] While the king suppressed both the Elamite and Babylonian revolts rather quickly, Babylon's establishment suffered severe religious penalties, as Darius conscripted the Babylonian priestly delegation as corvée labour in the building of his capital at Susa.[25] However, for a time, Darius maintained the existing Babylonian administrative system; it was not incorporated into a satrapy until the end of the sixth century, or possibly not even until the reign of Xerxes.[26] According to Darius himself, next came Persia, Media, Assyria, Egypt, Parthia, Margiana, Sattagydia, and Scythia (DB §21); in several instances, pretenders to the throne proclaimed themselves the true heir. In each case, Darius was successful in putting down the revolts, through a combination of personal military leadership and tactical balance from which he commanded the central position against rebels who had failed to coordinate in an organised fashion.[27] However, one of the places to which Darius never led an army during this period was Egypt. Instead, he dealt with this threat through a demonstrated respect for Egyptian religion (Polyaenus, *Strat.* 7.11.7; Diod. Sic. 1.95.4–5), and Diodorus (1.95.5) says that he was the first Persian king whom the Egyptians considered a living god. Darius also adopted Egyptian pharaonic titulary, and during his reign we can discern a distinct effort at creating a Egypto-Persian iconography.[28] Through these measures he was very successful at maintaining Persian control over Egypt. This was a similar strategy of rule-by-religious-tolerance that had been tried by Cyrus in his return of stolen gods to their home sanctuaries and in the reversal of the Babylonian Exile, which followed his capture of Babylon in 539 BCE (Cyrus Cylinder ls. 30–32).

In later inscriptions, we hear that India was brought under control of the Persians (DPe l. 11; DSab). But we do not hear much about the eastern territories

23 Root (1979).
24 For this stele, which was found broken in eleven different pieces, see Seidl (2009) 101–14.
25 Waerzeggers (2015) 200.
26 Kuhrt (1988) 131.
27 Young, Jr. (1980) 58–62.
28 Wasmuth (2015).

of the Persian Empire after the reign of Darius, although the lands east of Persepolis all the way to India were a significant part of the empire. However, that apparent disinterest may be a result of the bias of Greek authors, who actively sought to highlight the importance of their own territories in the context of Achaemenid strategy, resulting in the downplaying of any serious obligations elsewhere in the empire. From the reigns of Darius I until Darius III, there is very little documentation, and this only reoccurs as a result of the campaigns of Alexander the Great.[29] In actuality, the Persepolis Fortification Archive texts reveal that Persian interaction with the eastern provinces was as complex and organised as it was in the west. Although the tribute in these areas was often collected in kind or in uncoined money (whereas the west contributed minted coinage), nevertheless the eastern provinces were an integrated part of the centralised Achaemenid economy.[30] Additionally, Achaemenid administrative structures were actively employed as tools of control in these areas,[31] which also supplied important military bastions of the Achaemenid army.[32] The nature of the populations here (i.e., primarily nomadic) meant that the relationship between the Achaemenids and their subjects in the East was characterised by mutual dependencies, and it was necessary that power structures on both sides be acknowledged.[33] That reciprocal system seems to have worked well, as we see far fewer revolts in the eastern portions of the Persian Empire than we do in the western areas, and even when there was

29 Stark (2021) 696.

30 For example, King (2019) has shown that green chert objects were transferred from Arachosia to the central treasury in Persepolis as a form of payment. The eastern provinces were tax-paying tributary peoples of the empire (see Hdt. 3.93.3 and DSf ls. 36–40) and the communication between Central Asia and the centre of the empire was alive and well, although the extent to which the Persians imposed a concrete financial system in these satrapies is unclear. See Francfort (1988), 178–79.

31 For Bactria, see Henkelman (2013); for Arachosia, Henkelman (2017, 151–74), who shows that 'Achaemenid Arachosia had an institutional landscape and an administrative layout that a traveller from the heartland would not fail to recognize' (173); for the ways in which Achaemenid imperial structures inspired local municipalities in the Indus region of the empire, see Henkelman (2017) 174–86.

32 During the Persian Wars we hear most specifically of the Sakai; see Hdt. 6.113.1, at Marathon and Hdt. 9.31.3–5 at Plataia, where he also mentions the Bactrians and the Indians. Arrian *Anab.* 3.8.3–6 gives a roster of Darius III's army, which indicates that the main satrapal centres in the east at this point were Bactria, Arachosia, Areia, and Parthia (Vogelsang 1992, 221).

33 Matarese (2021, 45–55) notes the importance of the region of Bactria as a 'border zone' through which cooperation with the native peoples could aid the Persians in controlling the neighbouring nomadic populations.

unrest it seems to have resolved itself quickly.[34] But with the final defeat of the Achaemenid king Darius III, the reciprocal system collapsed.[35]

Darius was also at the helm of several sweeping reforms that aided in the consolidation of his massive empire. Herodotus (3.89) gives a detailed description of Darius' reforms, stating that he divided the Persian Empire into twenty satrapies, putting governors (satraps) in control of each one, and he appointed fixed yearly tribute payments to come from each of the territories 'according to *ethnos*' (πρὸς τοῖσι ἔθνεσι).[36] However, as previously mentioned, a skeleton of the satrapal system likely existed already in the reigns of Cyrus and Cambyses, as we have already seen, for instance, the satrap Gobryas in Babylonia and Bardiya in Bactra.[37] Furthermore, Xenophon (*An.* 7.4.2) states that Cyrus was receiving tribute from Cilicia and Cyprus, and Cambyses from Egypt, so it is likely that Darius was simply the first king to *fix* those payments.[38] Herodotus' numbers total an annual tribute of 14,560 talents a year, about thirty times more silver tribute than the Athenian Empire raised in its heyday.[39] However, although these contributions were certainly substantial, recent research has shown that tribute to the empire may not be quantifiable in monetary terms, as much of the tribute demanded by the Achaemenids were realised as physical labour and military service.[40] Importantly, tribute payments were actively used by the Achaemenids to draw off silver accumulations from potential adversaries, particularly Egypt and the territories in Asia Minor.[41]

Darius also ensured the accessibility and safety of all Persian territories through the construction of a massive 'royal' road network (described by Hdt. 5.52 as such) that connected all the major cities of the empire,[42] allowing

34 As evidenced, for instance, by the enrollment of Cadusian soldiers in Artaxerxes' army when fighting Cyrus at Cunaxa in 401 BCE (Plut. *Artax.* 9.1–2), despite their recent revolt.

35 Vogelsang (1992) 314.

36 For the satrapy list and their tribute payments (as attested by Herodotus), see Ruffing (2011, 86–94) as well as Bresson (2020, 225–34) for critical discussion.

37 Not to mention the fact that in the inscription at Bisotun, Darius lists satraps as already installed in important areas of the empire, such as Vivâna in Arachosia (DB §45 iii. 55–56).

38 Briant (2002) 64–69.

39 Tuplin (2011) 54. Vogelsang (1992, 201 table 15) provides a convenient tally for the tributes expected of each *nomoi* in a given year.

40 For the case of Babylonia see Jursa (2011); Dromard (2022) 204–209.

41 Blichert-Toft et al. (2022) 8.

42 In several places in Book 19, Diodorus gives us a great idea of the distance between major cities on the Royal Road; see Briant (2002) 358. Our best information, however, derives from the so-called Q texts from the Persepolis Fortification Archives, which are collected in Hallock (1969).

for a strongly centralised and deeply interconnected empire.[43] The Royal Road also allowed efficient and simple transport of tribute to the Persian royal capitals at Babylon, Susa, and Persepolis, as is most clearly illustrated by the contents of the Persepolis Fortification tablets and the Persepolis Treasury tablets.[44]

Despite the incredible complexity of their overland affairs, the control and management of waterways was also a top priority of the Achaemenids in the fifth century BCE. In Darius' inscriptions, the subjugation of Egypt was intimately connected to the mastering of the sea, as is apparent in his description of his construction of the Red Sea canal (DZc ls. 7–12).[45] Also mentioned in Darius' inscriptions are 'countries across the sea' (DPe ls. 14–15), which is likely a reference to Darius' campaigns against the Scythians. These campaigns, prosecuted between 514 and 512 BCE, represented the first time that the Persians had entered European soil. It entailed a performance of Darius' mastery over the sea when, en route to the Scythians, he made a bridge across the Bosporus (Hdt. 4.87). The result of the campaign was a swift subordination of a part of Thrace (Hdt. 4.93); it also entailed the conversion of Macedonia, whose kings swiftly and willingly offered earth and water to the king, into a Persian vassal state (Hdt. 5.17–18.1). Thraco-Macedonian coins discovered in a sixth-century horde in Central Asia indicate that already in this period the Achaemenids had good trade (rather than tribute) relationships with these areas in the reign of Darius;[46] furthermore, the area was of strategic importance, as it was a rich source for raw materials and manpower (Hdt. 5.11, 5.23). Importantly, our sources give us three separate stories as to the impetus behind the Scythian campaign, but importantly, all are focused on a revenge motif: Herodotus claims Darius wished to take revenge for the Scythian invasion of Media, the latter of whom had maintained power over them for almost a century (Hdt. 4.1.1); Ctesias mentions an exchange of hateful letters between Darius and the Scythian king Skytharches as the cause (*BNJ* 688 F13 20); and Justin notes that the campaign should be blamed on Darius' anger at being denied the

43 Colburn (2013) 29–52; see also Graf (1993) 149–68; Briant (2002) 357–77; Kuhrt (2014a) 122–31; Henkelman and Jacobs (2021) 719–35. The most detailed and lengthy contribution to the topic can be found in Henkelman (2017).

44 The former deal with the collection, distribution, and storage of foodstuff arriving from all corners of the empire, and the latter deal primarily with internal Persepolitan affairs. They also indicate the complex administrative structures with which the Achaemenid Empire conducted their affairs, particularly in the reign of Darius and his son Xerxes. See Briant (2002) 422–81.

45 See Haubold (2012) 5–24.

46 Kagan (2011) 231–32.

opportunity to marry the Scythian king's daughter (Just. 2.5.8–9). These explanations have left scholars dissatisfied, with interpretations varying between the desire for Scythian gold, the establishment of a new Persian boundary at the Danube, a preventive attack due to the Scythians' interest in lands south of the Danube, or a preamble to Darius' subjugation of Greece. Vasilev conjectures that the main intention was likely to establish a new Persian boundary at the Danube, 'for a good defense would turn the water into an obstacle for the warlike but fleetless Scythians'.[47] However, it should be noted that the motives mentioned by ancient authors are entirely consistent with previous and future Persian-led offensive campaigns, such as that of Cambyses against Egypt (see above), and that of both Darius and Xerxes against Greece (see below). Like other Persian-controlled areas, the Thracians now saw the imposition of military garrisons and Persian hyparchs (Hdt. 7.105.1–106.1), as well as the burden of taxes and tribute (Hdt. 3.96.1).[48] On the other hand, Persian occupation greatly increased trade in both directions, and the economic prosperity of the region was positively affected.[49]

These campaigns comprise only a small part of the Persian presence in Europe. In particular, the Greeks themselves are listed as subjects to the Persian Empire during the reign of Darius (e.g., DNa l. 28; DB i l. 15; DSf l. 48). Already in 517 BCE, the Persians held control over Samos, Lesbos, and Chios on the Ionian coast, and the king had been able to take advantage of their services in his campaigns against the Scythians. Their loyalty was secured through gifts and privileges: some faithful leaders were made tyrants of their cities, while others were granted large tracts of land.[50] But trouble began when, according to Hdt 5.73.2, in 507 BCE, an Athenian embassy approached Artaphernes, brother of Darius I. The Persians are represented as amenable to a treaty so long as the Athenians give earth and water to the king; otherwise, the embassy should 'be gone' (εἰ δὲ μὴ διδοῦσι, ἀπαλλάσσεσθαι αὐτοὺς ἐκέλευε). The embassy, according to Herodotus (5.73.3), agreed to terms. Athenian agreement to Persian terms would have entailed bearing tribute to the Persians, abiding by the Great King's directives and representing his interests, and providing support on Persian military campaigns. Therefore, it was understood that the Athenians were, at the very least, a tribute-bearing entity of the Persian Empire, which came with certain obligations – not least of which was acknowledging the sovereignty of the Persian king,[51] although the Athenians were still left to govern themselves

47 Vasilev (2015) 41–44 (quote at 44).
48 Vasilev (2015) 118.
49 Balcer (1995) 156–57.
50 Young, Jr. (1980) 67–68.
51 Waters (2016) 96.

how they saw fit. This was the way in which the Persians handled relationships with all Greek *poleis*.[52] From this survey it is clear that the first fifteen years of Darius' reign featured several avenues of approach to imposing Persian authority, including offensive military action; in many cases, however, we see a preference for diplomacy.[53] These strategies seem to have been effective, as until the beginning of the fifth century BCE, the tax incomes from Europe appear to have continued to reach the treasury of the Persian king, and Persian garrisons continued to function in these regions,[54] although Darius' European holdings do not appear to have ever constituted a Persian satrapy *per se*.[55]

The Ionian Revolt of 499–494 BCE is yet another example of a rebellion with which Darius was forced to contend. But this one receives quite a bit more attention in scholarship, as it was the catalyst to the Persians' greatest international conflict, a series of wars with Greece that would rage over the next 150 years. Although Persian policy was to maintain local governments 'if it was to the advantage of the political ambitions of the empire',[56] under Darius, the Ionians had grown weary of Persian rule. The greater activity in the western parts of the empire and across the Mediterranean had increased the demands on the local Greeks, who contributed to the Persian campaigns in Scythia. Together with the new expectation of military service also came a more significant presence of powerful satraps related to the royal family, a burdensome annual tribute in silver, and the disruption of normal trade patterns.[57] In 499 BCE, Aristagoras of Miletos led to mainland Greece an embassy on behalf of the Ionians to ask for assistance in separating themselves from Achaemenid authority. Having been refused by the Spartans, the Athenians promised twenty ships to the effort, with the Eretrians an additional five; for Herodotus, these ships were 'the beginning of evils between the Greeks and the barbarians' (αὗται δὲ αἱ νέες ἀρχὴ κακῶν ἐγένοντο Ἕλλησί τε καὶ βαρβάροισι, Hdt. 5.97). During the revolt, the Persians settled on a tactic of attacking each city in the resistance one by one; but this ultimately failed, and by 494 BCE, they were forced to attack the revolt at its source: Miletos.

52 So Tuplin (1987) 111: 'The Persians might interfere in the political matter of which individual or class controlled a particular city-state; they might visit enslavement, deportation, and destruction upon the population and buildings of a rebel *polis*, but they did not eradicate the institution (even in the latter cases), and did not necessarily object to the union of several cities in the league or the like'.

53 See Beckman in this volume.

54 Vasilev (2015) 124.

55 Zournatzi (2003) 9–10.

56 Brosius (2021) 78.

57 Murray (1988) 476–77.

After a difficult sea battle, the Persians defeated the Ionian naval contingent and besieged the city (Hdt. 6.18). Its people were reduced to slavery and its temple destroyed (Hdt. 6.19.3). The Persians had finally cut off the head of the resistance, which left the rest of the Ionians on an island. However, in the early days of the revolt, the Ionians had burned the great temple in the Lydian satrapy of Sardis, which the Persians would, according to Herodotus (Hdt. 5.102.1), cite as impetus for their retributive action while on campaign in Greece (see below). Although initial Persian reaction to the revolt was harsh, already in 493 BCE there is evidence that the territory of each Ionian city was assessed for fairer taxes and tribute, and tyrants were forcibly deposed in favour of traditional Greek 'democracies' (Hdt. 6.43.3).[58] This action was an effort to prevent dissatisfaction among the Greeks who were subject to the Persians, while also assuring the Persian king of their loyalty;[59] this policy aligned with the heretofore popular Persian policy of encouraging peace in the satrapies through religious and political toleration.[60] Although Balcer downplays the importance of these revolts, noting the amount of time it took the Persians to subdue it in comparison to rebellions in other areas,[61] the suppression of these discontents was incredibly important for Persian dominance over the Aegean. Hence, all measures were taken to ensure Ionia was pacified before further western imperialistic endeavours were tried.

In 492 BCE, after Darius' reestablishment of Persian power in Ionia, the Persian fleet began its campaign under Mardonios to Macedonia and Thrace. Persian strategy to this point had proved that a 'slow but safe penetration from the north under protection of the fleet', together with the Persian encouragement of Greek democratic governments, was both efficient and successful. Thasos was subjugated and Macedonia renewed its recognition of Persian overlordship. The Persians then quickly moved south to mainland Greece. Their push to the mainland was a retributive one, as official Persian rhetoric insisted that they should lead a campaign against Greece to punish the Athenians and Eretrians for their involvement in the Ionian revolt (Hdt. 6.43.4–44.1).[62] The Athenians, after all, had, from the Persian perspective, entered into an

58 Murray (1988) 489–90.

59 Dandamaev (1989) 169.

60 Wiesehöfer (2009) 86, notes that '... the Achaemenids were able to make cultural diversity serve the needs of imperial unity'. However, as was the case, for instance, of Egypt under Xerxes, they may also choose to assert their control over native religious affairs through the destruction of shrines, or through forced small changes in rituals or ceremonies (87).

61 Balcer (1995) 190.

62 Briant (2002) 158, also notes that other authors (for example, Ath. 14.652b–c and Plut. *Mor.* 173c) highlight Darius' obsession with revenge in other contexts as well. See also Konijnendijk (2021) 1145–46.

agreement with the Great King in 507 BCE (above). Therefore, the Athenians were viewed – most likely because of their involvement in the Ionian revolt – as having broken crucial diplomatic promises. Although Athens and Sparta rejected the demand for earth and water in recognition of the Great King, most Greek cities – and the islands – accepted Persian terms.[63] Despite the famous Athenian defeat of Persian forces at Marathon (Hdt. 6.111–117), the Persian loss was unlikely to have been but a blip on the imperial radar. The campaign had been launched with the pretext of vengeance for a broken treaty; the involvement of Athens and Eretria in the Ionian revolt had simply driven the Persians to attempt to extend their proximate buffer zone into the Aegean, but it was not a prerequisite for success.[64] Regardless, many cities on the mainland had given the King earth and water, and thus, Darius seems to have been satisfied that his objectives had been accomplished; although Ctesias (*BNJ* 688 F 13.22) claims that the Athenian general Miltiades defeated the 'barbarians' and killed Datis at Marathon, he prefaces his statement by noting that Datis 'laid waste to Hellas and the islands' (ἐπόρθει νήσους καὶ τὴν Ἑλλάδα). Additionally, in DSe ls. 27–29, the 'Greeks near and across the sea' were now listed as parts of Darius' Persian Empire.[65] Partly because of our abundance of source material, we can make firmer conclusions about the reign of Darius than any other Persian king: at all points, his main goals were the expansion of the imperial realm, control over military and trade routes, and the prevention of revolt. The rest of Persian imperial history would be characterised by a strategy of consolidation following the rule of Darius.

3 The Greco-Persian Wars

Shortly after the death of his father, Darius, and his subsequent accession to the Persian throne, the now-king Xerxes (r. 486–465 BCE) found himself overwhelmed with the prospect of an enormous military enterprise in the West. Opportunism was rampant in subjugated provinces of the empire, and Egypt was one of the first to test Xerxes' limits. The Greek sources provide very little information regarding these revolts, which have been dated to 484 BCE. The potential for Egypt to regain its navy and establish alliances with free Greek naval powers was a serious threat to Persia's position in the Mediterranean

63 See Waters (2016) 171.

64 Wiesehöfer (2004) sees the idea of Achaemenid world domination as a Herodotean invention, whereas Ruffing (2021) views it as a strategic expansionist policy critical to Achaemenid royal legitimacy during this period.

65 A sentiment echoed by Waters (2014) 91; Dandamaev (1989) 177; Jung (2013) 256.

environment, as it represented the potential loss of not only the province itself but also the secure possession of Syria-Palestine, Cilicia, Cyprus, and of the Persian position in the Aegean Sea.[66] According to Herodotus, the defeat of the revolt was harsh:

> And now, having thoroughly subdued them and made the whole of Egypt's servitude even harsher than it was under his father Darius, he entrusted the satrapy to his brother Achaemenes, son of Darius (Hdt 7.7.1).

The reign of Xerxes seems to have been a watershed in Egyptian-Persian relations, as he distanced himself from the strategies of his predecessors by installing his brother Achaemenes as governor in Egypt. Additionally, we see the (likely not coincidental) end to two demotic private papyrus archives around this time.[67] Similarities can be gleaned from Xerxes' treatment of Babylonia, which, like Egypt, was a perennial problem. Also in 484 BCE, there is evidence for two revolts in this satrapy.[68] Just as in the reign of his father, the repercussions of the revolts were widespread, with increased taxes and changes in temple personnel strictly enforced.[69]

It was the restoration of Persian control over Egypt that allowed Xerxes to begin preparations for a second invasion on Greece. Like his father, this campaign also required a display of dominance over the sea, presented to us in a famous Herodotean anecdote about his attempt to bridge the Hellespont (Hdt. 7.55–56).[70] However, we may glean a fundamental change in Achaemenid policy between Darius and Xerxes. Bridges points out that the list of subject nations is the same as that given in his father Darius I's inscription at Naqš-e Rostam (DNa), with one crucial difference in terminology: Darius says he 'seized' the lands on the list (DNa §15–22), whereas Xerxes simply says he 'ruled'

66 Wallinga (2005) 21.

67 Quack (2021) 558.

68 Waerzeggers (2003/2004) 151–56. Ctesias mentions a revolt of Babylon in the reign of Xerxes (*BNJ* 688 F 13 26) but gives no further details. Strabo (16.1.5) and Arrian (*Anab.* 3.16.4) are the only sources to mention Xerxes' destruction of sacred structures in Babylon as part of the events surrounding the revolt.

69 Waerzeggers (2015) 192–94. See also Kuhrt (2014b) 166–69, who argues that our sources demonstrate increasingly tight measures of control on imperial affairs during the reign of Xerxes, including in his management of the Babylonian revolts.

70 See the excellent discussion in van Rookhuijzen (2019) 78–86; the considerations of Rollinger (2013, 102–11), who sees reflections of Near Eastern ideas of the 'edge of the world' and Achaemenid rule over it in the Herodotean narrative; and Haubold (2012, 15–16), who argues that the Greeks misread Xerxes' act as a symbol of transgression, whereas in Persian imperial discourse it was rather a symbolic act of transcendence.

them as tribute-bearing countries (XPh §13–19).[71] With Herodotus (7.8β.1), Waters argues that, as a result of the confrontation between the Greeks and the Persians that occurred in 490 BCE at Marathon, Xerxes' plan to avenge his father was a 'fully justifiable, expected, even necessary campaign of retribution upon a rebellious state on the periphery', indeed a perquisite of Persian kingship as it had been formulated by his predecessors.[72] Ctesias (*BNJ* 688 F 13.25) even notes as such, when he says that Xerxes felt compelled to march against Greece partly because they destroyed the altar set up by his father and because the Athenians had failed to return Datis' corpse after the battle of Marathon. Therefore, this war, too, was one of vengeance; we should not consider it a campaign with primarily imperialist goals.[73] Regardless, we know that Xerxes was careful and methodical, ensuring the construction of storage centres, provisioning and outfitting the army, and opening communication and supply lines by land and sea.[74]

This is not the place to provide a lengthy summary of the Second Persian War: this task has been achieved in countless monographs and is the main subject of Herodotus' *Histories*. Xerxes, together with his land and naval military forces, approached mainland Greece via the pass at Thermopylai in 480 BCE, where the first Greco-Persian confrontation resulted in a stalemate at Artemision (by sea) and a Persian victory at Thermopylai (by land). The opening of the pass at Thermopylai allowed the Persians access to northern Attica, and the Persian army soon marched to Athens to besiege its Acropolis. Most of the citizens, however, had already been evacuated to the nearby island of Salamis. The island was the site for the second major clash, this time a purely naval affair, which the Persians approached in a weakened state.[75] After Xerxes left his general Mardonios in charge to prosecute the remainder of the war,

71 Bridges (2014) 90–96.

72 Waters (2016) 101–103 (quote at 101).

73 Briant (2002, 526) cites a passage in Herodotus (7.138.1) that proclaims: 'In name, the march of the king's army was undertaken against Athens, but truly it was against the whole of Greece' (ἡ δὲ στρατηλασίη ἡ βασιλέος οὔνομα μὲν εἶχε ὡς ἐπ᾽ Ἀθήνας ἐλαύνει, κατίετο δὲ ἐς πᾶσαν τὴν Ἑλλάδα). However, in Book 8 (102.3), Herodotus has the Mediser Artemisia remind Xerxes that he can feel free to go home after the burning of Athens, 'which was the whole purpose of the expedition' (τῶν εἵνεκα τὸν στόλον ἐποιήσαο). Additionally, see 'Conclusion' below.

74 Balcer (1995) 228–29. See also Ferrill (1966, 106), who argues that Hdt 7.8–10 gives, through the eyes of Xerxes, Artabanos, and Mardonios, evidence for a Persian strategic plan during the second Persian War (see also 7.49; 118–20).

75 Cawkwell (2005, 105–10) suggests that after Thermopylai, the Persian fleet was primarily used for supply convoys and Xerxes only fought the 'needless fight' at Salamis for its performative and symbolic potential.

the allied land forces of the Greeks met a large force of Persians (and their Greek allies, or 'Medisers') at the plain of Plataia in 479, where (not without great difficulty), the Persians were finally defeated and returned to the safety of the imperial centre. Certainly, the final repulsion of the Persians from the Greek mainland at Plataia left a strong impression in cultural memory which still dominates our modern interpretation of the Wars.[76]

However, to properly contextualise the Persian Wars from a Persian imperial point of view, it is important to step away from Herodotus and question whether other sources may problematise his account. The following passage from Dio Chrysostom's *Discourses* (11.149) has been a part of this larger discussion:

> I heard, for instance, a Mede declare that the Persians concede none of the claims made by the Greeks, but maintain that Darius dispatched Datis and Artaphernes against Naxos and Eretria, and that after capturing these cities they returned to the king; that, however, while they were lying at anchor off Euboea, a few of their ships were driven on to the Attic coast – not more than twenty – and their crews had some kind of an engagement with the inhabitants of that place; that, later on, Xerxes in his expedition against Greece conquered the Lacedaemonians at Thermopylai and slew their king Leonidas, then captured and razed the city of the Athenians and sold into slavery all who did not escape; and that after these successes he laid tribute upon the Greeks and withdrew to Asia. Now it is quite clear that this is a false account, but, since it was the natural thing to do, it is quite possible that the king ordered this story to be spread among the inland tribes in order to keep them quiet.[77]

According to Briant, Dio Chrysostom reveals the story that was likely closest to the 'official Persian version of the events' surrounding the First and Second Persian Wars of the early fifth century BCE.[78] Indeed, several pieces of evidence indicate that Dio's reported rumour is an appropriate representation of official Persian rhetoric. We have already indicated that the Persian view of events at Marathon was likely consistent with Dio's recounting of events.

76 That commentary has often been viewed as artificially dichotomising the Greeks against their Persian opponents. See, for instance, the excellent survey of Brosius (2013, 658–68), as well as that of LaForse (2013, 569–87), and Hornblower (2008, 45–48). However, Harrison (2011) argues that some Greek authors were informed by Persian sources, even if they 'moulded them to Greek models or even introduced elements from Greek predecessors' (20).

77 Cohoon (1932).

78 Briant (2002) 542.

But we have more information on the Second Persian War. In the so-called *daivā* inscription, Xerxes lists the Greeks, 'those who dwell on this side of the sea [Ionians] and those who dwell on that side of the sea [mainland Greeks]' as tribute-bearing subjects to his empire (XPh ls. 23–25).[79] Material evidence also corroborates Dio's 'Persian' accounting of events. The statue group of Harmodios and Aristogeiton, the 'tyrant-slayers' considered responsible for the foundation of the Athenian democracy in the late sixth century BCE, was discovered by Alexander the Great as displayed in the Persian capital of Susa (Arr. *Anab.* 7.19.2). The group was captured during the destruction of the Athenian Acropolis following the Persian defeat of Greek forces at Thermopylai in 480 BCE (Hdt. 8.52–53); this act of destruction was connected by Herodotus to the Persian desire for revenge for the burning of Sardis (Hdt. 5.102.2). Also stolen in the same raid was a bronze statue of a water carrier, originally dedicated by Themistocles when he was water commissioner in Athens; Themistocles would later view this piece in the Lydian city of Sardis, when he was in Persia as an exile (Plut. *Them.* 31.1).[80] Besides these minute pieces of evidence, there are no native Persian textual reminiscences – let alone impressive monuments – to commemorate the conflict with the Greeks.[81] From a Persian viewpoint, the war 'had the character of a minor set-back along the periphery of its realm'.[82]

The Persian losses in Greece did, however, initiate alterations in policy that would have a significant effect on Persian grand strategy thereafter. Immediately following the Persian loss at Plataia, Xerxes secured Daskyleion as a buffer zone against future Greek rebellion by placing his first cousin Artabazos in charge of the area. Furthermore, the Great King now hedged his bets in the east Aegean, reducing their military presence in favour of increased aid in areas like Egypt and Bactria; abandoning volatile areas such as Miletos; and allowing the East Greeks to remove themselves from Persian suzerainty (although garrisons were maintained in some areas on the coast, and the Persians continued to control military and inland routes in Asia Minor).[83] Persian strategy at this point seemed less concerned with expanding their territorial holdings into Greece and more interested in the ability of the Greco-Persian conflict to cause uprisings among the satrapies firmly in the empire's grasp.

79 Bridges (2014) 90–96.

80 On the discovery and repatriation of the statue group of Harmodios and Aristogeiton, see Finn (2014) 385–403, and Müller (2016) 173–202. On Xerxes' destruction of the Athenian Acropolis, see Garland (2017) 61–86; Kousser (2009) 261–69.

81 Compared to the plethora of such monuments in Greece. See, for instance, Custis West, III (1965), and more recently, Duffy (2018). However, Tuplin (2020) has done an interesting analysis of Persian seals that depict Persian victories over the Greeks in glyptic art.

82 Dandamae (1989) 226.

83 Cawkwell (2005) 126–38; Hyland (2018) 23–36.

4 The Peloponnesian Wars

The aftermath of the Greco-Persian Wars had left both sides in a place of reckoning: as Balcer notes, 'No dominant factor in ancient Greek history altered that society and its political subparts as did the long Persian wars in conquest of the Greeks'.[84] Following the war, the immense gulf between Athenian and Spartan military and cultural prerogatives, made ever starker by the Persian conflict, led directly to the so-called Peloponnesian Wars. It is through Persian interventions in this internecine war that we can glean a more accurate picture of Achaemenid relationships with Aegean states. After the formal end of Xerxes' invasion, in 478 BCE, the Athenians led the formation of the Delian League, a coalition of Greek city-states whose main goal was to prevent another Persian invasion of Greece. The Delian League was a threat to Persian supremacy over Ionia,[85] as the Athenians were able to coerce several city-states in Ionia, Caria, and Lycia to join the League; these states, although never relinquished by the Persians, were thus considered rebels by the Great King. The Athenians were particularly aggressive, even handing the Persians a naval and infantry defeat at Eurymedon in the early 460s.[86]

Closer to home, under Artaxerxes I, familiar troubles had also re-emerged: both Bactria and Egypt revolted (Thuc. 1.104–9; Diod. Sic. 11.71.3–4; *BNJ* 688 F14 35–37). The Egyptian revolt, led by Inaros, seems to have been spurred on by the Athenian successes in the Levant; alliance with them 'would give Egypt the enterprising, offensive capability the still intact Egyptian military system … did not alone provide'. Although the Persians were able to destroy Athens' Egyptian fleet, and hence quash the revolt, the threat of rebellion remained the same. The king's troubles deepened when Megabyzos, the satrap of Syria, also revolted (*BNJ* 688 F14 40); this appears to be the first instance – in what would become a long-running tradition – of Persian commanders utilising Greek mercenaries in battle (*BNJ* 688 F14 39–40). Megabyzos' son, Zopyros, did the same upon his father's death, and fled to the Athenians (*BNJ* 688 F14 45). In 450–449 BCE, these global concerns caused Artaxerxes to eschew any extension of Persian power in the West in favour of securing imperial hold over Egypt.[87] The culmination of these affairs was the signing of the Peace of Kallias

84 Balcer (1995) 331.

85 Which was under the aegis of Lydian satraps; see Klinkott (2005) 471.

86 Waters (2014) 132–33.

87 Ruzicka (2012) 30–33 (quote at 30). Egyptian officials were kept in place, with the caveat that they must agree not to rebel against the Persians, and they must also agree to send Egyptian marsh soldiers to the Great King. As Briant (2002, 576–77) notes, this system worked for the Persians for several decades.

in 449 BCE, the first agreement of its type struck by a Greek city-state (Athens) with the Persians. Plutarch (*Cim.* 13.4) refers to the peace as 'notorious' (περιβόητον), as it allowed all Greek cities to live under their own laws, prevented the Persian satraps from coming within three days' journey of the Ionian coast, and prohibited them from sailing warships west of the Kyaneian and Chelidonian islands. In return, the Athenians agreed not to send troops into territories over which the Persians ruled (Plut. *Cim.* 13.4; Diod. Sic. 12.4.5). Therefore, because of the treaty, Persian sea power was drastically weakened and a wide zone of influence in western Asia Minor was demilitarised.[88] However, it also prevented the Persians from having to continually finance their navy on the coast and allowed for increased merchant traffic. Furthermore, as Hyland argues, from a Persian ideological perspective, the war's end represented 'an imposition of order', with restraint in Ionia 'a voluntary grant rather than a forced concession'.[89] Perhaps most importantly, the treaty should be considered as consistent with a long-term Persian strategy that shifted the Greek focus onto wars with one another, thereby weakening their international impact and leaving the Persians to continue reinforcing their own position.[90]

As affairs within Greece intensified and the Peloponnesian War between the Athenians and the Spartans got underway, the Persians were yet involved in Greek affairs. From 423–413 BCE, Darius had avoided conflict with Athens; but their willingness to aid in revolts against the Persians (for instance, that of Amorges, a rebel in Caria – for which, see Thuc. 8.54.3) and their failures in Sicily in 415–413 BCE were cause for reconsideration at the Persian court. Concerns about the Greeks were heightened by the fact that internal revolts continued to plague the Persians. After Darius II's accession in 423 BCE, his brother Arsites contested his dominion in a revolt in which he was accompanied by Artyphios, son of Megabyzos, satrap of Syria (*BNJ* 688 F 15 52). In 420 BCE, Pissouthnes, satrap of Lydia, who had supported the Samian revolt against the Athenians, revolted against the Great King with the aid of Greek mercenaries. With Peloponnesian aid, the Persian general Tissaphernes was sent to quash the revolt, which ultimately ended in his appointment to the Lydian satrapy (*BNJ* 688 F 15 53, Thuc. 8.5.4–5). Similarly, Teritouchmes, the satrap of Armenia, also planned a revolt (*BNJ* 688 F 15 55). With the strain of internal affairs, Tissaphernes was able to reach an initial agreement with the Spartans in 412–411 BCE, with three major stipulations: that the Persians should retain whatever territories they already possessed, while preventing

88 Wallinga (1987) 73.

89 Hyland (2018) 31; see also the general discussion of the Peace of Kallias (16–36).

90 As argued by Beckman (2017).

the Athenians from profiting from these areas thereby; that the Persians would join with the Spartans in war against the Athenians, and would only make peace with them if that were the desire of the Spartans as well; and that a mutual protection agreement should exist between the two sides, whereby the enemy of one would be considered the enemy of the other (Thuc. 8.18). However, there were several iterations of this treaty, with the final one giving the Persian king rights over the country in his direct territorial control, Persian forbearance towards the Spartans and their allies, and payment for Spartan naval contingents until the arrival of the Persian fleet in Asia Minor. Crucially, it no longer required the Spartans to enter a mutual assistance agreement against internal rebellion and noted that if the two sides wanted to make peace with the Spartans, each would have a say in that peace (Thuc. 8.58). Even here, the wording of the treaty emphasises the importance of imperial messaging. It 'laid the foundations for the king to claim credit for final victory', allowing the king to 'promote his image as the world's ruler, capable of rewarding supporters, punishing enemies, and restoring peace across the sea'.[91] Ultimately, Tissaphernes was in no hurry to apply the terms of the treaty, and it amounted to very little in the grand scheme of things.

However, in the meantime, the Athenians had re-emerged in the war under the generalship of Alkibiades. It was at this point that Darius II decided to directly intervene in Greek affairs, abandoning the policy of his father. One of Darius' new policies was to reinstitute direct tribute payments from the Ionians. One additional measure was to eschew neutrality. Until 410 BCE, the Persian royal response to the Peloponnesian conflict was to refrain from taking one side or another. Finally, in 407 BCE Darius II committed to a definite stance, preferring to place his hand firmly with Sparta. A key turning point in the situation came in 408 BCE, when Cyrus (the Younger), son of Darius, assumed a super-satrapal command of Anatolia. The area had been recently plagued by competition between the satraps (most notably Tissaphernes and Pharnabazos, satrap of Hellespontine Phrygia); their unquenchable desire to impress the king (Thuc. 8.6.1), together with Ionian attacks on Persian garrisons, caused serious inefficiencies in this utterly important region of the empire.[92] To diffuse the situation, Darius assigned his son a general military command to take control over the war together with the Spartans (Xen. *Hell.* 1.4.13) and relieve the satrapal conflict. The impetus for Darius' change of heart was the potential for continued trouble in Asia Minor, but also in Egypt and Media (for which, see Xen. *Hell.* 1.2.19; 2.1.13).[93] That is, the heart of the empire remained of

91 Hyland (2018) 74–75, with general discussion of the treaty at 71–75.

92 Briant (2002) 594. See also Hyland (2018, 76–97) on this period.

93 Lewis (1977) 124–35.

utmost concern. The Persians were never interested in embroiling themselves directly in inter-Greek affairs; in fact, Darius failed to deliver the fleet promised to the Spartans in the treaty of 411 BCE (Diod. Sic. 13.46.6; Thuc. 8.87), likely expecting 'that the Greeks would wear each other out'.[94] Thus, as Wiesehöfer notes, Persian action, even in this tumultuous period, 'was always guided by what was in the best interests of the empire'.[95] Indeed, by 404 BCE, the Spartans had forced the Athenians into submission, largely with the help of Cyrus' aggressive aid to the Spartans. Nevertheless, for the Persians, the dynamic of the Mediterranean would never be the same. Now, political considerations had come to the fore in their strategic decision-making.

5 Fourth-Century Persia

In the same year as the conclusion of the Peloponnesian War, the Persians found themselves in a serious dynastic crisis. Upon Darius' death in 404 BCE, his son Artaxerxes II took the throne, despite the protests of Darius' wife, Parysatis, who had preferred Cyrus as crown prince (Plut. *Artax.* 2.3). Parysatis' intervention and Cyrus' possible attempt on his brother's life during the coronation ceremony at Pasargadae (Plut. *Artax.* 3.4; Xen. *An.* 1.3) were dead ends, however, and Cyrus remained at his post in Asia Minor. There, Cyrus began to raise an army to rebel against Artaxerxes. As it turned out, his aid to the Spartans in the last decade of the fifth century BCE had been self-motivated, as his aim in seeking a hurried conclusion to the Peloponnesian War was to collect a Greek mercenary army to help him overthrow his brother.[96] They were also useful for enforcing Cyrus' tribute claims and to deter resistance.[97] However, his campaigns came to an end a few years later upon his death at the battle of Cunaxa (401 BCE).[98] Ultimately, although Cyrus likely was the best Achaemenid candidate to extend Persian influence more permanently into

94 Briant (2002) 597. Hyland (2018, 86–91) proffers several probable reasons for the withdrawal, arguing that Darius did not desire to engage the Ionians in the autumn (due to the potential for debilitating storms); that he wished to conserve funds (a withdrawal of the fleet would put less pressure on the storehouses during a long encampment through the autumn and winter); and that he had lost his confidence in Tissaphernes' ability to supervise the fleet (given his previous failures to defend Ionian positions).

95 Wiesehöfer (2006) 662.

96 Waters (2014) 176.

97 Hyland (2018) 125, with bibliography.

98 Our main sources for the battle are Plut. *Artax.* 9–16, who retains the account of the Persian doctor Ctesias, who witnessed the battle; Diod. Sic. 14.22–24; and Xen. *An.* 1.8. A cogent visual reconstruction of the battle can be found in Hutchinson (2000) 256–57; see also the updated deployment plan in Rop (2019) 56.

Greece, his defeat and death seriously hampered Persian relations beyond the Aegean until a new war could restore it.[99] After long travails, the Greek mercenaries employed by Cyrus made their way home, but the battle was a major precipitating factor in the beginning of the Persian-Spartan Wars. That conflict would end in 387/6 BCE with the signing of the King's Peace (Xen. *Hell.* 5.1.131; Diod. Sic. 14.110.2–3), which brought the Ionian Greek city-states back firmly into Persian control and reestablished Persian dominance over much of the Aegean. It also provided autonomy for all Greek *poleis*, even those outside of Persia's border; that proclamation amounted to a statement of universal authority. The Greeks states were now viewed 'as beneficiaries of imperial order until the coming of Alexander'.[100]

Yet during this period it was not the Greeks that were Artaxerxes' main concern. Ruzicka provides a useful list of all the struggles faced by the new Persian king,[101] but the most time- and resource-consuming conflicts were with Egypt and in the suppression of satrapal revolts.[102] In 404 BCE, shortly after Artaxerxes' succession, the Libyan dynast Amyrtaios led a rebellion, and by 400 BCE, he was recognised as king even in Elephantine (*BNJ* 609 F3c).[103] The Egyptians had successfully seceded from the Persian Empire. From this point onwards, as Rop notes, 'The Achaemenid Empire ... was primarily concerned with the recovery of Egypt'.[104] Beginning in 389 BCE, the Persians embarked on an invasion of Egypt (Isoc. *Paneg.* 140), which ultimately failed.[105] However, it was not possible to remain idle; an independent Egypt was an incredible threat to Persia, as it would be a natural ally to anyone who wished to challenge the empire (even subordinates of the king himself).[106] Therefore, in 379 BCE, the Persians began preparations for yet another invasion of Egypt, which did not conclude until 373 BCE (Diod. 15.41.1–43.6). Simultaneously (and directly affecting Persian success in Egypt), the Persians faced difficult affairs in Cyprus, particularly with Evagoras of Salamis, who was supportive of the Egyptians in their war with the Persians (Diod. 15.4.3). The Persian king could not ignore

99 Hyland (2018) 127.
100 Hyland (2018) 166–68 (quote at 168).
101 Ruzicka (2006) 231.
102 The fullest treatment of Persia's conflict with Egypt during this period can be found in Ruzicka (2012).
103 Quack (2021) 560–61.
104 Rop (2019) 233.
105 Others date the rebellion to 385 BCE; see Quack (2021) 561.
106 Briant (2002) 652. However, see Agut-Labordère (2023, 750), who notes that the two decade-long delay in any attempt to decisively reconquer Egypt 'shows that the region's independence did not in itself pose a particular problem for the empire'.

this issue either, as Cyprus was an essential strategic stronghold for Persian power over the Aegean, and therefore both a logistical and military concern.[107] Ultimately, the situation was so desperate that Artaxerxes interfered in Hellenic politics to ensure the procurement of the naval and mercenary forces necessary to combat the Egyptian threat (Diod. 15.38.1). Only a year later, there were new preparations for another Egyptian campaign (Nep. *Datames* 3.5). And in 361 BCE, Artaxerxes faced yet another war with the Egyptian king Tachos (Diod. Sic. 15.90.2, 92.1–5), which saw Egypt lead a military expedition to Syria. However, internal affairs in Egypt (namely, the rebellion of Samaus, uncle of Tachos, who, in the king's absence, had placed Tachos' son Nectanebo II on the Egyptian throne) prevented any headway against the Persians in this encounter. Throughout the entire struggle between the two powers, both sides were aided by Greek mercenary troops.[108] However, this venture too had been made more difficult because of a revolt in the Phoenician city of Sidon (348/7 BCE), whose efforts were supplied by mercenary forces from Egypt.[109] The Egyptians were not formally reintegrated into the Persian Empire until the reign of Artaxerxes III (Ochus), who in 343 BCE was finally able to bring them back into the fold. This defeat inaugurated the second Persian rule over Egypt (Diod. Sic. 16.46.4–7; 16.51).[110]

Pressures in Egypt were exacerbated by additional stresses related to revolts of different satraps in the empire. The most prominent of these events is referred to as the 'Great Satraps' Revolt'.[111] Datames, satrap of Cappadocia, was the first to rise against the king (Diod. Sic. 15.90 dates the beginning of the revolt to 361 BCE). During his rebellion, Datames sought support from Athens and Sparta and was also allegedly aided by other satraps in Asia Minor; at the same time, he supported the Egyptian effort against Persia.[112] Ariobarzanes, satrap of Hellespontine Phrygia, followed suit (Dem. 15.9–10; Xen. *Ages.* 2.26), and Orontes, governor of Mysia (and formerly of Armenia), also joined in the unrest (Just. *prolog.* 10). Weiskopf has challenged the idea that the Satraps' Revolt was a well-organised, centralised threat against the empire, arguing that instead we should view it as a series of small disturbances, 'local instabilities'

107 A settlement was reached with Evagoras, which can most likely be dated to 381 BCE (Diod. 15.8.1–3; 15.9.1–2).

108 Rop (2019) 88–118.

109 On this revolt, see Wiesehöfer (2016) 93–112.

110 Quack (2021) 562. For the ways in which the politics of the Second Persian period in Egypt were manipulated by Alexander's successors, see Colburn (2015) 165–202.

111 A very useful discussion of the sources and problems relating to the Satraps' Revolt can be found in Briant (2002) 656–75.

112 This event is potentially referred to in Sachs and Hunger (1988) no. 367 lines 2–10.

in Asia Minor that never seriously threatened Artaxerxes' capacity to rule the empire.[113] Although little fighting was involved in these 'revolts', there was still yet a credible effort at amassing funds, hiring mercenaries, and creating connections with disaffected figures and peoples; therefore, the 'revolts' were meant simply to *appear* threatening, with the goal of establishing a bargaining position.[114] These events reveal imperial instability, in that the satraps clearly demonstrated the military and political capability of resisting the king's power, despite their lack of success in the end. Secondly, they indicate that, despite the dynamic situation on the Achaemenid periphery, the centre still maintained an ability to overcome these threats with either offensive military action or political repercussions, particularly under the direction of Ochus (later Artaxerxes III), who had succeeded in disbanding the mercenary armies of the coastal satraps (Schol. Dem. 4.19).[115] Thus, we should not view these revolts as a symptom of Persian decline in this period;[116] they were simply a consequence of its size and the dynamics of its political and administrative systems, which had succeeded remarkably for two centuries prior.

This observation brings us back to an important topic of consideration: Achaemenid expenditure on mercenary soldiers, which was a particularly acute concern in the fourth century BCE.[117] Their presence has often been interpreted with reference to the so-called 'Greek Thesis',[118] which argues that, due to the superiority of the Greek hoplite, the Persians became 'excessively dependent' on Greek mercenary soldiers as the empire descended into decadence and dysfunction. However, Tuplin has recently proffered the argument that we may view a structural link between the acquisition of mercenary troops and the possibility for satrapal disobedience and warlordism;[119] the latter issue, as we have seen, was one of the main causes for the collapse of Persian grand strategy, and hence, of the empire itself. Rop concurs, noting that at a fundamental level, the presence of Greek mercenaries at several important battles in the latter half of the life of the Achaemenid Empire also underscores the intricate and strategic political involvement of the Greeks in the structural web

113 Weiskopf (1989).
114 Ruzicka (2006) 234–35.
115 These are the observations of Briant (2002) 675.
116 In fact, as noted by Alram (2012) 74, money supply and circulation within the empire reached an enormous level under Artaxerxes III, indicating flourishing trade and a prosperous economy.
117 According to Plut. *Artax.* 20.4, these mercenaries were paid in darics.
118 For a summary of the 'Greek Thesis', see also Briant (2002) 783–787. Most recently, see Rop (2019, 1–29), who brilliantly exposes the fundamental flaws in the Greek thesis.
119 Tuplin (2017) 33.

of the Persian imperial system. However, he argues, the acquisition of Greek soldiers 'was not itself the primary purpose of the foreign policy of [Persia or Egypt] towards Greece'.[120] The extreme expense associated with their payment made the levying of a mercenary army a royal privilege, which thereby reduced potential threats abroad, and more importantly, at home.[121] Now, Greek mercenaries were being used as pawns in the internal Achaemenid political game, rather than serving as relatively harmless entities patrolling Aegean waters, as had been the case in the fifth century BCE.

Furthermore, the abundance of Achaemenid attention towards the satrapies of Asia Minor can reveal a good deal about their economic and strategic interests. The Persians ran an incredibly large and complex empire, buttressed by the richest economy the ancient world had yet seen. One estimate places the total precious metals available to the Greek and Achaemenid world somewhere around 2,500 talents per year.[122] Indeed, by its fall under Alexander, the total wealth of the Persian Empire is estimated to have been somewhere near 5200 metric tons of silver.[123] Numismatic evidence can reveal to us, in particular, the importance of the territories of Achaemenid Asia Minor. The first gold coins ('darics') in the Achaemenid monetary system were issued by Darius I (Hdt. 4.166.2), most likely following his return from the Scythian expedition in Europe.[124] These replaced the Lydian Croeseids that had until that time been predominant in Asia Minor. In the early periods, the coins were struck exclusively at Sardis. However, as Briant notes, these coins had no trade or compensatory function, but rather served as a symbolic tribute to Darius' unique power as founder of the empire.[125] He also instituted a silver coinage, known as a *siglos*, a type of coinage that would be used primarily to support Persian needs in Asia Minor.[126] This introduction of Persian imperial coinage in Asia Minor 'was a path-breaking financial reorganization and an important contribution to the further consolidation of the Achaemenid Empire in the west'.[127]

120 Rop (2019) 232–34 (quote at 233).
121 Blichert-Toft et al. (2022) 8.
122 Bresson (2020) 212.
123 These estimates are based on the amount of wealth accumulated by Alexander in his capture of the treasuries at Susa, Babylon, and Persepolis, following the defeat of Darius III. See Blichert-Toft et al. (2022) 1.
124 Root (1988) 1–12, argues for a date around 500 BCE, but does not favour pinpointing the European expedition as a cause for the reform. By the reign of Xerxes, one daric was equivalent to the monthly income of a soldier in the Persian army (Xen. *An.* 1.3.21).
125 Briant (2002) 408–409.
126 Alram (2012) 69.
127 Alram (2012) 65–66.

However, these royal coins coexisted in Asia Minor with local, satrapal coin strikes, as well as imported Greek coins. There were also special issues of coins, most notably an imitation of the Athenian tetradrachm that appears during the reign of the satrap Tissaphernes in the late fifth century BCE.[128]

The satrapies in Asia Minor constituted a significant source of revenue. Whereas Egypt, for instance, is reported by Herodotus to have contributed 700 talents, the three territories of Asia Minor contributed 1,260 talents, outweighing the numbers produced from Babylon and Assyria, which together produced 1,000 talents.[129] Furthermore, Diodorus (15.90.4) notes that the Satrap's Revolt of the 360s was so extensive that half of the king's revenues were cut off, leaving insufficient funds for the prosecution of war. It was also from this area that the oarsmen of the Persian navy were derived.[130] Additionally, together with Egypt, Asia Minor was far and away the hotspot for trade between Greek and Phoenician merchants in the Achaemenid period, particularly in the fifth century BCE.[131] Thus, Achaemenid use of mercenaries and their focus on the affairs of Asia Minor can provide a microcosmic example of Persian imperial policy: during this period, they were still keen on diligent resource management, peace in the provinces, and control over water and trade routes, but at this point in its history, the financial and political affairs of the empire were intimately intertwined, and of utmost priority.

6 Failures against Macedon

However, the final true test of the success of Achaemenid grand strategy was the rise of a newcomer in Europe: Macedon. As already discussed, Persian focus in the fourth century BCE was on Egypt, which meant that all other international players – and especially the Greeks – were treated as subsidiary issues. Because of this ambivalence, Greek pacification was aimed at securing a friendly hegemonic state or states. That disconnection from Greek politics meant that when Philip II came to the throne of Macedon in 360 BCE with the goal of becoming hegemon of all of Greece (Diod. Sic. 16.89.1), the Persians were quite unprepared for the potential consequences. It was only when the Athenians – far too late – sought aid from Persia in 343 BCE that the

128 Alram (2012) 71–72.

129 As noted by Bresson (2020) 232.

130 Briant (2002) 405, notes that the Achaemenids spent a considerable amount of money on building the Persian navy, which was to be manned by tributary coastal peoples.

131 See the discussion of Bresson (2020) 221–22.

Persians considered concluding an alliance with the Athenians against Philip (Dem. 10.31–34). The Persians had been aware of Philip's machinations; after all, some disgraced Persian aristocracy had sought refuge at his court (Diod. Sic. 16.52.3–4), and Artaxerxes III had supported Perinthos during Philip's siege of the city in 340/339 BCE (Diod. Sic. 16.75.1–2; Dem. 11.5–6).[132] After the battle of Chaeronea in 338 BCE, Philip became the hegemon of the Corinthian League and the Greek city-states, for the first time in their history, lost their independence. He used as a pretext for war against Persia the desire to avenge the Greeks for Persian profanities against their temples (Diod. Sic. 16.89.2; Just. 9.4), clearly a reference to the burning of the Acropolis during the Persian Wars of the fifth century BCE. What Philip really wanted, as Brosius has argued, was hegemony over all of Greece, and so, 'in order to prevent the development of the Persian-Athenian alliance, Philip "created" the enmity with Persia'. His aims went no further than the suppression of the satraps in Asia Minor, who presented the biggest threat through their interference in Greek affairs.[133] Greek memory served as a motivating tool for war; but perhaps more important was Philip's use of a popular fifth century Achaemenid pretext for conquest: revenge.

In the same year as Philip's death (336 BCE), the new – and last – Achaemenid king, Darius III would come to the throne. It was only after Alexander the Great's swift and ready assumption of the Macedonian kingship (and the leadership of the Greeks) that the Persians seem to have taken notice, and Darius quickly mustered his forces (Diod. Sic. 17.7.2). Alexander seemed more prepared to launch a full-fledged attack on the Persian Empire. Throughout his campaign, Alexander would invoke the revenge motif, as is apparent in his letter to Darius of 332 BCE, where he states: 'Your ancestors came to Macedonia and Greece and did great harm to us, although you had suffered no such harm yourselves' (Arr. *Anab.* 2.14.4).[134]

As with the Persian Wars of the fifth century BCE, there is a gigantic cache of scholarship on Alexander's campaigns in Persia, and therefore there is no need here to repeat it. The Persians would first meet Alexander in battle in 334 BCE, not far from Troy, where he initially beached his ships upon arrival in Asia. The battle at the Granikos river was the only engagement between the Macedonian

132 Kuhrt (2010) 419.

133 Brosius (2003) 235. Good overviews of the period can be found in Buckler (1994) 99–122, and Ellis (2008) 760–90.

134 An outstanding study of the trajectory of the vengeance theme during the reigns of Philip and Alexander can be found in Squillace (2010) 69–80, who finds that the Persian Wars were central to the self-definition and perception of both kings.

and Persian empires that would feature a solely satrapal army. Despite a close brush with death (Arr. *Anab.* 1.15.8) and an ugly encounter with the Greek mercenary forces fighting on behalf of the Persians (Arr. *Anab.* 1.16.6), the Macedonians were victorious. Following Alexander's victory at the Granikos, he moved his army down the Anatolian coast in a bid to disconnect the Achaemenids from their crucial naval bases and support on the Ionian coast. Understanding the emergent nature of the situation, Darius quickly gathered an army, with the hopes of presenting a real threat to Alexander's supply lines and communication. The Great King's quick movements to meet Alexander before he could subjugate the Anatolian coastline also gave him the possibility of linking up with the Achaemenid fleet at the Gulf of Issos. Therefore, Darius followed behind Alexander's army into Cilicia; this position forced a battle in 333 BCE.[135] Unfortunately for the Persians, the day spelled horrific carnage (Plut. *Alex.* 20.10 records 110,000 Persian soldiers dead), including many noble Persians (Diod. Sic. 17.34.5). Although Darius escaped, his family – including the next heir to the Persian throne – were captured in their tents (Plut. *Alex.* 21; Diod. Sic. 17.36.2–4). The battle resulted in the loss to Darius of the western half of his empire.[136]

After the battles of Granikos and Issos, and in preparation for a final decisive battle against Alexander, Darius introduced a major reform of the Persian army (Curt. 4.9); Darius now drew peoples from various satrapies together, gave them uniform arms, and increased the cavalry in the army.[137] The two sides met for their final set-piece battle at Gaugamela in 331 BCE. The Persians lost almost half of their men at Gaugamela.[138] The functional outcome of the battle, despite Darius' continued existence, was to make Alexander the ruler of the Persian Empire. But the war did not see a symbolic end until Alexander arrived at the Persian capital of Persepolis, first constructed by Darius I around 518 BCE. Alexander left Persepolis in ashes and again took up his pursuit of Darius into the territory of Media. Many have viewed the destruction of Persepolis as the final act in Alexander's 'revenge play', a symbolic burning of the most important Persian capital to represent the fulfillment of his promise

135 Farrokh (2007) 100.

136 Ashley (2004) 230.

137 Bittner (1987, 293) claims that Darius made 'Der alte taktische Irrtum des Xerxes' in his assumption that a multitude of men (and in Xerxes' case, ships, too) would be a guarantee of victory.

138 Diod. Sic. 17.61.3 lists 90,000 Persian infantry and cavalry dead, with 500 Macedonian dead and many more wounded. Curt. 4.16.26 lists 40,000 Persian and 300 Macedonian dead.

to the mainland Greeks that Persian wrongs done to their religious sites would be avenged in his campaigns.[139] If vengeance was so important to the Persian kings, we can hardly be critical of Alexander for giving them a taste of their own medicine.

The Persian defeat led to the total loss of their empire, which now fell into the hands of Alexander. However, although Alexander was indeed a great general, his opponent (as has often been argued) was not a weak or tactically inept king.[140] It is true that Alexander met with a Persian Empire that had been weakened over time by internecine warfare, constant revolts, and heavy taxes imposed on its subjects, all of which meant that those in the imperial purview followed the war with little interest and welcomed Alexander and his men as liberators.[141] But it was only now that the Persians had made a fatal mistake in their grand strategic plan: after centuries of involvement in intra-Greek affairs, they had failed to take the rise of Macedon seriously; the status quo of pitting the Greeks against one another had caused them to miss a new, innovative power player in the Mediterranean. Conversely, Alexander and his father had a firm grasp on Persian grand strategy, from logistics to politics to finances: control over the seaways, manipulation of the satrapal system (particularly in Asia Minor), protection of their financial holdings and tribute system, and maintenance of supply and communication lines had always been of utmost importance to their suzerainty. In their invasion, the Macedonians struck directly at these crucial areas of Achaemenid dominion and collapse quickly followed. Perhaps more interestingly, the Persians succumbed to a man from Macedon who understood – and expertly utilised – their favourite imperialistic pretext of vengeance. For that, they paid the ultimate price. But in the context of Achaemenid legacy, nothing could have been more fitting: the cycle had been completed.[142]

---

139 See, for instance, Seibert (1998) 5–58; Flower (2000) 96–135.

140 Dandamaev (1989, 331–32) argues that the Persians were not in fact inferior to the Macedonians during Alexander's campaigns, pointing out the multiple times when the Persians were able to push back the Macedonians in battle; the fact that their ranks included a large number of Greek mercenaries; the superiority of their fleet; and the disposal of an enormous economic potential. Furthermore, it is important to remember that the major reforms to the Achaemenid military machine came primarily in the reign of Darius III, debunking the idea that the last Achaemenid king was a weak or incompetent commander-in-chief, as is noted by Briant (2015) 425–48.

141 Dandamaev (1989) 331–32.

142 For a discussion of Alexander's revenge campaign in its Achaemenid context, see Bridges (2014) 119–25.

7 Conclusion: Persian Imperialism and the Provinces

For Murray, the Persian Empire survived on the performative nature of its military: 'An empire constantly threatened with national and religious revolts was held together essentially by the actions of the king in mobilising his subjects for war: it was only on expeditions that the power of the monarchy was exhibited and centrifugal tendencies were temporarily suspended'.[143] Others corroborate this assumption, claiming that Achaemenid royal ideology 'both legitimized and actively *demanded* war', further arguing that Xerxes, for instance, was compelled to make an attempt at the annexation of Greece at the beginning of his reign because he had to conquer *something*.[144] However, we must be careful to differentiate between expansionist aims and military success. The conception of *total* world domination speaks against what we know from Persian sources about their imperial intentions: never in their inscriptions do they aspire to such a goal, quite satisfied, as it were, with their rule over Asia and Egypt alone. Even for Xerxes, the conquering of Greece for the simple purpose of adding it to his empire made little historical sense.[145] As we have seen throughout this survey, at all points, the goal of Persian conquest was always control over the seaways and resources, consolidation of the empire (here, the continuation of tribute payments was not an insignificant matter[146]), and the establishment of peace in the provinces. It was perfectly acceptable if this control was mustered through military efforts or through diplomacy.[147] It was the appearance of control[148] – rather than expansionism – that we can consider the most crucial component of Achaemenid imperial ideology.[149] This prerogative helps explain the Persian rhetoric and iconography surrounding the Greco-Persian Wars and the consistent obsession with the reintegration of a rebellious Egypt.

143 Murray (1988) 465.

144 Konijnendijk (2021) 1144.

145 See the very convincing arguments of Wiesehöfer (2004) 214–18. Cataudella (1998, 51) has argued that the Persian punishment of rebellions, often presented in terms related to 'the will of Ahuramazda', should be viewed in terms of the restoration of imperial order, rather than a realisation of an imperial order *per se*. This is also the stance taken in Konijnendijk (2021) 1142–43.

146 See Tuplin (1987) 137–57. On tribute, see below.

147 See Beckman in this volume.

148 As discussed in Hyland (2018). He notes, for example, that the Achaemenid treaty with Athens under Artaxerxes I 'was both economically lucrative and illustrative of Persia's ability to bestow world order' (169). Control over the Sea in particular was not only strategically necessary for military and economic reasons, but it also became an important way for the Persians to symbolically express their power. See Haubold (2012) 17.

149 See Rung (this volume) on the importance of military success to Achaemenid royal ideology.

These considerations bring us directly to complications associated with the Persian preoccupation with the provinces. For the Achaemenids, the exercise of control in an area rested on several strategies, but often included a combination of the installation of military garrisons and the ready adoption of foreign governmental structures.[150] However, as has been noted, these strategies were not always successful or well received, and the Persians spent a significant amount of time and resources in quelling imperial rebellions. As we have seen, vengeance for the breaking of an oath or a treaty (which is often manifested by – or perceived as – rebellion) is a common pretext for Achaemenid conquest, at least until the reign of Xerxes. This theme appears in Persian inscriptions; Xerxes, for instance, is quite keen on noting the ferocity of his tendencies towards vengeance (XPh §2h), a conception that replicates that seen in the reign of his father, Darius.[151] Cataudella has made the interesting observation that the instances in which our sources cite revenge as a pretext for Achaemenid conquest are all in areas outside of Asia, external to the predetermined Persian dominion that was granted to the kings by the will of Ahuramazda.[152] However, Konijnendijk has convincingly argued that vengeance was an important motivation for military conquest in the Ancient Near East more generally (as well as in Persian inscriptions), and thus should be taken more seriously.[153] 'Vengeance' was part of the imperial messaging programme; this is certainly not unimportant (if perhaps ineffective), and it too was a concomitant part of Achaemenid grand strategy in different periods of the empire.

Further, Cataudella argues that there was a blanket policy of tolerance towards those territories in Asia – Israel, Babylonia, and Lydia, for instance – as opposed to those like European Scythia or the Greeks.[154] However, there was truly a range of Persian reaction to rebellion and we should view them on an *ad hoc* basis: Cambyses' expedition to Egypt, for instance, was predicated on vengeance and included heavy burdens of taxation on the population, and in the fifth and fourth centuries, the Persians actively encouraged some Greek democratic governments even after serious uprisings.[155] Throughout, the

150 For Egypt (one of the best-attested cases), see Colburn (2021) 303–11. As Tuplin (1987) 112: 'In short, it was the Persian way to use existing institutions and seek to harness the energies and interests of native dominant classes to their own ends'.

151 Bridges (2014) 82–83.

152 Cataudella (1998) 56–63.

153 Konijnedijk (2021) 1145–46. He rightly points out the comparandum of Assyrian vengeance for broken *adê* treaties.

154 Cataudella (1998) 62.

155 Further, Hyland (2018, 169–70) has made the important observation that, while fifth century interactions with the Greeks were largely based on principles of vengeance, these policies were altered in the fourth century. By then, efforts to resolve Greek conflicts

Achaemenids focused on policies of pragmatism, always considering primary the interests of their empire; in all periods, those prerogatives were tied up with controlling the Sea and ensuring peace in the provinces through the prevention of revolts and careful management of resources (political, economic, and logistic). The result is not a single Achaemenid grand strategy, but one that was flexible within all chronological and geographical boundaries. It was not until Alexander entered Persian territory that those strategic goals were hampered by a man who quite well understood their mortal weaknesses.[156]

Bibliography

Agut-Labordère, D. (2023) 'The Satrapies of the Persian Empire in Egypt', in Radner, Moeller, and Potts (2023) 737–83.

Alram, M. (2012) 'The Coinage of the Persian Empire', in Metcalf (2012) 61–88.

Ashley, J.R. (2004) *The Macedonian Empire: The Era of Warfare Under Philip II and Alexander the Great, 359–323 B.C.* Jefferson.

Balcer, J.M. (1995) *The Persian Conquest of the Greeks, 545–450 B.C.* Konstanz.

Beckman, D. (2017) 'King Artaxerxes' Aegean Policy'. *Journal of Persianate Studies* 10: 1–25.

Beckman, D. (2018) 'The Many Deaths of Cyrus the Great'. *Iran. Stud.* 51.1: 1–21.

Bichler, R. (2021) 'Kyros' letzte Schlacht und sein Tod', in Klinkott, Luther, and Wiesehöfer (2021) 49–92.

Blichert-Toft, J., de Callataÿ, F., Télouk, P., and Albarède, F. (2022) 'Origin and Fate of the Greatest Accumulation of Silver in Ancient History'. *Archaeological and Anthropological Sciences* 14.64.

Bittner, S. (1987). *Tracht und Bewaffnung des persischen Heeres zur Zeit der Achaimeniden.* Munich.

Boardman, J. (ed) (1988) *The Cambridge Ancient History Vol IV: Persia, Greece and the western Mediterranean, c. 525 to 479 B.C.* Cambridge.

Bosworth, A.B. and Baynham, E.J. (eds) (2000) *Alexander the Great in Fact and Fiction.* Oxford.

Bresson, A. (2020) 'Silverization, Prices, and Tribute in the Achaemenid Empire', in Tuplin and Ma (2020) 3: 209–48.

focused on expressions of universal authority, with direct military involvement a last resort.

156 My great thanks go to John Hyland and Khodadad Rezakhani for allowing me to contribute to this volume. I am especially indebted for the very careful editing and suggestions that greatly improved the quality of this contribution. For another take on Persian grand strategy, see this author's contribution to the forthcoming volume *Ancient Strategy* (eds. J. Lacey and D. Potter).

Briant, P. (1984) *L'Asie centrale et les royaumes proche-orientaux du premier millénaire (c. VIIIe–IVe siècles avant notre ère)*. Paris.

Briant, P. (2002) *From Cyrus to Alexander*. Winona Lake.

Briant, P. (2015) *Darius in the Shadow of Alexander*. Cambridge.

Briant, P. (2017) *Kings, Countries, Peoples: Selected Studies on the Achaemenid Empire*. Stuttgart.

Bridges, E. (2014) *Imagining Xerxes: Ancient Perspectives on a Persian King*. London.

Brosius, M. (2003) 'Why Persia became the Enemy of Macedon', in Henkelman and Kuhrt (2003) 227–38.

Brosius, M. (2013) 'Greek Sources on Achaemenid Iran', in Potts (2013) 658–66.

Brosius, M. (2021) *A History of Ancient Persia: The Achaemenid Empire*. Hoboken.

Buckler, J. (1994) 'Philip II, the Greeks, and the King 346–336 B.C.'. *Illinois Classical Studies* 19: 99–122.

Burn, A.R. (1984) *Persia and the Greeks: The Defense of the West, c. 546–478 B.C.* Stanford.

Cawkwell, G. (2005) *The Greek Wars: The Failure of Persia*. Oxford.

Callieri, P. (2023) 'The Southeastern Regions of the Persian Empire on the Indo-Iranian Frontier: Arachosia, Drangiana, Gedrosia, Sattagydia, Gandhara, and India', in Radner, Moeller, and Potts (2023) 837–86.

Campbell, B. and Tritle, L.A. (eds) (2013) *The Oxford Handbook of Warfare in the Classical World*. Oxford.

Carney, E. and Ogden, D. (eds) (2010) *Philip II and Alexander the Great: Father and Son, Lives and Afterlives*. Oxford.

Carradice, I. (ed) (1987) *Coinage and Administration in the Athenian and Persian Empires*. Oxford.

Cataudella, M. (1998) 'Vendetta e imperialism nella monarchia achaemenide', in Sordi (1998) 47–63.

Cohoon, J.W. trans. (1932) *Dio Chrysostom: Discourses 1–11, Loeb Classical Library 257*. Cambridge.

Colburn, H. (2013) 'Connectivity and Communication in the Achaemenid Empire'. *JESHO* 56: 29–52.

Colburn, H. (2015) 'Memories of the Second Persian Period in Egypt', in Silverman and Waerzeggers (2015) 165–202.

Colburn, H. (2021) '"The Spear of the Persian Man has Gone Forth Far:" The Achaemenid Empire and its African Periphery', in Daryaee and Rollinger (2021) 291–336.

Collins, J.J. and Manning, J.G. (eds) (2016) *Revolt and Resistance in the Ancient Classical World and the Near East: In the Crucible of Empire*. Leiden.

Dandamaev, M.A. (1989) *A Political History of the Achaemenid Empire*. Leiden.

Daryaee, T. and Rollinger, R. (eds) (2021) *Iran and Its Histories: From the Beginnings through the Achaemenid Empire: Proceedings of the First and Second Payravi Lectures on Ancient Iranian History, UC Irvine, March 23rd, 2018, & March 11th–12th, 2019*. Wiesbaden.

Dromard, B. (2022) 'Nippur and Its Region under the Neo-Babylonian and Achaemenid Empires: A Fiscal History', in Kleber (2021) 177–234.

Duffy, X. (2018) *Commemorating Conflict: Greek Monuments of the Persian Wars*. Oxford.

Dunsch, B. and Ruffing, K. (eds) (2013) *Herodots Quellen – Die Quellen Herodots*. Wiesbaden.

Dusinberre, E., Garrison, M.B., and Henkelman, W.F.M. (eds) (2020) *The Art of Empire in Achaemenid Persia: Studies in Honour of Margaret Cool Root*. Leiden.

Ellis, J.R. (2008) 'Macedonian Hegemony Created', in Lewis, Boardman, and Hornblower (2008) 760–90.

Ferrill, A. (1966) 'Herodotus and the Strategy and Tactics of the Invasion of Xerxes'. *AHR* 72: 102–15.

Finn, J. (2014) 'Alexander's Return of the Tyrannicide Statues to Athens'. *Historia* 63: 385–403.

Flower, M. (2000) 'Alexander the Great and Panhellenism', in Bosworth and Baynham (2000) 96–135.

Francfort, H.P. (1988) 'Central Asia and Eastern Iran', in Boardman (1988) 165–93.

Gaddis, J.L. (2018) *On Grand Strategy*. New York.

Garland, R. (2017) *Athens Burning: The Persian Invasion of Greece and the Evacuation of Attica*. Baltimore.

Gehler, M., and Rollinger, R. (eds) (2014) *Imperien in der Weltgeschichte. Epochenübergreifende und globalhistorische Vergleiche, 2 Bände*. Wiesbaden.

Graf, D.F. (1993) 'The Persian Royal Road System in Syria-Palestine'. *Recherches pluridisciplinaires sur une province de l'Empire Achéménide* 6: 149–68.

Hallock, R. (1969) *Persepolis Fortification Tablets*. Chicago.

Haubold, J. (2012) 'The Achaemenid Empire and the Sea'. *Mediterranean Historical Review* 27.1: 5–24.

Harrison, T. (2011) *Writing Ancient Persia*. London.

Henkelman, W.F.M. and Jacobs, B. (2017) 'Roads and Communication', in Jacobs and Rollinger (2021) 719–35.

Henkelman, W.F.M. and Kurt, A. (eds) (2003) *A Persian Perspective: Essays in Memory of Heleen Sancisi-Weerdenburg, Achamenid History* XIII. Leiden.

Henkelman, W.F.M. (2018) 'Bactrians in Persepolis – Persians in Bactria', in Lhuiller and Boroffka (2018) 223–55.

Henkelman, W.F.M. (2017) 'Imperial Signature and Imperial Paradigm: Achaemenid Administrative Structure and System Across and Beyond the Iranian Plateau', in Jacobs, Henkelman and Stolper (2017) 45–256.

Klinkott, H. (2005) *Der Satrap: Ein achaimenidischer Amtsträger und seine Handlungsspielräume* (*Oikumene. Studien Zur Antiken Weltgeschichte*). Berlin.

Klinkott, H., Luther, A., and Wiesehöfer, J. (eds) (2021) *Beiträge zur Geschichte und Kultur des alten Iran und benachbarter Gebiete Festschrift für Rüdiger Schmitt*. Stuttgart.

Holmes, N. (ed) (2011) *Proceedings of the XIVth International Numismatic Congress*. Glasgow.

Hornblower, S. (2008) 'Persia', in Lewis (2008) 45–96.

Hutchinson, G. (2000) *Xenophon and the Art of Command*. London.

Hyland, J.O. (2018) *Persian Interventions: The Achaemenid Empire, Athens, and Sparta, 450–386 BCE*. Baltimore.

Jacobs, B., Henkelman, W.F.M., and Stolper, M.W. (eds) (2017) *Die Verwaltung im Achämenidenreich – Imperiale Muster und Strukturen/Administration in the Achaemenid Empire – Tracing the Imperial Signature* (*Classica et Orientalia* 17). Wiesbaden.

Jacobs, B. and Rollinger, R. (eds) (2021) *A Companion to the Achaemenid Persian Empire*. Malden, MA.

Jung, M. (2013) 'Marathon and the Construction of the Persian Wars in Antiquity and Modern Times. Part I: Antiquity'. *BICS* 124: 255–66.

Jursa, M. (2011) 'Taxation and Service Obligations in Babylonian from Nebuchadnezzar to Darius and the Evidence for Darius' Tax Reform', in Rollinger, Truschnegg, and Bichler (2011) 431–48.

Kagan, J. (2011) 'Archaic Greek Coins East of the Tigris: Evidence for Circulation?', in Holmes (2011) 230–34.

Karageorghis, V. and Taifacos, I. (eds) (2004) *The World of Herodotus: Proceedings of an International Conference held at the Foundation Anastasios G. Leventis, Nicosia, September 18–21, 2003 and Organized by the Foundation Anastasios G. Leventis and the Faculty of Letters, University of Cyprus*. Nicosia.

King, R. (2019) 'Taxing Achaemenid Arachosia: Evidence from Persepolis'. *JNES* 78: 185–99.

Kleber, K. (2021) 'Babylonia' in Jacobs and Rollinger (2021) 905–22.

Kleber, K. (ed) (2021) *Taxation in the Achaemenid Empire. CleO 26*. Wiesbaden.

Konijnendijk, R. (2021) 'Legitimization of War', in Jacobs and Rollinger (2021) 1141–50.

Kousser, R. (2009) 'Destruction and Memory on the Athenian Acropolis'. *The Art Bulletin* 91: 263–82.

Kozuh, M., Henkelman, W.F.M., Jones, C., and Woods, C. (eds) (2014) *Extraction and Control: Studies in Honor of Matthew W.* Stolper (*Studies in Ancient Oriental Civilization* 68). Chicago.

Kuhrt, A. (1988) 'Babylonia from Cyrus to Xerxes', in Boardman (1988) 112–38.

Kuhrt, A. (2010) *The Persian Empire: A Corpus of Sources from the Achaemenid Period*. London.

Kuhrt, A. (2014a) 'State Communications in the Persian Empire', in Radner (2014) 112–40.

Kuhrt, A. (2014b) 'Reassessing the Reign of Xerxes in the Light of New Evidence', in Kozuh et al. (2014) 163–69.

Kuhrt, A. (2021) 'The Great Conquests', in Jacobs and Rollinger (2021) 403–13.

LaForse, B. (2013) 'Fighting the Other. Part I: Greeks and the Achaemenid Persians', in Campbell and Tritle (2013) 569–87.

Lewis, D.M. (1977) *Sparta and Persia: Lectures Delivered at the University of Cincinnati, Autumn 1976 in Memory of Donald W. Bradeen*. Leiden.

Lewis, D.M., Boardman, J., and Hornblower, S. (eds) (2008) *The Cambridge Ancient History Vol. 4: The Fourth Century BC*. Cambridge.

Lewis, S. (ed) (2006) *Ancient Tyranny*. Edinburgh.

Lhuiller, J., and Boroffka, N. (eds) (2018) *A Millennium of History: The Iron Age in Southern Central Asia (2nd and 1st Millennia BC)*. (*Archäologie in Iran und Turan 17 = Mémoires de la Délégation Archéologique Française en Afghanistan* 35). Berlin.

Matarese, C. (2021) *Deportationen im Perserreich in teispidisch-achaimenidischer Zeit.* (*Classica et Orientalia, 27*). Wiesbaden.

Metcalf, W. (ed) (2012) *The Oxford Handbook of Greek and Roman Coinage*. Oxford.

Morris, I. and Scheidel, W. (eds) (2009) *The Dynamics of Ancient Empires: State Power from Assyria to Byzantium*. Oxford.

Morrison, G. (2001) 'Alexander, Combat Psychology, And Persepolis'. *Antichthon* 35: 30–44.

Müller, S. (2016) 'Arrian, the Second Sophistic, Xerxes, and the Statues of Harmodios and Aristogeiton', in Svärd and Rollinger (2016) 173–202.

Murray, O. (1988) 'The Ionian Revolt: The Greek States', in Boardman (1988) 461–90.

Murray, W. (2011) 'Thoughts on Grand Strategy', in Murray, Hart Sinnreich, and Lacey (2011) 1–33.

Murray, W., Sinnreich, R.H. and Lacey, J. (eds) (2011) *The Shaping of Grand Strategy: Policy, Diplomacy, and War*. Cambridge.

Ñaco del Hoyo, T. and López Sánchez, F. (eds) (2017) *War, Warlords, and Interstate Relations in the Ancient Mediterranean*. Leiden.

Olmstead, A.T. (1948) *History of the Persian Empire: Achaemenid Period*. Chicago.

Potts, D. (ed) (2013) *The Oxford Handbook of Ancient Iran*. Oxford.

Quack, J.F. (2021) 'Egypt', in Jacobs and Rollinger (2021) 553–66.

Radner, K. (ed) (2014) *State Correspondence in the Ancient World: From New Kingdom Egypt to the Roman Empire*. Oxford.

Radner, K., Moeller, N. And Potts, D. (eds) (2023) *The Oxford History of the Ancient Near East Volume V: The Age of Persia*. Oxford.

Rollinger, R., Truschnegg, B. and Bichler, R. (eds) (2011) *Herodot und das Persische Weltreich. Akten des 3. Internationalen Kolloquiums zum Thema 'Vorderasien im Spannungsfeld klassischer und altorientalischer Überlieferungen', Innsbruck 24.–28. November 2008*. Wiesbaden.

Rollinger, R. (2008) 'The Median "Empire", the End of Urartu and Cyrus' the Great Campaign in 547 B.C. (Nabonidus Chronicle II 16)'. *Ancient West & East* 7: 49–63.

Rollinger, R. (2013) 'Dareios und Xerxes an den Rändern der Welt und die Inszenierung von Weltherrschaft. Altorientalisches bei Herodot', in Dunsch and Ruffing (2013) 95–116.

Rollinger, R. (2014) 'Das teispidisch-achaimenidische Großreich. Ein "Imperium" avant la lettere?', in Gehler and Rollinger (2014) 149–92.

Rollinger, R. (2021). 'From Assurbanipal to Cambyses', in Jacobs and Rollinger (2021) 365–88.

Root, M.C. (1979) *The King and Kingship in Achaemenid Art: Essays on the Creation of an Iconography of Empire*. Leiden.

Root, M.C. (1988) 'Evidence from Persepolis for the Dating of Persian and Archaic Greek Coinage'. *Num. Chron.* 148: 1–12.

Rop, J. (2019) *Greek Military Service in the Ancient Near East, 401–330 BCE*. Cambridge.

Ruffing, K. (2011) 'Herodot und die Wirtschaft des Achaimeniden-Reichs', in Rollinger, Truschnegg, and Bichler (2011) 75–102.

Ruffing, K. (2021) 'Westwards Bound ...: The Achaemenid Empire and the Mediterranean', in Daryaee and Rollinger (2021) 348–67.

Ruzicka, S. (2006) 'The Politics of Persian Autocracy, 424–334 BC', in Lewis (2006) 224–37.

Ruzicka, S. (2012) *Trouble in the West: Egypt and the Persian Empire, 525–332 BC*. Oxford.

Sachs, A.J. and Hunger, H. (1988) *Astronomical Diaries and Related Texts from Babylonia*. Wien.

Sancisi-Weerdenburg, H. (ed) (1987) *Sources, Structures and Synthesis: Proceedings of the Groningen 1983 Achaemenid History Workshop*. Leiden.

Schwinghammer, G. (2021) 'Imperial Crisis', in Jacobs and Rollinger (eds) (2021) 415–27.

Seibert, J. (1998) 'Panhellenischer Kreuzzug, Nationalkrieg, Rachefeldzug oder makedonischer Eroberungskrieg?-Überlegungen zu den Ursachen des Krieges gegen Persien', in Will (1998) 5–58.

Seidl, U. (2009) 'Ein Monument Dareios' I. aus Babylon'. *ZA* 89: 101–14.

Silverman, J.M. and Waerzeggers, C. (eds) (2015) *Political Memory in and after the Persian Empire*. Atlanta.

Sinnreich, R.H. (2011) 'Patterns of Grand Strategy', in Murray, Hart Sinnreich, and Lacey (2011) 254 70.

Sordi, M. (ed) (1998) *Responsabilità perdono e vendetta nel mondo antico* (*Contributi dell'Istituto di storia antica 24*). Milan.

Squillace, G. (2010) 'Consensus Strategies under Philip and Alexander: The Revenge Theme', in Carney and Ogden (2010) 69–80.

Stark, S. (2021) 'The Iranian East', in Jacobs and Rollinger (2021) 695–710.

Stökl, J. and Waerzeggers, C. (eds) (2015) *Exile and Return: The Babylonian Context*. Boston.

Svärd, S. and Rollinger, R. (eds) (2016) *Cross-Cultural Studies in Near Eastern History and Literature*. Münster.

Tuplin, C.J. (1987) 'The Administration of the Achaemenid Empire', in Carradice (1987) 109–66.

Tuplin, C.J. (2011) 'Managing the World: Herodotus on Achaemenid Imperial Organisation', in Rollinger, Truschnegg, and Bichler (2011) 39–63.

Tuplin, C.J. (2017) 'Mercenaries and Warlords in the Achaemenid Empire', in Ñaco del Hoyo and López (2017) 17–35.

Tuplin, C.J. (2020) and Ma, J. (eds) (2020) *Aršāma and his World: The Bodleian Letters in Context*. Oxford.

Tuplin, C.J. (2020) 'Sigillography and Soldiers: Cataloguing Military Activity on Achaemenid Period Seals', in Dusinberre, Garrison, and Henkelman (2020) 329–459.

van Rookhuijzen, J.Z. (2019) *Herodotus and the Topography of Xerxes' Invasion: Place and Memory in Greece and Anatolia*. Berlin.

Vasilev, M.I. (2015) *The Policy of Darius and Xerxes towards Thrace and Macedonia*. Leiden.

Vogelsang, W.J. (1992) *The Rise and Organisation of the Achaemenid Empire: The Eastern Iranian Evidence*. Leiden.

Waerzeggers, C. (2003/2004) 'The Babylonian Revolts against Xerxes and the "End of Archives"'. *Archiv für Orientforschung* 50: 150–73.

Waerzeggers, C. (2015) 'Babylonian Kingship in the Persian Period: Performance and Reception', in Stökl and Waerzeggers (2015) 181–208.

Wallinga, H.T. (1987) 'The Ancient Persian Navy', in Sancisi-Weerdenburg (1987) 47–77.

Wallinga, H.T. (2005) *Xerxes' Greek Adventure: The Naval Perspective*. Leiden.

Wasmuth, M. (2015) 'Political Memory in the Achaemenid Empire: The Integration of Egyptian Kingship into Persian Royal Display', in Silverman and Waerzeggers (2015) 203–38.

Waters, M.W. (2014) *Ancient Persia: A Concise History of the Achaemenid Empire, 550–330 BCE*. New York.

Waters, M.W. (2016) 'Xerxes and the Oath-Breakers: Empire and Rebellion on the Northwestern Front', in Collins and Manning (2016) 93–102.

West, W.C. III. (1965) *Greek Public Monuments of the Persian Wars*. https://chs.harvard.edu/book/west-william-custis-iii-greek-public-monuments-of-the-persian-wars/.

Wiesehöfer, J. (2004) '"O Master, Remember the Athenians": Herodotus and Persian Foreign Policy', in Karageorghis and Taifacos (2004) 209–21.

Wiesehöfer, J. (2009) 'The Achaemenid Empire', in Morris and Scheidel (2009) 66–98.

Weiskopf, M. (1989) *The So-called 'Great Satraps' Revolt', 366–360 B.C.: Concerning Local Instability in the Achaemenid Far West*. Wiesbaden.

Young, Jr., T.C. (1980) '480/479 BC – A Persian Perspective'. *IrAnt* 15: 213–39.

Young, Jr., T.C. 'The Consolidation of the Empire and Its Limits of Growth under Darius and Xerxes', in Boardman (1988) 53–111.

Zournatzi, A. (2003) 'The Apadana Coin Hoards, Darius I, and the West'. *American Journal of Numismatics* 15: 1–28.

CHAPTER 5

The Achaemenid Military System and Its Campaign Logistics

John Hyland

In the autumn of 522 BCE, the young Persian Empire was in turmoil, with two contenders struggling for the throne and provincial rebellions gaining momentum in Babylonia, Media, and other regions. Amidst the chaos, the would-be king Darius I summoned an army from Fars to join him at Babylon, where he wintered after crushing the first of two local usurpers. The Persian reinforcements responded to Darius' orders, but for reasons unknown, are said to have changed their allegiance at Babylon and professed support for Darius' rival Vahyazdata, who had meanwhile seized power in the Persian heartland. The Bisotun monument's terse narration hints at a vital linkage between military backing and royal authority, which temporarily threatened Darius' path to lasting power. Fortunately for Darius, the mutiny subsided, although Bisotun does not dwell on the details, leaving modern readers to guess whether he employed severity or concessions to regain the army's obedience. At the onset of spring, the 'hostile' soldiers accompanied the king to Media and participated in its reconquest, while more reliable troops proceeded to Fars to eliminate Vahyazdata (DB Pers. §40, Bab. §33–34).

The episode of Darius' recalcitrant army points not only to the military's vital role as a pillar of Achaemenid royal authority, but also to its stunning capacity for rapid long-distance travel. The mutineers' march of at least 1,100 miles in only half a year, regardless of their qualms about Darius' legitimacy, offers a striking example of the campaign mobility celebrated in his epitaph at Naqš-e Rostam – 'the spear of the Persian man has gone forth far; the Persian man has given battle far indeed from Persia' (DNa §4).

The extreme distances covered in military campaigns and the supply of armies along increasingly far-flung frontiers are among the most striking features of Achaemenid imperial expansion. Persia's first empire was a conquest state constructed with unprecedented speed over a geographic space of 2,500 miles from east to west, vastly exceeding the territorial extent of its Elamite, Assyrian, and Babylonian predecessors. It emerged through the long-distance campaigns of Cyrus II and Cambyses II, which overthrew the local rulers of Media in 550, Urartu and Lydia in the 540s, Babylon in 539, and Egypt in 526. Early geographic

 | DOI:10.1163/9789004710771_006

dispersal brought the potential for overextension and collapse, but despite Cyrus' death on campaign in Central Asia in 530, Cambyses' premature demise in 522, and the chaos that followed Darius' seizure of power, the succession wars of 522–521 restored central unity and consolidated control over key provinces. In the subsequent decade, Darius reiterated and expanded the empire's frontier claims through further expeditions to Central Asia, Egypt, India, and Thrace. Royal campaigns then grew rarer over time, with Xerxes' infamous Greek expedition in 480 as the last attested example in the expansionist period, but frontier warfare, delegated to subordinate generals, continued on a smaller scale.[1] The extensive road network and mounted messenger system facilitated rapid empire-wide communication, permitting the king to assemble and direct the movement of frontier forces while retaining the court at central locations.[2] The most substantial of these campaigns involved attempts to repeat the conquest of Egypt when this vital and insecure satrapy broke away from imperial control in 463–457 and 400–340; these repeated invasions involved substantial forces despite the lack of direct royal participation until the reconquest by Artaxerxes III. The fifth and fourth centuries also saw bursts of localised warfare, including responses to sporadic revolts by subjects and satraps, most dramatically but briefly attested in the confrontation between Artaxerxes II and Cyrus the Younger in 401, and low-intensity frontier conflicts against 'brigands' and nomadic raiders. But the empire's leaders retained the potential to assemble and supply massive campaign forces until the very end, when Persian organisational and logistical structures sustained Darius III's resistance to Macedonian conquest from 334–330, and subsequently provided important foundations for the military practices of the Diadochoi and Seleucids.

The close relationship between the Achaemenid state and military power makes it essential to understand how Persian rulers acquired, organised, and supplied their soldiers on campaign. The majority of studies, including chapter-length overviews in reference works, as well as a handful of short books for lay students of ancient warfare, have given less attention to the larger institutional framework than to matters of dress, armaments, and tactics.[3] But a more expansive structural approach, rooted in internal documentation, is

1 On the king as campaign leader, see Brosius (2005) 152–53; Tuplin (2017) 44–45. For a useful distinction between 'display campaigns' and '"low profile" regional or low intensity conflicts', formulated in reference to Neo-Assyrian warfare but equally applicable to the Achaemenids, see Parker (2014) 288.

2 See Briant (2012); Colburn (2013); Hyland (2019).

3 The monograph of Bittner (1985) concentrates on dress and equipment; for weaponry and its battlefield uses, see White (2020) as well as Rop in this volume. Chapter-length summaries include Shahbazi (1986); Tallis (2005); Hassan (2021); Tuplin and Jacobs (2021). Illustrated

becoming more feasible than it was in decades past. The recent appearance of the first full-length academic monograph on the Achaemenid armed forces exemplifies the potential for broader analysis, combining critical reassessments of Greek historiography, comparison with Neo-Assyrian and Neo-Babylonian military practice, and careful study of the temple and entrepreneurial archives of early Achaemenid Babylonia.[4]

The next step required in the study of Achaemenid military organisation and logistics is enhanced examination of the relevant content in the Persepolis Fortification Archive (PFA), the most important body of primary documentation for the wider field of Achaemenid studies. The Persepolis corpus, representing more than 15,000 original records issued by the satrapal bureaucracy of Fars in the years 509–493 BCE, sheds tremendous light on governance, economy, society, art, and religion at the height of Achaemenid power.[5] It has long been under-valued as a source for the Persian armed forces, because it does not contain explicit campaign records such as troop rosters or ration lists, but its numerous receipts for the issue of provisions to travelers speak to important aspects of military movement and logistics. The archive's soldiers have also begun to emerge from the shadows, due especially to the work of Wouter Henkelman and Christopher Tuplin. The discussion that follows will seek to build on recent advances, with particular focus on the Persepolis material in dialogue with Herodotus, to construct a new sketch of the organisation and supply systems that underpinned Persian military campaigns.

1 The Achaemenid Military System

The Achaemenids were heirs to earlier Near Eastern traditions of military organisation. The Elamites employed a force of court guards and assembled polyglot

---

works aimed at popular audiences include Sekunda (1989); Sekunda (1992); Head (1992); Farrokh (2007).

4 Manning (2021).

5 For initial publications of the Elamite tablets known respectively as PF and PFa, see Hallock (1969); Hallock (1978). The remaining Elamite tablets, labelled PF-NN and Fort., which have been subject to extensive collation and study by the Persepolis Fortification Archive Project under the leadership of Matthew Stolper, Wouter Henkelman, and Mark Garrison, will more than double the size of the published archive upon completion. Preliminary transliterations are available through the public OCHRE database (http://ochre.lib.uchicago.edu/PFA_Online/). The publication of the Aramaic tablets, although fewer in number, will add further depth to the corpus; see Azzoni (2008). For introductions to the archive, see Henkelman (2008b) 65–179; Garrison (2017) 15–77; Garrison and Henkelman (2020); Henkelman (2021); as well as the collected essays in Briant, Henkelman, and Stolper (2008), and Jacobs, Henkelman, and Stolper (2017).

campaign armies that also included contingents raised by regional governors and tribal levies; the Assyrians and Babylonians also employed a model including royal guards, provincial forces, and conscripts, broadly anticipating the Achaemenid system.[6] The Persians established a comparable model that drew soldiers for offensive campaigns from two principal categories, a standing guard force that accompanied the king and court, and non-elite troops assembled for temporary service through a wide variety of means, including tax-based conscription and mercenary recruitment. A third category, supportive and defensive in nature, comprised garrison soldiers assigned to major fortress centers and smaller rural outposts.[7] The following discussion will pay particular attention to the Persepolis evidence for each group in turn, before turning to the aspects of the military system that are visible in other parts of the empire.

The relevant Persepolis records do not deal explicitly with campaign contexts, and it is only within the last two decades that scholars have begun to grapple with the archive's military implications. Before examining the evidence in detail, it is necessary to start with a brief introduction to the relevant administrative terminology and its complexities. The Fortification tablets are intended to track the intake and dispensation of commodities at supply sites and the effectiveness of their managers; in this context, their scribes are usually uninterested in the precise activity of the ration recipients. References to soldiers are rooted in an initial distinction between free men, Elamite *šalup*, and dependent labourers, or *kurtaš*. Soldiers would typically fall into the former category, often receiving higher levels of compensation and supply, but the *šalup* are a broad enough social class to encompass many non-military personnel performing other duties on behalf of the empire. The more specific term employed for groups of soldiers, *taššup*, remains semantically broad. In the Elamite version of Darius' Bisotun inscription, it can designate either an army or a region's broader adult male populace; the same goes for its equivalents,

6 For the Achaemenid military debt to Near Eastern predecessors, see Tallis (2005) 214–15; Manning (2021) 65–114. On Elamite armed forces, see Álvarez-Mon and Wicks, this volume; for a guard force of spear-bearers at late Neo-Elamite Susa, see Henkelman (2002) 13–14. For Neo-Assyrian military organisation, see Malbran-Labat (1982); Fales (1990); Postgate (2000); Dezső (2012a); Dezső (2012b); Dezső (2016); Zaia (2022) 31–46. The Neo-Babylonian army is not as well documented, but the distinction between palace troops and short-term recruits probably retained a similar significance; see Sandowicz (2020) 164; Zaia (2022) 47–52.

7 The present chapter will comment on garrisons in the context of military service; due to the limits of space, it cannot address the extensive archaeological evidence for fortresses, or their roles within frontier defensive systems. Important discussions include Tuplin (1987); Briant (2002) 371–77; Dusinberre (2013) 84–113; Tuplin (2014).

Old Persian *kāra* and Babylonian *uqu*.[8] Accordingly, when *taššup* appear in Persepolis tablets dealing with contexts of stationary commodity storage and localised delivery, they are often difficult to identify as soldiers, which led Richard Hallock to employ a cautious translation as 'officials'; a number of examples refer more clearly to the local population from which agricultural produce might be collected by provincial authorities.[9] But when *taššup* travel long distances in large numbers, they are more plausibly viewed as military units.[10] Many of these groups are further designated by labels that hint at specific occupational categories within the military system, particularly the terms *hallinup*, *lin huttip*, and *halmarriš nuškip*, which will be discussed further below. The men who bear the archive's most explicit military title, the Spear-Bearers, while qualifying as *šalup*, are not described as *taššup* in surviving records, perhaps because they appear in documentary contexts as individuals or in very small groups rather than in larger units on the move. But they can be identified with greater confidence as members of the standing royal army, whose divisions would have marched in company with humbler *taššup* contingents when the king sent Persian men forth to give battle far from Persia.

1.1 *The Persepolis Spear-Bearers and the Standing Army*

The standing army may have played an important role in Darius' succession wars, despite Bisotun's semantic vagueness in describing military forces. The speed of the initial campaign against Babylon, less than two months after Darius seized power in Media, left little time for the recruitment and concentration of new conscripts. At least two military contingents associated with palace centers supported Darius' rivals, including the infamous group that gathered in Persia and mutinied at Babylon, as well as an army in Media that supported the rebel Fravartish (§24). The references to palaces appear

8 See Shahbazi (1986); Briant (2002) 103–104; Tuplin (2005) 230; Henkelman (2017a) 322; Manning (2021) 137–39. In the Aramaic copy found at Elephantine, the equivalent *ḥyl'* refers to armies, but the earlier sections of the text that might contain the broader meaning of 'people' have not survived.

9 See Hallock (1969) 62; Tuplin (2008) 369–72; Tuplin (2010) 136 n. 150; Stolper (2017) 781; Garrison and Henkelman (2020) 189 n. 45. Many of the ambiguous cases are situated at 'fortresses' (*halmarriš*), a term which in the Persepolis archive refers to sites of administrative storage and officials' country estates, and does not have to imply a context of armed defence; see Rossi (2010).

10 See Henkelman (2010a) 711; Tuplin (2014) 674–75; Henkelman (2017a) 322; Henkelman (2018b) 42–43; Henkelman (2020) 216.

suggestive of royal guards, although a possible alternative would involve the use of palaces as muster points for local levies.[11]

In the Persepolis Fortification archive, the personnel of the standing army come into clearer focus with the appearance of Spear-Bearers (Elamite *ištibarra* or *irtibarra*, transcribing Old Persian *ṛštibara*; also attested as the equivalent Elamite term *ripi kutira*, or *ŠI.KAK kutira*, *ŠI.KAK bakira*, or *šukurrum kutira*, Elamite phrases written with Akkadian logograms or incorporating an Akkadian loan-word).[12] As noted above, this organisation is an inheritance from the Neo-Elamite court of the early sixth century, attested in the Susa Acropole archive, as well as the Neo-Assyrian and Babylonian royal guards.[13] The Persepolis archive's Spear-Bearers circulate as individuals or in small teams of 'companions' (*akkayašbe*), almost always on direct authorisations from the king, whereas large groups of *taššup* more often carry travel passes from a satrap or satrapal deputy.[14] While they do not receive larger flour rations than other types of *šalup*, Spear-Bearers fulfil a wide range of obligations of an often administrative nature, including road survey and reconnaissance ahead of the moving court, and inspection of work personnel and *taššup* at forts, estates, and way stations. Prosopographic studies have shown that attested Spear-Bearers also performed other duties, overlapping with 'inspectors' (*pirrašaka*) and 'people-counters' (*karamaraš*), all of whom might be regarded as members of the same organisation, which Henkelman labelled the 'corps of lance-bearers'.[15] Based on the reappearances of personal names, it is possible that this large institution also included the men who served as 'professional guides' (*barrišdamap*) for travel parties on the royal roads and the imperial 'express messengers' (*pirradaziš*).[16]

---

11 See Briant (2002) 119; Manning (2021) 142–43.

12 See Henkelman (2002). Whereas Henkelman refers to 'lance-bearers' to emphasise the connotation of non-throwing weapons, 'spear' is preferred here for consistency with Anglophone discussions of thrusting weapons in other ancient military contexts, particularly the Greek hoplite spear (*dory*).

13 See note 6 above; for the Assyrian guards see Dezső (2012a) 115–42; Zaia (2022).

14 For the mechanics of Achaemenid travel authorization, see Briant (2012); Henkelman (2020).

15 See Henkelman (2002) 20–22; Henkelman (2017b) 72–77.

16 For example, the Ukbarauš who appears as a Spear-Bearer in NN 0904 may be the *pirradaziš* messenger who appears in Fort. 2292-101, and the investigator Manezza (NN 2062, NN 0257, NN 1807) may also reappear as a courier in NN 0570; see Henkelman (2017b) 159 n. 170. The name Rašdauka is associated with both a Spear-Bearer (NN 1788) and the guide of a travel party (PF 1508). See Hyland (2019) 156 n 27, for a list of ten individuals who appear to have served on both *barrišdama* and *pirradaziš* missions. Caution is required, though, as some cases may involve homonymy; patronymics are rare, while the

We might envision a model in which their commander(s), or other royal officials, selected men from a central pool of Spear-Bearers accompanying the court to detach on a range of assignments as individuals or in small groups. The campaign service of a royal army might seem remote from the inspectors and road-surveyors of the Persepolis tablets, but even the elite troops appearing in Greek histories are not immune from more mundane assignments, as attested by Herodotus' story of one Immortal's detachment for guard duty at the site of Xerxes' favourite plane tree (7.31). A similar engagement of royal guards in administrative tasks is attested for the Neo-Assyrian *qurbūtu*, who not only went out on periodic campaigns but also escorted travel parties, performed messenger service and diplomatic missions, supervised labourers, and even assisted in tax collection.[17]

Larger bodies of Spear-Bearers are not explicitly attested in the Persepolis archive, possibly because of their travel in company with the royal household establishment, which produced its own non-extant documentation, distinct from that of the Persepolis provincial bureaucracy.[18] Nevertheless, a sense of the Spear-Bearer organisation's magnitude can be obtained from the Fortification tablets that document provincial food outlays at locations visited by the royal entourage. Henkelman has shown that the enormous expenditures that appear in these tablets are not simply indicative of 'feasting' by the king and his intimates, but rather demonstrate the logistical needs of an enormous body of itinerant personnel.[19] Many of the flour amounts issued in these contexts reach into the tens of thousands of liters, and in one case at Persepolis itself, more than one hundred thousand (PF 701); the scale suggests the amplification of court manpower beyond civilian staff and dependents by large military forces. The Spear-Bearers' proximity to king and court is also reflected in the hundreds of soldier images that decorated the palace walls of Susa and Persepolis; variations in dress and equipment among the Persepolis reliefs, where some of the men carry bows or shields in addition to spears, may reflect distinctions between particular ranks or sub-units.[20]

---

seal impressions that would allow certain identification of individuals do not survive on some of the memoranda, and are absent from journal entries.

17 See Dezső (2012a) 123–41.

18 See Henkelman (2010) 675.

19 See Henkelman (2010) 674, 682–85.

20 See Brosius (2005) 147–48; Tuplin (2017) 36, 46; Tuplin and Jacobs (2021) 1173; Rop (this volume). For the original publication of the Susa glazed brick soldiers see Dieulafoy (1891); for recent discussion see Stronach (2011) 482–85. For the Persepolis reliefs see Schmidt (1953). Schmitt (2004) notes the need for caution regarding associations of the

This brings us to the high-status soldiers who accompany the king on campaign in Herodotus and the Alexander historians.[21] Describing Xerxes' departure from Sardis, Herodotus claims that units of 1,000 Spear-Bearers (*doryphoroi*) and 1,000 cavalry paraded in front of the royal chariot with spears reversed, and two more such units followed in its wake with spears pointing skywards; these Spear-Bearers may be equated with the so-called Apple-Bearers (*mēlophoroi*), named for their spear-pommels, who accompany Darius III in accounts of Gaugamela (Arr. *An.* 3.11.5, 3.13.1, 3.16.1).[22] The common ground in title, equipment, and royal proximity support a plausible identification between these royal guards and the Persepolis Spear-Bearer corps. It should also be noted that at least one known Persepolis Spear-Bearer, a man named Zamuš, also appears in Aramaic Persepolis texts recording rations for horses.[23] Many of the internal assignments associated with Spear-Bearer service required mounted travel, and the Persepolis Spear-Bearers may have comprised both Herodotus' infantry *doryphoroi* and his cavalry guards.

Alongside these elite of the elite, Herodotus describes a much larger permanent force that he famously calls the Immortals (*athanatoi*), glossing the term as a reference to maintenance at full strength through the prompt replacement of losses (7.83.1). The Persepolis records have not revealed an institutional equivalent for this title; a number of studies have attempted to connect it with Old Persian terms such as *anušiya*, but these associations remain uncertain.[24] Elsewhere in Herodotus, the Immortals seem to be the same as the 10,000 picked foot soldiers who follow the *doryphoroi* in Xerxes' parade and carry spears with decorated pommels, golden in the case of 1,000 men and silver for

---

Persepolis reliefs with the specific soldiers known as the Immortals; on their distinction from other guard units see further discussion below.

21 Cf. Henkelman (2017b) 76.

22 Sekunda (1992, 7) proposes that Herodotus erroneously doubled a single 1,000-man unit of *doryphoroi*; cf. Charles (2011) 121; Charles (2021) 177. On the *mēlophoroi*, see Charles (2011) 126–30; Charles (2021) 177–78.

23 Azzoni and Dusinberre (2014) 6.

24 For the theory of Greek confusion between *anušiya*, usually translated 'followers', and a postulated OP **anauša*, 'undying', see Pagliaro (1954) 146–51; Tallis (2005) 215. An alternative proposal gives a separate OP term **amrtaka*; see Sekunda (1989) 84; Sekunda (1992) 6. Schmitt (2004) is sceptical towards both interpretations. Daryaee (2021, 75–76) argues that *anušiya* could also mean 'deathless', with a direct linkage with Avestan and Middle Persian cognates (*anaoša* and *anōšag*), but Bisotun's usage is not suggestive of a heroic epithet unless it is applied ironically to the groups of prominent men executed for serving the false Bardiya (§13) and Darius' other defeated rivals (§32, 42–43, 47, 50); it appears only once in a positive context, designating the *kara* that served Darius' father Vištaspa against the rebels in Parthia (§35).

the rest (7.40–41); they are also accompanied by a column of 10,000 cavalry (7.41). Xenophon's *Cyropaedia* also claims that Cyrus established a royal guard corps of 10,000 *doryphoroi* (7.5.68). But the myriad may simply express a Greek conception of enormous size, and the recurrence of a conventional number may not be entirely reliable; exaggeration remains a possibility for the horsemen in particular.[25] Given Herodotus' hint of selection from among a larger body of Persian men and the similarity to the elite Spear-Bearers' armament, we should not rule out the possibility that members of this larger standing army appear in some of the Persepolis Spear-Bearer records alongside the guards proper.[26]

Unfortunately, the archives do not shed much light on how Persian authorities recruited men into the Spear-Bearers or associated units. Herodotus asserts that both the *doryphoroi* and Immortals were 'selected' from among the Persians, and that the former came from 'the best' backgrounds (7.40–41, 83).[27] He does not explain by whom and in what circumstances such hand-picking would have occurred, and there is no evidence that prior training or military experience outweighed socio-economic context.[28] Intriguingly, the tablets do indicate that elite service was not restricted to ethnic Persians, as a pair of Spear-Bearers assigned to the service of the future king Xerxes are identified as Parthians (NN 1657; see also NN 0516).[29] The record in question refers to the prince in a position of authority in Parthia, and suggests that some of the Spear-Bearers were stationed at provincial centers to support the activities of generals or satraps and facilitate their communications with court. This tradition might also lie behind Xenophon's later claim that the guards stationed in satrapal citadels were separate from other provincial soldiers and answered

25 Cf. Charles (2021) 179; see Tuplin (2010) for the absence of horsemen from Persepolis and Susa palace iconography (104–105) and the lack of concrete evidence for cavalry, as opposed to grooms and individuals associated with the feeding of horses, in the PFA (130–39).

26 For both Immortals and Spear-Bearers/Apple-Bearers as part of a larger court guard organisation, see Briant (2002) 261–62. For distinction between the Immortals, significant as a standing force, and the *doryphoroi* or *mēlophoroi* as more elite contingents, see Charles (2011) 130–32; Charles (2021) 174–75.

27 See Sekunda (1992) 11–12; Briant (1999) 116; Charles (2011) 119.

28 See Tuplin (2017) 46–47. The Assyrian king took a personal role in the selection of *qurbūtu* (Dezső (2012a) 125); Xen. *Cyr.* 7.5.67–68 claims that Cyrus selected 10,000 *doryphoroi* from among Persia's peasants (*autourgoi*), but the recruitment of forces on such a scale would have required considerable delegation.

29 See Henkelman (2010b) 28–31. Shahbazi (1986) already questioned the exclusive Persian origin of the Immortals based on doubts in the demographic capacity of early Achaemenid Fars.

directly to the king (*Cyr.* 8.6.9).[30] In theory, the practice would carry the benefit of surrounding ambitious regional administrators with soldiers whose primary loyalty was to the crown, as when the 1,000 Persian *doryphoroi* at Sardis eliminated the satrap Oroites on receipt of Darius' orders (Hdt. 3.127–128).

1.2 *The Persepolis* Taššup *and Non-Elite Military Service*

The Persepolis records suggest a distinction between Spear-Bearers and more generic *taššup* who receive supplies on the move. Numerous tablets document travel by hundreds or thousands of *taššup*, and sometimes similar groups described only as *šalup*, who are plausibly identifiable as armed personnel. The travelling groups referred to as *ḥyl'* in a small number of Aramaic tablets from Persepolis belong to the same category, as indicated by the equivalency of these terms in their respective versions of the Bisotun inscription, and the use of *ḥyl'* to designate both fortress garrisons and travelling military units in later documents from Achaemenid Egypt and Bactria.[31] While their travel is often authorised by satraps rather than the king, *taššup* and *ḥyl'* units can be commanded by high-status officials, in contrast with travelling *kurtaš* parties, which are led by imperial guides earning common soldiers' rations. Marching *taššup* units sometimes include both free men and unfree dependents, referred to as *libap* ('slaves/servants') or *puhu* ('boys', often referring to adults in a servile capacity).[32] Herodotus notes that the Immortals travelled in the company of 'concubines and numerous and well-equipped servants' (*pallakas kai therapēiēn pollēn te kai eu eskeuasmenēn*, 7.83.2), and humbler soldiers may have also taken enslaved persons with them or acquired them in the course of a campaign.

Marching *taššup* units are often designated with additional labels, at least three of which seem to refer to specific categories within the Persian armed

---

30 There may be an iconographic trace of provincial Spear-Bearers in the Persepolitan-style soldier reliefs which stood in the throne-room of the regional palace at Meydancıkkale in rough Cilicia; see Dusinberre (2013) 101.

31 For *ḥyl'* in the Persepolis Aramaic texts (PFAT 051, 054, 055), as well as their commanders (*rb ḥyl'*, PFAT 206, 210) and a cavalry officer (*rb swsn*, PFAT 181, 232), see Azzoni (2008) 261–62. In Egypt, references to *ḥyl'* occur in the Aršama archive (TAD A6.7, A6.8) and the Elephantine letters (TAD A4.5), which also make frequent reference to soldiers' association with a sub-unit or *dgl*; see Porten (1968) 29; Tuplin (2020a) 292–302. For attestations in Bactria, see Naveh and Shaked (2012) ADAB A2: 5–6; A4: 2, 5; A7: 1; B1: 3, 5.

32 For example, in NN 1375, a force of 2,700 *taššup* includes *libap* and *šalup*; NN 0271 records a group of 560 *taššup* including 460 *šalup* and 100 *puhu*; in PF 1397, a force of 180 *taššup* and 50 *puhu* set out together for India. The household entourage of Darius' father Vištaspa, which travelled to Persepolis in year 23, is labelled *taššup* in PF 1596 and broken down into 200 *šalup* and 1,194 *puhu* in NN 0124.

forces. The first, known as *hallinup*, often number in the thousands and carry out long marches between Fars and distant regions, including Karmania, Sagartia, Media, Bactria, and Gandhara.[33] Their attested commanders include the satraps Karkiš of Karmania (NN 2261: 16–18) and Harbamiššа of Areia (PF 1603, NN 1154, Fort. 0517-002: 1–2), as well as Unapa, son of Udana (Fort. 0472-101: 17–18), whom Herodotus names as the commander of the Elamite contingent in Xerxes' Greek expedition (7.62.2).[34] Henkelman suggests a translation as 'foot-men', in contrast with *telnup*, the term for the small groups of 'horsemen' who accompany fleeing rebel leaders in the Elamite Bisotun text.[35] Matthew Stolper has proposed a different etymology from Elamite *hal-*, 'country', making them '"rural officials/troops", that is, suppliers and other personnel stationed at villages and way-stations'.[36] The journal entry NN 2344: 3–5 appears to complicate the picture by applying the term to non-military personnel, recording the provision of seed grain to *kurtaš hallinup*.[37] But in other journals, conjunctions are sometimes omitted between adjacent but separate groups of ration recipients; for example, NN 1375 refers to '*libap šalup*' but does not indicate that these are one and the same. Alternatively, it is possible that some *kurtaš* might perform supporting duties in association with military units. This record need not prevent the identification of most travelling *hallinup* groups as soldiers, although it does make a comfortable fit with Stolper's hypothesis of extended rural deployments. Overall, the *hallinup* may have comprised a wide category of recruits who could be assigned to infantry service or supply duties in field armies or to stationary outposts supported by agricultural grants.

A second type, the *lin huttip*, were responsible for the long-distance movement and escort of supplies and property, including the baggage of the travelling royal court and presumably also of armies on the move.[38] Participants in such transport units were free men despite their engagement in demanding

33 PF 1603; PF-NN 2261: 16–18, 19–21, 26–29; PF-NN 0578 (PFa 31): 2–4; Fort. 0304-101: 5–7; Fort. 1301-101: 2–3, 4–5; Fort. 0472-101: 17–18; Fort. 2354-101: 7–9; PF-NN 2040: 4–6; Fort. 0517-002: 1–2, 3, 4, 5–6; PF-NN 1154; Fort. 1965-101: 11, 15, 19, 22; Fort. 2264-001: 1, 3, 4. For their probable military status, see Koch (1993) 15; Henkelman (2010) 707 n. 150; Henkelman (2018a) 228–29; Henkelman (2020) 216–17.

34 See Henkelman (2018) 228 n. 28.

35 See Henkelman (2020) 216 n. 39. For *telnup* at Bisotun, see DB El. §§19, 25, 34. They are rare in the Persepolis archive, where they cannot be definitely identified as soldiers. A single group led by a certain Bakadada and qualified by an otherwise unknown ethnic, Hapmatiš, receives supplies in PF 1179, NN 1875, and Fort. 03668; see Tuplin (2010) 132.

36 See Stolper (2017) 781.

37 See Stolper (2017) 781.

38 PF 1542; Fort. 02750; NN 1044. See Henkelman (2008b) 197 n. 423; Henkelman (2011) 136–37; Henkelman (2018b) 41–42 n. 30. The 1,500 *lin huttip* in PF 1542 travel in the same

manual labour; in NN 1044, the 1,060 *lin huttip* commanded by Ukama are described as both *šalup* and *taššup*. *Lin huttip* may also be involved in other texts that do not use the term, such as *taššup* carrying treasure from Media to Persepolis (Fort. 1316-101: 12–13), or *taššup* driving a herd of cattle to Persepolis (NN 0578 = PFa 31: 5–7). In the context of large-scale campaigns, involving much higher numbers than we find in the Persepolis archives, transporters played a key role supporting the movement of armies, as indicated by Herodotus' descriptions of Xerxes' marching columns. In the army's departure from Sardis, both '(human) baggage carriers and baggage animals' (*skeuophoroi te kai ta hypozygia*) preceded the troops (7.40.1); at the Hellespont, the soldiers crossed by one bridge while the 'baggage animals and servile contingent' (*ta hypozygia kai hē therapēiē*) took the other (7.55.1).

A third category, *halmarriš nuškip* or 'fortress-custodians', involves men deployed to service at prominent administrative locations.[39] Surviving journals record the travel of *halmarriš nuškip* between Persepolis and Ecbatana (Fort. 1316-101: 7–9, 10–11; Fort. 1301-101: 15–16, 17–18, 21–22, 23–24; Fort. 1206-102: 14–15, 18–19), and once between Karmania and Susa (Fort. 2173-102: 7'–8'); their movements suggest either new recruitment of troops to reinforce palace garrisons in other provinces, or exchanges of standing forces between them. These fortresses were not so much points of strategic defence as of storage and record-keeping, and small units of custodians may have carried out a variety of support tasks at these fixed locations, while the higher-status Spear-Bearers moved from palace to palace with the travelling court. The garrison system in more distant provinces, on the other hand, is more likely to have relied on the implantation of military colonies, supported by generous land grants, that could replenish their own manpower over multiple generations.[40]

The Persepolis tablets do not shed direct light on the mechanics of *taššup* conscription, although they do contain some references to the recruitment of workers and the 'summoning' of free men for unclear purposes.[41] Some

---

month and year, with the same destination and commander, as the 2,700 *libap* and *šalup* in NN 1375; they may constitute the *šalup* within the larger group.

39 See Henkelman (2017a) 322. For *halmarriš nuškip* as *taššup* see Fort. 1316-101: 7–9. A group of texts on a small squad at Persepolis use *halmarriš nuškip* interchangeably with the equivalent term *tiddabattišbe* (PF 1812; NN 0543; NN 0425; NN 1100; NN 1202; Fort. 0857-102).

40 For the ethnic diversity of garrison soldiers, see the cases of Deve Hüyük (Dusinberre (2013) 104–105) and Elephantine (Porten (1968) 29–35; Tuplin (2020a) 294–300).

41 The verb *rabba-*, which conveys coercion into service, appears in numerous texts involving *kurtaš* groups but none that mention *šalup*, *taššup*, or more specific groups of soldiers. Other texts record the issue of rations to *šalup* who have been 'summoned' (*kappaka*)

non-elite recruits may have been mustered through taxes associated with the rental or ownership of land, as in the Babylonian system; Henkelman has suggested a potential linkage between provisions of military supplies at rural estates and service obligations.[42] Tribal communities also might have provided quotas of men in the context of negotiated relationships with royal authority. In all cases, the duration of service is likely to have been stipulated in advance, and will have varied according to the nature of the soldiers' relationship with the central authorities and the purpose for which they were needed. For comparison, the Babylonian tablet Dar. 253 specifies that a group of cavalry soldiers has been called up to the royal campaign camp for a period of three years; the distances entailed in longer *taššup* marches, not to mention the journal entry on seed grain and those involving garrison service, certainly imply durations of at least a year and probably longer.

New recruits in Persia would have gathered at major regional centers, comparable to the muster points attested in other parts of the empire, such as Uruk in southern Babylonia and Kastolos in western Anatolia.[43] This might be an attractive context for the texts in which travelling *taššup* are labelled as men of Beziyamata (= Old Persian Paišiyauvada), a fortress in eastern Fars which had served as a base for Darius' rival Vahyazdata in 522–521, and was under the authority of the Karmanian satrap in the period of the Persepolis archives.[44] Travelling Beziyamatian units include both *hallinup* and *lin huttip*, and it seems unwarranted to postulate their membership in a unit of the Spear-Bearer corps named after the fortress; a more plausible solution would be to consider them contingents raised there for finite tours of service. New soldiers were likely counted and recorded on lists at the muster centers, a quantitative procedure better attested for *kurtaš* that would have been necessary for the effective control and supply of military *taššup*. Members of the Spear-Bearers, who helped to 'count' or 'write down' records of *taššup* estate personnel in seemingly

to Susa or Persepolis, and the same term appears in the Bisotun passage in which the military force 'summoned' from Anšan proves disloyal to Darius (DB El. §33); see Hallock (1969) 709. But the Persepolis *kappaka* travel over significant distances, in one case from India to Susa (PF 2051), making them incompatible with a context of initial conscription into military units which should have been gathered locally.

42 Henkelman (2018b) 43–44.

43 See Shahbazi (1986) on recruitment of non-elite levies. For the Uruk musters, see UC 9/3, 9/68, 10/61, 10/62; PBS 11/1:54, 162. For Kastolos, see Xen. *An.* 1.9.7; Xen. *Hell.* 1.4.3. One might compare the role of regional and royal military musters in the Neo-Assyrian Empire; see Dezső (2022).

44 PF 329; PF 330; PF 2027; Fort. 11811; R558; NN 1159; NN 1254; NN 1711; NN 1816. See Tuplin (2014) 674–75; Henkelman (2017b) 50 n. 6; Henkelman (2018b) 40–44.

non-military contexts, may have played a role in troops' enrollment and assignment to specific forces and commanders.[45]

Once assembled at muster points, individual soldiers may have been grouped into the ten-man squads and hundred-man companies that Herodotus mentions as subdivisions of military units (7.81); a comparable decimal division appears in non-military contexts in several Persepolis texts recording supply deliveries to 'centuries' (*satukbam*) and 'decuries' (*dasabam*).[46] When necessary, centuries could be combined into thousands as building blocks of garrisons or travelling armies; several of the travelling *taššup* groups numbered between 1,000 and 2,000, and the garrison *hyl'* stationed at Elephantine contained at least four smaller units called *dgl*, which may have been envisioned as forces of 1,000 but were certainly not kept up to full strength.[47] Herodotus' 'myriads' are doubtful as a regular organisational structure within Persian military forces, associated with the false assumption that Xerxes' army numbered in the millions; the Old Persian title **baivarabam* suggested by several modern scholars to refer to such enormous units is a linguistic postulate, nowhere attested in ancient sources.[48]

1.3 *Provincial Soldiers, Mercenaries, and Naval Crews*

Large-scale campaigns involved simultaneous assemblies of troops in multiple provinces; such regional contingents are understandably absent from the Persepolis archive but visible in the documentary sources from Babylonia and Egypt as well as Greek narratives. The most detailed primary evidence comes from the Babylonian entrepreneurial and temple documents involving the

---

45 PF 1280; PF 1620; Fort. 1081-104: 3; NN 2556. See Henkelman (2017b) 74 n. 44.

46 These texts mention Uštana's century (NN 2022; Fort. 2139-101), with its constituent decuries led by Mannanda (PF 138; PF 139; NN 2581) and Haturdada (NN 0227), and Zaratiya's century (PF 141; PF 142; PF 143; NN 1174; NN 1854; Fort. 2078-101; Fort. 1805-101), with one of its decuries led by Harripirtan (PF 140). Garrison and Henkelman (2020, 255) suggest that these texts might show a ranking system among mid-level provincial administrators 'borrowed from military context'; for the most detailed discussion, see Henkelman (2023).

47 For *taššup* group sizes on this scale, see the 1,060 men in NN 1044 and the mixed group of 2,700 *šalup* and *libap* in NN 1375; similar sizes are also implied by the daily ration amounts distributed in NN 2040: 4–6 and Fort. 0517-002: 1–4. For the *dgl* at Elephantine, see Porten (1968) 29–35; Tuplin (2020a) 299–300, n. 34. See also Manning (2021, 195–97; 2022, 162–63) on the need for caution in assumptions regarding a schematised decimal organisation.

48 The reconstruction of a military rank of **baivarapati*, 'myriad-commander', goes back to Marquart (1896) 19 n. 84, and is repeated by Shahbazi (1986). Sekunda (1989, 84) introduces a unit called the **baivarabam* as 'purely speculative', but Sekunda (1992, 5–6) gives the term as a standard 'division' of the Achaemenid army. For critique, see Schmitt (2004); Manning (2021) 196.

outfitting of small groups of soldiers in response to tax obligations. These are addressed in greater depth in Reinhard Pirngruber's chapter in this volume; to summarise briefly, they included service obligations for farmers or renters of the land parcels known as *hadru*, as well as manpower quotas that had to be met by the residents of designated urban wards, occupational groups, and temple institutions. The groups responsible for providing soldiers were also expected to contribute to their equipment and financial needs, and hiring of substitutes was expected, so that some men may have earned a living by fulfilling manpower demands on behalf of tax-payers.[49] Recruitment oversight was outsourced at the local level to entrepreneurs who gathered soldiers on behalf of the authorities before forwarding them to the appropriate muster points; a famous letter from Babylon illustrates the potential for competition over manpower resources among various local officials, as the city governor complains that his recruiting agent has cheated him of promised chariot troops and supplied them directly to a Persian fortress commander instead (BM 33077 = CT 22, 74).[50]

In Egypt, apart from colonies of garrison troops such as the Judaeans at Elephantine, several categories of local men were available for campaigns such as Xerxes' expedition, in which they performed both marine duties and later service in Mardonios' land army; Egyptian documentation shows them receiving land in exchange for the performance of internal policing and guard duties. Herodotus famously records the names of two prominent groups of Egyptian soldiers, the Kalasiries (*gl-šr.w*) and Hermotybies (*rmt-ḏm.w*), but errs in believing that they constituted hereditary classes of *machimoi* warriors, and exaggerates the scale of their landholdings.[51]

In western Anatolia, while Herodotus suggests the authorities conscripted men from the general population in preparation for Xerxes' campaign, Xenophon portrays a more complex recruitment system. Western Anatolian satraps could raise forces of several hundred cavalrymen, and even occasionally charioteers,

49 See Briant (2002) 597–99; MacGinnis (2010) 156–60; Jursa (2011) 435–42; Jursa (2015) 351–53; Jursa and Schmidl (2017), 719–27; Gombert (2018); Manning (2021) 159–203. For potential connections between the *hadru* system and excavated settlements in Achaemenid Palestine see Faust (2018); Kletter and Silverman (2020).

50 For text editions see Abraham (2004) no. 88; Hackl, Jursa, and Schmidl (2014) no. 61. Tolini (2011) 427–32, situates the letter alongside other references to mobilisation and hypothesises a possible connection with Persia's response to the Ionian Revolt; Gombert (2018) 460–64, argues for a different objective, situating the key toponym in southern Babylonia. See also Tuplin (2010) 124 n. 95; Jursa and Schmidl (2017) 731–32; Manning (2021) 167.

51 See Bovet-Fischer (2013) 210–19; Tuplin (2020a).

which may have been supported by landed estates.[52] In addition, they maintained the standing bodies of troops in urban garrisons, and were responsible for the upkeep of rural military colonies that originated in land grants by early Achaemenid rulers, whose manpower was allegedly inspected in annual musters.[53] For large campaigns, they could be augmented by allied warriors assembled by semi-autonomous figures such as the Lycian or Paphlagonian dynasts, and additional mercenary soldiers hired in peripheral regions, including but hardly limited to the Greek city-states. Xenophon, Diodorus, and the Alexander historians naturally emphasise those recruited in Greece by entrepreneurial commanders in the western regions of the empire, but recent scholarship has effectively deconstructed the literary trope of their qualitative dominance within late Achaemenid armies.[54]

Greek campaign narratives report the subdivision of large armies into regional or ethnic contingents. The specific details of Herodotus' army roster for Xerxes' expedition (7.61–80) have been called into question, and not all its units should not be taken as equal in size and significance; Pierre Briant has demonstrated that the assembly of men from the more remote and exotic groups in the list should be viewed as a token gathering of representatives from the empire's peoples, a 'parade army' separated from the actual campaign force.[55] Nevertheless, the overall idea of regional subdivisions within a field army is plausible enough, best attested at the end of the empire in the Persian contingent list at Gaugamela, which Aristoboulos claimed was derived from a document captured in the royal camp (Arr. *An.* 3.11.3). Over time, the empire's representatives may have engaged in military reform to standardise some aspects of recruitment and training between provincial and Persian conscripts, as suggested in Briant's hypothesis that the *kardakes* who appear in several fourth-century military contexts were part of a new type of infantry recruited throughout the provinces.[56]

A final aspect of recruitment in campaigns on the empire's western frontier zones involved the assembly of naval personnel to man trireme warships and supporting merchant craft. Some early expeditions depended on temporary

52 Cf. Xen. *An.* 1.2.4, 1.8.6; Xen. *Hell.* 3.4.10, 4.1.17–18. See Sekunda (1985); Sekunda (1988); Sekunda (1991); Tuplin (2010) 154–55; cf. Manning (2021) 215–16.

53 Xen. *Cyr.* 8.6.1–11; Xen. *Oec.* 4.5–11. See Tuplin (1987) 168–75; Dusinberre (2013) 87–89; Tuplin (2014) 676–77; Tuplin (2016) 17–19.

54 On mercenaries in the Achaemenid military system, see Rop (2021); Tuplin (2021). For critiques of Greek sources' overemphasis on Greek mercenaries, see Briant (2002) 791–800; Rop (2019).

55 See Briant (1999) 118–19; Tuplin (2017) 45.

56 See Briant (1999) 120–22; Charles (2012) 16–18; Tuplin (2014) 687–88.

requisition of vessels from coastal subjects (Hdt. 5.26), but reports of regular naval concentrations in Cilicia, Phoenicia, and other coastal regions may support the existence of more permanent naval structures during the era of Darius I and Xerxes, based on the maintenance of royally owned triremes and their required shipshed and dockyard facilities.[57] These combat vessels required regular replacement and the empire not only never returned to the ship numbers assembled for Xerxes' campaigns, but also seems to have let the fleet disappear in the mid-fifth century, before its reassembly and steady maintenance for the Egyptian, Aegean, and Cypriote conflicts of the fourth century.[58] The numbers required for fleets made up of triremes with crews of 200 men each, not counting the marines who served on the decks (Hdt. 7.184.1), pushed the authorities to gather naval manpower for individual campaigns and then demobilise them as swiftly as possible. This process occurred in the Mediterranean ports, from Egypt to western Anatolia, where each naval squadron gathered. It may have involved some conscription for the largest fleets, but otherwise will have shared much in common with the hiring of mercenaries for land forces.[59] An account of one fourth-century mission against rebels in Cyprus emphasises the role of promised wealth from looting, above and beyond regular wages, in attracting volunteers (Diod. Sic. 16.42.2). The Persian marines assigned to the decks of warships, attested in Xerxes' Greek expedition (Hdt. 7.96.1), were presumably selected and detached from the land army rather than men with permanent naval duties.[60]

2 Army Sizes and Campaign Logistics

The surviving documentary evidence does not testify to the sizes of full-scale armies or fleets, but illustrates the quantitative habit that characterised the empire's administrative system. One can assume that Achaemenid military officials gathered accurate data about the numbers of their forces, although such statistics were not accessible outside the administration and would not

57 For the theory of the standing navy, see Wallinga (1987) and Wallinga (2005); Cawkwell (2005, 257) argues that fleet mobilisations more likely supplemented small standing forces through naval requisitions from subjects such as the Phoenician city-states. For Cilicia's importance, see Wallinga (1991); Bouzid-Adler (2014).

58 For Achaemenid naval evolution after Xerxes, see Hyland (2018b) 23–28, 133–35, 143–45.

59 See Hyland (2018b) 26–28.

60 Provincial forces might be recruited directly for marine service, as in the case of Egyptian soldiers in Xerxes' fleet who later disembarked and served on land (Hdt. 9.32.2).

have made their way into the consciousness of the Greek authors.[61] It is undisputed that the empire was capable of assembling great numbers for the largest campaigns, although these figures rarely reached into the hundreds of thousands, despite the infamous exaggeration of numbers provided in Greek and Roman historians. The combination of land and naval forces was particularly significant as a force multiplier, as illustrated by the largest known Achaemenid military expedition, Xerxes' Greek invasion, for which even the lowest revisionist estimates require well over 100,000 personnel in the combined land army and combat fleet, not counting the additional manpower involved in the moving court, baggage train, and naval supply convoys.[62]

Campaigns on such a scale were unusual, but more typical field forces could still involve tens of thousands of soldiers, and in Mediterranean coastal contexts, tens of thousands of sailors and rowers to man fleets of 100 to 300 triremes.[63]

This brings us to the crucial matter of logistics. It is an old truism that armies march on their stomachs, and before Darius' ubiquitous Persian man battled the enemy far from Persia, his meals and fresh water access required careful preparation.[64] The empire's size meant that soldiers joining frontier expeditions might have to travel 1,500 miles or more from an original base or recruitment site just to reach the staging point for the campaign, before covering hundreds of miles on the actual expedition.[65] With this in mind, administrative capacities to feed moving forces were an essential factor in the empire's initial

61 Wallinga (2005, 130) makes the implausible claim that the ship loss figures for Salamis at Diod. Sic. 11.19.3 derived from imperial records left behind at Kyme, which found their way into Ephoros' fourth-century history.

62 Herodotus' manpower discussion (7.184–187), for all its infamously exaggerated statistics, usefully illustrates the complexity of the logistical issues. Modern efforts to estimate Xerxes' field army's size have generated an enormous bibliography; a useful starting point for critical examination is Tuplin (1997) 366–73. For introductory notes on Persian navies, see Tuplin and Jacobs (2021) 1175–77. Debate continues over the veracity of the 1,200 trireme total provided by Aeschylus and Herodotus. Wallinga (2005, 32–46), in attempting to justify the Classical figures, hypothesises that the total included large numbers of empty or under-manned replacement craft, but numerous studies have preferred to halve the number of active warships to 600. Cawkwell (2005, 263–67) makes an unconvincing effort to rehabilitate Hans Delbrück's minimalist estimate of 300, which would introduce drastic complications into the analysis of the campaign and Salamis in particular.

63 See Cawkwell (2005) 243–52, 255–67.

64 Maurice (1930), focusing on Xerxes' march across the Troad and Chersonese, remains a valuable study of the logistics of military hydration despite high estimates of the army's size.

65 Several Greek authors refer to Achaemenid marching rates or distances covered within a specific number of days, with resulting estimates ranging between 9 and 25 miles per day; see Hyland (2018a) 83 n. 37.

conquests and subsequent military activities. Modern studies have long recognised the impressive scale and reach of the logistical network implied by Greek descriptions of Achaemenid campaigns; George Cawkwell, for example, states on this basis that 'the Persians must rank as one of the greatest military powers of history'.[66] Again, while the Greek histories add important anecdotal testimony for the logistics of armies on campaign, the Persepolis Fortification texts and other internal documentation can offer more fine-grain detail on military movement within the empire.

Persian soldiers' staple ration was barley (or occasionally wheat), pre-ground into flour to avoid the necessity of milling on the march.[67] The Persepolis texts show average daily issues of 1.5 QA (liters/quarts) for *šalup*, *taššup*, and Spear-Bearers, whereas *kurtaš*, *libap*, and *puhu* often earn a lower ration of 1 QA; neither would have been sufficient on its own for full caloric maintenance, but soldiers are likely to have obtained additional foodstuffs on top of their base sustenance through private purchase and foraging.[68] Certain Persepolis journals record the dispensation of goats to travelling *taššup* (NN 2261: 16–18, 19–21, 26–29; Fort. 0472: 17–18), and Herodotus makes a categoric statement that the king brought animal herds along for provision whenever he went on campaign (1.188.1).[69] The cavalry mounts and pack animals that accompanied a military force had more complex dietary needs. When equids appear with travelling groups in Persepolis tablets, they typically receive 3 liters of grain fodder, but grazing remained an essential element of their diet, which varied according to activity and time of year; camels also require significant amounts of grazing for their primary nutrition, but this would have to be supplemented with solid grain-based rations for efficiency of long-distance movement.[70]

Daily consumption requirements for even moderate-sized military units thus required extensive quantitative planning, but the internal logistical network run by administrators in each imperial region was designed to provide

66 Cawkwell (2005) 237.

67 Barley flour (ŠE.BAR) is most common in the Persepolis archives. For the importance of milling in advance, see also Hdt 7.23.4 and 7.119.2 on Xerxes' campaign preparations.

68 Hdt. 7.187.2 estimates the average daily ration of Xerxes' troops as one Attic *choinix* of wheat, which may be calculated as 1.094 dry liters/quarts; see Moreno (2007) 325; O'Connor (2013) 333 n. 31. But the Persepolis Fortification tablets usually give rations in barley flour, which weighs less than wheat; for the equation of two *choinikes* of barley flour to one of wheat, which would weigh 1.52 lbs (.687 kg) before milling, see O'Connor (2013) 328–38.

69 See Henkelman (2008a) 310; Henkelman (2018a) 228 n. 28.

70 On the maintenance of horses at Persepolis, see Gabrielli (2006) 53–65, 134–35. Azzoni and Dusinberre (2014, 4–5) emphasise grain issues as a high-energy dietary supplement for especially active animals such as chariot teams or cavalry mounts. For camels' dietary needs, see Potts (2021) 232–37, 241–44.

exactly this kind of support. Massive granaries stored accumulated produce for redistribution as needed, and soldiers on the move could draw food from stockpiles at way stations on the royal road, small provincial forts, and officials' estates.[71] Commanders might also expect troops to purchase provisions in local markets, as Xerxes' officials arranged to supply the corvée workers on the Athos canal (Hdt. 7.23.4).

This regular access to resupply should have allowed troops to limit the amount of food transported on overland marches inside the empire.[72] Travelling contingents in the Persepolis archives most often stopped daily to pick up rations at estates and storehouses; exceptions, such as the 115 *šalup* and 50 *puhu* who received a month's supply of flour during a journey from Fars to India in the spring of 499 (Fort. 0313-103), probably reflect unique circumstances of temporary shortage along an anticipated route. Even the rebel army of Cyrus the Younger purchased food from city markets or requisitioned it from depots, estates, and villages for most of its march from Anatolia to the Euphrates; it was only late in the campaign that Cyrus assembled 400 wagons full of flour to sustain his soldiers through a temporary period in which impending battle was expected to interfere with its provisioning.[73] This is not to say that armies did not require significant baggage trains, but they were more likely used to haul tents, personal possessions, and military equipment rather than food supplies; the transport of water from the vicinity of Susa for the personal use of the king is the exception that proves the rule.[74] The pack animals used in such contexts included mules maintained in royal stables for the support of the travelling court, and Herodotus claims that the Immortals marched with their own column of camels and other baggage animals, separate from that of the provincial

71 For Persepolis and archaeological evidence for granary capacities, see Henkelman (2017b) 82–97; Briant (2019) 64. On the participation of estates in the logistical system, possibly as a form of regular taxation in fulfillment of service obligations, see Henkelman (2018b) 43–44.

72 The point is illustrated by several examples in the rebellion of Cyrus the Younger and the return march of his Greek mercenaries; see Descat (1995); Gabrielli (1995); Lee (2007) 134–37; O'Connor (2015) 135–36.

73 See Descat (1995) 101–105; for the wagons (Xen. *An.* 1.10.18), see O'Connor (2015). The 'Lydian market' of merchants that followed Cyrus' army sold flour to the troops at marked up prices in time of shortage (Xen. *An.* 1.5.6), but probably travelled with the purpose of buying slaves acquired by soldiers and selling other commodities.

74 Persepolis records of long-distance journeys with high ratios of equids and camels to men probably involve tribute convoys rather than strictly military movements, but also show the feasibility of stopping for daily food issues and thereby freeing the baggage animals to carry non-comestible supplies; see PF 2056 (100 mules, 18 horses, 588 men); NN 0431 (31 mules, 12 camels, 291 men); Fort. 0412-101 (15 mules, 10 horses, 108 men); NN 2569 (88 mules, 30 horses, 81 men). For royal water transport (Hdt. 1.188.2), see Briant (2017a).

forces (7.83.2); although the surviving Fortification tablets show camels only in contexts of travel by relatively small groups, their ability to carry more than twice the load of a mule would have added to their military value.[75]

Logistical stockpiling was most intensive in the concentration zones where armies and fleets assembled for expeditions beyond the frontier. Herodotus details Xerxes' use of naval transport to carry stores from multiple provinces to five major stockpiles between Perinthos and Doriskos, in order to support the army after its Hellespontine crossing (7.25.1). Expeditions into Egypt were facilitated by similar depots in the Levant, as well as the hydrological arrangements established to support military travel through the Sinai, with the help of local Arab clients who transported camel skins full of water to desert resupply posts (Hdt. 3.9.1).[76] In addition, provisions for court and military 'feasts' were requisitioned at least many months in advance from local communities that the army would pass on the march beyond the major stockpiles. Herodotus describes the significant burden this placed on the cities, reporting the quip of an Abderan citizen who thanked the gods that Xerxes did not order breakfast as well as dinner (7.119–120).

The most significant challenges began when an army or combined land and sea forces advanced beyond the frontier into regions outside the Achaemenid administrative reach. Not only were overland transport capacities limited, but the lack of comparable logistical infrastructure among their opponents meant that armies could not count on capturing major enemy grain stockpiles, as the Assyrians may have attempted in the famous invasion of Urartu by Sargon II.[77] The herds of meat on the hoof that accompanied Persian armies would have been depleted rapidly; troops may have attempted to capture local livestock to augment grain supplies, but their opponents sometimes managed to deny the army such moveable feasts, as when the men of the Greek fleet slaughtered flocks of sheep on Euboea to keep them out of Persian hands (Hdt. 8.19–20). In coastal regions, Persian commanders addressed supply challenges in part through the arrangement of naval food convoys, which played major supporting roles in Xerxes' Greek campaign (Hdt. 7.97, 185.3, 186.1, 190–91) and the fourth-century invasions of Egypt (Diod. Sic. 15.41.3, 16.40.6). But the

75 For royal mule and horse herds, see PF 1793; NN 1665; NN 1289; Fort. 1901A-101: 48. For camels in the Persepolis tablets and the comparison with mules, see Potts (2021) 237–39; for additional discussion and comparison with the presentation of camels in the Alexander historians, see Henkelman (2017b) 55–60.

76 Cf. Hassan (2021) 1158. On Cambyses' route, see Cruz-Uribe (2003); for interpretation of Achaemenid-period settlements in southern Palestine as a fortress and logistical network supporting Persia's fourth-century Egyptian campaigns, see Fantalkin and Tal (2012) 163–68.

77 See Marriott and Radner (2015) 133–34.

geography of the advance sometimes interrupted direct contact between land and sea forces, as in Xerxes' advance from Macedonia to Athens, and naval carrying capacity was not without its limitations. A Persian general had to raise the siege of Naxos in 499 when supplies ran out after a four-month blockade (Hdt. 5.34.2–3), and in a fourth-century expedition against Cyprian Salamis, enemy interception of supply convoys led to food shortage and provoked a mutiny in the Persian invasion force before a second food shipment alleviated the crisis (Diod. Sic. 15.3.1–3). Whenever possible, it was still preferable for imperial agents to arrange stockpiling by cooperative local communities in advance of an army's advance. This may have been one of the purposes of the 'earth and water' messengers Xerxes sent from Sardis to Greece many months before the start of his expedition (Hdt. 7.32); for the Persians in Greece no less than Caesar in Gaul, diplomatic client networks and logistics were closely interconnected.[78] Pre-assembly of stores in Thessaly can be assumed in light of Mardonios' extended winter encampment (Hdt. 8.113, 126), and the supplies of food gathered at Thebes before the summer campaign of 479 may have depended heavily on the contributions of the Persians' Greek allies (Hdt. 9.41.2).

Overall, the Persians were skillful at managing logistical challenges, even in the trans-Aegean campaigns where Aeschylus and Herodotus emphasise the difficulty of supplying a large army in rugged terrain. The destruction of grain transports in the storms off Magnesia (Hdt. 7.191.1) did not interrupt the subsequent advance on Athens; and claims of the army's starvation during Xerxes' retreat (Aesch. *Pers.* 482–491, Hdt. 8.115.2, 117.2) are contradicted by the extended winter operations and winter encampment of Artabazos' and Mardonios' forces in the same region (Hdt. 8.126–129).[79] Several modern studies have proposed that strains on the supply network pressured Persian generals into premature engagements, thereby contributing to battlefield defeats, but this critique depends in part on negative assessments of the tactical system (see Rop's chapter in this volume).[80] On the other hand, the loss of administrative centers, ports, and supply depots in Thrace and Ionia to Athenian counterattacks after 479 certainly imposed a negative impact on the Persians' regional power by removing the infrastructure required for a repetition of a campaign on a comparable scale.

Despite localised setbacks in the Aegean or the period of fourth-century Egyptian independence, the empire's internal logistical infrastructure continued to facilitate military activities on multiple frontiers throughout its

78 Cf. Engels (1978) 40–41, making a similar point regarding Alexander's invasion of Persia.
79 See Briant (2002) 530–31.
80 See Balcer (1995) 247–48, 255–56, 298; Cawkwell (2005) 106–107, 114–15, 123 nn. 54–55.

existence. In the context of the final conflict against Macedonian invaders, it supported the repetition of military assemblies and large-scale campaigns from Mesopotamian bases after initial defeats, and delayed the process of military collapse.[81] Ultimately, when the disaster of Gaugamela and the non-resistance of major fortified centers accelerated Persia's political and military breakdown, the remaining institutional framework empowered Alexander's army as it took it over and contributed to military continuities well into the Seleucid period.[82]

3 Conclusion

The discussion presented here remains preliminary, and is necessarily brief in its treatment of many important topics. Another aspect that could benefit from further attention is the development of military command in relation to the administrative system and the role of the regional generals referred to as **karanaya* (Greek *karanos*, Aramaic *krny*).[83] Our understanding of defensive networks in imperial frontier zones, already studied extensively by Tuplin, will doubtlessly continue to develop with new archaeological finds in areas ranging from western Anatolia to Central Asia, and the final publication of the complete Persepolis Fortification archive will naturally permit more comprehensive analysis of its military contents. Nevertheless, I hope to have offered a useful preliminary framework and illustrated the value of a systemic empire-wide approach to Persian military organisation and logistics with the Persepolis evidence at its center. The expanding study of the Achaemenid military in comprehensive structural terms will build a stronger foundation for future efforts to contextualise royal policy-making and campaigns on the imperial frontiers.

Bibliography

Abraham, K. (2004) *Business and Politics under the Persian Empire: The Financial Dealings of Marduk-nāsir-apli of the House of Egibi (521–487 B.C.E.)*. Bethesda.

Agut-Labordère, D., Boucharlat, R., Joannès, F., Kuhrt, A., and Stolper, M.W. (eds) (2021) *Achemenet. Vingt-ans après. Études offertes à Pierre Briant à l'occasion des vingt ans du Programme Achemenet*. Leuven.

81 See Briant (2002) 823–42, 864–65; Hyland (2018a).
82 See Briant (2002) 373; Henkelman (2017b) 79; Briant (2019) 64–67.
83 See Keen (1998); Hyland (2013).

Alcock, S., Bodel, J., and Talbert, R. (eds) (2012) *Highways, Byways, and Road Systems in the Pre-Modern World.* Oxford.

Álvarez-Mon, J. and Garrison, M. (2011) *Elam and Persia.* Winona Lake.

Azzoni, A. (2008) 'The Bowman MS and the Aramaic Tablets', in Briant, Henkelman, and Stolper (2008) 253–75.

Azzoni, A., and Dusinberre, E. (2014) 'Persepolis Fortification Aramaic Tablet Seal 002 and the Keeping of Horses', in Kozuh, Henkelman, Jones, and Woods (2014) 1–16.

Balcer, J. (1995) *The Persian Conquest of the Greeks, 545–450 BC*. Konstanz.

Bittner, S. (1985) *Tracht und Bewaffnung des persischen Heeres zur Zeit der Achaimeniden.* Munich.

Bouzid-Adler, F. (2014) 'Les marines des peuples d'origine louvite à l'époque achéménide'. *Res Antiquae* 11: 1–18.

Briant, P. (ed) (1995) *Dans les Pas des Dix-Mille. Peuples et Pays du Proche-Orient vus par un Grec* (*Pallas* 43). Toulouse.

Briant, P. (1999) 'The Achaemenid Empire', in Raaflaub and Rosenstein (1999) 105–28.

Briant, P. (2002) *From Cyrus to Alexander: A History of the Persian Empire.* Winona Lake.

Briant, P. (2012) 'From the Indus to the Mediterranean: The Administrative Organization and Logistics of the Great Roads of the Achaemenid Empire', in Alcock, Bodel, and Talbert (2012) 185–201.

Briant, P. (2017a) 'The Drinking Water of the Great King', in Briant (2017b) 286–304.

Briant, P. (2017b) *Kings, Countries, Peoples: Selected Studies on the Achaemenid Empire.* Stuttgart.

Briant, P. (2019) 'L'approvisionnement de l'armée Macédonienne: Alexandre le Grande et l'organisation logistique de l'Empire Achéménide', in Gondet and Haerinck (2019) 55–70.

Briant, P., Henkelman, W.F.M., and Stolper, M.W. (eds) (2008) *L'Archive des Fortifications de Persépolis: État des questions et perspectives de recherches* (*Persika* 12). Paris.

Brosius, M. (2005) 'Pax Persica: Königliche Ideologie und Kriegführung im Achämenidenreich'. in Meiβner and Schmitt (2005) 135–61.

Cawkwell, G. (2005) *The Greek Wars: The Failure of Persia.* Oxford.

Charles, M. (2011) 'Immortals and Apple-Bearers: Towards a Better Understanding of Achaemenid Infantry Units'. *CQ* 61: 114–33.

Charles, M. (2012) 'The Persian κάρδακες'. *JHS* 132: 7–21.

Charles, M. (2021) 'The Royal Elite of the Achaemenid Army', in Heckel, Naiden, Garvin, and Vanderspoel (2021) 173–84.

Clark, J. and Turner, B. (eds) (2018) *Brill's Companion to Military Defeat in Ancient Mediterranean Society*. Leiden.

Colburn, H. (2013) 'Connectivity and Communication in the Achaemenid Empire'. *JESHO* 56: 29–52.

Cruz-Uribe, E. (2003) 'The Invasion of Egypt by Cambyses'. *Transeuphratène* 25: 9–58.

Curtis, J. and Tallis, N. (eds) (2005) *Forgotten Empire: The World of Ancient Persia*. Berkeley.

Daryaee, T. (2021) '*Männerbund* aspects of Old Persian *anušiya*', in Agut-Labordère, Boucharlat, Joannès, Kuhrt, and Stolper (2021) 73–78.

Descat, R. (1995) 'Marché et tribut: l'approvisionement des Dix-Mille', in Briant (1995) 99–108.

Dezső, T. (2012a) *The Assyrian Army I. The Structure of the Neo-Assyrian Army. 1. Infantry*. Budapest.

Dezső, T. (2012b) *The Assyrian Army I. The Structure of the Neo-Assyrian Army. 2. Cavalry and Chariotry*. Budapest.

Dezső, T. (2016) *The Assyrian Army II. Recruitment and Logistics*. Budapest.

Dezső, T. (2022) *The Assyrian Army III. The Assyrian Army on Campaign. 1. Musters and Marching*. Budapest.

Dieulafoy, M. (1891) *L'Acropole de Susa III. Faiences et Terres Cuites*. Paris.

Düring, B. (ed) (2014) *Understanding Hegemonic Practices of the Early Assyrian Empire*. Leiden.

Dusinberre, E. (2013) *Empire, Authority, and Autonomy in Achaemenid Anatolia*. Cambridge.

Dusinberre, E., Garrison, M., and Henkelman, W.F.M. (eds) (2020) *The Art of Empire in Achaemenid Persia: Studies in Honour of Margaret Cool Root*. Leiden.

Engels, D. (1978) *Alexander the Great and the Logistics of the Macedonian Army*. Berkeley.

Fales, M. (1990) 'Grain Reserves, Daily Rations, and the Size of the Neo-Assyrian Army'. *State Archives of Assyrian Bulletin* 4: 23–34.

Fantalkin, A. and Tal, O. (2012) 'Judah and Its Neighbors in the Fourth Century BCE: A Time of Major Transformations', in Ro (2012) 134–97.

Farrokh, K. (2007) *Shadows in the Desert: Ancient Persia at War*. London.

Faust, A. (2018) 'Forts or Agricultural Estates?' *Palestine Exploration Quarterly* 150: 34–59.

Fagan, G. and Trundle, M. (eds) (2010) *New Perspectives on Ancient Warfare*. Leiden.

Fischer-Bovet, C. (2013) 'Egyptian Warriors: The "Machimoi" of Herodotus and the Ptolemaic Army'. *CQ* 63: 209–36.

Gabrielli, M. (1995) 'Transports et logistique militaire dans l'*Anabase*', in Briant (1995) 109–22.

Gabrielli, M. (2006) *Le cheval dans l'empire achéménide – Horse in the Achaemenid Empire* (*Studia ad Orientem Antiquum Pertinentia* 1). Istanbul.

Garrison, M., and Henkelman, W.F.M. (2020) 'Sigillophobe Suppliers and Idiosyncratic Scribes: Local Information Handling in Achaemenid Pārsa', in Dusinberre, Garrison, and Henkelman (2020) 167–286.

Garrison, M. and Henkelman, W.F.M. (2023) (eds) *The Persian World and Beyond: Achaemenid and Arsacid Studies in Honour of Bruno Jacobs*. Münster.

Gaspa, S., Greco, A., Bonacossi, D., Ponchia, S., and Rollinger, R. (eds) (2014) *From Source to History: Studies on Ancient Near Eastern Worlds and Beyond (Festschrift for Giovanni B. Lanfranchi)*. Münster.

Gombert, B. (2018) 'L'Armée en Babylonie du VIe au IVe Siècle av. N.È.', PhD Dissertation, Université Paris 1 Panthéon-Sorbonne.

Gondet, S. and Haerinck, E. (eds) (2019) *L'Orient est son Jardin: Hommages à Rémy Boucharlat*. Leuven.

Hackett, J. (ed) (1989) *Warfare in the Ancient World*. London.

Hackl, J., Jursa, M., and Schmidl, M. (2014) *Spätbabylonische Briefe 1. Spätbabylonische Privatbriefe*. Münster.

Hallock, R. (1969) *Persepolis Fortification Tablets*. Chicago.

Hallock, R. (1978) 'Selected Fortification Texts', *Cahiers de la Délégation Archéologique Française en Iran* 8: 109–36.

Hassan, C. (2021) 'Structure of the Army and Logistics', in Jacobs and Rollinger (2021), 1151–59.

Head, D. (1992) *The Achaemenid Persian Army*. London.

Hebblewhite, M. and Whately, C. (2022) *Brill's Companion to Bodyguards in the Ancient Mediterranean*. Leiden.

Heckel, W., Naiden, F., Garvin, E., and Vanderspoel, J. (2021) *A Companion to Greek Warfare*. Hoboken, NJ.

Henkelman, W.F.M. (2002) 'Exit der Posaunenbläser: On Lance-Guards and Lance-Bearers in the Persepolis Fortification Archive', *ARTA* 2002.007: 1–35.

Henkelman, W.F.M. (2008a) 'From Gabae to Taoce: The Geography of the Central Administrative Province', in Briant, Henkelman, and Stolper (2008), 303–16.

Henkelman, W.F.M. (2008b) *The Other Gods Who Are: Studies in Elamite-Iranian Acculturation based on the Persepolis Fortification Texts (Achaemenid History XIV)*. Leiden.

Henkelman, W.F.M. (2010a) 'Consumed before the King: The Table of Darius, that of Irdabama and Irtaštuna, and that of his Satrap, Karkiš', in Jacobs and Rollinger (2010) 667–775.

Henkelman, W.F.M. (2010b) 'Xerxes, Atossa, and the Persepolis Fortification Archive'. *Annual Report 2010 of the Netherlands Institute for the Near East*: 27–33.

Henkelman, W.F.M. (2011) 'Parnaka's Feast: šip in Pārsa and Elam', in Álvarez-Mon and Garrison (2011) 89–166.

Henkelman, W.F.M. (2017a) 'Humban & Auramazdā: Royal Gods in a Persian Landscape', in Henkelman and Redard (2017) 273–346.

Henkelman, W.F.M. (2017b) 'Imperial Signature and Imperial Paradigm: Achaemenid Administrative Structure and System across and beyond the Iranian Plateau', in Jacobs, Henkelman, and Stolper (2017) 45–256.

Henkelman, W.F.M. (2018a) 'Bactrians in Persepolis – Persians in Bactria', in Lhuillier and Boroffka (2018) 223–55.

Henkelman, W.F.M. (2018b) 'Precarious Gifts: Achaemenid Estates and Domains in Times of War and Peace', in Jullien (2018) 13–66.

Henkelman, W.F.M. (2020) 'Nakhthor in Persepolis', in Tuplin and Ma (2020a) 193–223.

Henkelman, W.F.M. (2021) 'Persia', in Jacobs and Rollinger (2021) 879–904.

Henkelman, W.F.M. (2023) 'Irdumartiya and the chiliarchies: The Contemporaneity of the Persepolis Fortification and Treasury Archives and Its Implications', in Garrison and Henkelman (2023) 235–300.

Henkelman, W.F.M. and Redard, C. (eds) (2017) *Persian Religion in the Achaemenid Period.* Wiesbaden.

Hillard, T., Kearsley, R., Nixon, C., and Nobbs, A. (eds) (1998) *Ancient History in a Modern University. Vol. 1: The Ancient Near East, Greece, and Rome.* Grand Rapids.

Hyland, J.O. (2013) 'Vishtaspa *krny*: An Achaemenid Military Official in 4th-century Bactria'. *ARTA* 2013.002: 1–7.

Hyland, J.O. (2018a) 'Achaemenid Soldiers, Alexander's Conquest, and the Experience of Defeat', in Clark and Turner (2018) 74–95.

Hyland, J.O. (2018b) *Persian Interventions: The Achaemenid Empire, Athens, and Sparta, 450–386 BCE.* Baltimore.

Hyland, J.O. (2019) 'The Achaemenid Messenger Service and the Ionian Revolt: New Evidence from the Persepolis Fortification Archive'. *Historia* 68: 150–69.

Jacobs, B., Henkelman, W.F.M., and Stolper, M.W. (eds) (2017) *Verwaltung im Achämenidenreich.* Wiesbaden.

Jacobs, B. and Rollinger, R. (eds) (2010) *Der Achämenidenhof.* Wiesbaden.

Jacobs, B. and Rollinger, R. (eds) (2021) *A Companion to the Achaemenid Persian Empire.* Hoboken, NJ.

Jullien, F. (ed) (2018) *Guerre et paix en monde iranien. Revisiter les lieux des rencontre.* Paris.

Jursa, M. (2011) 'Taxation and Service Obligations in Babylonia from Nebuchadnezzar to Darius and the Evidence for Darius' Tax Reform', in Rollinger, Truschnegg, and Bichler (2011) 431–48.

Jursa, M. (2015) 'Labor in Babylonia in the First Millennium B.C.'. in Steinkeller and Hudson (2015) 345–96.

Jursa, M., and Schmidl, M. (2017) 'Babylonia as a Source of Imperial Revenue from Cyrus to Xerxes', in Jacobs, Henkelman, and Stolper (2017) 715–40.

Keen, A.G. (1998) 'Persian *Karanoi* and their Relationship to the Satrapal System' in Hillard, Kearsley, Nixon, and Nobbs (1998) 88–95.

Kletter, R., and Silverman, J. (2020) '"Estates" or "Forts" in Persian Period Yehud'. *Palestine Exploration Quarterly* 152: 42–61.

Koch, H. (1993) *Achämeniden-Studien.* Wiesbaden.

Kozuh, M., Henkelman, W.F.M., Jones, C., and Woods, C. (eds) (2014) *Extraction and Control: Studies in Honor of Matthew W. Stolper.* Chicago.

Kuhrt, A. and Sancisi-Weerdenburg, H. (eds) (1988) *Achaemenid History III: Method and Theory*. Leiden.

Kuhrt, A. and Sancisi-Weerdenburg, H. (eds) (1991) *Achaemenid History VI: Asia Minor and Egypt*. Leiden.

Lee, G., Whittaker, H., and Wrightson, G. (eds) (2015). *Ancient Warfare: Introducing Current Research*. Vol. 1. Newcastle.

Lee, J.W.I. (2007) *A Greek Army on the March: Soldiers and Survival in Xenophon's* Anabasis. Cambridge.

Lhuillier, J. and Boroffka, N. (eds) (2018) *A Millennium of history: The Iron Age in Southern Central Asia (2nd and 1st Millennia BC)*. Berlin.

MacGinnis, J. (2010) 'Mobilisation and Militarisation in the Neo-Babylonian Empire' in Vidal (2010) 153–63.

Macuch, M., Weber, D., and Durkin-Meisterernst, D. (eds) (2010) *Ancient and Middle Iranian Studies: Proceedings of the 6th European Conference of Iranian Studies*. Wiesbaden.

Malbran-Labat, F. (1982) *L'Armée et l'Organisation Militaire de l'Assyrie*. Geneva.

Manning, S. (2021) *Armed Force in the Teispid-Achaemenid Empire: Past Approaches, Future Prospects*. Stuttgart.

Manning, S. (2022) 'The armies of the Teispids and Achaemenids: The Armies of an Ancient World Empire'. *Journal of Ancient Civilizations* 37: 147–92.

Marquart, J. (1896) *Untersuchungen zur Geschichte von Eran*. Vol. 1. Göttingen.

Marriott, J. and Radner, K. (2015) 'Sustaining the Assyrian Army among Friends and Enemies in 714 BCE'. *JCS* 67: 127–43.

Maurice, F. (1930) 'The Size of the Army of Xerxes in the Invasion of Greece 480 B.C.'. *JHS* 50: 210–35.

Meiβner, B. and Schmitt, O. (eds) (2005) *Krieg – Gesellschaft – Institutionen: Beiträge zu einer vergleichenden Kriegsgeschichte*. Halle-Wittenberg.

Moreno, A. (2007) *Feeding the Democracy: The Athenian Grain Supply in the Fifth and Fourth Centuries BC*. Oxford.

Naveh, J. and Shaked, S. (2012) *Aramaic Documents from Ancient Bactria from the Khalili Collection*. London.

O'Connor, S. (2013) 'The Daily Grain Consumption of Classical Greek Soldiers and Sailors'. *Chiron* 43: 327–56.

O'Connor, S. (2015) 'The Problem of the 400 Wagons: The Provisioning of the Ten Thousand on the March to Cunaxa', in Lee, Whittaker, and Wrightson (2015) 124–46.

Pagliaro, A. (1954) 'Riflesse di etimologie iraniche nella tradizione storiografica greca'. *Rendiconti dell'Accademia Nazionale dei Lincei. Classe di Scienze morali, storiche e filologiche*, 8th series, 9: 133–53.

Parker, B. (2014) 'Hegemony, Power, and the Use of Force in the Neo-Assyrian Empire', in Düring (2014) 287–99.

Porten, B. (1968) *Archives from Elephantine: The Life of an Ancient Jewish Military Colony*. Berkeley.

Postgate, J.N. (2000) 'The Assyrian Army in Zamua'. *JCS* 62: 89–108.

Potts, D.T. (2021) 'Camels and their Rations in the Persepolis Fortification Archive: An Enigma and Its Variations'. *Egitto e Vicino Oriente* 44: 231–47.

Ro, J. (ed) (2012) *From Judah to Judea: Socio-Economic Structures and Processes in the Persian Period*. Sheffield.

Rop, J. (2019) *Greek Military Service in the Ancient Near East, 401–330 BCE*. Cambridge.

Rop, J. (2021) 'Greeks in Foreign Service: The Case of the Achaemenid Empire', in Heckel, Naiden, Garvin, and Vanderspoel (2021) 160–69.

Raaflaub, K. and Rosenstein, N. (eds) (1999) *War and Society in the Ancient and Medieval Worlds: Asia, the Mediterranean, Europe, and Mesoamerica*. Cambridge, MA.

Rollinger, R., Truschnegg, B., and Bichler, R. (eds) (2011) *Herodot und das Persische Weltreich*. Wiesbaden.

Rossi, A. (2010) 'Élamite *halmarriš* – vieux-persan *didā* – est-elle vraiment une forteresse?' in Macuch, Weber, and Durkin-Meisterernst (2010) 205–18.

Sancisi-Weerdenburg, H. (ed) (1987) *Achaemenid History I: Sources, Structures and Synthesis*. Leiden.

Sandowicz, M. (2020) 'Companions of Nabonidus'. *ZA* 110: 161–75.

Scott, L. (2005) *Historical Commentary on Herodotus Book 6*. Leiden: Brill.

Schmidt, E. (1953) *Persepolis I. Structures – Reliefs – Inscriptions (OIP 69)*. Chicago.

Schmitt, R. (2004) 'Immortals'. *Encyclopaedia Iranica* XIII/1: 2–3, available online at http://www.iranicaonline.org/articles/immortals (accessed on 22 July, 2020).

Sekunda, N. (1985) 'Achaemenid Colonization in Lydia'. *Rev. Ét. Anc.* 87: 7–30.

Sekunda, N. (1988) 'Persian Settlement in Hellespontine Phrygia', in Kuhrt and Sancisi-Weerdenburg (1988) 175–96.

Sekunda, N. (1989) 'The Persians', in Hackett (1989) 82–103.

Sekunda, N. (1991) 'Achaemenid Settlement in Caria, Lycia, and Greater Phrygia', in Kuhrt and Sancisi-Weerdenburg (1991) 84–143.

Sekunda, N. (1992) *The Achaemenid Army 560–330 BC*. London.

Shahbazi, A.S. (1986) 'Army I. Pre-Islamic Iran'. *Encyclopaedia Iranica* II/5: 489–99, available online at http://www.iranicaonline.org/articles/army-i (accessed on 22 July, 2020).

Steinkeller, P. and Hudson, M. (eds) (2015) *Labor in the Ancient World*. Dresden.

Stolper, M.W. (2017) 'Investigating Irregularities at Persepolis', in Jacobs, Henkelman, and Stolper (2017) 741–823.

Stronach, D. (2011) 'Court Dress and Riding Dress at Persepolis', in Álvarez-Mon and Garrison (2011) 475–87.

Summerer, L., Ivantchik, A., and von Kienlin, A. (eds) (2016) *Kelainai-Apameia Kibotos: Développement urbain dans le contexte anatolienne*. Bordeaux.

Tallis, N. (2005) 'Transport and Warfare', in Curtis and Tallis (2005) 210–35.

Tolini, G. (2011) 'La Babylonie et Iran: les relations d'une province avec le coeur de l'empire achéménide (529–331 av. J.-C.)', PhD Dissertation, Université de Paris I.

Tuplin, C. (1987) 'Xenophon and the Garrisons of the Persian Empire'. *AMI* 20: 167–245.

Tuplin, C. (1997) 'Achaimenid Arithmetic: Numerical Problems in Persian History'. *Topoi Orient-Occident Supplément 1: Recherches récentes sur l'empire achéménide*: 365–421.

Tuplin, C. (2008) 'Taxation and Death: Certainties in the Fortification Archive?', in Briant, Henkelman, and Stolper (2008) 317–86.

Tuplin, C. (2010) 'All the King's Horse: In Search of Achaemenid Persian Cavalry', in Fagan and Trundle (2010) 101–82.

Tuplin, C. (2014) 'From Arshama to Alexander: Reflections on Persian Responses to Attack', in Gaspa, Greco, Bonacossi, Ponchia, and Rollinger (2014) 668–96.

Tuplin, C. (2016) 'The Persian Military Establishment in Western Anatolia', in Summerer, Ivantchik, and von Kienlin (2016) 15–27.

Tuplin, C. (2017) 'War and Peace in Achaemenid Imperial Ideology'. *Electrum* 24: 31–54.

Tuplin, C. (2020a) 'The Military Environment of Achaemenid Egypt', in Tuplin and Ma (2020b) 291–328.

Tuplin, C. (2020b) 'Sigillography and Soldiers: Cataloguing Military Activity on Achaemenid Period Seals', in Dusinberre, Garrison, and Henkelman (2020) 329–459.

Tuplin, C. (2021) 'Mercenaries', in Jacobs and Rollinger (2021) 1183–95.

Tuplin, C. and Ma, J. (eds) (2020a) *Aršāma and his World: The Bodleian Letters in Context. Volume II:* Bullae *and* Seals. Oxford.

Tuplin, C. and Ma, J. (eds) (2020b) *Aršāma and his World: The Bodleian Letters in Context. Volume III: Aršāma's World.* Oxford.

Tuplin, C. and Jacobs, B. (2021) 'Military Organisation and Equipment', in Jacobs and Rollinger (2021) 1161–81.

Vidal, J. (ed) (2010) *Studies on War in the Ancient Near East: Collected Essays on Military History*. Münster.

Wallinga, H.T. (1987) 'The Ancient Persian Navy and its Predecessors', in Sancisi-Weerdenburg (1987) 47–77.

Wallinga, H.T. (1991) 'Naval Installations in Cilicia Pedias: The Defence of the *Parathalassia* in Achaimenid Times and After'. *Anatolia Antiqua* 1: 277–82.

Wallinga, H.T. (2005) *Xerxes' Greek Adventure: The Naval Perspective*. Leiden.

White, J. (2020) 'Bows and Spears in Achaemenid Persia', PhD Dissertation, University of California Santa Barbara.

Zaia, S. (2022) 'Protecting the King in Mesopotamia in the First Millennium BCE: Perspectives from the Neo-Assyrian and Neo-Babylonian Empires', in Hebblewhite and Whately (2022) 31–58.

CHAPTER 6

The Persian Way of War: Infantry Tactics in the Achaemenid Empire

Jeffrey Rop

One of the most enduring explanations for the rise and fall of the Achaemenid Empire is premised on the idea that the Persian infantry was primarily composed of lightly equipped archers. This view asserts that the Persians avoided direct, hand-to-hand infantry battle whenever possible. Instead, they preferred to break enemy lines with arrows from massed bowmen firing from behind a single rank of spearmen carrying wicker shields, a formation commonly known today as the **sparabara*.[1] The Persian approach was sufficient to conquer the major powers of West Asia and Egypt, but floundered against Greek heavy infantry hoplites, whose superior armaments and disciplined phalanx formation allowed them to withstand Persian missile barrages from a distance and to annihilate the comparatively unarmoured Persian infantry at close range.[2] While a few scholars have challenged this view of Persian infantry combat,[3] it persists in both popular and academic histories: the superior equipment and tactics of the Greek hoplite and the Macedonian sarissa phalanx it inspired is seen as a major reason why the Persian invasions of Greece failed and Alexander's conquest of the Achaemenid Empire succeeded.[4]

1 *Sparabara* literally means 'shield-bearer', but is commonly used to refer to pair of soldiers consisting of a shield-bearer and an archer or to designate an infantry formation consisting of a single row of shield-bearers in front of several ranks of archers.

2 On the Persian preference for the *sparabara* formation, see Head (1992) 22–30 (depicted in fig. 11); Sekunda (1992) 16–18; Briant (2002) 537; Tallis (2005) 216–17; Fink (2014) 29; Rahe (2015) 151–52; Bardunias and Ray (2016) 9; Wrightson (2018) 70–72; Manning (2021) 303–307; Tuplin and Jacobs (2021) 1173–74. On its ineffectiveness against Greek hoplites, Rahe (1980) 79–82; Hanson (1989) 30; Briant (2002) 538; Cawkwell (2005) 251; Krentz (2010) 26–27; Butera and Sears (2019) 98–100.

3 Hyland (2011) 272–75; Charles (2012a); Konijnendijk (2012); White (2020) 163–66.

4 On the Macedonian adoption of military equipment and tactics from Greece, in particular from Thebes, see Markle (1978) 486–88; Borza (1990) 203–206; Worthington (2014) 32–36. On the use and improvement of these tactics to defeat the Persians, Hanson (1989) 222–27, (2001) 74–79, (2008) 34–43; Lendon (2005) 115–39. Ferrill (1985, 149–80) argued that Philip II 'wedded the best of the Greek and ancient Near Eastern military traditions' by combining Greek heavy infantry with Persian light infantry and cavalry.

 | DOI:10.1163/9789004710771_007

This chapter offers a revisionist history of the Persian royal infantry of the Achaemenid Empire.[5] The first half examines the written and archaeological evidence for their equipment and tactics through the invasion of Greece under King Xerxes in 480–479. I argue that the foot soldiers of the standing Persian army were primarily spearmen who were equipped and trained to fight as both hand-to-hand heavy infantry and light infantry archers, that the famous *sparabara* was a specialised rather than a preferred tactical formation, and that Persian spearmen were far more effective in close-quarters combat with Greek hoplites than is generally recognised. The second half addresses several misconceptions regarding Persian royal infantry from the aftermath of the Greco-Persian Wars to the Macedonian invasion of 334–330. I make the case that cross-training and equipping soldiers to fight as both heavy and light infantry was a Persian innovation that persisted to the end of the empire, and that the Greek experience of fighting both against and alongside the Persian royal infantry spurred the development of the hoplite phalanx and led to the invention of the Macedonian sarissa. In the end, I conclude that the adoption and refinement of the Persian approach to infantry warfare was a crucial factor in the Macedonian conquests of Greece and the Achaemenid Empire.

1 Persian Infantry Equipment and Tactics in the Greco-Persian Wars

Modern discussions of the equipment and tactics of Achaemenid Persian infantry typically adopt the perspective of ancient Greek authors. The Athenian tragedian and military veteran Aeschylus used the 'bow' and the 'spear' as a shorthand way to contrast Persian and Greek approaches to war in his *Persai*, written shortly after the Battle of Salamis of 480 BCE.[6] The late-fifth-century historian Herodotus of Halicarnassus reinforced this dichotomy in his *Histories*.

5 By 'royal' infantry, I mean the standing foot soldiers of the imperial army typically stationed in the core (see also Hyland's chapter in this volume). This includes the regular infantry, most famously known as the 10,000-strong Immortals but later referred to as *kardakes* (see discussion below), as well as the one or two 1,000-strong elite guards attached to the king. Although these forces may have included non-Persians such as Medes, for the sake of convenience I will simply refer to them as Persians. I do not include in this term conscripted levies drawn from Persians or imperial subjects, standing forces attached to satraps throughout the empire, or soldiers recruited from clients in areas like Greece that were not directly ruled by Persian officials.

6 E.g., '[Xerxes] leads against a people renowned for the spear (δουρικλύτοις) a warlike host of archers (τοξόδαμνον)' (85). Trans. Smyth. For one modern formulation, Cawkwell (2005, 251), 'Plataea was decisive in Persian military development, for it established the superiority of the Greek hoplite, of the spear over the bow, of Greek protective armor over the comparatively

He makes Aristagoras of Miletos deceptively dismiss the Persians as easy to defeat because of their short spears, bows, and lack of armour (5.49.3–4).[7] Elsewhere, Herodotus describes an early Persian battle against a Saka group called the Massagetae as beginning with an exchange of arrows followed by hand-to-hand combat only once both sides had spent all of their missiles (1.214). These earlier descriptions prepare the reader to accept the surprise of the Persians at the unexpected charge of the Athenians without archery support at Marathon (6.112), the ineffectiveness of the Persian short spears in close quarters fighting at Thermopylai (7.211), and the failure of the Persian *sparabara* formation at Plataia, where Herodotus adds that the Spartans' armour gave them a decisive advantage once the fighting moved to close quarters (9.62).

Taking Herodotus' descriptions at face value, until recently few scholars questioned the Greek characterisation of Persian infantry as primarily archers who were vulnerable to heavy infantry. Yet claims that the bow was the Persian weapon of choice are flatly contradicted by a closer investigation of the available Achaemenid and Greek evidence.[8] Early depictions and descriptions of Persian foot soldiers strongly associate their elite and regular corps with the spear, revealing the bow and different types of shields to be secondary armaments. There is no evidence in Persian sources to suggest that the *sparabara* was a standard tactical formation in open field battle, and Herodotus' battle narratives likewise reveal that it was used only under specific, siege-like conditions. The core of the Persian infantry was equipped and trained for flexibility, prepared both to engage enemies from a distance and to fight in close quarters, in the latter situation with underrated effectiveness against Greek hoplites.

1.1 *Early Descriptions and Depictions of Persian Infantry*

The most direct written description of Persian infantry equipment comes from Herodotus' *Histories*. In his account of the parade of soldiers who accompanied Xerxes during the invasion of Greece in 480 BCE, he writes that the Persians wore fabric caps, tunics with sleeves over iron scale armour, and trousers. They carried shields made of wicker (γέρρα), bows and arrows in a quiver, short spears (αἰχμὰς βραχέας),[9] and daggers (7.61). Notably, Herodotus' descriptions

ill-protected Persian'. For discussion of the role of the bow in Achaemenid royal ideology, see Rung in this volume.

7 As Konijnendijk (2016, 8–9) observes, Herodotus does not agree with Aristagoras' claim and is purposefully showing him to be ignorant or deceptive in much the same way he does with the Persian general Mardonios' characterisation of Greek warfare to Xerxes (7.9B1).

8 Hyland (2011, 272–75) and White (2020) anticipate and inform much of my own analysis in this chapter.

9 The controversy over the length of Persian spears is addressed in the following section.

of the 10,000-strong Immortals and the two 1,000-strong elite guard divisions emphasises their role as spearmen, since it was the adornment of their spears that distinguished them in status from the levied troops (7.41).[10]

While Herodotus observes that the Persian infantry fired arrows at Thermopylai (7.218.3, 225) and Plataia (9.61.3), outside of the speech of Aristagoras – which, again, is meant to be deceptive – he never calls the Persians primarily archers. His parade description of the Persian infantry presents them as equipped for both ranged and close quarters combat, and his battle narratives indicate a preference for the latter. The Persian infantry engage in hand-to-hand combat willingly in each of the four land engagements of the Greco-Persian Wars, but Persian archers do not appear at all at Marathon or Mykale.[11]

The sculptural reliefs of the Persian soldiers at the Achaemenid capitals of Susa and Persepolis give the same impression. In them, every single infantryman carries a spear.[12] At Susa, they have bows and quivers slung over their shoulders while carrying spears, which are decorated with a white glazed round butt reminiscent of the silver apples on the spears Herodotus says were carried by most of the Immortals.[13] The reliefs at Persepolis are greater in both number and variety, but the spear remains overwhelmingly predominant. Often the spear is their only weapon.[14] For example, 141 of the Persian soldiers carry only a spear on the east side of the entrance to the Apadana, the royal audience hall, another 132 carry a shield and a spear, and 169 carry a spear and a bow with quiver.[15]

Two types of shield appear at Persepolis. The most common is similar to the so-called Boeotian shield: oval-shaped, running from the chin to the thigh just above the knee, with a boss in the centre and two small holes on the edges to its left and right.[16] The boss of these shields was probably metal, and the rest

10 Charles (2011) 119.

11 Herodotus' comment that the Persians were surprised by the Athenian lack of archers is addressed below.

12 I include here all of the guard figures, some of whom wear traditional 'Median' or Elamite garments, since they appear interchangeably with the soldiers in what is more conventionally identified as Persian garb. It should be noted that the Median label itself reflects conjecture by the early archaeologists, in light of the lack of identifiable figural art from Media. Muscarella (1987). On the tradition of Elamite archery and its continuation by the Persians, see Álvarez-Mon and Wicks in this volume.

13 Muscarella, Caubet, and Tallon (1992) 226–28.

14 E.g., Schmidt (1953) pl. 50, 51, 58, 59.

15 Figures and description of the Apadana relief drawn from Roaf (1983) 30. Some of the guards occasionally also have a dagger or sword (e.g., Schmidt (1953) pl. 65).

16 E.g., Schmidt (1953) pl. 63. For the Boeotian shield, Rawlings (2007) 57. Echoing Schmidt (1953, n. 95), Head (1992, 22) reasons that because 'the shield-face, both inside and out,

of the shield was probably made of wood or woven strips of wood covered in hide. Soldiers with these shields appear in the palace reliefs at Persepolis in numerous prominent places, such as the central relief of the staircase into the Apadana. The second type of shield is rectangular and tall, stretching from the nose to the ground, with a striped surface pattern indicating it was probably made of wooden staves woven together by rawhide.[17] Soldiers with this shield appear on the doorjambs of entrances at the Apadana, the palace of Darius, and the palace of Xerxes.[18] In each case, the spearman carrying a rectangular shield is standing in front of a soldier bearing only a spear, and never with a bow and quiver.

Similarly, the Persepolis displays never show an oval shield-bearing spearman directly in front of a soldier with a bow and quiver. In fact, the other two types of soldier – the spearman and the shielded spearman – appear in locations that seem to indicate greater importance. For instance, the central reliefs on the façade of the Apadana show four pairs of two soldiers; in each pair, the first carries an oval shield and spear, and the second only a spear.[19] The soldiers with bows and spears are the smallest in size and positioned along the stairways,[20] whereas the spearmen with shields occupy the more prominent, central locations. In the upper register, the reliefs read like a sensible battle formation: shielded spearmen in the centre, flanked on each side by spearmen, with bow-carrying spearmen on the furthest wings.[21]

The Persepolis reliefs that come closest to showing a *sparabara* formation are found on the eastern and western door jambs of the eastern doorway into the Throne Hall at Persepolis.[22] The sculptures on these jambs are divided into six registers, with the top showing the king seated on his throne and the rest each depicting ten soldiers facing each other, five per side. In the top row of soldiers, two spearmen backed by three shielded spearmen face each other. In the bottom four rows, spearmen with bows and quivers face each other in alternating

is shown as smooth … it is not made of wicker, but of leather or leather-covered wood'. White (2020, 139–42) notes other instances of smaller, rounded *gerrha* of woven construction in Neo-Assyrian reliefs and in Xenophon's *Cyropaedia*.

17 White (2020) 144–49.

18 Schmidt (1953) pl. 94, 95, 136, 137, 151, 176, 177. White (2020, 141) also notes a depiction at the 'Harem', but I have not been able to find this image.

19 Schmidt (1953) pl. 53. Roaf (1983, 29) notes that the original scene here was of 'the king seated on his throne surrounded by his officials', but this was moved to the Treasury at some point and replaced with the alternating guards.

20 E.g., Schmidt (1953) pl. 56A and 56C.

21 Roaf (1983) 30. Neo-Assyrian military reliefs also appear to have been designed to reflect tactical formations and maneuvers; see Nadali (2005) and (2010); De Backer (2007).

22 Schmidt (1953) pl. 100 and 101.

styles of dress. Duncan Head suggested that these reliefs were intended to show a *sparabara* unit in action, with the top row of soldiers representing the front rank of an army and the rest showing the spearmen-archers stationed to their rear. Yet this is not quite the case: the central-most spearmen in the top row do not have shields, and the bottom rows are spearmen with bows – indicating, at least, that their primary purpose was not necessarily to fire arrows. Most importantly, the shields that appear are oval-shaped rather than the rectangular tower shields associated with the formation.[23] If these images do reflect a tactical reality, they indicate that the Persian infantry were cross trained to fight as both spearmen in close quarters and bowmen from a distance.

We must of course take care to avoid over-interpreting ancient art.[24] The artists who created these figures were not necessarily motivated to present a realistic portrait of Persian soldiers, and were undoubtedly informed by practical and ideological considerations. The smaller oval shields could appear more often because they were in fact more commonly used, or because the soldiers who used them were of a higher status, or because sculpting the larger rectangular shields was more difficult and so left to larger spaces like doorjambs rather than the more confined registers. Likewise, the decision to show only spearmen-archers at the palace of Darius in Susa must be for some ideological or artistic reason rather than a literal reflection of military reality, as the Apadana complex and Darius' palace at Persepolis contain soldiers with shields and spears and spears alone in roughly equal if not greater number to those with spears and bows.

Nonetheless, it is significant that the sculptural programs at Susa and Persepolis do not depict Persian elite soldiers as archers only or in anything that is unambiguously reflective of the *sparabara* formation. As far as I am aware, there is no practical limitation that would have prevented the sculptors from doing so, which means that the decision was based on contemporary military practice or ideological preference. If anything, the ease of depicting small, compact bows is reason to suspect that they may be overrepresented in the reliefs. At least, then, we can conclude that the Achaemenid elite did

23 Head (1992) 27. Overlooking the first two complications, Head dismissed the third by speculating that the oval shields were merely ceremonial and never used in combat. He does not provide a reason for rejecting the utility of the oval shields, giving the appearance that his dismissal of the oval shields is based solely on the fact that it does not fit preconceived notions about how Persian infantry fought. Still, it should be acknowledged that similar, unfounded dismissals (e.g., Snodgrass (1967, 55)) have been made regarding the similarly-shaped Boeotian shield's historicity in ancient Greece. See Rawlings (2007, 57) for discussion and further references.

24 Echeverría (2015) offers useful comments on this thorny issue.

not consider the *sparabara* in any sense – the spearman with the tower rectangular shield, the shieldman-archer unit, or the broader tactical formation of a single row of shieldmen in front of ranks of archers – as central to their military identity. The same pattern appears to hold for depictions of Persians fighting on seals, gems, and other artwork, from which the *sparabara* tactical combination in all forms is also entirely absent.[25]

The limited selection of Greek artwork depicting Persians close to the period of the invasions provides a similar view of the variety of their infantry equipment. They are shown carrying combinations of bows and swords,[26] shields and spears,[27] and, unlike in Persian art, shields and bows.[28] Additionally, Greek paintings depict Persians with bows and axes,[29] and sometimes wearing what appears to be scale armour.[30] The shields in art closest to the invasions are nearly all rectangular tower shields with patterning that indicates they were of woven construction.[31] The bow is certainly more prominent among Persian soldiers than the Greeks, but it is one among many weapons they carry.[32] Greek artists present Persian infantry as versatile adversaries able to fight both at close range and a distance.

The *sparabara* shield-bearer and archer pair and larger formation is also absent from Greek depictions of Persian infantry. The Basseggio Cup, which shows several separate scenes of fighting between Greek hoplites and Persians, probably comes the closest.[33] In one place, a single Persian soldier swings a sword in one hand and holds a bow in another while standing behind a tower

25 No such images are shown in Head (1992) Sekunda (1992) or in Tuplin (2020), which offers the most complete survey of Achaemenid military imagery on seals available.

26 Bovon (1963) fig. 7 (with shield), fig. 13; Head (1992) fig. 9c (with shield), 12a, 12c, 12f, 12g.

27 Bovon (1963) fig. 2; Head (1992) fig. 9b.

28 Also with a sword. Bovon (1963) fig. 7; Head (1992) fig. 9c.

29 Head (1992) fig. 9c and 12h.

30 Head (1992) fig. 9b; for identification of the armour as a scale cuirass, see Barrett and Vickers (1978) 22.

31 Bovon (1963) fig. 14; Head (1992) fig. 9d. Later Greek art shows Persians with rounded crescent *peltē* style shield (discussed below); the oval, Boeotian style shield from the Persepolis reliefs is absent.

32 Head (1992, fig. 12b) has one figure armed solely with a bow, while fig. 12d has only a bow as well, but the figure, which appears on the famed Eurymedon vase, has his arms raised in a (mocking?) surrender-like pose. Fig. 12c is missing his right hand, which is raised and likely bears another weapon or shield.

33 Bovon (1963) fig. 7. Head (1992, fig. 9c) and Sekunda (1992, 23) include clearer images, but do not show the entire scene. The cup also has two other scenes of Persians and Greeks in combat; none of the four Persians shown has any type of shield. Bovon (1963, fig. 2) depicts a spearman with a tower shield, i.e., a literal *sparabara*, but he is not accompanied by an archer or archers.

shield affixed to the ground. Elsewhere on the same register is a pair of Persian soldiers: the first stands, firing a bow in one hand while holding an axe in the other, the second crouches, swinging his sword while holding his bow in his other hand. Neither has a shield, and there is no tower shield stationed close enough to be associated with them. Thus the cup contains all of the individual elements of a *sparabara* unit – a tower shield, swordsman (rather than spearman), archers, and a swordsman/archer pair – but they are not depicted in a combination that represents it, that is, with a shieldman protecting an archer.

In brief, the sculptural reliefs from the Achaemenid imperial core indicate that the royal Persian infantry were spearmen who were also trained to use shields and bows. Herodotus' descriptions of Persian equipment reinforce this image, and Greek artistic depictions of Persian foot soldiers similarly show them wielding a wide array of hand-to-hand weapons along with bows and the occasional large rectangular shield. Based on this evidence, there is no reason to doubt that the Persians could have deployed their infantry in shieldbearer/archer pairs or in the larger *sparabara* formation with a front rank of shieldmen protecting several rows of archers. At the same time, these descriptions and depictions give no reason to believe that this was a favoured or even common tactic. If anything, the prevalence of spearman and of shielded spearman pairs at Persepolis indicates an ideological preference for close range fighting, a conclusion that is reinforced by examining the descriptions of Persian battle tactics in the *Histories*.

1.2 *Persian Infantry Tactics and Effectiveness during the Invasions of Greece*

Despite its prominence in modern literature as a standard Persian tactic, Herodotus' account of the Battle of Plataia contains the only historical description of the *sparabara* formation in action.[34] In this engagement, the Persian royal infantry shot arrows from behind a fence of shields (9.60–61), and then put away their bows and fought at close quarters along the wall of shields when the Spartans and Tegeans charged (9.62). Herodotus reports a similar tactic in his account of the Battle of Mykale, where the Persians fought against the Athenians and their allies at a barricade made of their shields, but here he makes no mention of arrows at all (9.99). Still, these two reports along with the visual representations of tower shields in Persian and Greek art mentioned in the foregoing section provide sufficient evidence to state with confidence that 1) the Persians did at least sometimes use their tower shields to create fences, and 2) that they could and did fight with both spears and bows from behind them.

34 On the possible reports of the *sparabara* tactic in Xenophon's *Cyropaedia* and *Anabasis*, see discussion below.

What the available evidence does not support is the widespread belief that fighting from a distance in this formation was a common or preferred method of combat for the Persian royal infantry. Herodotus makes no mention of the *sparabara* tactic or the presence of both wicker shield-bearers and archers at all in the other two major battles of the Persian invasions, Marathon and Thermopylai. The famous charge of the Greeks at Marathon is often reasonably thought to have been made to avoid Persian arrows, but the use of ranged weapons and shield walls is not mentioned in the battle narrative (6.112–114).[35] Similarly, at Thermopylai Persian shields and archers are absent from the fighting at the Gates; they appear only to warn the Phocians off the Anopaia path (7.218.3), in an anecdotal aside about Persian arrows providing shade for the Greek defenders (7.226), and in the description of the final slaughter of the Greeks on the Colonus hill (7.225). Herodotus' description of the battle otherwise places great stress on the shorter Persian spears and their over-eagerness to pursue the Spartans during their feigned retreats (7.211).

While the Persian infantry were equipped and prepared to fight from a distance, the weight of the evidence from the two Persian invasions of Greece indicates that their preferred tactic was to engage in decisive, hand-to-hand combat.[36] All three major battles on the Greek mainland involved the Greeks seizing defensible positions and refusing for days to march into battle.[37] At Marathon, the Athenians held the southern pass and refused to enter the plain for days, waiting to charge until they had the advantage of surprise (6.109–111).[38] At Thermopylai, the Spartan-led defenders held a narrow, fortified position where they received Persian assaults and did not venture forth until learning that their position had been turned (7.210–212, 223). At Plataia, they first camped in the foothills of Cithaeron (9.19), then moved further down nearer to the plain to a position that was still in the lower foothills (9.25). Although the Persians harassed the Greeks in both of their encampments (9.20, 40, 49) and presented for battle by marching to the Asopos River at the centre of the plain (9.31),[39] the Greeks refused to descend to meet them (9.36). Ultimately

35 Krentz (2010, 152) suggests that the Greek charge was made to prevent the Persian cavalry from deploying, but had the added benefit of reducing the harm of Persian arrows as they came into range.

36 On the Persian preference for pitched battle in Herodotus and the reticence for the same, respectively, see Konijnendijk (2016); Konijnendijk (2021).

37 Rawlings (2007) 88–89.

38 Krentz (2010) 118–21, 142–52.

39 The Greeks' inability to draw water from the Asopus river in the plain for fear of Persian attacks reveals that the Persians controlled the plain (9.49), and Herodotus claims that Mardonios' attempt to goad the Spartans into agreeing to fight was ignored (9.48).

it was the Persians who crossed the river and attacked upon seeing the Greeks attempting a disorganised retreat to higher ground (9.58–60).

The practice of planting a shield wall in the ground was only utilised under specific circumstances that did not favour direct assault during pitched battle.[40] The Persians were on the defensive at Mykale and, given that they faced both frontal and flanking attacks (9.102.1), perhaps outnumbered. They constructed a makeshift wall around their ships (9.96–97), and their infantry additionally used their shields to form an exterior perimeter outside of that camp wall (9.99). At Plataia, they used their shields as a counter-fortification opposite the Spartans and Tegeans, who had taken up a defensible position on broken ground in the lower slopes of Cithaeron (9.56.2), near a stream and a shrine of Demeter (9.57.2).[41] Rather than conduct a direct assault against an entrenched foe, they fired arrows on the defenders to draw them out and then engaged in close quarters. In this approach the Persians were likely carrying on a tactic popularised by the Neo-Assyrians, whose reliefs most frequently depict *sparabara*-like shield-bearer and archer pairs in siege contexts.[42]

Much has been made of Herodotus' assertion that that the Persians were like naked (γυμνῆτες) men when facing the Greeks at close quarters (9.63.2), but in fact this appears to mean only that they did not carry large, round, wooden Greek-style *aspides* shields. The importance of this specific type of shield was so central to the identity of a hoplite in the days of Herodotus that for many Greeks not having one was akin to not having any protection at all.[43] As we have seen, however, the Persian infantry did carry shields of a different type: at a minimum, some carried *gerrha* of the tall, rectangular tower shape that could be planted in the ground. Furthermore, Herodotus describes all of the Persian soldiers as carrying *gerrha* in his review of Xerxes' troops. While we should be cautious about accepting his description at face value, it is noteworthy that nothing he writes precludes the possibility that many of the Persian Immortals and elite guards, especially those not arrayed in the front ranks, carried smaller, oval shaped *gerrha* like those depicted on the reliefs at Persepolis.

---

40 Thermopylai could also be likened to a siege; Herodotus makes no mention of Persian shield walls, but van Wees (2018, 32–37) makes a reasonable argument for their use against the Phocian Wall. This may be due to the availability of flanking attacks against the Greek position on sea and land, the latter of which succeeded and made attempts to engage directly at the Phocian Wall unnecessary. For the Persian plan of attack, see Rop (2019b).

41 Plut. *Arist.* 11.6–7 identifies this location as near the village of Hysiae and offering excellent natural protection from cavalry.

42 Barron (2010) 128–34; Dezső (2012) 83.

43 Van Wees (2004) 48; Rawlings (2007) 47–48; Konijnendijk (2012) 5–6.

The quality of these shields has been greatly underrated in modern scholarship. While *gerrha* is usually rendered into English as 'wicker', primarily because the shields were made of woven strips of wood and hide, one recent experimental recreation demonstrated that this translation does not adequately convey the composition of these shields.[44] The wooden staves used were much thicker than the type of material used in modern wicker furniture sets, and the hides that bound them together tightly created a sturdy mass that remained light and easy to manipulate. While not as solid or heavy as an *aspis*, this type of shield – whether in tower or oval shape – offered its own advantages in terms of lightness and manoeuvrability, giving reason to believe that a soldier using one was far from helpless against a hoplite. Indeed, Herodotus never directly states that the *gerrha* themselves were incapable of stopping Greek spear thrusts or were otherwise ineffective against hoplites.

There is also much more to close-quarters combat than shields. For protection, Herodotus reports that the Persian infantry wore iron armour that looked like fish scales (7.61),[45] and notes that the scale armour of the Persian cavalry commander Masistios was so effective that he could only be killed with a blow to the eye (9.22.2). Archaeological finds from this period reveal that the use of this type of scale armour was widespread both in the Achaemenid Empire and in Greece itself; perhaps not coincidentally, a number of Attic vase paintings dating to the 470s show hoplites in scale armour.[46] Most importantly, Herodotus' claims that the Persians were at a disadvantage due to their shorter spears has not withstood closer scrutiny. There is no evidence that the length of Greek hoplite spears was standardised by the time of the Persian invasions, and neither Persian nor Greek iconography confirms that one side used a longer spear than the other.[47]

44 White (2020) 144–56.

45 Charles (2012a) 260.

46 De Backer (2012). Konijnendijk (2012, 4) also observes that the Greeks were no more likely than the Persians to have been fully armoured, as 'the argument that the finest equipment would have only been available to the wealthy few applies to both sides'. In fact, given the wealth disparities between Persians and Greeks at this time, it seems more probable that the professional Immortals and elite guards of the royal infantry were uniformly equipped with body armour than their Greek militia adversaries. In addition to their comparatively vast wealth, the Persians possessed administrative organisation and palace arsenals capable of large-scale production and storage on a scale unimaginable for the archaic *poleis* of Greece. For a recent survey of the surviving material evidence of military equipment throughout the empire, see Manning (2021) 223–59. Hyland's chapter in this volume also addresses the logistical capabilities of the Persians.

47 Hyland (2011) 273; Konijnendijk (2012) 5; Manning (2021) 280–82. White (2020, 55–91) offers the most extensive survey of Persian spears to date, concluding that 'despite what

The historical record of Persian infantry in combat likewise gives no indication of inferiority to the Greek hoplite. Prior to Xerxes' invasion of Greece, Persian forces defeated armies with Greek and Carian hoplites in Egypt at the Battle of Pelusium (3.11), during the conquest of Ionia (1.169), and during the suppression of the Ionian Revolt (e.g., 5.102, 120).[48] Despite losing at the Battle of Marathon in 490, Herodotus' own account of the fighting acknowledges that the battle was won on the wings first, where the Persians had stationed their unidentified subject-allies, probably including Ionian Greeks.[49] The Persian infantry in the centre drove back the Athenians who charged them in the first phase of the fighting, and were only defeated when their exposed flanks were attacked in the second phase. Herodotus makes no comment on the inferior weapons or equipment of the Greek infantry in the centre, attributing their setback instead to the thinness of the line there and, in turn, their victory on the wings to their greater depth (6.112–113).

The Persian infantry performed reasonably well against Greek hoplites during the invasion of Xerxes, too. At Thermopylai they used a multi-pronged attack plan to outflank and annihilate their entrenched Greek foes in a matter of days. Here their versatility was on full display, as they used arrows to warn the Phocians off the Anopaia path, defeated the Greek soldiers at the Gates in hand-to-hand combat, and then once again used their ranged weapons to finish off the final group of defenders once they had retreated to a nearby hill.[50] Herodotus' claim that the Persians suffered heavy casualties over three days of fighting is belied by his later report that visitors from the fleet witnessed only 1,000 Persian to 4,000 Greek dead during their tour of the battleground (8.24–25).

It is noteworthy that Thermopylai is the only land engagement of the Greek and Persian Wars at which the full Persian infantry was mustered. The Greek

many modern scholars and ancient authors believe, Persian spears were not much smaller or weaker than those of the Greeks' (91–92).

48 Since Herodotus does not provide tactical details of these battles, other than to say that the Persian cavalry struck the decisive blow at Malene (6.28–29), it is impossible to know how the Persian infantry specifically fared vis-à-vis the Greeks, other than to say that they were at least not so inferior as to make the combined Persian military suffer defeat. At the same time, it is telling that the only hoplite victory during the Ionian Revolt was accomplished by the Carians in an ambush (5.121).

49 For the possibility of Ionian Greeks in the Persian army at Marathon, see Krentz (2010) 140–42, 159–60; Hyland (2011) 275; Rop (2021) 163.

50 For a recent reconstruction of the Persian attack plan and the possibility of Phocian betrayal at Thermopylai, see Rop (2019b).

coalition refused to meet the Persian army in the open field during the first year of Xerxes' campaign, and the victories at Plataia and Mykale in 479 were against partial elements of the Persian ground forces.[51] Even so, the Persian infantry harried the Spartans at Plataia after their retreat to higher ground. They then resisted stoutly both at the shield fence and for some time after it fell all around the sanctuary, falling back only upon the death of their commander Mardonios (9.62–63). In the final stage of the fighting, the Persians withstood the Spartan assault against their walled camp until Athenian reinforcements arrived (9.70). Similarly, the Persian infantry at Mykale – drawn from a mixture of land and naval forces, and with Greek subject-conscripts of questionable loyalty – defended the Greek frontal assault at their temporary shield wall for some time before withdrawing into their fortified camp (9.102.2–3), where they resisted until the arrival of the Spartans on their flank (9.103.1).

All told, the record of Persian infantry against Greek hoplites through 479 BCE reveals more victories than defeats. While the Greeks certainly told themselves otherwise, even from the evidence included in the battle narratives of Herodotus it is not clear that the Persians faced a disadvantage in close quarters combat. Defeats can occur for many reasons other than tactics and equipment,[52] and tellingly the Greeks never met either the full or partial Persian royal infantry muster in the open environment that they preferred. In fact, Herodotus never reports that the Greeks faced either of the two elite 1,000-strong infantry guard units in combat during any of the land battles he describes. The strategic need for a decisive battlefield victory during Xerxes' invasion forced the Persians to attack in less than favourable circumstances at the Thermopylai pass and in the foothills of Plataia, but the flexibility of their equipment and training allowed them to adapt to these environments with different tactics including the *sparabara*. But while it was effective in forcing the Spartans and Tegeans out of their defensive position within the sanctuary of Demeter near Plataia, the *sparabara* was not the signature or standard Persian infantry formation. Much like the Roman *testudo*, it was a memorable and impressive tactic utilised only under specific conditions.

---

51 Xerxes was in Sardis at the time, presumably with both of his elite guard divisions (9.108), and the Immortals had reportedly remained with Mardonios in Greece, albeit without at least their commander Hydarnes (8.113).

52 Balcer (1995) explores the Persian military errors and logistical challenges that led to the failure of their invasions of Greece.

2 Continuities, Reforms, and Innovations in Persian Infantry Warfare

It is frequently stated that military defeats in Greece spurred the Persians to reform their infantry equipment, tactics, organisation, and recruitment strategies. The most common narrative goes like this: Persian infantry proved incapable of fighting in close quarters against the Greeks, which led them to abandon the Immortals and the *sparabara* formation and eventually in the fourth century to hire hoplites as mercenaries in order to compensate for their own heavy infantry deficiencies. These Greeks were the best soldiers in the Eastern Mediterranean and mainstays of Persian armies for nearly seventy years until Philip II of Macedon invented the sarissa pike phalanx. This innovation upon the hoplite phalanx empowered Philip to subdue Greece and his son Alexander to conquer the Achaemenid Empire.

Once again, however, the conventional view does not withstand scrutiny. Closer inspection of battles involving the Persian royal infantry in the late fifth and fourth centuries reveals that the same adaptable approach to equipment and tactics persisted to the end of the Achaemenid Empire. There is a stronger case to be made that it was the Greeks who changed their approach to infantry combat in the aftermath of the Persian invasions. Furthermore, the source of inspiration for Philip II's Macedonian sarissa phalanx is to be found in Greek participation in the Persian campaigns against Egypt in the fourth century. If there are any major innovations to infantry warfare attributable to the Persians, they represent a refinement in the approaches of earlier West Asian armies and took place before the invasions of Greece.

2.1 *Persian Infantry after the Invasions of Greece*

As we have seen, the ineffectiveness of the royal Persian infantry in Greece and its reliance on the *sparabara* tactical formation have been greatly overstated. I have made the case elsewhere against the longstanding view that the Persians depended upon superior Greek hoplites in the fourth century.[53] Two additional arguments that Persian infantry was reformed as a result of the failure of the invasions of Greece will be addressed in this section. One is that Persian soldiers in Greek art in the decades following the invasions are frequently shown carrying a new type of shield, which has been explained as an attempt to improve their ability to engage with more heavily armoured hoplites. Another is that later narratives of Persian royal infantry at the Battles of Cunaxa in 401, Issos in 333, and Gaugamela in 331 do not mention the Immortals or describe

53 Rop (2019a). See also Briant (2002) 783–800; Bushman (2017).

tactics that resemble the *sparabara*. To the contrary, the Hellenocentric search for Persian reforms in response to the defeat of three expeditionary armies on the imperial frontier has led many to overlook the evidence for continuity within the Persian infantry tradition.

In the period closer to the Persian invasions of 490 and 480–79, Persian soldiers are usually either shown unshielded or with larger, rectangular tower shields.[54] Starting roughly in the 460s, Greek depictions of Persians begin to incorporate a new crescent shaped shield, much like the *peltē* used by Greek light infantry peltasts. While peltasts traditionally fought with javelins, the Persian soldiers carrying these shields are depicted with many types of weapons, including spears, battle-axes, swords, and bows. Several of the figures using other weapons are shown with a bow and quiver attached at their hip. When firing arrows, the shield is strapped onto the back of the archer.[55]

One popular explanation for this new shield is that the Persians felt the need to better equip their infantry after their experience during the invasions of Greece.[56] This view ignores the numerous earlier Persian successes against Greek hoplites in Ionia and Egypt, and is based on the errant assumption that the Persians themselves recognised that infantry equipment and tactics were the reason for their defeats at Marathon, Plataia, and Mykale.[57] A more likely explanation for the change in depictions of Persians is that these images reflect the equipment of soldiers mustered from the western regions of the empire whom the Greeks fought in the decades after the invasion of Xerxes. There is no reason to believe or evidence to suggest that units from the standing Persian royal infantry remained in western Anatolia following the events of 479, and so logically it makes sense that Greek depictions over time changed to reflect the troops that they had first-hand experience fighting.[58]

Mainland Greek soldiers did not face elements of the Persian royal infantry for quite some time after the invasion of Xerxes. The next major land battle

---

54 Head (1992) fig. 23, 25.

55 Head (1992) fig. 26, 27.

56 Sekunda (1992) 18–19; Charles (2021) 176–77. Head (1992, 39–44) concludes, 'having shown their inability to stand up to the Greek hoplites, the *sparabara* may have begun to be replaced before the end of the Greek wars, and have disappeared soon after'. Cawkwell (2005, 251–52) suggests that, while the Persians reformed their infantry in response to hoplite superiority, their major innovations came in the realm of cavalry tactics.

57 Manning (2021, 310) accepts that Persian infantry combat changed from the fifth into the fourth centuries, but rightfully cautions against uncritically adopting the views of Greek sources or understanding any reforms as a response to the Greek wars.

58 Head (1992, 40) makes a similar observation, but only a few pages later (p. 44) oddly concludes that the Persian royal infantry must have made reforms due to their ineffectiveness in Greece. See also Sekunda (1992, 24).

after Mykale took place in the 460s at Eurymedon, where Athenian-led Delian League forces defeated what was probably a hastily mustered levy of local conscripts under the command of nearby satraps.[59] Our sources provide no details regarding the makeup of the two Persian armies dispatched to suppress the rebellion of Inaros in the 450s. It is not unreasonable to think that either or both included troops drawn from the Immortals, but at the same time the lack of a royal presence on the campaign must have prevented the full royal infantry from being mustered. Even so, it is not clear to what extent the Athenian allies of Inaros participated alongside Egyptian soldiers in the land battles that first defeated the Persian general Achaemenes and then were defeated by his replacement Megabyzos.[60]

The first confirmed open field battle involving both Persian royal infantry and Greek hoplites after the invasion of Xerxes took place near the village of Cunaxa not far from Babylon in 401. Xenophon offers an eyewitness account of the encounter (*An.* 1.8), but his description of the royal army has provoked a great deal of confusion. Gone are the Immortals and the king's elite guard, replaced with 'Egyptian hoplites' who appear neither to be from Egypt – which was in rebellion at the time – nor to be equipped with weapons and armour found in contemporary Egypt (1.8.9).[61] The size he gives for the royal army – 1.25 million with 900,000 present – is beyond plausibility, and the only numbered contingent is the 6,000-strong cavalry stationed in front of King Artaxerxes II's position (1.7.11–12). The composition of individual units is unclear otherwise, with no distinctions made between regular royal forces and militia levies or Persians and non-Persians other than the so-called Egyptians.

Similar challenges also appear in the next set-piece battles involving Persian royal infantry, which took place during the Macedonian conquest at Issos in 333 and Gaugamela in 331.[62] The Immortals are missing from narratives of these battles, replaced at Issos with the *kardakes*, whom ancient authors variously compare

59 Meyer (2018).

60 Thucydides (1.104–105, 109–110) does not report that the Athenians participated in any ground battles: he omits entirely the defeat of the first Persian response force under Achaemenes, writing only that the Athenians arrived after Inaros had secured Egypt, and adds that the successful Persian army under Megabyzos first defeated the Egyptian army before driving the Greeks from Memphis. Meanwhile, Diodorus (11.74–75, 77) claims that the Athenians helped defeat the ground forces of Achaemenes and then were prepared to do the same against Megabyzos until the Egyptians abandoned them.

61 Shannahan (2014) 60–68.

62 It is possible that royal infantry were among the reinforcements dispatched by the king to Tissaphernes against a Spartan army under Agesilaos in 395, but Persian infantry were absent from both small skirmishes against this Spartan army (Xen. *Hell.* 3.4.6, 11–24). For more on this campaign, see Rop (2018).

to heavy infantry hoplites and to more lightly equipped peltasts. Confoundingly, these *kardakes* are themselves absent two years later at Gaugamela, where the Persian infantry is once again not described in detail.[63]

Still, there are some strong hints at continuity from the fifth century. In his semi-fictionalised *Cyropaedia*, Xenophon reports that Cyrus II established an infantry corps of 10,000 soldiers that still existed and accompanied the king in the fourth century (7.5.68–70). While he does not identify this group with any specific name or title, the size and function imply an obvious connection to the Immortals of Herodotus. The Immortals do reappear in a highly literary parade description of King Darius III's army as it marched from Babylon toward Issos in Curtius Rufus (3.3.13). Finally, at Gaugamela Arrian mentions that King Darius III was guarded by the *mēlophoroi*, one of the two 1,000-strong elite guard units whom Herodotus identified as protecting Xerxes (3.11.5).[64] Attempts to explain the variations in the terms Greek and Latin writers use to identify the standing units of Persian royal infantry have resulted in frustration due to inconsistency and lack of evidence.[65] Nonetheless, it seems reasonable to believe that Persian kings from Xerxes onward at least always fielded 1) a larger standing infantry force akin to the Immortals, and 2) a smaller elite guard akin to the *mēlophoroi*. It should not be surprising if the sizes and names for these units changed over the centuries, or that Greek and Latin authors writing from the imperial periphery and/or centuries after the fall of the Achaemenid Empire would fail to provide consistent and precise details concerning their composition.

Focusing on the puzzling names and varying sizes of royal Persian infantry contingents over time has obscured that many continuities in equipment and tactics did persist from the invasions of Greece to the Macedonian conquest. The Persian foot soldiers opposite the Greeks at Cunaxa were equipped with *gerrha* shields (1.8.9), and Xenophon's comment that they fled before all but one of the Greeks entered range of their arrows indicates they also carried bows (1.8.19–20). Xenophon does not mention the shape of the *gerrha*, but his later distinction between them and the smaller *peltai* – often made in the same woven or wicker composition as tower *gerrha*[66] – that the Greeks scavenged from the battlefield may indicate that they were rectangular tower

63 Charles (2011, 124–30; 2021, 175–77) surveys the (lack of) evidence for the Immortals and other elite Persian infantry units during the invasion of Alexander.

64 Also in the centre according to Arrian were several other groups, including Carians, Babylonians, and Mardian archers (3.11).

65 Charles (2011) 133.

66 Best (1969) 3–5; van Wees (2004) 62.

shields (2.1.6). In that case, these *peltai* might reasonably be the very same oval Boeotian or dipylon style shields carried by most of the Persian guards on the reliefs at Persepolis.[67] If these soldiers were also equipped with spears, just as the Persian infantry of the early fifth century were,[68] then we have all the same elements carried by the Herodotean Immortals.

Tactically, the Persian infantry retreated before the charging Greek hoplites could reach them, and in fact even before all but one of the individual Greek soldiers came within range of their arrows. I have argued elsewhere that this retreat was a feint intended to draw the Greeks into a position that exposed their rear and flank, as Xenophon himself suggests the charge did (1.8.24).[69] Such a tactic required the discipline of well-trained professionals, and indeed Xenophon's observation that the Persians approached the Greek line marching in silence and in step (1.8.11), contrary to Cyrus' earlier prediction (1.7.4), might be a clue that they were arrayed against a division of the royal infantry rather than a motley collection of conscripts.[70]

While feigned withdrawal is not mentioned in the few extant battle descriptions of Persian infantry in action prior to Cunaxa, there is reason to think that Greek familiarity with and perhaps the expectation of the *sparabara* tactic played a role in making the manoeuvre a success. Consider that, as we have seen, the victories at Mykale, Plataia, and potentially Marathon involved a charge through arrows, a charge against a shield wall, or both. We might speculate then that the Greeks at Cunaxa anticipated that they would need to make a similar charge, a belief that would only be confirmed if the formation opposite them set up their *gerrha* as a fence and then fired a round of arrows as the Greeks approached. Once the ten thousand hoplites began their run, it would be almost impossible for their commanders to stop them even in the event that they did suspect the sudden, premature flight of the Persians was

67 Another possibility is that they were the shields carried by the peltasts on the Greek right wing (*An.* 1.8.5), although Xenophon claims they did not suffer harm during the charge of the Persian cavalry opposite them (1.10.7).

68 Note that Xenophon also does not mention the weapons carried by the Egyptian hoplites, which were undoubtedly thrusting spears or even perhaps pikes (see discussion below).

69 Rop (2019a) 52–63; Rop (2020). For the argument that it was not an intentional retreat, see Shannahan (2014) 68–76.

70 As Wrightson (2019, 174) notes regarding Philip's victory at Chaeronea, 'the pretend withdrawal is one of the most difficult maneuvers to accomplish in the chaos of an ancient battle and Philip's reliance on it when drawn up en echelon, and its perfect execution here, shows both his own tactical genius and the superb training of the Macedonian army'.

intentional.[71] This would explain why the Greeks collected arrows and *gerrha* along with other shields and wagons – but not spears – for firewood the next day (2.1.6), and why Xenophon, perhaps having encountered it in action, made his own Cyrus invent a *sparabara*-like formation in the *Cyropaedia* (6.3.23–27).

Just as it makes sense that King Artaxerxes II would entrust the challenging tactic of the feigned retreat to his Immortals or their equivalent in 401, so too is it logical to identify the densely packed infantry defending his own personal position in the centre as his elite royal guards (1.8.13). Xenophon does not use the same terminology as Herodotus and Arrian to describe these soldiers, but importantly his description is not mutually exclusive of theirs, either. As we have seen, these Egyptian hoplites (1.8.9) do not make sense as ethnic Egyptians; either Xenophon was simply confused or, as I believe, he identified these soldiers by their equipment. Much as 'Cretan archers' were a specialised, skirmishing type of light infantry who were not necessarily ethnic Cretans,[72] Xenophon's Egyptian hoplites may simply be any foot soldiers equipped with long spears or pikes. Notably, while he does not identify the weapons carried by the Egyptians in the *Anabasis*, in the *Cyropaedia* he includes Egyptian hoplites armed with the same kind of shield along with 'large spears, just like those they carry today' (δόρασι μεγάλοις, οἷάπερ καὶ νῦν ἔχουσι, 6.2.10; reiterated at 7.1.33).[73] This would not be the first time that the Persians had adopted military gear from the Egyptians; Herodotus claims Persian infantry had also borrowed the Egyptian thorax (1.135).

The wooden tower shields and pikes of the Egyptian-style infantry may represent an innovation in the Persian infantry since the fifth century. Herodotus and other early sources do not mention the use of either pikes or wooden tower shields, while the reliefs at Persepolis are inconclusive. They do show rectangular tower shields and spears, but in neither case can the composition or length be fully confirmed, although the patterning on the tower shields does suggest a woven rather than solid construction. Nonetheless, both implements have a long history predating the Achaemenid Empire in West Asia. Neo-Assyrian

71 Rop (2020, 160–62) discusses the lack of discipline in the Greek soldiers at Cunaxa, in contrast to a similar scenario in the *Cyropaedia*, evidenced by their wild charge and inability to halt.

72 Craven (2017) 43–46, 123–37.

73 Observed by Ueda-Sarson (2002) n. 13. Xenophon may not have felt it necessary to include their weapons in the *Anabasis* since it was the pikes that made them 'Egyptian' in his mind, but he did describe the tower shields because these were not. Shannahan (2014, 60–68) makes clear that the Egyptians of Xenophon's day had not used tower shields like this for centuries. The significance of Persians and Egyptians using pikes or long spears is addressed further in the following section.

infantry used many types of tower shields, including both wicker and wooden, and it should not be surprising to learn that the Persians might have carried on this tradition of adapting their military shields as circumstances required.[74] Likewise, one of the earliest depictions of military activity in ancient Sumer, on the so-called Vulture Stele of Lagash, includes a phalanx of pikemen marching behind rectangular tower shields that appear to be made of solid material.[75]

We cannot rule out the possibility that the Egyptian hoplite phalanx at Cunaxa represents a new Persian infantry unit or at least a new way of equipping the king's guards. However, neither the spears nor the shields of these soldiers were new to warfare in West Asia or Egypt, and, as we have seen, the Persians themselves were no strangers to heavy infantry tactics. We must remember that Greek sources only discuss a small number of the battles fought by Persian royal infantry in the history of the empire, and of these only a few include details regarding military equipment. When our sources mention the Persians using an item that has a long prior history in the ancient Near East, such as the wooden tower shield, it is by some degree far more likely to be a carryover not previously noticed by the Greeks rather than a Persian re-invention.

After the Egyptians at Cunaxa, the next Persian royal infantry whose equipment and tactics are described in Greek and Latin sources are the *kardakes*. Ancient authors variously associate them with light-infantry bowmen, javelin-throwing peltasts, heavy-infantry hoplites, and even slingers.[76] Confusing matters further is that the peltasts of fourth-century Greece – that is, soldiers who carried *peltai* shields – expanded from being lightly equipped javelineers to include more flexible infantry who could fight hand-to-hand with thrusting spears and, eventually under the Athenian general Iphikrates, with even longer pike-length spears. The *kardakes* at Issos are described by Arrian as hoplites arrayed in a phalanx (*Anab.* 2.8.6, 8), and accordingly they held the left centre of the royal battle line next to the Greek hoplites in the right centre.[77]

Michael Charles has reasonably speculated that the *kardakes* were 'general purpose infantry' equipped with thrusting spears and small shields in the manner of the better-known Iphikratean peltasts.[78] While he suggests that they occupied a lighter role than the Immortals of the fifth century, who were replaced by Greek heavy infantry hoplites, his description closely matches the

74 Dezső (2012) 89–93; see especially the lists of different types of regular Assyrian infantry and shield types under Ashurbanipal in fig. 3. See also White (2020) 132–44.

75 Nadali (2014) fig. 1.

76 For survey of the different sources and descriptions of the *kardakes*, see Charles (2012b) 8–16.

77 Charles (2012b) 12–14; (Rop 2019a) 215–16.

78 Charles (2012b) 16–18.

images of the spearmen with smaller oval-shaped shields on the sixth and fifth century Persepolis reliefs. Given these similarities, it seems overzealous to discredit the appearance of the Immortals in Curtius' description of the king's campaign preparations.

Additionally, Arrian's comment that the king's forces set up makeshift fortifications at some places along their line could be taken as another instance of Persian infantry entrenching themselves much as at Plataia and Mykale (2.10.1).[79] I previously suggested that the *kardakes*' position along the riverbank and use of fortifications may suggest that King Darius III considered them less suited to a heavy infantry role than his Greek hoplites. At the same time, it is also true that they were the only group that Alexander chose to avoid directly assaulting, taking them in both flanks instead.[80] This itself could be interpreted as an indication of their effectiveness: shielded spearmen stationed behind fortifications along a riverbank would have posed a formidable threat, especially if they were equipped with javelins and bows like the Persian infantry depicted in other sources and images.

The *kardakes* go unmentioned in the accounts of Gaugamela in 331. As mentioned above, however, the heavy infantry unit carrying spears with golden apples on their butt ends (*mēlophoroi*), earlier described by Herodotus, reappears in Arrian's description of the Persian line. They are stationed in the royal centre guarding King Darius III, with Greek hoplites on either side of them. Arrian also mentions a mixed contingent of Persian cavalry and foot soldiers – presumably equipped as light infantry runners – as part of the right wing, but otherwise the Persian army here is composed of contingents from all over the empire. Most of Arrian's narrative concerns cavalry fighting, irrelevant to our purposes here, and so we receive little tactical information on the Persian infantry other than that they were deployed as both heavy and light infantry.[81]

Organisational changes in the royal Persian infantry corps following the failed invasions of Greece in the early fifth century do not appear to have led to reforms in their armament and tactics. In the extremely limited sample available to us, the regular and elite Persian foot soldiers who served in the standing royal army were armed primarily with spears, bows, and *gerrha* shields

79 Although the specific word he uses to describe this stockade (χάρακα) is more associated with a stake palisade than a shield wall, whereas Herodotus uses different language to describe the shield walls – φράξαντες γὰρ τὰ γέρρα (9.61) and συνεφόρησαν τὰ γέρρα ἕρκος (9.99) – Arrian's wording does not rule out the possibility of shields forming part of these defences.

80 Rop (2019a) 214–20.

81 For a recent reconstruction of Gaugamela, see Rop (2019a) 220–26.

of rectangular and oval shapes. Like the troops at Plataia, their training and equipment enabled them to operate both in a heavy infantry phalanx formation and as light infantry skirmishers. The few innovations attributed to them during this period, supposedly in response to Greek hoplite superiority, were in fact part of a longer tradition of warfare in West Asia. The large wooden tower shield and two-handed spear predated the Achaemenid Empire by centuries if not millennia, and the crescent shaped *peltē* that appears in mid-fifth century Greek art reflects the traditional equipment of Persian subject forces in western Anatolia. The flexibility, training, and professionalism of the royal infantry was critical to the maintenance of the empire over several centuries. From that perspective, it should come as little surprise that they were effective in combat against Greek hoplites and were not overwhelmed by the Macedonian sarissa phalanx at either Issos or Gaugamela.

2.2 *Innovations in Persian and Greco-Macedonian Infantry Warfare*

In many respects, the Persians refined longstanding traditions of infantry warfare in ancient West Asia and, apart from possibly the scythed chariot, do not appear to have introduced new, innovative weapons or equipment.[82] Nonetheless, one Persian innovation may have been the cross-training and equipping of their royal infantry. The Neo-Assyrians seem to have typically deployed specialised units of heavy infantry, light infantry, shield bearers, and archers, and to this end Neo-Assyrian reliefs show only the king's mounted bodyguards carrying both bow and spear.[83] By contrast, the reliefs at Susa and Persepolis and Herodotus' *Histories* give strong evidence that this was common among the Persian royal infantry. With their scale body armour and many different types of shields, the Persian Immortals in particular were ready to face the enemy in hand-to-hand combat, to rain arrows upon any foe who refused to come into close quarters, and to engage in siege and counter-siege operations using tactics like the *sparabara*.

This versatility would have been advantageous not only because of the wide range of environments in which they operated, but also because it allowed

82 On whether the scythed chariot was a Persian or earlier invention, see Nefiodkin (2004); Nefiodkin (2014); Rop (2013).

83 Desző (2012) 147: 'Those officers who are equipped with a spear, bow, quiver, sword, and wear boots were most probably officers of the cavalry bodyguard, since with the exception of the bodyguard cavalry not a single arm of the army of Sennacherib and Assurbanipal was equipped with both spears and bows'.

them to seamlessly coordinate with local subject and client allies and against different types of enemies. It is further reflected in the royal propaganda and reports concerning the education of Persian nobles. In his funerary inscription at Naqš-e Rostam, Darius I proclaims his abilities as a bowman and spearman on foot and horseback.[84] Xenophon echoes the importance of training with both the bow and the spear or javelin for Persian youth in his *Anabasis* (1.9.5) and *Cyropaedia* (1.2.8–9).[85] The later writer Strabo paints a similar picture: Persian nobles were trained with bow and spear on foot and horseback from the ages of 5 to 24, adding that from 20 to 50 they served as *kardakes* armed with *gerrha*, swords, bows, and iron scale armour (*Geog.* 15.3.18–19).[86]

For all the unending talk of Persian efforts to address the superiority of the Greek hoplite, there is a stronger case to be made that it was the Greeks who adapted their own approach to warfare following the fifth-century Persian invasions. The date of the emergence of the hoplite phalanx itself remains controversial, but one prominent position is that it developed in the aftermath of the Greco-Persian Wars.[87] Proponents of this view note that Herodotus reports that the Athenians charged at Marathon all together, and – other than a contingent of Athenian archers – his description of Plataia gives no indication that the Greek infantry was divided into separate divisions based on their equipment. It is therefore not implausible to suspect that the experience of fighting infantry units with coordinated, uniform equipment and tactics like the Immortals helped spur the Greeks to more deliberately standardise their own infantry equipment and tactics.[88] Herodotus credits Cyrus II with inventing the division of troops by their equipment (1.103); he is mistaken on this point, but the very fact that he believed it lends some weight to the suggestion that the Greeks learned this lesson from the Persians.

84 DNb 9 (Lecoq (1997) 223).

85 As White (2020, 88–89) notes, elsewhere Xenophon praises the Persian *palton* spear over the Greek *dory* for its strength and its versatility as both a thrusting and throwing weapon, and because Persian troops armed with two *palta* could discharge one from a distance and fight hand-to-hand with the other (Xen. *Eq.* 12.12). While these comments refer to the weapons used by Persian cavalrymen, it is not out of the question that Persian foot soldiers might have used the *palton* in the same way.

86 Charles (2012b) 8–9.

87 Van Wees (2004) 177–83 and (2009); Konijnendijk (2012) 8–11; Konijnendijk (2018) 95–116; Krentz (2013). For a recent examination of hoplite origins and tactics, see Kagan and Viggiano (2013).

88 This would hardly be the first time that inhabitants of Greece adopted military concepts from West Asia. E.g., Hale (2013) makes the argument that the hoplite was born out of Greek military involvement with the Assyrians and Egyptians.

It is likely that the single most important infantry innovation in fourth-century Aegean warfare – the Macedonian sarissa phalanx – derived from Greek involvement in Persian campaigns during this period. While Philip II of Macedon has been credited with inventing the pike phalanx, it is probably more accurate to say that he invented a specific type of this formation.[89] In West Asia, the Vulture Stele of Lagash dating to ca. 2600 BCE depicts a front row of tower shield-bearers marching in front of soldiers carrying spears with two hands, and on another panel shows the king on his war cart bearing a long lance.[90] In the art of the Akkadian Empire only a few centuries later, 'the two-handed long spear or pike makes a frequent appearance', including on the Victory Stele of Naram-Sin and the fragmentary Stele of Rimush.[91] Pike-length thrusting weapons do not appear in Neo-Assyrian art, but it is possible that a few particularly large spearheads found by archaeologists were designed for atypically long spears.[92]

As we have seen, Xenophon appears to have associated long thrusting spears with the Egyptian-style Persian infantry he saw at Cunaxa (*Cyr.* 6.2.10; 7.1.33). Before him, Herodotus notes that the Egyptian sailors in Xerxes' navy were equipped with long spears and pole-axes (7.89.3). The use of such weapons was not at all novel in Egypt: several Old Kingdom tombs include reliefs of figures on boats in ritualistic combat wielding long poles with two hands.[93] It is not difficult to understand how long spears might be particularly useful in naval and marine combat, allowing soldiers to extend their reach from ship to ship or ship to shore. Not coincidentally, then, the first reported Greek use of a long spear comes about as a result of Greeks fighting for Persians during a maritime invasion of Egypt.

Following the rebellion of the Egyptian Amyrtaios in 404, the Persians made repeated attempts to regain control over their former satrapy throughout the fourth century. The rebellion of Cyrus the Younger in 401 forestalled the first of these and gave the Egyptians time to fortify the eastern Delta region extensively, adding to the significant geographic and environmental challenges already present for would-be conquerors.[94] Greeks fought for the Persians in

89 See Wrightson (2019) 164 for discussion and further bibliography.

90 Nadali (2014) 142.

91 Hamblin (2006) 86–88.

92 Barron (2010) 84–85, 92. Muscarella (2013, 257–58) notes the finding of several iron pikes from the site of Hasanlu, which was destroyed in the late ninth century probably by rivals of Assyria from Urartu.

93 Bolshakov (1993); Trimm (2017) 264–65.

94 On the challenges of conquering Egypt, see Kahn and Tammuz (2008); Fischer-Bovet (2014). On the Egyptian defensive preparations under Amyrtaios and his successors, see Ruzicka (2012) 41–42; (Rop 2019a) 98–100.

several of these invasions, which involved ship-borne beach assaults and necessitated a departure from the traditional phalanx-based hoplite form of combat. In the successful 343–342 campaign, for instance, Theban hoplites opened the assault against the Egyptian fort at Pelusium with a charge across a deep and narrow canal (Diod. Sic. 16.46.8). The next stage of the fighting involved a joint Persian and Argive strike force sailing past Pelusium to capture a key strategic position near Daphnae (16.44.2–3, 47.3, 48.3–5).[95]

The Athenians were particularly well suited for this type of warfare due to their long experience as a maritime power, but even so the peculiarities of invading the Nile Delta required specialised training. According to Diodorus, the Athenian Iphikrates instituted a series of reforms to the hoplites under his command as a direct result of his preparation and campaigning with the Persians over a period of several years in the 370s. He replaced the hoplite shield with a smaller shield,[96] doubled the length of their spears and swords, and provided them with a new style of boots (Diod. Sic. 15.44). Cornelius Nepos adds that he further lightened their equipment by giving them linen body armour (*Iph.* 1), in doing so probably matching the Egyptian thorax reportedly adopted by the Persians already in the fifth century.[97]

The obvious similarities between the long spears and small shields of the new Iphikratid peltasts and the Macedonian sarissa pike phalanx has led several scholars to suggest that the former inspired the latter. Indeed, after returning from Egypt, Iphikrates' next two campaigns were against Sparta at Corcyra and then in Thrace, where he operated in support of his adoptive brother and the future king of Macedon, Philip. While no source directly states that Philip adopted and refined the innovations of Iphikrates, the circumstantial evidence for this transmission scenario is compelling.[98]

Another plausible source of inspiration is Thrace. Thracian peltasts equipped with what seem to be long thrusting spears appear in Greek art from the fifth century, and Iphikrates spent a long time campaigning with the Thracians prior to his Persian service. Macedon's proximity to and frequent clashes against the Thracians would have made Philip familiar with their style of warfare.[99] This type of armament was probably not limited to Thrace, either, as Xenophon

95 Ruzicka (2012) 186; Rop (2019a) 161.

96 Shannahan (2014, 65–68) notes that Egyptian troops are frequently depicted with smaller shields during this period. Herodotus reports that the Egyptian marines in Xerxes' fleet were equipped with hollow shields (ἀσπίδας κοίλας 7.89.3). Sekunda (2014, 131–32) suggests that Iphikrates adopted smaller Persian shields made of leather and wood.

97 For full discussion with more references, see Ruzicka (2012) 106–11.

98 Ueda-Sarson (2002); Anson (2010) 65–66; Sekunda (2014); Wrightson (2019) 164.

99 Against this position, however, see Bertosa (2014).

mentions that the Ten Thousand confronted soldiers armed with *peltai* and long spears on their march from Cunaxa to the Black Sea.[100] Whether they came to Macedon via northern Aegean groups or from Iphikrates or a combination thereof, it seems likely that infantry armed with two-handed thrusting spears and small, *peltē*-like shields were more common in Eastern Mediterranean and West Asian warfare than is typically recognised.

Finally, Macedonian infantry under Philip II appear to have been cross-trained much like their Persian counterparts. In situations when the use of a long two-handed pike was not advantageous, such as assaulting cities and on the march, Philip's infantry corps was also able to fight with swords, shorter spears, and even javelins that could be either thrust or thrown much like the *palton* that Xenophon associated with the Persians.[101] The Macedonians were therefore prepared to fight hand-to-hand as heavy infantry or at a distance as light infantry, whatever the situation demanded. Philip's capacity to train a large standing army rather than a small elite guard to fight with two different weapons requiring different tactics and formations was as effective as it was difficult.[102] It should not be surprising that the royal Persian infantry were similarly cross-trained, that Philip imitated the wealthiest and most powerful imperial army in his world when reforming the Macedonian armed forces,[103] or that his signature victory at Chaeronea made use of the same feigned retreat tactic which won the day for Artaxerxes II at Cunaxa sixty-three years earlier.

3 Conclusion

The widespread belief that the royal infantry of the Achaemenid Persian Empire were bowmen who preferred to fight from a distance behind a single row of wicker tower shields does not withstand scrutiny. The evidence, from the reliefs at Susa and Persepolis to Greek art and historical narratives, reveals instead that they were primarily spearmen who both preferred and excelled in hand-to-hand combat. They were protected by scale body armour, carried large tower or smaller oval shields made of strong, flexible wooden staves woven

100 Chaldean hoplites in the king's employ appear with long wicker shields and spears (Xen. *An.* 4.3.4), and another group carries spears measuring six cubits or roughly nine feet, a little longer than the traditional Greek spear (5.4.12). For the translation of cubits to feet, see Stone (2014) 5.

101 Markle (1982) 98–104.

102 Ueda-Sarson (2002).

103 There is a better case to be made for Persian attempts to reform cavalry between Issos and Gaugamela; sometimes this effort is confused as an attempt to implement a Persian sarissa infantry phalanx. For recent discussion, see Heckel (2017).

together with rawhide, and fought with spears as well as battle axes, swords, and pikes. Unlike the Greek hoplite to whom they are often negatively compared, the Persian royal infantry were also archers. This cross-training granted the Persian army an unprecedented degree of tactical versatility and may have represented a significant innovation in West Asian warfare.

The Greeks themselves witnessed the extent and effectiveness of this versatility over the course of the fifth century. At Malene during the Ionian Revolt, the Persian foot soldiers stood toe-to-toe with the Greek infantry until their support cavalry struck the decisive blow; at Marathon, they defeated the Greek centre in direct, close quarters combat; and at Cunaxa, Persian infantry with pikes and tower shields protected King Artaxerxes in the centre while Persian spearmen on the wing used their superior mobility to defeat their slower hoplite adversaries without coming to blows at all. When the Spartans and their allies refused to engage in the open field at Thermopylai and Plataia, the Persians harried them from a distance with arrows. Under such circumstances, the *sparabara* formation could be deployed to provide their archers with a degree of protection behind a standing tower-shield wall. And as Mykale and the second phase of the fighting at Plataia reveal, the shield wall could also be used as a makeshift defensive fortification for close quarters combat.

Of course, the Persians themselves also experienced defeat at the hands of Greek infantry, most famously at Marathon, Plataia, and Mykale. The full royal infantry was not present at any of these battles, however, and these defeats did not spur the Persians to reform their equipment or tactics. The Persian royal infantry remained unrivalled in West Asia and the Eastern Mediterranean until the middle of the fourth century, when Philip II of Macedon created his own standing, cross-trained infantry force. While the Macedonians were not equipped in the exact same manner as their Persian counterparts, their ability to fight with javelins, spears, and pikes as circumstances demanded meant that they too could deploy as light or heavy infantry. Much like the Persians, moreover, the Macedonians were able to make use of combined arms tactics by complementing their royal infantry with a strong cavalry force. In many respects, the armies that Philip II and Alexander III used to conquer Greece and then the Achaemenid Empire represent the adoption and refinement of not the Greek, but the Persian way of war.

Bibliography

Anson, E.M. (2010) 'The Introduction of the "Sarisa" in Macedonian Warfare'. *Anc. Soc.* 40: 51–68.

Balcer, J.M. (1995) *The Persian Conquest of the Greeks 545–450 BC*. Konstanz.

Bardunias, P., and Ray, F. (2016) *Hoplites at War: A Comprehensive Analysis of Heavy Infantry Combat in the Greek World, 750–100 BCE*. Jefferson, NC.

Barrett, A. and Vickers, M. (1978) 'The Oxford Brygos Cup Reconsidered'. *JHS* 98: 17–24.

Barron, A. (2010) 'Late Assyrian Arms and Armour: Art versus Artifact', PhD Dissertation, University of Toronto.

Barr-Sharrar, B. and Borza, E. (eds) (1982) *Macedonia and Greece in Late Classical and Early Hellenistic Times*. Washington.

Bertosa, B. (2014) 'Peltast Equipment and the Battle of Lechaeum', in Sekunda and Burliga (2014) 113–25.

Best, J. 1969. *Thracian Peltasts and Their Influence on Greek Warfare*. Groningen.

Bolshakov, A. (1993) 'The Scene of the Boatmen Jousting in Old Kingdom Tomb Representations'. *Société d'Égyptologie, Genève* 17: 29–39.

Borza, E. (1990) *In the Shadow of Olympus: The Emergence of Macedon*. Princeton.

Bovon, A. (1963) 'La représentation des guerriers perses et la notion de Barbare dans la première moitié du Ve siècle'. *BCH* 87: 579–602.

Briant, P. (2002) *From Cyrus to Alexander: A History of the Persian Empire*. Winona Lake.

Bushman, A. (2017) 'We're Kind of a Big Deal: Greek Mercenaries in Achaemenid Persian Armies'. *Ancient World* 48: 3–24.

Butera, C.J. and Sears, M.A. (2019) *Battles and Battlefields of Ancient Greece: A Guide to Their History, Topography and Archaeology*. Barnsley.

Cawkwell, G. (2005) *The Greek Wars: The Failure of Persia*. Oxford.

Charles, M.B. (2011) 'Immortals and Apple Bearers: Towards a Better Understanding of Achaemenid Infantry Units'. *CQ* 61: 114–33.

Charles, M.B. (2012a) 'Herodotus, Body Armour, and Achaemenid Infantry'. *Historia* 61: 257–69.

Charles, M.B. (2012b) 'The Persian Kardakes'. *JHS* 13: 7–21.

Charles, M.B. (2021) 'The Royal Elite of the Achaemenid Army', in Heckel, Naiden, Garvin, and Vanderspoel (2021) 173–84.

Chlup, J. and Whately, C. (eds) (2020) *Greek and Roman Military Manuals: Genre and History*. London.

Clark, J. and Turner, B. (eds) (2018) *Brill's Companion to Military Defeat in Ancient Mediterranean Society*. Leiden.

Craven, S.P. (2017) 'The Mercenaries of Hellenistic Crete', PhD Dissertation, University of Texas at Austin.

De Backer, F. (2007) 'Some Basic Tactics of Neo-Assyrian Warfare'. *Ugarit-Forschungen* 39: 69–119.

De Backer, F. (2012) 'Scale-armour in the Mediterranean Area during the Early Iron Age: A) From the IXth to the IIIrd Century BC'. *Revue Des Études Militaires Anciennes* 5: 1–38.

Dezső, T. (2012) *The Assyrian Army I: The Structure of the Neo-Assyrian Army. 1: Infantry.* Budapest.

Dusinberre, E., Garrison, M., and Henkelman, W.F.M. (eds) (2020) *The Art of Empire in Achaemenid Persia: Studies in Honour of Margaret Cool Root.* Leiden.

Echeverría, F. (2015) 'Heroic Fiction, Combat Scenes, and the Scholarly Reconstruction of Archaic Greek Warfare'. *BICS* 58: 33–60.

Ferrill, A. (1985) *The Origins of War: From the Stone Age to Alexander the Great.* London.

Fink, D.L. (2014) *The Battle of Marathon in Scholarship: Research, Theories and Controversies Since 1850.* Jefferson, NC.

Fischer-Bovet, C. (2014) 'Est-il facile de conquérir l'Égypte? L'invasion d'Antiochos IV et ses conséquences', Feyel and Graslin, (2014) 209–59. Nancy.

Hale, J.R. (2013) 'Not Patriots, Not Farmers, Not Amateurs: Greek Soldiers of Fortune and the Origins of Hoplite Warfare', in Kagan and Viggiano (2013) 176–93.

Hamblin, W.J. (2006) *Warfare in the Ancient Near East to 1600 BC: Holy Warriors at the Dawn of History.* London.

Hanson, V.D. (1989) *The Western Way of War.* New York.

Hanson, V.D. (2001) *Carnage and Culture: Landmark Battles in the Rise of Western Power.* New York.

Hanson, V.D. (2008) 'Genesis of the Infantry 600–350 BCE', in Parker (2008) 12–31.

Harper, P., Aruz, J. and Tallon, F. (eds) (1992) *The Royal City of Susa: Ancient Near Eastern Treasures in the Louvre.* New York.

Head, D. 1992. *The Achaemenid Persian Army.* Reddish.

Heckel, W. (2017) 'Dareios III's Military Reforms Before Gaugamela and the Alexander Mosaic: A Note'. *AHB* 31: 65–69.

Heckel, W., Naiden, F.S., Garvin, E.E., and Vanderspoel, J. (2021) *A Companion to Greek Warfare.* Hoboken.

Hyland, J.O. (2011) 'Contesting Marathon: Billows, Krentz, and the Persian Problem'. *CPhil* 106: 265–77.

Jacobs, B. and Rollinger, R. (eds) (2021) *A Companion to the Achaemenid Persian Empire.* Chichester.

Kagan, D. and Viggiano, G.F. (eds) (2013) *Men of Bronze: Hoplite Warfare in Ancient Greece.* Princeton.

Kahn, D. and Tammuz, O. (2008) 'Egypt Is Difficult to Enter: Invading Egypt – A Game Plan (Seventh–Fourth Centuries BCE)'. *Journal for the Society for the Study of Egyptian Antiquities* 35: 37–66.

Konijnendijk, R. (2012) 'Neither the Less Valorous nor the Weaker: Persian Military Might and the Battle of Plataia'. *Historia* 61: 1–17.

Konijnendijk, R. (2016) 'Mardonius' Senseless Greeks'. *CQ* 66: 1–12.

Konijnendijk, R. (2018) *Classical Greek Tactics: A Cultural History.* Leiden.

Konijnendijk, R. (2021) 'Playing Dice for the Polis: Pitched Battle in Greek Military Thought'. *TAPA* 151: 1–33.

Krentz, P. (2010) *The Battle of Marathon*. New Haven.

Krentz, P. (2013) 'Hoplite Hell: How Hoplites Fought', in Kagan and Viggiano (2013) 134–56.

Lecoq, P. (1997) *Les inscriptions de la Perse achéménide. Traduit du vieux perse, de l'élamite, du babylonien et de l'araméen (L'aube des peuples)*. Paris.

Lendon, J.E. (2005) *Soldiers and Ghosts: A History of Battle in Classical Antiquity*. New Haven.

Manning, S. (2021) *Armed Force in the Teispid-Achaemenid Empire: Past Approaches, Future Prospects*. Stuttgart.

Markle, M.M. (1978) 'Use of the Sarissa by Philip and Alexander of Macedon'. *AJA* 82: 483–97.

Markle, M.M. (1982) 'Macedonian Arms and Tactics under Alexander the Great', in Barr-Sharrar and Borza (1982) 113–22.

Meyer, E. (2018) 'Cimon's Eurymedon Campaign Reconsidered'. *AHB* 32: 25–43.

Muscarella, O.W. (1987) 'Median Art and Medizing Scholarship'. *JNES* 46: 109–27.

Muscarella, O.W. (2013) *Archaeology, Artifacts and Antiquities of the Ancient Near East: Sites, Cultures, and Proveniences*. Leiden.

Muscarella, O.W., Caubet, A., and Tallon, F. (1992) 'Susa in the Achaemenid Period', in Harper, Aruz, and Tallon (1992) 215–52.

Nadali, D. (2005) 'Assyrians to War: Positions, Patterns, and Canons in the Tactics of the Assyrian Armies in the VII Century BC'. *Contributi E Materiali de Archaeologia Orientale* 10: 167–207.

Nadali, D. (2010) 'Assyrian Open Field Battles: An Attempt at Reconstruction and Analysis', in Vidal (2010) 117–52.

Nadali, D. (2014) 'How Many Soldiers on the "Stele of the Vultures"? A Hypothetical Reconstruction'. *Iraq* 76: 141–48.

Nefiodkin, A.K. (2004) 'On the Origin of the Scythed Chariots'. *Historia* 53: 369–78.

Nefiodkin, A.K. (2014) 'Once More on the Origin of the Scythed Chariot'. *AHB* 28: 112–18.

Parker, G. (ed) (2008) *The Cambridge Illustrated History of Warfare: The Triumph of the West. Revised and Updated*. Cambridge.

Rahe, P.A. (1980) 'The Military Situation in Western Asia on the Eve of Cunaxa'. *AJPhil* 101: 79–96.

Rahe, P.A. (2015) *The Grand Strategy of Classical Sparta: The Persian Challenge*. New Haven.

Rawlings, L. (2007) *The Ancient Greeks at War*. Manchester.

Roaf, M. (1983) 'Sculptures and Sculptors at Persepolis'. *Iran* 21: 1–164.

Rop, J. (2013) 'Reconsidering the Origin of the Scythed Chariot'. *Historia* 62: 167–81.

Rop, J. (2018) 'The Assassination of Tissaphernes: Royal Responses to Military Defeat in the Achaemenid Empire', in Clark and Turner (2018) 51–73.

Rop, J. (2019a) *Greek Military Service in the Ancient Near East, 401–330 BCE*. Cambridge.

Rop, J. (2019b) 'The Phocian Betrayal at Thermopylae'. *Historia* 68: 413–35.

Rop, J. (2020) 'Refighting Cunaxa: Xenophon's Education of Cyrus as a Manual on Military Leadership', in Chlup and Whately (2020) 153–71.

Rop, J. (2021) 'Greeks in Foreign Service: The Case of the Achaemenid Empire', in Heckel, Naiden, Garvin, and Vanderspoel (2021) 160–70.

Ruzicka, S. (2012) *Trouble in the West: Egypt and the Persian Empire, 525–332 BC*. Oxford.

Schmidt, E. (1953) *Persepolis 1: Structures – Reliefs – Inscriptions* (*OIP* 69). Chicago.

Sekunda, N. (1992) *The Persian Army 650–330 BC*. Oxford.

Sekunda, N. (2014) 'The Chronology of the Iphicratean Peltast Reform', in Sekunda and Burliga (2014) 126–44.

Sekunda, N. and Burliga, B. (eds) (2014) *Iphicrates, Peltasts and Lechaeum*. Gdańsk.

Shannahan, J. (2014) 'Two Notes on the Battle of Cunaxa'. *AHB* 28: 61–81.

Snodgrass, A. (1967) *Arms and Armour of the Greeks*. Ithaca.

Stone, M. (2014) 'The Cubit: A History and Measurement Commentary'. *Journal of Anthropology* 2014: 1–11.

Tallis, N. (2005) 'Transport and Warfare', in Tallis and Curtis (2005) 210–35.

Tallis, N. and Curtis, J. (2005) *Forgotten Empire: The World of Ancient Persia*. London.

Trimm, C. (2017) *Fighting for the King and the Gods: A Survey of Warfare in the Ancient Near East*. Atlanta.

Tuplin, C. (2020) 'Sigillography and Soldiers: Cataloguing Military Activity on Achaemenid Period Seals', in Dusinberre, Garrison, and Henkelman (2020) 329–459.

Tuplin, C. and Jacobs, B. (2021) 'Military Organization and Equipment', in Jacobs and Rollinger (2021) 1161–81.

Ueda-Sarson, L. (2002) 'The Evolution of Hellenistic Infantry, Part 1: The Reforms of Iphicrates'. *Slingshot* 222: 30–36. http://lukeuedasarson.com/Iphikrates1.html (accessed on 20 August, 2021).

van Gils, L.W., de Jong, I.J.F., and Kroon, C.H.M. (eds) (2018) *Textual Strategies in Ancient War Narrative: Thermopylae, Cannae and Beyond*. Leiden.

van Wees, H. (2004) *Greek Warfare: Myths and Realities*. London.

van Wees, H. (2009) 'The Development of the Hoplite Phalanx: Iconography and Reality in the Seventh Century', in van Wees (2009) 125–66.

van Wees, H. (ed) (2009) *War and Violence in Ancient Greece*. Swansea.

van Wees, H. (2018) 'Thermopylae: Herodotus versus the Legend', in van Gils, de Jong, and Kroon (2018) 19–53.

Vidal, J. (ed) (2010) *Studies on War in the Ancient Near East: Collected Essays on Military History*. Münster.

White, J. (2020) 'Bows and Spears in Achaemenid Persia', PhD Dissertation, University of California, Santa Barbara.

Worthington, I. (2014) *By the Spear: Philip II, Alexander the Great, and the Rise and Fall of the Macedonian Empire*. Oxford.

Wrightson, G. (2019) *Combined Arms Warfare in Ancient Greece: From Homer to Alexander the Great and His Successors*. London.

CHAPTER 7

Achaemenid Diplomacy in Historical Perspective

Daniel Beckman

The Achaemenid Persian Empire is famed for its meteoric expansion, the explosion of military conquest that swept aside no fewer than three venerable kingdoms (Lydia, Babylon, Egypt) and absorbed many more smaller states in just a few decades. No less importantly, the Achaemenids also employed non-violent diplomatic strategies in pursuit of their imperial goals. As befitting a chapter in a volume on war, Achaemenid diplomacy was not about avoiding warfare or creating peace; instead, as I will argue, these diplomatic strategies were nothing short of 'war by other means', aimed at accomplishing the same ends as military violence, but used when a military campaign was deemed to be undesirable.

In this chapter I will survey the development of diplomatic activity through the course of the Achaemenid period. I define 'diplomacy' as any formal discourse between autonomous entities concerning peace, territory, trade, or other such affairs; this may include letters sent between rulers or official representatives, face-to-face negotiations, or treaties and other agreements. I define a 'treaty' as any formal, binding political agreement between two or more parties, regardless of the relative power of the parties; this would include, but is not limited to, non-interference in commercial and economic activity, the maintenance of local political forms or constitution, and recognition of the territory and borders of the opposite party.[1] Unfortunately, very little evidence of Achaemenid diplomacy has survived, and much of what is available is concerned almost exclusively with the Aegean world, produced by authors of doubtful reliability. Thus I must emphasise that it is impossible to state with certainly what ideological and strategic concerns motivated the Achaemenid approach to diplomacy, and how this approach might have varied chronologically and geographically.

As a result, in order to build a more complete image of Achaemenid diplomacy, this chapter relies on a comparison between Achaemenid and Neo-Assyrian

1 For introductions to Near Eastern diplomacy, see McCarthy (1978); Magnetti (1978); Tadmor (1982); Parpola and Watanabe (1988); Bederman (2001); Weeks (2004); Altman (2010); Kitchen and Lawrence (2012); Lauinger (2013); Radner (2019). For Hellenic perspectives, see Adcock and Mosley (1975); Wolpert (2001); Rhodes (2008); Hunt (2010).

 | DOI:10.1163/9789004710771_008

diplomacy. The Neo-Assyrian Empire (hereafter, simply 'the Assyrians') ruled much of the Ancient Near East from roughly the tenth through the end of the seventh centuries BCE using, among other strategies, a well-developed programme of treaties. My goal is not to argue that the Achaemenids were aware of or influenced by Assyrian diplomatic practices. Instead, comparison with the Assyrians will help to emphasise the most important elements of Achaemenid diplomacy. By framing Achaemenid diplomacy as a part of Near Eastern history, this approach also helps counterbalance the Greek bias of our main narrative sources.

1 Historical Overview

The paragraphs below offer a brief summary of the most important developments of Achaemenid diplomatic practice. The reader is encouraged to use this section as a guide to a more thorough exploration of the topics introduced here.

The years of the expansion of the Persian Empire, ca. 540–479, offer only a few examples of diplomacy; in general, the Persians were able to simply absorb other states through conquest, and the new subjects and many of their social and political institutions were incorporated into the machinery of the empire. However, in a few cases, foreign states submitted to the Persians without violence. During Cyrus' conquest of Asia Minor through the mid-540s, he negotiated the peaceful submission of Miletos (Hdt. 1.141). Between ca. 513 and 480, Darius I and Xerxes I secured the submission of Macedon and a number of Greek states by demanding 'earth and water'. The only positive evidence for this demand comes from Herodotus, in reference to events in the Aegean region.[2]

Following the Persian invasions of Greece (492–79), the Achaemenids made no further attempts to conquer large swathes of new territory. Assaults on imperial holdings by the Athenian-led Delian League, conflicts among the imperial elites, and the rebellion of Egypt may have tested the Achaemenids' ability to finance and carry out new campaigns of expansion. However, this picture of an empire under extreme stress is due to our sources, almost all of which are focused on the western regions of the empire, and are afflicted by their anti-Persian bias. There is little evidence for endemic strife in the eastern satrapies, but given the minimal evidence for this region on any matter, this cannot be used to draw any firm conclusions.

2 E.g., Hdt. 4.143–4; 5.18; 5.73–5; 7.131–2.

The first possible example of a negotiated peace treaty between the Achaemenids and another party is the Peace of Kallias. The date, terms, and even the very existence of this treaty have been contested for over a century, but today there appears to be a scholarly consensus that, sometime in the mid-fifth century, the Achaemenids and Athenians agreed to cease aggressive actions against one another. If one accepts the historicity of the Peace of Kallias, it stands as the first step in a series of negotiated agreements between the Achaemenids and several Greek states (especially the Athenians, Spartans, and Thebans). The mechanisms of the negotiations, as well as their ideological and strategic ramifications, will be addressed in greater detail below. Regardless of precisely why and how the Achaemenids decided to enter into treaties with the Greek states, the effect of these agreements was to limit the number of Greek incursions into imperial territory, especially post-387.

I must reiterate that our sources are overwhelmingly concerned with the western portions of the empire, so it is impossible to ascertain the nature of Achaemenid diplomacy elsewhere. The regions that possessed well-developed diplomatic traditions prior to the Achaemenid conquests (especially Mesopotamia, Asia Minor, and Egypt) had been totally incorporated into the empire, and no longer had any independent institutions with which the Achaemenids could exchange embassies or form interstate agreements. Outside of the empire, there were not many other polities that possessed the literary, political, and/or artistic traditions that would create evidence of this diplomacy for us to discover. All of this is to say that, while there is no reason to doubt that the Achaemenids might have maintained diplomatic relations with a variety of powers, it is highly unlikely that evidence of such relations will ever come to light.

2 Methodology

As emphasised above, any attempt to study Achaemenid diplomacy is hamstrung by the fact that little relevant evidence has come down to us today, and most of this betrays a strong Greek bias, as well as a general ignorance of Achaemenid affairs. Although this means that our study will never be as thorough and definitive as one might wish, nevertheless there are a number of steps we can take to improve the quality of our conclusions.

First, it is necessary to consider Achaemenid diplomacy in its historical context. The Achaemenid Empire did not arise from a vacuum. Parsa, the Achaemenid homeland in southwest Iran, stood within the orbit of two previous regional powers, Elam and Media, and was intimately connected to

Mesopotamia and the broader Near East. A review of the diplomatic culture of the Near East prior to the rise of the Achaemenids will reveal a world criss-crossed by ambassadors, traveling between states connected by marriage alliances and treaties. Whoever the first Persians were and whenever they first appeared,[3] it is evident that they took part in a culture which included a diplomatic tradition going back millennia.

Next, we must examine the testimonies, whatever their origin and however problematic they may be, for Achaemenid diplomatic activity. I have selected three cases for which we have substantive evidence, and concerning which modern scholars can agree on the basic facts. While this will, by its nature, limit the types of conclusions that can be drawn, nevertheless it will allow us to sketch a picture of the types of diplomatic activity undertaken by the Achaemenids, which will then serve as an object of comparison with the Assyrian examples.

The purpose of this comparison is not to establish a causal link between the Achaemenids and their predecessors; to say that a trait of Assyrian diplomacy is very similar to an Achaemenid trait is not a claim that the Achaemenids 'inherited' this trait from the Assyrians, or were 'influenced' by them. While it may well be true that the Achaemenids were aware of a given Assyrian (or Babylonian, or Median, or Elamite) practice, demonstration of a direct connection between the two is nearly impossible. By leaving open the question of the source of shared practices, we allow for the possibility of influence or inheritance of a prior Near Eastern diplomatic culture, without disregarding the possibility of convergent evolution, resulting from shared interests, capabilities, and circumstances experienced by the two states. This allows us to focus on a precise and accurate description of Achaemenid diplomacy – a difficult but achievable goal – rather than the impossible task of identifying every strand of its heritage.

3 Historical Predecessors

When Cyrus the Great came to power in the mid-sixth century BCE, Mesopotamia and Western Iran were largely under the control of the Neo-Babylonian

3 On the Persian ethnogenesis, see Briant (1984) 80–100; de Miroschedji (1985a), (1985b), (1990); Amiet (1992) 92–4; Potts (1999) 306–307; Waters (1999); Álvarez-Mon (2006) 267–68, (2010) 191–316; Tavernier (2018) 164–72; Henkelman (2011) 582–95; Henkelman (2017) 290–303; Henkelman (2018) 804–15; Rossi (2018) 854–58.

Empire and the Median kingdom,[4] respectively. Also significant were the remnants of the Assyrian Empire and Elamite kingdom. All four of these states left deep impressions on the Achaemenids.[5] These four states also participated in a very active diplomatic culture with roots as far back as the third millennium.[6] However, for the first millennium, all of the primary texts come from Assyrian contexts, and thus it is the Assyrian perspective that will dominate our analysis.

The corpus of Assyrian diplomatic texts is difficult to define; the Aramaic word *adê* is often translated as 'treaty' but can also mean 'loyalty oath' (an agreement made by a subordinate party to their superior).[7] As the surviving texts all come from Assyrian contexts, they portray the *adê*-relationship as one of absolute submission of the other party to the Assyrian king, which can make it difficult to determine to what extent the non-Assyrian party existed as politically distinct from Assyria. Hence, the number of extant treaty texts could be as low as four or five,[8] or as high as fourteen.[9] One must also decide how to interpret references to agreements or oaths in other texts, where the treaty itself is lost.

The Assyrians made frequent use of *adê*-agreements in the administration of their empire. They appear across the Near East, from Egypt[10] to Scythia,[11] and seem to have been fundamental to the structure of the empire. Broadly

4 While most scholars reject the Herodotean image of a Median 'empire', there is no consensus on its size or structure. For attempts to synthesise the linguistic, archaeological, and literary data, see Genito (1986); Sancisi-Weerdenburg (1988); Schmitt (2003); Tuplin (2004); Rossi (2010); Waters (2011); Radner (2003); Radner (2013); Rollinger (2020).

5 The Sumero-Akkadian and Aramaic scribal traditions survived into the Achaemenid period (Folmer (2011), (2012)); the Babylonian scribes may have influenced Achaemenid royal titulature (Shayegan (2012) 247–91). For Achaemenid adoptions of Mesopotamian art and iconography see Stronach (1978) 51–52, 74–77; Stronach (2001) 99–102; Álvarez-Mon (2011) 334–56; Garrison (1991) 7–10; Garrison (2017) 341–47, 366–73; Garrison and Root (2001) 16–18. For deities of Mesopotamian origin in the Persepolis pantheon, see Henkelman (2012) 305–23. The Cyrus Cylinder was unequivocally Babylonian in its language and ideology; see Kuhrt (1983); Beckman (2018) 154–57; Pongratz-Leisten (2018) 98–103; Schaudig (2018) 70–88.

6 The oldest surviving Elamite text is a fragmentary treaty between Narām-Sîn (r. ca. 2254–2218) and an uncertain Elamite king; see Hinz (1967) 91–93. Steinkeller (2018), 189, prefers to call the document a 'record of privileges bestowed by Naram-Suen upon the chief gods of Susa and of their recognition of Naram-Suen as their servant and protector' rather than a 'treaty'.

7 Tadmor (1982) 142–43; Grayson (1987) 128; Parpola (1987) 180–83; Brinkman (1990) 88–105.

8 Altman (2012) 189–92.

9 Parpola and Watanabe (1988) xv–xxiii.

10 E.g., Ashurbanipal's *adê* with Necho, RINAP 5.1 = Novotny and Jeffers (2018) 2, v.11–12.

11 E.g., reference to a potential marriage alliance between Esarhaddon and Bartatua, king of Scythia (*Iškuza*), SAA 4 = Starr (1990), 20 obv. 1–9.

speaking, these agreements were either bilateral or unilateral, depending on the degree to which Assyria gained control over the other entity, but in neither case was the *adê* primarily intended to bring peace. It was a tool of imperial expansion, either by making the conquered subject complicit in their own exploitation by the Assyrians, or by giving the Assyrian king a *casus belli*, which would often result in (re-)conquest and new, harsher treaty. As a treaty-oath was sworn in the name of the gods, to violate the *adê* was a grave sin, and would be punished severely. The subject or vassal was forced by the stipulations to do or not do certain things. If the oath were to be broken, the treaty then functioned as a way to legitimise the inevitable retaliation by the Assyrian king.[12]

Conversely, there is no evidence that the Neo-Babylonian Empire used treaties or other formalised diplomatic arrangements in the administration of their empire. This may be an artefact of their literary and ideological interests, rather than their diplomatic practices; the Neo-Babylonian king's legitimacy derived more from his proper treatment of the Babylonian gods and his related building projects than from the expansion of the empire, and this dictated the focus of the royal inscriptions.[13]

With the rise of the Assyrian Empire in the first millennium, we find increasing evidence of diplomats at work. In 784, an Elamite ambassador was in Kalhu;[14] around this same time, Tiglath-pileser III established the first in a series of Assyrian provinces in Iran, run by Assyrian eunuchs. Local Median dynasts, called 'city-lords', were maintained and strengthened alongside Assyrian administrative structures, and were bound by eternal loyalty oaths to the Assyrian king.[15] Similar oaths are recorded in Esarhaddon's Succession Treaty, dated to ca. 672.[16] Among those who took the oath were a number of Median city-lords, who had most likely provided men to serve in Ashurbanipal's royal bodyguard.[17] Of course, the Assyrians were not party to every diplomatic development. The Babylonian king Nabû-šuma-iškun (r. 760–748) sent gifts to Elam, implying an anti-Assyrian alliance. In 720, the Elamite king Humban-nikash joined the Babylonian Marduk-apla-iddina against Sargon II at Der.[18]

12 Radner (2019) 309–10, 319–25.

13 Da Riva (2008) 26–31; Jursa (2014) 137–42.

14 Dalley and Postgate (1984) 145 iv 26.

15 Ponchia (2014) 504–10; Radner (2013) 443–49; Waters (2011).

16 The *editio princeps* is Wiseman (1958). See also SAA 2 = Parpola and Watanabe (1988) 6. Lauinger (2015, 293) emphasises that it is a composite text, an abstraction that refers to the text that served as a model for all of the copies.

17 Liverani (1995). This should not be taken as evidence that others who swore the oath also served as bodyguards.

18 Luckenbill (1927) 14–15, 26; Potts (1999) 263–64.

What did this type of diplomacy actually look like? According to Esarhaddon's Nineveh A inscription, such was the reputation of what 'the might of the god Aššur, my lord, had done among all of (my) enemies', that 'fear and terror poured over them'. As a result, in ca. 674 the Elamite king Urtak(u) and an unnamed Gutian king sent ambassadors to negotiate 'so that there would be no trespassing on the borders of their countries' (RINAP 4 = Leichty (2011), Esarhaddon 1 v 26–33a). While we would be right to doubt that the Elamites and Gutians were simply so terrified that they begged for a treaty, their border agreement does come on the heels of Esarhaddon's campaigns in the Zagros where he 'imposed the tribute and payment of my lordship' on the Medes, forcing them to send resources like horses, camels, and lapis lazuli to Nineveh (RINAP 4 = Leichty (2011), Esarhaddon 1 iv 32–52). An inquiry sent to the oracle of Šamaš confirms that the impetus for the agreement came from the Elamite:

> Šamaš, great lord, [give me a firm positive an]swer to what I am asking you! If Urtak, king of Elam has se[nt this proposal for making peace] to Esarhaddon, king of Assyria, [has he honestly sent] true, sincere words of re[conciliation to Esarhaddon, king of Assyria]? (SAA 4 = Starr (1990) 74.1–4)

This treaty was the culmination of a diplomatic exchange wherein Urtak and Esarhaddon 'constantly listened to each other' (SAA 18 = Reynolds (2003), 7 obv.3–4), that is, exchanged messengers. While the Elamite and Gutian representatives swore oaths to uphold whatever border agreement had been reached, Esarhaddon does not mention swearing a reciprocal oath himself.[19] However, in a letter to Urtak, Esarhaddon refers to him as 'my brother', suggesting a level of parity that implies that the treaty was bilateral, and not imposed by the Assyrians on the Elamites.[20] There is also evidence that Assyria was on the defensive and would have happily taken advantage of a peace offering from her enemies. The Elamite king Huban-haltaš II had defeated the Assyrian army at Sippar in 675, and supported a rival claimant for control of Sealand, nominally held by an Assyrian client.[21] The Assyrians were also struggling to conquer Egypt at this same time.[22] An Assyrian inventory, which may date from 674, includes a list of treasures sent to Elam, echoing Nabû-šuma-iškun's gifts of the previous century.[23]

19 RINAP 4 = Leichty (2011), Esarhaddon 1 v 26–33a, 22.
20 SAA 16 = Luukko and Van Buylaere (2002) 1 1–13; Álvarez-Mon (2011) 356.
21 Waters (1997) 62–65.
22 ABC = Grayson (1975) 1 iv 16–18.
23 Waters (1997) 69; Frame (1992) 83–100.

Without the treaty text itself, we can only guess at the terms, but since Esarhaddon specifically mentioned borders, we must assume that boundaries were agreed upon. A later letter also shows that Urtak promised not to aid Assyria's enemies:

> Last year when the Palace Superintendent and the Magnates went down to Chaldea, the brothers of the king of Elam tried to incite their brother the king, saying: 'Let us gather an army, cross over to Chaldea and take it away from Assyria'. The king of Elam, however, did not sin, but refused to listen to them, saying: 'I shall not violate the treaty'. (SAA 18 = Reynolds 2003, 202)

The peace agreement survived for a decade, with a continual exchange of envoys reestablishing the agreement as needed. When a local Assyrian official detained an Elamite party, the crown prince, Ashurbanipal, intervened and sent the visitors to Esarhaddon, presumably to ensure that the insult would not create an international conflict (SAA 18 = Reynolds 2003, 7). Such was the relationship between Assyria and Elam that Ashurbanipal, Esarhaddon's successor, sent food to Elam during a famine, and allowed Elamite refugees to settle in Assyrian territory until the situation improved (RINAP 5.1 = Novotny and Jeffers 2018, 3 iv 17–24, 35–38).

In Assyrian texts, ambassadors are generally specialised messengers, often 'mounted' (*rakbâšu*). In the Biblical account of Sennacherib's siege of Jerusalem in 701 (2 Kings 18:17), three Assyrian officials called 'Tartan', 'Rabsaris', and 'Rabshakeh' approached the city walls to convince the residents to open the gates. The identity of these men is not certain, although possibly the names are actually Akkadian titles. *Tartan* may be the Hebrew rendering of *turtānu*, 'commander-in-chief of the army'. *Rabsaris* may be from *rab ša-rēsi*, a senior official of unknown function.[24] *Rabshakeh* may be from *rab šaqu*, a civil official and a sort of personal servant of the king.[25] Whether this represents the makeup of a typical Assyrian embassy can only be guessed.

After Ashurbanipal's sack of Susa in 647, a decentralised Elamite kingdom continued to exist for the next century. A letter testifies to the movement of ambassadors from Elam to Assyria,[26] and Nabopolassar's return of plundered cult statues to Susa suggests strong contacts between Elam and Babylon (ABC = Grayson (1975), 2 16–17). Unfortunately, little more is known about Elam's relations with other peoples beyond this point.

24 Tadmor (1983) 279–81.
25 Levin (2015) 327.
26 Gorris (2018).

Likewise, the Medes are absent from the surviving textual record until the end of the seventh century. By 614, the Medes had allied with the rebel and eventual founder of the Neo-Babylonian dynasty, Nabopolassar, and were able to capture the city of Aššur by themselves, before the Babylonians arrived on the scene (ABC = Grayson (1975), 3 23–30). There is no indication that the Medes received or were interested in any share of the Assyrian Empire, so the treaty of alliance may have guaranteed the independence of Media in the post-Assyrian world. Berossos states that Amytis, Cyaxares' daughter, was married to the son of Nabopolassar, but this is a confused and unreliable passage.[27]

The Medes also made at least one treaty with a non-Mesopotamian party. According to Herodotus, the Median king Cyaxares came into a long conflict with Alyattes, king of Lydia. In the sixth year of the war (585), an eclipse occurred during a battle. Both sides took this as an omen. Syennesis of Cilicia and Labynetus of Babylon represented Lydia and Media, respectively, in the negotiations. They agreed to make a truce (*horkion*) and a marriage alliance between the son of Cyaxares and the daughter of Alyattes (Hdt. 1.73–74). When Cyrus the Great came to power in ca. 550, the Median-Lydian alliance was still intact.

4 Achaemenid Case Studies

4.1 *Case Study 1: Darius in Scythia*

In ca. 513 BCE, Darius the Great invaded European Scythia. Our Greek sources report that Darius was attempting to punish the Scythians for a perceived crime (Hdt. 4.1.1; Ctes. *FGrH* 688 F 13.20–1; Justin 2.5.8–12); modern scholars tend to emphasise economic or security motivations.[28] Herodotus reports that the Scythians stubbornly refused to submit, or even to fight. Frustrated, Darius sent a message to Idanthyrsos, the Scythian king, demanding that he give the 'gifts' of earth and water, and come to terms with the Persians (Hdt. 4.126). Although Idanthyrsos continued his resistance, later the same demand met with great success in Macedonia. Megabazos, Darius' commander, sent seven Persians who, 'besides himself, were the most honorable in his army' (Hdt. 5.17) as messengers to demand earth and water from Amyntas, king of Macedonia, on behalf of Darius. According to Herodotus, prince Alexander became so incensed at the impropriety of the Persians that he organised the murder of the entire embassy. Afterward, he bribed Boubares, the Persian commander

27 Burstein (1978) 25, §6a.

28 Gardiner-Garden (1987) 343–45; Balcer (1995) 148; Kuhrt (2007) 183.

sent to investigate their disappearance, with a large sum of money and his sister, Gygaia, in marriage (Hdt. 18–21).

As this is the first known instance of submission to the Achaemenid king via the gift of earth and water, it is worth investigating the purpose of the ritual, as well as its broader diplomatic implications. Darius makes it clear to Idanthyrsos that to give these gifts is to acknowledge his weakness in the face of Persian military power, and requires him to engage in some sort of negotiated submission. While the semantics of this gift are debated, it is generally accepted that it was a ritual of submission by a weaker party to the Achaemenid king, and not a bilateral agreement between two equal parties.[29]

Examining the account of Amyntas' submission, some scholars have suspected that Herodotus cannot be taken at his word. His tale of the demise of the seven Persian noblemen and their retinue may have been influenced by Macedonian propaganda, as well as his own literary motives.[30] We might therefore assume that Megabazos and his seven noblemen negotiated the submission of Amyntas. His gift of earth and water was followed by a celebratory banquet, and Gygaia was married to Boubares to further cement the new arrangement.[31]

For the next three decades, the Argead kings were Persian vassals. This meant that they owed tribute to the Achaemenids, as well as contributions of material and manpower for imperial military campaigns – Macedonians were 'added to the slaves' of Mardonios' army in 492 (Hdt. 6.44), and would fight with his forces at Plataea in 479 (Hdt. 9.31). In that same campaign, Alexander, now king himself, acted as Xerxes' ambassador to the Athenians (Hdt. 8.136–144). Beyond these obligations, it is not certain what subjugation to Persia meant for Macedonia. Macedonia may have been incorporated into the satrapy of Skudra, and may have been connected with the country-name *yaunā takabarā* in the Old Persian inscriptions; however, it is not certain that 'Skudra' was ever a satrapy, rather than an ethnic label used by the Persian administration to describe a disparate collection of peoples.[32] There is no evidence that any Persian commander took up a permanent residence in Macedonia, nor that any garrisons were established there. While it risks overstating the case to

29 For various interpretations, see Kuhrt (1988) 88–96; Balcer (1995) 155, 203; Nenci (2001) 38; Herrenschmidt and Lincoln (2004) 269–72; Munn (2006) 225; Waters (2014); Rung (2015); Klinkott (2016).

30 Erbse (1992) 101–4; Badian (1994); Fearn (2007) 101–27.

31 Sprawski (2010) 134–38; Harrison (2019).

32 Balcer (1988); Rollinger (2006); Henkelman and Stolper (2009) 288–99; Olbrycht (2010) 345–46.

call the Macedonians 'willing and helpful allies' of the Persians,[33] it does not appear that their vassalage was overly burdensome.

As noted above, the only references to this gift occurred in reference to the Scythians, Macedonians, and, later, many Greek *poleis* (Hdt. 7.131–2), all in the Aegean region. Certainly this might be an artefact of the incompleteness of our sources. However, it may be relevant that the rest of the empire had been built through conquest, while the demand for earth and water specifically took place in the absence of conquest; recall that Darius made the demand of Idanthyrsos only after the latter refused to give battle, and that Megabazos sent his seven nobles to Amyntas before hostilities broke out. Later, when Artaphernes demands earth and water of two Athenian ambassadors, it was once again prior to hostilities between Athens and Persia (Hdt. 5.73–75).[34] The gift of earth and water, then, would not be a ritual of submission in all cases, but of submission only when the Great King is unwilling or unable to militarily conquer a foreign entity. The ritual gift was demanded in the Aegean, as this region was far from the centers of imperial power, and offered fewer potential benefits for the empire, in comparison with places like Egypt and Mesopotamia.

4.2 *Case Study 2: Spartan-Persian Treaties, 412–411 BCE*

Greco-Persian relations went through dramatic changes in the century after Amyntas' gift of earth and water. First, there was the Ionian Revolt (499–493), during which many Greek cities of Asia Minor attempted to gain independence from the empire. As a result of Athenian and Eretrian participation in the revolt, Darius and Xerxes sent a total of three invasions to punish the meddlesome Greeks, and to enforce a degree of order on the northwestern corner of the empire. Ultimately, these campaigns had the opposite effect: the Athenians emerged a regional power, ideologically prepared to fight any and all who would threaten their autonomy. For the next thirty years, Athenian fleets disrupted Achaemenid control over Cyprus, Asia Minor, and Egypt.

Throughout this turbulent period, instances of Greco-Persian diplomacy are few. During the Athenian interference in Egypt, Artaxerxes sent a Persian named Megabazos with fifty talents to bribe the Spartans to invade Attica, with

33 As does Borza (1990), 115.

34 Note that Artaphernes only made this demand after the Athenians requested Persian assistance against Spartan aggression. There is also debate as to whether or not the Athenians actually concluded a formal agreement with Artaphernes. See Raubitschek (1964); Schachermeyr (1973); Kramer (2004); West (2011); Waters (2016).

the intention of forcing an Athenian withdrawal (Thuc. 1.109).[35] Even though the Spartans refused, the incident shows that the king was well aware of Greek goings-on, and was ready to exploit them. Arthmios of Zeleia was likewise sent with gold to the Peloponnese,[36] although the specific context of his mission is lost to us.

In the mid-fifth century we find (perhaps) our first Greco-Persian treaty. Among those scholars who accept the reality of the Peace of Kallias, it is generally dated to ca. 449/8,[37] although it has also been placed back in the 460s.[38] If the treaty was in fact concluded, it most likely established a non-aggression pact, delineated a boundary between Athenian and Persian territory, and allowed for relative – not total – autonomy of the Greek cities within the empire.[39] With the Peace of Epilykos, Darius II confirmed the agreement in ca. 423, after his chaotic succession (Andoc. *De Pace*, 3.28–29; *IG* I^3 227).[40]

Peace between Athens and Persia did not last, and there were likely at least two causes of its demise. First, at some point between 420 and 415, Pissouthnes, the satrap of Lydia, went into rebellion. After his capture and execution (Ctes. *FGrH* 688 F 15.53), his son, Amorges, continued the fight, and the Athenians decided to give him assistance (Andoc. *De Pace*, 3.29; Thuc. 8.5, 19, 28, 54). It is also possible that Lykon, the Athenian in command of Pissouthnes' Greek mercenaries, was acting in accordance with his city's wishes,[41] which might help explain their support for Amorges. If they had hoped to use the rebellion as leverage against Darius, their plans evaporated when Amorges, too, was defeated and captured (Thuc. 8.28).

Second, the diplomatic contacts between the Spartans and the Persians had been growing more frequent, if not particularly fruitful. Unlike the Spartans who, in the 450s, sent Megabazos away without taking his gold, in 432 Archidamus was ready to publicly declare the necessity of reaching out for financial or naval assistance from both Greek and barbarian allies; in the

35 Lewis (1989).

36 Dem. 9.42; Cary (1935); Walbank (1978) 67–75.

37 Recent examples include: Hornblower (1991) 179–81; Lewis (1992) 121–27; Cawkwell (1997) 115–22; Rung (2008) 31–33; Hyland (2018) 16–17.

38 Walsh (1981) 31–41; Meister (1982) 28–30; Badian (1987) 2–11; Badian (1993) 2–17.

39 Dem. *De Falsa Legatione* 273; Isoc. *Panath.* 12.59–61; Lycurgus *Leoc.* 73; Diod. Sic. 12.4.5; Plut. *Cim.* 13.4. See Sealey (1955), 331; Wade-Gery (1958) 214–20; Thompson (1981) 169–77; Badian (1987) 33–38; Cawkwell (1997) 22–29.

40 Wade-Gery (1958) 207–21; Andrewes (1961) 1–5; Thompson (1971) 119–24; Blamire (1975); Lewis (1977) 76–77; Descat (1991) 33–39; Rhodes (2008) 12–14; Hyland (2018) 42–43. The main attacks on the existence of this treaty are in Harris (1999) and (2000) 479–506.

41 Hyland (2018) 43.

following year, Spartan embassies did indeed go out to 'the king and other barbarians (Thuc. 1.82, 2.7)'. In 430, a Spartan embassy seeking money from the king was captured in Thrace; they were brought to Athens and executed (Thuc. 2.67). In 425, the Athenians intercepted the Persian Artaphernes on his way from Artaxerxes to Sparta. He was bearing a letter written in Aramaic in which the king complained that, despite numerous embassies sent to him, the Spartans seemed to him to be unable to adopt a single, clear policy. However, Artaxerxes was still willing to negotiate with them, if they were prepared to 'speak plainly' with him (Thuc. 4.50).[42]

Ultimately, the break between Athens and Persia was more the result of a change in Persian, rather than Spartan, policy.[43] In 413, Darius ordered Tissaphernes, satrap of Lydia, to submit tribute payments, for which he was in arrears. Tissaphernes attempted to collect this from the Greek cities of Ionia, but the Athenians prevented him from doing so. Tissaphernes had also been tasked with putting down Amorges' rebellion in Caria. According to Thucydides, Tissaphernes sought to forge an alliance with the Spartans and their allies, whose troops might help him accomplish both goals (Thuc. 8.5.5).

As the first step in this plan, Tissaphernes sent an ambassador to Sparta as part of a joint embassy with representatives from Chios and Erythrai, Athenian allies who wished to revolt (Thuc. 8.5.4–5). At the same time, two more ambassadors arrived in Sparta: Kalligeitos of Megara and Timagoras of Kyzikos, exiles living in Daskyleion, the court of Pharnabazos, satrap of Hellespontine Phrygia. Their mission was to convince the Spartans to send a fleet not to Tissaphernes, but to Pharnabazos, so that the latter could achieve the reassertion of Darius' control over the Asian Greeks. Kalligeitos and Timagoras brought twenty-five talents to help pay for the fleet, should they sail to Pharnabazos. Although Thucydides does not state how much Tissaphernes' man brought, he does say that the satrap offered to furnish provisions for any Spartan troops sent to him. The Spartans eventually decided to support the Chians and Erythraians (Thuc. 8.6).[44] This agreement,[45] although initially unfruitful because the

42 Lewis (1977) 65–67; Hyland (2018) 41–42.

43 Persian policy certainly included an analysis of the Athenian disaster on Sicily, but a discussion of this would take us too far from our main subject.

44 Thucydides states that the Spartans made an alliance (σύμμαχος) with the Chians and Erythraeans, not yet with Tissaphernes.

45 Consensus holds that the first two agreements were only drafts or temporary agreements, and that only the third version was the formally ratified 'treaty'. See Westlake (1979) 36 n. 7; Andrewes (1981) 40, 143; Cartledge (1987) 266; Cawkwell (2005) 149–50; Hyland (2018) 55–56. Contra Lévy (1983, 229) who sees three treaties with a meaningful progression of correction and specification.

Athenians intercepted the fleet (Thuc. 8.11), initiated the alliance between the Spartans and Achaemenids. Immediately after the revolt of Miletos from Athens, in the summer of 412, Tissaphernes concluded an alliance with the Spartans, represented by Chalkideus, the Spartan commander in Ionia:[46]

> Whichever lands and cities the king has, or the king's ancestors had, shall be the king's; and however much money or other things came to the Athenians from these cities, the king and the Spartans and their allies shall in common prevent the Athenians from acquiring either money or any other thing.
>
> The king and the Spartans and their allies shall make war on the Athenians in common; and it shall not be lawful to bring an end to the war with the Athenians, unless it is deemed right by both the king and the Spartans and their allies.
>
> If any should revolt from the king, they shall be the enemies of the Spartans and their allies. And likewise, if any should revolt from the Spartans and their allies they shall be the enemies of the king. (Thuc. 8.18)

Tissaphernes had declared a bold reassertion of Persian imperial power. The first and third terms indicated that Darius claimed the right to intervene in, and indeed to conquer and hold, the parts of Greece that had been made subject to Darius I and Xerxes between 513 and 480.[47] Even if Darius was unlikely to make good on this claim, it strongly reinforced his claim to authority, both to the Greeks and to his satraps in the region. After the successful capture of Amorges in 412, Tissaphernes paid the fleet one month's wages at the rate of one drachma per day, for a total of fifty-five talents. However, he said that, going forward, he would only pay half that rate (twenty-seven and a half talents for the fleet), until he received permission from Darius to give the full drachma. In response, the Spartans negotiated him up to a total of thirty talents (Thuc. 8.29). Although Darius did not participate in the negotiations, leaving that task to Tissaphernes, the satrap's statements about the pay increase show that Darius was still very much aware of and in charge of the final diplomatic agreement.[48]

That winter, continued Spartan dissatisfaction with the treaty led to renewed negotiations. The resulting new treaty made the relationship between the

46 For general commentary and literature, see Andrewes (1981) 39–42; Hornblower (2008) 800–802.

47 Amit (1974) 62; Andrewes (1981) 58–59; Hyland (2018) 55.

48 Hyland (2018) 60–62.

Spartans and Achaemenids more equitable.[49] Where the first only defended the territories of the king, the second treaty provided for mutual defence of each party's territory, in addition to mutual aggression against the Athenians; this second treaty also specified that Darius would cover the expenses for all troops serving in his territory. This was no act of generosity; by extending his largess to the Spartans, Darius made them his clients, and effectively claimed their war and their (potential) victory as his own.[50] Finally, while the first treaty was officially concluded between 'The Spartans and their allies' and 'the King and Tissaphernes', this second was between 'The Spartans and their allies' and 'King Darius and the sons of the King, and with Tissaphernes' (Thuc. 8.37).[51] By bringing in his sons, Darius affirmed that he intended his successors to maintain the agreement in the future.[52]

At the same time that Tissaphernes negotiated this treaty, his rival Pharnabazos was still at work. Putting their master's twenty-five talents to good use, the exiles Kalligeitos and Timagoras sailed out from the Peloponnese with twenty-seven ships. Thucydides' wording makes clear that the Spartan commanders were ordered to send some, but not necessarily all of these ships to the Hellespont, 'if they thought it best' (ἢν δοκῇ), and no treaty formalised their relationship with Pharnabazos (Thuc. 8.39). What is more, among the fleet was Lichas, a Spartiate advisor on the war effort in Asia. One of his first actions upon meeting Tissaphernes on Knidos was to disavow the two previous agreements and demand a new one, which would not oblige the Spartans to recognise Persian claims on the islands and the mainland (Thuc. 8.43).

Still in the winter of 412–11, an Athenian embassy met with Tissaphernes to discuss a possible treaty. The Athenians were willing to cede control of all of Ionia to Darius, but refused to consent to the creation of an imperial fleet in the Aegean and to grant it total freedom of operation along the coast (Thuc. 8.56). Having come with the intention of ending Tissaphernes' support for the Spartans, the Athenians were yet unwilling to achieve this at the cost of allowing a second hostile fleet into the region. Immediately after the failed

49 For general commentary and literature, see Andrewes (1981) 79–82; Hornblower (2008) 854–57.

50 Lévy (1983) 235. See Hyland (2018), 74, in reference to the third and final treaty.

51 This contradicts the claim that the Great King could never allow himself to be bound by an oath on equal terms to a mere Greek *polis*, contra Martin (1963) 230–31; Badian (1987) 27–28; Rhodes and Osborne (2003) 86–87. See also Andrewes (1981) 140; Hornblower (2008) 923–24.

52 Lévy (1983) 227. Waters (2016, 95 n. 8) calls this reference to the king's sons an 'echo' of Assyrian *adê*, which often demand loyalty to the king and to his successors.

talks with the Athenians in spring 411, Tissaphernes returned to the Spartan delegation and made a final treaty:[53]

> In the thirteenth year of the reign of Darius, while Alexippidas was ephor in Sparta, a treaty was concluded in the plain of the Maeander[54] by the Spartans and their allies with Tissaphernes, Hieramenes, and the sons of Pharnakes, concerning the affairs of the king and of the Spartans and their allies.
>
> The king's lands in Asia shall be the king's, and the king shall do as he wishes concerning his lands.
>
> The Spartans and their allies shall not invade or injure the king's lands; nor shall the king invade or injure that of the Spartans or of their allies. If any of the Spartans or their allies should invade or injure the king's lands, the Spartans and their allies shall hinder it; and if any from the king's country invade or injure the lands of the Spartans or of their allies, the king shall hinder it.
>
> Tissaphernes shall provide for the ships now present, according to the agreement, until the arrival of the king's vessels; but the Spartans and their allies may provide for their own ships after the arrival of the king's ships, if they should wish to. If, however, they choose to receive the pay from Tissaphernes, Tissaphernes shall furnish it; and the Spartans and their allies shall repay him at the end of the war as much money as they will have received.
>
> After the king's ships have arrived, the ships of the Spartans and of their allies and those of the king shall carry on the war in common, as Tissaphernes and the Spartans and their allies should see fit. If they should wish to end the war with the Athenians, they shall also end it in common. (Thuc. 8.58)

The use of both Achaemenid and Spartan dating formulas, and the description of this treaty as 'concluded' (σπένδειν; literally, 'poured', a reference to the ritual offering of libations),[55] suggests that both parties were finally satisfied with the terms and ready to formalise the relationship. It names the parties as the Spartans and their allies and 'Tissaphernes, Hieramenes, and the sons of Pharnakes'. Hieramenes may have been Darius' brother-in-law (Xen.

53 For general commentary and literature, see Andrewes (1981) 138–46, Hornblower (2008) 924–31.

54 See Hyland (2018) 70–73 on the logistics of the negotiations.

55 Karavites (1984).

Hell. 2.1.8–9),[56] while Pharnakes was Pharnabazos' father and predecessor as satrap of Hellespontine Phrygia. We know only one of Pharnabazos' brothers (either 'Magaios', Plut. *Alc.* 39, or 'Bagaios', Xen. *Hell.* 3.4.13), but it is clear that the treaty is meant to include the entire Pharnakid dynasty, which had a hereditary hold over the satrapy.[57] Darius and his heirs are not parties to this treaty, but of course that does not mean that the king was absent. Instead, Tissaphernes, Hieramenes, and the Pharnakids would be expected to represent the king's interests. Darius may have intended that each of the three parties would serve as a balance to the others, preventing any one party from achieving dominance over the others. As the wording indicates, all of the military and financial responsibilities fall on Tissaphernes. Pharnabazos is bound to not violate the terms of the treaty, but the text places no further limits or demands on his activities, a fact that the satrap would exploit to the fullest extent.[58] The structure of the treaty reflects the regional strategy (the attempt to enforce stability by assisting in a Spartan victory), as well as the internal imperial strategy (the king's manipulation of competing factions within the satrapal system to prevent rivals to his own authority).

But it did not last. The Spartans were still upset about their wages, the Athenians were still attempting to draw Tissaphernes to their side, and Pharnabazos was still in competition with Tissaphernes (e.g., Thuc. 8.80, 8.99; Xen. *Hell.* 1.1.27–31). What is more, despite all of the effort that Tissaphernes put into the Spartan treaties, within a few months he broke his side of the agreement. Although he went to Aspendos to fetch the fleet of Phoenician ships there, Tissaphernes never brought them into play; furthermore, he failed to pay the Spartans and their allies while they waited for the fleet to arrive (Thuc. 8.83, 87, 99).[59] Tissaphernes' apparent neglect of the Spartans and their allies caused them to forcibly expel Persian garrisons from Miletos, Knidos, and Antandros (Thuc. 8.84, 104–109).

Several hypotheses have been put forward to explain Tissaphernes' actions: the fleet may have been needed to put down unrest in Egypt, or elsewhere

56 Hieramenes may be mentioned in the Xanthos stele: Keen (1998a) 138.

57 The family may have had only *de facto* control over the satrapy, reconfirmed by the king upon the succession of each new inheritor of the position. Balcer (1984) 315; Briant (1987) 25–28; Briant (2002) 339–40. Hornblower (2008, 928) regards the inclusion of Pharnabazos' patronymic as 'a small sign of the integrated narrative character of the present treaty' within the context of Thucydides' work.

58 Lewis (1977) 103–107.

59 Hyland (2018, 79–81) argues that Tissaphernes was likely paying a reduced rate due to the inefficiencies of Achaemenid financial disbursement.

in the region;[60] Darius may have ordered the fleet to remain at Aspendos in response to the riots at Miletos, Knidos, and Antandros;[61] or, as Thucydides claims, Tissaphernes may have intended to harm the Spartan cause, to ensure that the Athenians and Spartans would wear each other out.[62] The overall effect of the breakdown of Spartan-Persian cooperation meant that, over the next few years, the Athenian-Spartan conflict played itself out ever more on Asian shores (e.g., Xen. *Hell.* 1.2.4–17). This helps to explain why, in 408, Pharnabazos made an agreement (συγχωρεῖν) with the Athenians to give them twenty talents and to escort their embassy to the king. More significantly, Argive and Spartan representatives joined the Athenians (Xen. *Hell.* 1.3.12–13). If neither the Spartans nor Athenians could find outright victory in battle, Persian intervention on only one side would likewise prove insufficient. This first 'common' embassy was, then, a sign of things to come.

Before moving to the post-Peloponnesian War period, let us pause to consider how Greco-Persian diplomacy had developed between the Peace of Kallias and the arrival of Cyrus the Younger on the scene. The first development is the participation of the Spartans, giving the Great King two opposing parties to choose from. This is mirrored by the second development, that of the rival satraps, Pharnabazos and Tissaphernes. There was an essential contradiction in the administrative structure of the northwestern edge of the Achaemenid Empire: in order to contain Greek forces, the satraps had to be sufficiently powerful, wealthy, and autonomous to respond rapidly to unpredictable threats; but if any one man were invested with such power, he would pose a potential threat against the regional stability of the empire.[63]

If Darius II was aware of this risk, he may have allowed Pharnabazos and Tissaphernes to persist in their rivalry as a way to ensure that, should one gain too much power, the other would leap at the opportunity to destroy him. This interpretation helps to explain an apparent paradox. The Achaemenids, despite their reputation for limitless wealth, were very concerned about maximising

60 Lewis (1958); Andrewes (1981) 289–93; Keen (1998c) 100–101.

61 Heitch (2007) 94–96, 134–36, 154–66; Hyland (2004) 89–96; Hyland (2018) 85–91.

62 Lateiner (1976) 274–90; Hawthorn (2014) 214–17. Cawkwell (2005, 153–55) argues that this idea originated with Darius, not Tissaphernes. Westlake (1985, 52–54) discusses Thucydides' possible sources for information on Tissaphernes.

63 Relevant examples include Oroites (ca. 515; Hdt. 3.120–28), Pissouthnes (ca. 420; Thuc. 8.5.5, 8.28.3), Cyrus the Younger (401; Plut. *Artax.* 6–14; Xen. *An.* 1), Datames (ca. 362, Nep. *Datames* 5–11; Diod. Sic. 15.91), Artabazos (ca. 356; Diod. Sic. 16.34, 52). Jacobs (2003), 246–60, uses these cases to argue for the existence of a regularised response to rebellions. See Weiskopf (1989) and Beckman (2020) for alternative analyses.

income and minimising expenditures.[64] Since peace would accomplish both, simultaneously, Darius should have wanted to achieve a stable peace as soon as possible. Why, then, do we see Tissaphernes and the Spartans squabbling over a few extra talents? Why was Tissaphernes allowed to squander the costs of the mission to Aspendos? Why did Darius tolerate Tissaphernes' failure to maintain stability in his own region and to work in harmony with Pharnabazos, if Darius had been steadfast in his desire for a rapid Spartan victory? Darius spent time and money, so that he could keep both the Greeks and his satraps from growing dangerously powerful.[65]

But Darius did not allow this situation to last indefinitely. His decision to send Cyrus as *karanos* was likely a response to the Spartans' inability to bring the war to a close after the battle of Kyzikos in 410 (Xen. *Hell.* 1.1.11–26). This neatly illustrates the point: as *karanos*, Cyrus was uniquely able to oversee a successful conclusion of the war. Because Cyrus was his younger son, Darius probably expected that it was safe to endow him with power over the entire region; when Darius died in 404, however, Cyrus used his position to raise an army and challenge his brother, Artaxerxes II, for the throne in 401.

4.3 *Case Study 3: The King's Peace*

One of the reasons for Darius' support for the Spartans may have been their traditional aversion to overseas adventures. Yet, his successor Artaxerxes II soon found that, having made the defence of the Asian Greeks the keystone of their wartime propaganda, the Spartans were unwilling to quit that position after 404. This attitude was likely reinforced by the Ionians' fear of Tissaphernes after Cyrus' failure in 401 (Xen. *Hell.* 3.1.3–4).[66] Artaxerxes made Tissaphernes the overall commander (στρατηγὸς τῶν πάντων: Xen. *Hell.* 3.2.13; Diod. Sic. 14.35.2–3), and compelled Pharnabazos to recognise Tissaphernes' authority in military matters; their rivalry was subdued, but not forgotten.

Two events forced the Great King to make a radical change. First, in 396, the Spartans concluded an alliance with the pharaoh Nefaarud I (Nepherites: Diod. Sic. 14.79.4). The next year, the Spartan king Agesilaos led a raid on the outskirts of Sardis. Although this attack did not cause terrible damage, Artaxerxes

64 See *Hell. Oxy.* 19.2.

65 Hamilton (1979, 101) argues that an Achaemenid king might allow his satraps to feud, as long it did not get too out of hand, because he feared that royal intervention might precipitate a full-scale rebellion. Ruzicka (1985, 208) and Rop (2019b, 60–68) discussing a later context, suggest that Artaxerxes was well aware of the potential and actual conflicts between the various officials in Asia Minor, and may have counted on them to restrain Cyrus the Younger.

66 Lewis (1977) 110–17; Hyland (2018) 128–30.

had Tissaphernes executed for this failure, and replaced him with Tithraustes (Xen. *Hell.* 3.4.21–25).[67] About this time, Timokrates of Rhodes was sent (either by Tithraustes or Pharnabazos) with fifty talents to encourage resistance to the Spartans in Greece.[68] The resulting Corinthian War, coupled with Konon's successes at sea (namely at Rhodes and Knidos[69]), forced Agesilaos to abandon Asia in 394 (Xen. *Hell.* 4.8.8–10; Diod. Sic. 14.39, 79).

In light of Persian support for their enemies, in 392 the Spartans looked to cut off Pharnabazos' aid (Xen. *Hell.* 4.8.12). The Spartans sent Antalkidas to Tiribazos, the 'king's *stratēgos*'[70] at Sardis, intending to reach a private agreement. When the Athenians found out about Antalkidas' mission, they and their allies sent embassies as well. However, the talks fell apart when the Spartans insisted on the autonomy of all the islands; as the Athenian goal seems to have been to reconstitute some form of their empire, autonomy would only serve to weaken them (Xen. *Hell.* 4.8.14–15).[71] After the 392 talks failed, Tiribazos gave money to the Spartans, but hid this from the king, and imprisoned Konon. He then went to consult the king, who replaced him with Strouthas. Strouthas threw his weight behind the Athenians, provoking the Spartans to raid royal territory (Xen. *Hell.* 4.8.16–17).

The failure of the talks in 392, and the conflicting objectives of Tiribazos and Strouthas, led to renewed warfare in Ionia and the islands, as both the Spartans and Athenians attempted to reassert their right to intervene in the affairs of the Asian Greek *poleis* (Diod. Sic. 14.99; Xen. *Hell.* 4.8.18–19). By 388, Tiribazos had returned from his consultation with Artaxerxes, who had agreed to support the Spartans against the Athenians, but only after giving them an opportunity to accept a new peace agreement. The Athenians balked at first, but after Antalkidas' fleet blocked grain shipments to Athens, they finally relented. In 387 Tiribazos called the various embassies to Sardis and read out a royal edict summarising the so-called 'King's Peace' or 'Peace of Antalkidas', reflecting the Spartans' key role in the proceedings.

67 Westlake (1981); Rop (2018) 64–71.

68 Xen. *Hell.* 3.5.1, 4.2.1–4; Paus. 3.9.8; Plut. *Artax.* 20.3–4, *Hell. Oxy.* 7.2. The details of Timokrates' mission are extremely contentious. See the recent analyses of Cook (1990); Rung (2004); Bleckmann (2006) 91–100; Schepens (2012).

69 Hyland (2018) 133–37, 145–47, 149–50.

70 Tiribazos had been a subordinate (*hyparchos*) to the satrap of Armenia and an advisor of Artaxerxes during the rebellion of Cyrus the Younger in 404 (Xen. *An.* 4.4.4), but there is no information on when he was sent to the coast. Keen (1998b) suggests that Tiribazos may have been *karanos*.

71 See Wilker (2012); Hyland (2018) 156–61.

> King Artaxerxes thinks it right that the cities in Asia should be his, as well as Klazomenai and Cyprus among the islands, and that the other Greek cities, both small and large, should be left autonomous, except Lemnos, Imbros, and Skyros; and these should be Athenian, as of old. But whichever of the two[72] does not accept this peace, I will make war against them, with those who desire this arrangement, both on land and by sea, and with ships and with money. (Xen. *Hell.* 5.1.31)

This is an epitome of the original, which likely would have included a preamble, as in the third Spartan-Persian treaty in 411, and appropriate rituals.[73] Isocrates states that the original text established an 'eternal alliance' between the king and the Greeks, and that the Greeks were required to display an inscription of the peace in their temples (*Paneg.* 4.128, 180), but the rhetorical intent of this speech means we cannot take it at face value. Xenophon reports that 'each of the ambassadors returned to their own cities ... and all the cities swore that they would observe the articles of the peace, except the Thebans' (Xen. *Hell.* 5.1.32). Shortly thereafter, Agesilaos forced Thebes into swearing.

The King's Peace solidified the trends that began in the previous decades. For the first time, an agreement was made between all the parties concerned, not just the king and his favoured Greek side. The inability of either the Spartans or Athenians to maintain a stable hegemony, and the overwhelming impact of Achaemenid financial intervention, meant that there was no realistic alternative to this new system. This treaty codified the inviolability of Asia. It also reflected the king's inability to prevent Greeks from disturbing his realm on his own (due mainly to his problems in Egypt and Cyprus[74]). But the guarantee of autonomy for all *poleis*, even those outside his realm, would go far towards preventing the formation of any anti-Persian coalitions in the future.

The year 387 in no way marked the end of Greco-Persian diplomatic interaction, and neither the Greek world nor the empire were internally stable. But future developments would take place within the framework of the King's Peace. The clearest testimony to the impact of the King's Peace is the enthusiasm with which the Greeks would comply with it. For one example, in 384/3, Athens and Chios formed an alliance. Not wishing to appear to violate the

72 ὁπότεροι, i.e., the Spartans and the Athenians and their respective allies.

73 Cawkwell (1981); Badian (1991) 35–48; Buckler and Beck (2008) 71.

74 On Cyprus, Evagoras of Salamis went into revolt in ca. 391 and made an alliance with Akoris of Egypt in the early 380s; see Diod. Sic. 14.98; Tuplin (1983) 176–80; Hyland (2018) 160–1. According to Shrimpton (1991, 10), 'the reduction, or at least neutralization, of Evagoras would be an unavoidable preliminary' for the reconquest of Egypt.

autonomy clause of the King's Peace, this treaty text specified that the alliance preserved 'the peace, the friendship, the oaths and the existing agreement, which the king, the Athenians, the Spartans and the other Greeks swore', and heralded 'good things for the People of Athens, for all of Greece, and for the king' (*IG* II^2 34 9–15).[75] Similar statements were made at the Athenian peace conference in 371 (Xen. *Hell.* 6.3.12; 6.5.1–2).

This was not empty rhetoric. The Greeks had learned from hard experience that the King would not hesitate to enforce the terms of his treaties, by paying one side to attack the other. This allowed him to place controls on Greek military energy. In 378, Pharnabazos, threatening the loss of Artaxerxes' favour, insisted that the Athenians recall the general Chabrias from Egypt, who had joined the rebels there as a mercenary; not only did the Athenians comply, they sent Iphikrates to serve alongside Pharnabazos (Diod. Sic. 15.29.1–4). In 375, Artaxerxes encouraged peace between the Athenians and Spartans to ensure that he would have a reliable stock of mercenaries for yet another Egyptian campaign (Diod. Sic. 15.38.1–2; Xen. *Hell.* 6.2.1–2).[76] In 367, following the rise of Theban power in Greece, the Thebans, Spartans, and Athenians travelled to Susa for a conference with the king. The resulting peace, once again made in the form of a royal proclamation, brought an end to Spartan and Athenian hegemony: it guaranteed the autonomy of Messene, and forced the Athenian fleet to return to port. The King would make war on either in the event of a violation, and on any city that refused to join a retributive campaign (Xen. *Hell.* 7.1.34–6).

5 Conclusion

We have seen that, even as the form of Achaemenid diplomacy changed over time (from earth and water, to bilateral treaties, to multilateral treaties), diplomatic efforts always worked in concert with the empire's military activities. The demand for earth and water was intended to avoid or minimise open warfare in situations where outright victory was uncertain. Written treaties were intended to bring stability to a turbulent region, so that the king could focus his martial energies on other, more important regions. Earth and water and

75 See also the charter of the Second Athenian League, *IG* II^2 43 9–25.

76 Rop (2019a, 88–118) emphasises that both the Persians and the rebellious Egyptians called on Greek mercenaries not because they were dependent on the martial prowess of Greeks, but in order to 'establish or to reinforce alliances between foreign leaders and Greek city-states' (89).

the treaties both clearly established the demands the king, and thus provided a *casus belli* in the event of future instability.

Diplomacy had an important function within the empire as well. The king used diplomacy not only to solve external conflicts, but also to redirect the energies of his satraps and commanders outward, potentially lowering the number and severity of rebellions. In addition, the king was able to use these diplomatic ventures to bolster his own prestige, even if he was not himself personally involved. By positioning himself as the final arbiter of interstate relations, and as the ultimate paymaster of both his satraps and his foreign allies, the king intended that no significant decisions could be made against his will. If, in its language, a treaty placed limits on the king's power, it would be framed as a confirmation of his power and righteousness: his subjects came to him so that he might govern them through his law, a duty bestowed upon him by Ahuramazda. According to Darius the Great, 'When Ahuramazda saw this earth in turmoil, at that time he bestowed it upon me ... By the greatness of Ahuramazda I put it back in its place (DNa 30–36)'.

Most of the features of Achaemenid diplomacy have Near Eastern precedents. Where Darius demanded earth and water, Esarhaddon collected loyalty oaths from the Median city-lords. Where Artaxerxes I acknowledged Athenian control over the Aegean, Esarhaddon recognised Urtak(u) as his 'brother' and formalised a border agreement with him. Boubares' marriage to the Macedonian princess echoes the marriage alliance between the son of Cyaxares and the daughter of Alyattes. There are, of course, important differences: the Assyrians established formalised diplomatic agreements across the Near East, while the Achaemenids seem to have limited the practice to the Aegean region. While the Assyrians are not known to have financed the wars of their allies, the Achaemenids made great use of 'daric diplomacy'. This would seem to be a function of the Achaemenids' traditional manipulation of local traditions within the satrapal system; in this case, their financing of Greek wars was made possible by their mastery of Greek economic and military (mercenary) traditions.

In the absence of more information about Achaemenid diplomacy, these comparisons allow us to create a better sense of the role of diplomacy in the Achaemenid Empire. Diplomacy was a never a goal in and of itself; it was a tool for carrying out the king's will. Nor was it a particularly favoured tool; the king preferred to conquer territory through military action, and then rule over that territory through a satrap or a client king. Diplomacy's subordinate ideological position relative to military conquest would explain the absence of the former from the empire's visual and inscriptional record. Like, for example, bureaucratic record keeping, diplomacy was simply one of the necessary but unglamorous realities of empire. But the Achaemenids nevertheless were committed to the long-term success of their diplomatic ventures. They expended a great deal of money and energy on them, not only in the conduct of negotiation and fulfillment of treaty terms, but in learning and adapting to Greek notions of diplomatic conduct.

Bibliography

Adcock, F. and Mosley, D. (1975) *Diplomacy in Ancient Greece*. New York.

Altman, A. (2010) 'How Many Treaty Traditions Existed in the Ancient Near East?', in Cohen, Gilan, and Miller (2010) 17–36.

Altman, A. (2012) *Tracing the Earliest Recorded Concepts of International Law: The Ancient Near East (2500–330 BCE)*. Legal History Library, Studies in the History of International Law 8. Leiden.

Álvarez-Mon, J. (2006) 'The Arjan Tomb: At the Crossroads of the Elamite and the Persian Empires', PhD Dissertation, University of California, Berkeley.

Álvarez-Mon, J. (2010) *The Arjan Tomb: At the Crossroads of the Elamite and the Persian Empires*. Acta Iranica 49. Leuven.

Álvarez-Mon, J. (2011) 'The Golden Griffin from Arjan', in Álvarez-Mon and Garrison (2011) 299–373.

Álvarez-Mon, J. and Garrison, M. (eds) (2011) *Elam and Persia*. Winona Lake, IN.

Álvarez-Mon, J., Basello, G.P., and Wicks, Y. (eds) (2018) *The Elamite World*. London.

Amiet, P. (1992) 'Sur l'histoire Élamite'. *IrAnt* 27: 75–94.

Amit, M. (1974) 'A Peace Treaty Between Sparta and Persia'. *RSA* 4: 55–63.

Andrewes, A. (1961) 'Thucydides and the Persians'. *Historia* 10: 1–18.

Andrewes, A. (1981) *A Historical Commentary on Thucydides. Vol. 5: Book VIII*. Oxford.

Badian, E. (1987) 'The Peace of Callias'. *JHS* 50: 1–39.

Badian, E. (1991) 'The King's Peace'. *BICS* 37: 25–48.

Badian, E. (1993) *From Plataea to Potidaea: Studies in the History and Historiography of the Pentecontaetia*. Baltimore.

Badian, E. (1994) 'Herodotus on Alexander I of Macedon: A Study in Some Subtle Silences', in Hornblower (1994) 107–30.

Balcer, J. (1984) *Sparda by the Bitter Sea: Imperial Interaction in Western Anatolia*. Chico, CA.

Balcer, J. (1988) 'Persian Occupied Thrace (Skudra)'. *Historia* 37: 1–21.

Balcer, J. (1995) *The Persian Conquest of the Greeks, 545–450 B.C.* Konstanz.

Beckman, D. (2017) 'King Artaxerxes' Aegean Policy'. *JPS* 10: 1–25.

Beckman, D. (2020) 'Law, Mercy, and Reconciliation in the Achaemenid Empire'. *JAH* 8: 127–51.

Binder, C., Börm, H., and Luther, A. (eds) (2016) *Diwan – Studies in the History and Culture of the Ancient Near East and the Eastern Mediterranean*. Duisburg.

Bederman, D. (2001) *International Law in Antiquity*. Cambridge.

Blamire, A. (1975) 'Epilycus' Negotiations with Persia'. *Phoenix* 29: 21–26.

Bleckmann, B. (2006) *Fiktion als Geschichte: Neue Studien zum Autor der Hellenika Oxyrhynchia und zur Historiographie des vierten vorchristlichen Jahrhunderts*. Göttingen.

Borza, E. (1990) *In the Shadow of Olympus*. Princeton.

Briant, P. (1984) 'La Perse avant l'Empire (un état de la question)'. *IrAnt* 19: 71–118.

Briant, P. (1987) 'Pouvoir central et polycentrisme culturel dans l'Empire achéménide (Quelques réflexions et suggestions)', in Sancisi-Weerdenberg (1987) 1–31.

Briant, P. (2002) *From Cyrus to Alexander: A History of the Persian Empire*. Winona Lake.

Brinkman, J. (1990) 'Political Covenants, Treaties, and Loyalty Oaths in Babylonia and between Assyria and Babylonia', in Canfora, Liverani, and Zaccagnini (1990) 81–111.

Buckler, J. and Beck, H. (eds) (2008) *Central Greece and the Politics of Power in the Fourth Century BC*. Cambridge.

Burstein, S. (1978) *The 'Babyloniaca' of Berossus*. Vol. 1. Sources from the Ancient Near East. Malibu, CA.

Canfora, L., Liverani, M., and Zaccagnini, C. (eds) (1990) *I Trattati Nel Mondo Antico: Forma, Ideologia, Funzione*. Rome.

Cartledge, P. (1987) *Agesilaos and the Crisis of Sparta*. London.

Cary, M. (1935) 'Arthmius of Zeleia'. *CQ* 29: 177–80.

Cawkwell, G.L. (1981) 'The King's Peace'. *CQ* 31: 69–83.

Cawkwell, G.L. (1997) 'The Peace between Athens and Persia'. *Phoenix* 51: 115–30.

Cawkwell, G.L. (2005) *The Greek Wars: The Failure of Persia*. Oxford.

Cereti, C. (ed) (2010) *Iranian Identity in the Course of History: Proceedings of the Conference Held in Rome, 21–24 September 2005*. Rome.

Clark, J. and Turner, B. (eds) (2018) *Brill's Companion to Military Defeat in Ancient Mediterranean Society*. Leiden and Boston.

Cohen, Y., Gilan, A., and Miller, J. (eds) (2010) *Pax Hethitica: Studies on the Hittites and Their Neighbours in Honour of Itamar Singer*. Wiesbaden.

Collins, J. and Manning, J. (eds) (2016) *Revolt and Resistance in the Ancient Classical World and the Near East*. Leiden.

Cook, M.L. (1990) 'Timocrates' 50 Talents and the Cost of Ancient Warfare'. *Eranos* 88: 69–97.

Da Riva, R. (2008) *The Neo-Babylonian Royal Inscriptions: An Introduction*. Münster.

Dalley, S. and Postgate, J.N. (1984) *The Tablets from Fort Shalmaneser*. Oxford.

Delnero, P. and Lauinger, J. (eds) (2015) *Texts and Contexts: The Circulation and Transmission of Cuneiform Texts in Social Space*. Boston.

Descat, R. (1991) 'Colophon et la Paix d'Epilycos', in Malay (1991) 33–39.

Erbse, H. (1992) *Studien zum Verständnis Herodots*. Berlin.

Fearn, D. (2007) 'Narrating Ambiguity: Murder and Macedonian Allegiance (5.17–22)', in Irwin and Greenwood (2007) 98–127.

Flensted-Jensen, P., Nielsen, T., and Rubinstein, L. (eds) (2000) *Polis and Politics: Studies in Ancient Greek History*. Copenhagen.

Folmer, M. (2011) 'Old and Imperial Aramaic', in Gzella (2011) 128–59.

Folmer, M. (2012) 'Imperial Aramaic as an Administrative Language of the Achaemenid Period', in Weninger (2012) 587–97.

Frame, G. (1992) *Babylonia 689–627 B.C.: A Political History*. Istanbul.

Frame, G., Leichty, E., Sonik, K., Tigay, J., and Tinney, S. (eds) (2011) *A Common Cultural Heritage: Studies on Mesopotamia and the Biblical World in Honor of Barry l. Eichler*. Bethesda, MD.

France, J. and de Souza, P. (eds) (2008) *War and Peace in Ancient and Medieval History*. Cambridge.

Gabriella, M., Bertinelli, A., and Piccirilli, L. (eds) (2001) *Linguaggio e terminologia diplomatica dall'antico oriente all'impero Bizantino (atti del Convegno nazionale, Genova, 19 novembre 1998)*. Serta antiqua et Mediaevalia 4. Rome.

Gardiner-Garden, J. (1987) 'Dareios' Scythian Expedition and Its Aftermath'. *Klio* 69: 326–50.

Garrison, M. (1991) 'Seals and the Elite at Persepolis: Some Observations on Early Achaemenid Persian Art'. *Ars Orientalis* 21: 1–29.

Garrison, M. (2017) *The Ritual Landscape at Persepolis: Glyptic Imagery from the Persepolis Fortification and Treasury Archives*. Studies in Ancient Oriental Civilization 72. Chicago.

Garrison, M. and Root, M.C. (2001) *Seals on the Persepolis Fortification Tablets, Volume I: Images of Heroic Encounter. Part I: Text*. Chicago.

Gaspa, S., Greco, A., Bonacossi, D., Ponchia, S., and Rollinger, R. (eds) (2014) *From Source to History: Studies on Ancient Near Eastern Worlds and Beyond (Festschrift for Giovanni B. Lanfranchi)*. Münster.

Gehler, M. and Rollinger, R. (eds) (2014) *Imperien und Reiche in Der Weltgeschichte: Epochenübergreifende und Globalhistorische Vergleiche*. Wiesbaden.

Genito, B. (1986) 'The Medes: A Reassessment of the Archaeological Evidence'. *East and West* 36: 11–81.

Gorris, E. (2018) 'Crossing the Elamite Borderlands: A Study of Interregional Contacts Between Elam and "Kingdom" of Hara(n)', in Tavernier, Gorris, Abraham, and Boschloos (2018) 313–44.

Grayson, A.K. (1975) *Assyrian and Babylonian Chronicles*. Locust Valley, NY.

Grayson, A.K. (1987) 'Akkadian Treaties of the Seventh Century B.C.'. *JCS* 39: 127–60.

Gzella, H. (ed) (2011) *Languages from the World of the Bible*. Boston.

Hamilton, C.D. (1979) *Sparta's Bitter Victories: Politics and Diplomacy in the Corinthian War*. Ithaca.

Harris, E. (1999) '*IG* I³ 227 and the So-called Peace of Epilykos'. *ZPE* 126: 123–28.

Harris, E. (2000) 'The Authenticity of Andocides *De Pace*: A Subversive Essay', in Flensted-Jensen, Nielsen, and Rubinstein (2000) 479–506.

Harrison, T. (2019) 'A Persian Marriage Feast in Macedon? (Herodotus 5.17–21)'. *CQ* 69: 507–14.

Hawthorn, G. (2014) *Thucydides on Politics: Back to the Present*. Cambridge.

Heitsch, E. (2007) *Geschichte und Personen bei Thukydides: eine Interpretation des achten Buches*. Berlin.

Henkelman, W. (2011) 'Cyrus the Persian and Darius the Elamite: A Case of Mistaken Identity', in Rollinger, Truschnegg, and Bichler (2011) 577–634.

Henkelman, W. (2012) 'The Achaemenid Heartland: An Archaeological-Historical Perspective', in Potts (2012) 931–62.

Henkelman, W. (2017) 'Humban & Auramazdā: Royal Gods in a Persian Landscape', in Henkelman and Redard (2017) 273–346.

Henkelman, W. (2018) 'Elamite Administrative and Religious Heritage in the Persian Heartland', in Álvarez-Mon, Basello, and Wicks (2018) 803–28.

Henkelman, W. and Redard, C. (eds) (2017) *Persian Religion in the Achaemenid Period / La Religion Perse à l'époque Achéménide*. Wiesbaden.

Herrenschmidt, C. and Lincoln, B. (2004) 'Healing and Salt Waters: The Bifurcated Cosmos of Mazdaean Religion'. *History of Religions* 43: 269–83.

Hillard, T. and Judge, E. (eds) (1998) *Ancient History in a Modern University: Proceedings of a Conference Held at Macquarie University, 8–13 July, 1993: To Mark Twenty-Five Years of the Teaching of Ancient History at Macquarie University and the Retirement from the Chair of Professor Edwin Judge*. New South Wales.

Hinz, W. (1967) 'Elams Vertrag Mit Narām-Sîn von Akkade'. *ZA* 58: 66–96.

Hobden, F. and Tuplin, C. (eds) (2012) *Xenophon: Ethical Principles and Historical Enquiry*. Leiden.

Hornblower, S. (1991) *A Commentary on Thucydides. Vol. 1: Books I–III*. Oxford.

Hornblower, S. (ed) (1994) *Greek Historiography*. Oxford.

Hornblower, S. (2008) *A Commentary on Thucydides. Vol. 3: Books 5.25–8.109*. Oxford.

Hunt, P. (2010) *War, Peace, and Alliance in Demosthenes' Athens*. Cambridge.

Hurowitz, V. and Yonah, S. (eds) (2015), *Marbeh Ḥokmah: Studies in the Bible and the Ancient Near East in Loving Memory of Victor Avigdor Hurowitz*. Winona Lake.

Hyland, J. (2004) 'Waiting for Tissaphernes: Athens and Persia in Thucydides VIII'. *Syllecta Classica* 15: 71–101.

Hyland, J. (2018) *Persian Interventions: The Achaemenid Empire, Athens, and Sparta, 450–386 BCE*. Baltimore.

Irwin, E. and Greenwood, E. (eds) (2007) *Reading Herodotus: A Study of the Logoi in Book 5 of Herodotus' Histories*. Cambridge.

Jacobs, B. (2003) 'Mechanismen der Konfliktbewältigung in der Verwaltungsorganisation Kleinasiens zur Achämenidenzeit', in Kuhrt and Henkelman (2003) 239–62.

Jursa, M. (2014) 'The Neo-Babylonian Empire', in Gehler and Rollinger (2014) 121–48.

Karavites, P. (1984) '*Spondai-Spendein* in the Fifth Century B.C.'. *Ant. Class.* 53: 60–70.

Keen, A.G. (1998a) *Dynastic Lycia: A Political of History of the Lycians and Their Relations with Foreign Powers: C. 545–362 B.C.* Leiden.

Keen, A.G. (1998b) 'Persian Karanoi and Their Relationship to the Satrapal System', in Hillard and Judge (1998) 88–95.

Keen, A.G. (1998c) 'Persian Policy in the Aegean, 412–386 B.C.'. *JAC* 13: 93–110.

Kitchen, K. and Lawrence, P. (2012) *Treaty, Law and Covenant in the Ancient Near East*. Wiesbaden.

Klinkott, H. (2016) '"Dem König Erde und Wasser bringen". Persisches Unterwerfungsritual oder herodoteisches Konstrukt?', in Binder, Börm, and Luther (2016) 133–82.

Kozuh, M., Henkelman, W., Jones, C., and Woods, C. (eds) (2014) *Extraction & Control: Studies in Honor of Matthew W. Stolper*. Chicago.

Kramer, N. (2004) 'Athen: keine Stadt des Grosskönigs!' *Hermes* 132: 257–70.

Kuhrt, A. (1983) 'The Cyrus Cylinder and Achaemenid Imperial Policy'. *JSOT* 8: 83–97.

Kuhrt, A. (1988) 'Earth and Water', in Kuhrt and Sancisi-Weerdenburg (1988) 87–99.

Kuhrt, A. (2007) *The Persian Empire: A Corpus of Sources from the Achaemenid Period*. London.

Kuhrt, A. and Henkelman, W. (eds) (2003) *A Persian Perspective: Achaemenid Essays in Memory of Heleeen Sancisi-Weerdenburg*. Leiden.

Kuhrt, A. and Sancisi-Weerdenburg, H. (eds) (1987) *The Greek Sources: Proceedings of the Groningen 1984 Achaemenid History Workshop*. Leiden.

Kuhrt, A. and Sancisi-Weerdenburg, H. (eds) (1988) *Method and Theory: Proceedings of the London 1985 Achaemenid History Workshop*. Leiden.

Lanfranchi, G., Roaf, M., and Rollinger, R. (eds) (2003) *Continuity of Empire (?): Assyria, Media, Persia*. Padova.

Lanfranchi, G., Mattila, R., and Rollinger, R. (eds) (2019) *Writing Neo-Assyrian History. Sources, Problems, and Approaches*. State Archives of Assyria Studies 30. Helsinki.

Lateiner, D. (1976) 'Tissaphernes and the Phoenician Fleet (Thucydides 8. 87)'. *TAPA* 106: 267–90.

Lauinger, J. (2013) 'The Neo-Assyrian *Adê*: Treaty, Oath, or Something Else?' *ZAR* 19: 99–116.

Lauinger, J. (2015) 'Neo-Assyrian Scribes, "Esarhaddon's Succession Treaty", and the Dynamics of Textual Mass Production', in Delnero and Lauinger (2015) 285–314.

Leichty, E. (2011) *The Royal Inscriptions of Esarhaddon, King of Assyria (680–669 BC)*. RINAP 4. Winona Lake.

Levin, Y. (2015) 'How Did Rabshakeh Know the Language of Judah?', in Hurowitz and Yonah (2015) 323–37.

Lévy, E. (1983) 'Les trois traités entre Sparte et le Roi'. *Bulletin de Correspondance Hellénique* 107: 221–41.

Lewis, D.M. (1958) 'The Phoenician Fleet in 411'. *Historia* 7: 392–97.

Lewis, D.M. (1977) *Sparta and Persia: Lectures Delivered at the University of Cincinnati, Autumn 1976 in Memory of Donald W. Bradeen*. Leiden.

Lewis, D.M. (1989) 'Persian Gold in Greek International Relations'. *REA* 91: 227–34.

Lewis, D.M. (1992) 'The Thirty Years' Peace', in *The Cambridge Ancient History*, 2nd edn. Cambridge. 121–46.

Liverani, M. (1995) 'The Medes at Esarhaddon's Court'. *JCS* 47: 57–62.

Luckenbill, D. (1927) *Ancient Records of Assyria and Babylonia*. Vol. 2. Chicago.

Luukko, M. and van Buylaere, G. (2002) *The Political Correspondence of Esarhaddon*. SAA 16. Helsinki.

Magnetti, D. (1978) 'The Function of the Oath in the Ancient Near Eastern International Treaty'. *The American Journal of International Law* 72: 815–29.

Malay, H. (ed) (1991) *Erol Atalay memorial (Arkeoloji dergisi özel sayi 1)*. Izmir.

Martin, V. (1963) 'Quelques remarques à l'occasion d'une nouvelle édition des "Staatsverträge des Altertums"'. *Museum Helveticum* 20: 230–33.

McCarthy, D. (1978) *Treaty and Covenant: A Study in Form in the Ancient Oriental Documents and in the Old Testament*. Analecta Biblica 21A. Rome.

Meister, K. (1982) *Die Ungeschichtlichkeit des Kalliasfriedens und deren historische Folgen*. Wiesbaden.

Meyers, C. and O'Connor, M. (eds) (1983) *The Word of the Lord Shall Go Forth: Essays in Honor of David Noel Freedman in Celebration of His 60. Birthday*. Winona Lake.

Miroschedji, P. de (1985a) 'Du royaume élamite d'Anshan et de Suse au royaume perse d'Anshan'. *Paléorient* 11: 57–61.

Miroschedji, P. de (1985b) 'La Fin Du Royaume d'Anšan et de Suse et La Naissance de l'Empire Perse'. *ZA* 75: 265–306.

Miroschedji, P. de (1990) 'La fin de l'Elam'. *IrAnt* 25: 47–95.

Munn, M.H. (2006) *The Mother of the Gods, Athens, and the Tyranny of Asia: A Study of Sovereignty in Ancient Religion*. Berkeley.

Nenci, G. (2001) 'La formula della richiesta della terra e dell'acqua nel lessico diplomatico achemenide', in Gabriella, Bertinelli, and Piccirilli (2001) 33–42.

Novotny, J. and Jeffers, J. (2018) *The Royal Inscriptions of Ashurbanipal (668–631 BC), Aššur-Etal-Ilāni (630–627 BC), and Sîn-Šarra-Iškun (626–612 BC), Kings of Assyria*. Vol. 1. RINAP 5. University Park, PA.

Olbrycht, M. (2010) 'Macedonia and Persia', in Roisman and Worthington (2010) 342–69.

Parpola, S. (1987) 'Neo-Assyrian Treaties from the Royal Archives of Nineveh'. *JCS* 39: 161–89.

Parpola, S. and Watanabe, K. (1988) *Neo-Assyrian Treaties and Loyalty Oaths*. SAA 2. Helsinki.

Ponchia, S. (2014) 'The Neo-Assyrian *Adê* Protocol and the Administration of the Empire', in Gaspa, Greco, Bonacossi, Ponchia, and Rollinger (2014) 501–25.

Pongratz-Leisten, B. (2018) '"Ich Bin Ein Babylonier": The Political-Religious Message Of the Cyrus Cylinder', in Shayegan (2018) 92–105.

Potts, D. (1999) *The Archaeology of Elam: Formation and Transformation of an Ancient Iranian State*. Cambridge.

Potts, D. (ed) (2012) *A Companion to the Archaeology of the Ancient Near East*. Chichester.

Potts, D. (ed) (2013) *The Oxford Handbook of Ancient Iran*. Oxford.

Radner, K. (2003) 'An Assyrian View on the Medes', in Lanfranchi, Roaf, and Rollinger (2003) 37–64.

Radner, K. (2013) 'Assyria and the Medes', in Potts (2013) 442–56.

Radner, K. (2019) 'Neo-Assyrian Treaties as a Source for the Historian: Bonds of Friendship, the Vigilant Subject and the Vengeful King's Treaty', in Lanfranchi, Mattila, and Rollinger (2019) 309–28.

Raubitschek, A. (1964) 'The Treaties between Persia and Athens'. *GRBS* 5: 151–59.

Rhodes, P.J. (2008) 'Making and Breaking Treaties in the Greek World', in France and de Souza (2008) 6–27.

Rhodes, P.J. and Osborne, R. (2003) *Greek Historical Inscriptions: 404–323 BC*. Oxford.

Roisman, J. and Worthington, I. (eds) (2010) *A Companion to Ancient Macedonia*. Malden, MA.

Rollinger, R. (2006) 'Yaunā takabarā und maginnāta tragende "Ionier". Zum Problem der "griechischen" Thronträgerfiguren in Naqsch-i Rustam und Persepolis', in Rollinger and Truschnegg (2006) 365–400.

Rollinger, R. (2020) 'The Medes of the 7th and 6th c. BCE: A Short-term Empire or Rather a Short-term Confederacy?', in Rollinger, Degen, and Gehler (2020) 187–211.

Rollinger, R., Degen, J., and Gehler, M. (eds) (2020) *Short-term Empires in World History*. Wiesbaden.

Rollinger, R. and Truschnegg, B. (eds) (2006) *Altertum und Mittelmeerraum: Die antike Welt diesseits und jenseits der Levante*. Stuttgart.

Rollinger, R., Truschnegg, B., and Bichler, R. (eds) (2011) *Herodot und das Persische Weltreich = Herodotus and the Persian Empire: Akten Des 3. Internationalen Kolloquiums zum Thema 'Vorderasien im Spannungsfeld Klassischer und Altorientalischer Überlieferungen', Innsbruck, 24.–28. November 2008*. Wiesbaden.

Rop, J. (2018) 'The Assassination of Tissaphernes: Royal Responses to Military Defeat in the Achaemenid Empire', in Clark and Turner (2018) 51–73.

Rop, J. (2019a) *Greek Military Service in the Ancient Near East, 401–330 BCE*. Cambridge.

Rop, J. (2019b) 'The Outbreak of the Rebellion of Cyrus the Younger'. *GRBS* 59: 57–85.

Rossi, A. (2010) 'Elusive Identities in Pre-Achaemenid Iran: The Medes and the Median Language', in Cereti (2010) 289–329.

Rossi, A. (2018) 'Elam in Achaemenid Studies', in Álvarez-Mon, Basello, and Wicks (2018) 851–61.

Rung, E. (2004) 'Xenophon, the Oxyrhynchus Historian and the Mission of Timocrates to Greece', in Tuplin (2004b) 413–25.

Rung, E. (2008) 'War, Peace and Diplomacy in Graeco-Persian Relations from the Sixth to the Fourth Century BC', in France and de Souza (2008) 28–50.

Rung, E. (2015) 'The Language of the Achaemenid Imperial Diplomacy towards the Greeks: The Meaning of Earth and Water'. *Klio* 97: 503–15.

Ruzicka, S. (1985) 'Cyrus and Tissaphernes, 407–401 B.C.'. *CJ* 80: 204–11.

Reynolds, F. (2003) *The Babylonian Correspondence of Esarhaddon and Letters to Assurbanipal and Sin-Šarru-Iškun from Northern and Central Babylonia*. SAA 18. Helsinki.

Sancisi-Weerdenberg, H. (ed) (1987), *Sources, Structures and Synthesis: Proceedings of the Groningen 1983 Achaemenid History Workshop*. Leiden.

Sancisi-Weerdenburg and A. Kuhrt, eds, *The Greek Sources: Proceedings of the Groningen 1984 Achaemenid History Workshop*. Leiden.

Sancisi-Weerdenburg, H. (1988) 'Was There Ever a Median Empire?', in Kuhrt and Sancisi-Weerdenburg (1988) 197–212.

Schachermeyr, F. (1973) 'Athen als Stadt des Großkönigs'. *Grazer Beiträge* 1: 211–20.

Schaudig, H. (2018) 'The Magnanimous Heart of Cyrus: The Cyrus Cylinder and Its Literary Models', in Shayegan (2018) 67–91.

Schepens, G. (2012) 'Timocrates' Mission to Greece – Once Again', in Hobden and Tuplin (2012) 213–41.

Schmitt, R. (2003) 'Die Sprache der Meder – eine grosse Unbekannte', in Lanfranchi, Roaf, and Rollinger (2003) 23–36.

Shayegan, M.R. (2011) *Arsacids and Sasanians: Political Ideology in Post-Hellenistic and Late Antique Persia*. Cambridge.

Shayegan, M.R. (ed) (2018) *Cyrus the Great: Life and Lore*. Boston.

Shrimpton, G. (1991) 'Persian Strategy against Egypt and the Date for the Battle of Citium'. *Phoenix* 45: 1–20.

Sprawski, S. (2010) 'The Early Temenid Kings to Alexander I', in Roisman and Worthington (2010) 127–44.

Starr, I. (1990) *Queries to the Sungod: Divination and Politics in Sargonid Assyria*. State Archives of Assyria 4. Helsinki.

Steinkeller, P. (2018) 'The Birth of Elam in History', in Álvarez-Mon, Basello, and Wicks (2018) 177–202.

Stronach, D. (1978) *Pasargadae: A Report on the Excavations Conducted by the British Institute of Persian Studies from 1961 to 1963*. Oxford.

Tadmor, H. (1982) 'Treaty and Oath in the Ancient Near East: A Historian's Approach', in Tucker and Knight (1982) 127–52.

Tadmor, H. (1983) 'Rab-Saris or Rab-Shakeh in 2 Kings 18', in Meyers and O'Connor (1983) 279–85.

Tavernier, J. (2018) 'Elamites and Iranians', in Álvarez-Mon, Basello, and Wicks (2018) 163–74.

Tavernier, J., Gorris, E., Abraham, K., and Boschloos, V. (eds) (2018) *Topography and Toponymy in the Ancient Near East*. Louvain-la-Neuve.

Thompson, W. (1971) 'Notes on the Peace of Callias'. *Classical Philology* 66.1: 29–30.

Thompson, W. (1981) 'The Peace of Callias in the Fourth Century'. *Historia* 30.2: 164–77.

Tucker, G. and Knight, D. (eds) (1982) *Humanizing America's Iconic Book: Society of Biblical Literature Centennial Addresses 1980*. Chico, CA.

Tuplin, C. (1987) 'The Treaty of Boiotios', in Kuhrt and Sancisi-Weerdenburg (1987) 133–53.

Tuplin, C. (2004a) 'Medes in Media, Mesopotamia, and Anatolia: Empire, Hegemony, Domination or Illusion?' *Ancient West & East* 3.2: 223–51.

Tuplin, C. (ed) (2004b) *Xenophon and His World: Papers from a Conference Held in Liverpool in July 1999*. Wiesbaden.

Wade-Gery, H.T. (1958) *Essays in Greek History*. Oxford.

Walbank, M. (1978) *Athenian Proxenies of the Fifth Century B.C.* Toronto.

Walsh, J. (1981) 'The Authenticity and the Dates of the Peace of Callias and the Congress Decree'. *Chiron* 11: 31–63.

Waters, M. (1997) 'A Survey of Neo-Elamite History', PhD Dissertation, University of Pennsylvania.

Waters, M. (1999) 'The Earliest Persians in Southwestern Iran: The Textual Evidence'. *Iran. Stud.* 32.1: 99–107.

Waters, M. (2011) 'Notes on the Medes and Their Empire from Jer. 25.25 to Hdt. 1.134', in Frame, Leichty, Sonik, Tigay, and Tinney (2011) 243–53.

Waters, M. (2014) 'Earth, Water, and Friendship with the King: Argos and Persia in the Mid-Fifth Century', in Kozuh, Henkelman, Jones, and Woods (2014) 331–36.

Waters, M. (2016) 'Xerxes and the Oathbreakers: Empire and Rebellion on the Northwestern Front', in Collins and Manning (2016) 91–102.

Weeks, N. (2004) *Admonition and Curse: The Ancient Near Eastern Treaty/Covenant Form As a Problem in Inter-Cultural Relationships*. London.

Weiskopf, M. (1989) *The So-Called 'Great Satraps Revolt', 366–360 B.C.: Concerning Local Instability in the Achaemenid Far West*. Stuttgart.

Weninger, S. (ed) (2012) *The Semitic Languages: An International Handbook*. Berlin.

West, S. (2011) 'A Diplomatic Fiasco: The First Athenian Embassy to Sardis (Hdt. 5,73)'. *Rh. Mus.* 154.1: 9–21.

Westlake, H. (1979) 'Ionians in the Ionian War'. *CQ* 29: 9–44.

Westlake, H. (1981) 'Decline and Fall of Tissaphernes'. *Historia* 30: 257–79.

Westlake, H. (1985) 'Tissaphernes in Thucydides'. *CQ* 35: 43–54.

Wilker, J. (ed) (2012a) *Maintaining Peace and Interstate Stability in Archaic and Classical Greece*. Mainz.

Wilker, J. (2012b) 'War and Peace at the Beginning of the Fourth Century: The Emergence of the Koine Eirene', in Wilker (2012a) 92–117.

Wiseman, D. (1958) 'The Vassal-Treaties of Esarhaddon'. *Iraq* 20: 1–99.

Wolpert, A. (2001) 'The Genealogy of Diplomacy in Classical Greece'. *Diplomacy & Statecraft* 12: 71–88.

CHAPTER 8

Impact of Empire: Aspects of Warfare in Babylonia, Achaemenid to Parthian Periods

Reinhard Pirngruber

1 Introduction: Chronology and the Written Documentation

Attempts at discussing any aspect of the history of Achaemenid Babylonia encounter a fundamental difficulty. The roughly two hundred years can be divided into two distinct periods characterised by very different socio-economic frameworks; moreover, the types of documentation extant from these periods are quite divergent. While the first fifty-plus years display a fundamental continuity with the preceding Neo-Babylonian Empire, there is a watershed in 484 BCE, the second regnal year of Xerxes, when after two abortive revolts under local anti-kings, retaliatory measures resulted in an upsetting of the country's power balance. Among other consequences, we witness a displacement of the hitherto socially and politically dominant local elites in northern Babylonia and a much stronger direct exploitation of the resources of the province by the crown. This process is furthermore accompanied by a sudden break in the written documentation at our disposal.[1] It has consequently become customary in Ancient Near Eastern studies to study the 'long sixth century BCE', that is, the period between the establishment of the Neo-Babylonian dynasty in 626 BCE and the 'end of archives' in Xerxes' second year, as one coherent unit. The last two decades have seen a proliferation of studies on this period, mainly socio-economic in outlook, while the Late Achaemenid period, on the other hand, has been comparatively neglected.[2]

This 'long sixth century BCE' is one of the best documented periods in Ancient Near Eastern History. It has yielded tens of thousands of archival cuneiform tablets, mainly administrative records from the large temple households,

1 The seminal article is Waerzeggers (2003/04); see most recently the contributions in Waerzeggers and Seire (2018). Jursa (2014) provides an in-depth analysis of the changing institutional framework.

2 For an exhaustive economic history of Babylonia in this period see Jursa (2010) (and the literature quoted therein as well as later works of the contributors to this monograph). For Late Achaemenid Babylonia see Pirngruber (2017, 47–70, 95–107) and Hackl (2018).

 | DOI:10.1163/9789004710771_009

in particular the Ebabbar temple of the sun god Šamaš in Sippar and the Eanna temple, abode of the goddess Ištar in Uruk as well as legal documents from dozens of private archives of businessmen from all over Babylonia.[3] The largest of these archives comprises more than 1,700 tablets and belongs to the Egibi family, an exceptionally wealthy clan of agricultural entrepreneurs from the city of Babylon. Other archives – of priestly families in Borsippa, small-scale merchants active in the quay of Sippar, etc. – are smaller in size, but their interplay provides us with detailed insight into Babylonian society during the 'long sixth century BCE'.

Characterisations of the later Achaemenid period on the other hand, largely rely on the evidence of a single archive, of the Murašû family from the city of Nippur.[4] Additionally, after ca. 400 BCE archival texts can be supplemented by data culled from chronographic texts including chronicles and Astronomical Diaries (AD). The latter are observational records yielding information on celestial as well as terrestrial phenomena, including descriptions of military campaigns and notations of commodity prices, which in turn can be shown to have been influenced by the course of political history.[5] They extend into the Parthian period and are thus among the latest extant cuneiform records.

This contribution will investigate the relevance of the cuneiform documentation just described on the topic of warfare on different time scales, in a Braudelian sense.[6] In a first step, I will consider the *histoire événementielle*

3 Jursa (2005) provides an overview of the extant text material as well as a succinct description of the main text genres, including promissory notes, sales records and rental agreements. The number of tablets form abovementioned temple archives amounts to ca. 8,000 texts for the Eanna and exceeds 30,000 in case of the Ebabbar (but note that the latter contains a much higher proportion of ephemeral preliminary memoranda that are difficult to contextualise).

4 Stolper (1985); Pirngruber (2017) 47–66. Other archives from the Late Achaemenid period (the Tattannu and Kaṣr archives as well as the Late Achaemenid Esangila archives) have not yet been adequately published. Note, however, that they mostly confirm – and expand – the insights obtained from the Murašû archive and make a case for the latter's representativeness for patterns of land tenure in Babylonia after the second year of Xerxes.

5 For a detailed analysis of the price data in the ADs, see Pirngruber (2017). Diaries from the Late Achaemenid period are edited in Hunger and Sachs (1988).

6 Achaemenid warfare and armed forces have received considerable scholarly interest in recent times. In addition to several papers on the topic by J. MacGinnis (2010 *et passim*), there are two recent doctoral dissertations dedicated to the subject matter. B. Gombert (2018) provides a very detailed overview of the archival cuneiform documentation of the Neo-Babylonian and Achaemenid empires pertaining to military matters as well as extensive (if not always innovative) discussions of key terms and institutions. S. Manning (2021), building on a dissertation dating to 2018, focuses on the Achaemenid army only; there is some overlap with Gombert's work, particularly in Manning's chapter four, *Commoners at War: The Perspective of Letters and Documents*, which deals with the evidence from cuneiform tablets,

and discuss episodes of armed conflict in Babylonia and beyond, as well as their immediate impact. Next, I will look at structural components, hence the *moyenne* and *longue durée*. A prime example are the systems of land tenure and taxation in (not only) Achaemenid Babylonia. At least on a conceptual level – though maybe not as much in actual practice, as we will see below – the exigencies of warfare left an imprint in the country's socio-economic framework. This much is indicated by the terminology employed: the division of agricultural plots into bow land (*bīt qašti*), horse estates (*bīt sisê*) and chariot domains (*bīt narkabti*), as well as the central service obligation of the 'royal soldier' (*ṣab šarri*) for urban elites. Finally, I will discuss structural transformations brought about by war, with a focus on the kings' deportation policies.

2 Episodes of Warfare in Babylonia, Achaemenid to Parthian Periods

Both the beginning and end of Achaemenid rule over Babylonia were marked by rather peaceful takeovers. A. Kuhrt pointed out the similarities between Cyrus' 'own' account of his entry in Babylon in the famed Cylinder inscription – 'When I entered Babylon in peace, I took up my lordly abode in the royal palace amidst jubilation and rejoicing' – and the descriptions of Greco-Roman historiographers of Alexander the Great's arrival.[7] The underlying literary template reflecting the carefully staged triumphant entries of the conquerors has an inherent tendency to downplay the amount of violence accompanying any military endeavour. Yet it is true that in both instances the decisive battles took place elsewhere (the towns of Opis and Gaugamela, respectively) and Babylon itself escaped relatively unscathed.[8] While this peaceful capture is not a unique event in the city's rich history – the Parthian takeover in 141 BCE was also relatively quiet according to the evidence of the Astronomical Diaries, where the only indicator of political change was a new dating formula (rather than accounts of battles and social disruption, as was the case for example

but Manning makes use of non-cuneiform primary sources – accounts of Greek historiographers but also archaeological evidence – to a greater extent. The two works thus complement each other quite well.

7 See Kuhrt (1990). The most recent edition of the Cyrus Cylinder is Schaudig (2019a) (whence the translation above originates); see also Schaudig (2019b) for a literary analysis of the text. Among Greco-Roman historiographers, the accounts of Arrian (Anab. 3.16.2–5) and Curtius Rufus (3.1.17–24) stand out; see also AD -330A + B.

8 See Tolini (2005), who points out that repair work on Babylon's city walls carried out in the first years of Achaemenid reign may be one indicator of damage caused by Cyrus' army; also Jursa (2007) 74–75.

during the war of the Diadochi)[9] – it was certainly not the way things normally went. The most traumatic event in this regard was certainly Sennacherib's destruction of the city in 689 BCE.[10]

The transition from the Neo-Babylonian to the Achaemenid period, however, is thought to have been unproblematic overall. In particular, the local temple elites seem to have found mutually satisfactory arrangements with the new rulers quickly, and there is not much turnover, for example, in the chief temple officials (e.g., the *šatammu* of the Eanna and the *šangû* of the Ebabbar) between the reigns of Nabonidus and Cyrus.[11] Yet much of the amicable relationship between conquerors and conquered can be considered mere decorum and as soon as the opportunity presented itself, the Babylonians attempted to defect from the Persian Empire. Two longer episodes are known from the sources. First, as commemorated by the Bisotun inscription, two rebel kings – both adopting the throne name of Nebuchadnezzar – arose in Babylonia during the succession crisis following Cambyses' death; date formulas in contemporary cuneiform documents show that both found widespread recognition in the country before the revolts were put down by Darius.[12] Second, likely as a consequence of rising tax pressures during Darius' reign that further alienated the Babylonian elites, two more usurpers (Šamaš-erība and Bēl-šimânni) revolted in Xerxes' second regnal year, 484 BCE. We have no account of these rebellions other than possibly a brief remark in Ctesias mentioning the suppression of a revolt against Xerxes by the king's son-in-law Megabyzos (FGrH 688 F13 26), but they had important consequences, as their suppression resulted in the 'end of archives', a sudden break in the documentation at our disposal mentioned above.[13]

Not much is known about the military operations resulting in the defeats of these local pretenders to the throne. Only for the campaign to subdue

9 Pirngruber (2017, 107–22) dealt with the troubled last quarter of the fourth century BC in Babylonia from the perspective of economic history; see also his discussion (175–209) on the economic effects of warfare.

10 Brinkman (1973).

11 Jursa (2007). For C. Waerzeggers (2015a, 184), the account of the Cyrus cylinder 'probably represents a consensus opinion within literate Babylonian society at the time'.

12 The pertinent cuneiform texts are collected in Lorenz (2008). For the Bisotun inscription, see the remarks in Briant (2002) 114–21. A new comprehensive edition of the inscription under the direction of W. Henkelman is in preparation; see for now the translation of Kuhrt (2007) 141–58.

13 Jursa (2011) provides a succinct description of the Achaemenid system of taxation and its development over time; see also Pirngruber (2018) for the gradual tightening of the crown's grip of the countries resources (which started already in the final decades of the preceding Neo-Babylonian Empire). For the 'end of archives', see above, footnote 2.

Nebuchadnezzar III does the Bisotun inscription record some details. Darius claims to have led the army in person, and describes the crossing of the Tigris, partly on horses' and dromedaries' backs, as well as the fact that the rebel king escaped twice after defeats of his troops.[14] The latter battle resulted in a widespread massacre of the remnants of the rebel army, which were driven into the Euphrates to drown. Ultimately, Nebuchadnezzar III was seized in Babylon and put to death. To quell the second Babylonian revolt (of Nebuchadnezzar IV), Darius sent out his trusted vassal Intaphernes rather than engaging in battle in person; the rebel and his allies were subsequently impaled. Consequences for the civilian population – such as deportations of suspected collaborators – or damage to the city are not reported in both instances.

Hence, not even two decades after the Achaemenids' conquest of Babylon, we witness a quick succession of revolts that were not suppressed easily – both Nebuchadnezzar III and IV, as well as later Šamaš-erība, were recognised as kings of Babylonia for about three months, according to the cuneiform documentation.[15] This turmoil belies the visions of cosmic order conveyed by Achaemenid royal inscriptions and peaceful reliefs of tribute-bringers in the palaces of Persepolis, where in the words of one scholar the empire was conceived of as 'eine auf Ewigkeit ausgerichtete und von der Gottheit sanktionierte statische Einrichtung'.[16]

For the remaining ca. 150 years of Achaemenid rule over Babylonia, sources are few and often fragmentary. From about 400 BCE onwards, the Astronomical Diaries regularly mention events involving the king and his military troops; yet, usually, these events occur elsewhere in the empire. AD -381C[17] seems to mention the conflict between Artaxerxes II and the Cypriote king Evagoras, a reference to 'royal troops' in AD -373B is tentatively connected to the ultimately abortive attempt at reconquering Egypt, and AD -369 refers to the rallying of troops for a campaign against the land of Razaunda.[18] AD -366 mentions fighting involving the royal troops in northern Mesopotamia; according to

14 The crossing is analysed in Rollinger (2013) 64–67.

15 The reign of Bēl-šimânni seems to have been locally confined (in Borsippa and Dilbat) and lasted for only a couple of weeks. It also overlapped in time with that of Šamaš-erība, who established himself in Sippar and Babylon and later in some of the cities previously held by Bēl-šimânni; see Waerzeggers (2003/04).

16 Rollinger (2017) 214.

17 The number in Hunger and Sachs (1988) is AD -440, but the tablet has since been re-dated. The discussion of the historical passage in Van der Spek (1998, 240–51) already refers to the new data and siglum.

18 Van der Spek (1998, 253) connects this passage with an expedition against the Cadusians mentioned by Plutarch, but see the doubts raised by Binder (2008) 316–20.

Van der Spek, this passage possibly describes the conflict between the central authority and the satrap of Cappadocia, Datames, in the context of the Great Satraps' Revolt.[19] The most interesting fragment is AD -362 (first published by Hunger and Van der Spek 2006). Below I give a translation of the historical section, which also conveys an idea of the fragmentary state of much of the source material at our disposal. Note that about ten signs are missing to the left in each line, and more to the right:

Rev 1: [(broken)]
r2: [...] his sons, tr[oops? ...]
r3: [...] in the month of *Arahsamnu* (VIII) to Si[ppar ...]
r4: [...] upstream and downstream (lit.: above and below the wind) they encamped [...]
r5: [...] they took. The crown prince (*mār šarri*) who to/for the guard [...]
r6: [...] their defeat he brought about; their booty [...]
r7: [... the troops of] the king which for battle in/to the mountains [...]
r8: [...] they (or: he) conquered. They (or: He) captured the town [...]

Hence, all that is certain is that in the winter of 363/2 BCE military activities took place in the vicinity of Sippar, involving the victorious crown prince and royal troops. Whether fighting took place in the immediate vicinity of Sippar (i.e., in the Babylonian heartland) or whether the town was merely a rallying point for a campaign into the hill country to the north of Babylonia remains unclear. Booty was taken and a town was conquered. The editors noted at least two possibilities for contextualising this fragment – if we are not dealing with an event unknown to Classical historiographers.[20] The first option would be to assume a relationship to the Great Satraps' Revolt, and hence either to date the insurgency of Datames later than postulated above (in which case the question of the historical context of -366 would present itself), or to attribute the fighting in the present instance to Orontes, the satrap of Armenia. Alternatively, the 'sons' mentioned in line 2 may have been the sons of Artaxerxes II, who rose up against their father but were defeated by their brother and heir apparent Ochus (the alter Artaxerxes III). Indeed, according to Plutarch (*Art.* 26–29)

19 Van der Spek (1998, 253–55); on the revolt, likely a loosely connected succession of local uprisings rather than an imperial crisis, see Weiskopf (1989).

20 Hunger and Van der Spek (2006), 8–13.

Artaxerxes II's oldest son Darius revolted against his father but was killed in the event.[21]

To sum up, the Achaemenid period was characterised by frequent military activity in Babylonia. The repression of local revolts under Darius and especially Xerxes seem to have stifled the natives' capacity to organise armed resistance against the Persian overlords. Thereafter, army campaigns mentioned in cuneiform records usually take place elsewhere in the empire; for Babylonia, no further attempts at secession are recorded after 484 BCE. This does not mean that warfare left no imprint on the country, as military assemblies (e.g., as for the campaign against Razaunda in AD -369) constituted a significant drain on the country's resources, as will be discussed below. The only instance of warfare in Babylonia attested in cuneiform sources was brought about by external forces, maybe even the royal family itself. The conflict between Cyrus the Younger and his brother, the reigning king Artaxerxes II, culminating in 401 BCE at Cunaxa on the Euphrates (to the north of Babylon), left no imprint in local sources (as far as the latter are extant).[22] The troubled accession to the throne of Artaxerxes II's father and predecessor Darius II a quarter of a century earlier, on the other hand, may be reflected in the Murašû archive. According to M. Stolper's interpretation of the evidence, the fact that debt notes cluster in the first months of Darius II's accession year is indicative of a large-scale military levy that strained the fief holders' economic capacity, causing widespread indebtedness. Ultimately, these developments caused a collapse of the system of land tenure (at least temporarily, and on the regional level) and led to the displacement of the Murašûs by the Persian price Arsames.[23]

Overall, then, the interplay of succession crises, frequent campaigning on part of the king, the nature of Achaemenid rule after 484 BCE with its focus on direct rent extraction and reliance on local minions, as well as occasional natural disasters, had a negative impact on the country's welfare. This is shown *inter alia* by a rather high price level for basic commodities such as barley and dates, and considerable price volatility in this period.[24] Another factor contributing to the observed market instability in the Late Achaemenid period was the city of Babylon's role as recruitment base of the royal army. Greek

---

21 In Plutarch's account, Darius conspired upon instigation of the powerful courtier Tiribazos to kill his father while the latter was asleep in his bed, but a eunuch informed the king of this plan. While this anecdote may well preserve a historical kernel – the revolt of a son of Artaxerxes II – its historicity is doubtful at best; see Binder (2008) 347–50.

22 Briant (2002, 615–30) provides a detailed analysis of this episode.

23 Stolper (1985) 104–24; Tuplin (2020a) 49–52.

24 Pirngruber (2017, 95–107) provides a concise price history of the Late Achaemenid period. For the changed institutional framework see Pirngruber (2017) 47–70; Jursa (2014).

and Roman historiographers agree that in the run-up to the decisive battles between Darius III and Alexander the Great at Issus and Gaugamela, the army was convened at Babylon (a practice continued by the victorious Alexander).[25]

After the havoc wrought upon Babylonia during the war of the Diadochi following Alexander the Great's premature death in the last quarter of the fourth century BCE, Babylon briefly played a key role – both on the political and the ideological plane – during the emergence of the Seleucid dynasty.[26] However, the city was soon supplanted first by Seleucia-on-the-Tigris and then, decisively, by Antioch-on-the-Orontes as the focus of royal attention. The main conflicts of this era, the Syrian Wars between the Seleucids and Ptolemaic Egypt, took place in the Levant and western Syria; they affected Babylonia only occasionally, most clearly in 246/5 BCE, when the Ptolemaic army invaded the country and briefly laid siege to Babylon itself.[27]

In the Parthian period, this situation was quite similar. The scribes of the Astronomical Diaries lamented the frequent absence of the king, who campaigned in distant lands while delegating the exercise of power in Babylonia to his satraps and generals-in-command.[28] However, contrary to the Seleucid period, Babylonia was frequently afflicted by unrest and armed conflicts on a local and regional scale in the first decades after the rather peaceful Parthian takeover in 141 BCE. This holds particularly true for the 130s and 120s BC, which saw an abortive attempt at Seleucid re-conquest by Antiochus VII Sidetes, bellicose activities of the Elamite kings Kamnaskires and Pittit and Hyspaosines of Mesene in southern Babylonia, as well as recurrent incursions of nomadic Arab tribes.[29] Most of these events are best explained in terms of power politics. Hence, Antiochus VII's chief concern was his own rather precarious position as his predecessor as king, Demetrius II, was still alive in Parthian captivity, while Hyspaosines strove to expand his own sphere of influence into central Babylonia – with considerable success, as he was even briefly recognised as

25 Briant (2002) 797. Pirngruber (2017, 204–207) discusses the economic impact of army presence by means of a demand-centred approach.

26 Capdetrey (2007) 35–38. On the economic impact of the war between the successors, see Pirngruber (2017) 107–22.

27 The relevant text, chronicle BCHP 11, is currently being prepared for publication in print by R.J. van der Spek; see preliminarily the translation on livius.org (https://www.livius.org/sources/content/mesopotamian-chronicles-content/bchp-11-invasion-of-ptolemy-iii-chronicle/; last accessed 25.11.2020) and Clancier (2012).

28 Haubold (2019) 276–80.

29 Shayegan (2011) 60–150 provides a very detailed account of the political history of these rather chaotic years.

king of Babylonia both according to the Astronomical Diaries and the contemporary archival documentation.

Other phenomena were more ephemeral. It has been observed that the recurrent raids of the Babylonian countryside by Arab tribes correlate with a period of worsening climatic conditions. These invasions are thus maybe best characterised as expeditions geared at the procurement of necessities in times of hardship. Taking advantage of the power vacuum prevailing in Babylonia in these years, they plundered the countryside or coerced local authorities into buying them off.[30] The troubled decades after the Parthian takeover also left their mark on the domestic atmosphere in Babylonia, which is repeatedly characterised with terms such as *hattu* and/or *gilittu*, 'fear' and 'terror'.[31] There are a number of mostly fragmentary instances of strife and hostilities involving military officials, such as the killing of a *rab maṣṣarti* (chief of the guard) in AD -123A, r7f. Perhaps the most emblematic episode is the rise of a prophet of the goddess Nanāya, attracting a large crowd of followers and stirring unrest in Babylon and Borsippa in 133 BC, which is indicative of widespread dissatisfaction of the country's inhabitants with the religious-political establishment in these troubled times.[32]

In the Achaemenid and Parthian periods, then, warfare was a regular occurrence for Babylonians, and in different guises. The province was occasionally afflicted by armed conflict, sometimes in the context of the suppression of local revolts. At other times, Babylon was the place where army contingents met for expeditions to other regions of the empire, and sometimes, military events abroad were simply registered by the chronographers compiling the Astronomical Diaries.

3 Warfare and Empire: a Structural Perspective

3.1 *Deportations*

In addition to immediate economic repercussions, as visible for example in the level of commodity prices, warfare brought about structural transformations in the country as well. In the Near East during the first millennium BCE,

30 Pirngruber (2016) gives a detailed discussion of all pertinent attestations. Note that with the exception of two late attestations, all recorded invasions date between 130 and 118 BCE. Several years record more than one invasion, in which case repeated activities by smaller bands can be assumed.

31 ADs -137A, r6; -136C, 18 and -122E, r7 (*gilittu*), and -140D, 11; -137D, 12 and r3; -124B, 21 and -123A, r7 (*hattu*).

32 Dirven (2019) 163–74.

the deportation of conquered peoples on a large scale organised by victorious kings since the Neo-Assyrian period fundamentally altered the social landscape of both imperial core and periphery. The Babylonian chronicle ABC 9 attests to the persistence of this practice also under the Achaemenids in the context of Artaxerxes III's activities in the Levant in the 340s BCE, which preceded his re-conquest of Egypt:[33]

> The fourteenth year of Umasu, who was called Artaxerxes (III). In the month of *Tašrītu* (month VII), the prisoners whom the king had captured at Sidon [were brought] to Babylon and Susa. On the thirteenth day of the same month, a few of these troops entered Babylon. On the sixteenth day the women remaining among the prisoners of Sidon, whom the king had sent to Babylon – on that day they entered the palace of the king.

A few females from among of these deportees, likely members of the Sidonian elite, stayed as privileged hostages in the royal palace. A similar arrangement is famously attested for the captured Judean king Jehoiachin and his sons under Nebuchadnezzar II, who appear in the textual record as recipients of food rations from the palace administration.[34] Alongside them, a number of skilled craftsmen from Levantine cities (and Ionia), including carpenters and sailors, as well as musicians, are mentioned.

The far greater number of deportees, however, were settled in the Babylonian countryside. In a centripetal movement, their labour power was to be exploited in the empire's central region, to the detriment of more peripheral regions. They were integrated into an economic system based on the allotment of agricultural land in exchange for labour service and taxes in silver and kind. Two text groups in particular inform us about the living conditions of these deportees in the Achaemenid period, the Murašû archive and the tablets from Judean exiles settled in and near Āl-Yahūdu.[35] Note that Babylonia was not the only region where deportees were settled in the Achaemenid period. This much seems to emerge from the evidence from the Fortification Archive from Persepolis, where more than two dozen ethnonyms are attested. At least some

33 Briant (2002) 682–85; Ruzicka (2012) 172–74. The *locus classicus* on Neo-Assyrian deportation practice is Oded (1979); for Achaemenid period Babylonia see also MacGinnis (2014) 121–22.

34 See most recently Alstola (2019, 63–77), who points out the presence of further rulers of small kingdoms (Ashkelon and Lydia) in the palace archive from Babylon.

35 For the latter, see Waerzeggers (2015b).

of these may refer to deportees, who joined the ranks of the (semi-)dependent *kurtaš*-workers.[36]

3.2 *Taxation*

Let us move, then, from an event-centred overview to a structural synopsis of the archival sources from Babylonia. Warfare permeated society also on the socio-economic level. The Achaemenids adopted and expanded the land-for-service scheme already in evidence under the Chaldeans.[37] In essence, the crown gave plots of agricultural land in fief, thereby obliging the holders of this fief to the performance of corvée service as well as the payment of various dues in kind and/or silver to the crown in turn. The obligations on these lands, known in their totality as *ilku*, usually consisted of:[38]

- *ṣāb šarri* the 'royal soldier', hence a service obligation
- *qēmû ša šarri* ('flour of the king'), a delivery in kind
- *bāra*, maybe an obligation to deliver produce to the royal household (note that the term is a Persian loanword)
- further unspecified duties and fees, summarised as *mimma nadnātu ša bīt šarri* ('whatever else is owed to the royal palace')

Superficial similarities to the Medieval European feudal system have been repeatedly noted, and the settlers, primarily producers of agricultural staples, are sometimes referred to as 'military colonies'. The terminology employed to describe these landed holdings betrays indeed a military origin, as the main elements were the 'bow land' (*bīt qašti*), the 'horse land' (*bīt sīsê*), and the 'chariot land' (*bīt narkabti*). There is thus an obvious connection to the army division to be staffed and outfitted from these lands. Holdings were further grouped into administrative units called *hadru*, overseen by a state official called *šaknu*, who was responsible for the collection of the associated dues. Also, some – though by no means all – designations for these *hadru* units are

36 On the *kurtaš*, see Briant (2002) 433–39; Tamerus (2018). Henkelman and Stolper (2009) provide an extensive discussion of ethnonyms in Persepolis and some specific case studies. Although the evidence is difficult to interpret, and there is no definite proof for such an assertion, all authors allow for the possibility that deportees and prisoners of war may have been among the *kurtaš*, whose status remains elusive.

37 The most exhaustive discussion of this system is Van Driel (2002) 226–73. For pre-Achaemenid attestations see Jursa (2011) 436–37.

38 Van Driel (2002) 254–59, 268–70. A suitable text example is PBS 2/1 193 from year 2 of the reign of Darius II, a receipt for *ilku*. The text starts: 'One mina of silver, the complete *ilku* – *ṣāb šarri, qēme ša šarri, bāra* and anything else that is owed to the royal palace from month I to the end of month XII of year 2 of King Darius (II.), incumbent on orchards and fields on settled as well as empty bow plots [...] in the *hadru* of the clay-workers'.

military in origin; pertinent examples include the *hadru* of the 'scouts of the left flank' and the *hadru* of the 'shield-bearers of the right flank'.[39]

The system is primarily in evidence in rural, comparatively marginal zones, for example around the central Babylonia town of Nippur. There, a local family of agricultural entrepreneurs, the Murašû clan, made a fortune by acting as intermediaries between the cultivators and the crown agents.[40] The land-for-service scheme also served the purpose of facilitating the integration of foreigners, often war deportees, into the Babylonian economy. Consequently, a number of *hadru*s were named after the ethnicity of their members, including Carians (*Bannēšāya*), Arameans (*Arumāya*), and Cimmerians (*Gimirrāya*).[41] A peculiar set of documents that has received much attention in recent years are the tablets associated with the community of deportees from the kingdom of Judah, who had been resettled after Nebuchadnezzar II's capture of Jerusalem in the central Babylonian countryside. Tellingly, the central town of the archive is called Āl-Yahūdu, 'Judah-town'. The tablets are earlier than the evidence from the Murašû archive, dating to the sixth and early fifth century BCE, but confirm central aspects that are in evidence in the latter, such as settling of exiles in ethnic communities and especially the prominence of the land-for-service sector in the Babylonia countryside and its importance in the economic integration of forcibly settled foreigners.

In addition to the rural population, urban elites and the temple households were liable to contributions to the Achaemenid state, including its army. Among the variety of obligations due from the residents of Babylonian cities, evidence from the city of Borsippa shows that a bow tax (*qaštu*) existed in an urban context, too. The precise nature of this tax, and in particular whether it implied tenure of a bow plot (*bīt qašti*), is unclear. It has been suggested that we are rather dealing with a standardisation of tax-related terminology during the reign of Darius I, which applied the language of the land-for-service system to other types of arrangements, such as the *hanšû*-allotments of land and orchards in the city's immediate hinterland to urban notables.[42] Also, the *ṣāb šarri* is attested in Borsippa and was likely part of the *qaštu* obligation. As members of the elites were reluctant to perform corvée services themselves, silver paid under designation of *ṣāb šarri* was usually received by the substitutes carrying out the labour duties in lieu of the targeted individuals. For example,

39 Examples from Stolper (1985) 72–79.

40 Stolper (1985); Pirngruber (2017) 47–70.

41 Examples from Stolper (1985) 72–79; note also the generic *hadru* of the refugees (*maqtūtu*).

42 Jursa (2009), 246–47, 265; see also 255–57 on the bow tax in general. The *hanšû*-scheme, dating to the late seventh century BCE is discussed in Still (2019) 65–88.

in the accession year of Cambyses, the small-scale merchant Iššar-tarībi and his brother jointly hired a third person, Nabû-bēlšunu, to perform their *ṣāb šarri*-service, which was part of their *qaštu* obligation. Nabû-bēlšunu had to perform his duty in the town of Lahīru, an important way station on the Transtigridian road connecting Babylonia with Ecbatana, and one may surmise that he was posted there on guard duty to safeguard the imperial road system.[43] Part of the 'royal soldier' obligation was the outfitting of those doing the actual service with suitable equipment, known under the title of *rikis qabli*, and the provisioning with food rations (*ṣidītu*).[44]

Another peculiar obligation of wealthy citizens was the (temporary?) maintenance of foreigners. This practice is best attested for Carians during the reign of Darius I. It has been hypothesised that these persons were former members of Cambyses' army at the time of his conquest of Egypt, who were at first stationed in Egypt and later moved to other parts in the empire. Possibly, they were to be settled in the land-for-service system, but until a definite arrangement was concluded, their upkeep was devolved on Babylonian citizens.[45]

3.3 *Institutional Households*

Not only private persons, be they urban elites or settlers (and deportees) in the countryside, but also the Babylonian temple households were liable to provide personnel for the royal army in the context of taxation.[46] According to one scholar, the Ebabbar temple from Sippar contributed at least fifty archers (i.e., infantry) in a regular year, and potentially more in times of high demand. These archers were organised in decuries of eight soldiers, further divided into two units comprising four persons. Each decury was under the command of a *rab eširti*, and the *rab qašti* oversaw the whole of the archers.[47] Importantly,

43 Pirngruber (2020), text no. 2. On Lahīru, see Kleber (2008) 186; Tuplin (2020b) 151–52.

44 On *rikis qabli*, see Van Driel (2002) 249–53.

45 Waerzeggers (2006). The cuneiform texts indeed identify these persons occasionally as 'Egyptians'. Also, the word for 'Carian' in this dossier is not the usual term (*Bannēšāya*) but rather *Karšāya*, maybe a calque from the Egyptian designation of these people, who are thus identified as 'Egypto-Carians'.

46 Kleber (2008, 198–236) discusses the evidence from the Eanna temple in Uruk, with due consideration of information gleaned from Greek authors, in particular Xenophon. For the Sipparean Ebabbar temple, see MacGinnis (2010a); MacGinnis (2010b); MacGinnis (2012). Note that in the case of the usually undated memoranda, it is not always possible to establish whether a text dates to the Neo-Babylonian or Achaemenid period, which is, however, unproblematic in the light of the fundamental administrative continuity between these two empires, at least in Babylonia.

47 MacGinnis (2010a) 497–98; MacGinnis (2012) 5–11. Bongenaar (1997, 131–32) gives a prosopography of the *rab qašti* of the Ebabbar.

these archers were not professional soldiers but temporarily conscripted from the ranks of the temple serfs (*širku*), who ordinarily were active in the temple's agricultural sector as farmers (*ikkaru*) or in other professions. MacGinnis (2012) no. 2 from the reign of Darius, a simple list of temple dependents that have been recruited to perform bow service, designates them in its heading as 'the troops of the temple farmers and shepherds of the bow'.[48] Texts from the Eanna archive indicate that the town of Opis, located on the eastern bank of the Tigris river, was one important rallying point for conscripts from Uruk. The city was strategically important and located along the most important axis of travel.[49] Evidence from the Ebabbar points in a similar direction. A telling example is CT 56 555, according to which a certain Tattannu, who was attached to the Ebabbar's poultry-rearing sector (*bīt iṣṣurī*), was outfitted with a bow and sent there.[50]

Dozens of texts furthermore attest to the manufacture of equipment for archers, and the issue of these items as well as provisions of foodstuff as travel rations (*ṣidītu*). Occasionally, silver was paid instead, with a specification of the purpose of the payment. In some instances, the troops are explicitly said to be (on campaign) with the king. According to MacGinnis (2012) no. 33 dating to year 16 of Darius, for example, an official of the Ebabbar received more than 3.5 minas of silver for travel rations (*ṣidītu*) for the troops 'who are with the king' by order of the temple's chief administrator, the *šangû* Gūzānu. Other texts document the receipt of equipment by the temple's craftsmen for repair. The standard armament of an archer consisted of bow and bow case, arrows and an iron dagger; sometimes, spears are also recorded.[51] General equipment – from military caps to jerkins to sandals – also occurs in the record in the context of outfitting archers, which clearly was the temple's responsibility.

---

48 Similar texts include MacGinnis (2012) nos. 1, 3–5, and 43, the latter likely dating to the reign of Darius. Homologous evidence from the Eanna is discussed in Kleber (2008) 214–23.

49 Jursa (2010) 80–84.

50 Unfortunately, the aim of his mission is in a lacuna. Pinches in CT 56 reads *ana* U[GU], commenting that the *muhhi*/UGU ('for the purpose of …') was written over other characters. Alternatively, one may tentatively suggest reading *a-na* IGI! LU[GAL!?], hence, '(he went) before the king'. In any case, the fact that the official was outfitted with a bow indicates some kind of venture for which arming was considered necessary. On the temple's bird house, see Janković (2004) 50–61. Tattannu is a rather well-attested official from the reigns of Nabonidus and Cyrus, who most frequently occurs in the context of the feeding/fattening of ducks and other birds; see Janković (2004) 65–67.

51 See, e.g., MacGinnis (2012) no. 27. The text lists a number of recipients, each being issued one bow, bow case and dagger as well as forty or fifty arrows. For the Akkadian terminology, see MacGinnis (2012) 50.

Cavalry forces and chariot units are poorly documented beyond the mere fact that they existed, and there is a distinct possibility (in particular in case of the latter) that they were under direct supervision of the royal palace and hence fell beyond the competences of the temple administration.[52] One relevant document in this regard is CT 57 82, an administrative memorandum listing miscellaneous silver expenses of the Ebabbar temple in the fourth year of Darius (518/7 BCE) from its income from house rentals. In addition to an expense of one mina and fifty shekels of silver (almost three times the yearly income of an unskilled labourer!) for blankets and jerkins for archers, the text registers the issue of half a mina and eight shekels of silver to a one 'Šamaš-iddin and the cavalry troops who have returned from Egypt'.[53] The text is not only one of the very few texts referring to cavalry troops maintained by a Babylonian temple, but the passage is also one of the rare instances of an administrative document that can be connected with political history on the imperial scale. The return of Šamaš-iddin and the troops likely under his command coincided with the pacification of Egypt, which had revolted at Darius' contested accession to the throne under the usurper Petubastis IV.[54]

Evidence from the Eanna temple in Uruk helps to round off our knowledge of the tasks of armed forces in Babylonia during the long sixth century BCE. Also here, we find mentions of temple serfs (*širku*) performing their military service in the king's camp (*madāktu*). A local peculiarity is the service as *musahhiru*; likely, these persons were responsible for the logistics of the military forces.[55] The Eanna archive also yields information on the activity of 'archers' within Babylonia during times of peace. Those of the *širku*s who were summoned for the 'royal guard duty' (*maṣṣartu ša šarri*) were most frequently stationed in guard posts in the countryside; these *kādānu* can likely be identified as the *phylaktēria* of Greek historians.[56] Archers recruited from the ranks of the shepherds (of the Lady-of-Uruk) occur with particular frequency, and this is likely less of a coincidence in the extant documentation than a reflection of the importance of animal husbandry and the sale of wool in the Eanna temple's economy. Correspondingly, and as these shepherd-archers were posted

52 MacGinnis (2012) 12–18.

53 See Bongenaar (1997, 133) for this text and further references to cavalry troops; also MacGinnis (2010a, 495), and MacGinnis (2012, no. 12, a list of eight persons 'on horses' (*ša ina* ANŠE.KUR.RA)). For more texts showing a connection between income from house rentals and military expenses, see Zawadzki (2019) 141; for an edition of CT 57 82 see *ibid.* no. 53.

54 Wijnsma (2018).

55 Discussed in Kleber (2008) 223–26.

56 Kleber (2008) 204–208, citing Hdt. 5.52 and Xen. *Cyr.* 8.6.

in *kādānu* along the Tigris, it is clear that one central task was the guard over Eanna's herd grazing in the pasture between the Tigris and the foothills of the Zagros mountains.[57] GC 2, 102 (lines 8–9) serves as explicit confirmation. In this text, the highest administrators of the Eanna, the *šatammu* and the *bēl piqitti* give the order to the supervisors in charge to station 'the archers (recruited) from the shepherds, as every year, in the *kādānu* on the great river (i.e., the Tigris)'.

Escort and custody was another area of activity of these troops recruited from the temples. J. MacGinnis (2012), 10–11 lists several related tasks in this regard in his overview of attested duties, including protection of long-distant merchants and their cargo, safeguarded by troops dispatched to perform labour service on behalf of the temple, as well as protection of temple property. During one occasion, for example, rations were issued to 'troops' accompanying the sacred chariot of Ebabbar's main deity, the sun-god Šamaš from Sippar to Babylon, likely for participation in the New Year's Festival. These 'troops' are interpreted as armed guardians of the valuable cultic paraphernalia during their journey.[58]

A final point that deserves attention is the length of the military service. In an institutional context, a period of service of one year may have been the norm for archers, but longer periods of up to three years are attested as well, in particular in the case of mounted troops.[59]

3.4 *Summary: Warfare and Taxation*

In spite of the militaristic terminology of the system of taxation in place in Achaemenid Babylonia, it should be borne in mind that the state's main concern was the mobilisation of troops for labour service in a broader sense. State-sponsored building projects included military structures – a prime example being Nebuchadnezzar II's Median Wall; designed as protection against invasion from (semi-)nomadic pastoralists from the north – but more often concerned canal digging and building work on palaces. Military service was but a small niche within this overarching framework. From an economic perspective, the delegation of labour and military service to substitutes and their provisioning with silver rations facilitated the ongoing monetisation of Babylonia's economy. This, then, was maybe the most important impact of any military aspect on the country.

57 Kleber (2008) 213. On the importance of Eanna's trade in wool see Kleber (2008) 237–53; Kleber in Jursa (2010) 595–615.

58 The text is edited and discussed in MacGinnis (2000/01).

59 Bongenaar (1997) 133; MacGinnis (2012) 40; Kleber (2008) 230.

The texts at our disposal allow thus a rather detailed reconstruction of how subject populations in the Achaemenid Empire were integrated into the state and its military apparatus. For Babylonia, we see an adoption of pre-existing mechanisms that were only gradually altered. Notable is the usually local deployment of troops; however, caution is warranted, as the army on campaign is beyond the purview of our sources.

3.5 *The Higher Ranks*

The highest ranks of the army are hardly visible in the cuneiform sources at our disposal from the long sixth century BCE, which is centred on Babylonia's urban elites and their central institution, the temple of the city god. The decurion (*rab eširti*) had above him the 'chief of one-hundred' (*rab meʾati*), and still higher ranked the 'chief of one-thousand' (*rab limmi*), but their competences remain in the dark. Directly below the king in the hierarchy was maybe the *rab ummu*.[60] This is the only official who is still attested after the break in the documentation in 484 BCE. In addition to some attestations in the Murašû archive, where he appears in the context of the land-for-service system in control of large estates, the title also occurs in an Astronomical Diary (AD -366A).[61] The text contains the passage that has been interpreted as a reference to the invasion of northern Mesopotamia by Datames (see above). However, the context for the mention of the *rab ummu* is likely a purely administrative matter, as the preceding lines refer to the appointment of the *mašennu* (another high official in the land-for-service system, usually translated as 'canal inspector') to the position of satrap.

The situation is different in the Parthian period, when the Astronomical Diaries provide insight into the high command of the army. The supreme commander in the country was the 'general (of Babylonia) who is above the four generals', ^{lú}gal erín (kur uri^{ki}) *ša ana* ugu 4 ^{lú}gal erín^{meš} (and variants). There is broad agreement that the title corresponds to the Greek *stratēgos* and that this commander-in-chief was the highest military and civil authority of later Seleucid and Parthian Babylonia. However, there are also several aspects that are obscure; in particular, the identity of the four generals serving under this commander-in-chief is still elusive.[62]

---

60 See the reconstruction of MacGinnis (2012) 31–37.

61 The official is discussed at length in Stolper (2001/02). He suggests deriving the previously elusive element *ummu* from the word for 'quiver'.

62 Shayegan (2011, 208–21) provides a detailed discussion of the title in the context of Parthian military authorities; see n. 677 on relevant earlier literature. His scheme (221, table 9) that identifies the four minor generals as *rab uqān* (*māt Akkad*), *mumaʾir* (*māt Akkad*), *pāhāt Bābili* (maybe Greek *epistatēs*) and *rab*/*muhhi maṣṣarti*, thus, general,

An interesting passage that illustrates several aspects of this office at once is found in AD -140C, lines 37–39, which reads:

> An(tiochus) the general who is above the 4 generals, who was representing king Arsaces, went out from Sel[eucia which is on] the Tigris towards the Elamite for fighting; from the river Kabari he departed, and the numerous troops [...] went out for battle.

First, the terminology in the diaries continues unbroken from the Seleucid to Parthian periods, and this continuity is also reflected in the identity of the office-holder. The first attested commander-in-chief of Parthian Babylonia was a Greek by the name of Antiochus, and it is likely that he was in office already before the Parthian conquest the year before. Second, like their Seleucid predecessors, the Parthian commanders-in-chief resided in Seleucia-on-the-Tigris rather than Babylon. Travels of high officials between the two cities, and the performance of sacrifices to the gods on their arrival, are a recurring *topos* in the diaries. Third, the important position of Antiochus is rendered explicit by the scribe of this diary, clarifying that the general-in-command was entitled to acted in the king's stead in the latter's absence (which, as has been pointed out above, was the normal but bemoaned state of affairs in the Parthian period). And finally, the passage underscores the general-in-command's military competences, and illustrates the protracted conflicts with Elam (and other local powers) in the first two decades of the Parthian takeover in Babylonia. Incidentally, AD -140C continues on the reverse (lines r30–r34) with a rather unique story of treason, as Antiochus is reported to have defected from Parthian authority and joined the Elamite side (after which he disappears from the record).

To conclude this section, it should be pointed out that at least for the Parthian period, there is clear evidence that all higher-ranking military officials were appointed by the king himself. AD -126A lines 7–8, for example, mentions one Timarchos, who had been appointed as citadel commander (*rab maṣṣarti*) by order of the king. Again, a Greek was selected for an important office some fifteen years after the Parthian conquest. The Parthians thus drew heavily on local elites and expertise from the preceding empire in military matters.

---

satrap, 'governor' and citadel commander, is appealing at first glance. Yet, as he admits himself, these are simply the only other military titles known from the ADs, and other suggestions (e.g., that identity these generals as commanders of four important cities in Babylonia) cannot be discarded. Also, his suggested identification sits ill with the fact that the office of satrap was still suspended when this title was conceived (a suggestion by Hermann Bengtson, still upheld, e.g., in Capdetrey (2007) 285; see also Capdetrey (2007, 283–93) for a discussion of the title of 'general who is above the four generals').

4 Conclusions

To sum up, this contribution drew on two fundamentally different corpora to shed light on aspects of warfare in Babylonia during the second half of the second millennium BCE. For the first fifty years or so of Achaemenid rule, the uniquely dense coverage in terms of archival documents – both business records of private individuals and administrative records of the large temple households – provides us with detailed insight into the mechanism of mobilisation of manpower for both civic and military purposes in the context of taxation. On the other hand, chronographic documents such as Astronomical Diaries and Chronicles record minutiae of local and regional political history, and give occasional insight into political structures on the imperial level for the Late Achaemenid, Hellenistic and Parthian periods. Conversely, due to the lack of archival sources after 484 BCE, we are largely in the dark as regards the integration of the region into the state apparatus of the Seleucid and Parthian empires.

Recent research has shown that the standardisation and reformation of the system of taxation during the reign of Darius I went hand in hand with increasing tax pressure, continuing a trend rooted in the final years of the Neo-Babylonian Empire. This development seems to have been the *causa prima* for a series of local revolts that ultimately changed Babylonia's socio-economic configuration to a significant extent. Uprisings against foreign rule are hardly an exceptional event in Ancient Near Eastern history. A case in point is the history of Babylonia during the Neo-Assyrian period, when the city regularly defected from its overlords, be it upon instigation of regional (Chaldean tribal) leaders such as Marduk-apla-iddina II (the Biblical Merodach-Baladan) in the last quarter of the eighth century BCE, under the guidance of local puppet kings like Bēl-ibni in the final years of the same century, or under Assyrian vice-kings like Šamaš-šum-ukīn, the brother of the reigning king Ashurbanipal in the mid-seventh century. The unique feature of Achaemenid-period Babylonia is that the rich and heterogeneous primary sources from these roughly two centuries provide uniquely detailed insight into the origins of warfare on a local scale beyond programmatic statements of victorious kings.[63]

---

63 Research for this article took place under the auspices of the Austrian Science Fund's project P 30311, "Astronomical Diaries Digital". The *siglum* 'AD -xxx' refers to Astronomical Diaries published in Hunger and Sachs (1988; ADs -652 to -262) and (1996; ADs -161 to-62); 'ABC' to texts in Grayson (1975). Transcriptions and translations of Astronomical Diaries are available at http://oracc.museum.upenn.edu/adsd/. Abbreviations follow Assyriological conventions.

Bibliography

Alstola, T. (2019) *Judeans in Babylonia: A Study of Deportees in the Sixth and Fifth Centuries BCE*. CHANE 109. Leiden.

Binder, C. (2008) *Plutarchs Vita des Artaxerxes: Ein historischer Kommentar*. Göttinger Forum für Altertumswissenschaften, Beihefte N.F. 1. Berlin.

Bongenaar, A.C.V.M. (1997) *The Neo-Babylonian Ebabbar Temple at Sippar: Its Administration and Its Prosopography*. PIHANS 80. Istanbul.

Briant, P. (2002) *From Cyrus to Alexander: A History of the Persian Empire*. Winona Lake.

Briant, P. and Chauveau, M. (eds) (2009) *Organisation des pouvoirs et contacts culturels dans les pays de l'empire achéménide*. Persika 14. Paris.

Brinkman, J. (1973) 'Sennacherib's Babylonian Problem: An Interpretation'. *JCS* 25.2: 89–95.

Brosius, M. and Kuhrt, A. (eds) (1998) *Achaemenid History XI: Studies in Persian History. Essays in Memory of David M. Lewis*. Leiden.

Capdetrey, L. (2007) *Le pouvoir séleucide: Territoire, administration, finances d'un royaume hellénistique (312–129 av. J.-C.)*. Rennes.

Clancier, P. (2012) '"Le *rab sikkati*" de Babylone contre "l'homme de renom venu d'Égypte": le troisième guerre syrienne dans les rues de Babylone', in Goukowksy and Feyel (2012) 9–31.

Curtis, J. and Simpson, St.J. (eds) (2010) *The World of Achaemenid Persia: History, Art and Society in Iran and the Ancient Near East. Proceedings of a Conference at the British Museum 29th September–1st October 2005*. London.

Dirven, L. (2019) 'The Astronomical Diaries and Religion in Seleucid and Parthian Babylon: The Case of the Prophet of Nanāya', in Haubold, Steele, and Stevens (2019) 154–79.

Garcia-Ventura, A. (ed) (2018) *What's in a Name? Terminology Related to Work Force and Job Categories in the Ancient Near East*. AOAT 440. Münster.

Gombert, B. (2018) 'L'armée en Babylonie du VIè au IVè siècle av. n. è.', PhD Dissertation, Université Paris 1 Panthéon-Sorbonne.

Goukowksy, P. and Feyel, C. (eds) (2012) *Folia Greca in honorem É. Will*. Paris.

Grayson, A.K. (1975) *Assyrian and Babylonian Chronicles*. TCS V. New York.

Graziani, S. (ed) (2000/2001) *Studi sul Vicino Oriente antico dedicati alla memoria di Luigi Cagni*. Vol. 2. Rome.

Hackl, J. (2018) 'The Esangila Temple during the Late Achaemenid Period and the Impact of Xerxes' Reprisals on the Northern Babylonian Temple Households', in Waerzeggers and Seire (2018) 165–87.

Haubold, J. (2019) 'History and Historiography in the Early Parthian Diaries', in Haubold, Steele and Stevens (2019) 269–93.

Haubold, J., Steele, J. and Stevens, K. (eds) (2019) *Keeping Watch in Babylon: The Astronomical Diaries in Context*. CHANE 100. Leiden.

Henkelman, W. and Stolper, M. (2009) 'Ethnic Identity and Ethnic Labelling at Persepolis: The Case of the Skudrians', in Briant and Chauveau (2009) 271–327.

Hunger, H. and Sachs, A. (1988) *Astronomical Diaries and Related Texts from Babylonia. Vol. 1: 652–262 BC*. Vienna.

Hunger, H. and Sachs, A. (1996) *Astronomical Diaries and Related Texts from Babylonia. Vol. 3: 164–61 BC*. Vienna.

Hunger, H. and van der Spek, R.J. (2006) 'An Astronomical Diary concerning Artaxerxes II: military operations in Babylonia'. *ARTA* 2006.002: 1–16.

Janković, B. (2004) *Vogelzucht und Vogelfang in Sippar im 1. Jahrtausend v. Chr.* AOAT 315. Münster.

Jursa, M. (2005) *Neo-Babylonian Legal and Administrative Documents. Typology, Contents and Archives*. GMTR 1. Münster.

Jursa, M. (2009) with contributions by C. Waerzeggers 'On Aspects of Taxation in Achaemenid Babylonia: New Evidence from Borsippa', in Briant and Chauveau (2009) 237–69.

Jursa, M. (2010) *Aspects of the Economic History of Babylonia in the First Millennium BC. Economic Geography, Economic Mentalities, Agriculture, the Use of Money and the Problem of Economic Growth*. AOAT 377. Veröffentlichungen zur Wirtschaftsgeschichte Babyloniens im 1. Jahrtausend v. Chr., Band 4. With contributions by J. Hackl, B. Janković, K. Kleber, E.E. Payne, C. Waerzeggers and M. Weszeli. Münster.

Jursa, M. (2011) 'Taxation and Service Obligations in Babylonia from Nebuchadnezzar to Darius and the Evidence for Darius' Tax Reform', in Rollinger, Truschnegg, and Wiesehöfer (2012) 431–48.

Jursa, M. (2014) 'On the Existence of Factor Markets in Babylonia from the Long Sixth Century to the End of Achaemenid rule'. *JESHO* 57: 173–202.

Kleber, K. (2008) *Tempel und Palast: Die Beziehungen zwischen dem König und dem Eanna-Tempel im spätbabylonischen Uruk*. AOAT 358. Veröffentlichungen zur Wirtschaftsgeschichte Babyloniens im 1. Jahrtausend v.Chr., Band 3. Münster.

Kuhrt, A. (1990) 'Alexander and Babylon', in Sancisi-Weerdenburg and Drijvers (1990) 121–30.

Kuhrt, A. (2007) *The Persian Empire: A Corpus of Sources from the Achaemenid Period*. 2 volumes. London.

Lorenz, J. (2008) *Nebukadnezzar III/IV: die politischen Wirren nach dem Tod des Kambyses im Spiegel der Keilschrifttexte*. Dresden.

Luciani, M. (ed) (2016) *The Archaeology of North Arabia: Oases and Landscapes*. OREA 4. Vienna.

MacGinnis, J. (2000/01) 'The Chariot of Šamaš Goes to Babylon', in Graziani (2000/2001) 621–30.

MacGinnis, J. (2010a) 'The Role of Babylonian Temples in Contributing to the Army in the Early Achaemenid Empire', in J. Curtis and St.J. Simpson (2010) 495–502.

MacGinnis, J. (2010b) 'Mobilisation and Militarisation in the Neo-Babylonian Empire', in Vidal (2010) 153–63.

MacGinnis, J. (2012) *The Arrows of the Sun: Armed Forces in Sippar in the First Millennium* BC. With Copies of the Cuneiform Texts by Cornelia Wunsch. BAr 4. Dresden.

MacGinnis, J. (2014) 'Minions to the Rule of Erra: The Impact of War on the Civilian Population of Babylonia in the First Millennium BC', in Nadali and Vidal (2014) 113–25.

Manning, S. (2021) *Armed Forces in the Teispid-Achaemenid Empire: Past Approaches, Future Prospects*. Oriens et Occidens 32. Stuttgart.

Nadali, D. and Vidal, J. (eds) (2014) *The Other Face of the Battle: The Impact of War on Civilians in the Ancient Near East*. AOAT 413. Münster.

Oded, B. (1979) *Mass Deportations and Deportees in the Neo-Assyrian Empire*. Wiesbaden.

Pirngruber, R. (2016) '"Arabs" in Late First Millennium BCE Babylonia', in Luciani (2016) 345–55.

Pirnguber, R. (2017) *The Economy of Late Achaemenid and Seleucid Babylonia*. Cambridge.

Pirngruber, R. (2018) 'Towards a Framework for Interpreting Social and Economic Change in Babylonia during the Long 6th Century BCE', in Waerzeggers and Seire (2018) 19–33.

Pirngruber, R. (2020) 'Minor Archives from First Millennium BCE Babylonia: The Archive of Iššar-tarību from Sippar'. *JCS* 72: 165–98.

Rebenich, S. (ed) (2017) *Monarchische Herrschaft im Altertum*. Schriften des Historischen Kollegs, Kolloquien 94. Oldenburg.

Rollinger, R. (2013) *Alexander und die großen Ströme. Die Flussüberquerungen im Lichte altorientalischer Pioniertechniken*. Wiesbaden.

Rollinger, R (2017) 'Monarchische Herrschaft am Beispiel des teispidisch-achaimenidischen Großreichs', in Rebenich (2017) 189–215.

Rollinger, R., Truschnegg, B., and Wiesehöfer, J. (eds) (2011) *Herodot und das Persische Weltreich. Akten des 3. Internationalen Kolloquiums zum Thema "Vorderasien im Spannungsfeld klassischer und altorientalischer Überlieferungen", Innsbruck, 24.–28. November 2008*. Wiesbaden.

Ruzicka, S. (2012) *Trouble in the West: Egypt and the Persian Empire 525–332 BCE*. Oxford.

Sancisi-Weerdenburg, H. and Drijvers, W. (eds) (1990) *Achaemenid History v: The Roots of the European Tradition*. Leiden.

Schaudig, H. (2019a) 'The Text of the Cyrus Cylinder', in Shayegan (2019) 16–25.

Schaudig, H. (2019b) 'The Magnanimous Heart of Cyrus: The Cyrus Cylinder and Its Literary Models', in Shayegan (2018) 67–91.

Shayegan, M.R. (2011) *Arsacids and Sasanians: Political Ideology in Post-Hellenistic and Late Antique Persia*. Cambridge.

Shayegan, M.R. (ed) (2019) *Cyrus the Great: Life and Lore*. Ilex Series 21. Boston.

Still, B. (2019) *The Social World of the Babylonian Priest*. CHANE 103. Leiden.

Stökl, J. and Waerzeggers, C. (eds) (2015) *Exile and Return: The Babylonian Context*. BZAW 478. Berlin.

Stolper, M. (1985) *Entrepreneurs and Empire: The Murašû Archive, the Murašû Firm, and Persian Rule in Babylonia*. PIHANS 54. Istanbul.

Stolper, M. (2001/02) 'Texts of the Murašûs and from Their Surroundings'. *JCS* 53: 83–132.

Tamerus, M. (2018) 'Labour in the Achaemenid Heartland', in Garcia-Ventura (2018) 467–93.

Tolini, G. (2005) Quelques éléments concernant la prise de Babylone par Cyrus le Grand (octobre 539 av. J.-C.). *ARTA* 2005.003.

Tuplin, C. (2020a) 'Aršāma: Prince and Satrap', in Tuplin and Ma (2020b) 3–72.

Tuplin, C. (2020b) 'The Bodleian Letters: Commentary', in Tuplin and Ma (2020a) 61–283.

Tuplin, C. and Ma, J. (eds) (2020a) *Aršāma and his World: The Bodleian Letters in Context. Volume I: The Bodleian Letters*. Oxford.

Tuplin, C. and Ma, J. (eds) (2020b) *Aršāma and his World: The Bodleian Letters in Context. Volume III: Aršāma's World*. Oxford.

Van der Spek, R.J. (1998) 'The Chronology of the Wars of Artaxerxes II in the Babylonian Astronomical Diaries', in Brosius and Kuhrt (1998) 239–56.

Van Driel, G. (2002) *Elusive Silver: In Search for a Market in an Agrarian Environment. Aspects of Mesopotamia's Society*. PIHANS 95. Istanbul.

Vidal, J. (ed) (2010) *Studies on War in the Ancient Near East: Collected Essays on Military History*. AOAT 372. Münster.

Waerzeggers, C. (2003/04) 'The Babylonian Revolts against Xerxes and the "End of Archives"'. *AFO* 50: 150–73.

Waerzeggers, C. (2006) 'The Carians of Borsippa'. *Iraq* 68: 1–22.

Waerzeggers, C. (2015a) 'Babylonian Kingship in the Persian Period: Performance and Reception', in Stökl and Waerzeggers (2015) 181–222.

Waerzeggers, C. (2015b) 'Review Article of Laurie E. Pearce and Cornelia Wunsch, Documents of Judean Exiles and West Semites in Babylonia in the Collection of David Sofer. CUSAS 28'. *STRATA* 33: 179–94.

Waerzeggers, C. and Seire, M. (eds) (2018) *Xerxes and Babylonia: The Cuneiform Evidence*. Leuven.

Weiskopf, M. (1989) *The So-called "Great Satrap's Revolt", 366–360 BC: Concerning Local Instability in the Achaemenid Far West*. Historia Einzelschriften 63. Stuttgart.

Wijnsma, U. (2018) 'The Worst Revolt of the Bisitun Crisis: A Chronological Reconstruction of the Egyptian Revolt under Petubastis IV'. *JNES* 77: 157–73.

Zawadzki, S. (2018) *The Rental of Houses in the Neo-Babylonian Period. (VI–V Centuries BC)*. Warsaw.

CHAPTER 9

Alexander and the Iranian Satraps in War and Peace

Waldemar Heckel

A short essay, in a volume titled *How to Lose a Battle: Foolish Plans and Great Military Blunders*, reaches the following conclusion about the Battle of Arbela (Gaugamela):

> Why did Darius III lose? The battle was fought where he wanted, how he wanted (at least at first), and he had massive superiority in numbers. But Alexander based his plan on one fact: when they battled the year before at Issus, the conflict had been decided when Darius III fled the field. *Simply put*, it appears that the Persian emperor was unable to face personal danger. Darius III lost two battles and the largest empire in the world because he was a coward.

This observation is neither correct with respect to Alexander's strategy nor to Darius' character and actions, but since the view of the latter has often been extended to include Iranian leaders in general, it is important to provide a more balanced view.[1]

1 Fawcett (2006) 10 (emphasis added). 'Simply put' is an apt way of describing Fawcett's conclusion, which I cite only because it typifies the persistent view of modern writers, both laymen and scholars, that the Persians were poor fighters and cowardly by nature. For the fifth-century origins of the 'unmanning' of Persians see the excellent work of Hall (1989); Hall (1993). The notion of the 'cowardly Darius' has been debunked on several occasions by competent scholars (Seibert (1987); Nylander (1993); Badian (2000b); Briant (2010); Briant (2015)), whose work Fawcett and his ilk have not bothered to read, and this corrective view must apply also to the actions of other Persian satraps and commanders. See also Heckel (2008); Heckel (2020). I wish to thank Professor Maxim Kholod for sending me, in advance of publication, his chapter on 'The Administration of Alexander's Empire', which will appear in *The Cambridge Companion to Alexander the Great* (edited by Daniel Ogden).

 | DOI:10.1163/9789004710771_010

1 The First Clash with the Invader

In the last year of his life, Philip II, who had brought the Greek states under his control and forged the League of Corinth, from which he planned to requisition forces to fight in his Persian campaign, sent an advance force of 10,000 men to Asia Minor under the command of Attalos, Parmenion and a certain Amyntas.[2] Despite Philip's assassination in autumn 336, it was clear that a full-scale invasion of the Persian Empire was still on the drawing board – though Philip's successor, Alexander, was temporarily preoccupied with affairs in Europe[3] – and the efforts of this smaller expeditionary army continued up to the time of Alexander's crossing of the Hellespont in May 334. What is particularly noteworthy is that this Macedonian contingent made very little headway in Asia Minor and, by the time of Alexander's arrival, had been all but neutralised by Memnon of Rhodes, acting on the authority of the satrap of Hellespontine Phrygia. Greek sources, predictably, depicted Memnon not only as the reason for this success but even as the man who could, even after Alexander's initial victories in Asia Minor, turn the tide against the invader and turn the war back upon Greece. Death from illness ended that dream, just as Agesilaos' recall from Asia had allegedly saved Persia from imminent defeat.[4] At least, that was how Greek historians chose to imagine events. In the Greek imagination, Memnon's potential to halt Alexander in his tracks at the Granikos was squandered by the jealousy and distrust of the Persian commanders who lacked both his strategic and tactical brilliance.[5]

Although the experience of the Ten Thousand had exposed to Greeks and Persians alike some of the deficiencies of the Persian Empire, it was also the limited impact of both these mercenaries and the forces of Agesilaos that explains Darius' failure to marshal a larger army to deal with the Macedonians – whether or not he was distracted by other events or unable to bring his naval forces to the Hellespont. Clearly, the Great King thought that a coalition of satraps,

2 Just. 9.5.8 (the force was sent *initio veris*); Diod. Sic. 16.91.2; cf. 16.93.9. Polyaenus 5.44.4 says it numbered 10,000. Amyntas appears to have been the son of Arrhabaios; Heckel no. 78 (all prosopographical references are to Heckel (2021a) unless otherwise indicated).

3 For the years 336–334 see Heckel (2020) 20–40.

4 See Diod. Sic. 15.31.3; Plut. *Ages.* 15.1.

5 See McCoy (1989); Kholod (2018). For Memnon's exaggerated role in the Greek historians see Briant (2002) 820–1. Memnon appears to have been a mercenary leader in the service of Arsites of Hellespontine Phrygia. It was not until after the disaster at the Granikos and the suicide of Arsites that Memnon was appointed ὕπαρχος τῆς κατὰ τὴν Ἀσίαν παραλίας (cf. Diod. Sic. 17.23.5–6; 17.29.1). See Heckel no. 701.

with their satrapal armies, would be sufficient to repel the invader, though the presence of leaders such as Mithrobouzanes of Cappadocia, Pharnakes, and Mithridates (the last two relatives of Darius) shows that he did not take the invasion lightly. The rulers of Asia Minor came out in full force and fought gallantly, many of them dying in defence of their lands.[6] The flight of the cavalry was not the cause of the Persian defeat but rather a consequence of it. Furthermore, it is worth noting that Arsites committed suicide after his satrapy was lost and both Arsames of Cilicia and Atizyes, satrap of Greater Phrygia, joined Darius to fight again at Issos, where they died noble deaths.[7]

Despite the voluntary surrender of Sardis by its garrison commander, the Persians continued to put up stubborn resistance in the Aegean and at Miletos and Halikarnassos, where Orontobates of Caria and Hydarnes, along with several members of the family of Artabazos, distinguished themselves. After Issos (333), where again the Persian opposition was more determined than the Greek sources let on,[8] remnants of the army took refuge in Cappadocia and Paphlagonia, preparing to fight behind Macedonian lines.[9] By contrast, those who surrendered and collaborated were few and relatively insignificant in the years 334–332. Hence, we may regard Alexander's readiness to reach an accommodation with the defeated Iranian leaders (along with the local dynasts) in these years as more than an enlightened gesture to the vanquished. It was, in fact, a necessary move in light of the failure of the empire to collapse in the aftermath of two decisive defeats.

Alexander had grown up in a kingdom strengthened by conquest and assimilation of the defeated; for Macedon was less xenophobic than the Greek *poleis* to the south. Philip II had offered land, citizenship, and even membership in aristocratic circles, particularly at the court, in the body of the *hetairoi* or Companions. Though many outsiders continued to be identified by place of

6 References to Arrian are to his *Anabasis of Alexander* unless otherwise specified. Arr. 1.12.8 names the commanders: Arsames, Rheomithres, Petenes, Niphates, Spithridates (also his brother Rhoisakes: Arr. 1.15.7–8), Arsites. Of these all but Arsames and Arsites died (for Rheomithres see note on Issos below), along with Mithrobouzanes, Pharnakes, Mithridates, Arboupales and Omares.

7 Arsites' suicide: Arr. 1.16.3. Arsames' death at Issos: Arr. 2.11.8. Atizyes was almost certainly at the Granikos, despite Arrian's failure to name him. Diod. Sic. 17.21.3 says he died there; but his death is again reported by Diod. Sic. 17.34.5 at Issos (Atizyes and Antixyes are almost certainly the same individual). Rheomithres also fought a second time at Issos and met his end there (Diod. Sic. 17.34.5).

8 Typically the Greek sources regard the death of Memnon (from illness) as the deciding factor (Diod. Sic. 17.29.4).

9 Curt. 4.1.34–40. See also Ruzicka (1988).

birth, they were nevertheless designated *Makedones* and proved loyal to the state that had adopted them and awarded both wealth and status.[10] From the royal house down, intermarriage was common and the offspring of these unions enjoyed the same privileges as those whose parents were both native Macedonians. The neighboring Thracians, Illyrians, Triballians, and Paionians were, to the Greek world in general, no less barbarian than the Persians; indeed, in the eyes of many who dwelt in the south, the Hellenicity of the Macedonians themselves was at best questionable and their leaders often derided as barbarian.[11] The inclusion of members of aristocratic or royal families from Asia in the body of the *hetairoi* and the later intermarriage of Macedonians and Iranians were natural extensions of Philip II's practices and not a sign of a sudden cultural enlightenment on Alexander's part.[12]

John D. Grainger claims that 'Alexander had assumed that he could simply take over the Persian Empire from Darius, as he had taken over Macedon from Philip, needing only to replace a few satraps', adding that 'it was not possible to run the empire alone; he needed a system, a bureaucracy and dependable officials'.[13] But Grainger's view of Alexander's conquest as a failure is based on a superficial reading of the sources and an apparent belief that the king's failure to micro-manage affairs points to a lack of interest in ruling. Richard A. Billows shares this view of Alexander: 'he showed no interest in organising his conquests into a coherent, functioning empire, merely continuing the Persian system for the sake of convenience'.[14] In truth, when one considers the immensity of the undertaking and the brevity of Alexander's life, one must reach the opposite conclusion. 'Continuing the Persian system' ought not to be regarded as a simple avoidance of the ruler's duties but a very sensible adoption of a system that had proved flexible, and thus durable. The Persian Empire, for its Achaemenid rulers as much as the Macedonian conquerors, was not conducive to a 'one size fits all' approach to its administration.

10 Nearchos is named as a citizen of Amphipolis, but also identified as Cretan; cf. the Mytilenaian sons of Larichos, Erigyios and Laomedon.

11 Alexander I of Macedon, although he had provided the Greeks with sufficient proof of his Greek ancestry (Hdt. 5.22), was nevertheless known by the curious *epiklēsis* 'Philhellene'; for the Greek view of Macedonian rulers as barbarians see Cawkwell (1978) 22.

12 See Abdalonymos of Sidon (Diod. Sic. 17.47.6); Nikokles of Soloi and Nithaphon of Salamis (Arr. *Ind.* 18.8); Bagoas son of Pharnouches (Arr. *Ind.* 18.8). Cf. Arr. 7.6.2–5. Intermarriage: Diodorus, contents of Book 17; Metz Epitome 31; Arr. 7.4.4–8; Ael. *VH* 8.7; Just. 12.10.9–10.

13 Grainger (2007) 85.

14 Billows (2018) 181.

2 The First Steps toward Integration

In 334, soon after his victory over Persian forces at the Granikos River, Alexander sent a force under Amyntas son of Andromenes against Sardis, where its Iranian *phrourarchos*, Mithrenes, surrendered the citadel.[15] He would be the first of many, but it is significant that already at this very early stage of the campaign Alexander kept the defector in his entourage with a view to making use of him when the opportunity arose. If there had been any doubts about whether the young conqueror would attempt to 'win the hearts' of his opponents, these were already dispelled when he buried the Persian dead after the Granikos battle.[16] In the same campaigning season, Alexander reinstated the ousted Carian queen (or dynast) Ada to her rightful position, going so far as to accept her has his adoptive mother; full control of Halikarnassos was not achieved until 333.[17] But, upon closer examination, this act, although designed to win over the Carian population, was actually anti-Persian, and, like the favourable treatment of Sardis and Daskyleion, intended to weaken the power of Darius III. It does, however, show once again that Alexander's so-called orientalising policy – I would not go so far as to call it 'fusion' or *Verschmelzungspolitik* – was not simply a reaction to the events of 330 that led up to and followed the death of the Achaemenid ruler.[18]

Alexander's propaganda depicted the conquest of Asia Minor as an act of liberation, which left no room for the retention of Iranian satraps, hyparchs, or even Greek tyrannies and oligarchies, on the Aegean coast and in Anatolia.[19] Nevertheless, he kept Iranian defectors or collaborators in his entourage with a view to employing them elsewhere. Elsewhere was not far to seek. Regions such as Cappadocia and Armenia, which stubbornly resisted the Macedonian advance, had populations with rulers loyal to Darius III. To win them over, Alexander sought to install Sabiktas in Cappadocia and, somewhat later, Mithrenes in Armenia.[20] Their apparent failure to establish themselves does

15 Arr. 1.17.3–7. For Mithrenes, see Heckel no. 739. *Phrourarchos*: Arr. 1.17.3; Diod. Sic. 17.21.7 calls him satrap, clearly an error, since Arrian makes a distinction between the leading citizens of Sardis, who surrendered the city, and Mithrenes who turned over control of the citadel and the treasures. For Hydarnes and Orontobates, see Heckel nos. 542, 820.

16 Arr. 1.16.6.

17 Ada (no. 7). Arr. 1.23.7–8; Diod. Sic. 17.24.2. See also Sears (2014); Howe (2021).

18 For the question of Alexander's policy of fusion see especially Berve (1938); Bosworth (1980b); Hamilton (1988).

19 For detailed discussion see Heckel (2020).

20 Mithrenes (no. 739). His appointment as satrap of Armenia dates to 331 (Arr. 3.16.5; Diod. Sic. 17.64.6; Curt. 5.1.44; cf. Julien (1914) 27–8). Sabiktas (no. 1030) was sent to

not alter the fact that the conqueror adapted his policies to suit the needs of different regions and their populations. In southwestern Asia Minor, he supported a native dynast, just as he would retain trustworthy Sidonian and Cypriot rulers in 332.[21] The administration of Syria was assigned to Menon son of Kerdimmas; its satrap, Mazaios, remained with Darius, who was recruiting new forces to redress the defeat at Issos.[22]

Egypt presented Alexander with an unusual problem, leading to a hybrid administrative solution. For over half a century, Egypt had remained independent until its reconquest by Artaxerxes III in 343/2, and even then resistance continued in the form of an uprising by a certain Khabbabash, perhaps between 342 and 338 or between 338 and 336.[23] By the time of Alexander's invasion of Asia Minor, the satrapy was again securely in Persian hands. The Iranian Sabakes (or Sauakes) brought a contingent of troops to Issos, where he died on the battlefield. His replacement, Mazakes, may have been the man who ruled the satrapy in Sabakes' absence, unless Darius sent a new official to Egypt in 332; at any rate, we know that he governed with the Great King's approval.[24] Soon after Issos, Mazakes defeated the forces of Amyntas son of Antiochos, a Macedonian exile who had fought for Darius and escaped from Issos with a force of 4,000 mercenaries.[25] Initially, some of the Egyptians rallied to his side, intent on overthrowing the Persian garrisons,[26] but the mercenaries harmed their cause when they turned to plunder the countryside and were defeated by Mazakes. Hence, despite the prompt surrender of Mazakes, who learned that the Macedonians had reached Pelousion, Alexander knew that he could neither confirm the defector as satrap nor appoint a single native governor. Arrian (3.5.2–5) tells us:

Cappadocia in the summer of 333 (Arr. 2.4.2; cf. Julien (1914) 18–19; Baumbach (1911) 59 n. 2). Curt. 4.1.34–5, in this context, calls him Abistamenes, which is clearly a corruption of Sabiktas; Hedicke in his edition of Curtius emends the name to *Sabistamenes* (cf. Schachermeyr (1973) 194 n. 206). Neither Mithrenes nor Sabiktas is heard of again. Cappadocia was ruled by Ariarathes (no. 179) at the time of Alexander's death in 323.

21 For the Sidonian rulers, see further discussion under Straton (no. 1084) and Abdalonymos (no. 1).

22 Menon (no. 722, almost certainly identical with the Arimmas of Arr. 3.6.8). Mazaios (no. 685) and the Syrian troops (Arr. 3.8.6) remained with Darius (Curt. 4.12.12; Arr. 3.11.4). See also Briant (2002) 713–14.

23 Briant (2002) 718; Burstein (2000).

24 Sabakes (no. 1029); for his death see Curt. 3.11.10; Arr. 2.11.8. Mazakes (no. 686) is called *praetor Darei* by Curt. 4.7.4 and satrap by Arr. 3.1.1.

25 Curt. 4.1.27–33; Diod. Sic. 17.48.1–5; Arr. 2.13.2–3. Amyntas (no. 77). See also Heckel (2020) 116–19.

26 Curt. 4.1.30; cf. 4.7.1.

> Alexander appointed two Egyptians as nomarchs of Egypt, Doloaspis and Petisis, dividing between them the country of Egypt; when Petisis declined the power, Doloaspis took it all. *Phrourarchoi* were appointed from the Companions, Pantaleon of Pydna at Memphis, and Polemon son of Megakles of Pella at Pelousion; Lykidas the Aitolian was to command the mercenaries and Eugnostos son of Xenophantes, one of the Companions, was made secretary in charge of the mercenaries, while Aischylos and Ephippos son of Chalkideus were appointed overseers (*episkopoi*). Neighbouring Libya was given to Apollonios son of Charinos to govern; and Arabia around Heroönpolis to Kleomenes from Naukratis. The latter was told to allow the nomarchs to govern the nomes under their control according to the ancient practices, but to exact tribute from them himself; they in turn were ordered to pay it to him. As generals of the army which he left in Egypt he appointed Peukestas son of Makartatos and Balakros son of Amyntas, and as admiral of the fleet, Polemon son of Theramenes.

These measures show that Alexander once more retained existing structures, assigning control over Upper and Lower Egypt to Egyptians and leaving in place the lower-level native nomarchs, who continued to function in the traditional manner that went back to age of the pharaohs. It has been noted that 'two principles were applied: the favour of the natives was safeguarded, and power was (as far as possible) divided'.[27] At the same time, in the east and west, Greco-Macedonian officials were appointed, and these were supported by commandants in Memphis and Pelousion as well as by an army of mercenaries and a fleet.[28] The former Persian satrap, Mazakes, along with a certain Amminapes, remained in Alexander's entourage; the latter was appointed satrap of Parthia and Hyrcania after the death of Darius III.[29]

27 Badian (1965) 171.

28 Bosworth (1980a, 275) calls the appointment of the Egyptians 'primarily a public-relations exercise'. Hölbl (2001, 12) believes that Doloaspis was Iranian – certainly the name suggests this – but if this is so he was probably a Persianised member of a local aristocratic family. Briant (2002, 869) notes: 'As soon as the royal military was conquered, local leaders found themselves with a simple choice (which their ancestors had known since the conquests of Cyrus): to negotiate with the victor a way to maintain their dominant position within their own society'.

29 On Amminapes see below. What became of Mazakes is unknown.

3 Darius' Final Battle and Its Aftermath

In late summer and early autumn 331,[30] Darius organised what was to be his final military campaign against Alexander. Having failed to make headway on the diplomatic front, where he offered territorial concessions and kinship by marriage – clearly an attempt to neutralise his enemy in the short term with the hope recovering his power in the future – he was forced to risk everything in a single, full-scale engagement.[31] Having learned from his mistakes at Issos, where he had abandoned the northern Mesopotamian plain at Sochoi for the narrows on the Gulf of Issos,[32] Darius was determined to fight on the battlefield of his own choosing, not only selecting the location but grooming the terrain to suit his needs.[33] The Persian army was certainly vastly superior in numbers, and particularly strong when it came to the horsemen from the upper satrapies, but the figures found in Greek sources have been deliberately exaggerated to enhance the prestige of Alexander's victory; similarly, they make Darius' flight seem more disgraceful. The Great King is alleged to have marshalled a force of anywhere from 500,000 to more than a million against a Macedonian force of under 50,000.[34] His aim was to bring about a double envelopment of his enemy, placing large numbers of cavalry on each wing. Many of these were now armed with the *xyston*, the thrusting spear used so effectively by the Macedonians, others with shields and longer swords.[35] Even the scythed chariots achieved a measure of success, though the evidence of Diodorus and

30 The encounter was preceded by desultory skirmishing with the enemy who marched north from Phoenicia (Diod. Sic. 17.55; Curt. 3.9.1–10.15; Arr. 3.7.1–8.2); according to a Babylonian astronomical diary (AD -330 = Kuhrt (2007) 447), the battle of Gaugamela was fought on October 1, 331.

31 For these diplomatic exchanges see Mikrojannakis (1970); Berhardt (1988); Bloedow (1995); Badian (2000b) 257–58. Note that Briant (2002, 832–40) challenges their historicity.

32 Cf. Curt. 4.14.4.

33 For the location of the battle see Sushko (1936); Zouboulakis (2015). For the battle, Marsden (1964); Fuller (1960) 163–80; English (2011) 129–47. The battlefield was levelled in places (Arr. 3.8.7) and spikes were planted to counter the Macedonian cavalry (Curt. 4.13.36).

34 For the numbers see Diod. Sic. 17.39.4; Just. 11.12.5; Arr. 3.8.6; Curt. 4.12.13 claims Darius had 45,000 cavalry and 200,000 infantry, contradicting his earlier claim that his army was double the size of that at Issos (4.9.3), which he numbered as 250,000 infantry and 62,200 cavalry (3.2.4–9).

35 Diod. Sic. 17.53.1. Curt. 4.9.3–4 says that Darius armed his cavalry with shields and sword (*scuta gladiique*) in place of javelins (*iacula*). Contrary to the view of some scholars, the sources do not attest the use of the *sarissa* by Persian forces; nor does the Alexander Mosaic support their arguments. See Heckel (2017).

Curtius – apparently at odds with Arrian's account – has been misunderstood.[36] But Alexander's army caught the Persian ruler between the anvil of the sarissa-bearing phalanx and the hammer of the Companion cavalry, breaking the Persian formation. Although a gap in the phalanx had allowed Mazaios and his cavalry to put pressure on the Macedonian left under the command of Parmenion, while the Scythian and Persian horse plundered the baggage camp,[37] Darius was forced to flee, thus completing the Macedonian victory.[38] Nevertheless, by letting the king escape his grasp, Alexander had achieved only a stalemate, one that prolonged the conquest and forced him to seek victory by other means.

After Gaugamela, Darius fled in the direction of the upper satrapies, abandoning Babylonia, Susa, and Persis to face the Macedonian conquerors and planning his next moves from Ecbatana in Media. The satrap of Babylonia, Boupares, was probably among those who died in the battle.[39] What remained of his contingent was taken to Babylon by Mazaios, who recognised the impact of the Persian defeat and surrendered the city to Alexander. What assurances, if any, he had received from the conqueror are unknown. Mazaios himself was allowed to retain the satrapy. Despite Alexander's treatment of Mithrenes, Sabiktas, Mazakes, and Amminapes, Mazaios was the first of Darius' appointees to be confirmed as satrap under the new regime. His importance is reflected in the prominent position he occupies in the Alexander historians, particularly those belonging to the so-called Vulgate tradition. If the Alexander Sarcophagus, found in the royal necropolis at Sidon, was commissioned by him (or for him), the sculptures on the panels and gables of the sarcophagus can

36 Arr. 3.13.5–6 describes an attack on Macedonian light-armed troops who killed the drivers with javelins, whereafter the Macedonian grooms (*hippokomoi*) brought the horses under control. But Diod. Sic. 17.58.2–5 and Curt. 4.15.14–17 describe the chariots exploiting the gap in the Macedonian infantry line on the left. See Heckel, Willekes and Wrightson (2010).

37 Arr. 3.14.5. Diod. Sic. 17.59.5 and Curt. 4.15.12 say they were Scythians and Cadusians.

38 Arr. 3.15.2; Diod. Sic. 17.61.3; Curt. 4.16.32 all mention the wounding of prominent Macedonian commanders, but we do not hear of Persian satraps who died on the battlefield (as they did at the Granikos and Issos).

39 Boupares (no. 308) is known only as the leader of Babylonian troops at Gaugamela (Arr. 3.8.5), after which he is no longer mentioned (but see Bosworth (1980a, 291), following Leuze (1935, 410–12)). His identification as satrap is thus a reasonable assumption, especially since Mazaios' presence as the political authority in Babylon implies that the previous satrap had either died or been removed from office (cf. Briant (2002) 849). That Mazaios took control of the city without Darius' approval is highly unlikely, especially in light of the actions of the *gazophylax* Bagophanes (no. 279), who made no attempt to hold out against the Macedonian forces (Curt. 5.1.7–20).

be regarded as evidence of a reconciliation between the conqueror and the defeated. The Macedonian king and his *hetairoi* are depicted conducting a lion hunt in a game park (*paradeisos*) in Babylonia in conjunction with Mazaios and his noble entourage.[40] This shows that Alexander went beyond the simple step of retaining Iranian rulers for the sake of convenience. Not much later, Aboulites of Sousiana was also confirmed as satrap after his surrender. In both satrapal capitals, the king left Macedonians in charge of the military forces and the collection and protection of revenues.[41] This process may suggest that the Iranian satrap was little more than a figurehead, but this cannot be the case. Aboulites was later executed by Alexander for his failure to supply the army during the Gedrosian campaign, which suggests that he had the authority to requisition and dispatch what the king required. The Macedonian officials were there to defend the satrapy against insurrection but also to serve the satrap when he issued legitimate orders.[42] Conversely, the abuses of Macedonian officials and generals in Media were punished by the king, but no charges were brought against the Iranian satrap.[43]

The pattern of appointing Iranians to office continued through 330 and 329, but the criteria for selecting the specific individual to hold the office varied. This can be seen in the next three satrapal appointments. In Persis, Alexander appointed Phrasaortes son of Rheomithres in place of Darius' satrap, Ariobarzanes, about whose fate the ancient sources disagree: either he was killed defending the city against the invader or he fled after a last-ditch

40 See Heckel (2006). The traditional view, going back to Studniczka (1894), is that the sarcophagus belonged to Abdalonymos of Sidon, the evidence for which is limited to the fact that it was found in the royal necropolis and the imaginative interpretation of what is called a 'murder scene' on one of the gables (Stewart (2014) 258 n. 2); Palagia (2017, 181–82) also reiterates her earlier support for the Abdalonymos theory (Palagia (2000)). But German scholars of the late nineteenth century recognised that the depiction of a Persian – whose features and dress are totally inappropriate for a Sidonian king who, it may be added, neither fought at Issos or Gaugamela nor hunted in the Persian game-park – pointed to an Iranian occupant (or intended occupant) of the sarcophagus: see Reinach (1892); Judeich (1895); Willrich (1899).

41 For the *stratēgoi* and *phrourarchoi*, see below. Kallikrates (no. 563) is named by Curtius as treasurer or *gazophylax* (5.2.17). He may, in fact, have been responsible for the collection of tribute.

42 Cf. Badian (1965) 174.

43 Death of Aboulites (no. 4): Plut. *Alex.* 68.7; Arr. 7.4.1 says he was mismanaged affairs in Susa. Agathon, Kleandros, and Sitalkes were all stationed in Media and were punished for their crimes there, but no charges were brought against Atropates (no. 261). On this see Hyland (2013) 132–37.

skirmish before Persepolis.[44] We are not told what it was that made Phrasaortes acceptable to the king, though he appears to have come from a prominent family.[45] Tiridates was allowed to remain in Persepolis as *gazophylax*.[46] Then, in Ecbatana, Alexander selected Oxydates to rule Media; its satrap, Atropates, remained with Darius as the Persian forces fled eastward. Oxydates himself had been arrested and incarcerated in Susa on capital charges by Darius, and it was for this opposition to the monarch that Alexander considered him trustworthy.[47] Amminapes, who was installed as satrap of Parthia (and Hyrcania) soon after the appointment of Oxydates, was a native Parthian and a man who had spent time in exile at the court of Philip II. But that exile appears to belong to the time of the Satraps' Revolt. He had fled from Artaxerxes III Ochos, and his presence with Mazakes in Egypt suggests that he had no personal quarrel with Darius III.[48] Whether Alexander had already met him in Macedonia, when he still a young man, we do not know, but clearly Amminapes' early defection spoke in his favour.

After the death of Darius, at the hands of Bessos, Nabarzanes, and Barsaentes, Alexander received a flood of defectors: Artabazos and his sons, Kophen, Ariobarzanes, and Arsames; Autophradates; Phrataphernes; and even Nabarzanes himself, who surrendered only after prior negotiations. We may assume that Atropates, Darius' satrap of Media, submitted at this time, but there is no specific mention of this. Of these, only Autophradates was reinstated immediately, not only resuming his position as satrap of the Tapourians but being given responsibility for the Mardians as well.[49] Why did Alexander

44 The identity of Ariobarzanes is uncertain. He may have been a member of a noble family from Cappadocia (see no. 182). Whether he is identical with no. 183, the son of Artabazos (no. 223), depends on whether we accept Curtius' claim that the satrap of Persis was killed in the fighting before the city (Curt. 5.4.34). Arr. 3.23.7 tells us that the son of Artabazos surrendered along with his father in Hyrcania. If both statements are correct, the satrap of Persis cannot have been the son of Artabazos. But, if Curt. 5.4.34 is incorrect, and the satrap was the son of Artabazos, his opposition to Alexander may explain why he lapses into obscurity.

45 Phrasaortes (no. 957) seems to have been the son of that Rheomithres (no. 1026) who died at Issos (Arr. 2.11.8; Curt. 3.11.10; Diod. Sic. 17.34.5).

46 Tiridates (no. 1148). Curt. 5.6.11. Bagophanes (no. 279), by contrast, was deposed in Babylon and told to accompany the king (Curt. 5.1.44).

47 Oxydates (no. 835). His appointment: Arr. 3.20.3; Curt. 6.2.11. Badian (1965) 171 observes that, in Media, 'no *strategus* is mentioned'. But it was not long before Parmenion, Kleandros and others were stationed there (cf. Arr. 3.26.3; Curt. 7.2.19–32).

48 Amminapes (no. 71) surrenders with Mazakes: Arr. 3.22.1. Curt. 6.4.25 adds that he had been an exile at Philip's court during the reign of Ochos.

49 Curt. 6.4.25; Arr. 3.23.7. The Mardians were placed under his authority not much later (Arr. 3.24.3).

reinstate Autophradates but not Phrataphernes? The answer may be as simple as the fact that, by the time Phrataphernes surrendered, Amminapes had already been appointed and this act could not easily be undone.[50] Perhaps the king believed that the Parthians would be more amenable to a native satrap. Parthia and Hyrcania appear to have formed an administrative unit and it may be that Alexander was reluctant to destabilise the area by detaching Hyrcania and giving it back to Phrataphernes. But the sources are confusing rather than helpful: Curtius (6.4.25) names Amminapes[51] as satrap of Hyrcania, with no mention of Parthia, and when Phrataphernes is finally reinstated, Arrian calls him satrap of Parthia on two occasions (3.28.3; 4.18.1) and then, describing the events of 326, satrap of Parthia and Hyrcania (5.20.7). This is precisely the reverse of what one might expect. Furthermore, we are not told what became of Amminapes, whether he died while in office or was deposed. It is possible, however, that Alexander was uncertain of Phrataphernes' loyalty, preferring to keep him in his entourage before deciding his political future. It was not long before Phrataphernes was reinstated and he proved to be one of the most loyal satraps in Alexander's empire, but this could not be predicted.[52] In 330, he had surrendered in tandem with Nabarzanes, who did so reluctantly; for he had played a leading role in the murder of Darius III and feared punishment.[53]

This raises the question of Alexander's attitude toward the regicides. In 329, he made a great show of punishing Bessos for his betrayal of Darius, though in fact his true crime was that he assumed the upright tiara and styled himself Artaxerxes (V).[54] By doing so he challenged Alexander's claim to be king of Asia. But the matter is obscured by contemporary propaganda and subsequent justification of the king's actions. For years the Great King and the conqueror had waged a diplomatic and propaganda war. But the Alexander historians would have us believe that Darius in his final days had praised his opponent. The popular (and more sensational) Alexander tradition preserves a story that,

50 Cf. Badian (1985) 450.

51 The MS reading is *manapin*.

52 He is already named as satrap of Parthyaia by Arr. 3.28.2, unless this designation is anachronistic, referring to his position under Darius. Phrataphernes in 323: Diod. Sic. 18.3.3; Just. 13.4.23 (with Yardley, Wheatley, and Heckel (2011, 114–15), where my reference to Phrataphernes 'as satrap of Areia' is an error for Parthyaia); Dexippos, *FGrH* 100 F8.6, where the name is given as Rhadaphernes. Phrataphernes and Oxyartes are the only Iranians who remained in charge of their satrapies after Alexander's death; Atropates was not deposed but his authority was limited to Lesser Media (Media Atropatene, as it was later called).

53 See Heckel (2018) 104: 'For Phrataphernes, there must have been guilt by association, but he was able to redeem himself, and did so in the coming year'.

54 Arr. 3.25.3.

as he lay dying, Darius asked that his conqueror avenge his death: 'it was a question of precedent and the common cause of kings; for Alexander to ignore this would be dishonorable' (11.15.12).[55] Alexander could thus place himself in the legitimate line of succession and avenge, as duty required, the murder of his predecessor. In truth, the Macedonian king was prepared to do what was expedient. Of the regicides, Barsaentes had fled, first to Arachosia and thence to India, only to be brought to Alexander and executed.[56] Bessos could not be forgiven, since he had not only failed to surrender but also claimed the kingship for himself. By contrast, he was willing to forgive Nabarzanes, who was equally guilty. He had made his submission long before Bessos was arrested. Bosworth believes that 'Alexander in late 330 seems consciously to have placed satrapies in the hands of men of proven disloyalty to Darius'.[57] With the exception of Oxydates, this would be true only of Satibarzanes, if one assumes that Arr. 3.21.10 is not an error for Nabarzanes;[58] in both cases he was disappointed. He would later be forced to remove Oxydates from office on the charge of shirking his official responsibilities (ἐθελοκακεῖν).[59] Satibarzanes surrendered to Alexander at Sousia (Tus)[60] and was allowed to remain as satrap of Areia, but the man wasted no time in rebelling against his new master, killing a force of forty mounted javelin-men (*hippakontistai*) who had accompanied him back to his satrapal capital, along with their leader, Anaxippos.[61] The treachery forced Alexander, who was hurrying to Bactria via the Merv oasis, to turn back and deal with the traitor. But Satibarzanes fled, only to be pursued by a Macedonian contingent led by Erigyios, who killed him in single combat.[62] In

55 Heckel and Yardley (2003) 67.

56 Barsaentes: no. 286. His death Arr. 3.25.8; cf. Curt. 8.13.3–4.

57 Bosworth (1980a) 339.

58 Bosworth (1980) 344 regards the text as correct, noting that Σατιβαρζάνης is the *lectio difficilior*. But this does not rule out an error by the author himself or a later scribe; cf. Heckel (1981) 66. In fact, Arr. 3.21.10 contradicts 3.21.1, where those who arrested Darius are named as Nabarzanes, Bessos, and Barsaentes.

59 Arr. 4.18.3. See also Bosworth (1981) 21–22; Bosworth (1995) 123. For full discussion of the satraps of Media and their conduct see Hyland (2013).

60 Arr. 3.25.1. Curt. 6.6.13 says that it was Satibarzanes who brought him word of Bessos' usurpation of the kingship (but Arr. 3.25.3, says this information came from other Persians).

61 Satibarzanes (no. 1037). Arrian (3.21.10) is the only source to name him as a regicide; cf. Heckel (1981). For the fate of Anaxippos (no. 95) see Arr. 3.25.2, 5; cf. Diod. Sic. 17.78.1.

62 See Briant (1999) 111. I am inclined to believe (perhaps naïvely) that single combat in battle was a common feature (and not just a historiographical motif), whether this was done by prearrangement or in the heat of battle, designed to set the tone and, in the case of the victor, to demoralise the enemy.

spite of the betrayal, Alexander continued to rely on Iranians, assigning Areia to Arsakes.[63]

4 Nabarzanes

If Satibarzanes was not one of the regicides but merely a satrap who surrendered out of necessity and then rebelled on the news of Bessos' usurpation, we are left with only one person who had murdered Darius and gained a reprieve: Nabarzanes. This man had been Darius' chiliarch[64] and was thus also an influential member of the court. Despite his importance, Arrian seems to know little about him except that he conspired to murder Darius and later surrendered himself to Alexander in the company of Phrataphernes. In order to flesh out his career we must turn to Curtius who gives a full account of his role in the conspiracy and his subsequent dealings with Alexander. After Darios' death, Bessos had returned to Bactria, leaving Nabarzanes in control of Hyrcania. But, upon Alexander's approach, Nabarzanes despaired of his situtation and submitted to the conqueror, but not before writing him a letter that defended his actions against Darius and asked for a pledge of safe conduct.[65] To this Alexander agreed, and Nabarzanes surrendered to him in Hyrcania, bringing with him the eunuch Bagoas, whose charms persuaded the king to pardon the former chiliarch. Nabarzanes was, however, not given a satrapy or any higher office and the text of Arrian, as it has come down to us, gives the impression that he simply disappeared from history.[66] But the fate of Nabarzanes is to be found in a corrupt passage of Arrian (4.7.1), where it is reported that 'at this time, Phrataphernes, the satrap of Parthia, came to him, along with Stasanor, who had been sent to Areia to arrest Arsakes. They brought Arsakes in chains and also Brazanes, whom Bessos had installed as satrap of Parthia, along with some others who had rebelled with Bessos'. This clearly refers to Nabarzanes' earlier agreement with Bessos, which saw the latter return to Bactria, while he himself occupied Hyrcania and Parthia. We may conclude that, after failing

63 Arsakes (no. 215): Arr. 3.25.7. Nothing is known about his family or earlier career. Death of Satibarzanes: Arr. 3.28.3; Curt. 7.4.33–7; Diod. Sic. 17.83.4–6.

64 Curt. 6.5.22–3. For the position of chiliarch and its adoption by Alexander see Schachermeyr (1970) 31–7; Heckel (2016) 294–7; Collins (2001); Collins (2012); Meeus (2009).

65 Nabarzanes (no. 762). In charge of Hyrcania in support of Bessos: Curt. 6.3.9. Letter to Alexander: Curt. 6.4.8–14.

66 Badian (1958) 146; Schachermeyr (1973) 308. Berve (1926, 2.269) implausibly suggests that Nabarzanes may have been executed by the king. That, it seems, was yet to come.

to secure an office from Alexander, Nabarzanes continued to support Bessos by wreaking havoc in the Caspian region. Phrataphernes, who had been reinstated in his old satrapy, was sent to arrest him and bring him to Alexander.[67] Hence, Alexander's policy was not to reward men who were disloyal, but rather to rely upon those who respected the sanctity of kingship.

5 Artabazos and His Family

Of all the Iranian satraps who reached an accommodation, Mazaios (see above) and Artabazos are by far the most important.[68] They provide convenient bookends for the discussion of Alexander's Iranian policy. Artabazos belonged to the ruling house of Hellespontine Phrygia (Daskyleion) and was the son of the famous Pharnabazos and Apame, a daughter of Artaxerxes II.[69] Because of his involvement in the Great Satraps' Revolt, he and his family had spent some years in exile at the Macedonian court, and it is likely that Alexander met him and members of his family at that time.[70] Arrian says that he held Artabazos and his sons in honour on account of their loyalty to Darius; he omits to mention that Artabazos' daughter, Barsine, was Alexander's mistress.[71] It was to Artabazos that Alexander entrusted the administration of Bactria, adding Sogdiana after the capture of the so-called Sogdianian Rock.[72] The choice was, however, not a happy one. This was an unruly region, where local hyparchs, men such as

67 See discussion in Heckel (1981) 66–67; cf. Heckel (2020) 166–70. In Heckel (2018, 102–103), I have argued that the full extent of Nabarzanes' betrayal of Darius will have been reported to Alexander by Artabazos, who had remained loyal to the Great King.

68 On Artabazos see now Heckel (2021b), refining the arguments of Heckel (2018). For the sake of thoroughness and clarity many of the views expressed in these two studies are repeated here.

69 For Pharnabazos' role in the events of the late fifth and early fourth centuries, see Hyland (2018). His marriage to Apame: Plut. *Artax.* 27.7; cf. *Alex.* 21.9; Xen. *Hell.* 5.1.28; see also Brunt (1975). The date of the wedding is 387, or possibly 391.

70 Artabazos (no. 223). For his exile, in the late 350s, see Curt. 6.5.2.

71 Arr. 3.23.7; cf. Curt. 5.9.1. Curt. 6.5.1, 4 says that he was accompanied by nine of his sons. In 333, Barsine (no. 287) was captured at Damascus (Curt. 3.13.14); for her role as Alexander's mistress, and their son Herakles (no. 522), see Just. 11.10.2–3; Plut. *Eum.* 1.7; cf. Just. 13.2.7; 15.2.3; Paus. 9.7.2; Curt. 10.6.10–13; Diod. Sic. 20.20.1, 28.1. Müller (2021) considers Alexander's sexual relations with Barsine a (deliberate) insult to her family, especially her brother Pharnabazos (no. 889), who had replaced Memnon (no. 701) as commander in the Aegean.

72 Curt. 7.11.29. Lane Fox (1973) 318 claims that Artabazos 'was retired to the governorship of the first Sogdian rock in place of the baron Ariamazes who had been crucified'. This is arrant nonsense. Not only would such a move have been considered an insulting

Ariamazes and Sisimithres, exercised control from their mountain fortresses.[73] The other nobles had either served with Bessos in the Great King's armies or supported him at home. Those who had been with Bessos at the time of Darius' death were known to Artabazos and, doubtless, many feared his influence with Alexander.[74] It is hard to find fault with Artabazos' performance of his duties: he had campaigned with Alexander's forces against those who continued the resistance,[75] and his son, Kophen, was involved in negotiating the surrender of Ariamazes in late 329.[76] It appears that another son, Arsames, was appointed satrap of Drangiana.[77] In order to reject the evidence of Curtius Rufus, we must assume that he made two separate errors on two separate occasions (7.3.1; 8.3.17), not only confusing Arsakes and Arsames but also the satrapies of Areia and Drangiana. And, if we consider that in the second passage (8.3.17) he mentions both Arsames and Arsakes in the same sentence, the likelihood

---

demotion, but it is contradicted by the fact that several months later Artabazos was still satrap. His appointment as satrap of Bactria: Arr. 3.29.1; Curt. 7.5.1; cf. 8.1.19.

73 Ariamazes (no. 178); Sisimithres (no. 1049).

74 The account given by Curt. 5.9.1–12.20 depicts Artabazos as a faithful supporter and adviser of Darius and Nabarzanes, supported by Bessos, as the voice of the opposition. Despite Curtius' dramatic touches, the story must have been based on a reasonably factual account, which most likely came from Artabazos himself, or possibly one of his sons. After the arrest and extradition of Bessos, many of his supporters will have attempted to distance themselves.

75 According to one source, Artabazos campaigned with Hephaistion (Curt. 8.1.10; cf. 8.1.1); Arr. 4.16.2–3 says he campaigned with Koinos against the Scythians. He had earlier accompanied Erigyios against Satibarzanes. These activities disprove Badian's assumption (2000a) 84 that Artabazos' absence from Gaugamela was due to his age; cf. Badian (1985) 460.

76 Curt. 7.11.5, 22–3; cf. Metz Epit. 17, who calls him Dares. For his career see no. 618.

77 Arsames (no. 218). Arrian does not specify who was to govern the Drangians (Zarangians), saying only that Alexander won over the Drangians and Gedrosians; then he appointed Menon (no. 723) satrap of the Arachosians (Arr. 3.28.1). Curt. 7.3.1 makes it clear that the governing of Drangiana went to Arsames, and that Menon received Arachosia (Curt. 7.3.5). Drangiana cannot be a mistake for Areia, since Stasanor, when he was already satrap of Areia, was sent to take over Drangiana (see note below). Menon's appointment was the first departure from the king's policy of appointing Iranians (noted by Badian (1985) 455 n. 1): he had appointed twelve Iranians in succession up to this point, fifteen if we include Astaspes (no. 245) of Carmania, who was probably confirmed when Alexander was in Persis; Amedines (no. 69), who was placed in charge of the Ariaspians (Curt. 7.3.4 calls him *scriba regis*; hence he probably surrendered after Darius' death); and Tiridates (no. 1149), who is called governor of the Gedrosians and Euergetai (Diod. Sic. 17.81.2). Since the Euergetai and the Ariaspians are the same people (Arr. 3.27.4), Diodorus may have omitted Amedines and assigned both peoples to Tiridates.

.at these two men are the same individual becomes remote.[78] But, if Arsames was installed as satrap of Drangiana, he was soon removed from office either for some fault of his own or, what is more likely, because Alexander found it necessary to part ways with Artabazos and his family. The third son named by Arrian, Ariobarzanes, is not heard of again; this strengthens the case that he is identical with the former satrap of Persis and that Alexander looked upon him with suspicion. In all likelihood, Artabazos was regarded as an outsider, with no roots in Central Asia. His rule was relatively short, not much more than a year: in the late summer or autumn of 328 he relinquished his office, citing old age, although he was only in his early sixties. It appears that his resignation for personal reasons was a face-saving measure.[79]

---

78 Curt. 8.3.17: *Arsami, Drangarum praefecto, substitutus est Stasanor, Arsaces in Mediam missus ut Oxydates inde discederet*. Now we know that Stasanor arrested Arsakes in Areia and brought him to Alexander, who was in Bactra/Zariaspa in 329/8 (Arr. 4.7.1). In the following year, he is described as the satrap of Areia (Arr. 4.18.1) and, at that time, he was instructed to assume control of Drangiana (Arr. 4.18.3). If we believe that Arsames and Drangiana are mistakes for Arsakes and Areia, we must then also accept that *Arsaces in Mediam missus* is an error for *Atropates in Mediam missus*. The phrase *ut Oxydates inde discederet* is a peculiar way of describing the deposing and arrest of a delinquent satrap. I suspect that some information has dropped out of this summary of administrative changes, which originally recorded that Arsakes was replaced by Stasanor (cf. Bosworth (1995) 121), that Atropates was sent to Media to arrest and replace Oxydates, and that in the following year Stasanor was to assume control of Drangiana so that Arsames could leave. Arsakes and Arsames are clearly different people, as we can see from (1) their names, (2) their satrapies, and (3) the time at which they were replaced (one in 329, the other in 328). Arsames was to accompany his father and his sister back to their ancestral territories in the West.

79 Bosworth (1995) 118: 'his old age is likely to have been a diplomatic excuse'. Curt. 6.5.4 says he was in his ninety-fifth year, but he could not have been born any earlier than 390; 386 is the most likely year, in which case he was only 56 when he surrendered to Alexander. On the other hand, dissatisfaction on the king's part must be ruled out for three reasons. (1) At this time, Barsine became pregnant with Alexander's child, Herakles. Diod. Sic. 20.20.1 says that Herakles, the son of Alexander and Barsine, was 17 in 310, in which case he was born in 328/7 (Just. 15.2.3 claims he was 15, which is far less likely, in light of Alexander's marriage to Rhoxane in the winter of 328/7). Justin has probably confused Alexander IV and Herakles (see Yardley, Wheatley, and Heckel (2011) 67–68, 207–208). Barsine was already resident in Pergamon when Alexander died, and it is likely that she and her family returned to the West in 328 or 327. Artabazos relinquished his satrapy in the summer of 328 (Curt. 8.1.19). (2) In 324, at the mass marriage ceremony in Susa, where ninety-two of Alexander's *hetairoi* took Persian brides (Chares, *FGrH* 125 F4 = Athen. 12.538b–539a), Ptolemy married Artakama (no. 224; Arr. 7.4.6), sometimes called Apame (Plut. *Eum.* 1.7), and Eumenes Artonis (no. 227; Arr. 7.4.6; Plut. *Eum.* 1.7); Nearchos' wife, unnamed (Heckel (2021a) A39–41; Arr. 7.4.6), was a daughter of Barsine and Mentor. (3) Kophen son of Artabazos was enrolled in a Persian cavalry *agema* along with two sons of Mazaios, Hydarnes (no. 543) and Artiboles (no. 226); two sons of

6 Artabazos and Oxyartes

What, then, might have been the true reason for his removal from office? Given the nature of the evidence, any conclusion must be regarded as speculative. Barsine's pregnancy may, however, have been a factor. The sources agree that Alexander's marriage to Rhoxane was a love-match, recounting how the young king met Oxyartes' daughter at a banquet, was smitten by her beauty, and decided to make her his bride.[80] The union of Alexander and Rhoxane had the effect of stabilising affairs in Bactria-Sogdiana, and this was surely no coincidence. There can be little doubt that the marriage served a political purpose, which increases the likelihood that it was planned and not spontaneous. Hence, we must attempt to determine the role of Oxyartes in all this. In 329, Oxyartes was among those leaders who fled across the Oxos into Sogdiana along with Bessos. Whether he had accompanied Bessos to Gaugamela and returned with him after Darius' death is unknown.[81] Nor do we know if he participated in the betrayal of Darius. But he was one of the local barons who continued to resist the Macedonians after the extradition of Bessos. For Arrian says that Oxyartes placed his family on the Sogdianian Rock 'when he himself rebelled against Alexander'.[82] The Rock of Sogdiana, which was the fortress of Ariamazes, was captured early in 328, but all the other Alexander historians place the marriage of Alexander and Rhoxane on the Rock of Chorienes; this fortress was taken almost a year later, some time after the Kleitos affair at Marakanda in the late summer or autumn of that year. Certainly, during the banquet at which Kleitos raised objections to Alexander's conduct, no mention was made of his marriage to a barbarian. Indeed, all the available evidence, with the exception of Arrian's faulty account, dates Oxyartes' surrender to Alexander before the capture of his family, who were left for safekeeping on the Rock of Chorienes (that is, the fortress of Sisimithres). Our first mention of Oxyartes in Alexander's

---

Phrataphernes, Sisines (no. 1051) and Phradasmenes (no. 957); and Itanes son of Oxyartes (no. 556); see Arr. 7.6.4–5. This does not point to a family in disgrace.

80 Rhoxane (no. 1028). Curt. 8.4.21–30 (with textual problems); Metz Epit. 28–9; Plut. *Alex*. 47.7–8. The banquet was hosted by Chorienes, also known as Sisimithres. Arr. 4.19.4–5 says that the family of Oxyartes, including Rhoxane, were captured at the Sogdianian Rock in the spring of 327 (for the date, see Arr. 4.18.4). But is clearly an error on Arrian's part. The Sogdianian Rock was the fortress of Ariamazes, which Alexander captured in late 329; Strabo 11.11.4 C517 says that the Sogdianian Rock and place of Rhoxane's capture were two different places. Curt. 7.11.28 tells us that Ariamazes and his noble adherents were whipped and crucified at the foot of the Rock. See the detailed discussion in Heckel (2020) 186–200; also Holt (2005) 85–93.

81 Arr. 3.28.9–10.

82 Arr. 4.18.4. For the revolt of the Bactrian barons see Arr. 4.1.5.

entourage comes when the king sends him, upon Sisimithres' own request, to negotiate that baron's surrender.

We cannot determine the exact date of Oxyartes' surrender. All we know is that he was with the king in late 328, when he besieged the fortress of Sisimithres,[83] and that, if Curtius reports the events of that year in chronological sequence, it occurred before the death of Spitamenes at hands of the Massagetai and the arrest of Dataphernes by the Dahai prompted his surrender.[84] We do not know whether Oxyartes submitted to Alexander before Artabazos relinquished his position as satrap, shortly before the Kleitos affair.[85] But, if Oxyartes surrendered soon after the fall of Ariamazes' fortress, he may have impressed upon the king the unwillingness of the local barons to serve under Artabazos, whom they feared as a political enemy. Uncertain of how he would be received, Oxyartes left his family in the secure fortress of Sisimithres. In the event, he made a favourable impression on the king and the two may have discussed the possibility of a political marriage, as a means of bringing the insurgency to an end. Such an arrangement left no room for Artabazos and his pregnant daughter at Alexander's court, and it was announced that Artabazos was relinquishing his position for personal reasons. The entire family returned to western Asia Minor: father, daughter, and sons, including Arsames, who left his post in Drangiana. The marriage of Alexander to Rhoxane on the Rock of Chorienes was prearranged, and the story of the conqueror smitten by the captive maiden (who just happened to offer great political advantages) is consigned to the world of fairy tales.[86] But, if Artabazos resigned before Oxyartes' surrender, our picture of cause and effect is reduced to sheer coincidence.

83 Curt. 8.2.25–27, 30–31 mentions a certain Oxartes (clearly a corruption of Oxyartes). Cf. Plut. *Alex.* 58.3–4. Curtius and Plutarch report that Oxyartes was sent to Sisimithres; Arr. 4.21.7–8 calls him Chorienes, but clearly Sisimithres and Chorienes are the same person: Chorienes was Sisimithres' official title: thus Berve (1926) 2.354; Heckel (1986).

84 Death of Spitamenes (no. 1072): Arr. 4.17.7 and Strabo 11.11.6 C518 (killed by the Massagetai); Curt. 8.3.1–10 and Metz Epit. 20–1 (murdered by his own wife). Dataphernes (no. 342): Curt. 8.3.16 says he was arrested by the Dahai and turned over to Alexander; cf. Metz Epit. 23.

85 This assumes that Curt. 8.1.19 is correct in claiming that Kleitos was designated as Artabazos' successor. Doubted by Badian (1965, 177 n. 3), who thinks Curtius intended 'to heighten the tragedy of Clitus' death'. Arrian does not mention this, but he does put the appointment of Amyntas son of Nikolaos in place of Artabazos (4.17.3) before the death of Spitamenes (4.18.7).

86 Bosworth (1995) 131: 'She may have been in Alexander's entourage for several months before the king noticed and married her'. But Bosworth's reconstruction of events is very different from mine. See especially Heckel (2020) 199 n. 104.

Once the turmoil in Bactria-Sogdiana had subsided, Alexander crossed the Hindu Kush a second time and moved through Gandhara in the direction of the Indus River. In Parapamisos, he removed from office Proëxes, whom he had installed as satrap in 330/29, replacing him with another Iranian, Tyriespis.[87] But from this point on he found himself among the Indians, who would not look favourably on Iranian overlords.[88] It marked the end of such appointments in the East. Oxyartes himself appears not to have accompanied his son-in-law to India. But there is no explicit evidence of a satrapal appointment. In 325 he rejoined Alexander at the confluence of the Akesines and the Indus, probably bringing Tyriespis, whose conduct of affairs was not in order, with him. Tyriespis was deposed, and Oxyartes now assumed control of the Parapamisadai.[89] It was the last time the king installed an Iranian satrap.

7 Iranian Satraps and Troops in the Service of Alexander

The reinstatement of Iranian satraps, or their replacement by other Iranians, did not automatically translate into military cooperation. In fact, the normal practice of installing garrison troops and appointing Macedonian or Greek *stratēgoi* alongside the satraps was clear indication that king did not entirely trust his Iranian officials and a warning against rebellion in his absence.[90]

87 Tyriespis (no. 1154); Proëxes (no. 994). Appointment of Proëxes: Arr. 3.28.4. Replaced by Tyriespis: Arr. 4.22.4–5.

88 Badian (1965, 177) misses the point when he says, 'Alexander had no reliable Orientals left'.

89 Arr. 6.15.3; Curt. 9.8.9–10, adding that Alexander put Tyriespes (whom Curtius calls Terioltes) to death. We have no record of an earlier satrapal appointment for Oxyartes, but both Arrian and Curtius imply that he had one. Arr. 6.15.3 says that Parapamisos was given to him 'in addition'. Brunt's suggestion in the Loeb Arrian (2.144 n. 4) that this may mean 'in addition to being Alexander's father-in-law' is unconvincing. Curt. 9.8.10 calls him *praetor Bactrianorum*. Since his arrival followed the uprising in Bactria, in which Alexander's appointee, Amyntas son of Nikolaos, may have been killed, it is possible that Oxyartes was encouraged by the Macedonians stationed there to assume control of the region. He may have come to Alexander to report the news of the unrest there.

90 Babylon: Apollodoros (no. 146) of Amphipolis was appointed *stratēgos*, Asklepiodoros (no. 239) son of Philon as tax or tribute collector (Arr. 3.16.4; cf. Diod. Sic. 17.64.5). Curt. 5.2.16–17 says that Susa was entrusted to Archelaos (no. 159) with a force of 3,000 troops (cf. Arr. 3.16.9: *straēegos*); Xenophilos (no. 1165) was appointed *phrourarchos* (apparently replacing Mazaros (no. 687): Arr. 3.16.9; for the problem of Mazaros see Bosworth (1980a) 319). At Persepolis Nikarchides (no. 787) was installed as *phrourarchos*, again with 3,000 Macedonian troops (Curt. 5.6.11). If the forty *hippakontistai* under the command of Anaxippos (no. 95) were intended to keep Satibarzanes in check, they were clearly not up to job (Arr. 3.25.2, 5).

This was certainly the case while Darius lived. But soon after his death, the Macedonian conqueror began to employ those whom he regarded as most loyal, particularly against Iranian rebels. Hence, Phrataphernes and Artabazos must have led contingents of native and mercenary soldiers on several occasions.[91] Early attempts to introduce Iranians into the court and military were met with a predictable measure of hostility, but the king prevailed nevertheless.[92] Alexander appointed the most prominent Iranians as *doryphoroi*, among them Oxyathres, the brother of Darius; at court he introduced Persian *rhabdouchoi* (περὶ τὴν αὐλὴν εἶχε ῥαβδούχους Ἀσιαγενεῖς); but it is unlikely that the *taxis* of *hippakontistai* included native troops.[93] In Bactria and Sogdiana, contingents of Bactrians and Sogdianians served with the general Koinos and the satrap Amyntas.[94] Whatever the ethnicity of the mounted javelin-men, it is clear that the mounted archers were Scythians. In India, the value of the *hippotoxotai* became clear, especially when confronting charioteers.[95] They appear to have replaced the *hippakontistai* and the Macedonian *sarissophoroi*, and they excelled in the battle against Poros and its preliminaries. Specially recruited Iranians, trained in the Macedonian art of war in their home satrapies, were eventually summoned to serve their new king, but both the occasions on which they were employed and their actual military contributions are uncertain.[96] Eventually, in 324, Iranian fighters were added to existing units of Alexander's

91 Phrataphernes: Arr. 4.7.1, 18.1. Two of his sons were later enrolled in the *agēma* of the cavalry (Arr. 7.6.4). Artabazos: Arr. 3.28.2; 4.16.2–3; Curt. 7.3.2; 8.1.10; cf. 8.1.1. His son Kophen also joined the *agēma* (Arr. 7.6.4). Other prominent Iranians whose sons served in Alexander's cavalry guard were Mazaios, Oxyartes, and Hystaspes; all had proved themselves trustworthy. Since Koinos and Artabazos served together in the campaign against Spitamenes (Arr. 4.16.3) their force no doubt included native troops.

92 Hostility to Alexander's 'orientalising' policy: Just. 12.3.8–4.12; Plut. *Alex.* 45.1–2; Diod. Sic. 17.77.4–6; Curt. 6.6.2; Arr. 4.7.4. This played no small part in the conspiracy of Demetrios the Bodyguard (no. 358), the news of which Philotas failed to report to Alexander. There is an unmistakable connection between Alexander's integration of the Persian nobility, at court and in the army, and the plot that was discovered at Phrada; see Heckel (2020) 203–208.

93 See Olbrycht (2007) 312, with references. *Hippakontistai*: Olbrycht (2007, 321) assumes an Iranian 'core of the formation', but English (2009, 50–51) is rightly sceptical of the view that the *hippakontistai*, mentioned by Arr. 3.24.1 in the context of the Mardian campaign, were Iranian troops. Arrian does not say that they were a new addition to Alexander's army, but adds that they had by now formed their own *taxis*. This does not rule out the possibility that the existing force was augmented by new recruits from the ranks of the defeated, but one would expect Arrian to comment on such integration. See Bosworth (1980a, 352) against Berve (1926, 1.151), who thinks they were Persian deserters.

94 Arr. 4.17.3 distinguishes between the *hippakontistai* and the Bactrians and Sogdianians.

95 Arr. 5.14.3, 16.4. On the *hippotoxotai* see Olbrycht (2007) 316; English (2009) 51.

96 The so-called *epigonoi*. Just. 12.4.11 is confused; Curt. 8.5.1; Plut. *Alex.* 71.1; Arr. 6.6.1; Diod. Sic. 17.108.1–3. See Bosworth (1980b) 17–18; Briant (1972) 51–60. Olbrycht (2015); Olbrycht

army,[97] and, although this brought the Macedonians to the point of mutiny, the new formations appear to have endured; for we find native troops in the armies of all the Successors and the later Hellenistic kings. Greek sources tended to focus on the achievements of Greco-Macedonian troops and the proportion of native fighters and their value to the success of the final stages of the campaign – to say nothing of their intended role in Alexander's future campaigns – are mostly hidden from us. Just as Alexander had adapted his court to include Iranians in a variety of roles, so he now extended this merger to the army.

8 The So-Called Reign of Terror

When he returned from India, having marched through the Gedrosian desert, Alexander deposed Astaspes, the satrap of Carmania, on the charge of having plotted rebellion.[98] Soon afterwards, the king made administrative changes in response to charges of malfeasance. The numerous charges brought against the officials of the empire have been viewed by many scholars as a purge or, in the words of Ernst Badian, a 'reign of terror'.[99] But this author rightly observes, 'One or two of the Iranian satraps were punished for attempted treason: thus Astaspes and Autophradates. This, of course, had been happening ever since Alexander began the practice of appointing Iranian satraps. But it is worth noting how remarkably *few* such cases there were during his long absence'.[100] Those Greeks and Macedonians who were removed from office and, in many cases, executed were not victims of Alexander but rather criminals who merited punishment, often for crimes against the native population. Of those Iranians appointed by Alexander, several either died while they were in office or relinquished their satrapies.[101] Others were removed for abusing their

(2007, 316) believes they joined Alexander in India, though the more compelling evidence suggests that they did not arrive until the king returned to Sousiana.

97 Persian units in Alexander's army: Olbrycht (2007); Olbrycht (2010); Olbrycht (2011); Olbrycht (2015); Bosworth (1988) 271–73; Bosworth (2010); Hamilton (1988) 476–84.

98 Curt. 9.10.21, 29. Arr. 6.27.1–2 mentions the changes in Carmania without naming Astaspes. His account is confused and inaccurate. See Heckel (2017). Hyland (2013, 129 n. 39) rightly notes that there is no evidence for his engaging in 'open defiance of Macedonian authority'.

99 Badian (1961); Higgins (1980) provides a useful corrective.

100 Badian (1961) 20, original emphasis.

101 Died while in office: Mazaios (Arr. 4.18.3), Phrasaortes (Arr. 6.29.2) and perhaps also Amminapes. Artabazos was excused for personal reasons (Curt. 8.1.19), as was (apparently) his son Arsames. Mithrenes and Sabiktas may have been killed attempting to gain control of Armenia and Cappadocia respectively.

powers.[102] Only Satibarzanes and Autophradates were openly rebellious.[103] Other 'rebels' were arrested during the course of the campaign but these were not Alexander's appointees.[104] In fact, Alexander was confronted by 'rebels' no more frequently than the earlier Achaemenids, and far less often than in the time of Artaxerxes II. It should be noted that rebellion came from those at or near the top of the socio-political ladder, from men who sought power for themselves, rather than from the grassroots level. And, indeed, Alexander punished those – both Greco-Macedonians and Iranians – who abused the native population. The idea that Alexander was met with wholesale resistance is simply untrue.

102 Abuse of power, poor performance, or shirking responsibilities: Oxydates, Proëxes, Tyriespis, Aboulites, Oxathres (son of Aboulites and ruler of Paraitakene: Arr. 3.19.2; see no. 830). Also Arsakes, who was not fulfilling his obligations (Arr. 3.29.5: ὅτι ἐθελοκακεῖν αὐτῷ) – presumably in connection with Satibarzanes' uprising – and was arrested by Stasanor but we do not know why (Arr. 4.7.1). Orxines (no. 828; Curtius calls him Orsines) was of a distinguished family (descended from one of the Seven and from Cyrus the Great: Curt. 4.12.8; 10.1.23). He took over the government of Persis, after Phrasaortes' death, without Alexander's approval (Arr. 6.29.2, 30.1), but the charges against him were that he plundered temples and tombs and put innocent Persians to death (Arr. 6.30.2; Curt. 10.1.22–38 depicts him as an innocent victim of Bagoas, on which see Badian (1958) 147–9). For the undoing of Alexander's policy of appointing Iranians see Lerner (2018) 121–23.

103 Execution of Autophradates (Curt. 10.1.39, who calls him Phradates).

104 Baryaxes (no. 288) was a Mede who was charged with usurping the kingship (Arr. 6.29.3; see Hyland (2013) 128–32, who thinks that Baryaxes' actions may have been motivated by the abuses of Alexander's generals). Clearly, he believed that he was more entitled to rule than the satrap Oxydates or Alexander himself. Badian (2000a, 93) believes he may have been descended 'from the old Median kings'. Barzanes or Brazanes (Arr. 4.7.1; cf. 4.18.1) was probably identical with Nabarzanes. Like Satibarzanes, who may or may not have conspired against Darius, he was a supporter of Bessos. Satibarzanes' revolt: Arr. 3.25.5; 3.28.2; Curt. 6.6.20–4; Curt. 7.4.33–7; Diod. Sic. 17.78.1, 83.4–6. Bessos (no. 295) did not rebel but rather continued to resist, assuming the upright tiara and calling himself Artaxerxes: Diod. Sic. 17.74.2; Curt. 6.6.13; Arr. 3.25.3; Metz Epit. 3. Other rebels (or 'hold-outs') were Zariaspes (no. 1166) and Ordanes (no. 818, also called Ozines), who were arrested by Krateros on his return from India (Curt. 9.10.19; cf. Arr. 6.27.3); Hyland (2013, 129 n. 39) suggests that Zariaspes and Tyriespis may be the same individual, but Curtius speaks of both Zariaspes and Terioltes (= Tyriespis) and mentions the execution of Terioltes (9.8.9) two chapters before he reports the arrest of Zariaspes (9.10.19). Among those who rose up against Macedonian authority we must also include a certain Boxos (no. 310) who is alleged to have been involved in the uprising led by Athenodoros (no. 257) and Biton (no. 302). Curt. 9.7.3 says that Athenodoros declared himself 'king' for the purpose of leading the rebellious mercenaries home, adding that he was joined by Boxos, 'a Bactrian'. Why a Greek mercenary would take the title of king in order to lead his troops home is incomprehensible (cf. Iliakis (2013) 188–9); it is more likely that Boxos led a local uprising, emulating Bessos in claiming the kingship, and that the Greek mercenaries supported him. Boxos then arranged a banquet where Athenodoros was murdered, for which he was subsequently put to death (Curt. 9.7.4–8). See Heckel (2021b) 69–71.

9 A Final Note on *Verschmelzungspolitik*

I would conclude with the observation that, although a full-scale mixing of populations – a fusion or *Verschmelzung* – was not Alexander's intention (unless we accept Diodorus' claim that he planned to transplant populations between Europe and Asia),[105] the evidence shows that the conqueror made a serious effort at integrating Iranians and other subject elites into the governing class, as well as into his court and his army. He adopted the dress and practices of the Persian court to the extent that it was acceptable to his *hetairoi* and his troops, drawing back somewhat when he met with resistance. After the death of Darius, he adopted Iranian dress (except for trousers and the pointed hat),[106] added Persian ushers and guardsmen,[107] and included prominent Persians among his courtiers. By the end of his career, there were numerous Iranians in his armed forces, both new additions and those units that had fought with him since his campaigns in the upper satrapies. He himself married two Iranian princesses and he arranged the marriages of over ninety of his *hetairoi* to the daughters of Iranian nobles.[108] He had, in the past, encouraged such unions, both officially and unofficially. He died too soon to complete the organization of his empire, but there is no good evidence to suggest that he did not care about ruling but was only interested in further conquest. The instability of his empire, after Alexander's death, was the fault of his lieutenants, those who had accepted military commands and administrative offices from the king (or as a result of the division of power in Babylon in 323) but chose to work for their own ends rather than the welfare of the state. If anything, the Macedonians

105 Diod. Sic. 18.4.4.

106 Plut. *Alex.* 45.2: he did not adopt the Median fashion – the trousers, the sleeved vest or the tiara – but rather wore a blend of Persian and Median styles. Diod. Sic. 17.77.4–6; Just. 12.3.8–11; Curt. 6.6.4–5.

107 Diod. Sic. 17.77.4; Metz Epit. 2.

108 Alexander's marriage to Rhoxane has already been discussed. Metz Epit. 31 says that Alexander's *hetairoi* also took barbarian brides at this time (cf. Diod. Sic. contents of Book 17: ὡς Ἀλέξανδρος ἐρασθεὶς Ῥωξάνης τῆς Ὀξυάρτου ἔγημεν αὐτὴν καὶ τῶν φίλων πολλοὺς ἔπεισε γῆμαι τὰς τῶν ἐπισήμων βαρβάρων θυγατέρας). Baynham (1995) 70 thinks that this applied to garrison troops who took native wives; this will have been the case, but Diodorus specifies that Alexander encouraged his *philoi* (= *hetairoi*) to follow his example. Alexander's marriages to Stateira (no. 1079) and Parysatis (no. 849): Arr. 7.4.4; cf. Plut. *Alex.* 77.6. For the mass marriage in Susa see Diod. Sic. 17.107.6; Just. 12.10.9; Plut. *Alex.* 70.3. It is unlikely that the primary aim of this was racial/ethnic blending, although this may have been a factor; for the offspring of these unions would form a pool of future governors and administrators. I have suggested elsewhere that the brides brought with them lands and wealth: see Heckel (2008) 138–39, comparing the unions of conquistadors and native Mesoamerican women.

who inherited empire were more to blame for its disintegration than the Iranians who had, for the most part, made their peace with the conqueror.

Bibliography

Alonso Troncoso, V. and Anson, E.M. (eds) (2013) *After Alexander: The Time of the Diadochi (323–281 BC)*. Oxford.

Badian, E. (1958) 'The Eunuch Bagoas: A Study in Method'. *CQ* 8: 144–57.

Badian, E. (1961) 'Harpalus'. *JHS* 81: 16–43.

Badian, E. (1965) 'The Administration of the Empire'. *G & R* 12: 166–82.

Badian, E. (1985) 'Alexander in Iran', in Gershevitch (1985) 420–501.

Badian, E. (2000a) 'Conspiracies', in Bosworth and Baynham (2000) 50–95.

Badian, E. (2000b) 'Darius III'. *HSCP* 100: 241–67.

Baynham, E.J. (1995) 'An Introduction to the Metz Epitome: Its Traditions and Value'. *Antichthon* 29: 60–77.

Baumbach, A. (1911) *Kleinasien unter Alexander dem Grossen*. Weida.

Berve, H. (1926) *Das Alexanderreich auf prosopographischer Grundlage*. 2 vols. Munich.

Berve, H. (1938) 'Die Verschmelzungspolitik Alexanders des Grossen'. *Klio* 31: 135–68.

Billows, Richard A. (2018) *Before and After Alexander. The Legend and Legacy of Alexander the Great*. London.

Bosworth, A.B. (1980a) *A Historical Commentary on Arrian's* History of Alexander, *Books I–III*. Oxford.

Bosworth, A.B. (1980b) 'Alexander and the Iranians'. *JHS* 100: 1–21.

Bosworth, A.B. (1981) 'A Missing Year in the History of Alexander'. *JHS* 101: 17–39.

Bosworth, A.B. (1988) *Conquest and Empire: The Reign of Alexander the Great*. Cambridge.

Bosworth, A.B. (1995) *A Historical Commentary on Arrian's* History of Alexander, *Books IV–V*. Oxford.

Bosworth, A.B. (2010) 'The Argeads and the Phalanx', in Carney and Ogden (2010) 91–102.

Bosworth, A.B. and E.J. Baynham (eds) (2000) *Alexander the Great in Fact and Fiction*. Oxford.

Briant, P. (2002) *From Cyrus to Alexander: A History of the Persian Empire*. Winona Lake.

Briant, P. (1999) 'The Achaemenid Empire', in Raaflaub and Rosenstein (1999) 103–28.

Briant, P. (2015) *Darius in the Shadow of Alexander*. Cambridge, MA.

Brunt, P.A. (1975) 'Alexander, Barsine and Heracles'. *RFIC* 103: 22–34.

Carlsen, J., Due, B., Due, O., and Poulsen, B. (eds) (1993) *Alexander the Great: Reality and Myth*. Rome.

Carney, E.D. (2019) *Eurydice and the Birth of Macedonian Power*. Oxford.

Carney, E.D. and Ogden, D. (eds) (2010) *Philip II and Alexander the Great: Father and Son, Lives and Afterlives*. Oxford.

Cawkwell, G. (1978) *Philip of Macedon*. London.

Cawkwell, G. (2005) *The Greek Wars: The Failure of Persia*. Oxford.

Collins, A.W. (2001) 'The Office of Chiliarch under Alexander and the Successors'. *Phoenix* 55: 259–83.

Collins, A.W. (2012) 'Alexander and the Persian Court Chiliarchy'. *Historia* 61: 159–67.

D'Agostini, M., Anson, E.M., and Pownall, F. (eds) (2021) *Affective Relations and Personal Bonds in Hellenistic Antiquity*. Oxford.

Droß-Krüpe, K. and Fink, S. (eds) (2021) *Perception and (Self)Presentation of Powerful Women in the Ancient World*. Münster.

Elayi, J. (2013) *Histoire de la Phénicie*. Paris.

English, S. (2009) *The Army of Alexander the Great*. Barnsley.

English, S. (2011) *The Field Campaigns of Alexander the Great*. Barnsley.

Fawcett, B. (ed) (2006) *How to Lose a Battle: Foolish Plans and Great Military Blunders*. New York.

Fuller, J.F.C. (1960) *The Generalship of Alexander the Great*. New York.

Galewicz, C., Pstrusinska, J., and Sudyka, L. (eds.) (2007) *Understanding Eurasia: From Ancient Times to the Present Day*. Krakow.

Gershevitch, I. (ed) (1985) *The Cambridge History of Iran. Vol 2: The Median and Achaemenian Periods*. Cambridge.

Grainger, J.D. (2007) *Alexander the Great Failure: The Collapse of the Macedonian Empire*. London and New York.

Hall, E. (1989) *Inventing the Barbarian: Greek Self-Definition through Tragedy*. Oxford.

Hall, E. (1993) 'Asia Unmanned: Images of Victory in Classical Athens', in Rich and Shipley (1993) 108–33.

Hamilton, J.R. (1988) 'Alexander's Iranian Policy', in Will and Heinrichs (1988) 167–86.

Heckel, W. (1981) 'Some Speculations on the Prosopography of the *Alexanderreich*'. *LCM* 6: 63–70.

Heckel, W. (2006) 'Mazaeus, Callisthenes and the Alexander Sarcophagus'. *Historia* 55: 185–96.

Heckel, W. (2008) *The Conquests of Alexander the Great*. Cambridge.

Heckel, W. (2016) *Alexander's Marshals. A Study of the Makedonian Aristocracy and the Politics of Military Leadership*. London.

Heckel, W. (2017). 'Was Sibyrtios Ever Satrap of Karmania?' *Anabasis* 8: 38–43.

Heckel, W. (2018) 'Artabazos in the Lands beyond the Caspian'. *Anabasis* 9: 93–109.

Heckel, W. (2020) *In the Path of Conquest: Resistance to Alexander the Great*. Oxford.

Heckel, W. (2021a) *Who's Who in the Age of Alexander and his Successors: From Chaironeia to Ipsos (338–301 BC)*. London.

Heckel, W. (2021b) 'Notes on Alexander in Central Asia'. *AHB* 35: 55–72.

Heckel, W., Müller, S., and Wrightson, G. (eds.) (2015) *The Many Faces of War in the Ancient World*. Newcastle upon Tyne.

Heckel, W., Willekes, C., and Wrightson, G. (2010) 'Scythed Chariots at Gaugamela: A Case Study', in Carney and Ogden (2010) 103–109, 272–75.

Heckel, W. and Yardley, J.C. (2003) *Alexander the Great: Historical Texts in Translation*. Oxford.

Higgins, W.E. (1980) 'Aspects of Alexander's Imperial Administration: Some Modern Methods and Views Reviewed'. *Athenaeum* 48: 131–52.

Hölbl, G. (2001) *A History of the Ptolemaic Empire*. Translated by T. Saavedra. London.

Holt, F.L. (2005) *Into the Land of Bones. Alexander the Great in Afghanistan*. Berkeley.

Howe, T. (2021) '(Re)taking Halikarnassos: Ada, Alexander the Great and Karian Queenship', in Droß-Krüpe and Fink (2021) 241–67.

Howe, T., Müller, S., and Stoneman, R. (eds.) (2017) *Ancient Historiography on War and Empire*. Oxford.

Hyland, J.O. (2013) 'Alexander's Satraps of Media'. *JAH* 1: 119–44.

Hyland, J.O. (2018) *Persian Interventions: The Achaemenid Empire, Athens and Sparta, 450–386 BCE*. Baltimore.

Iliakis, M. (2013) 'Greek Mercenary Revolts in Bactria: A Re-appraisal'. *Historia* 62: 182–95.

Judeich, W. (1895) 'Der Grabherr des Alexandersarkophages'. *JdI* 10: 165–82.

Julien, P. (1914) *Zur Verwaltung der Satrapien unter Alexander dem Grossen*. Weida.

Kholod, M. (2018) 'The Macedonian Expeditionary Corps in Asia Minor (336–335 BC)'. *Klio* 100: 407–46.

Kuhrt, A. (2007) *The Persian Empire: A Corpus of Sources from the Achaemenid Period*. London.

Lerner, J.D. (2018) 'Alexander's Settlement of the Upper Satrapies in Policy and Practice'. *Anabasis* 9: 110–28.

Leuze, O. (1935) *Die Satrapieneinteilung in Syrien und im Zweistromlande von 520–320*. Halle.

McCoy, J.W. (1989) 'Memnon of Rhodes at the Granicus'. *AJP* 110: 413–33.

Meeus, A. (2009) 'Some Institutional Problems concerning the Succession to Alexander the Great: *Prostasia* and Chiliarchy'. *Historia* 58: 289–310.

Müller, S. (2013) 'The Female Element of the Political Self-fashioning of the Diadochi: Ptolemy, Seleucus, Lysimachus, and Their Iranian Brides', in Alonso Troncoso and Anson (2013) 199–214.

Müller, S. (2021) 'Barsine, Antigone and Macedonian War', in D'Agostini, Anson, and Pownall (2021) 81–96.

Nylander, C. (1993) 'Darius III – the Coward King: Point and Counterpoint', in J. Carlsen, Due, Due, and Poulsen (1993) 145–59.

Olbrycht, M.J. (2007) 'The Military Reforms of Alexander the Great during his Campaigns in Iran, Afghanistan and Central Asia', in Galewicz, Pstrusinska, and Sudyka (2007) 309–21.

Olbrycht, M.J. (2008) 'Curtius Rufus, the Macedonian Mutiny at Opis and Alexander's Iranian Policy in 324 BC', in Pigon (2008) 231–52.

Olbrycht, M.J. (2011) 'First Iranian Military Units in the Army of Alexander the Great'. *Anabasis* 2: 67–84.

Olbrycht, M.J. (2015) 'The Epigonoi: The Iranian Phalanx of Alexander the Great', in Heckel, Müller, and Wrightson (2015) 196–212.

Palagia, O. (2000) 'Hephaestion's Pyre and the Royal Hunt of Alexander', in Bosworth and Baynham (2000) 167–206.

Palagia, O. (2017) 'Alexander's Battles against Persians in the Art of the Successors', in Howe, Müller, and Stoneman (2017) 177–87.

Pigon, J. (ed) (2008) *The Children of Herodotus: Greek and Roman Historiography and Related Genres*. Newcastle upon Tyne.

Raaflaub, K. and Rosenstein, N. (eds) (1999) *War and Society in the Ancient and Medieval Worlds: Asia, the Mediterranean, Europe, and Mesoamerica*. Cambridge, MA.

Reinach, T. (1892) *Une nécropole royale à Sidon*. Paris.

Rich, J. and Shipley, G. (eds) (1993) *War and Society in the Greek World*. London.

Ruzicka, S. (1988) 'War in the Aegean, 333–331 B.C.: A Reconsideration'. *Phoenix* 42: 131–51.

Schachermeyr, F. (1970) *Alexander in Babylon und die Reichsordnung nach seinem Tode*. Vienna.

Schachermeyr, F. (1973) *Alexander der Große: Das Problem seiner Persönlichkeit und seines Wirkens*. Vienna.

Sears, M.A. (2014) 'Alexander and Ada Reconsidered'. *CP* 109: 211–21.

Seibert, J. (1987) 'Dareios III', in Will and Heinrichs (1987) 437–56.

Stewart, A. (2014) *Art in the Hellenistic World*. Cambridge.

Will, W. and Heinrichs, J. (eds.) (1987) *Zu Alexander d. Gr. Festschrift G. Wirth*. Vol. 1. Amsterdam.

Will, W. and Heinrichs, J. (eds.) (1988) *Zu Alexander d. Gr. Festschrift G. Wirth*. Vol. 2. Amsterdam.

Willrich, H. (1899) 'Krateros und der Grabherr des Alexandersarkophags von Sidon'. *Hermes* 34: 231–50.

Yardley, J.C., Wheatley, P., and Heckel, W. (2011) *Justin: Epitome of the* Philippic History *of Pompeius Trogus. Volume II: Books 13–15. The Successors to Alexander the Great*. Oxford.

Zouboulakis, K. (2015) 'The Topography of the battle of Gaugamela. Contributing to an old Controversy'. *Pharos* 21: 31–62.

CHAPTER 10

The Military Organisation of the Upper Satrapies Under the Seleucids

Omar Coloru

In 311 BCE, Seleucus I defeated Nikanor, the officer appointed by Antigonus I Monophthalmos as satrap of Media and *stratēgos* of the Upper Satrapies (Diod. 19.91–92). This military success allowed Seleucus to gain control over the majority of the territory of ancient Media and Persia together with a series of unspecified neighbouring countries. More than Babylon, it was the possession of the Iranian centres of Ecbatana, Persepolis, and Susa that formed the real core on which Seleucus would lay the foundation of his empire.[1] In fact, these territories provided Seleucus with the financial and military resources that between 308 and 305 allowed him to conquer the satrapies East of Media and later to defeat Antigonus. Nevertheless, Seleucus had to renounce the regions south of the Hindu Kush, which were already under the sway of Chandragupta Maurya.[2] In the following pages, we are going to explore the way the Seleucids arranged the defence of this wide geographical space and how they made use of its military resources. Since 1938, when Elias Bickerman devoted part of his *Institutions des Séleucides* to the study of the Seleucid army, a few major works have provided useful insights in the field of Seleucid military studies. Bezalel Bar-Kochva's *The Seleucid Army* (1976) explored the organisation of the army and its practical use on the occasion of some celebrated battles, while Nicholas Sekunda's *Hellenistic Infantry Reform in the 160s BC* (2001) focused on the military reform that, according to his view, would have led both the Seleucid and the Ptolemaic infantries to abandon the Hellenistic phalanx and adopt Roman manipular tactics and units in order to compete with the more effective and powerful legions of Rome. On the other hand, Laurent Capdetrey's *Le pouvoir séleucide* (2007) contributed to our understanding of the role of the army and military officers in the administrative organisation of the vast and diverse Seleucid Empire.[3] However, our knowledge of the structure of the Seleucid

1 Capdetrey (2007) 28–29, 271. In the years between 311 and 308/7 the Seleucid possession of Babylon was far from being assured.

2 See Capdetrey (2007) 39–50; Kosmin (2014) 31–61.

3 For a general overview on Seleucid rule in the eastern satrapies, see Plischke (2014).

 | DOI:10.1163/9789004710771_011

military is still suffering from the slow pace at which fresh evidence emerges, a problem that affects the eastern sector of the Seleucid world more than its western counterpart. In this situation, I think that a reappraisal of the military organisation of the Seleucid Upper Satrapies in the light of the available sources may prove useful to put the question in perspective and provide new lines of research.

1 The Military Landscape of the Upper Satrapies

Like the other sectors of the empire, the Upper Satrapies underwent a phase of urbanisation and military control of territory primarily under Seleucus I and his son Antiochus I, but later interventions by their successors are equally attested. In order to enhance the appropriation of the royal space, the Seleucids engaged in a programme of foundation – sometimes refoundation – of a series of settlements of different typology, from forts to proper cities. These settlements were often named after the king, his family and Macedonian toponyms. For instance, Susa was renamed Seleucia-on-the-Eulaeus,[4] the new Seleucid foundation on the Persian Gulf was called Antioch in Persis (possibly modern Bushire), Rhagae (modern Ray) became Europos and a neighbouring settlement Apamea.[5] The Achaemenid site of Erk Kala in Margiana, which a very old tradition still identifies with the remains of the former settlement of Alexandria in Margiana, was included within the urban plan of the new city of Antioch in Margiana (site of Gyaur Kala, 30 km east of modern Merv) and became its citadel.[6] In the oasis of Bactra the existence of a settlement called Amphipolis near Karelote has been recently discovered.[7] Some important Achaemenid centres like Bactra kept their original name while others like Ecbatana and Susa were renamed later.[8]

4 It is probable that the settlement received its new name under Seleucus II in the second half of the third century rather than under Seleucus I; see Martinez-Sève (2011) 57–60.

5 For references on these sites, see Cohen (2013) 185–89, 194–99, 204, 209–10.

6 Actually, Alexander never passed from the Merv oasis during his campaign and even if we admit that an Alexandria in Margiana may have been founded by delegation to one of the Alexander officials (Ptolemy? Craterus?) we lack evidence to test this hypothesis; see Coloru (2016) 38–40. On the settlement of Antioch in Margiana see Zavyalov (2007) 323–29.

7 Clarysse and Thompson (2007) 275–77.

8 Pliny the Elder (*HN* 6.17) relates that Seleucus I was the founder of Ecbatana, but this information has to be interpreted as a refoundation. In any case, Ecbatana kept its name until the reign of Antiochus IV when it was renamed *Epiphaneia* after the royal epithet of the king, *Epiphanes*, 'God manifest'. As for Susa, it was renamed Seleucia-on-the-Eulaeus during the third century; see Martinez-Sève (2011) 57–63.

When possible, the urban plan of the cities followed quite a widespread pattern among the settlements of the Hellenistic East: proximity to a water source and a fertile area, a subdivision between a lower city and an upper city protected by a fortifications encircling a citadel which also contained the buildings of political and religious institutions. These features are more or less observable, for instance, at Ai Khanum, Marakanda/Samarkand, Susa, and Ecbatana. As for the last of these, we still have to rely on the description handed down by Polybius (10.27.6) when he reports that Ecbatana did not have walls but its citadel was nevertheless protected by an impressive system of fortifications of the length of seven *stadia* (ca. 1.4 km).[9] This information was later echoed in Ammianus Marcellinus (23.6.39), when he relates that some settlements in Media that were the object of Seleucid (re)foundation, namely Ecbatana, Europos, and Heraclea, were celebrated both for their wealth and the size of their fortifications.[10]

In general, the defences of the settlements of the Upper Satrapies display Hellenistic innovations while at the same time adopting solutions that are deeply rooted in the local tradition;[11] in any case, the choice of preferring a particular feature over another was ultimately dictated by practical reasons. In this regard, the walls of Marakanda/Samarkand provide a very good example.[12] The ramparts are built with square bricks, a typical component of the Hellenistic military architecture introduced in Bactria and Sogdiana but made out of mud instead of stone according to a widespread tradition spanning Mesopotamia to Central Asia. Unlike the Greek tradition, which privileges thick walls, the Hellenistic ramparts of Marakanda present galleries according to Sogdian military architecture. This choice is motivated by the fact that the fortifications are located in an elevated position, thus safe from the attacks of a besieging army that could use rams or mine tunnels to make a breach in thinner walls.[13] In the case of Ai Khanum, however, the conception of a wall circuit protecting all the strategic points within a vast perimeter finds parallels in

9 Archaeological excavations at Hamadan have not yet brought to light the strata corresponding to the Seleucid occupation of the site. Achaemenid and Seleucid Ecbatana should be looked for on Tepe Mosallah; see Boucharlat (2018) 1–34. On Media under the Seleucids, see Coloru (2023) 201–16.

10 For a commentary see Feraco (2004) 192–94.

11 See Engels (2017), 78–80.

12 Rapin and Isamiddinov (1994) 551–56.

13 Recent excavations on the citadel of Marakanda brought to light an early Hellenistic granary where the supplies of the Greek garrison were stored; see Baratin and Martinez-Sève (2013) 373–89.

several Mediterranean settlements such as Syracuse, Messene, and Iasos.[14] The city walls, whose foundation dates back to Antiochus I (281–261),[15] embraced a considerable surface of ca. 150 hectares while the acropolis hosted a citadel protected by ramparts and towers measuring 10 metres in height. Likewise, the exceptional thickness of the walls (from 6 to 8 metres) with walkways and massive rectangular towers and bastions suggests that the fortifications of Ai Khanum were designed according to the principle of static defence. This allowed the city to resist the attacks of siege engines and to be defended from atop the walls by a limited number of soldiers. However, this strategic model itself did not take inspiration from the Greek world but rather from Mesopotamia, particularly the defensive system of Babylon. As for Antioch in Margiana, the fortifications dating to the Seleucid period, always built with mudbricks, present a striking example of a three-storey defensive wall, which seems to apply the principles of military architecture described by Philo of Byzantium (third century BCE) in his treatise.[16] In addition, other features such as embrasures, merlons, and arched chambers find remarkable parallels in the fortifications of Hellenistic cities of Asia Minor like Kaunos, Perge, and Side.[17] Nevertheless, there are also differences ascribable to the building material and local tradition, which means that, unlike the stone fortifications of the Mediterranean area, the base of the walls was battered instead of having vertical sides, and wide platforms extended for some distance beyond the line of the walls.[18]

Although the Classical tradition celebrated the high degree of urbanisation of regions such as Media and Bactria, the typology of the *polis* represented a modest minority when compared to the plethora of smaller-sized settlements throughout the diverse landscape of the Upper Satrapies. Like larger cities, these smaller settlements served to both practically and symbolically appropriate nearby territory. In addition, they served to concentrate the presence of colonists in these lands. If we have to establish a hierarchy,[19] we may surmise that the *polis* controlled the main axes of communication in a given

14 On the fortifications of Ai Khanum, see Leriche (1986); Leriche (2007) 141–44; Martinez-Sève (2014a) 270–74.

15 Martinez-Sève (2014a) 270–71.

16 Apparently, Philo's principles were used also in the construction of the Hellenistic walls of Barikot (ancient Bazira) in Swat; see Olivieri (1996) 48; Callieri (2007a) 142.

17 Zavyalov (2007) 317–19, 325–28. Philo's description of the arches used in curtain walls took the fortifications of Rhodes as a model.

18 Zavyalov (2007) 327.

19 I employ here the schematisation of Kosmin (2014) 190–92.

region. Next came a category of smaller establishments called *katoikiai*, located along the region's secondary axes. For a long time a *katoikia* was understood almost exclusively as a military colony inhabited by veterans or troops (with their families) who were entrusted with the defence of the territory. Recent research, however, has challenged this view by demonstrating that the term *katoikia* was also applied to agricultural colonies, and more generally to any kind of community – including indigenous villages.[20] In any case, the soldiers settled in the Seleucid *katoikiai* would receive an allotment (*klēros*) from the royal land and also comprise (though not exclusively) the recruitment pool for the army.[21] The general picture suggests that the inhabitants of the Seleucid *katoikiai* were mostly civilians, while soldiers were garrisoned elsewhere.[22]

The lower step of the regional hierarchy was represented by the *phrourion*, a fortress occupied by a garrison entrusted with guarding small areas in proximity to strategic points such as mountain passes and fords. Evidence from Asia Minor shows that soldiers serving in a *phrourion* were assigned a land lot like the settlers of a *katoikia*.[23] Yet again, we lack information for the eastern part of the empire and we can only speculate by means of analogy with what occurred in the west. One of the earliest examples of a *phrourion* in the Seleucid Upper Satrapies seems to be the cave complex of Karafto, which between the end of the fourth and the early third century was occupied by a military garrison guarding the borders of northern Media.[24] On numismatic grounds, it is possible to assume that during the reign of Antiochus III the territory east of Ecbatana bordering with Parthia was filled with garrisons and guard posts in order to protect the Seleucid eastern frontier against possible attacks by the Parthians.[25]

As for the Upper Satrapies, our sources mention the inhabitants of a Seleucid *katoikia* only in relation to two bloody episodes in Persis: the first is set during the reign of Seleucus (in all likelihood Seleucus I) and refers to 300 Macedonian and Thracian horsemen under the command of a certain Seiles

---

20 The idea that a *katoikia* was a military colony was developed for the first time by Radet (1890) 6; for *katoikia* as agricultural colony see Robert (1985) 483; Cohen (1991) 43; for *katoikia* in the general sense of community and more specifically of indigenous village (*kōmē*) see Schuler (1998) 33–34. For a recent analysis see Capdetrey (2012) 322–26.

21 Capdetrey (2007) 162–66.

22 Fischer-Bovet (2014) 296–97.

23 Capdetrey (2007) 164–65.

24 Bernard (1980) 301–24; Callieri (2007b) 68–69.

25 It seems that Antiochus III opened three new mints in Media in areas close to Parthia in order to serve the needs of the garrisons located at the eastern frontier; see Houghton *et al.* (2008) 694–98.

who ambushed and killed 3,000 Persians because they were involved in a revolt (Polyaenus *Strat.* 7.39).[26] Possibly, Seiles was a Macedonian governor of Persis during the last years of Seleucus I, and the episode related by Polyaenus may refer to the support provided by the Seleucid forces to the *frataraka* Baydad of Persis against his rival Ardaxsher.[27] The second episode reports the massacre of 3,000 *katoikoi* by the order of Oborzos – the *frataraka* Wahabarz, early third century – because they were conspiring against him (Polyaenus, *Strat.* 7.40).[28] In the present state of knowledge, it is difficult to draw a clear distinction between *katoikia* and *phrourion* in the archaeological sites studied so far. The status of *katoikia* has been attributed to the fortified site on the island of Ikaros (modern Failaka) in the Persian Gulf and for Kampyr-tepe in Bactria, a fortified settlement controlling a crossing point on the Oxus (modern Amudarya) built around the third century.[29] These identifications, however, are highly conjectural, and while nothing certain can be determined in this matter, nevertheless the general picture suggests that many of these settlements had a military function.[30]

2 The Military Administration of the Upper Satrapies and Its Officers

In the eastern sector of the Seleucid kingdom, the satraps represented the highest-ranking officials on whom the administration and control of the various satrapies depended. From this point of view, the Seleucids were in continuity with the Achaemenid past. The governors of the Upper Satrapies were powerful figures holding regional authority with wide margins of autonomy. They performed both administrative and military functions, and exercised

26 Bar-Kochva (1976, 50) supposes that these Thracians were military *katoikoi*. This hypothesis seems acceptable when considering that Thracian settlers in the Upper Satrapies were already mentioned by Diodorus Siculus (19.27.5) in the context of the battle of Gabiene (312) between Antigonus Monophthalmos and Eumenes. Tuplin (2007) 231 has a similar view. On the Thracian settlers in Persis, see the critical remarks in Dumitru (2013) 78–79.

27 For the Macedonian origins of Seiles, see Sekunda (2007) 231; the Seleucid intervention for Baydad is posited by Müseler (2018) 94. Engels (2018, 177–85) dates this episode and that concerning Oborzos to the reign of Seleucus II.

28 For the chronology of Oborzos and the other *frataraka* of Persis, see Müseler (2018) 84–106. A different chronology is proposed by Wiesehöfer (2010) 107–21; Wiesehöfer (2013) 718–27.

29 For Ikaros as a *katoikia* see Salles (1992) 224–25; for Kampyr Tepe as a *katoikia* see Rtveladze (1995) 24. See Dvurechenskaya (2019) for the Uzundara fort in Bactria.

30 Leriche (2007) 147–48; the policy of founding small military settlements was applied in the whole of the Seleucid East, from Mesopotamia to Bactria.

strong control over the troops stationed in their territories.[31] An astronomical diary from Babylon records that in 273 Antiochus I received twenty elephants sent to Syria by the satrap of Bactria in order to use them against Ptolemy II during the First Syrian War (274–271).[32] Their status as regional leaders with a satrapal court explains why in some moments of weakness of the Seleucid authority a few satraps managed to detach themselves from the central power for a limited time, as in the case of Molon and Timarchos, or definitively as in the case of Andragoras of Parthia and Diodotus I of Bactria.[33] Molon could raise a large army both drawing on the conspicuous military resources of Media and of the neighbouring officials and dynasts, in particular Artabarzanes of Media Atropatene, whose kingdom provided brave warriors – especially horsemen – and armaments (Polyb. 5.43, 55). The same strategy was adopted by Timarchos, who used Media and the surrounding areas as a recruitment pool and obtained the military support of king Artaxias of Greater Armenia (Diod. Sic. 31.27a).

Some regions were administered by governors bearing the title of *stratēgos*. Despite the name, they did not exert exclusively military functions and one should be aware that the title and the functions associated with it were not uniform throughout the kingdom and very seldom used by the office-holders themselves. In some cases, the title of *stratēgos* was also assigned to dynasts who were integrated into the Seleucid sphere of control through that rhetorical expedient. This is the case of the two Armenian kings Artaxias and Zariadris appointed by Antiochus III as *stratēgoi* of Greater Armenia and Sophene respectively, where they 'ruled by delegation' of Antiochus (Strabo 11.14.5, see also 11.14.14).[34] From Antiochus III onwards, officers at the head of certain administrative units like *eparchoi* and *hyparchoi* were replaced by others bearing the title of *stratēgos*. As for the Upper Satrapies, we know that after the revolt of Molon in 222 Antiochus III removed Pythiades, the *eparchos* of the Erythrean Sea, from his office and appointed Tychon as *stratēgos* 'of the region

31 Capdetrey (2007) 286.

32 Astronomical diary dated to the 24th day of the month of Addaru, year 38 of the Seleucid Era (= March 27th, 273); see Boiy (2004) 141.

33 On the usurpers in the Seleucid kingdom, see Chrubasik (2016).

34 On the titles of these Armenian dynasts, see Traina (2017) 373–84. For Central Asia, we have the slightly later example of Aspiones (second century) who administered one of the satrapies of the Greek kingdom of Bactria; see Strabo 11.11.2. His Iranian onomastics may suggest that he was a local dynast in charge of a region between Parthia and Bactria (probably close to Areia) that Mithridates I took away to Eucratides I; see Coloru (2009) 217–19. Former Achaemenid satraps managed to keep their power under Alexander and beyond; see Hyland (2013) 119–44.

of the Erythrean Sea' (Polyb. 5.46.7 and 54.12). It is worth noting that before taking on his new charge, Tychon was performing the function of chief secretary of the army (*archigrammateus*), thus a position related to the military sphere. This suggests that even his office of *stratēgos* was both military and administrative. While the extension of the district under Tychon remained the same, it is possible that this change in terminology was not accidental as it aimed to diminish the ideological dimension of the satrapal power and create a stronger bond between the king and his officers.[35] Moreover, an inscription from Akko-Ptolemais records the dedication made by another *archigrammateus* of the army to Antiochus VII, his wife Cleopatra Thea and their children (*SEG* 19: 903). It is interesting to note that this military officer also bears the aulic title of 'one of the *prōtoi philoi*' of the king. Again, the evidence does not allow us to understand how this and other centralised hierarchical positions related to provincial commands, but the inscription suggests that this military officer ranked among the most important figures at the Seleucid court and in the administrative system as well. What we can say is that an *archigrammateus* of the army could be a valuable functionary because of his double competence in bureaucratic and military matters. This made him effective in strategic areas of the empire such as the Persian Gulf and Coele-Syria. Finally, an inscription from Susa, probably composed between the reigns of Antiochus III and Antiochus IV, provides the only epigraphic witness of a *stratēgos* of Susiana named Arrheneides.[36]

Beginning with the coregency of Antiochus I in 294 the eastern half of the empire was placed under the control of a supra-satrapal authority. Sometimes this charge was the responsibility of the crown princes, while in the third century we see that an officer could control what was left of the eastern Seleucid territories with the titles of *stratēgos* and 'governor of the Upper Satrapies' (ὁ ἐπὶ τῶν ἄνω σατραπειῶν). After Antiochus I, the authority over the Seleucid East was possibly exerted by his sons Seleucus and Antiochus II, then by Antiochus III and his namesake son.[37] As for the latter two, our sources are

35 Capdetrey (2007) 293.

36 Mentioned in a votive inscription to a deity whose name is lost; see Rougemont (2012) no. 6. Arrheneides is honoured by Pythagoras, son of Aristarchus, who is labelled as *sōmatophylax*, literally 'bodyguard'. Despite the military origin of this title, it is quite probable that in this and other contexts the charge of *sōmatophylax* is simply an honorific title; see Rougemont (2012) n. 101, 47.

37 Our evidence cannot confirm that all these princes held the office of governor of the Upper Satrapies, even if it is not improbable. Seleucus, son of Antiochus I, was co-ruler, a charge that he shared with his younger brother Antiochus II beginning at least from 268, as witnessed by an inscription of Stratonicea in Caria (*IStratonikeia* 1030, ll. 1–2). The

controversial: Polybius (5.40.6) reports that Molon and his brother Alexander were appointed the governors of the Upper Satrapies by Seleucus III, but his younger brother Antiochus (III) was in the east when the king died in 223. Bengtson interpreted this passage as a hint that the Seleucid prince was governor of the eastern satrapies,[38] but at first sight it seems odd that the young Antiochus III could hold the same office as Molon.[39] However, this pattern seems to repeat itself in a later phase of the reign of Antiochus III. In fact, in the spring of 193 the king sent his son Antiochus back to Syria to watch over 'the farthest regions of the kingdom' (*ultimae partes regni*, Livy 35.13.4–5) and to prevent the risk of revolts while he was occupied with the conflict against the Pisidians. Presumably, that expression alludes to the Upper Satrapies. In a period chronologically close to this event Menedemos was satrap of Media and *stratēgos* of the Upper Satrapies. More precisely, the letters he received from Antiochus III in the summer of 193 refer to his capacity as satrap of Media, while an undated honorary inscription from Laodicea in Media praises him as governor of the Upper Satrapies, which was a more prestigious title.[40] If these inscriptions were contemporary, then Menedemos would already have been governor of the Upper Satrapies when the young Antiochus was entrusted the task of supervising the Seleucid East by his father.[41] The honorary inscription mentions the commitment of Menedemos 'toward the kings' (ll. 3–4). It is probable that we are dealing with a reference to Antiochus III and his namesake son rather than to Antiochus III and his elder brother Seleucus III, especially because Molon was governor of the Upper Satrapies under Seleucus III and the beginning of the reign of Antiochus III. If this hypothesis is correct, the honorary inscription for Menedemos can also imply that the young Antiochus was still alive when Menedemos was governor of the Upper Satrapies. The son of Antiochus III died in Syria in the late summer of the same year 193 before his journey to the East (Livy 35.15.3), but his appointment had already been decided in the spring and doubtless transmitted to the eastern Seleucid satrapies. It is possible that Menedemos received this information together with the announcement of the appointment of Laodike, sister and wife of the young

joint rule of Antiochus I, Seleucus, and Antiochus II is explicitly recorded in a Babylonian tablet dated on 4 Abu 46 Seleucid Era (= August 14th, 266); see Del Monte (1997a) 433–44; Del Monte (1997b) 228–30.

38 Bengtson (1964^2) 84–85.

39 Frye (1984, 169) points out that it does not necessarily mean that Antiochus III was in the east in his capacity of governor. The same scepticism is shared by Strootman (2011) 72.

40 Rougemont (2012) no. 66–68.

41 Rougemont (2012, 182) admits that on stylistic grounds the royal epistle of Antiochus III to Menedemos as well as the honorary inscription for the latter appear closely contemporary.

Antiochus, as priestess of the cult for Queen Laodike in Media and possibly the Upper Satrapies.[42] If this historical reconstruction is consistent, we again find ourselves in the conflicting situation of a Seleucid prince exerting a function of control over the Seleucid East at the same time as another subordinate officer. As the testimonies of Polybius and Livy seem to be reliable, we may resolve this apparent contradiction by positing that both Antiochus III and his son had exerted some sort of control/supervision over the governors of the Upper Satrapies. Of course, we could also posit that the princes might have been sent on a journey of inspection without taking over a long-term administrative command, perhaps in the context of the itinerant royal court where a prince could gain experience in the oversight of a royal province through the opportunity to manage one without the presence of the king. However, Livy (35.13.5) states that the young Antiochus was sent by his father to watch over (*ad custodiam*) the farthest regions of the kingdom, which in my opinion seems to suggest a task extending beyond the simple tour of inspection. Perhaps in these situations, the conflict of powers was settled following the old principle that 'where the greater is there, the lesser ceases'.

The loss of Parthia and Bactria in ca. 246–245 required the administrative reorganisation of the extant area – in essence, Media, Persis, Susiana and possibly Karmania. The satrapy of Media was now the eastern Seleucid frontier and its satrap, as we have seen above, was invested with a wider authority over the extant Upper Satrapies. Apparently, we are dealing with a renewal of the administrative situation in place during the wars of the Diadochs, more precisely the period between 315 and 311, when Antigonus I Monophthalmos appointed Hippostratos and then Nikanor as satraps of Media and *stratēgoi* of the Upper Satrapies (Diod. Sic. 19.92.1–5, 100.3; App. *Syr.* 55, 57). The main difference now was the reduced extension of the territory controlled by the Seleucids in the second half of the third century. The earliest reference to a satrap of Media and *stratēgos* of the Upper Satrapies dates back to the reign of Seleucus III (ca. 226/5 or 225/4–222) with the appointment of Molon as governor of Media and 'of the inner sectors of the kingdom' (τὰ δ' ἄνω μέρη τῆς βασιλείας, Polyb. 5.40.6). This charge was shared with Molon's brother, Alexander, governor of Persis. At this time, it seems that the nature of Molon's authority was more military than administrative.[43] Polybius' words seem to imply that sometimes the government of the Upper Satrapies could be assigned to a couple of officers, probably in cooperation with the satrap of Persis, who nevertheless appears as subordinated to the satrap of Media.

42 Rougemont (2012) 179–80; see also Ma (2004) 407.
43 Capdetrey (2007) 98.

Although the available sources do not provide explicit information about the primacy of Media over Persis, it is safe to assume that the reason for this is the fact that Media was exposed to Arsacid raids, while Persis lacked a comparable military threat. From what we can grasp from the scanty evidence from the region, it seems that the *frataraka* never claimed an independent authority, and their very title – as well as that of *krny* (Greek *karanos*) – shows that they remained officers subordinated to the Seleucids.[44] Another example of this double government is attested under Demetrius I with the Milesian brothers Apollonius *junior* and Lachares, who according to their funerary epigram were commanders-in-chief of 'Greater Asia'.[45] The inscription suggests that the two brothers were appointed governors of the Upper Satrapies after the victory of Demetrius I over the usurper Timarchos.

It is only during the reign of Antiochus III that we learn of the existence in Media of a *phylakē*, a military district controlled by an officer.[46] In this particular case, a letter written by the aforementioned Menedemos informs us that in the area close to modern Kermanshah (Iran) there was a *phylakē* commanded by a certain Thoas. As Menedemos transmits his orders to Thoas without intermediary agents, one may assume that Thoas was a high-ranking official. We do not know the exact extent of the territory under Thoas' command even if on the basis of the information provided by the letter of Menedemos it is possible to argue that it was large enough to include a certain number of shrines. It was located in a strategic position along the road linking Ecbatana to Seleucia-on-the-Tigris in order to protect the main communication route against the raids of the mountain people of the Zagros.[47]

3 The Eastern Troops and Their Officers

Our knowledge of the military troops stationed in or originating from the Upper Satrapies mainly draws on the description of the composition of the Seleucid army during major battles such as Raphia (217) and Magnesia (190) as well as the military parade organised by Antiochus IV at Daphnae in 167 or

44 Engels (2013) 79; Müseler (2018) 96–98.

45 Herrmann (1987), 175–79. Sources on Apollonius and Lachares in Savalli-Lestrade (1998), no. 64 and 67.

46 Rougemont (2012) no. 68.

47 An interesting parallel is offered by the Cappadocian *phylakē* of Tyana; see Iossif and Lorber (2010a) 432–47.

165.[48] This information is complemented by a few other passages of Greco-Roman historiography as well as by epigraphic evidence. The image conveyed by the ancient sources tends to present the Seleucid army as composed of heterogeneous troops both in an ethnic sense and due to the diverse kinds of armament and combat styles of the soldiers.[49] The wide geographical extension of the Seleucid Empire doubtless made a large number of different typologies of soldiers available, which contributed to such an exotic perception from a Greek and Roman point of view. In this regard, Polybius (5.79) says that before the battle of Raphia, Antiochus III as well as Ptolemy IV addressed the troops with the assistance of interpreters. In addition, during the war between Antiochus III and Rome, the Romans used the diversity of the Seleucid army to present it as an avatar of the Persian army, and Antiochus was depicted as a new Xerxes aiming to deprive Greece of its freedom.[50] The Upper Satrapies alone provided a large number of troops of which we will try to provide a general review. Unlike the Ptolemies, the Seleucids constantly kept their troops on duty in garrisons or engaged in warfare, while only a minority of soldiers could have the status of military colonist after the assignment of a *klēros*.[51]

In many cases, the issue concerning the presence of soldiers of Greco-Macedonian background from the eastern Seleucid settlements is difficult to assess because of the ambiguity of the ethnic by which some of the corps are labelled. For instance, it is not easy to tell if a regiment of 'Macedonians' was formed by men of actual Macedonian descent or by soldiers armed and fighting in the Macedonian style. Still during his reign, Alexander had young Iranian men trained in the Macedonian style and named them *Epigonoi*, 'the descendants', and it is reasonable to think that the Seleucids followed the same practice.[52] In fact, the settlement of communities from Greece and the eastern Mediterranean in Central Asia must have already stopped in the first half of the third century as far as the easternmost satrapies are concerned because of the loss of Parthia and Bactria. Soldiers labelled as Macedonians are active in Persis under Seleucus I (Polyaenus, *Strat.* 7.39), while at Ai Khanum the

48 For a general description of the Seleucid army, see Bickerman (1938) 51–105; Bar-Kochva (1976); Sekunda (1994).

49 Eumenes' troops had a very diverse ethnic composition and this fact is restated in the account of the battle of Gabiene (317/6) where these soldiers are said to fight in the Macedonian manner; see Diod. Sic. 18.30.5 and 19.27.6.

50 See, e.g., Flor. 1.24.12. On the association between Xerxes and Antiochus III, see Tsitsiou-Chelidoni (2007) 23–44; Primo (2009) 93, 129, 234.

51 Aperghis (2004) 194–97; Fischer-Bovet (2014) 82.

52 Curt. 8.5.1; Plut. *Alex.* 47.6, 71.1; Diod. Sic. 17.108.1 and 17.110.1–2; Just. 12.4.2–10; Arr. *Anab.* 7.6.1., 7.23.1–4. On the *Epigonoi*, see Olbrycht (2015) 196–212; Olbrycht (2016) 66–69.

descendants of a military officer connected with the Thracian tribe of the Triballians are still active in the second century. If we take into account that in Ptolemaic Egypt the Greek population represented only 5% of the overall population (ca. 200,000 individuals) while the remaining 95% was formed by native Egyptians (ca. 4,000,000 individuals),[53] we have to imagine that the situation was not that different in the much wider Seleucid Empire. On the other hand, the difficulty in recruiting troops from the Greek world underpins the architectural choices of several fortifications of the Seleucid East: as we have seen, the ramparts of Ai Khanum were designed to be defended by a limited number of soldiers and restored by a larger amount of local manpower. However, we may posit that for the area of modern Iran a certain number – very limited, though – of individuals was still settled between the third century and the first half of the second century as a consequence of the intense activity in that area of rulers such as Antiochus III and Antiochus IV.[54] For example, an Athenian officer named Soteles is attested in the Seleucid garrison on Ikaros in the first half of the third century.[55] Epigraphic evidence from Susa/Seleucia-on-the-Eulaeus provides information on officers and troops in Susiana under the Seleucids. Between the end of the third and the first half of the second century, Leon, chief of a military unit, together with the officials (*hēgemones*) and the soldiers under his command honoured Arete, the daughter of Timon and Atheno.[56] Timon has the title of 'chief of the court (*aulē*) of the king', which probably means that he was in charge of the ward of the royal palace at Susa. In 183/2 the horseman Kalliphon, son of Diodorus, serving in the cavalry unit commanded by Alexander, freed his slave and consecrated him (or her) to Apollo and Artemis *Daittai* for the safety of Seleucus IV, the queen mother Laodike, and his wife Laodike.[57] A few years later, maybe under Antiochus IV, we have another act of manumission by a certain Bacchios serving in an unknown unit (again in the cavalry, perhaps?) commanded by Evander.[58] His slave girl Mikra is freed and consecrated to the goddess Nanaia for the safety of king Antiochus (IV?) and his queen Laodike. The epigraphic dossier from Susa under the Parthians concerning soldiers is not substantially different from that of the Seleucid period. In general, we catch a glimpse on

53 See Fischer-Bovet (2014) 170.

54 According to the demographic studies of Fischer-Bovet (2011, 135–54; 2014, 169–71), Macedonian immigration to Egypt was irregular and the large-scale immigration of Greeks stopped in the years preceding the battle of Raphia.

55 For the epigraphic dossier concerning Soteles, see Gatier (2007) 75–79.

56 Rougemont (2012) no. 7.

57 Rougemont (2012) no. 13.

58 Rougemont (2012) no. 17.

everyday life of the troops stationed in town: honorific inscriptions attest the popularity of certain officers among their troops, while the acts of manumission show the preference of the soldiers for traditional Seleucid deities such as Apollo and Artemis or for the Mesopotamian goddess Nanaia, the main deity of Susa, whose cult spread from Greece to Central Asia.

The presence of soldiers of Iranian or other eastern background in the Seleucid troops is underrepresented in our sources and, as in the case of the Macedonians, one cannot always assign precise ethnicities. The *sympoliteia* treaty (ca. 245) between Smyrna and Magnesia ad Sipylum attests an officer bearing the Iranian name of Omanes, commander of a group of Persians in the fort of Palaimagnesia (*OGIS* 229, ll. 104–109). The absence of any other detail prevents us from telling when they were stationed there and by whom. At Raphia a contingent of 3,000 units commanded by Menedemos of Alabanda was formed by 2,000 soldiers including Agrianians and Persian archers and slingers together with 1,000 Thracians (Polyb. 5.79).[59] Bar-Kochva is probably right in supposing that this contingent was raised in Persis because the troops were under the command of the same officer, which means that they were recruited in the same area.[60] In this case, we would be dealing with a national contingent possibly provided by the local *frataraka* dynasts, while the Agrianian and Thracian troops would be units without a defined ethnic composition. Contingents of national troops could also serve under officers from the same cultural and linguistic area as in the case of Aspasianos the Mede, who commanded 5,000 Medes, Kissians, Kadousians, and Karmanians (Polyb. 5.79), or the Arab chief Zabdibelos, who had under him a contingent of 10,000 Arabs and the neighbouring tribes (Polyb. 5.79). Dionysos the Mede, satrap or general in Mesopotamia, supported the royal family against the usurper Diodotos Tryphon (Diod. Sic. 33.28). Although isolated, the case of these three commanders suggests that the eastern Seleucid contingents relied also on non-Greek individuals who not only had military skill but were also bilingual so that they could act as linguistic intermediaries between the native soldiers and officers under their command and the Hellenophone general staff.[61]

59 According to Grainger (1997, 104), Menedemos of Alabanda could be identical with the officer who in 193 was appointed *stratēgos* of the Upper Satrapies by Antiochus III. Although he does not provide any element to justify this suggestion, this eventuality is not improbable at all: Menedemos' experience with Iranian troops and the Iranian world may have represented a strong point in his appointment as governor of the Upper Satrapies by Antiochus III.

60 Bar-Kochva (1976) 50.

61 See also Engels (2017) 73–78.

It is often agreed that Iranians formed the backbone of the Seleucid cavalry (and more generally the army):[62] this opinion seems reasonable when we think that the Seleucids had direct access to the vast recruiting pool of the Iranian Plateau and Central Asia. Upon closer analysis, however, it is important to nuance this view and make a distinction between Iranian cavalry techniques and ethnic composition of those troops. In the first case, the Seleucids employed resources and tactics that doubtlessly belong to the Iranian tradition: the higher proportion of cavalry to infantry (1:10) is a legacy from the late Achaemenid army, notably at Gaugamela, as is the use of the scythed chariots. Although the latter was already an old-fashioned weapon, and not always effective, scythed chariots were attested under Seleucus I at the battle of Ipsos (Diod. Sic. 20.113; Plut. *Demetr.* 28, 48), the usurper Molon (Polyb. 5.53), and Antiochus III at Magnesia (Livy 37.41).[63] However, Alexander also employed substantial cavalry-to-infantry ratios, for example at Gaugamela where he deployed 7,000 cavalry and 40,000 infantry (Arr. *Anab.* 3.12.5), and one must bear in mind that Tuplin's in-depth analysis has downplayed earlier assumptions of cavalry dominance within Achaemenid forces.[64] In addition, the Seleucids continued to maintain the Median stud farms (Polyb. 5.44) that provided these celebrated horses. The privileges bestowed by Seleucus I, and confirmed by Seleucus II, on a group of Median villages involved in horse breeding[65] witness a clear will to preserve these farms, which might have been depleted during the campaigns of Alexander and the wars of the Diadochi (cf. Diod. Sic. 17.110.6, Arr. *Anab.* 7.13.1). Other cavalry units such as the Dahae mounted archers and the cataphracts are indeed Iranian but belong to the world of the steppe. On the other hand, it is safe to assume that the adoption of Iranian war techniques and weapons does not imply that the soldiers of these troops always stemmed from this ethnic background. As Michele Minardi points out, even under the Achaemenids the deployment of these troops was not necessarily organised according to our idea of 'nationality' but rather military purposes.[66] Only in a few cases may we infer the presence of 'national' contingents, for instance the abovementioned Persians serving under Menedemos of Alabanda, the *agēma* of 1,000 horsemen from Media who fought at Magnesia (Livy 37.40.5–6) and the mounted archers from the Dahae nomads from the eastern shores of the Caspian. The latter are attested in the army of Antiochus III at Raphia in 217

62 See, e.g., Sherwin-White and Kuhrt (1993) 53–59.
63 Bickerman (1938) 60; Tuplin (2014) 245–76.
64 Tuplin (2010) 100–82.
65 Coloru (2018) 9–12.
66 Minardi (2023) 804.

(Polyb. 5.79.3) and at Magnesia in 190 (Livy 37.40; App. *Syr*. 32). The presence of Dahae at these two different dates is interesting in that it suggests that the Seleucids were still able to recruit troops in territories that had been lost to their control for several decades. In fact, in 217 the regions east of Media were under the Parthians and presumably not on peaceful terms with their Seleucid neighbours. However, in 208 Antiochus managed to make an alliance with Arsaces II after a fierce battle (Just. 41.5), so we may posit alternatively that the Dahae at Magnesia were sent by the Parthians as part of the clause of the treaty of 208 – which seems to me the most plausible scenario – or that the Seleucids were free to recruit mercenary contingents in an area near to or under Parthian rule. The troops from the Upper Satrapies could also rely on the mountain tribes of the Zagros and the Parachoatras Range in northern Media: again, the reign of Antiochus III provides our extant evidence for their recruitment. According to Polybius (10.29–30), mountain tribes provided excellent javelin and stone throwers that were employed by Antiochus against the Parthians when crossing the Elborz Range in order to reach Hyrcania. In particular, Strabo (11.12.5) praises the Kadousians as extraordinary javelineers (*akontistai*) and they are also employed at the battle of Raphia in the same contingent including Medians, Karmanians, and Kissians. The Kissians, who provided cavalry units, served together with Median contingents (Polyb. 5.79.7; 5.82.1) but this tactical disposition was already in place under the Achaemenids, as the two units are mentioned together by Herodotus at the battle of Thermopylai in 480 (Hdt. 7.210.1). However, in Herodotus the Kissians seem to stand in for Elamites in general, whereas at the time of the Seleucids we may not exclude that this name had lost its former ethnic connotation. As for the Kadousians, they fought under Darius III at Gaugamela in a contingent that was at least geographically homogenous, as it included troops from Media and bordering regions (Arr. *Anab*. 3.8.4; 11.3; 19.3; Diod. Sic. 17.59.5; Curt. 4.12.12). Troops, money, and military equipment as well as pack animals were provided by other rulers who were connected to the Seleucid house as vassals or were included in the Seleucid sphere of influence through treaties. While the kings of Media Atropatene may have supplied contingents of *akontistai* and horsemen (but the extant sources lack precise information on their numbers and military contribution),[67] we know that in ca. 212 Xerxes of Sophene gave Antiochus III 300 talents, 1,000 horses, and 1,000 equipped mules after his act of submission to the Seleucid king (Polyb. 8.23). Arsaces II of Parthia may have acted similarly in compliance with the treaty of 208 with Antiochus III (Just. 41.5), while

67 Bar Kochva (1976) 50.

Euthydemus I of Bactria and the Indian king Sophagasenos offered elephants in 206 (Polyb. 11.34).

As for war elephants, the Seleucids got an upper hand over the other Hellenistic dynasties thanks to their good diplomatic relations with the Mauryan kings. In fact, the peace treaty of 303 with Chandragupta brought Seleucus I the impressive total of 500 elephants and it seems that in the following years – until the loss of the Upper Satrapies – these animals would have left India and reached Syria via Bactria. In actuality, the elephants and the personnel responsible for their care and training were mainly stationed at Apamea in Syria (Strabo 16.2.10) but one cannot rule out the possibility that elephant stations existed in other sectors.[68] Unlike their Achaemenid predecessors,[69] the Seleucid army made greater use of Indian elephants in battle: Seleucus I deployed them at the battle of Ipsos (301)[70] and Antiochus I would make use of them at the alleged 'Battle of the Elephants' against the Galatians (275/268) in Asia Minor,[71] while Antiochus III used elephants against the usurper Molon in 222,[72] at Raphia (217),[73] during his *Anabasis* (212–205),[74] at Panion (200),[75] and at Magnesia (190).[76] After Antiochus III, the supply of Indian elephants must have come to an end because the Seleucids were cut off from India by the Parthian expansion on the one hand and the conquest of northwestern India by Demetrius I of Bactria and his successors on the other. The elephants deployed by Antiochus IV represented what was left of his father's herd and after 162 the elephants disappeared from the Seleucid army.[77] The association between elephants and the Seleucids was a strong one in the official iconography as an allusion to the Indian exploits of Seleucus I as well as symbol of kingship.[78] However, the frequent military use of those animals by the Seleucids could also

68 Epplett (2007, 224) suggests Babylon, Seleucia-on-the-Tigris, and Bactra. The latter could be the most probable location, as cuneiform evidence shows that elephants were sent to Antiochus I by the governor of Bactria.

69 See Briant (1997) 177–90; Tuplin (2014) 271. On the extensive use of elephants by the Seleucids see Bar-Kochva (1976) 75–83.

70 Diod. Sic. 21.1.2; Plut. *Demetr.* 28–29.

71 Luc. *Zeux.* 8–11; doubts on the historicity of this battle have been raised by Coşkun (2012, 57–73), with discussion of sources and bibliography.

72 Polyb. 5.53.4.

73 Polyb. 5.79–87.

74 The presence of a contingent of elephants during this long campaign is attested by Polyb. 11.34.10.

75 Polyb. 16.18–19.

76 Livy 37.40.

77 Bar-Kochva (1976) 81.

78 Iossif and Lorber (2010b) 147–164. See also Kosmin (2014) 3–4, 33–37.

be the object of parody by their rivals as can be seen from a passage of Plutarch (*Demetr.* 25.4) relating that the flatterers of Demetrius Poliorcetes would mock the king's opponents by toasting to Seleucus as *elephantarch*, to Ptolemy as admiral (*navarch*), to Lysimachus as treasurer, and to Agathokles as governor of Sicily.[79] This pun aimed at the royal pretensions of his rivals by assigning them the titles of subordinate officers while the underlying message was that only Demetrius was worthy of the title of king. In fact, the title of *elephantarch*, 'master of elephants', designated the officer in command of the corps of elephants, which was a high rank position in the Hellenistic armies, but was obviously not related to the concept of kingship.

4 The Navy

Although it is not the first aspect that comes to mind when dealing with the Upper Satrapies and sometimes the latter are approached only in their terrestrial dimension, the navy played an important role in the military organisation of the Seleucid East, even if with different outcomes. The two maritime spaces that saw the involvement of the Seleucid navy were the Caspian Sea and the Persian Gulf. The first one was certainly the least exploited: as far as we know, the admiral Patrokles made a periplus of the Caspian Sea between 285 and 282 by order of Seleucus I and Antiochus I (the two kings may have taken part to the exploration according to Pliny, *HN* 6.58.7–9). The report of Patrokles was an authority for the geography of the Caspian area and the hydrography of Sogdiana and Bactria. In addition, Patrokles was convinced that the Caspian Sea was not a closed sea, but opened on the Boreal Ocean at a distance of 6,000 stadia, thus allowing the circumnavigation of the Indian coast. It seems that after this mission the Seleucids lost interest in the Caspian: the explorations in the north had a political purpose in order to gain a better knowledge of a world that the Seleucids mastered only partially.[80] This exploration could also have economic goals: citing a passage of Aristobulus who in turn took the information by Patrokles himself, Strabo (11.7.3) reports the existence of a trade route originating from India which included navigation along the river Oxus and the Caspian Sea. This aspect is nevertheless addressed when Strabo (11.7.2) deplores the fact that the navigation has not been exploited, although the Hyrcanian (Caspian) Sea had many inhabitable islands among which there

79 On this pun, see Hauben (1974) 105; Coloru (2014) 16–17; on humour on Alexander and the Hellenistic kings, see Muccioli (2018).

80 Callieri (1999) 31–43.

were some with gold-bearing deposits. This lack of interest for the Caspian Sea is equally shared by the Achaemenids, the Seleucids, and the Parthians, even if for different causes. As for the Persians and the Parthians, Strabo explains their neglect for the development of a Caspian navy with an ethnic prejudice: they are barbarians – thus, they have no competence in maritime affairs – and the lands surrounding the Caspian teem with brigands and nomads, two categories that are often equivalent in the Hellenic vision of the world. On the other hand, the Seleucids were too occupied with wars in other parts of their empire and what is more their rule over the Caspian shores was short-lived – which is true if we think that territories overlooking the Caspian Sea such as Hyrcania were lost to the Seleucids by the mid-third century.

If the development of a Seleucid navy on the Caspian Sea was prevented by logistical and political difficulties, the Persian Gulf on the contrary received special attention. In fact, the Persian Gulf represented a strategic position connecting Mesopotamia to India and thereby provided access to a thriving trade network. The Seleucids were only the latest in a series of authorities exerting control over that geographical space dating back to at least the third millennium and the Achaemenids had also organised the Gulf islands as a satrapy.[81] A Seleucid district of the Erythrean Sea encompassing the mouths of the Tigris and the Euphrates as well as the Shatt al-Arab was in place since the third century (under the reign of Seleucus II or Antiochus III) when its governor Ikadion wrote to the Seleucid officer Anaxarchos in charge of the settlement on the island of Ikaros.[82] As Anaxarchos refers to the policy followed by the king Seleucus/Antiochus and his ancestors in relation to the island, it is possible to posit that the Seleucids were already in control of the Persian Gulf area in the first half of the third century.[83] Ikaros hosted a garrison but provided also a naval base for the Seleucids as well as the island of Tylos, modern Bahrain.[84] In addition, the Seleucid presence in the Gulf is attested by the settlements of Alexandria Charax (later renamed Antioch), Antioch in Persis, and Seleucia-on-the-Erythrean Sea. Probably, the administrative capital of the district of the Erythrean Sea and consequently the seat of the *stratēgos* were

81 On the control exerted by the first dynasty of the Sealand in the second millennium, see Potts (2009). On the Achaemenid control over the Persian Gulf, see Salles (1996) 255–57; Briant (2002) 758, 1028–29; Kosmin (2013) 69; Coloru (2017) 98, 102–103.

82 Jeppesen (1989) 83–92; Sherwin White and Kuhrt (1993) 174–78; Martinez-Sève (2011) 59–60.

83 See Capdetrey (2007) 251–52.

84 Kosmin (2013) 67. A funerary inscription dating to the second century mentions a *kybernētēs* ('helmsman') bearing the Babylonian name of Abidistaras; see Gatier et al. (2002) 229.

located at Antioch Charax at least since its re-foundation by Antiochus IV, but also Antioch in Persis hosted an important naval base, as suggested by the iconography of its bronze coinage whose main feature is the prow of a warship.[85]

Albeit fragmentary, our sources attest several military operations led by the Seleucids in the Gulf and prove the importance they attached to the area. The Seleucids most likely had a fleet stationed in the Persian Gulf that assured the control as well as the protection of the area. In 205/4 Antiochus III ended his *Anabasis* with a sojourn at Antioch in Persis and a naval campaign to exact tribute from Gerrha (Thaj). On the same occasion his fleet sailed to Thylos and possibly to Ikaros, where the archaeological evidence suggests the existence of a conflict for the reaffirmation of the Seleucid rule on the island.[86] Antiochus IV would be the first Seleucid king to have organised an exploration of the Persian Gulf (Pliny *HN* 6.147)[87] and his satrap Noumenios would obtain in the same day a double victory by sea and land in the Strait of Hormuz area against 'Persians' (Pliny *HN* 6.152). The latter may be identified with populations of the Karmanian littoral devoted to piracy.[88] Of a more difficult interpretation is the reference to military operations led by a king Demetrius against the Arabs because of the fragmentary condition of the tablet.[89]

5 Conclusions

Titus Quinctius Flamininus disparaged the army of Antiochus III by saying that despite the apparent number and variety of its troops and units, the soldiers of the king were only Syrians, thus implying that they were unwarlike and easily conquered opponents (Plut. *Flam.* 17). Paradoxically, such a distorted perspective shows the high degree of integration reached by the Seleucid army in which Greco-Macedonian and Eastern features coexisted at the same time. Like the western half of the empire, the military organisation of the Upper Satrapies was defined by taking into account the regional features of every single territory and the Hellenistic tradition was shaped according to those

85 Hyspaosines, the Seleucid governor of the Erythrean Sea district, had his seat in Charax in the last years of the reign of Antiochus IV; see Schuol (2000) 291–300; on the bronze coinage of Antioch in Persis, see Houghton, Lorber, Hoover (2008) no. 1527.

86 The local population may have gained control of Ikaros, taking advantage of the turmoil caused by the usurpation of Molon; see Martinez-Sève (2011) 58–60.

87 The presence of Antiochus in 165 is deduced by an astronomical diary; see Martinez-Sève (2014b) 366–67.

88 Martinez-Sève (2014b) 382–83.

89 BM 34433.

peculiarities. Likewise, the difficulty in assigning precise ethnic identities to the soldiers is indeed due to the dearth of sources, but I cannot help but think that since the beginning of the Seleucid rule the army had become a place where ethnic background lost a great deal of its importance in favour of an ethnicity chosen by the soldiers on the basis of the corps to which they belonged. Thus, the contingents from the Upper Satrapies were part of a wider supranational army: Greek, Iranian, Macedonian, Thracian were ethnic labels no more or less than that of Syrian applied by Flamininus to the Seleucid soldiers.

Bibliography

Aperghis, M. (2004) *The Seleukid Royal Economy: The Finances and Financial Administration of the Seleukid Empire*. Cambridge.

Baratin, C. and Martinez-Sève L. (2013) 'Le grenier grec de Samarkand'. *Cahiers d'Asie Centrale* 21/22: 373–89.

Bar-Kochva, B. (1976) *The Seleucid Army: Organization and Tactics in the Great Campaigns*. Cambridge.

Baumer, Ch. and Novák, M. (eds) (2019) *Urban Cultures of Central Asia from the Bronze Age to the Karakhanids*. Wiesbaden.

Bearzot, C. and Landucci, F. (eds) (2016) *Alexander's Legacy: Atti del Convegno Università Cattolica del Sacro Cuore Milano 2015*. Rome.

Bengtson, H. (1964²) *Die Strategie in der hellenistischen Zeit*. Munich.

Bernard, P. (1980) 'Héraclès, les grottes de Karafto et le sanctuaire du mont Sambulos en Iran'. *StIr* 9: 301–24.

Bernardini, M., Bonora, G., and Traina, G. (eds) (2016) *Turkmenistan: Histories of a Country, Cities and a Desert*. Torino.

Bickerman, E. (1938) *Institutions des Séleucides*. Paris.

Boiy, T. (2004) *Late Achaemenid and Hellenistic Babylon*. Leuven.

Boucharlat, R. (2018) 'Les traces archéologiques des palais achéménides de Hamadan'. *ARTA* 2018.002: 1–34.

Briant, P. (2002) *From Cyrus to Alexander: A History of the Persian Empire*. Winona Lake.

Callieri, P. (1999) 'L'esplorazione geografica dell'Iran in epoca ellenistica e romana: Il contributo della documentazione archeologica'. *Ocnus* 7: 31–43.

Callieri, P. (2007a) 'Barikot: An Indo-Greek Urban Center in Gandhāra', in Srinivasan (2007) 133–64.

Callieri, P. (2007b) *L'archéologie du Fars à l'époque hellénistique*. Paris.

Capdetrey, L. (2007) *Le pouvoir séleucide: Territoire, administration, finances d'un royaume hellénistique (312–129 avant J.-C.)*. Rennes.

Capdetrey, L. (2012) 'Fondations, diasporas et territoires dans l'Asie hellénistique au IIIᵉ siècle', in Martinez-Sève (2012) 319–44.

Chrubasik, B. (2016) *Kings and Usurpers in the Seleukid Empire: The Men Who Would Be King*. Oxford.

Clarysse, W. and Thompson, D. (2007) 'Two Greek Texts on Skin from Hellenistic Bactria'. *ZPE* 159: 273–79.

Cohen, G.M. (1991) 'Katoikia, Katoikoi and Macedonians in Asia Minor'. *AncSoc* 22: 41–50.

Cohen, G.M. (2013) *The Hellenistic Settlements in the East from Armenia and Mesopotamia to Bactria and India*. Berkeley.

Coloru, O. (2009) *Da Alessandro a Menandro. Il regno greco di Battriana*. Pisa.

Coloru, O. (2014) 'Rex ridens, rex cavillatus: L'umorismo e la regalità ellenistica', in Coloru and Minunno (2014) 20–36.

Coloru, O. (2016) 'From the Achaemenids to Alexander', in Bernardini, Bonora and Traina (2016) 33–41.

Coloru, O. (2017) 'Potere e territorio: Gli Achemenidi nei *Geographikà* di Strabone'. *Electrum* 24: 93–105.

Coloru, O. (2023) 'Seen from Ecbatana: Aspects of Seleucid Policy in Media', in Daryaee, Rollinger, and Canepa (2023) 201–16.

Coloru, O. and Minunno, G. (eds) (2014) *L'Umorismo in prospettiva interculturale: Immagini, Aspetti e Linguaggi/Crosscultural Humour: Images, Aspects, and Languages*. Parma.

Cribb, J. and Herrmann, G. (eds) (2007) *After Alexander: Central Asia before Islam*. Oxford.

Daryaee, T, Rollinger, R., and Canepa, M. (eds) (2023) *Iran and the Transformation of Ancient Near Eastern History: The Seleucids (ca. 312–150 BCE)*. Wiesbaden.

Del Monte, G. (1997a) 'Antioco I Soter e i figli Seleuco e Antioco: Un nuovo testo da Babilonia'. *SCO* 45: 433–44.

Del Monte, G. (1997b) *Testi dalla Babilonia ellenistica*. Pisa.

Duindam, J., Kunt, M., and Artan, T. (eds) (2011) *Royal Courts in Dynastic States and Empires: A Global Perspective*. Leiden.

Dumitru, A. (2013) 'Les Séleucides et les Balkans: les Thraces dans l'armée séleucide', in Rufin Solas, Parissaki, and Kosmidou (2013) 69–89.

Dvurechenskaya, N.D. (2019) 'The Hellenistic Fortress of Uzundara', in Baumer and Novák (2019) 153–62.

Engels, D. (2017) 'The Achaemenid and Seleucid Court: Continuity or Change?', in Erskine, Llewellyn-Jones, and Wallace (2017) 69–100.

Engels, D. (2018) 'Iranian Identity and Seleucid Allegiance; Vahbarz, the Frataraka and Early Arsacid Coinage', in Erickson (2018) 173–96.

Epplet, C. (2007) 'War Elephants in the Hellenistic World', in Heckel, Tritle, Lawrence, and Wheatley (2017) 209–32.

Erickson, K. (ed) (2018) *The Seleukid Empire, 281–222 BC. War within the Family*. Swansea.

Erskine, A., Llewellyn-Jones, L., and Wallace, S. (eds) (2017) *The Hellenistic Court*. Swansea.

Feraco, F. (2004) *Ammiano geografo: la digressione sulla Persia (23, 6)*. Naples.
Feyel, C. and Graslin, L. (eds) (2014) *Le projet politique d'Antiochos IV*. Nancy.
Feyel, C. and Graslin-Thomé, L. (eds) (2017) *Antiochos III et l'Orient. Actes de la rencontre franco-allemande tenue à Nancy du 6 au 8 juin 2016*. Nancy.
Fischer-Bovet, C. (2011) 'Counting the Greeks in Egypt: Immigration in the First Century of Ptolemaic Rule', in Holeran and Pudsey (2011) 135–54.
Fischer-Bovet, C. (2014) *Army and Society in Ptolemaic Egypt*. New York.
Frye, R.N. (1984) *The History of Ancient Iran*. Munich.
Gatier, P.-L. (2007) 'Sôtélès l'Athénien'. *Arabian Archaeology and Epigraphy* 18: 75–79.
Gatier, P.-L., Lombard, P., and al-Sindi, K. (2002) 'Greek Inscriptions from Bahrain'. *Arabian Archaeology and Epigraphy* 13: 223–33.
Grainger, J.D. (1997) *A Seleukid Prosopography and Gazetteer*. Leiden.
Hauben, H. (1974) 'A Royal Toast in 302 B.C.'. *AncSoc* 5: 105–17.
Heckel, W., Tritle, L., Lawrence, A., and Wheatley, P. (eds) (2007) *Alexander's Empire. Formulation to Decay: A Companion to Crossroads of History*. Claremont, CA.
Heckel, W., Müller, S., and Wrightson, G. (eds) (2015) *The Many Faces of War in the Ancient World*. Newcastle upon Tyne.
Herrmann, P. (1987) 'Milesier am Seleukidenhof'. *Chiron* 17: 171–92.
Hoffmann, F. and Schmidt, K.S. (eds) (2014) *Orient und Okzident in hellenistischer Zeit. Beiträge zur Tagung 'Orient und Okzident – Antagonismus oder Konstrukt? Machtstrukturen, Ideologien und Kulturtransfer in hellenistischer Zeit', Würzburg 10.–13. April 2008*. Vaterstetten.
Holeran, C., and Pudsey, A. (eds) (2011) *Demography and the Graeco-Roman World: New Insights and Approaches*. Cambridge.
Houghton, A., Lorber, C.C., and Hoover, O. (2008) *Seleucid Coins: A Comprehensive Catalogue. Part II: Seleucus IV through Antiochus XIII*. New York.
Hyland, J. (2013) 'Alexander's Satraps of Media'. *JAH* 1: 119–44.
Iossif, P. and Lorber, C.C. (2010a) '*Hypaithros*: A Numismatic Contribution to the Military History of Cappadocia'. *Historia* 59.4: 432–47.
Iossif, P. and Lorber, C.C. (2010b) 'The Elephantarches Bronze of Seleucos I Nikator'. *Syria. Archéologie, art et histoire* 87: 147–64.
Jeppesen, K. (1989) *Ikaros. The Hellenistic Settlements III: The Sacred Enclosure in the Early Hellenistic Period*. Aarhus.
Kosmin, P. (2013) 'Rethinking the Hellenistic Gulf: The New Greek Inscription from Bahrain'. *JHS* 133: 61–79.
Kosmin, P. (2014) *The Land of the Elephant Kings: Space, Territory, and Ideology in the Seleucid Empire*. Cambridge, MA.
Leriche, P. (1986) *Fouilles d'Aï Khanoum. Vol. 5: Les remparts et les monuments associés*. Mémoires de la Délégation archéologique française en Afghanistan 29. Paris.
Leriche, P. (2007) 'Bactria, Land of a Thousand Cities', in Cribb and Herrmann (2007) 121–53.

Ma, J. (2004) *Antiochos III et les cités de l'Asie Mineure occidentale*. Paris.

Martinez-Sève, L. (2011) 'Suse et les Séleucides au IIIe siècle avant J.-C.'. *Electrum* 18: 41–66.

Martinez-Sève, L. (ed) (2012) *Les diasporas grecques du VIIIe à la fin du IIIe siècle av. J.-C.* (*Pallas* 89). Toulouse.

Martinez-Sève, L. (2014a) 'The Spatial Organization of Ai Khanoum, a Greek City in Afghanistan'. *AJA* 118: 267–83.

Martinez-Sève, L. (2014b) 'Antiochos IV en Susiane, dans le Golfe Persique et en Élymaïde', in Feyel and Graslin (2014) 363–93.

Minardi, M. (2023) 'The Northeastern Regions of the Persian Empire: Bactriana, Sogdiana, Margiana, Chorasmia, Aria, Parthia, the Sakas, and the Dahae', in Radner, Moeller, and Potts (2023) 784–836.

Muccioli, F. (2018) *Le orecchie lunghe di Alessandro Magno. Satira del potere nel mondo greco (IV–I secolo a.C.)*. Rome.

Müseler, W. (2018) 'The Dating and the Sequence of the Persid Frataraka Revisited'. *Koinon. The International Journal of Classical Numismatic Studies* 1: 84–106.

Olbrycht, M. (2015) 'The Epigonoi – The Iranian Phalanx of Alexander the Great', in Heckel, Müller, and Wrightson (2015) 196–212.

Olbrycht, M. (2016) 'Alexander the Great at Susa (324 B.C.)', in Bearzot and Landucci (2016) 61–72.

Olivieri, L.M. (1996) 'Notes on the Problematic Sequence of Alexander's Itinerary in Swat: A Geo-Historical Approach'. *East and West* 46: 45–78.

Plischke, S. (2014) *Die Seleukiden und Iran. Die seleukidische Herrschaftspolitik in den östlichen Satrapien*. Wiesbaden.

Potter, L. (ed) (2009) *The Persian Gulf in History*. New York.

Potts, D. (2009) 'The Archaeology and Early History of the Persian Gulf', in Potter (2009) 27–56.

Primo, A. (2009), *La storiografia sui Seleucidi: Da Megastene a Eusebio di Cesarea*. Pisa.

Radet, G. (1890) *De Coloniis a Macedonibus in Asiam Cis Taurum Deductis*. Paris.

Radner, K., Moeller, N., and Potts D.T. (eds) (2023) *The Oxford History of the Ancient Near East. Volume V: The Age of Persia*. New York.

Rapin, C. and Isamiddinov, M. (1994) 'Fortifications hellénistiques de Samarcande (Samarkand-Afrasiab)'. *Topoi. Orient-Occident* 4: 547–65.

Reade, J. (ed) (1996) *The Indian Ocean in Antiquity*. London.

Robert, L. (1985) 'Documents d'Asie Mineure. XXXV. Retour à Pergame. 2. Strabon et la *katoikia* de Pergame'. *BCH* 109: 481–84.

Rougemont, G. (2012) *Inscriptions grecques d'Iran et d'Asie centrale. Corpus inscriptionum Iranicarum, Part II: Inscriptions of the Seleucid and Parthian Periods of Eastern Iran and Central Asia. Vol. 1: Inscriptions in Non-Iranian Languages, 1*. London.

Rtveladze, E.V (1995) 'Découvertes en numismatique et épigraphie gréco-bactriennes à Kampyr-Tepe (Bactriane du Nord)'. *RN* 150: 20–24.

Rufin Solas, A., Parissaki, M.-G., and Kosmidou, E. (eds) (2013) *Armées grecques et romaines dans le nord des Balkans*. Gdánsk.

Salles, J.-F. (1992) 'L'Arabie sans Alexandre. D.T. Potts, *The Arabian Gulf in Antiquity. Volume II. From Alexander the Great to the Coming of Islam*, 1990'. *Topoi* 2: 201–35.

Salles, J.-F. (1996) 'Achaemenid and Hellenistic Trade in the Indian Ocean', in Reade (1996) 251–67.

Savalli-Lestrade, I. (1998), *Les* philoi *royaux dans l'Asie hellénistique*. Geneva.

Schuler, C. (1998) *Ländliche Siedlungen und Gemeinden im hellenistischen und römischen Kleinasien*. Munich.

Schuol, M. (2000) *Die Charakene: Ein mesopotamisches Königreich in hellenistisch-parthischer Zeit*. Stuttgart.

Sekunda, N. (1994) *Seleucid and Ptolemaic Reformed Armies 168–145 BC, Volume 1: The Seleucid Army*. Stockport.

Sekunda, N. (2007) 'Boxus the Persian and the Hellenization of Persis', in Tuplin (2007) 225–35.

Sherwin-White, S. and Kuhrt, A. (1993) *From Samarkhand to Sardis: A New Approach to the Seleucid Empire*. Berkeley.

Srinivasan, D.M. (ed) (2007) *On the Cusp of an Era: Art in the Pre-Kuṣāṇa World*. Leiden.

Strootman, R. (2011) 'Hellenistic Court Society: The Seleukid Imperial Court under Antiochos the Great, 223–187 BC', in Duindam, Kunt, and Artan (2011) 63–89.

Traina, G. (2017) 'Rois ou dynastes? Les territoires arméniens à l'époque d'Antiochos III', in Feyel and Graslin-Thomé (2017) 373–84.

Tsitsiou-Chelidoni, C. (2007) 'Kleinasien zwischen Ost und West. Titus Livius' Bericht über den Kampf zwischen Antiochos dem Grossen und den Römern (192–188 v.Chr.)', in Urso (2007) 23–44.

Tuplin, C.J. (ed) (2007) *Persian Responses: Political and Cultural Interaction with(in) the Achaemenid Empire*. Swansea.

Tuplin, C.J. (2014) 'The Military Dimension of Hellenistic Kingship: An Achaemenid Inheritance?', in Hoffmann and Schmidt (2014) 245–76.

Urso, G. (ed) (2007) *Tra Oriente e Occidente. Indigeni, Greci e Romani in Asia minore. Atti del convegno internazionale, Cividale del Friuli, 28–30 settembre 2006*. Pisa.

Zavyalov, V.A. (2007) 'The Fortifications of the City of Gyaur Kala, Marv', in Cribb and Herrmann (2007) 313–29.

CHAPTER 11

Patricide, Fratricide, and Betrayal: the Role of Parthia in the First War with Rome

Nikolaus Overtoom

Compared to the scholarly attention paid to the other major pre-Islamic empires of the Middle East, the empires of the Achaemenids, the Seleucids, and the Sasanians, the Arsacids receive little appreciation.[1] Archaeological research on the Arsacid era only began to elicit comprehensive attention in the past half-century, severely handicapping Parthian studies prior to the 1970s.[2] Moreover, interest in much later Roman and Parthian interactions and the agency of the Romans in those interactions still dominates the field.[3] These modern obstacles minimise and restrict the Parthians' agency, accomplishments, reputation, and presence in the study of antiquity.

Unfortunately, no account of the Parthians' history survives from their perspective, and therefore, in reconstructing Parthian history, we must manoeuvre through Greek and Roman literary sources that dominate the narrative and often treat the Parthians from a mostly hostile or cursory foreign perspective.[4] Although Armenian, Arab, Chinese, and Jewish literary sources offer some evidence of Parthian history and culture, they also often suffer from the aforementioned limitations.[5] Meanwhile, scholars must supplement the historical record of the Parthians with material evidence from archaeology, numismatics, and cuneiform records, like the Babylonian *Astronomical Diaries*, to help recreate acceptable and plausible reconstructions of Parthian history.[6] Where

1 Colledge (1967) 176; Hauser (2013) 728–30; Gregoratti (2015) 203–204; Sampson (2015) xvi, 32; Daryaee (2015) 286.

2 Keall (1967); Keall (1975); Keall (1977a); Keall (1977b); Keall (1982); Sherwin-White and Kuhrt (1993) 88; Dąbrowa (2012) 25; Callieri and Chaverdi (2013) 690; Hauser (2013) 728–30.

3 Schlude (2009); Sheldon (2010); Poirot (2014); Sampson (2015); McLaughlin (2016); Gregoratti (2017b); Barnett (2017); Syvänne (2017); Bishop (2018); Fratantuono (2019); Schlude (2020).

4 Wiesehöfer (1998); Dąbrowa (2010b) 123; Dąbrowa (2012) 21–25; Overtoom (2020a): 27–64.

5 Hackl, Jacobs, and Weber (2010b) 175–472, 482–512; Tao (2007); Gruen (2017); Overtoom (forthcoming).

6 Widengren (1983) 1261–69; Boyce (1983); Sachs and Hunger (1988); Sachs and Hunger (1989); Sachs and Hunger (1996); Wiesehöfer (1998); Wolski (2003); Lerouge-Cohen (2007) 26–37; Hackl, Jacobs, and Weber (2010a) 1–175; Dąbrowa (2010b) 123; Dąbrowa (2012) 21–25;

 | DOI:10.1163/9789004710771_012

possible, this chapter strikes a balance between the surviving literature and material evidence to understand the Parthians on their own terms.

This chapter breaks from the dominant Rome-centric, anti-Crassus traditions concerning the geopolitical developments before, during, and after the Carrhae campaign (54–53 BCE) to reconstruct these events as much as possible from the Parthians' perspective.[7] The Roman focus of our sources and centuries of anti-Crassus propaganda have led most scholars to discount or overlook the critical agency of the Parthians and the crucial developments within the Parthian Empire that eventually led to Parthia's first war with Rome. This chapter challenges the misguided tradition that describes the Parthians as feeble, passive, and parasitic and that places agency in the hands of the Parthians' rivals, blaming the incompetence or weakness of the Parthians' enemies to explain their actions and successes.[8] Indeed, the Parthians were important, active, and powerful agents in the origin of their rivalry with Rome, something often obscured or neglected because of the inherent Roman focus and bias of the Greco-Roman literary tradition and modern scholarship.

1 Parthian Kings and the Aristocracy

In the middle second century, the Parthians established an empire that stretched from Babylonia to the Hindu Kush at the expense of the Greco-Bactrians and Seleucids.[9] A generation later the military and diplomatic successes of Mithridates II against the Armenians and Seleucids in the 110s–90s meant that Parthian lands and allied territories came to share an extensive frontier with Pontus, Cappadocia, and Syria, lands in which the Romans had a growing

---

Ellerbrock and Winkelmann (2012); Hauser (2013) 729–30; Sampson (2015) 186–99; Rezakhani (2016); Wiesehöfer and Müller (2017); Ellerbrock (2021); Olbrycht (2021c).

7 Regling (1899); Regling (1907); Debevoise (1938) 78 n. 36; Garzetti (1944); Timpe (1962); Ward (1977) 1–5; Keaveney (1981); Keaveney (1982); Sherwin-White (1984) 279; Schippmann (1987a); Mielczarek (1990); Zadorojniy (1997); Arnaud (1998); Frendo (2003/2007); Weggen (2011); Schlude (2012); Muccioli (2012); Sampson (2015); Goldsworthy (2016) 331; Overtoom (2016b); Overtoom (2017a); Overtoom (2017b); Sampson (2020); Schlude (2020). All dates are BCE.

8 Lindsay (1852) 15–16; Mommsen (1856) iii 288; Rawlinson (1876) 24, 33–34, 41, 97–98, 105, 118–19; Vogt (1955b) 74; Wolski (1956–1957); Syme (1958) 47, 493–97; Ward (1977) 287, 289; Keaveney (1981); Keaveney (1982); Yarshater (1983) xlix–l; Campbell (1993); Warry (1995) 157; Potter (2010b) 157–60; Sheldon (2010) 1–2, 6, 10, 21–23, 36–40, 61, 177; Edwell (2013) 191, 194, 206; Grainger (2013) 185–86; Grainger (2014) 199, 210; Sampson (2015) 44–45; Grainger (2016).

9 Overtoom (2019a); Overtoom (2020a); Overtoom (2020b); Overtoom (2021a); Overtoom (2022).

interest.[10] Yet from the middle 90s to early 50s, the Parthians and Romans only interacted directly on four occasions, and the two world powers remained distant from one another, largely concerned with foreign conflicts and internal strife.[11]

Although the Parthian Empire reached its widest extent under Mithridates II and his immediate successors, beginning toward the end of Mithridates' reign in the late 90s, a series of civil wars hamstrung the Parthians' wider imperial ambitions.[12] The cultural flexibility of the Parthians had helped encourage indigenous communities to support the early Arsacid kings, which in turn allowed the Parthians to occupy large sections of the Middle East quickly under capable and charismatic leaders.[13] Indigenous aristocrats and communities had heightened regional autonomy and power under the Parthians with greater access to authority and participation within the Parthian imperial system.[14] The early Arsacid kings embraced regional aristocracies and incorporated them into a more accommodating structure of empire.[15] The Parthian state from its conception built upon the social flexibility of a multi-cultural union of regional leaders under the ultimate authority of the early Arsacid kings. This made the high command of the Parthian state versatile and eclectic, allowing the Parthians to develop a more inclusive system of administration that emphasised, utilised, and internalised the capabilities and strengths of local leaders.[16] Yet it also made aristocratic families increasingly influential and powerful, and invested in their regional loyalties and concerns.

Unfortunately, the end of Mithridates II's reign marks the beginning of what scholars call the 'Parthian Dark Age' (late 90s to early 50s), which was a period of geopolitical turmoil and dynastic instability and conflict. Almost no written

10 Overtoom (2019b); Overtoom (2019c); Overtoom (2023).

11 Gregoratti (2012); Overtoom (2016c); Schlude (2020) 22–59; Overtoom (2021b).

12 Simonetta (1953–1957) 111–21; Simonetta (1966) 15–40; Sellwood (1976) 2–25; Mørkholm (1980) 33–47; Simonetta (2001) 69–108; Assar (2006b) 55–96; Assar (2009); Assar (2011) 121–28.

13 Dąbrowa (1994); Frye (2000) 18; Dąbrowa (2008/2009); Dąbrowa (2011a); Grajetzki (2011) 14; Overtoom (2016a); Olbrycht (2016a); Olbrycht (2017); Dąbrowa (2017); Lerner (2017); Overtoom (2019a); Overtoom (2019b); Overtoom (2019c); Overtoom (2020a).

14 Wolski (1989); Olbrycht (2010b) 147; Grajetzki (2011) 11, 15; Dąbrowa (2011b); Dąbrowa (2012); Graslin-Thomé (2012); Stevens (2016); Haubold (2016); Wiesehöfer (2016) 217–18, 227; Gregoratti (2016a) 5; Olbrycht (2017b).

15 Dąbrowa (2011a); Wiesehöfer (2016) 227; Gregoratti (2016b) 86–89; Gregoratti (2017a); Overtoom (2019a).

16 Keall (1994) 271–72; Dąbrowa (2013); Wiesehöfer (2016) 217–18; Gregoratti (2017a); Olbrycht (2017b).

sources of this period have survived, so scholars mostly rely on coinage to reconstruct events and, unsurprisingly, disagree widely on the details.[17] Yet it is possible that by the end of Mithridates' reign, separate, rival groups within the Parthian aristocracy, based in the eastern and western territories of the empire, helped initiate this period of crisis through civil war.[18] Never before had the Parthians fought a civil war; however, it became common over the next forty years, as the heightened factionalism and ambition of the Parthian aristocracy facilitated the dynastic chaos and warfare of this period.[19] Factions within the Parthian aristocracy challenged one another and the king for power and influence, and by the 50s, the power of these factions was at its height.[20] No Arsacid candidate could hope to rule the empire unchallenged without their support. Within this context of crisis, the murder of a father and a war amongst brothers facilitated Parthia's first war with Rome.

2 The Early Civil Wars of Parthia

Mithridates II came to the Parthian throne in ca. 121 in a time of geopolitical chaos.[21] In the 120s, invading Central Asian tribes, like the Saka and Tochari, ravaged the Iranian plateau and killed no less than two Parthian kings.[22] Mithridates campaigned against these incursions with determination and success over the first several years of his reign, restoring and then expanding the eastern frontier of the Parthian Empire (Strabo 11.8.2; Just. 42.2.2–5; Trog. *Prol.* 41).[23] His victory in the east was decisive, allowing him to stabilise the eastern territories of the empire and return them to prosperity.[24] The

17 Simonetta (1953–1957) 111–21; Simonetta (1966) 15–40; Dobbins (1974); Sellwood (1976) 2–25; Mørkholm (1980) 33–47; Simonetta (2001) 69–108; Vardanyan (2006); (Assar 2006b); Assar (2009); Fowler (2013) 5073.

18 Assar (2005a) 53; Assar (2005b); Assar (2006a) 145; Assar (2006b) 56–62; Assar (2011) 121–24; Valverde (2017).

19 Olbrycht (1997); Assar (2006b) 56–96; Assar (2009); Assar (2011) 121–28; Overtoom (2016c); Overtoom (2020a) 246–76.

20 Dąbrowa (2013); Gregoratti (2013).

21 Assar (2006a) 129–34; Olbrycht (2010b); Assar (2011) 120.

22 Enoki, Koshelenko, and Haidary (1999); van de Mieroop (2002); Benjamin (2007); Bivar (2009); Rtveladze (2011) 150; Olbrycht (2012); Mair (2014) 17, 19; Lerner (2015) 311–18; McLaughlin (2016) 27, 41–91, 98–99, 101–2, 179, 188–90, 192–93, 199; Overtoom (2019c); Overtoom (2020a) 148–49, 172, 216–17, 230–31, 240–41, 244–45, 252–54.

23 Overtoom (2019b); Overtoom (2020a) 247–55.

24 Bivar (1983a) 41; Olbrycht (1998) 96–100; Assar (2005a) 52; Olbrycht (2010b) 151–55; Rtveladze (2011); Schmitt (2011a); Rezakhani (2013) 769; Lerner (2015) 312; Callieri (2016); Overtoom (2019b).

aristocratic families living in these eastern territories benefitted greatly from Mithridates' protection and strong eastern focus early in his reign, and they supported him with considerable resources, including 20,000 horsemen.[25] Yet Mithridates' attention shifted suddenly and drastically to the west in the late 110s, as expansion into northern Mesopotamia, Armenia, Anatolia, and Syria became the focus of the remainder of his reign.[26] Mithridates increasingly favoured western nobles and lands to the mounting frustration of the eastern families, who no longer enjoyed their privileged status.[27] Although some scholars do not recognise this conflict, it is possible that Mithridates' prolonged western emphasis and the relegation of the eastern families finally resulted in rebellion and civil war in 94/93.[28] This conflict would mean that for the first time in Parthian history, a faction of the Parthian aristocracy chose to support a usurper to the throne who would champion their interests better than the king, and they chose perhaps the only candidate worthy enough to challenge Mithridates II, an Arsacid prince named Sinatruces (93–70/69).

Sinatruces had a strong claim to the throne. He was the son of a former king, Mithridates I, the brother of a former king, Phraates II, and the cousin of Mithridates II.[29] When his brother died in battle fighting against rebellious Central Asian mercenaries in 127, it is possible Sinatruces was captured and spent much of his young adulthood as a captive amongst the Saka (Just. 42.2.1).[30] After his return to Parthia, likely following the victories of Mithridates II along the eastern fronter in the early 110s, Sinatruces maintained a cooperative relationship with the eastern families and the Central Asian tribes for decades.[31] Once Mithridates' marginalisation of the eastern families had alienated them to the point of rebellion, Sinatruces appears to have emerged as their champion.

In this reconsideration of the reconstruction of the nebulous 'Dark Age' period, Sinatruces, with the support of the eastern families and his Central Asian allies, quickly seized control of the eastern territories, overran Media,

25 For the interaction of the Chinese envoy, Zhang Qian, with Mithridates' army in the eastern territories of the Parthian Empire, see Mair (2014) 158; Overtoom (2020a) 256.

26 Overtoom (2020a) 255–76.

27 Sachs and Hunger (1996) no. -119A2 + B1, no. -119C, no. -111B, no. -111C, no. -110; no. -109A, no. -99B, no. -95A, no. -95C, no. -94A; Del Monte (1997) 147–48; Olbrycht (2010b) 148–49, 152–53; Shayegan (2011) 224; Dąbrowa (2013); Finkel and van der Spek (2020) BCHP 20; Overtoom (2020a) 258–63; Overtoom (2023).

28 Sachs and Hunger (1996) no. -93A; Assar (2011) 121; Overtoom (2016c). Compare Simonetta (2001) 86; Curtis (2012) 69; Shayegan (2011) 235; Dąbrowa (2011b) 169; Olbrycht (2016b) 23–24.

29 Overtoom (2020a) xxxiii.

30 Assar (2006a) 109–12; Assar (2011) 119.

31 Assar (2011) 121; Overtoom (2019b); Olbrycht (2021a).

and even occupied Susiana in southern Iran in 93/92.[32] With the support of the western families, Mithridates II marched against Sinatruces but likely died in battle in 91.[33] With Sinatruces threatening Babylonia, the western families elected to support Mithridates' son, Gotarzes I (91–87), against Sinatruces, and the civil war continued.[34] Gotarzes took his father's title of 'King of Kings' to distinguish his authority against the usurper.[35] Although the evidence is fragmentary and mostly circumstantial, it appears Gotarzes and Sinatruces fought a back-and-forth war over Babylonia and Susiana from 91–87 until Gotarzes finally defeated Sinatruces in battle, regained the eastern lands of the empire, and forced Sinatruces into exile on the Central Asian steppe.[36] Yet Gotarzes' success was short-lived because he suddenly died after expelling Sinatruces from the empire. Further, it appears that the succession was poorly planned, with tension emerging between two candidates, Mithridates and Orodes.[37] These men either were related as uncle and nephew or as sons of Gotarzes, but in any case, Mithridates took this opportunity to seize the throne.[38] Mithridates III (87–80) used the specific and rare epithet of 'Autocrat', which suggests that he was not Gotarzes' chosen successor.[39] Thus, another usurpation further undermined the stability of the empire.

Mithridates III's reign was chaotic and violent. His generals defeated the Seleucid king Demetrius III and established Philip I as a vassal ruler in Antioch.[40] Yet he lost several western lands to the rising imperial ambitions of Tigranes II of Armenia in the middle and late 80s.[41] The weakness of Mithridates' position allowed Orodes I (80–75) to seize much of Babylonia with the help of a faction within the western families in 80, and the two men appear to have fought over Babylonia for several months before Mithridates fled to the eastern territories and died.[42] Yet at this time Sinatruces likely came out of exile to renew his claim to the throne with further support from his Central Asian allies (Lucian, *Macr.* 15).[43]

32 Assar (2005a) 53; Assar (2006a) 145–49; Assar (2006b) 55–62; Assar (2011) 121; Valverde (2017).
33 Assar (2006a) 147–49; Assar (2006b) 62; Assar (2011) 121; Overtoom (2020a) 275.
34 Sachs and Hunger (1996) no. -90; Assar (2006b) 62–66; Shayegan (2011) 196–97, 225–26.
35 Assar (2006b) 65; Assar (2011) 122.
36 Sachs and Hunger (1996) no. -87C; Assar (2006b) 60–61, 66–68; Assar (2011) 122.
37 Assar (2011) 122–23; Shayegan (2011) 226.
38 Assar (2020) 84–90.
39 Simonetta and Sellwood (1978) 108; Assar (2006b) 70–71.
40 Overtoom (2020a) 270–71.
41 Strabo 11.14.15; Plut. *Luc.* 14.5, 26.1–4; Eutr. 6.8; Oros. 6.4.9; Isid. 6; Cass. Dio 36.14.2. See Schmitt (2011b); Overtoom (2016c).
42 Sachs and Hunger (1996) no. -79; Assar (2006b) 75–77; Assar (2011) 123.
43 Assar (2006b) 71–72, 77, 84–85; Valverde (2017); Olbrycht (2021a).

Orodes I's reign was equally chaotic and violent. Armenia had annexed northern Mesopotamia; the king of Media Atropatene had abandoned the Parthians and married into Tigranes II's family; Elymais in southwestern Iran rebelled; and civil war continued to destabilise the eastern territories (Cass. Dio 36.14.2).[44] In 78/77 it is possible a new usurper, simply known as Arsaces XVI (78/77–62/61) and who may have been a young son of Mithridates II, seized Susa in rebellion against Orodes.[45] It appears that Arsaces even formed an alliance against Orodes with the king of Elymais, Kamnaskires III, in 76.[46] This alliance proved successful as Arsaces quickly occupied most of the eastern territories and then appears to have defeated and killed Orodes in Babylonia in early 75.[47] Arsaces' coinage from Ecbatana, Rhagae, Nisa, and Margiane (S30.14–17, 20, 23, 26–27) illustrates that, despite the harassment of Sinatruces' raids from the Central Asian steppe, he maintained control over much of the Iranian plateau throughout the middle and late 70s.[48] Yet with Sinatruces' death in 69/68, his son, Phraates III (69/68–58/57), took up his father's cause, campaigned through northern Iran, struck at Babylonia, and ousted Arsaces from the western territories in 67/66 (Cass. Dio 36.45.3).[49]

As the new ruler of the western territories, Phraates III acted aggressively against a now vulnerable Armenia in 66–64, reclaiming northern Mesopotamia and Media Atropatene.[50] The Romans had defeated Armenia in 66, but they were uninterested in conflict with the Parthians.[51] Meanwhile, Phraates was satisfied with his gains in northern Mesopotamia because he did not want conflict with Rome and he had internal disputes in Media Atropatene, Elymais, and the eastern territories to overcome.[52]

Although the campaign was difficult, from 64 to 61 Phraates III finally brought an end to the ongoing Parthian civil war. In 63 there was a major calamity in Babylon likely connected to the civil war.[53] Moreover, it is possible

44 Sachs and Hunger (1996) no. -77B; Del Monte (1997) 180; Hansman (1998); Assar (2006b) 77, 79–80, 84; Assar (2011) 123; Shayegan (2011) 325; Dąbrowa (2014) 64.

45 Assar (2006b) 81–84, 87; Assar (2011) 124.

46 Assar (2006b) 79, 82; Dąbrowa (2014) 64.

47 Sachs and Hunger (1996) no. -75; Assar (2006b) 80–81.

48 Sachs and Hunger (1996) no. -72; Assar (2006b) 80–85.

49 McEwan (1986) 93; Del Monte (1997) 256; Assar (2006b) 85–91; Assar (2011) 124–25.

50 Cass. Dio 36.45.3–5, 37.5.2–7.4; Plut. *Pomp.* 33.6–8, 36.1–2; App. *Mith.* 16.106; Strabo 11.13.2. See Assar (2011) 125; Shayegan (2011) 326; Overtoom (2016c).

51 Plut. *Pomp.* 33.2–5; App. *Mith.* 15.104–105; Cass. Dio 36.52.1–53.2; Vell. Pat. 2.37.4–5; Strabo 11.14.10; Cic. *Sest.* 27.58–59.

52 Schippmann (1987b); Assar (2006b) 93; Assar (2011) 125.

53 McEwan (1986) 93; Sachs and Hunger (1988) 34; Sachs and Hunger (1996) no. -62; Del Monte (1997) 182, 257; Assar (2006b) 91.

Kamnaskires IV of Elymais seized Susa in 63 and held the city for two years.[54] Yet Arsaces XVI's coinage ends in 62/61, marking Phraates' final triumph along the eastern frontier.[55] Phraates then returned west, at which point he either defeated Kamnaskires in battle or forced him to submit. Seemingly, Kamnaskires remained king of Elymais after the war, and the Elymeans retained their vassal status after the rebellion but lost control of Susa.[56] A Babylonian colophon from 61/60 makes no distinction between rival Parthian royal candidates, confirming the end of the civil war and Phraates' victory.[57]

3 Patricide and Renewed War with Armenia

For the first time since the reign of Mithridates II, one man ruled the Parthian Empire uncontested. Phraates III's victories were substantial and important, re-establishing Parthia as the dominant power throughout the Middle East at least temporarily, and he celebrated his triumph by adding the epithet 'God' to his coinage (S.39.1). Yet the period of peace that followed did not satisfy his ambitious sons, Mithridates and Orodes, who murdered their father in winter 58/57 (Cass. Dio 39.56.2).[58] Never had an Arsacid king been assassinated, demonstrating further the continued deterioration of dynastic stability and family cohesion in the first century. It is unclear if the influential aristocratic families had encouraged the actions of the brothers; however, with Phraates' death tensions between the western and eastern families re-emerged.

Mithridates, now known as Mithridates IV (58/57–55), initially gained the throne (Just. 42.4.1–4; App. *Syr.* 8.51).[59] Although he favoured the western territories (particularly Mesopotamia and Media), early in his reign he issued

54 Assar (2006b) 93–95; Shayegan (2011) 325.

55 Assar (2006b) 95.

56 Hansman (1998); Dąbrowa (2014) 64–65.

57 George (1992) 71, Tablet 1, pl. 4; Assar (2006b) 93; Assar (2011) 125–26.

58 Bivar (1983a) 48–56; Karras-Klapproth (1988) 104–109; Wolski (1993) 127–40; Olbrycht (1998) 115–17; Lerouge-Cohen (2007) 63–98; Assar (2011) 125–26; Olbrycht (2021a).

59 Sachs (1955) 177, LBAT 1184; Assar (2006b) 95–96; Assar (2011) 126; Valverde (2017). Some argue Justin confuses this Mithridates with Mithridates III (ca. 87–80) or Mithridates II (ca. 121–91); see Manandyan (1940) 162 n. 1; Marshall (1976) 142; Sherwin-White (1984) 279 n. 19. Yet Justin twice states that Mithridates and Orodes were brothers, and Appian confirms this fraternal relationship (*Syr.* 8.51). Mithridates II did not have a brother named Orodes, and Mithridates III's relationship with Orodes I is unclear; see Assar (2006b) 75, 96 n. 200; Assar (2011) 120–28. Moreover, Justin emphasises that Orodes, after killing his brother, fought Crassus in 54–53. It is impossible that Justin confused Orodes II with the obscure Orodes I (ca. 80–75); see Assar (2011) 123–24.

coinage throughout the entirety of the Iranian plateau (S40.1–20, S41.2–23), indicating that he ruled over the eastern territories with authority and unopposed, at least initially.[60] As the new king, Mithridates pursued a war against Armenia to help legitimise his otherwise tenuous position as ruler. Ironically, his father had pursued a similar policy in the middle 60s.

In 64 Phraates III and Tigranes II had agreed to end a border dispute along the northern Tigris, where the Parthians regained northern Mesopotamia, but the Armenians retained control of Gordyene (Cass. Dio 37.6–7; App. *Mith.* 16.106; Plut. *Pomp.* 39.3). Years of peace between Parthia and Armenia followed; however, Tigranes died of illness at the advanced age of 85 in ca. early 56, leaving the throne to his last remaining son, Artavasdes II.[61] Although Mithridates IV did not pursue a conflict against Tigranes, possibly renewing his father's peaceful relationship with Armenia at least initially, the geopolitical situation suddenly shifted in Parthia's favour with Tigranes' death.[62] Tigranes had been a great rival and menace for the Parthians since the 80s, but now the untested Artavasdes sat on the Armenian throne.[63] From the perspective of Mithridates, Armenia was vulnerable, and therefore he renewed aggression in 56 (Just. 42.4.1).

Despite Justin's statement that Mithridates IV concluded a war against Armenia, there is little scholarly consensus on the matter. Although Mommsen exaggerated the conflict as 'a declaration of war against Rome', he accepted Justin's account.[64] Rawlinson also accepted Justin's account and connected the war to Orodes' brother.[65] Debevoise and Sherwin-White ignored the issue completely.[66] Manandyan's efforts to discredit Justin's account assumed erroneously that it referred to Mithridates II in 42.4.1.[67] Moreover, Assar recently has tried to validate the passage by assigning the war to Mithridates III in the 80s; however, it is highly unlikely that Justin here refers to the obscure Mithridates III instead of the more recognisable Mithridates IV, brother of Orodes II.[68]

60 Assar (2006b) 96 n. 200; Assar (2011) 126–27.

61 Cass. Dio 37.7.1–4; App. *Mith.* 16.106; Lucian *Macr.* 15; Cic. *Sest.* 27.59. When one considers the actions of the Parthians, Tigranes' death likely occurred in early 56, not in 55; see Sherwin-White (1984) 224–25, n. 102; Garsoian (2005); Shayegan (2011) 326.

62 Cicero states that the Parthians remained peaceful in late 57, further supporting that Mithridates did not immediately attack Armenia after becoming king (*De Dom.* 21.55, 23.60, 47.124).

63 Schmitt (2011c); Overtoom (2016c).

64 Mommsen (1856) v 151.

65 Rawlinson (1876) 82–83.

66 Debevoise (1938) 76; Sherwin-White (1984) 272, 279 n. 19.

67 Manandyan (1940) 162 n. 1.

68 Assar (2006b) 75; Assar (2011) 122–23; Assar (2020).

Although the Romans could have considered Parthian aggression against Armenia a cause for war since Pompey had accepted Armenia's submission in 66, they seemingly did not. Gabinius, who was the governor of Syria and a close supporter of Pompey, appears to have shared Pompey's detached and reserved policy toward the ongoing rivalry of the Parthians and Armenians. The evolving political crisis at Rome, which culminated in the election of Pompey and Crassus as co-consuls in 55, took precedence.[69] It is notable that Gabinius did not interfere in Armenia, and he did not use this incident as an excuse for war. The Romans were not looking simply for any excuse to bait Parthia into open conflict, and more importantly, the Parthians still did not associate a war against Armenia with a declaration of war against Rome. Even at this late stage, both powers remained distant and isolated. Ultimately, Mithridates IV wanted to return to the aggressive western policies of Mithridates II and his father.[70] A quick war against a vulnerable Armenia in 56 was an opportunity he eagerly seized.[71]

The objective of the war likely was the recapture of Gordyene, which was a prize Mithridates II had secured, and one Phraates III had failed to regain (Strabo 11.14.15; Cass. Dio 37.7.1–4; App. *Mith.* 16.106).[72] The conquest of Gordyene was militarily and politically important to the Parthians since control of the region protected the important but exposed lands of northern Mesopotamia. Moreover, a successful military action against Armenia would have been attractive to the vulnerable Mithridates IV. He compensated for his inexperience and scandalous rise to power in part by quickly adopting numerous bold titles, like 'Great King', 'Just', 'Illustrious', 'Beneficent', 'Admirer of Greeks', and 'Well-Born of a God' (S40.1–20). Yet he later adopted the even bolder titles 'King of Kings' and 'Founder' for stronger propagandistic value (S41.2–23, S44.1), likely to celebrate his victory in Armenia.[73] He wanted to distance himself from the murder of his father by embracing and surpassing his father's legacy, while positioning himself as a legitimate heir to Mithridates II's much greater legacy. The war against Armenia, in part, allowed him to accomplish these goals and forge an imperial identity of his own.

In 56 the Parthians put pressure on the southern and eastern frontiers of Armenia, forcing Artavasdes II to make concessions to Mithridates IV in

69 Plut. *Cat. Min.* 42.1; Plut. *Crass.* 15.4; App. *B Civ.* 2.18; Cass. Dio 39.31.1–2.

70 Overtoom (2019b).

71 Overtoom (2021b).

72 Marciak (2017); Overtoom (2020a) 246–76.

73 Sellwood (1980) 122, 131; Sellwood (1983) 290; Chaumont (1986); Assar (2006b) 96–97; Assar (2011) 126–27; Callataÿ and Lorber (2011) 426.

exchange for a new peace treaty (Just. 42.4.1; Cass. Dio 39.56.2). Like the concessions made by Tigranes II to Phraates III nearly a decade prior, Artavasdes' decision was a compromise to preserve himself and his vulnerable kingdom. He was in no position to resist the full force of the Parthians without Roman support, and with Rome on the edge of political chaos, no Roman support came to Armenia in 56.[74] Rome was a nonfactor in this conflict and its resolution. Meanwhile, the Parthians' attack was sudden, but the brevity of the conflict indicates that Mithridates' goals were limited. Both sides proved agreeable to negotiations in late 56 to avoid a protracted conflict. Ultimately, Artavasdes lost the brief war but retained his most important lands in Greater Armenia, and Mithridates expanded his western frontier at minimal cost and gained the quick glory he sought.

4 The War of the Brothers

Although Mithridates IV concluded a favourable war against Armenia, he remained vulnerable as king. His western focus afforded him support in Media and Mesopotamia; however, his harsh policies toward the Parthian aristocracy, especially toward the eastern families, created hostility (Cass. Dio 39.56.2; Just. 42.4.3–4; Plut. *Crass.* 21.7). Moreover, Mithridates had not addressed the ambitions of his younger brother.

Orodes had helped kill their father to usurp the throne, but he had been passed over in favour of his older brother in the succession. It does not appear that Orodes held any office of significance in his brother's regime, and his lack of prospects encouraged him to scheme against Mithridates. He did so, in part, by marrying into eastern families and making alliances with Central Asian tribes.[75]

One of the only aristocratic families within the Parthian Empire mentioned in the surviving Greco-Roman literature by name also happened to be one of the strongest, oldest, and most influential. The Suren family controlled immense resources, commanded thousands of soldiers, and held the privilege of crowning the Arsacid kings at their coronations (Plut. *Crass.* 21.6–7). Although the extent of their vast lands is unclear, there are important clues available as to their location. The Suren's privileged status indicates that they were an important family in eastern Parthia when Arsaces I seized the region

74 Artavasdes submitted to Parthia again in 53. Plut. *Crass.* 21.5, 22.2, 33.1–2.

75 His sons were the product of these eastern marriage alliances. Olbrycht (2015); Olbrycht (2016b); Olbrycht (2021a); Olbrycht (2021b).

from the rebellious Seleucid satrap, Andragoras, in the early 230s.[76] They must have been fundamental supporters of the early Arsacids in their various efforts to expand the early Parthian state throughout the Iranian plateau, and they likely were rewarded generously for that support with expanded lands.[77] For example, early Islamic sources connect the family to Sakastan (Sistan) in the eastern Iranian plateau, although the connection of the Suren to these lands is difficult to confirm.[78] In the late 120s and early 110s, the Suren likely aided Mithridates II in his defeat of the Central Asian invaders, helped settle these conquered peoples into their lands, and eventually intermarried with the re-settled invaders.[79] By the 50s, their power and influence was undiminished. The strength and seniority of the Suren make it clear that they remained the leading family in the eastern territories.

The Suren were positioned to play kingmaker like no other aristocratic family in the empire, and when the focus of the empire shifted to the west under Mithridates II, threatening their primacy, they likely spearheaded the rebellion of Sinatruces and facilitated the dynastic and geopolitical chaos that followed. The Suren must have agreed with the other aristocratic families to support Mithridates IV's uncontested ascension to the throne in 58; however, it appears that Mithridates overlooked the Suren for prominent positions within his regime, favouring statesmen from western families such as Orsanes (Joseph. *AJ* 14.103; *BJ* 1.175).[80] Mithridates' further neglect and abuse of the eastern families encouraged the Suren to search for a new, more accommodating king.[81] Late in 56 the head of the Suren family, a man known only by his family name, Surena, threw his support behind Orodes II to challenge his brother for the throne.[82] The other eastern families and allied Central Asian tribes joined Orodes' cause, and Mithridates quickly lost control of the eastern territories.[83]

76 Overtoom (2016a); Overtoom (2020a) 65–93; Overtoom (2021).

77 Balakhvantsev (2017); Overtoom (2020a); Ellerbrock (2021); Olbrycht (2021c).

78 Olbrycht (1998) 117; Gazerani (2015) 20; Olbrycht (2021a).

79 Bivar (2007) 28; Overtoom (2019b); Overtoom (2019c); Overtoom (2020a) 246–76; Olbrycht (2021a).

80 Karras-Klapproth (1988) 111–12; Olbrycht (2021a).

81 Cass. Dio 39.56.2; App. *Syr.* 8.51; Just. 42.4.1–4; Plut. *Crass.* 21.7.

82 Karras-Klapproth (1988) 165–71; Shahbazi (1990); Bivar (2007); Assar (2006b) 96–97; Assar (2011) 124–27; Suren-Pahlav (2019).

83 Inscriptions on Parthian coinage minted in the east at this moment became irregular and noncommittal to either faction (S41.6, S41.7, S41.17). Assar (2011) 127. Mithridates perhaps retained the support of the Saka, whom Orodes' allies defeated in the east early in the conflict; see Olbrycht (2021a).

As he established his authority throughout the east, Orodes II began to put pressure on Mithridates IV in Media (Cass. Dio 39.56.2). Mithridates clearly retained some support amongst the western families but not enough to counter the now considerable resources of Orodes.[84] He needed to acquire more resources, especially soldiers; however, Orodes blocked access to mercenaries from the Central Asian steppe, Artavasdes II and the hostile tribes of the Caucasus blocked access to mercenaries from the Pontic-Caspian steppe, and the Romans controlled access to the resources to the west. Mithridates could not negotiate with his rivals, Orodes and Artavasdes; however, negotiation with the Romans was possible. Mithridates abandoned Media, began consolidating what forces he had in Babylonia, and journeyed to Syria to solicit the aid of Gabinius in winter 56/55.[85] Mithridates' decision to seek Roman support in a Parthian conflict was unprecedented and forever changed Parthian and Roman relations.[86]

Gabinius, who had served as consul in 58, successfully transferred his proconsular governorship of Cilicia to Syria to stabilise the region, receiving a large military force and grant of money, control of operations against the various Arab tribes threatening Syria, and a command of at least three years.[87] With Mithridates IV's arrival, the two statesmen met to discuss the deteriorating situation to the east. Mithridates would have stressed the illegitimacy of his brother, the plans of his brother to attack Mesopotamia, and the threat his brother potentially posed to Roman Syria if he succeeded. At least initially, Gabinius became convinced that his intervention in Mesopotamia was merited and necessary, accepting Mithridates' friendship and preparing an abortive expedition.[88] Despite his lack of follow-through, Gabinius' decision effectively declared war on Orodes II, who now reigned over most of the Parthian Empire.[89]

Although Gabinius' initial objective likely was to help install Mithridates IV as a pro-Roman candidate on the Parthian throne, as Gabinius crossed into northern Mesopotamia in early 55, Ptolemy XII arrived in Syria with letters from Pompey, who was Gabinius' political ally and an active consul, instructing

84 Joseph. *AJ* 14.103; Joseph. *BJ* 1.175; Just. 42.4.1–4; Plut. *Crass.* 21.7.

85 Cass. Dio 39.56.2; App. *Syr.* 8.51; Joseph. *AJ* 14.103; Joseph. *BJ* 1.175; Moses 2.16.

86 Overtoom (2021b).

87 Cic. *Sest.* 10.24, 25.55, Cic. *Dom.* 9.23–4, 21.55, 23.60, Cic. *Pis.* 21.49, 23.55, 89, Cic. *Prov. cons.* 1.1–2.3, 4.9, 7.17, Cic. *Red. pop.* 11; Plut. *Cic.* 30.1; Aur. Vict. 81.4.

88 Cass. Dio 39.56.1–3; App. *Syr.* 8.51; Joseph. *AJ* 14.98; Joseph. *BJ* 1.175; *V Maccabees* 40.17–18; Moses 2.16; Strabo 12.3.34, 17.1.11.

89 Overtoom (2021b).

Gabinius to help restore Ptolemy to his throne in Egypt instead.[90] Gabinius' change of focus frustrated Mithridates, who soon abandoned hope of immediate Roman military aid and returned to his remaining allies in Mesopotamia (Joseph. *AJ* 14.103, *BJ* 1.175). From the perspective of the Romans, their involvement in the Parthian conflict remained open-ended, freeing the triumvirs and the senate to assign the ongoing Parthian conflict to Crassus.[91] Yet the perception of the Parthians would have been quite different. For Mithridates, the gamble of seeking aid from the Romans had been necessary but ultimately unsuccessful. He needed to return to Mesopotamia as quickly as possible to rally what support remained there for his cause. Meanwhile, for Orodes II, Mithridates' desperate actions further emphasised the need to eliminate his brother as a threat to the empire, and Gabinius' interference in a Parthian conflict justified retaliation against Roman aggression.

Mithridates IV returned to Mesopotamia by summer 55, where he took command of a small army of supporters based in Babylonia (Just. 42.4.2; Plut. *Crass.* 21.7). From Seleucia he began minting tetradrachms that emphasised his right to rule (S41.1). On these coins Mithridates stressed only one title, 'Philhellene', illustrating his desire to acquire and maintain the support of the large Greek population of the region.

It appears that Mithridates IV's sudden return and the loyal support he mustered in Babylonia surprised Orodes II. Plutarch states that 'Orodes was driven out of Parthia' before being restored by his head general, Surena (Plut. *Crass.* 21.7). Although Plutarch exaggerates Mithridates' success in driving out his brother, this passage indicates that, during Mithridates' absence in Syria, Orodes likely advanced through Media into Babylonia with a modest force to put pressure on Mithridates' remaining western supporters. Although he was unable to take Babylonia by force without more soldiers, Orodes could have hoped that his presence would encourage Mithridates' generals in the region to surrender. When this did not materialise, he likely began negotiations with these men to secure the region. Yet Mithridates' sudden return in summer 55 alongside his most fervent supporter, Orsanes, solidified the support of the generals in Babylonia, allowing Mithridates to strike abruptly at a now vulnerable

90 Plut. *Cat. Min.* 42.1, Plut. *Crass.* 15.4; App. *B Civ.* 2.18; Cass. Dio 39.31.1–2, 56.2–3; App. *Syr.* 8.51; Joseph. *AJ* 14.98; Joseph. *BJ* 1.175. See Sherwin-White (1984) 273–74; Williams (1985); Siani-Davies (1997); Seager (2002) 112–16.

91 Cic. *Att.* 4.13.2, Cic. *Fam.* 1.9.20, Cic. *Div.* 1.29–30; Caes. *BGall.* 6.1.2; Vell. Pat. 2.46.3; Luc. 3.126–27; Plin. *HN* 15.83; Plut. *Cic.* 26.1, Plut. *Crass.* 16.3–6, Plut. *Pomp.* 52.5; Flor. 1.46.1–3; App. *B Civ.* 2.18; Cass. Dio 39.39.1–7; Strabo 12.3.34. See Arnaud (1998); Lerouge-Cohen (2007) 70–75.

Orodes (Joseph. *AJ* 14.103, *BJ* 1.175). Orodes found himself outnumbered and outmanoeuvred, retreating into the eastern territories of the empire.

Yet Mithridates IV's sudden success was short-lived. He was unable to pursue and defeat his brother, who joined forces with the main army under the command of Surena.[92] It is likely that this combined force marched west from Parthia proper through Media and then south through the Zagros Mountains into Babylonia along the main northern route (Isid. 1–19).[93] Mithridates did not have the resources to confront this large army in the field, so he withdrew into Babylonia's fortified cities. The mobile Parthian army was not well-suited for offensive or defensive siege warfare (Just. 41.2.7).[94] Yet Mithridates had far fewer cavalry than his brother, so siege warfare was to his advantage. Additionally, within these important urban centres Mithridates could rely on the favour of their friendly populations and their abundant resources.[95] Mithridates' decision forced Orodes into a difficult series of sieges.

Mithridates IV left part of his force in Seleucia, presumably alongside the Greek garrison, and he fortified himself in Babylon (Plut. *Crass.* 21.7; Just. 42.4.2). Meanwhile, it appears Orodes II and Surena split their forces, with Surena putting pressure on Seleucia along the Tigris and Orodes marching to Babylon on the Euphrates. Although it is not mentioned by name, it seems likely that Mithridates also fortified Ctesiphon across the river from Seleucia at this time.[96] Thus, Surena took on the more difficult task of reclaiming the well-defended northern cities in Babylonia as the more accomplished commander, and Orodes followed his brother southward and settled into a siege at Babylon.

Surena decided to avoid a protracted siege at Seleucia, preferring instead to take the walls by force (Plut. *Crass.* 21.7). At the head of his soldiers, Surena assaulted the battlements. He apparently was the first to mount the walls, and he fought off the defenders in a valiant display of courage and skill. His brave actions galvanised his troops, and Seleucia quickly fell.

To the south, Babylon did not fall to an assault, but rather the siege carried on for months (Just. 42.4.2). Eventually famine began to take hold in the city, encouraging its surrender. It is unclear if Surena reinforced Orodes II and took part in the siege of Babylon. It is possible he spent this time stabilising Seleucia and reducing Ctesiphon. Yet by late 55 Mithridates IV realised he could no

92 Surena had been busy stabilising the eastern territories; see Olbrycht (2021a).

93 Kawami (2013) 752.

94 Overtoom (2017b).

95 Overtoom (2023).

96 Kröger (2011); Iossif (2013); Lendering (2019).

longer resist his brother, and instead of taking his own life, Mithridates chose to rely on the mercy of his brother by surrendering (Just. 42.4.3).

Unfortunately for Mithridates IV, his brother was not a merciful man. Orodes II considered Mithridates untrustworthy and dangerous, and he did not desire to reconcile. He had Mithridates brought before him as a prisoner, at which point he had him executed (Just. 42.4.4). This fratricide brought a violent end to the war of the brothers in late 55. Orodes even tried to erase his brother's memory throughout the region by overstriking Mithridates' late coinage (S41.1).[97] The reverse of these overstruck coins breaks with the traditional imagery of the seated Parthian archer, instead displaying Orodes enthroned accepting the submission of Tyche and helping her rise (S45.1–8). The inscription hails him as 'King of Kings', 'Beneficent', 'Just', 'Illustrious', and 'Admirer of Greeks', which became a new standard on Parthian coinage.[98]

Orodes II wanted everyone to know that he was unrivalled, the benefactor of fortune, justified in his actions, glorious in his rule, and a friend of the Greeks, even those who had shown favour to his brother. He wanted to make it clear to his western rivals that the war was over and that he had won. To this point, Orodes also minted victory coinage for wide distribution (S42.1–3), which, again, rather uniquely features Nike crowning Orodes to celebrate a great victory, likely his triumph over Mithridates.[99] In the inscription, Orodes adopted the epithet 'Philopator', which served to distance him from the murder of his father and to foist the stain of patricide squarely on the memory of his dead and disgraced brother as much as possible.

The murder of Phraates III and the civil war between Mithridates IV and Orodes II had major ramifications for the Arsacids and the Parthian aristocracy. Never had family members so close in relation acted so violently towards one another. In fact, the Arsacid dynasty traditionally had a history of brothers working closely together to rule the empire.[100] It is notable that Mithridates did not kill Orodes once he became king, and at the end of the civil war, he elected to surrender himself to Orodes, confident that his brother would spare him because of their bond as siblings (Just. 42.4.3–4). Yet with the assassination of Phraates (58/57), followed by Orodes' execution of his brother (55), the betrayal and murder of even close family members became solidified as an

97 Dressel (1922) 156–77; Sellwood (1971) 111; Sellwood (1980) 123; Assar (2006b) 97.

98 Assar (2011) 127.

99 Dilmaghani (1986) 223; Assar (2011) 128.

100 Notably Phraates I and Mithridates I, Mithridates I and Bagasis, and Mithridates II and Artabanus II (Just. 41.5.9–10, 6.6–7). Note Sachs and Hunger (1996) no. -118A; Del Monte (1997) 55–57, 150; van der Spek (2001) 453–54; Assar (2005a) 41, 48, 52; Assar (2006a) 89, 140; Olbrycht (2010a) 239; Assar (2011) 116; Shayegan (2011) 72–73, 114; Overtoom (2019b); Overtoom (2019c); Overtoom (2020a).

acceptable path to power within the Arsacid dynasty (Just. 42.4.16). In fact, the most notorious offender was Orodes' son, Phraates IV (38/37–3/2), who learned well from his father, murdering Orodes, his twenty-nine brothers (and their families), and one of his own sons.[101]

One of the greatest strengths of the early Parthian state had been the capable leadership and dynastic stability of the Arsacids.[102] Yet the violent fracturing of the Arsacid dynasty during the 'Dark Age' and in its aftermath, even within its nuclear units, severely damaged the Parthians' momentum in expanding their empire beyond the Euphrates.[103] Although they were able to influence Armenia and the surrounding minor kingdoms along their western frontier and they were able to defend their territory against major Roman invasions in the 50s and 30s, their efforts to act aggressively in Syria, Anatolia, and the Levant in this period were disastrous.[104] Meanwhile, the war of the brothers caused 'a lasting split' in the Parthian aristocracy between the western and eastern families that further destabilised and restricted the empire.[105] Finally, it set another dangerous precedent, namely that the Parthians and Romans began looking to one another to interfere in times of internal crisis.[106] Mithridates IV's arrogance, his combative relationship with the Parthian aristocracy, and the successful scheming of his brother had allowed for the renewal of civil war, but the most wide-reaching consequence of the war of the brothers, which sets it most apart from the previous cycles of civil war, is that it facilitated Parthia's first war with Rome.[107]

5 Parthia Defeats Rome

Although Mithridates IV's death in late 55 ended the civil war in Parthia, Gabinius' offer of support to Mithridates in winter 56/55 had made Orodes II an enemy of Rome. Throughout the middle 50s, the conflict between Orodes

101 Just. 42.5.1–2; Trog. *Prol.* 42; Cass. Dio 49.23.2–4; Plut. *Ant.* 37.1. See Olbrycht 2021b.

102 Overtoom (2020a) 7, 30, 161.

103 Overtoom (2020a) 264–76.

104 Cass. Dio 40.28.3–29.3, 48.24.3–26.2, 41.1–4, 49.19.1–21.2; Cic. *Att.* 5.18.1, 20.2–3, 21.2, 6.1.14; Cic. *Fam.* 2.10.2, 3.8.10, 8.10.1–2, 15.1.1–2, 3.1, 4.2, 7; Cic. *Phil.* 11.35; Vell. Pat. 2.46.4; Frontin. *Str.* 1.1.6, 2.2.5, 5.35, 37; Just. 42.4.5, 7–13; Ruf. Fest. 17.4, 18.1–2; Aur. Vict. 83.2; Eutr. 6.18.2, 7.5.1; Oros. 6.13.5; Moses 2.18–20; Joseph. *BJ* 1.248–68, 317; Joseph. *AJ* 14.330–64, 404, 465, 469, 15.323, 20.245–46; Plut. *Crass.* 33.5; Plut. *Ant.* 28.1, 34.2–3; Flor. 2.19.3–7; App. *B Civ.* 5.10; Hor. *Carm.* 3.6.9–12; Livy, *Per.* 127; Ampel. 31.4; Tac. *Hist.* 5.9; Strabo 16.751.

105 Dąbrowa (2013) 57.

106 Schlude (2012); Nabel (2017); Schlude (2020); Sampson (2020); van Wijlick (2021).

107 Overtoom (2021b).

and Rome remained an open issue, and both sides needed to decide upon a solution.[108] For the Romans, Crassus inherited the war in Parthia and pursued a military solution from his new base in Syria in 54–53, temporarily seizing portions of northern Mesopotamia (Plut. *Crass.* 17.2–4; Cass. Dio 40.12.1–13.4).[109] Meanwhile, for the Parthians, in 54–53 Orodes' generals stabilised Babylonia and conducted raids and reconnaissance missions into northern Mesopotamia with part of the army, while Orodes prepared to invade Armenia (Plut. *Crass.* 18.3–5, 21.3–5, 22.2–3; Cass. Dio 40.16.1–3).

The sources universally paint Crassus as an unrealistic, greedy fool who sought personal advancement and wealth above all else.[110] Crassus is an avaricious aggressor, making mistakes at every turn, and the Parthians are the reactive victims of his aggression until they eventually overcome him due to his mismanagement, selfishness, and irreverence for the gods.[111] For decades scholarship has mirrored the biased pro-Roman, anti-Crassus sentiments of the sources.[112] Yet Crassus' expedition was not a surprise to the Romans or Parthians, nor were the Parthians passively awaiting Roman aggression. Roman anger and shame over the disaster of Crassus at Carrhae in 53 encouraged the Romans to ignore the legitimate causes and objectives of the war against Parthia in favour of passing full blame upon one disgraced and dead man.[113] Meanwhile, the Romans' obsession with blaming Crassus for the disaster at Carrhae has obscured the agency of the Parthians in the campaign and their eagerness to pursue a successful war against Rome. By 54 Orodes II had just as much to gain from a widescale war along his western frontier, and he prepared his forces to reclaim hegemony over northern Mesopotamia, Armenia, and Syria.

There is good reason to accept Plutarch and Dio's claims that Crassus and his men expected the expedition into Mesopotamia to be quick and successful

108 Strabo 12.3.34; *HA Clod.* 13.6; Ruf. Fest. 17.1; Eutr. 6.18; Oros. 6.13.1.

109 Arnaud (1998); Sampson (2015) 92–93; Overtoom (2016b); Overtoom (2017a).

110 Cic. *Att.* 4.13.2; Cic. *Fam.* 1.9.20; Cic. *Div.* 1.29–30; Caes. *BGall.* 6.1.2; Vell. Pat. 2.46.3; Luc. 3.126–127; Plin. *HN* 2.147, 15.83; Plut. *Cic.* 26.1; Plut. *Crass.* 16.3–6. 17.4–6, 19.3–6; Plut. *Pomp.* 52.5; Flor. 1.46.1–3; App. *B Civ.* 2.18; Cass. Dio 39.39.1–7, 40.17.3–19.4; *HA Clod.* 13.6; Val. Max. 1.6.11; Joseph. *AJ* 14.105–111; Joseph. *BJ* 1.179; Joseph. *Ap.* 2.82.

111 Plut. *Crass.* 18.1–2; Cass. Dio 40.12.1–16.3; Ruf. Fest. 17.1; Oros. 6.13.1; Livy *Per.* 106. See Arnaud (1998) 31–32; Sampson (2015) 79–80, 83–84; Overtoom (2016b); Overtoom (2017a); Overtoom (2017b).

112 Colledge (1967) 37; Keaveney (1981); Keaveney (1982); Sherwin-White (1984) 279; Sartre (2005) 48–49; Sampson (2015) 94, 98; Goldsworthy (2016) 331.

113 Sampson (2015) 95; Overtoom (2016b); Overtoom (2017a); Overtoom (2017b); Schlude (2020) 60–74.

against ineffective Parthian forces.[114] Few of the Parthians' military endeavours in the past three decades would have convinced the Romans otherwise, since the Parthian army had struggled against the Armenians in the 80s–70s and since a seemingly unending series of civil wars had sapped Parthia's unity and strength for decades.[115] Moreover, the Romans failed to appreciate that the army composition and asymmetrical fighting style of the Parthians was unique in the east because of its nomadic influences.[116] This major oversight proved a critical factor in the Carrhae campaign.

In 54 Crassus gathered supplies, prepared his men, and established a forward base of operations in northern Mesopotamia (Cass. Dio 40.12.1–13.4; Plut. *Crass.* 17.2–6). To do this, Crassus occupied the towns of southwestern Osrhoene between the Euphrates and Balikh rivers with 7,000 infantry and 1,000 cavalry (Plut. *Crass.* 17.4). The Romans even forced the Parthian commander of northern Mesopotamia, Silaces, to abandon his fortress, Ichnae, and retreat eastward to inform Orodes II of Crassus' movements (Cass. Dio 40.12.2). Although the first military engagement between the Parthians and Romans had been minor, the result further encouraged Crassus to pursue the campaign deeper into Mesopotamia against the seemingly weak and vulnerable Parthians (Plut. *Crass.* 20.1). At the close of 54, Crassus had established a strong forward position in northern Mesopotamia while he awaited the arrival of his son Publius in Syria with an elite force of 1,000 Gallic cavalry (Plut. *Crass.* 17.4–6, 19.1–3, 21.7; Just. 42.4.2–4). Crassus had every reason to feel confident in his plans to strike at Babylonia the following year to force the submission of Orodes.

Yet the Parthians had not remained idle since the death of Mithridates IV. After regaining Babylonia, Orodes II's generals collected resources, prepared defences, and stabilised the western territories, for instance, by issuing overstruck coinage with Orodes' image (Plut. *Crass.* 17.4; Cass. Dio 40.13.3).[117] From Babylonia, Silaces regained control of northern Mesopotamia in ca. winter 55/54, while Orodes prepared an invasion of Armenia and perhaps countered disturbances along the eastern frontier (Cass. Dio 40.12.2, 16.2; Plut. *Crass.* 21.5).[118] A Parthian envoy sent to Crassus in early 53 demonstrates immense confidence, forcefulness, and aggression in his negotiations with the Romans. Orodes' message was brief but powerful, threatening 'war without truce and without treaty'

114 Plut. *Crass.* 18.3–4, 20.1, Plut. *Nic. and Crass.* 4.1, Plut. *Luc.* 36.6–7; Cass. Dio 40.12.1.

115 Overtoom (2016c).

116 Overtoom (2017b).

117 Orodes had his brother's tetradrachms (S41.1) overstruck with his image and propaganda (S45.1–8); see Assar (2011) 127.

118 Olbrycht (1998) 114; Olbrycht (2021a).

if Crassus did not admit blame for the war and abandon the campaign (Plut. *Crass.* 18.1). When Crassus insisted that he would only negotiate from a position of victory after capturing Babylonia, the senior Parthian emissary, Vagises, laughed in his face and threatened him with violence.[119] The Parthians were not willing to back down from the Romans. Orodes dispatched his generals with an elite force of cavalry to harass, delay, and if possible, defeat Crassus in northern Mesopotamia, and he marched against Artavasdes II to keep Armenia neutralised.[120]

Crassus' campaign early in 53, the decisive Battle of Carrhae, and the subsequent Roman retreat were a disaster for the Romans thanks to the skill, ingenuity, and determination of the Parthians. Crassus assumed Orodes II would defend the important cities of Babylonia, allowing him the opportunity he needed to take the important cities within the region by siege, where the Roman army had an advantage (Just. 41.2.7; Cass. Dio 40.15.2–6). The recent argument that Orodes wanted to sacrifice Surena and his army at Carrhae is unconvincing.[121] In reality, both forces under Surena and Orodes played a crucial role in winning the campaign. It was quite common for Parthian kings to turn over the defence of Mesopotamia to subordinates, while they campaigned elsewhere.[122]

The Parthians likely mustered 20,000–50,000 soldiers for active combat duty throughout most of their history.[123] With the unification of the empire under Orodes II and the strong support of the Suren and Central Asian allies, it is likely the Parthian military in the middle 50s numbered around 50,000 soldiers.[124] Numbers on this scale were sufficient for splitting the Parthian army into two forces, with Orodes leading the main force against Armenia and Surena leading an elite force of cavalry against Crassus in northern Mesopotamia. By doing so, the Parthians wanted either to force Crassus to withdraw back into Syria by threatening his flank or, more likely, to isolate and trap Crassus between their two aggressive forces in a classic pincer movement, something they had accomplished time and again in their wars against the Seleucids.[125] To

119 Plut. *Crass.* 18.1–2; Flor. 1.46.4–5; Cass. Dio 40.16.1–3; Ruf. Fest. 17.1; Oros. 6.13.2. See Courtieu (2018).

120 Overtoom (2020a) 37–64.

121 Sampson (2015) 112–13, 144.

122 Overtoom (2020a) 180–83, 202, 258.

123 Potter (2010b) 157; Olbrycht (2016c) 295; McLaughlin (2016) 223–24; Overtoom (2020a) 230–31.

124 Olbrycht (2021a); Olbrycht (2021b).

125 Just. 36.1.5, 38.9.2, 10.7–10, 41.4.9–10; Trog. *Prol.* 35–36; Diod. Sic. 33–34/35; Joseph. *AJ* 13. 186, 218–19, 253; *I Maccabees* 14.2–3; *V Maccabees* 21.23–24; App. *Syr.* 11.67–68; Amm.

accomplish this objective, Orodes successfully subdued Armenia, threatening Roman Syria to the west and Crassus's army to the south by way of the bountiful lands of Sophene and Gordyene along the northern Tigris.[126] Meanwhile, Surena moved aggressively into northern Mesopotamia to harass, delay, and, if possible, challenge the Romans with his mobile army on the battlefield.[127]

The swift and total annihilation of Crassus' army in 53 is a testament to the cunning and aggressiveness of the Parthians' unique mode of asymmetric warfare at both strategic and tactical levels.[128] The Parthian cavalry used scouting missions and skirmishes to test Crassus and his army. By implementing feigned retreats, they lulled the Romans into a false sense of superiority and security (Plut. *Crass.* 18.4, 20.1; Cass. Dio 40.12.2). The perceived unwillingness or inability of the Parthians to defend northern Mesopotamia convinced the Romans that Babylonia was vulnerable (Plut. *Crass.* 18.2; Cass. Dio 40.13.3, 16.3). This deception proved crucial when Artavasdes II offered Crassus considerable military support if he marched through Armenia, but Crassus refused, preferring the more direct approach against Babylonia (Plut. *Crass.* 19.1–3). Artavasdes could not abandon his kingdom in the face of Orodes II's invasion, leaving Crassus without thousands of much needed cavalry in Mesopotamia (Cass. Dio 40.12.2, 16.2; Plut. *Crass.* 21.5). Thus, the Parthians' strategy and tactics encouraged the enemy forces to remain divided. With the bulk of the Parthian military, Orodes then invaded and overwhelmed Armenia, knocking Artavasdes out of the conflict (Plut. *Crass.* 22.2–3). This left the Romans isolated, and the Parthians began to harass and assault Roman detachments, damaging Roman morale (Plut. *Crass.* 18.2–4; Cass. Dio 40.16.1).

In these skirmishes, the Parthians learned much about their new enemy. The Romans appeared easily manipulated by speed and manoeuvrability; they relied almost entirely on dense heavy-infantry formations that could do little to counter the Parthian cavalry; and their arms and armour were made of inferior

---

Mar. 23.6.3; Mal. 8.198; Athen. 5.38; Posid. 16 (F.H.G. III.258). See Overtoom (2017b); Overtoom (2020a) 131–245.

126 Marciak (2012a); Marciak (2012b); Marciak (2013); Marciak (2017). Orodes could have used the route through Sophene suggested by Artavasdes II to Crassus, except in reverse, to penetrate Syria or northern Mesopotamia. Plut. *Crass.* 18.2–19.3. Although much diminished after the Romans' sack of the city in 69, Tigranocerta could have served as a useful, centralised supply depot for Orodes. Strabo 11.532, 12.539; App. *Mith.* 67, 86; Quadratus 18; Cic. *Leg. Man.* 23; Memnon 38.6; Plut. *Luc.* 29.3–5; Cass. Dio 36.2.3–4; Ruf. Fest. 15.3; Eutr. 6.9.1.

127 Plut. *Crass.* 23.1–27.8; Plut. *Luc.* 36.5–7; Plut. *Pomp.* 55.1; Cass. Dio 40.21.1–25.5, 49.21.2; Cic. *Div.* 2.22; Ov. *Fast.* 5.583–586, 6.465–468; Vell. Pat. 2.46.4; Val. Max. 1.6.11; Plin. *HN* 6.47; Flor. 1.46.8, 10; App. *B Civ.* 2.18; Just. 42.4.4; Ruf. Fest. 17.2; Eutr. 6.18.1, 7.5.1; Oros. 6.13.3. See Overtoom (2016c) 246–76; Overtoom (2017a).

128 Overtoom (2017b).

metal (Plut. *Crass.* 18.2–4, 20.1–2; Oros. 6.13.2). The Parthians' access to abundant supplies of high-quality mounts, powerful composite bows, and superior steel afforded them a major advantage on the battlefield.[129] Although Orodes II had most of the Parthian army in Armenia, his generals in Mesopotamia commanded elite mobile forces of light and heavy cavalry (Plut. *Crass.* 21.4–6). These forces were highly adaptable, well-supplied, and synchronised. They exploited every Roman weakness as Crassus advanced through Mesopotamia, and the effectiveness of Parthian tactics and weapons increasingly surprised and worried the Romans (Plut. *Crass.* 18.2–4).

With a good sense of their advantages and the limitations of the Roman army, the Parthians next utilised superior military intelligence and espionage to manipulate the Roman army into favourable terrain. Surena, and his subordinate Silaces, picked a perfect battle site near Carrhae. The open plain was ideal for the swift movement of the Parthian cavalry, leaving the Romans isolated and vulnerable, and the surrounding hills allowed for cover to disguise and protect the Parthian horsemen from Roman assaults (Plut. *Crass.* 23.4–7; Cass. Dio 40.21.1–3). The Parthians also used a network of scouts and spies to track and manipulate the Roman army, encouraging Crassus to fight near Carrhae.[130]

The Battle of Carrhae in May 53 was an expertly planned and executed ambush. The Romans never held the initiative or an advantage in the engagement. The Parthian horse archers excelled in hit-and-run engagements with the Roman infantry, while the Parthian heavy cavalry protected the horse archers and exploited weak points in the Roman formation.[131] The Roman army was surrounded and picked apart by the superior mobility of the Parthian cavalry, the exceptional power of Parthian arrows and lances, and the outstanding battleplan of Surena (Plut. *Crass.* 23.1–27.8; Cass. Dio. 40.21.1–25.5).[132] In the aftermath of the battle, the Parthians harassed the Roman retreat, even killing Crassus during a meeting of the two generals that turned violent.[133]

It is unclear if the Parthians actively attempted to capture or kill Crassus at this meeting or if the violence that erupted was an accident (Plut. *Crass.* 31.1–6).

129 Strabo 3.4.15, 11.13.7; Plin. *HN* 34.41; Hor. *Carm.* 3.2; Plut. *Crass.* 18.2–4, 24, 25.7, 27.1–2. See Olbrycht (2001); Sheldon (2010) 36; Ellerbrock and Winkelmann (2012); McLaughlin (2016) 4–5, 154–61; Farrokh *et al.* (2017); Olbrycht (2021a).

130 Plut. *Crass.* 20.1–22.6; Cass. Dio 40.20.1–4; Flor. 1.46.6–7; Ruf. Fest. 17.1.

131 Overtoom (2020a) 37–64.

132 Overtoom (2017b).

133 Plut. *Crass.* 28.1–33.4; Plut. *Pomp.* 53.8–10; Cass. Dio 40.26.1–27.4; Liv. *Per.* 106; Strabo 16.747; Luc. 1.103–111, 8.431–439; Flor. 1.46.9–10; Polyaenus, *Strat.* 7.41.1; Ath. 6.252d–e; Ruf. Fest. 17.2–3; Oros. 6.13.4.

Yet the death of Crassus enhanced the Parthian strategy to protect Mesopotamia and to pursue the war aggressively in Syria, and it served to heighten the international standing of Orodes II. While Crassus remained in command, the Roman army remained a threat that concerned Surena (Cass. Dio 40.26.1). Yet Crassus' death broke the remaining cohesion of the Roman army, allowing the Parthians to kill and capture thousands more soldiers (Plut. *Crass.* 31.6–7; Cass. Dio 40.27.4). The Parthians used the captured Roman soldiers as new muscle to bolster the garrisons along their ever-vulnerable eastern frontier.[134]

Crassus' death also threw the geopolitical situation in Syria into chaos, making it vulnerable to Parthian attacks.[135] Meanwhile, it meant that Roman gains in northern Mesopotamia were untenable, allowing the Parthians to reclaim the region quickly (Cass. Dio 40.28.1). Finally, control of Crassus was valuable to Parthian propaganda. Surena sent Crassus' head and hand to Orodes II in Armenia as a symbol of Parthia's triumph and renewed power, before celebrating his victory at Seleucia (Plut. *Crass.* 32.1–5). In Armenia, the delighted king honoured his triumphant officers at a great banquet that now celebrated not only Parthia's triumph over the Armenians, but also Parthia's triumph over the Romans (Plut. *Crass.* 33.1–4).

6 Conclusion: Patricide, Fratricide, and Betrayal Re-Emerge in the Aftermath of Carrhae

Crassus' failed invasion initiated a drastic change in the relations and policy between Parthia and Rome as a major hegemonic rivalry emerged between the two world powers.[136] Moreover, the failure of Crassus and the success of the Parthians did not tell 'Rome that it could never occupy the lands beyond the Euphrates'; rather, it made further conflict between the two powers inevitable because it made Parthia a legitimate and direct threat to Roman Syria.[137] The meddling of the Romans in the Parthian civil war and the campaign of Crassus encouraged the Parthians to retaliate against Rome along the Parthians' western frontier, while encouraging the Romans to attempt to recover their reputation through further efforts to gain the submission of Parthia.[138] The

134 Plin. *HN* 6.47; Hor. *Carm.* 3.5.5; Solinus 48.3; Vell. Pat. 2.82; Flor. 2.20.4.

135 Plut. *Brut.* 43.7–8, Plut. *Crass.* 19.4; Cass. Dio 40.28.2; Vell. Pat. 2.46.4; Joseph. *BJ* 1.180; Joseph. *AJ* 14.119; App. *B Civ.* 4.59; Ruf. Fest. 17.4; Eutr. 6.18.2; Oros. 6.13.5.

136 Just. 41.1.1, 6–9; Strabo 1.2.1, 11.6.4, 9.2, 16.1.28; Phot. *Bibl.* 58; App. *B Civ.* 2.83, 4.32–33; Vell. Pat. 2.101; Tac. *Ann.* 15.13. See Schlude (2012); Schlude (2020) 60–115; Sampson (2020).

137 Chahin (1987) 243.

138 Sheldon (2010); Sampson (2015); Sampson (2020); Schlude (2020); van Wijlick (2021).

Carrhae campaign was a catastrophe for Crassus and his reputation; however, betrayal and family intrigue quickly re-emerged in Parthia, derailing the Parthians' momentum in the campaign's aftermath.

In early 53, Orodes II had overwhelmed Armenia with an impressive show of force (Plut. *Crass.* 22.2). It is unclear if a decisive battle took place; however, Artavasdes II did not think he could resist the Parthians, even with an army of 30,000 infantry and 16,000 cavalry (Plut. *Crass.* 19.1). Orodes' army must have been considerable, making it likely that he commanded most of the Parthian forces and a large contingent of Central Asian mercenaries secured through war and diplomacy.[139] He accepted Artavasdes' submission and began making new alliances to secure his northern and western frontier, even marrying his eldest son to Artavasdes' sister (Plut. *Crass.* 33.1).[140] His victory in Armenia, which had been the focus of the Parthians' strategy in 54–53, made Crassus' campaign in Mesopotamia untenable even before the battle of Carrhae in May 53. Had the Romans avoided the disaster at Carrhae, Crassus could not have continued toward Babylonia with Orodes and his network of new allies gaining momentum in his rear.

Orodes II's determination to end the conflict with Artavasdes II quickly with an important marriage alliance in early 53 demonstrates his eagerness to move against Crassus in northern Mesopotamia, something the Romans feared during their retreat to Syria (Cass. Dio 40.26.2). Yet the sudden and decisive victory of Surena without the direct help of Orodes and the main Parthian army was unexpected. The arrival of Silaces with the head of Crassus at Orodes' banquet in Armenia was met with joy and celebration. Armenia was subdued, the frontiers were stabilised, and the Roman invasion was destroyed. Thus, the closing question of this chapter is, why did the Parthians not annex the mostly undefended Roman Syria following Carrhae, and the answer yet again lies in betrayal, murder, and court intrigue.

After his victory over Crassus, Surena reoccupied northern Mesopotamia (Cass. Dio 40.28.1). Yet he did not press the advantage into Syria, instead returning to the site of his other great triumph, Seleucia (Plut. *Crass.* 21.7). Although Surena sent the head of Crassus to Orodes II in Armenia, he sent messengers to Babylonia claiming he had captured the general (Plut. *Crass.* 32.1). Surena then celebrated his victory over Rome with a mock Roman triumph, which included the procession of a fake Crassus in humiliation (Plut. *Crass.* 32.2–5).

---

139 Olbrycht (2021a).

140 Orodes made alliances with Armenia, Commagene, Osrhoene, and tribes within Southern Caucasia; see Olbrycht (2021a).

Such a self-aggrandising celebration by a general was unprecedented in Parthian history, and it was an aggressive display of power and wealth. The news of this event angered Orodes II greatly, who had Surena captured and executed in 53/52 (Plut. *Crass.* 33.5). Rather than Orodes acting from jealousy as Plutarch claims, it is more likely Surena's brazen public display of his authority and his deception regarding Crassus necessitated his removal. Orodes felt betrayed and threatened by his head general, and therefore he did not hesitate to betray Surena in turn. As a usurper, Orodes naturally feared usurpation his entire reign.[141] By late 53, Surena was more dangerous than useful to Orodes, so he had him killed.

Although Surena's execution did not create chaos, it destabilised Parthia enough to distract from a full-scale invasion of Syria. Orodes II severed his ties with the Suren family, who declined in favour and position for a generation; however, he retained enough favour amongst the eastern families and his Central Asian allies to avoid another civil war.[142] The fall of Surena created opportunities for others, and Orodes favoured statesmen, like Silaces, Pomaxathres, Ornodapates, and Osaces, to replace him (Plut. *Crass.* 33.3–4; Cass. Dio 40.29.3, 30.2; Oros. 6.13.5). Thus, Orodes further secured his position by removing Surena from power, but he had to keep a watchful eye on the internal developments of his kingdom. He could not afford to campaign in Syria personally, nor could he afford to commit considerable resources to the venture, since the bulk of the Parthian army was needed to stabilise the northern and eastern frontiers.[143] Instead, he commissioned others to attack Syria.

Throughout 52 the Parthians raided Syria, scouting the Roman position, and probing it for weaknesses; however, Cassius had taken up command of the region and organised its defences (Cass. Dio 40.28.1–2). Then in 51, Orodes II ordered a larger invasion and gave command to his young son, Pacorus, but since Pacorus was not yet an adult, Orodes appointed Osaces as his guardian on the campaign and the true commander, perhaps upsetting Ornodapates, who began plotting against Orodes (Cass. Dio 40.28.3, 30.2).[144] Osaces swept through northern Syria and Cilicia with a cavalry force mostly unchecked, but Cassius had fortified Antioch, and the Parthian horsemen were unable to assault the city (Cass. Dio 40.28.3–29.2; Cic. *Att.* 5.18.1, 20.2, 6.1.14; Cic. *Fam.* 3.8.10, 8.10.1–2, 15.1.1–2, 3.1, 4.2). It is possible that soldiers from Armenia were meant to support this invasion, but this support did not materialise in

141 Cass. Dio 40.30.1–3, 49.23.2–3; Just. 42.4.5, 4.14–5.2; Plut. *Crass.* 33.5; Plut. *Ant.* 37.1.
142 Bivar (1983b) 196–97; Olbrycht (2021a).
143 Olbrycht (2021a).
144 Assar (2011) 128.

time (Cic. *Att.* 5.21.2). Cassius ambushed the Parthians near Antioch, and in the defeat Osaces was killed.[145]

With the death of Osaces, the invasion stalled. In 50 the new Parthian commander, the ambitious Ornodapates, and the new Roman commander, Bibulus, worked together to manipulate young Pacorus, whom Ornodapates propped up as a rival to his father for the throne.[146] Yet Ornodapates did not command enough support, and his rebellion accomplished little. Orodes II counteracted the plot quickly, recalling Pacorus from Syria, with whom he reconciled, and presumably he then killed Ornodapates. Yet this new development of betrayal was troubling enough for Orodes to decide to continue to consolidate his position within his kingdom, putting his hopes of gaining Syria aside for a decade.[147]

Unfortunately for Orodes II, this later campaign, although initially successful, ultimately failed to claim Syria for the Parthians as well.[148] Pacorus, who had become the favorite son of Orodes and his leading commander, lost his life during the campaign in 38, and the news devastated the king.[149] In his grief, Orodes gave another of his sons, Phraates, access to power, and as Orodes had done prior, Phraates murdered his father and butchered his brothers.[150] Phraates IV reigned for thirty-five years (38/37–3/2), but he endured usurpations and civil wars before being overthrown and murdered by his son, Phraates V (Joseph. *AJ* 18.39–42).[151]

Phraates IV's remarkable but turbulent reign epitomised the enduring power of the Parthian state and military, despite the sudden, intense instability and violence of the Arsacids in the first century.[152] The Parthians established

145 Cass. Dio 40.29.3; Cic. *Att.* 5.20.3, 21.2; Cic. *Fam.* 2.10.2, 15.4.7; Cic. *Phil.* 11.35; Vell. Pat. 2.46.4; Frontin. *Str.* 2.5.35; Just. 42.4.5; Ruf. Fest. 17.4; Aur. Vict. 83.2; Eutr. 6.18.2; Oros. 6.13.5; Moses 2.18.

146 Cass. Dio 40.30.2; Just. 42.4.5; Cic. *Att.* 6.8.5; Cic. *Fam.* 12.19.2; Caes. *BCiv.* 3.30.3; Livy *Per.* 108; App. *Syr.* 8.51.

147 Assar (2011) 128; Schlude (2020) 75–91; Sampson (2020); van Wijlick (2021).

148 Cass. Dio 48.24.3–26.2, 41.1–4; Just. 42.4.7; Joseph. *BJ* 1.248–68; Joseph. *AJ* 14.330–64, 404, 465, 469, 15.323, 20.245–46; Plut. *Ant.* 28.1; Flor. 2.19.3–5; App. *B Civ.* 5.10; Hor. *Od.* 3.6.9–12; Liv. *Per.* 127; Ampel. 31.4; Ruf. *Fest.* 18.1; Tac. *Hist.* 5.9; Moses 2.19; Strabo 16.751; Frontin. *Str.* 2.5.37.

149 Strabo 16.751; Joseph. *BJ* 1.317; Cass. Dio 49.19.1–21.2; Just. 42.4.7–13; Frontin. *Str.* 1.1.6, 2.2.5; Tac. *Hist.* 5.9; Plut. *Crass.* 33.5; Plut. *Ant.* 34.2–3; Flor. 2.19.5–7; Ruf. Fest. 18.2; Eutr. 7.5.1; Moses 2. 20.

150 Plut. *Ant.* 37.1; Cass. Dio 49.23.2–4; Just. 42.4.14–5.2; Trog. *Prol.* 42. See Assar (2011) 128–29; Olbrycht (2021a).

151 Assar (2011) 130–35; Olbrycht (2021b).

152 Overtoom (2020a); Schlude (2020) 75–115; Sampson (2020); Olbrycht (2021b).

a military reputation that concerned the Romans and made them take note of the power of Parthia.[153] Yet patricide, fratricide, and betrayal remained defining characteristics of the Arsacids for generations. Indeed, the Arsacids' ability to survive their attempted self-destruction, beginning in the 90s and carrying into the first century CE, and their ability to thrive in spite of it for more than three centuries are subjects worthy of future scholarly attention.

Bibliography

Arnaud, P. (1998) 'Le guerres parthiques de Gabinius et de Crassus et la politique occidentale des Parthes Arsacides entre 70 et 53 av. J.-C.'. *Electrum* 2: 13–34.

Assar, G.R.F. (2005a) 'Genealogy and Coinage of the Early Parthian Rulers II: A Revised Stemma'. *Parthica* 7: 29–63.

Assar, G.R.F. (2005b) 'The Genealogy of the Parthian King Sinatruces'. *Journal of the Classical & Medieval Numismatic Society* 6.2: 16–33.

Assar, G.R.F. (2006a) 'A Revised Parthian Chronology of the Period 165–91 BC'. *Electrum* 11: 87–158.

Assar, G.R.F. (2006b) 'A Revised Parthian Chronology of the Period 91–55 BC'. *Parthica, Incontri di culture nel mondo antico* 8: 55–104.

Assar, G.R.F. (2009) 'Some Remarks on the Chronology and Coinage of the Parthian "Dark Age"'. *Electrum* 15: 195–234.

Assar, G.R.F. (2011) 'Iran under the Arsacids, 247 BC–AD 224/227', in Nelson (2011) 113–71.

Assar, G.R.F. 'The Gold Variety of a Silver Drachm of Mithradates III of Parthia (87–80 BC)', in Faghfoury (2020) 63–112.

Atkinson, K. (ed) (Forthcoming) *The Oxford Handbook of Josephus*. Oxford.

Babelon, J. and Lafaurie, J. (eds) (1953–1957) *Actes du congrès international de numismatique*. Vol. 2. Paris.

Bagnall, R., Broderson, K., Champion, C., Erskine, A., and Huebner, S. (eds) (2013) *The Encyclopedia of Ancient History*. Malden, MA.

Balakhvantsev, A. (2017) *Political History of Early Parthia* (*Политическая история ранней Парфии*). Moscow.

Barnett, G. (2017) *Emulating Alexander: How Alexander the Great's Legacy Fuelled Rome's Wars With Persia*. Barnsley.

Bemmann, J. and Schmauder, M. (eds) (2015) *Complexity of Interaction Along the Eurasian Steppe Zone in the First Millennium CE*. Bonn.

153 Just. 41.1.1, 6–9; Strabo 1.2.1, 11.6.4, 9.2, 16.1.28; Photius *Bib.* 58; App. *B Civ.* 2.83, 4.32–33; Vell. Pat. 2.101; Tac. *Ann.* 15.13. See Rezakhani (2013) 770; Overtoom (2016b); Overtoom (2017a); Engels (2017) 55–57.

Benjamin, C. (2007) *The Yuezhi: Origin, Migration, and the Conquest of Northern Baktria*. Turnhout.

Benjamin, C. (ed) (2015) *Cambridge World History*. Vol 4. Cambridge.

Bishop, M. (2018) *Lucius Verus and the Roman Defence of the East*. Barnsley.

Bivar, A. (1983a) 'The Political History of Iran under the Arsacids', in Yarshater (1983) 21–99.

Bivar, A. (1983b) 'The History of Eastern Iran', in Yarshater (1983) 181–217.

Bivar, A. (2007) 'Gondophares and the Indo-Parthians', in Curtis and Stewart (2007) 26–36.

Bivar, A. (2009) 'Kushan Dynasty i: Dynastic History'. *EncIr*. Last accessed 4/2021. <https://iranicaonline.org/articles/kushan-dynasty-i-history>.

Boyce, M. (1983) 'Parthian Writings and Literature', in Yarshater (1983) 1151–65.

Brice, L. and Donahue, J. (ed) (2023) *Brill's Companion to Diet and Logistics in Greek and Roman Warfare*. Leiden.

Burton, P. (ed) (2013) *Culture, Identity and Politics in the Ancient Mediterranean World*. Canberra.

Callataÿ, F. and Lorber, C. (2011) 'The Pattern of Royal Epithets on Hellenistic Coinages', in Iossif, Chankowski, and Lorber (2011) 417–55.

Callieri, P. (2016) 'Sakas: In Afghanistan'. *EncIr*. Last accessed 4/2021. <https://iranicaonline.org/articles/sakas-in-afghanistan>.

Callieri, P. and Chaverdi, A. (2013) 'Media, Khuzestan, and Fars between the End of the Achaemenids and the Rise of the Sasanians', in Potts (2013) 690–717.

Campbell, B. (1993) 'War and Diplomacy: Rome and Parthia, 31 BC–AD 235', in Rich and Shipley (1993) 213–40.

Chahin, M. (1987) *The Kingdom of Armenia*. New York.

Chaumont, M. (1986) 'Armenia and Iran II: The pre-Islamic Period', *EncIr*. Last accessed 4/2020. <http://www.iranicaonline.org/articles/armenia-ii>.

Colledge, M. (1967) *The Parthians*. New York.

Courtieu, G. (2018) 'Vagises' Virtuous Hand: An Unforeseen Note on Plutarch, Life of Crassus 18, 3'. *Iran. Stud.* 51: 633–42.

Curtis, E. (ed) (2012) *Altertum und Gegenwart: Gesammelte Reden und Vorträge*. Innsbruck.

Curtis, J. (ed) (2000) *Mesopotamia and Iran in the Parthian and Sasanian Periods*. London.

Curtis, V. and Stewart, S. (eds) (2007) *The Age of the Parthians*. London.

Curtis, V. (2012) 'Parthian Coins: Kingship and Divine Glory', in Wick and Zehnder (2012) 67–83.

Curtis, V., Pendleton, E., Alram, A., and Daryaee, T. (eds) (2016) *The Parthian and Early Sasanian Empires: Adaptation and Expansion*. Oxford and Philadelphia.

Dąbrowa, E. (1994) 'Dall'autonomia alla dipendenza. Le città greche e gli Arsacidi nella prima metà del I secolo d.C.'. *Mesopotamia* 29: 185–98.

Dąbrowa, E. (2002) '"... *ostentasse Romana arma satis* ..." The Military Factor in Roman-Parthian Relations under Augustus and Tiberius', in Freeman, Bennett, Fiema, and Hoffmann (2002) 275–80.

Dąbrowa, E. (2006) 'Marc Antoine, les Parthes et l'Arménie', in Traina (2006) 343–52.

Dąbrowa, E. (2008/2009) 'The Political Propaganda of the First Arsacids and Its Targets: From Arsaces I to Mithridates II'. *Parthica* 10: 25–31.

Dąbrowa, E. (ed) (2010a) *Hortus Historiae: Studies Dedicated to Prof. Jozef Wolski*. Krakow.

Dąbrowa, E. (2010b) 'The Parthian Kingship', in Lanfranchi and Rollinger (2010) 123–34.

Dąbrowa, E. (2011a) *Studia Graeco-Parthica: Political and Cultural Relations between Greeks and Parthians*. Wiesbaden.

Dąbrowa, E. (2011b) 'The Arsacid Empire', in Daryaee (2011) 187–207.

Dąbrowa, E. (2012) 'The Arsacids and Their State', in Curtis (2012) 21–52.

Dąbrowa, E. (2013) 'The Parthian Aristocracy: Its Social Position and Social Activity'. *Parthica, Incontri di culture nel mondo antico* 15: 53–62.

Dąbrowa, E. (2014) 'A Troublesome Vassal? Elymais and Parthia (141 BC–AD 228)'. *Parthica, Incontri di culture nel mondo antico* 16: 61–68.

Dąbrowa, E. (2017) 'Sacral Kingship in Parthia'. *Studi Ellenistici* 31: 287–96.

Dalziel, N. and MacKenzie, J.M. (eds) (2016) *The Encyclopedia of Empire*. Hoboken.

Dani, A.H. and Harmatta, J. (eds) (1999) *History of Civilizations of Central Asia*. Delhi.

Daryaee, T. (ed) (2011) *The Oxford Handbook of Iranian History*. Oxford.

Daryaee, T. (2015) 'Western and Central Eurasia', in Benjamin (2015) 271–99.

Daryaee, T. (ed) (2017) *King of the Seven Climes*. Irvine.

Debevoise, N. (1938) *A Political History of Parthia*. Chicago.

Del Monte, G. (1997) *Testi dalla Babilonia Ellenistica*. Vol. 1. Pisa.

Dilmaghani, J. (1986) 'Parthian Coins from Mithridates II of Parthia to Orodes II'. *Num. Chron.* 146: 216–24.

Dobbins, K. (1974) 'Mithridates II and His Successors: A Study of the Parthian Crisis'. *Antichthon* 8: 63–79.

Dressel, H. (1922) 'Ein Tetradrachmon des Arsakiden Mithridates III'. *Zeitschrift für Numismatik* 33: 156–77.

Edwell, P. (2013) 'The Euphrates as a Boundary between Rome and Parthia in the Late Republic and Early Empire', in Burton (2013) 191–206.

Ellerbrock, U. (2021) *The Parthians: The Forgotten Empire*. Abingdon.

Ellerbrock, U., and Winkelmann, S. (2012) *Die Parther: Die vergessene Großmacht*. Darmstadt.

Engels, D. (2017) *Benefactors, Kings, Rulers: Studies on the Seleukid Empire between East and West*. Leuven.

Enoki, K., Koshelenko, G., and Haidary, Z. (1999) 'The Yueh-chih and Their Migrations', in Dani and Harmatta (1999) 171–90.

Farrokh, K., Karamian, G., Kubic, A., and Oshterinani, M. (2017) 'An Examination of Parthian and Sasanian Military Helmets (2nd century BC–7th century CE)', in Maksymiuk and Karamian (2017) 121–63.

Feyel, C., Fournier, J., Graslin-Thomé, L., and Kirbihler, F. (eds) (2012) *Communautés locales et pouvoir central dans l'Orient hellénistique et romain*. Nancy.

Finkel, I. and van der Spek, R. (2020) 'BCHP 20 (Euphrates Chronicle)'. *Babylonian Chronographic Texts from the Hellenistic Period*. Last accessed 4/2021. <https://www.livius.org/sources/content/mesopotamian-chronicles-content/bchp-20-euphrates-chronicle/>.

Faghfoury, M. (ed) (2020) *Ancient Iranian Numismatics in Memory of David Sellwood*. Leiden.

Fowler, R. (2013) 'Parthians, Rulers', in Bagnall et al. (2013) 5071–75.

Fratantuono, L. (2019) *Roman Conquests: Mesopotamia and Arabia*. Barnsley.

Freeman, P., Bennett, J., Fiema, Z.T., and Hoffmann, B. (eds) (2002) *Limes XVIII: Proceedings of the XVIIIth International Congress of Roman Frontier Studies, Held in Amman, Jordan (September 2000)*. Vol. 1. Oxford.

Frendo, D. (2003/2007) 'Roman Expansion and the Graeco-Iranian World: Carrhae, Its Explanation and Aftermath in Plutarch'. *Bulletin of the Asia Institute* 17: 71–81.

Frye, R. (2000) 'Parthian and Sasanian History of Iran', in Curtis (2000) 17–22.

Garsoian, N. (2005) 'Tigran II'. *EncIr*. Last accessed 4/2020. <http://www.iranicaonline.org/articles/tigran-ii>.

Garzetti, A. (1944) 'M. Licinio Crasso. VI: La campagna contro i Parti'. *Athenaeum*. NS. 32: 35–61.

Gazerani, S. (2015) *The Sistani Cycle of Epics and Iran's National History: On the Margins of Historiography*. Leiden.

George, A. (1992) *Babylonian Typographical Texts*. Leuven.

Goldsworthy, A. (2016) *Pax Romana: War, Peace and Conquest in the Roman World*. New Haven.

Grainger, J. (2013) *Rome, Parthia, India: The Violent Emergence of a New World Order 150–140 BC*. Barnsley.

Grainger, J. (2014) *The Rise of the Seleukid Empire (323–223 BC): Seleukos I to Seleukos III*. Barnsley.

Grainger, J. (2016) *The Fall of the Seleukid Empire 187–75 BC*. Barnsley.

Grajetzki, W. (2011) *Greeks and Parthians in Mesopotamia and Beyond 331 BC–224 AD*. London.

Graslin-Thomé, L. (2012) 'La Babylonie hellénistique: le point de vue des "dominés"', in Feyel, Fournier, Graslin-Thomé, and Kirbihler (2012) 237–58.

Gregoratti, L. (2012) 'Between Rome and Ctesiphon: The Problem of Ruling Armenia', in Ivanov and Zulumyan (2012) 134–40.

Gregoratti, L. (2013) 'The Journey East of the Great King: East and West in the Parthian Kingdom'. *Parthica, Incontri di culture nel mondo antico* 15: 43–63.

Gregoratti, L. (2015) 'A Tale of Two Great Kings: Artabanus and Vologaeses', in Krasnowolska (2015) 203–10.

Gregoratti, L. (2016a) 'Parthian Empire', in Dalziel and MacKenzie (2016).

Gregoratti, L. (2016b) 'Legendary and Real Wealth in the Arsacid Kingdom', in Santangelo and Bissa (2016) 83–92.

Gregoratti, L. (2017a) 'Sinews of the Other Empire: The Parthian Great King's Rule Over Vassal Kingdoms', in Terigon and Seland (2017) 95–104.

Gregoratti, L. (2017b) 'The Arsacid Empire', in Daryaee (2017) 125–53.

Gruen, E. (2017) 'Josephus' Image of the Parthians', in Wiesehöfer and Müller (2017) 223–40.

Hackl, U., Jacobs, B., and Weber, D. (eds) (2010a) *Quellen zur Geschichte des Partherreiches: Textsammlung mit Übersetzungen und Kommentaren*. Vol. 2. Göttingen.

Hackl, U., Jacobs, B., and Weber, D. (eds) (2010b) *Quellen zur Geschichte des Partherreiches: Textsammlung mit Übersetzungen und Kommentaren*. Vol. 3. Göttingen.

Hansman, J. (1998) 'Elymais'. *EncIr*. Last accessed 04/2020. <http://www.iranicaonline.org/articles/elymais>.

Haubold, J. (2016) 'Hellenism, Cosmopolitanism and the Role of Babylonian Elites in the Seleucid Empire', in Lavan, Payne, and Weiswiler (2016) 89–102.

Hauser, S. (2013) 'The Arsacids (Parthians)', in Potts (2013) 728–50.

Iossif, P., Chankowski, A., and Lorber, C. (eds) (2011) *More than Men, Less than Gods: Studies on Royal Cult and Imperial Worship*. Leuven.

Iossif, P. (2013) 'Seleukeia', in Bagnall *et al.* (2013) 6125–26.

Ivanov, V. and Zulumyan, B. (eds) (2012) *Armenia-Iran: History. Culture. Modern Development Prospects*. Moscow.

Jayaswal, V. (ed) (2012) *Glory of the Kushans*. New Delhi.

Karras-Klapproth, M. (1988) *Prosopographische Studien zur Geschichte des Partherreiches auf der Grundlage antiker literarischer Überlieferung*. Bonn.

Kawami, T. (2013) 'Parthian and Elymaean Rock Reliefs', in Potts (2013) 751–65.

Keall, E. (1967) 'Qal'eh-i Yazdigird, A Sasanian Palace Stronghold in Persian Kurdistan'. *Iran* 5: 99–121.

Keall, E. (1975) 'Parthian Nippur and Vologases' Southern Strategy: A Hypothesis'. *Journal of the American Oriental Society* 95: 620–32.

Keall, E. (1977a) 'Political Economic and Social Factors on the Parthian Landscape of Mesopotamia and Western Iran: Evidence from Two Case Studies', in Levine and Young (1977) 81–89.

Keall, E. (1977b) 'Qal'eh-i Yazdigird: The Question of Its Date'. *Iran* 15: 1–9.

Keall, E. (1982) 'Qal'eh-i Yazdigird: An Overview of the Monumental Architecture'. *Iran* 20: 51–72.

Keall, E. (1994) 'How Many Kings did the Parthian King of Kings Rule?'. *IrAnt* 29: 254–72.

Keaveney, A. (1981) 'Roman Treaties with Parthia circa 95–circa 64 B.C.'. *AJPhil* 102.2: 195–212.

Keaveney, A. (1982) 'The King and the War-Lords: Romano-Parthian Relations Circa 64–53 B.C.'. *AJPhil* 103.4: 412–28.

Krasnowolska, A. (ed) (2015) *Studies on the Iranian World I, Before Islam*. Krakow.

Kröger, J. (2011) 'Ctesiphon'. *EncIr*. Last accessed 4/2021. <https://iranicaonline.org/articles/ctesiphon>.

Lanfranchi, G. and Rollinger, R. (eds) (2010) *Concepts of Kingship in Antiquity*. Padova.

Lavan, M., Payne, R., and Weiswiler, J. (eds) (2016) *Cosmopolitanism and Empire*. Oxford.

Lendering, J. (2019) 'Ctesiphon (Parthian: Tyspwn), The Capital of the Parthians and the Sassanids'. *Iran Chamber Society*. Last accessed 4/2021. <https://www.iranchamber.com/history/ctesiphon/ctesiphon.php>.

Lerner, J. (2015) 'Regional Study: Baktria – The Crossroads of Ancient Eurasia', in Benjamin (2015) 300–24.

Lerner, J. (2017) 'Mithridates I and the Parthian Archer', in Rubin and Schlude (2017) 1–24.

Lerouge-Cohen, C. (2007) *L'image des Parthes dans le monde gréco-romain: Du début du Ier siècle av. J.-C. jusqu'à la fin du Haut-Empire romain*. Stuttgart.

Levine, L., and Young, T.C. (eds) (1977) *Mountains and Lowlands: Essays in the Archaeology of Greater Mesopotamia*. Malibu.

Lindsay, J. (1852) *A View of the History and Coinage of the Parthians*. Cork.

Mair, V. (ed) (2014) *A Concise Commentary on Memoirs on the Western Regions in the Official Histories of the Western and Eastern Han, Wei, Jin, and Southern and Northern Dynasties*. Trans. Y. Taishan. Beijing.

Maksymiuk, K. and Karamian, G. (eds) (2017) *Crowns, Hats, Turbans and Helmets: The Headgear in Iranian History, Volume I: Pre-Islamic Period*. Siedlce and Tehran.

Manandyan, H. (1940) *Tigranes II and Rome: A New Interpretation Based on Primary Sources*. Trans. G. Bournoutian. Costa Mesa.

Marciak, M. (2012a) 'The Historical Geography of Sophene'. *Acta Ant. Hung*. 52: 295–338.

Marciak, M. (2012b) 'The Historical Geography of Gordyene. Part 1: Classical Sources'. *Anabasis* 3: 173–213.

Marciak, M. (2013) 'The Historical Geography of Gordyene. Part 2: Oriental Sources'. *Anabasis* 4: 149–59.

Marciak, M. (2017) *Sophene, Gordyene, and Adiabene: Three Regna Minora of Northern Mesopotamia Between East and West*. Leiden.

Marshall, B. (1993) *Crassus: A Political Biography*. Amsterdam.

Masson, V. (ed) (2001) *The Role of Ahalteke Horse in the Formation of World Horse-Breeding*. Ashgabat.

McEwan, G. (1986) 'A Parthian Campaign against Elymais in 77 B.C.'. *Iran* 24: 91–94.

McLaughlin, R. (2016) *The Roman Empire and the Silk Routes: The Ancient World Economy and the Empires of Parthia, Central Asia and Han China*. Barnsley.

Mielczarek, M. (1990) 'Die parthische Panzerreiterei bei Carrhae: Aus den Studien über Plutarchus, Crassus XXIV–XXVII'. *Fasciculi Archaeologiae Historicae* 4: 31–38.

Mommsen, T. (1856) *The History of Rome*. Vols. 3–5. W. Dickson (trans.). New York.

Mørkholm, O. (1980) 'The Parthian Coinage of Seleucia on the Tigris, c. 90–55 B.C.'. *Num. Chron.* 20: 33–47.

Muccioli, F. (2012) 'La testa mozza di Crasso (Plut., Crass. 32–33): A proposito di un libro recente sulla battaglia di Carre'. *Electrum* 19: 167–78.

Müller, S. (2017) 'Das Bild der Parther bei Trogus-Justin', in Wiesehöfer and Müller (2017) 241–58.

Nabel, J. (2017) 'The Arsacids of Rome: Royal Hostages and Roman-Parthian Relations in the First Century CE', PhD Dissertation, Cornell University.

Nelson, B. (ed) (2011) *Numismatic Art of Persia*. Lancaster.

Olbrycht, M. (1997) 'Parthian King's Tiara – Numismatic Evidence and Some Aspects of Arsacid Political Ideology'. *Notae Numismaticae/Zapiski Numizmatyczne* 2: 27–65.

Olbrycht, M. (1998) *Parthia et ulteriores gentes. Die politischen Beziehungen zwischen demarsakidischen Iran und den Nomaden der eurasischen Steppen*. Munich.

Olbrycht, M. (2001) 'The Origins of the Arsacid Parthian Cavalry: Some Remarks', in Masson (2001) 108–11.

Olbrycht, M. (2010a) 'Mithradates I of Parthia and His Conquests up to 141 B.C.', in Dąbrowa (2010b) 229–45.

Olbrycht, M. (2010b) 'The Early Reign of Mithradates II the Great in Parthia'. *Anabasis* 1: 144–58.

Olbrycht, M. (2012) 'Eastern Baktria under Da Yuezhi Hegemony', in Jayaswal (2012) 79–86.

Olbrycht, M. (2015) 'Arsacid Iran and the Nomads of Central Asia – Ways of Cultural Transfer', in Bemmann and Schmauder (2015) 333–90.

Olbrycht, M. (2016a) 'The Sacral Kingship of the Early Arsacids I: Fire Cult and Kingly Glory'. *Anabasis* 7: 91–106.

Olbrycht, M. (2016b) 'Dynastic Connections in the Arsacid Empire and the Origins of the House of Sāsān', in Curtis, Pendleton, Alram, and Daryaee (2016) 23–36.

Olbrycht, M. (2016c) 'Manpower Resources and Army Organisation in the Arsakid Empire'. *Ancient Society* 46: 291–338.

Olbrycht, M. (2017) 'Greeks in the Arsacid Empire', in Wiesehöfer and Müller (2017) 3–28.

Olbrycht, M. (2021a) 'Orodes II'. *EncIr.* Last accessed 4/2021. <https://referenceworks.brillonline.com/entries/encyclopaedia-iranica-online/orodes-ii-COM_336463>.

Olbrycht, M. (2021b) 'Phraates IV'. *EncIr.* Last accessed 4/2021. <https://referenceworks.brillonline.com/entries/encyclopaedia-iranica-online/phraates-iv-COM_336464>.

Olbrycht, M. (2021c) *Early Arsakid Parthia (ca. 250–165 B.C.): At the Crossroads of Iranian, Hellenistic, and Central Asian History*. Leiden.

Overtoom, N.L. (2016a) 'The Power-Transition Crisis of the 240s and the Creation of the Parthian State'. *International History Review* 38: 984–1013.

Overtoom, N.L. (2016b) 'The Rivalry of Rome and Parthia in the Sources from the Augustan Age to Late Antiquity'. *Anabasis* 7: 137–74.

Overtoom, N.L. (2016c) 'Challenging Roman Domination: The End of Hellenistic Rule and the Rise of the Parthian State from the Third to the First Centuries', PhD Dissertation, Louisiana State University.

Overtoom, N.L. (2017a) 'The Parthian Rival and Rome's Failure in the East: Roman Propaganda and the Stain of Crassus'. *Acta Ant. Hung.* 57: 415–35.

Overtoom, N.L. (2017b) 'The Parthians' Unique Mode of Warfare: A Tradition of Parthian Militarism and the Battle of Carrhae'. *Anabasis* 8: 99–122.

Overtoom, N.L. (2019a) 'The Power-Transition Crisis of the 160s–130s BCE and the Formation of the Parthian Empire'. *JAH* 7: 111–55.

Overtoom, N.L. (2019b) 'A Reconsideration of Mithridates II's Early Reign: A Savior Restores the Eastern Frontier of the Parthian Empire'. *Parthica* 21: 9–21.

Overtoom, N.L. (2019c) 'Considering the Failures of the Parthians against the Invasions of the Central Asian Tribal Confederations in the 120s BCE'. *Stud. Iran.* 48: 77–111.

Overtoom, N.L. (2020a) *Reign of Arrows: The Rise of the Parthian Empire in the Hellenistic Middle East*. Oxford.

Overtoom, N.L. (2020b) 'The Parthians' Failed Vassalage of Syria: The Shortsighted Western Policy of Phraates II and the Second Reign of Demetrius II (129–125 BCE)'. *Acta Acta Ant. Hung.* 60: 1–14.

Overtoom, N.L. (2021a) 'A Fight to Reclaim the Central Asian Frontier: The Seleucid and Parthian Rivalry in the 230s BC'. *Studia Hercynia* 26: 105–18.

Overtoom, N.L. (2021b) 'Reassessing the Role of Parthia and Rome in the Origins of the First Romano-Parthian War (56/5–50 BCE)'. *JAH* 9: 238–68.

Overtoom, N.L. (2022) 'Phriapatius'. *EncIr.* Last accessed 8/2023. (http://dx.doi.org/10.1163/2330-4804_EIRO_COM_363949).

Overtoom, N.L. (forthcoming) 'The Parthian Empire in Josephus's Writings', in Atkinson (Forthcoming).

Overtoom, N.L. (2023) 'Logistics and Strategy in Hellenistic Greece: Parthians and Seleucids', in Brice and Donahue (2023) 258–86.

Poirot, J. (2014) 'The Romano-Parthian Cold War: Julio-Claudian Foreign Policy in the First Century CE and Tacitus' Annales', PhD Dissertation, Louisiana State University.

Potter, D. (ed) (2010a) *A Companion to the Roman Empire*. Malden, MA.

Potter, D. (2010b) 'The Transformation of the Empire: 235–337 CE', in Potter (2010a) 153–74.

Potts, D. (ed) (2013) *Oxford Handbook of Ancient Iran*. Oxford.

Rawlinson, G. (1876) *The Seven Great Monarchies of the Ancient Eastern World*. Vol. 3. Piscataway.

Regling, K. (1899) *De belli Parthici Crassiani fontibus*. Berlin.

Regling, K. (1907) 'Crassus' Partherkrieg'. *Klio* 7: 357–94.

Rezakhani, K. (2013) 'Arsacid, Elymaean, and Persid Coinage', in Potts (2013) 766–78.

Rezakhani, K. (2016) 'From Aramaic to Pahlavi: Epigraphic Observations Based on the Persis Coin Series', in Curtis, Pendleton, Alram, and Daryaee (2016) 69–75.

Rich, J. and Shipley, G. (1993) *War and Society in the Roman World*. London.

Rtveladze, E. (2011) 'Parthians in the Oxus Valley: Struggle for the Great Indian Road'. *Anabasis* 2: 149–78.

Rubin, B. and Schlude, J. (eds) (2017) *Arsacids, Romans and Local Elites: Cross-Cultural Interactions of the Parthian Empire*. Oxford.

Sachs, A. (1955) *Late Babylonian Astronomical and Related Texts*. Providence.

Sachs, A. and Hunger, H. (1988) *Astronomical Diaries and Related Texts from Babylonia: Diaries from 652 B.C. to 262 B.C.* Vienna.

Sachs, A. and Hunger, H. (1989) *Astronomical Diaries and Related Texts from Babylonia: Diaries from 261 B.C. to 165 B.C.* Vienna.

Sachs, A. and Hunger, H. (1996) *Astronomical Diaries and Related Texts from Babylonia: Diaries from 164 B.C. to 61 B.C.* Vienna.

Sampson, G. (2015) *The Defeat of Rome: Crassus, Carrhae and the Invasion of the East*. Barnsley.

Sampson, G. (2020) *Rome and Parthia: Empires at War – Ventidius, Antony and the Second Romano-Parthian War, 40–20 BC*. Barnsley.

Santangelo, F. and Bissa, E. (eds) (2016) *Studies on Wealth in the Ancient World*. London.

Sartre, M. (2005) *The Middle East Under Rome*. C. Porter and E. Rawlings (trans.). Cambridge.

Scheidel, W. and Bang, P. (eds) (2016) *Oxford Handbook of the State in the Ancient Near East and Mediterranean*. Oxford.

Schippmann, K. (1987a) 'Arsacids ii. The Arsacid Dynasty'. *EncIr*. Last accessed 4/2020. <http://www.iranicaonline.org/articles/arsacids-ii>.

Schippmann, K. (1987b) 'Azerbaijan iii. Pre-Islamic History'. *EncIr*. Last accessed 4/2020. <http://www.iranicaonline.org/articles/azerbaijan-iii>.

Schlude, J. (2009) 'Rome, Parthia, and Empire: The First Century of Roman-Parthian Relations', PhD Dissertation, University of California, Berkeley.

Schlude, J. (2012) 'The Parthian Response to the Campaign of Crassus'. *Latomus* 71/1: 11–23.

Schlude, J. (2015) 'The Early Parthian Policy of Augustus'. *Anabasis* 6: 139–56.

Schlude, J. (2020) *Rome, Parthia, and the Politics of Peace: The Origins of War in the Ancient Middle East*. Abingdon.

Schmitt, R. (2011a) 'Arachosia'. *EncIr*. Last accessed 4/2021. <https://iranicaonline.org/articles/arachosia>.

Schmitt, R. (2011b) 'Armenia and Iran I. Armina, Achaemenid province'. *EncIr*. Last accessed 4/2021. <https://iranicaonline.org/articles/armenia-i>.

Schmitt, R. (2011c) 'Artavasdes'. *EncIr*. Last accessed 4/2021. <https://iranicaonline.org/articles/artavasdes-old-iranian-male-personal-name-attested-as-greek-artaousdes-artabzes-artbazos-artozos-latin-artava>.

Seager, R. (2002) *Pompey the Great: A Political Biography*. Malden, MA.

Sellwood, D. (1971) *An Introduction to the Coinage of Parthia*. London.

Sellwood, D. (1976) 'The Drachms of the Parthian "Dark Age"'. *JRAS* 108: 2–25.

Sellwood, D. (1980) *An Introduction to the Coinage of Parthia*. London.

Sellwood, D. (1983) 'Parthian Coins', in Yarshater (1983) 279–98.

Siani-Davies, M. (1997) 'Ptolemy XII Auletes and the Romans'. *Historia* 46: 306–40.

Shahbazi, A. (1990) 'Carrhae'. *EncIr*. Last accessed 4/2020. <http://www.iranicaonline.org/articles/carrhae-harran-town-in-mesopotamia-where-in-may-53-b>.

Shayegan, M.R. (2011) *Arsacids and Sasanians: Political Ideology in Post-Hellenistic and Late Antique Persia*. Cambridge.

Sheldon, R. (2010) *Rome's Wars in Parthia: Blood in the Sand*. London.

Sherwin-White, A. (1984) *Roman Foreign Policy in the East 168 B.C. to A.D. 1*. Norman.

Sherwin-White, S. and Kuhrt, A. (1993) *From Samarkhand to Sardis: A New Approach to the Seleucid Empire*. Berkeley.

Simonetta, A. (1953–1957) 'Notes on the Parthian and Indo-Parthian Issues of the First Century B.C.', in Babelon and Lafaurie (1953–1957) 111–21.

Simonetta, A. (1966) 'Some Remarks on the Arsacid Coinage of the Period 90–57 B.C.'. *Num. Chron*. 7.6: 15–40.

Simonetta, A. (2001) 'A Proposed Revision of the Attributions of the Parthian Coins Struck During the So-called "Dark Age" and Its Historical Significance'. *East & West* 52: 69–108.

Simonetta, A. and Sellwood, D. (1978) 'Again on the Parthian Coinage from Mithradates II to Orodes II'. *Quaderni Ticinesi di Numismatica e di Antichità Classiche* 7: 95–119.

Stevens, K. (2016) 'Empire Begins at Home: Local Elites and Imperial Ideologies in Hellenistic Greece and Babylonia', in Lavan, Payne, and Weiswiler (2016) 65–88.

Suren-Pahlav, S. (2019) 'General Surena, The Hero of Carrhae'. *Iran Chamber Society*. Last accessed 4/2021. <https://www.iranchamber.com/history/surena/surena.php>.

Syme, R. (1958) *Tacitus*. Oxford.

Syvänne, I. (2017) *Caracalla: A Military Biography*. Barnsley.

Tao, W. (2007) 'Parthia in China: A Re-examination of the Historical Records', in Curtis and Stewart (2007) 87–104.

Terigon, H. and Seland, E. (eds) (2017) *Sinews of Empire*. Oxford.

Timpe, D. (1962) 'Die Bedeutung der Schlacht von Carrhae'. *MH* 19: 104–29.

Traina, G. (ed) (2006) *Studi sull'eta di Marco Antonio*. Rudiae 18. Lecce.

Valverde, L. (2017) 'Algunas novedades sobre numismática parta desde Fraates II a Orodes II (132–38 a.C.)'. *Gaceta Numismática* 193: 111–28.

van de Mieroop, M. (2002) 'Gutians'. *EncIr*. Last accessed 4/2021. <https://iranicaonline.org/articles/gutians>.

van der Spek, R. (2001) 'The Theatre of Babylon in Cuneiform', in Veenhof (2001) 445–56.

van Wijlick, H. (2021) *Rome and the Near Eastern Kingdoms and Principalities, 44–31 BC: A Study of Political Relations during Civil War*. Leiden.

Vardanyan, R. (2006) 'Some Remarks on the Arrangement of the Parthian "Dark Age" Coin Series'. *Parthica, Incontri di culture nel mondo antico* 8: 105–29.

Veenhof, K. (ed) (2001) *Veenhof Anniversary Volume: Studies Presented to Klaas R. Veenhof on the Occasion of His Sixty-Fifth Birthday*. Leiden.

Vogt, J. (ed) (1955a) *Gesetz und Handlungsfreiheit in der Geschichte. Studien zur historischen Wiederholung*. Stuttgart.

Vogt, J. (1955b) 'Die Tochter des Grosskönigs und Pausanias, Alexander, Caracalla', in Vogt (1955a) 55–80.

Ward, A. (1977) *Marcus Crassus and the Late Roman Republic*. Columbia.

Warry, J. (1995) *Warfare in the Classical World*. Norman.

Weggen, K. (2011) *Der lange Schatten von Carrhae: Studien zu M. Licinius Crassus*. Hamburg.

Wick, P. and Zehnder, M.P. (eds) (2012) *The Parthian Empire and Its Religions: Studies in the Dynamics of Religious Diversity*. Gutenberg.

Widengren, G. (1983) 'Sources of Parthian and Sasanian History', in Yarshater (1983) 1261–83.

Wiesehöfer, J. (ed) (1998) *Das Partherreich und seine Zeugnisse*. Stuggart.

Wiesehöfer, J. (2016) 'Iranian Empires', in Scheidel and Bang (2016) 199–233.

Wiesehöfer, J. and Müller, S. (eds) (2017) *Parthika: Greek and Roman Authors' Views of the Arsacid Empire*. Wiesbaden.

Williams, R. (1985) 'Rei Publicae Causa: Gabinius' Defense of His Restoration of Ptolemy Auletes'. *CJ* 81: 25–38.

Wolski, J. (1956–1957) 'The Decay of the Iranian Empire of the Seleucids and the Chronology of the Parthian Beginnings'. *Berytus* 12: 35–52.

Wolski, J. (1989) 'Die gesellschaftliche und politische Stellung der großen partischen Familien'. *Tyche* 4: 221–27.

Wolski, J. (1993) *L'Empire des Arsacides (Acta Iranica 32) – Textes et Mémoires XVIII*. Leuven.

Wolski, J. (2003) *Seleucid and Arsacid Studies: A Progress Report on Developments in Source Research*. Krakow.

Yarshater, E. (ed) (1983) *Cambridge History of Iran. Vol. 3.1–2: The Seleucid, Parthian, and Sasanid Periods*. Cambridge.

Zadorojniy, A. (1997) 'Tragedy and Epic in Plutarch's "Crassus"'. *Hermes* 125: 169–82.

CHAPTER 12

The Arsacids vs. Rome (First to Third Centuries CE): Observations on Parthian Tactics and Warfare

Leonardo Gregoratti

The emperor Trajan's invasion of both Arsacid Armenia and Parthian Mesopotamia in 114 CE, Trajan's Parthian War in modern scholarship, the *Bellum Parthicum* of Latin inscriptions,[1] is the most famous and best known of the many open conflicts between Roman legionaries and Arsacid armies.[2] It is not by coincidence that the war, which in many aspects marked a turning point in the relationships between Parthia and the Roman state, was the only conflict between the Roman *Urbs* and its eastern neighbours chosen as the subject of two monographs. Julien Guey's *Essai sur la guerre parthique de Trajan (114–117)*, published in 1937, was followed not many years later by Frank Lepper's *Trajan's Parthian War* (1948). Even after the publication of more recent studies on the subject, these volumes still represent the two main and most comprehensive studies on Trajan's Armenian and Parthian invasions.[3]

Nonetheless, in browsing the pages of these two authors, a striking element, perhaps not so evident to a historian of the Roman world, would surely draw the attention of the contemporary scholar of the Parthians – always struggling, tied to the fragile ship mast of the available sources, against the sirens' song of a too Romano-centric approach. This element is the absolutely secondary role played by the Arsacids in the historical narration, corresponding with an almost complete absence of the Parthian army from the theatre of war.[4] The original flaw is, of course, not to be ascribed to Guey and Lepper: their

1 E.g., IK-17.4112; IDRE 2.363; CIL 10.1202, 3733; 11.3198.

2 For an excellent synthetic overview on the Parthian side, see Fabian (2020). I am, of course, completely aware that such a significant topic that covers the relationship between two major powers spanning for more than two centuries cannot be dealt with adequately in the space available in this chapter. Therefore, to avoid a dull and summary narration of events, I decided to focus on a particular question of Parthian warfare. This topic, rooted in first-century BCE history, will be used as a sort of *fil rouge* in order to introduce the various historical events and tie them together.

3 See also Lightfoot (1990); Angeli Bertinelli (2000); Gonzales (2000).

4 Guey (1937) deals with Trajan's adversaries in his three-page appendix, 'Les Arsacides' (123–25).

 | DOI:10.1163/9789004710771_013

discussions of the conflict simply mirror the elusiveness of Parthian political and military power in the surviving Roman sources. After the loss of Arrian's *Parthica*, the best-known work explicitly dedicated to the later Arsacid kingdom, which left only a few dozen floating fragments, what remains, except for a few passages by Tacitus, is a series of historical chronicles whose authors were clearly more interested in the feats of the Roman leaders than in those of their eastern counterpart (Plutarch, Cassius Dio, Herodian, and the *Historia Augusta*).

If a subordinate presence of the Parthians in the relationship between the two superpowers or discussions of Rome's eastern policy is a common feature, considering the situation of the ancient sources, this circumstance assumes particularly striking characteristics in the case of a large-scale armed conflict where Roman legions were supposed to fight a comparable opponent at least in terms of political size or power. Yet, through the pages of these ancient authors, the Roman emperors and their loyal generals and soldiers seem to be capable of wandering within the trans-Euphratean theatre of operations, the battlefields, cities, and open spaces of Mesopotamia, free to march, manoeuvre and conquer, fighting nothing more than an evanescent presence of the enemy, a shadow which seldom would take the shape of something concrete and tangible, the shadow of the Parthian army.

This is even more astonishing if we think about the attention given by ancient historians to the Arsacid army and strategies put into action during the ill-fated campaign of Crassus.[5] The two main sources, Plutarch and Cassius Dio, both writing in the centuries after Crassus' war, are much more generous in providing details about Arsacid warfare on that (and for Cassius Dio, only that) occasion.[6] But of course, the main focus remains on the Romans.[7] This certainly depends on the considerable amount of analysis of the defeat already present in their sources, which gathered all the elements that might help to explain and justify the devastating defeat.[8] The same richness of information is the basis of the modern studies on the events which took place in northern Mesopotamia in the spring of 53 BCE.

5 Major studies include Regling (1907); Smith (1916); Debevoise (1938) 78–93; Timpe (1962); Brizzi (1983a); Wolski (1993) 128–34; Frendo (2003); Lerouge (2007) 67–75; Traina (2009) and to a lesser extent Traina (2010); Schlude (2020) 60–74; Brizzi (2022) 41–55; and more relevantly, Overtoom (2017b) and of course Overtoom's chapter in this volume.

6 Plut. *Crass.* 16–33; Cass. Dio 40.12–30. See Gabba (1966) 54 and 67; Brizzi (1983a) 18; Frendo (2003) 73.

7 Plut. *Crass.* 21, 23–27; Cass. Dio 40.15–16, 22–24.

8 Traina (2011) 211–216; Sheldon (2010) 39; Overtoom (2017b).

Unfortunately, this interest in the Arsacid way of fighting did not last long, and it seems to remain limited to Crassus' episode. The sources regarding the later invasions of the first century BCE, Pacorus' and Labienus' war and Antony's invasion of Media, still provide some information on the topic, but they give way to narrations of later events, sometimes by the same authors, where this occurs more rarely.[9] It seems that Parthian warfare was in the spotlight and continued to be of interest a long time after the events, only concerning Crassus' disaster, primarily as an element able to shed light on Rome's astonishing defeat at Carrhae. Beyond that particular event, after the defeat was discussed and thoroughly investigated, no genuine interest remained for the way the Parthians behaved in battle. From what can be read in the available sources concerning later conflicts, it seems the attention was focused elsewhere.

The relatively detailed descriptions of Parthian military organisation, strategy, and tactics opposed to Crassus' legions were therefore instrumental in describing nobody else but the Romans themselves. The ancient authors reflected specifically on the Romans' mistakes (or better on those of one Roman in particular), in order to explain to the imperial audience how the most formidable military power of the time could have been severely defeated by Rome's eastern barbarian neighbours and their rudimentary military skills.

Following these considerations, since it is evident that what the sources report the Parthians did at Carrhae played a relevant role in the analysis of the defeat, and therefore in the whole apparatus of explanations Roman historians and Roman propaganda managed to collect, it seems legitimate to look with caution at the unusual richness of information concerning the battle. The authenticity of what we are told about it is hard to discuss due to the lack of comparably rich Parthian sources on the specific event and warfare in general. On the other side, with the scarcity of sources on Parthia, the Carrhae-related information undeniably constitutes a fundamental and essential treasure of knowledge.[10]

What should be questioned is the opportunity of considering what happened at Carrhae, an example of the usual Parthian way of fighting and tactics applicable always and everywhere in the over four centuries of history of the empire. It would be risky to give a general value to the war tactics employed in

9 In the case of Cassius Dio's *Roman History*, it should be pointed out that second-century events are available only through later epitome or through excerpts dealing only with specific aspects: Lepper (1948) 1–12.

10 For example, Mielczarek (1990); Wolski (1967); Wolski (1969); Wolski (1981); Olbrycht (2016).

the field against Crassus only because a relatively detailed description of them is available. In other words, there is the definite possibility that that Arsacid tactical behaviour at Carrhae could be the result of a combination of exceptional circumstances like the extraordinarily competent and skilled leadership of Surena, the fortunate availability of the necessary troops effectively supplied for the task, and the repeated and continuous errors on the Roman part.[11]

This contribution will therefore go through the principal military events that occurred in the history of the relationship between the Roman and Arsacid empires, from Augustus to the fall of the latter, paying specific attention to the information provided by the sources concerning Parthian warcraft to establish whether the tactics employed by the Arsacids in Crassus' campaign could be considered a valid example of how they otherwise acted in the field.

1 Rome vs. Parthia in the Late Republic

Of course, the natural beginning of this investigation is that day in the late spring of 53 BCE on the plains of northeastern Mesopotamia, where Crassus met his fate.[12] The topic is analysed in detail elsewhere in this volume by an expert on the subject,[13] and it will be enough here to recall what the two primary sources say about Parthian field tactics. Cassius Dio explains in a short paragraph the key to Surena's victory: the combined use of heavily armoured mounted lancers and mounted archers.[14] He explicitly claims that after the loss of Publius Crassus' Gallic cavalry, thanks to a classic feigned-retreat tactic, the legionary infantry, eager for revenge, could not 'accomplish anything worthy of themselves because of the enemy numbers and tactics'.[15] Cassius Dio goes on to explain how the situation of the poor infantry units was desperate: when they chose to fight close together, shield against shield, they were smashed to pieces by the charges of the lancers who scattered those who escaped the deadly impetus. If they chose to extend their ranks to avoid being an easy target for the Parthian heavy cavalry, they were surrounded and targeted by volleys

11 The author agrees with Overtoom's opinion (2017a) that Carrhae was not unique (in the sense that the tactics, strategies, and units employed on that occasion constituted a repertoire indicative of Parthian methods of combat). However, what is exceptional about Carrhae is the unusual combination of those units and tactics.

12 For the actual day, see Groebe (1907).

13 See Overtoom's chapter in this volume.

14 Cass. Dio 40.22.2–3.

15 Cass. Dio 40.22.1; Sheldon (2010) 36–37; Overtoom (2017a) 110–11; Brizzi (2022) 45–48.

of arrows. Thus, they could neither move nor stay put.[16] Plutarch's account is slightly different: the cataphract/archer combination emerges only when Roman detachments attempted a counterattack. Concerning the main army, the disruptive action of the bowmen was complemented by manoeuvres of the heavy cavalry aimed at control and encirclement.[17] It is therefore interesting to note that only in the later source, Cassius Dio, do the combined and perfectly synchronised tactics involving cataphract charges and mounted bowmen's deadly arrows play a vital role in defeating the Roman army, along with the client king Abgar's sudden and unforeseen change of sides, which forced the soldiers to defend their backs as well.[18]

In the military episodes that followed Carrhae, things developed differently. At the battle of the Cilician Gates (39 BCE), where Antony's lieutenant, P. Ventidius, got the upper hand over Q. Labienus, the former supporter of the Caesaricides who had joined the Great King's service,[19] the Parthian heavy cavalry seems to have acted alone.[20] The Arsacid armoured cavalry, after arrogantly avoiding joining forces with Labienus' Roman deserters, charged without archers' support. Therefore the Roman leader's advantage of higher ground granted him the victory.[21] In the decisive battle at Gindaros the following year, the situation seems to mirror Carrhae with reversed roles: the overconfident Parthian units charged uphill again with the same disastrous results,[22] while the Roman slingers kept the enemy at a distance while battering his lines as the Parthian archers did in Mesopotamia.[23] Again, no mention is made of bowmen in Cassius Dio, but in Frontinus' *Strategemata*, a didactic dissertation dealing specifically with the expedients which gained victory for Roman generals, Ventidius refers to the choice of a hilly battleground to better protect his soldiers from enemy archers.[24] At the same time, Frontinus ascribes to him the tactic of letting the enemy get closer before ordering a rapid attack, thus forcing the Parthians into close combat immediately, neutralising their

16 Cass. Dio 40.22.4–5.

17 Plut. *Crass.* 24–25; Regling (1907) 383–84; Smith (1916) 253–58; Gabba (1966) 55; Brizzi (1983a) 19–22; Traina (2010) 78–80; Brizzi (2022) 48.

18 Cass. Dio 40.23.1–2. See Gabba (1966) 66–67; Coulston (1986) 61, 68; Overtoom (2017a). For Brizzi (2016, 744), the role of the cataphracts was secondary in comparison with that of the mounted archers in winning the battle.

19 Cass. Dio 48.24.5; Flor. 2.19.4.

20 Cass. Dio 48.39–41; Debevoise (1938) 108–20; Linz (2009) 47–48; Sheldon (2010) 57–60; Schlude (2012) 13–14; Schlude (2020) 84–85; Brizzi (2016) 745–46; Brizzi (2022) 66–67.

21 Cass. Dio 48.40.1–2.

22 Cass. Dio 49.20.1–2.

23 Cass. Dio 49.20.3; Brizzi (2022) 64–66.

24 Frontin. *Str.* 1.1.6.

devastating long-distance attacks.[25] It is interesting that for this latter stratagem no specific context is given; it is simply labelled as Ventidius' tactic against the Parthians, maybe used at some point of the battle at Gindaros, in contrast with Cassius Dio's account. Other passages by Frontinus refer to the tactic of hiding units before staging a feigned retreat and luring the Parthian mounted troops into an ambush; again, no mention is made of the famous Arsacid archers.[26] As is evident, the Parthian tactics in these references consist of a head-on charge against the Romans; bows and arrows are seldom mentioned, and certainly there is no reference to the combination of cataphracts and bowmen that won the day at Carrhae.

The situation changes again with Antony's disastrous invasion of Media (36 BCE).[27] Here Frontinus mentions constant attacks of Parthian mounted archers, starting early in the morning at the beginning of each marching day,[28] against which the Roman warlord employed the *testudo* formation until the enemy supply of arrows was exhausted,[29] a problem the Parthian bowmen did not face at Carrhae thanks to Surena's far-sightedness.[30] In Plutarch, the Arsacids harassed the Roman marching units on every side. Only Antony's surprise counter-attacks were able to win the day by forcing the enemy into a close-quarters fight and, as Ventidius did, neutralising the archers' long-distance attacks.[31] On one occasion, if we decide to believe Plutarch, the Parthians took the *testudo* deployment for a sign of fatigue and exhaustion of the Romans;[32] this convinced them to put aside their bows and charge into the Roman ranks with spears. According to these passages of Plutarch and Cassius Dio, the same light-armed bowmen conducted the charges. No reference is given to cataphracts or a combined attack of heavy cavalry and *toxotai*.

From this rapid examination, it seems evident that in the historical accounts of the Romano-Parthian conflicts that followed Carrhae, still relatively rich in information about Parthian tactical methods in comparison with those of the first and second century CE, there is no trace of the combined employment

25 Frontin. *Str.* 2.2.5; Flor. 2.19.

26 Frontin. *Str.* 2.5.37.

27 Kromayer (1896); Debevoise (1938) 121–32; Bengston (1974); Schieber (1979); Dabrowa (2006); Sheldon (2010) 70–73; Van Wijlick (2015); Patterson (2015) 87–93; Jones (2017); Brizzi (2022) 71–72.

28 Frontin. *Str.* 2.13.7; Cass. Dio 49.29.3.

29 Plut. *Ant.* 45.2; Frontin. *Str.* 2.3.15.

30 Plut. *Crass.* 25.1; Overtoom (2017a) 99–100.

31 Plut. *Ant.* 39.4; 41.5. Other strategies involve the employment of slingers whose firepower was superior: Plut. *Ant.* 41.4–5, 42.1, 45.1, 49.1; Cass. Dio 49.26.2–3.

32 Cass. Dio 49.9.3–4; Plut. *Ant.* 45.3.

of mounted archers and cataphract units typical of Crassus' defeat. In the invasion of Syria by Pacorus and Labienus (40–38 BCE), a few years after the deaths of Marcus and Publius Crassus, the primary role on the field is played by charging mounted warriors. Archers are not significantly employed on the battlefield, possibly due to the anti-archer tactics conceived by Ventidius.[33] In Antony's invasion of Media, on the contrary, the heavily armoured warriors have vanished while bowmen and light-armed mounted spearmen have the task of pinning down, encircling and harassing the enemy. In neither case do the two kinds of unit appear together to operate in the field as an ensemble.

We cannot exclude that the anti-archer tactics possibly employed by Ventidius and Antony, deprived the cataphracts of the bowmen units instrumental in suppressing the enemy and helping them to perform an effective charge, eventually forcing the Parthian generals to employ those corps separately. But this is simply nowhere stated. From the texts it seems instead that when the mounted warriors charged Ventidius, the archers were not there, or were not deemed relevant enough to the operations to deserve a mention. On the other side, it looks as the cataphracts were entirely absent from Antony's Parthian wars, or better, they do not seem to have been deployed in battle and remained in the background during the fighting, since Pompeius Trogus clearly states that at least 400 free men, members of the Parthian aristocracy and therefore in all probability heavily armed knights, were part of Phraates IV's army.[34]

2 Rome vs. Parthia: from Augustus to Nero

Antony's eagerness for revenge against the Parthian Great King after the failed invasion of Media died with him in Cleopatra's arms following the defeat at Actium.[35] It will be necessary to wait more than ninety years to see Roman legionary formations fighting Parthian mounted ranks again. Although Rome's 'Golden Age' propaganda reaffirmed the need to avenge the Roman pride wounded at Carrhae and Phraaspa as soon as possible,[36] Augustus' political shrewdness allowed him to realise that the situation was ripe for the adoption of a different strategy against Parthia.[37]

---

33 Overtoom (2017a) 98–99; Brizzi (2022) 64–66.

34 Just. 41.2 6; Olbrycht (2016) 292–93.

35 Plut. *Ant.* 77.

36 Lerouge (2007) 93–97; Linz (2009) 44; Overtoom (2017b) 14–16.

37 In general for the diplomatic activity, see Nabel's chapter in this volume.

Rome needed time to consolidate its presence in the Near East after the storm of the Civil Wars. In addition, the Parthian Empire was shaken by internal troubles and friction with the nobility, probably originating with the bloodbath in the royal house that granted the throne to Phraates IV.[38] Between 31 and 26 BCE Tiridates, perhaps a general who gained a solid reputation fighting the Romans, exploited this inner tension to gain some aristocratic support in his attempt to usurp the throne.[39] His plan failed, and he was forced to leave Parthia, but not before managing to lay hands on one of the Arsacid princes to back his claims to the throne and convince Augustus, during a meeting in Hispania, to support him. The *princeps* denied him Rome's help, but it is evident that through that episode, the Roman leadership witnessed how vital and powerful was the possibility of controlling a member of the royal Arsacid family to consolidate the Parthian aristocratic opposition to the king around a pretender to the throne, and therefore have some chance of success in taking control of all of Parthia.

What Augustus learned on that occasion proved itself useful some years later. The climate of reciprocal trust and collaboration between Rome and Parthia, inaugurated by the return of the standards lost to the Arsacids by Crassus and Anthony in 20 BCE,[40] was strengthened a decade later by the decision made by the Great King Phraates IV to give custody of all his sons, except the one born to Queen Musa, and their families, to the Roman emperor.[41] The chosen heir, Phraates V, having obtained the throne and shown a collaborative attitude in Armenia,[42] undoubtedly understanding the relevant political advantage Rome

38 Just. 42.5.1–2; Plut. *Ant.* 37.1; *Crass.* 33.8–9; Cass. Dio 49.23.5.

39 Just. 42.5.4–9; *Res Gestae* 32; Isid. Char. *Mans. Parth.* 1; Cass. Dio 51.18, 53.33.1–2, 54.8.1; Tarn (1932) 832–36; Debevoise (1938) 136–38; Oltramare (1938) 122–23; Ziegler (1964) 45–47; Timpe (1975) 155–58; Schippmann (1980) 20, 105–106; Bivar (1983) 65–66; Dąbrowa, (1983) 40; Frye (1984) 235; Karras-Klapproth (1988) 171–76; Wolski (1993) 146; Verstandig (2001) 211–16; Wiesehöfer (2010) 187–88; Schulde (2015) 148–49; Schlude (2020) 97–99; Schlude and Rubin (2017a) 69.

40 Just. 42.5.11; *Res Gestae* 29; Cass. Dio 54.8.1–2; Liv. *Per.* 141; Strabo 6.4.2; 16.1.28; Vell. Pat. 2.91.1; Flor. 2.34; Eutr. VII, 9; Oltramare (1938) 131–132; Angeli Bertinelli (1979) 52; Dąbrowa (1983) 41; Bivar (1983) 66–67; Barzanò (1985) 211; Wolski (1993) 147–48; Lerouge (2007) 99–104; Gregoratti (2020).

41 Just. 42.5.12; *Res Gestae* 32; Joseph. *AJ* 18.42; Tac. *Ann.* 2.1–2; Vell. Pat. 2.94.4; Suet. *Aug.* 21.3; 43.4; Oros. 6.21.29; CIL 14.2216 = ILS 843; CIL 6.1799 = ILS 842; Ziegler (1964) 51–52; Dąbrowa (1987) 64–65; Schippmann (1980) 47–48; Frye (1984) 230; Nedergaard (1988) 106–108; Wolski (1993) 148; Gregoratti (2012a) 184–86; Nabel (2017) 36; Schlude (2020) 100–101.

42 Vell. Pat. 2.101.1–2; Cass. Dio 55.10.18–19; Suet. *Tib.* 12.2; Oros. 7.3.4; Ziegler (1964) 53–54; Schippmann (1980) 48–49; Dąbrowa (1983) 43–44; Schlude and Rubin (2017a) 70; Linz (2009) 73–74; Schlude (2020) 102–103.

had gained, and foreseeing the consequences of his father's decision, asked for the restitution of his brothers, to no avail.[43] These princes, guests of the Roman Senate in Rome, and their descendants, who maintained only a vague memory of the motherland or were born in the Urbs, constituted one of the two principal instruments of Julio-Claudian strategy against Parthia in the following decades.[44] The terms of the new approach became clear once the unpopularity of Phraates V's policy strengthened the aristocratic opposition.

At the beginning of the first century CE, the Arsacids found themselves in a condition of deep social and institutional instability. The king's leadership had in fact been overruled. Arsacid monarchs were maintained on the throne by aristocratic factions, and royal family members were induced to rebel against the ruling king with the sole intent of providing formal legitimacy to the power gained and wielded by one or the other group of the aristocracy. Parthia was torn apart by competition between the main noble factions embodying the different political orientations coexisting inside the kingdom. They had a common interest in weakening the crown to enlarge their power, based on territorial control and private militias, thus enhancing their independence from the king.[45]

Phraates and his mother were forced to renounce the throne, plunging the Parthian empire into a political crisis.[46] With no direct relatives available at court, between 6 and 8 CE a delegation of Parthian dignitaries reached Rome to ask Augustus for an Arsacid prince as the new Great King.[47] The elderly Vonones was chosen, a choice that, because of his inclination towards Roman culture and customs and the dependence on Rome that it demonstrated, displeased several aristocratic groups, who in those years wielded the real power in the empire.[48] These chose Artabanus II, a member of the ruling dynasty of Media with strong relationships with the eastern satrapies. Vonones, after some successes, was finally defeated and forced to find refuge in Armenia (12 CE). Although the first attempt to put a 'friendly' king on the throne at Ctesiphon ended in failure, the events showed the Roman leadership that an Arsacid prince could quickly gain enough aristocratic support and military forces to contest the throne.

43 Cass. Dio 55.10a.4. Dąbrowa (1987); Dąbrowa (1989a) 315; Frézouls (1995) 487.

44 Gregoratti (2015a).

45 Wolski (1981); Wolski (1981/1984); Gregoratti (2015b).

46 Joseph. *AJ* 18.42–44; Ziegler (1964) 56; Dąbrowa (1983) 64.

47 Joseph. *AJ* 18.46; Suet. *Tib.* 16.1; *Res Gestae* 33; Tac. *Ann.* 2.1–3; 6.36 and 42; Cass. Dio 40.15.3–4; Arr. *Parth.* fr. *31; Debevoise (1938) 151; Kahrstedt (1950) 17–18; Ziegler (1964) 56; Angeli Bertinelli (1979) 53; Schippmann (1980) 49; Dąbrowa (1983) 44–45; Frye (1984) 237; Wolski (1993) 150.

48 Tac. *Ann.* 2.1–2; Sellwood (1980^{2}) typ. 60.; Dąbrowa (2017) 173–74; Gregoratti (2018) 25.

The approach, anticipated decades before by Tiridates' coup, continued to be adopted by the Julio-Claudians to weaken their rival by taking advantage of Parthian internal political strife and fragmentation with relatively limited costs and political involvement. Augustus' successors continued to draw from the reserve of Arsacid princes at their disposal in Rome, even though this strategy, now that Artabanus ruled from a consolidated position, had to be integrated with other means. Through the years, the government of Artabanus, eager to get rid of the influence of the aristocratic factions and seeking alternative powers in its territory, deterred the rise of a strong opposition to the king.[49] The nobility waited for a moment of royal weakness to make its move. This occurred in 35 CE when Artabanus tried to put his son Arsaces on the throne of the long-contended kingdom of Armenia.[50] Emperor Tiberius reacted politically to the Parthian military offensive, offering the Armenian crown to Mithridates, brother of Pharasmanes, monarch of Caucasian Iberia, a minor state on the Caucasus already dependent on Rome.[51]

Thanks to Roman financial support, Mithridates was able to take control of the kingdom quickly. The Great King then dispatched a second Arsacid prince, Orodes, after acquiring the help of Sarmatian mercenaries from the tribes living north of the Caucasus through gifts and a recruiting campaign. However, thanks to his royal brother's solid control over the main Caucasus passes, Mithridates managed to prevent these contingents of Orodes' allies from reaching the theatre of the military operations and supporting the Parthians, while nonetheless allowing his own allied Sarmatians to get through.[52]

Aware of being outnumbered, Orodes wisely avoided giving battle. Pharasmanes then threatened the Parthian camps, ravaged the foraging grounds, and began to encircle isolated strongholds. Only after these many provocations and after the protests of his officers, who were not intentioned to tolerate such acts of insolence, Orodes finally decided to take the field.[53] Tacitus provides a short account of the battle.[54] The Roman historian states clearly that while

49 Gregoratti (2015b); Gregoratti (2017b) 130–31; Gregoratti (2018) 25–28.

50 Tac. *Ann.* 6.31; Kahrstedt (1950) 80; Ziegler (1964) 60; Schippmann (1980) 51; Linz (2009) 95; Olbrycht (2012) 217–19.

51 Tac. *Ann.* 6.32.2; Joseph. *AJ* 18.96–97; Cass. Dio 58.26; Plin. *HN* 6.17; Braund (1997) 219–20; Dąbrowa (2017) 175–77; Dąbrowa (2021) 50–52.

52 Tac. *Ann.* 6.33.2–3; Bachrach (1973) 6–7; Chaumont (1976) 89; Olbrycht (1998) 146–149; Ash (1999); Gregoratti (2013c).

53 Tac. *Ann.* 6.34.1.

54 The reliability of Tacitus' account of this battle, where no Roman units were involved, has been rightly put in doubt due to the number of stereotypes employed concerning both Iberians and Parthians: Ash (1999). Nonetheless, concerning the tactics employed, the information provided should be taken into account.

the Parthians employed only cavalry units, the Iberians and their allies, the Albanians and the Sarmatians, were skilled warriors both on horseback and on foot.[55] Apart from the interesting stereotypical representation of their enemies made by the leaders of the two armies addressing their troops, a probable fictitious reconstruction by Tacitus, some tactical elements seem relevant here. Several Sarmatian wand-bearers (*sceptuchi*) and many others among the Iberian ranks all agreed that a battle of archers was to be avoided. The Sarmatian tactics then consisted of anticipating the clash by rushing towards the enemy to force him into hand-to-hand combat.[56] The Parthians instead tried their usual cavalry tactics, manoeuvring the horse squadrons back and forth in other to keep enough distance between them and the savage Sarmatians and let the volley of arrows win the day. According to Tacitus, the Parthian plan failed, as the trans-Caucasian horsemen, who had dropped their useless short-range bows early in the battle, managed to engage the Arsacids in a close-quarters fight, eliminating the advantage of mobility and archery. The situation became unbearable once the Iberian and Albanian infantry entered the fray, unsaddling and wounding from below the horsemen already engaged in the fight.[57] A wound suffered by Orodes concluded the day.[58] What is worth observing here is that even if Parthian mounted bowmen were able to engage in close-quarters fighting against a formation of both cavalry and infantry, there is no mention of charges by the famous cataphracts or the combination of the two contingents, which brought about success at Carrhae.

On this occasion Tiberius used monarchs designated as 'friends and allies of the Roman people' for his foreign policy, but also the already tested strategy of sending an Arsacid candidate as a symbolic leader of the king's opposition. While the situation in Armenia was still to be decided by arms, a clandestine delegation reached Tiberius. It was formed by the eunuch Abdus and the young nobleman Sinnaces, representing the circles of the nobility opposed to Artabanus, and particularly the influential aristocratic leader Abdagaeses, Sinnaces' father.[59] Rome decided to support the rebel nobles and sent the young Tiridates, a grandson of Phraates IV, to Parthia. Only when the Roman political and economic help materialised and the aristocratic leaders got the Arsacid candidate they needed they took the field openly against the monarch.

55 Tac. *Ann.* 6.34.1.
56 Tac. *Ann.* 6.35.1
57 Tac. *Ann.* 6.35.2.
58 Tac. *Ann.* 6.33–35; 38; Cass. Dio 58.26.1–4.
59 Tac. *Ann.* 6.31; Karras-Klapproth (1988) 13–14, 164–65; Olbrycht (2012) 219–21; Schlude (2020) 120.

In the first months of 35 CE, while the Armenian situation was quickly escalating, L. Vitellius, the Roman governor of Syria, led the young Tiridates to the Euphrates, the border between Rome and the Arsacid Empire.[60] The king's opponents met him there with their noble militias, their private aristocratic armies.[61] They led the prince in a quick and triumphal ride through the cities and towns of Parthian Mesopotamia[62] as far as the royal court in Ctesiphon, where Tiridates was crowned the new king of the Parthians. Artabanus, after witnessing the defection of most of his men,[63] quickly abandoned the field and looked for shelter in the eastern satrapies.[64]

Rome's coordinated action in both Parthia and Armenia proved itself successful: the Roman candidate managed to sit on the throne of Parthia. Still, not a few Arsacid nobles remained neutral and gained time. Among them, there were the satraps Phraates and Hiero who were the holders of the two most important satrapies (*validissimas praefecturas obtinebant*). According to Tacitus, the two were not enthusiastic supporters of Artabanus' rule. Still, they were even more afraid of their rival Abdagaeses, who controlled the new puppet-king and therefore had risen to leadership within the new regime.[65] The two powerful officers decided to defect to the old king's side and supported him financially in gathering a new army from the population of the east as well as Saka and Dahae nomad mercenaries.[66]

The political and military resurrection of Artabanus and his hasty and unexpected advance towards Mesopotamia revealed all the weaknesses of the new ruling faction. Tiridates hesitated to strike before his opponent could gain enough followers. Abdagaeses preferred to withdraw to northern Mesopotamia, defending the line of the Tigris and relying on troops about to come from Elymais and Armenia. The planned strategic withdrawal soon turned into a disordered rout. Defections multiplied exponentially, and Abdagaeses' army simply vanished. The few remnants of the army gained the western bank of the Euphrates. Between 37 and 38 CE, a diplomatic meeting mediated by Herod Antipas of Judaea between Vitellius and Artabanus acknowledged

60 Tac. *Ann.* 6.37.
61 Wolski (1967); Wolski (1981); Wolski (1985).
62 Tac. *Ann.* 6.41–42.
63 Tac. *Ann*, 6.36; Joseph. *AJ* 18.100.
64 Schippmann (1980) 52; Bivar (1983) 73–74; Dąbrowa (1983) 90–91; Dąbrowa (1989a) 317–18; Wolski (1993) 161; Olbrycht (1998) 151–55.
65 Tac. *Ann.* 6.43.
66 Tac. *Ann.* 6.44; Joseph. *AJ* 18.100; Angeli Bertinelli (1976) 135; Olbrycht (2012) 226–27.

the restoration of the previous situation.[67] As a result, Armenia was given to Mithridates, entering the area of Roman control.

Artabanus' death once again plunged the kingdom into a period of civic strife during which two pretenders to the throne, Vardanes and Gotarzes II, fought for supremacy. It was only a few years after the final triumph of the latter in 49 CE that, for the last time, a delegation by the Parthian aristocracy asked emperor Claudius to send another candidate to the throne to oppose the new great king.[68] With a less powerful aristocratic opposition after Artabanus' countermeasures, and new players on the scene like the Parthian client kings, and in the absence of an effort by the emperor to coordinate Rome's candidate strategy with indirect attacks through Roman allied kings, this operation obtained meagre results. Meherdates, another grandson of Phraates, was escorted by the governor Cassius Longinus and delivered to the leaders of the Karen household, who at this time had risen to the leadership of the royal opposition.[69] The kings of Osrhoene and Adiabene, who officially supported the Arsacid prince, actively operated to slow down his march towards the core of the empire. Gotarzes avoided confrontation and moved his troops rapidly from Mesopotamia to gain time to corrupt the rebel nobles.[70] Meherdates witnessed, like his predecessor a few years before, the rapid evaporation of both his alliances and his army a long time before even being able to reach the capital.[71] Both Meherdates and the Karenas fell in the final battle when Gotarzes decided to wipe out the remnants of their army.

These events demonstrated to the Roman leadership that the strategy of feeding the instability of the Arsacid Empire through interfering in its inner conflicts, in other words, the idea of winning the confrontation between the two empires through indirect conflict, was no longer effective due to the increasing solidity of the Parthian kingship. Therefore, the time was ripe for more direct intervention.

---

67 Cass. Dio 59.17.5, 27.2–3; Suet. *Vit.* 2.4; *Gaius* 14.3; Garzetti (1956) 217–29; Ziegler (1964) 62–63; Chaumont (1976) 90; Olbrycht (1998) 156.

68 Tac. *Ann.* 12.10–14.

69 Abruzzese (1911) 396–400; Dąbrowa (2017) 178–79; Schlude (2020) 124.

70 Marciak (2014a) 113–14.

71 Kahrstedt (1950) 61–62, 69–70; Bivar (1983) 76–77; Dąbrowa (1983) 121–22.

3 Rome vs. Parthia: Nero and the War for Armenia

The first direct military clash between the two superpowers after Antony found its origins again in the instability of the Armenian kingdom. The Iberian prince Rhadamistus, the son of the king mentioned above, Pharasmanes, managed to get control of the kingdom in 51 CE first through the bribery of his uncle's officers, Armenian nobles, and their Roman 'protectors', and then with an army.[72] Vologases I, recently risen to the throne of Great King of Parthia, considered this to be a coup by an illegitimate usurper,[73] and a good chance to give the throne to his little brother Tiridates.[74] The reform project conceived by Vologases entailed the assignment/bestowing of the major client thrones on his younger brothers to create a solid shared base of power for the Great King. The price for this new ruling system was a direct war with Rome.[75]

Vologases' intervention in Armenia took the form of a large-scale invasion in 52/53 CE. The Parthians took control of the major cities, Tigranocerta and Artaxata, but the frigid winters slowed down the planned rapid conquest. Then, in 54 CE, the Armenian aristocracy revolted against Rhadamistus, considering him the main responsible for their recent troubles, forced him to leave and asked for help from Rome, where a young Nero had just begun to rule.

The precarious situation of the Roman influence on Armenia, which risked disappearing entirely after years of dormancy under Claudius, called for a resolute and direct intervention. Bridges were built on the Euphrates and the allied border kings alerted, while Nero, under the influence of his tutors Burrus and Seneca, appointed Cn. Corbulo, a veteran general of the wars in Germania, supreme military authority for the east.[76]

Due to internal Parthian instability and the lack of preparation of Corbulo's troops, both the contenders resorted to diplomatic means during 55–58 CE, postponing any military operation.[77] On the Parthian side, the rise of the usurper

---

72 Tac. *Ann.* 12.44–51; Abruzzese (1911) 400–408; Linz (2009) 103; Dąbrowa (2017) 179–80.

73 Tac. *Ann.* 12.48.1–3, 15.50.1; Gregoratti (2012b).

74 Tac. *Ann.* 12.44–47; Cass. Dio 61.6.6; Chaumont (1976) 97–98; Dąbrowa (1983) 133–34; Braund (1994) 221–24; Heil (1997) 59–61.

75 Gregoratti (2013a) 47–48; Gregoratti (2017a), 101–102; Gregoratti (2017b) 131–32; Gregoratti (2018) 28–30.

76 Tac. *Ann.* 13.35.1–4; Abruzzese (1911) 408–29; Debevoise (1938) 179–80; Ziegler (1964) 67–69; Chaumont (1976) 98–101; Schippmann (1980) 54–57; Dąbrowa (1983) 136–38; Frye (1984) 239; Wolski (1993) 165–67; Heil (1997) 63; Dąbrowa (1989b) 70.

77 Tac. *Ann.* 13.34.2, 37.4. 55–57 CE: Henderson (1901) 205–206, 209–10, 266–74; Chaumont (1976) 100; Wheeler (1997) 383–85, 392–97. On Corbulo: Schur (1925); Sanford (1937);

filius Vardanis, perhaps an anonymous son of the previous Great King, followed by the upheaval of the Hyrcanians and the eastern satrapies, forced the Great King to move his troops elsewhere, breaking the brothers' front and forcing young Tiridates to deal with the Romans with only his own Arsacid troops.

In 58 CE, Corbulo was ready to launch a lightning invasion of Armenia. Artaxata[78] and Tigranocerta,[79] the main cities, and the other strongholds capitulated immediately or surrendered after a short siege. Artaxata was plundered and burned down to strike terror in the Armenians and force Tiridates into action. The Arsacid employed against Corbulo the same guerrilla tactics he had used to take control of the region from the Iberians. Carefully avoiding a field battle, he used his highly mobile cavalry units to attack Rome's allies, plunder, make raids, and spread terror, forcing Corbulo to divide his troops and distribute them across the territory. The only notable fighting occurred when Tiridates showed up at the siege of Artaxata, harassing the Roman marching army from many directions through volleys of arrows. Avoiding getting pinned down in a siege battle, he manoeuvred his mounted warriors to encircle the legions, harassing them with minor attacks, probably with long-distance weapons, and employing fake-flight tactics in the hope of breaking the Roman ranks and eliminating isolated pursuers.[80]

Although Tiridates' tactics are reminiscent of those effectively employed against the marching legionnaires of Crassus and Antony, Tacitus explicitly states that Corbulo was not taken by surprise: the army was arranged to be able to defend itself even during the march.[81] The presence of cavalry units, archers, and slingers commanded to stay in line and fend off the attacks without leaving the line, and the fact that hills sheltered one side of the army, were countermeasures that determined the failure of Tiridates' stratagems.[82] Even if the circumstances were different, the impression from Tacitus's account is that under Corbulo, who probably studied the tactics that the 'barbarians of the east' had employed in the past, the imperial legions began to use on the field the improvements that rendered Parthian ways of fighting ineffective against large units of infantry.[83]

Syme (1960); Momigliano (1975); Vervaet (1999a); Vervaet (1999b); Vervaet (2002); Vervaet (2003).

78 Tac. *Ann.* 13.39–41; Cass. Dio, 62.19–20; Sheldon (2010) 107; Gregoratti (2017c).

79 Tac. *Ann.* 14.24–25; Frontin. *Str.* 2.9.5; 4.1.21, 28. Henderson (1911) 210–12; Schur (1923) 89–93; Wheeler (1997) 386–92; Brizzi (2022) 109–10.

80 Tac. *Ann.* 13.37; Overtoom (2017a).

81 Tac. *Ann.* 13.38–39.

82 Tac. *Ann.* 13.40.

83 In addition to this, for the improvement of Roman weapons and armour see the detailed explanation by Brizzi (2022) 101–108.

The candidate Rome put on the throne left vacant by Rhadamistus and Tiridates, Tigranes VI, grew bold quickly and invaded the Parthian client kingdom of Adiabene.[84] The protests of Monobazus, the influential monarch of that kingdom east of the Tigris, found an echo in those of the deposed Tiridates at Vologases's court. The accusation of the two kings threatened to undermine the relationship based on trust and protection between the Great King and his loyal royal servants and, therefore, to destroy Vologases' new ruling structure.[85] Vologases had to act decisively and immediately.[86]

The Great King ordered his personal guard, led by Monaeses, to join Monobazus and his Adiabenes and retake Armenia, which was garrisoned only by a few Roman units. While this northern army found itself bogged down in the difficult siege of Tigranocerta,[87] Vologases personally led an army southward, threatening the province of Syria,[88] but found himself soon short of supplies. The Parthian double offensive, although militarily unfruitful, rebalanced the situation and produced a stalemate, soon exploited by the diplomacy of the two empires which agreed on a truce in 61 CE, during which both armies had to depart from Armenia.[89]

Corbulo's request for a shared command was answered when in 62 CE L. Caesennius Paetus, more in Nero's favour than his colleague, was appointed as governor of Cappadocia.[90] Corbulo retained the command of the Syrian province legions. The newcomer, responsible for the Roman province facing the enemy, aimed at pursuing military glory and honours through the annexation of Armenia.[91] Refusing to adopt the cautious attitude of his colleague and to follow his example in giving precedence to diplomatic methods, he launched a rushed and poorly planned invasion of the kingdom.[92] He advanced rapidly, failing to take any precautions to protect his run towards Tigranocerta. Vologases decided to change his strategy completely;[93] he left only a screen of troops on the Euphrates to watch Corbulo's movements and concentrated

84 Schur (1923) 19–20; Wheeler (1997) 387.

85 Tac. *Ann.* 15.1.2; Cass. Dio 62.20.2; Hammond (1934) 92; Chaumont (1976) 107–109; Angeli Bertinelli (1979) 55; Schippmann (1980) 55; Dąbrowa (1983) 140; Frye (1984) 239; Wolski (1993) 167; Gilmartin (1973) 605–607; Gregoratti (2017c).

86 Hammond (1934) 92–94; Kahrstedt (1950) 23, 36, 82; Dillemann (1962) 268–72; Ziegler (1964) 68–81; Chaumont (1976) 109–11; Dąbrowa (1983) 141–42.

87 Tac. *Ann.* 15.4.3.

88 Tac. *Ann.* 13.37.5, 14.26.2, 15.2.4, 15.3.1; Cass. Dio 62.20.2.

89 Tac. *Ann.* 15.5.4, 6.1–2; Cass. Dio 62.20.4.

90 On Paetus in Armenia, Carter (2004) 370–79; Gregoratti (2017c) 115–18.

91 Tac. *Ann.* 15.6.3–4; Patterson (2020) 769–70 with bibliography.

92 Tac. *Ann.* 15.8.1–2.

93 Schur (1923) 28–29; Chaumont (1976) 111–14, 116–18; Dąbrowa (1983) 143–46; Frézouls (1995) 494–95; Dąbrowa (1989b) 70–71.

the bulk of his army against Paetus.[94] After leading his legions deep into the enemy's territory instead of remaining in the easily defendable camp, and after losing a few soldiers sent in advance, Paetus panicked. He ordered retreat, scattered his forces, and weakened his army, detaching troops that would have successfully hampered Parthian rapid manoeuvres.[95]

In the south, Corbulo built bridges, strongholds, and a legionary camp on the eastern bank of the Euphrates after driving off the Parthian mounted archers who were defending it. He used catapults and ballistae mounted on turrets that fired from pontoons floating in the middle of the river. The more extended range of such devices compared with the Parthian bow forced the Parthians to give up. That accomplished with all due precautions, Corbulo slowly moved northwards to rescue his colleague.[96] Pressed by Vologases' troops besieging the camp, Paetus lost control and surrendered. Cassius Dio gives some more detailed elements concerning the siege.[97] Unlike Vologases, Paetus had plenty of food at his disposal and could have resisted until the arrival of Corbulo. Furthermore, the Parthians had no heavy infantry that could get close and assault the fortifications. All that the Parthians could do was to get closer to the defenders' positions and target them with volleys of arrows.[98] Dio states explicitly that Vologases had a vast number of warriors at his disposal.[99] They seemed to be everywhere harassing the defending positions; furthermore, their aim was so accurate that they could successfully hit any target also in the heart of the enemy camp. The fact that Paetus' soldiers refused to abandon the shelter of the camp and organise a sally to keep Parthian archers away from the wall allowed Vologases' troops to get extraordinarily close to the defendants and the fortifications, making them capable of keeping the whole camp under the fire of their deadly arrows. Watching the Roman legions trapped in their encampment, reduced to little more than sitting ducks under lethal fire from every direction, broke Paetus' morale.[100]

On his route westwards, Paetus met Corbulo's rescue force and proposed to join their armies to reconquer Armenia and save the situation. Corbulo's answer seems particularly relevant for this study:

---

94 Tac. *Ann.* 15.12.1, 17.2.
95 Tac. *Ann.* 15.10.2–3.
96 Tac. *Ann.* 15.9.
97 Tac. *Ann.* 15.13–14, 16.1; Cass. Dio 61.211.
98 Tac. *Ann.* 15.13.1–2.
99 Cass. Dio 61.11–13.
100 Cass. Dio 61.20.3–4, 21.1–2.

> Corbulo said that ... he should return to Syria, and, even as it was, he must pray for fortune under her most favourable aspect in order that the infantry, worn out by long marches, might keep pace with the enemy's untiring cavalry, certain to outstrip him on the plains, which facilitated their movements.[101]

Parthian mounted ranks were still a formidable and deadly opponent for an infantry army marching through the plains, even more so if tired by the hardships of days of forced marches like Corbulo's legions. The example of Antony's retreat from Media was still vivid.

This description matches other Tacitean ones concerning Parthian armies and tactics: Tiridates' force before Artaxata, Monaeses' army at Tigranocerta, Vologases' army at Rhandeia, and finally the border garrison facing Corbulo's crossing of the Euphrates. All Parthian armies appearing in Tacitus' narration of the Armenian wars consist of highly mobile units of mounted archers, constantly attacking the enemy with a hit-and-run tactic to induce enemy units to leave the main formation in pursuit.[102] Again, no explicit mention is made of cataphracts and of the combined arms tactics of which they were an integral part. In Tacitus, 'the Parthian is an enemy whose battles are decided by missiles'.[103]

Thanks to Paetus' surrender, Vologases could take part in the resulting peace agreements from a position of advantage;[104] therefore, he obtained the evacuation of Roman units from Armenia and the eastern bank of the Euphrates[105] and the crown for his brother Tiridates,[106] confirmed by emperor Nero in Rome in 66 CE during a solemn ceremony.

4 Rome vs. Parthia: Trajan's War

Decades of peace between the empires followed, but the political plans for direct Roman rule in Armenia that led Paetus to his fate were never forgotten. These same ideas inspired the first major invasion of the Parthian empire by

---

101 Tac. *Ann.* 15.17.1–2.
102 Overtoom (2017a).
103 Tac. *Ann.* 15.7.2.
104 Tac. *Ann.* 15.14.1, 24.1–2, 25.2–3; Cass. Dio 62.21.3–4, 22.2–3; Ruf. Fest. 20; Suet. *Ner.* 39.1.
105 Tac. *Ann.* 15.14; Cass. Dio 62.21.
106 Kovacs (2016) 32–33 n. 195; Schottky (2019) 97–98.

Trajan, fifty years after the agreements at Rhandeia. The Roman Empire and its presence in the Near East had changed significantly in those years: the system of client kingdoms had been gradually dismantled and substituted by direct provincial rule over most of the eastern imperial territories, while large military bases moved close to the border with Parthia, indicating that the only possible major threat to provincial life could come only from there.[107] With Trajan, the strategy that determined Julio-Claudian actions in the east was dismissed entirely. A new era was inaugurated, a period of direct clash between the empires. Roman emperors starting from Trajan did not think anymore in terms of appointing easily controllable monarchs in Parthia or Armenia but aimed at destroying the Arsacid Empire, launching their attacks at the core of its state system: the royal court in central Mesopotamia.

In order to get the war he wanted, Trajan needed two things: a formidable invasion army on a scale never seen before,[108] and a *casus belli*, able to justify the aggression against a state, the Arsacid Empire, that never failed to fulfil the conditions of the peace agreement signed decades before. The chance he was looking for came of course from the Armenian kingdom, where one of the two pretenders to the throne of the Great King of Parthia, in the context of internal struggle for the control of the Arsacid client kingdoms, appointed between 108 and 112 CE a certain Axidares without asking Trajan's consent first. The Great King Osroes soon replaced him with Parthamasiris in 113.[109] These changes on the throne of Armenia made by the Arsacids gave Trajan the chance he needed to denounce the end of the Rhandeia agreements and move into the Near East, where the massive army he had been gathering for years was waiting for its leader.[110]

In the autumn of 113 CE at Athens, on the way to the legions' bases on the Euphrates, Trajan rejected Osroes' peace requests, gifts, and the offer of approving Axidares' substitution.[111] Refusing any form of compromise and dialogue, the emperor decided not to follow the path of diplomacy employed extensively by Nero and Corbulo before him. For the veteran of many wars, the time had come when arms, and only arms, would decide the fate of Armenia and the rivalry between Rome and Parthia. This new approach was immediately

107 For a summary with bibliography, see Gregoratti (2016).

108 At least fourteen legions were involved plus *vexillationes* from several others, probably about 180,000 armed men, four or five times the size of Crassus' army.

109 Schlude (2020) 157–165.

110 Guey (1937) 15–43; Lepper (1948) 28–30, 156–204.

111 Cass. Dio 68.17.2–3; Arr. *Parth.* fr. 37–*38 and 40 = FGrH 156, F 51, 120; Mal. 11.270–272; Longden (1931); Guey (1937) 39–42; Debevoise (1938) 218–19; Ziegler (1964) 99–100; Angeli Bertinelli (1979) 73–74; Schippmann (1980) 60; Wolski (1993) 178–80.

evident from the emperor's behaviour towards local kings and lords: only those willing to appear in person in front of Trajan offering submission had the chance to survive; all others were considered enemies and treated as such.

Many local rulers of northern Mesopotamia like Abgar VII of Osrhoene, whose kingdom lay on the main highway towards the Parthian core cities, were forced *de facto* into an unwilling alliance to survive.[112] Those who still hesitated in 114 CE were quickly persuaded by the fate of the king of Armenia once the huge army that had crossed the Euphrates from the bases of Melitene and Satala stormed the kingdom, taking control of the most important cities and strongholds.[113]

Parthamasiris, the defeated Arsacid king of Armenia, asked for a meeting at Elegeia, looking for a way to maintain the throne through submission.[114] Explicitly referring to the Rhandeia treaty of 63 CE, he laid the crown at Trajan's feet, hoping he would pick it up to place it on his head as first Corbulo, then Nero had done with Tiridates. Instead, the crown remained on the ground, much to the surprise of the Arsacid prince, a surprise not shared by those who knew Trajan's plans for the Near East, which he made clear and publicly announced through a solemn address to all those present.[115]

That was the end of the kingdom of Armenia and the beginning of the Roman province of Armenia. No one would wear that crown anymore; no one could bear the title of king of Armenia anymore; the Parthians were allowed to leave the meeting, but the Armenians had just become subjects of the emperor. Perhaps not by chance, Parthamasiris was eliminated in dubious circumstances when a skirmish arose between Romans and Parthians while he was leaving the meeting.[116]

After getting control of Armenia, Trajan and his army descended into northern Mesopotamia in the autumn of 114 CE. Establishing his headquarters in Edessa,[117] he spent the whole of the following year, according to the most popular chronology of Trajan's Parthian Wars, consolidating his control

112 Cass. Dio 68.18.1; Ruf. Fest. 15 e 20; Guey (1937) 43–49; Lightfoot (1990) 116.

113 Cass. Dio 68.19–20; Front. *Princ. Hist.* 15; Eutr. 8.3.1; Cass. Dio 68.18.2–3; 19.2; Longden (1931) 1–4, 8–12; Guey (1937) 28–29, 32–35, 50–65; Debevoise (1938) 221–26; Lepper (1948) 88–96; Ziegler (1964) 100–103; Angeli Bertinelli (1976) 11–16; Chaumont (1976) 130–39; Angeli Bertinelli (1979) 74–75; Schippmann (1980) 61; Bivar (1983) 88–89; Lightfoot (1990) 117.

114 Arr. *Parth.* fr. 5 = FGrH 156, F 36.

115 Cass. Dio 68.19.3–4, 20.3–4.

116 Arr. *Parth.* fr. 40 = FGrH 156, F 51; Ruf. Fest. 20.

117 Cass. Dio 68.21–22; Arr. *Parth.* fr. 42–*48, *54–*55 = FGrH 156, F 142, 171; Longden (1931) 11–15; Guey (1937) 67–76; Lepper (1948) 54–62; Dillemann (1962) 281–83; Lightfoot (1990) 118.

of the region and its cities,[118] Nisibis, Singara, and Batnae, while his general Lusius Quietus conducted raids east of the Tigris in Adiabene.[119] In this first phase of the invasion, a supreme Parthian authority able to organise a coordinated resistance or reaction was absent. Weakened by internal struggles, the central authority deserted the field *de facto*, delegating the task of organising the defence to local rulers. Predictably, most of these lords and kings switched sides in front of an overwhelming and unstoppable invasion force.[120] The only exception seemed to be Mebarsapes, king of Adiabene, who dared to take the field against Roman legions before being forced to withdraw to his realm.[121] In the spring of 116 CE, according to the 'slow chronology', after the winter in Antioch, Trajan ordered the construction of a pontoon bridge across the Tigris and ordered ships to be filled with archers and soldiers and used to protect the works from the attacks of the enemy deployed on the eastern bank.[122] The necessary vessels were brought there by the army, mounted on the riverbank, and set afloat. Not expecting such a large mass of boats filled with soldiers, the Parthian defenders had to give up and withdraw to avoid being overwhelmed.[123] Trajan personally led his troops in the invasion of Babylonia, marching downstream along the Euphrates with a river fleet in support.[124] They took all the towns and cities along the river and reached Ctesiphon, which had already been left to its fate by the Great King Osroes,[125] before descending as far as the Persian Gulf, where Trajan was welcomed by Attambelus VII, king of the Arsacid vassal kingdom of Characene.[126]

The most populated and economically relevant areas of the Parthian Empire thus fell into Roman hands. Armenia and northern Mesopotamia were to become the first Roman provinces beyond the Euphrates.[127] In the Persian Gulf region, it is possible that the Arsacid client kingdom system was to be maintained under Roman suzerainty. In the surprisingly scanty accounts left

118 Cass. Dio 68.23.2; 28.2; Longden (1931) 10–12.
119 Cass. Dio 68.22.2.
120 Brizzi (2022) 152–53.
121 Cass. Dio 68.21–22; Marciak (2014b) 244.
122 Cass. Dio 68.24–25; Mal. 11.275; Longden (1931) 7; González (2000a) 210.
123 Arr. *Parth.* fr. 11–12 = FGrH 156, F 41, 43; Guey 1937, 95–120.
124 Cass. Dio 68.26; 28; Arr. *Parth.* fr. 8–10; 14–17; 36; 57–63; 67–*69 FGrH 156, F 39–40, 42, 45–48, 127, 139, 146, 151–152, 154, 164–165; Longden (1931) 12–13; Debevoise (1938) 229–33; Wirth (1963) 288–300; Ziegler (1964) 103; Angeli Bertinelli (1976) 15–17; Angeli Bertinelli (1979) 74–75; Schippmann (1980) 61–62; Bivar (1983) 89–91; Lightfoot (1990) 120.
125 Cass. Dio 68.28.3; González (2000a) 212; González (2018) 16–17.
126 Cass. Dio 68.28.3–29.1; Arr. *Parth.* fr. *70–*71; González (2000a) 215; Gregoratti (2011) 219.
127 Mal. 11.274; Jer. *Chron.* 223, 196–197; Cagnat (1927); Guey (1937) 78–80; Dillemann (1962) 196–97, 287–89; Angeli Bertinelli (1979) 74–75.

by ancient sources on this war, the Arsacid army is seldom mentioned, and no pitched battle took place: the Arsacid generals, aware of the superiority of the legions on the field, simply decided to evacuate Mesopotamia, avoiding fighting, and leaving Roman armies free to move far away from their bases, stretching their supply lines.

The Parthian Empire was not only Parthian Mesopotamia. The cities and satrapies of the Iranian plateau and Central Asia constituted for the Arsacids a protected base of operations, where they might organise a counter-attack. The Arsacid army was intact, and more soldiers could be recruited among the noble houses of the east. However, the strategic initiative was still firmly in the hands of Trajan, whose vanguard units were already penetrating south-west Iran.

What the Great King needed was something capable of stopping the Roman advance until he could be ready to react. The revolt of the major cities of northern Mesopotamia behind Roman legions' backs and along their supply lines took Trajan by surprise and offered the Parthians the chance they had wished for. Between 116 and 117 CE, probably with the involvement of the local Jewish communities, Roman garrisons left in the major cities of North Mesopotamia and Babylonia were annihilated with a substantial risk that Trajan would remain trapped in south Mesopotamia at the head of his troops.[128] Leaving the Gulf in a rush, the emperor sent his generals to deal with the hotspots of the uprising: Lusius Quietus took and plundered Nisibis and Edessa, and Erucius Clarus and Julius Alexander retook Seleucia-on-the-Tigris and set it in flames.[129] At the same time Appius Maximus Santra met his fate facing the first serious Arsacid counterattack in northern Mesopotamia, led by a certain Sanatruces. The Parthian armies began to retake the initiative, putting pressure on Armenia and middle Mesopotamia. Trajan's attempt to break the Parthian resistance by appointing an Arsacid prince, Parthmaspates, as Great King at Ctesiphon, thus *de facto* returning to the old client kingdom strategy, proved useless.[130] In 117 CE, with the emperor's health rapidly deteriorating, even though the revolt of the cities had been quashed and the Parthian offensive contained, the Roman columns began to retrace their steps back into Roman territory. After unsuccessfully besieging the city of Hatra,[131] Trajan reached Syria, where he set sail towards Rome. Death reached him on the journey back in August 117 CE.

---

128 Guey (1937) 126–28; Lepper (1948) 88–96; Neusner (1976) 57–59; Pucci (1979a); Pucci (1979b); Pucci (1981); Neusner (1986) 74–88.

129 Cass. Dio 68.30.1–2.

130 Cass. Dio 68.30.3; SHA *Hadr.* 5.4; Mal. 11.274.

131 Gregoratti (2013b) 51–52 with bibliography.

The new emperor Hadrian signed a peace with both Great Kings Osroes and Vologases II, renouncing his predecessor's grandiose expansion plans.

Unfortunately, despite the relevance and length of the conflict, no information is given by the ancient authors concerning the Parthian armies' authentic way of fighting, apart from some general references to mounted warriors shooting arrows on the eastern bank of the Tigris. This can be partly explained by the lack of contact with the enemy in the first phase of the war and the character of urban war for the control of the cities that the war assumed later, when very few field battles took place. Still, the lack of any details concerning the Parthian reaction, including the counterattack targeting Roman supply lines, and particularly the defeat suffered by Santra, prevents any remarks on Arsacid field tactics.

5 Rome vs. Parthia: Wars of the Late Second and Third Centuries

Ironically, scholars have slightly more information on warfare at their disposal for the major conflict that followed, although it is usually considered far less documented than Trajan's. In 161 CE, the Great King Vologases III, who managed to get control of the empire in 147 CE, was consolidating his position and looking for military glory against the western arch-enemy to cement his consensus among his subjects, noble families, royal officers, court dignitaries, and client kings. Considering the co-rule of Marcus Aurelius and Lucius Verus as a sign of weakness, he launched an invasion of Armenia with the purpose of including it again in the system of the Arsacid vassal kingdoms.[132] The attack, as it seems, took Roman leadership in the Near East completely by surprise.[133] The *legatus* of Cappadocia, C. Sedatius Severianus, was defeated at Elegeia and found his death along with many of his legionnaires. After these successes, which allowed the Arsacid general Osroes to threaten Syria, the invasion lost momentum.[134] In 162 CE, Verus rushed to Antioch and assigned generals to

132 SHA *Marc.* 8.6, 12.7; Dodds (1911).

133 Cass. Dio 71.2–3; Lucian, *Hist. conscr.* 18, 20–21, 25; Oros. 7.15.2; SHA *Verus* 6.9, 7.1; SHA *Marc.* 8.6, 9.1; Eutr. 8.10.2; Fronto, *Ep. ad Verum imp.* 2.1; Debevoise (1938) 246–54; Ziegler (1964) 113–16; Angeli Bertinelli (1976) 23–29; Chaumont (1976) 147–51; Angeli Bertinelli (1979) 76–77; Schippmann (1980) 65–67; Frye (1984) 243; Sheldon (2010) 156–57; Hartmann (2019); Schlude (2020) 165–67. For a new interpretation, Juntunen (2020). For recent considerations on these events, see Hartmann (2019) and Wheeler (2022).

134 Brizzi (2022) 202–203.

lead a counteroffensive that lasted for the next four years.[135] Armenia was quickly retaken. Avidius Cassius then descended along the Euphrates, clearing his way to Babylonia with a series of major battles on the banks, at Sura and Dura-Europos,[136] which resulted in disastrous defeats for the Parthians. Seleucia was plundered and destroyed, sharing the fate of the royal palace in Ctesiphon.[137] This time an epidemic of smallpox spread among the troops and prevented the Roman legions from heading further east into the Iranian plateau and Adiabene.[138]

The following peace agreement granted Roman control over the middle Euphrates and Mesopotamia north of the Khabur River. From the information available, it seems clear that the Parthian army, as in Trajan's times, was no match for the imperial legions. The attempts to stop the Roman juggernaut resulted in disastrous defeats if we decide to believe the eccentric Lucian.[139] The transformation of the Roman army gave birth to units capable of neutralising Parthian tactics based on horse manoeuvres and distance fighting. The standard Arsacid tactics of surrounding the enemy to launch multiple attacks based on mounted archery were only successful against the small army that the unprepared Severianus managed to deploy at Elegeia.[140]

The end of the third century CE saw the third Roman invasion of Parthian Mesopotamia. In the spring of 195 CE Septimius Severus decided to launch a military expedition to punish the lords of Edessa and Adiabene for helping his rival, Pescennius Niger, and occupying northern Mesopotamia during the Roman civil war.[141] Between 195 and 198/9 CE, in two separate campaigns, the Romans reoccupied the northern Mesopotamian cities and were left free to raid and plunder Babylonia, which was abandoned by the great king and his army. Severus created the provinces of Osrhoene and Mesopotamia in the north, while in the south, Ctesiphon and Seleucia, quickly deserted by their

135 Cass. Dio 71.2.1–2; SHA *Verus* 6.9, 7.1, 7.6–7, 8.12; Sillar (2002); Brizzi (2022) 211–14.

136 Hartmann (2022).

137 Cass. Dio 71.2.3; SHA *Verus* 8.3–4; Amm. Marc. 23.6.24; Zon. 12.2.

138 Gilliam (1961); Littman and Littman (1973).

139 Lucian, *Hist. conscr.* 20–21, 24, 28–29. The purpose of Lucian's treatise was to satirise the work of the contemporary historians of Verus' Parthian war. In the absence of any other information on the actual fighting, his scattered references to major battles (while criticising other authors' methods) are priceless.

140 Cass. Dio 71.2.1; 'officers were shot down'; SHA *Verus* 6.9; Lucian, *Hist. conscr.* 16; Sillar (2002) 16–17; Linz (2009) 210–11; Sheldon (2010) 156; Brizzi (2022) 203–204.

141 Cass. Dio 75.1–3; 9–12; Hdn. 2.8.8; 3.1.2; SHA *Sev.* 9.9–11, 18.1, 15–16; Debevoise (1938) 255–61; Ziegler (1964) 129–32; Angeli Bertinelli (1979) 78–79; Wolski (1993) 187–91; Verstandig (2001) 337–45; Gradoni (2007/2009); Linz (2009) 221–24; Sheldon (2010) 164–70; Schlude (2020) 166–69; Brizzi (2022) 224–30.

garrisons, were taken and plundered. Great King Vologases IV decided to leave Mesopotamia and avoid any military operation. He was mindful of the defeats inflicted on the Euphrates thirty years before and probably aware that the structural crisis which affected the empire, with the weakening of the central authority and the increased autonomy of the client kings, rendered it particularly difficult to put together a large army. Only the city of Hatra and its inhabitants resisted Septimius.

Concerning the last offensive against Parthia by emperor Caracalla (216–217 CE), even the most reliable sources do not provide much detail on military aspects.[142] It seems that Caracalla ravaged Adiabene, using the justification that the Great King Artabanus had rejected his request of marrying one of his daughters.[143] Herodian, a third-century writer with a penchant for novelistic narrations, provides some more details. According to him, the entire marriage affair was a ruse to penetrate Parthian territory, take the enemy by surprise, and massacre the unarmed warriors. Caracalla refers to the skilled Parthian horse archers, who, combined with Roman infantry prowess, would make Partho-Roman armies invincible once the two empires were united through his marriage, to convince his rival.[144] Again, during the onslaught at the marriage ceremony, Herodian stresses the extreme vulnerability of the Arsacid warriors, who laid down their bows and got off their horses before joining the celebrations.[145]

Herodian's story goes on pointing out how the outrageous betrayal by Caracalla[146] allowed the enraged Artabanus to gather a massive army of mounted archers and mail-clad warriors riding dromedaries and hurling spears.[147] After Caracalla's assassination, his successor Macrinus had to face such a danger. The whole story seems fabricated to blame Caracalla's behaviour for Artabanus' military reaction. The Parthian king invaded Mesopotamia and gained a decisive victory at Nisibis, forcing Macrinus to come to terms and ask for peace and a safe passage into Roman territory in exchange for money.[148] Again, Herodian is the only one providing some detail about the battle at Nisibis.

On this occasion, differently from their custom and surely in connection with Artabanus' desire for revenge, the Parthians looked for close combat.

142 Cass. Dio 79.1–5; Hdn. 4.10–15; SHA *Car.* 6.1–7.1; Debevoise (1938) 263–65; Ziegler (1964) 132–35; Wolski (1993) 191–92; Linz (2009) 234–41; Sheldon (2010) 172–75; Schlude (2020) 170–71.

143 Vogt (1952); Motta (2020).

144 Hdn. 4.10.3–4.

145 Hdn. 4.11.6–7.

146 Hdn. 4.14.3, 6, 8.

147 Hdn. 4.14.3.

148 Cass. Dio 79.26–27.

After more than two and a half centuries, Herodian once again puts in the field the fearsome mail-clad heavy horsemen (now riding dromedaries), and makes them perform their deadly charge into the Roman ranks. This is the first battle description after that of Carrhae by Plutarch and Cassius Dio, where a charge of heavy-armed cavalry is launched in combination with the usual Parthian tactics of arrow shooting.[149] Concerning this aspect, it will be wise to act with the same caution generally adopted when dealing with Herodian.

But the Roman army in the third century was not like Crassus' army anymore. Anti-Parthian measures to withstand the impact of the charge and keep at bay the mounted bowmen were put into practice: cavalry units and Moors armed with javelins were placed on the wings while light-armed footmen filled the gaps in the ranks.[150] The Romans thus managed to resist the impact, then used caltrops to break the charge and force helpless cataphracts to fight on foot and succumb.[151] Made stubborn by their rage, on the third day of fighting, the Parthians resorted to their usual tactics of encirclement and multi-directional long-distance attacks to which the Romans answered by over-extending their front to render encirclement impossible.[152] When Artabanus was informed of Caracalla's death, he put an end to the last military confrontation between Rome and Parthia after almost three centuries. Many specific elements beyond Herodian's reputation suggest taking his account of the battle and the entire campaign with caution.[153] Nevertheless, it cannot be excluded that, stressing Caracalla's responsibility for how things developed in the east, he decided to adapt Parthian tactics to Artabanus' resentment, introducing at the same time some elements reminiscent of Crassus' defeat like the still-feared mail-clad horsemen. In the absence of other detailed accounts, if we take Herodian's narration for genuine, one fact is undeniable. At the beginning of the third century, only the rage of their Great King and their desire to avenge the offence forced the noble Parthian cataphracts, who had avoided a direct confrontation with the legionnaires for centuries, to charge again into the Roman ranks as they did at Carrhae, but with entirely different results.

Except for Nisibis in 217 CE, as narrated by Herodian, on no occasion other than Carrhae did the combination of charging mail-clad cavalry and archers delivering deadly arrows at a distance appear on the battlefield. When the terrific cataphracts were present, as in the invading army of Pacorus and Labienus, the archers are not mentioned; when these were employed, as in the other

149 Hdn. 4.15.1–2.
150 Hdn. 4.15.1.
151 Hdn. 4.15.2–3.
152 Hdn. 4.15.4.
153 Hartmann (2009) 249.

conflicts, the cataphracts are missing from the accounts. This does not mean that those specific units were not present in the Parthian armies, but judging from the available sources, if present, they did not play a sufficient tactical role to deserve mention.

Caution would then suggest treating this combined-arms tactic as a unique occurrence employed by Surena against Crassus and never again by the Parthians, at least against the Romans. It is possible that the tactic was conceived by Surena himself. In this case, perhaps continuous supply of arrows attributed to him could have been only one of the arrangements made by the famous general to make the combination of units so effective. Perhaps that fighting method was familiar among the private militias of the Suren clan, where the aristocratic cataphracts were trained to fight in combination with mounted archers. If so, Surena was not the inventor of the tactic; he only understood that at Carrhae, the mistakes made by Crassus in deploying the units on the field determined the perfect conditions for its use, conditions that evidently never occurred again, perhaps due to Roman countermeasures.

It is nonetheless tempting, considering Surena's execution after Carrhae, to see a connection between his premature demise, which deprived Parthia of a skilled tactician, and the disappearance in western sources of the combined use of armour-clad knights and mounted bowmen on the field. What is certain is that such a tactic appears in the sources only at Carrhae; therefore, it cannot be considered an example of the way the Parthians usually fought pitched battles. Carrhae was an exceptional battle where exceptional field tactics were employed under exceptional circumstances by an exceptional military leader.

Bibliography

Abruzzese, A. (1911) 'Le relazioni politiche fra l'Impero Romano e l'Armenia da Claudio a Traiano'. *Bessarione*: 389–434.

Angeli Bertinelli, M.G. (1976) 'In I Romani oltre l'Eufrate nel II secolo d.C. (le province di Assiria, Mesopotamia e Osroene)', in Temporini and Haase (1976) 3–45.

Angeli Bertinelli, M.G (1979) *Roma e l'Oriente: Strategia, economia, società e cultura nelle relazioni politiche fra Roma, la Giudea e l'Iran*, Problemi e Ricerche di Storia Antica 7. Rome.

Angeli Bertinelli, M.G. (2000) 'In Traiano in oriente. La conquista dell'Armenia, della Mesopotamia e dell'Assiria', in González (2000b) 25–54.

Ash, R. (1999) 'An Exemplary Conflict: Tacitus' Parthian Battle Narrative (Annals 6.34–35)'. *Phoenix* 53: 114–35.

Aßkamp, R. and Esch, T. (eds) (2010) *Imperium. Varus und seine Zeit*. Münster.

Bachrach, B.S. (1973) *A History of the Alans in the West: From Their First Appearance in the Sources of Classical Antiquity through the Early Middle Ages*. Minneapolis.

Barzanò, A. (1985) 'Roma e i Parti tra pace e guerra fredda nel I secolo dell'impero', in Sordi (1985) 211–22.

Bengston, H. (1974) *Zum Partherfeldzug des Antonius*. Sitzungsberichte der Bayerischen Akademie der Wissenschaften. Philosophisch-historische Klasse. Munich.

Bivar, A.D.H. (1983) 'The Political History of Iran under the Arsacids', in Yarshater (1983) vol. 3.1, 21–99.

Braund, D. (1994) *Georgia in Antiquity: A History of Colchis and Transcaucasian Iberia 550 BC–AD 562*. Oxford.

Brizzi, G. (1983a) 'Note sulla battaglia di Carre', in Brizzi (1983b) 9–30.

Brizzi, G. (1983b) *Studi militari romani*. Bologna.

Brizzi, G. (2016) 'Cassio Dione e le campagne d'Oriente', in Fromentin, Bertrand, Coltelloni-Trannoy, Molin, and Urso (2016) 741–71.

Brizzi, G. (2022) *Roma contro i Parti. Due imperi in guerra*. Rome.

Cagnat, R. (1927) 'Inscription romaine du Sindjar au nom de Trajan'. *Syria* 8: 53–54.

Carter, M. (2004) 'The "*Lex Portorii*" and the Armenian Campaign of Caesennius Paetus'. *Latomus* 63.2: 370–79.

Chaumont, M.-L. (1976) 'L'Arménie entre Rome et l'Iran I. De l'avènement d'Auguste à l'avénement de Dioclétien', in Temporini and Haase (1976) 71–194.

Coulston, J.C. (1986) 'Roman, Parthian and Sassanid Tactical Developments', in Freeman and Kennedy (1986) 59–75.

Dąbrowa, E. (1983) *La politique de l'état parthe à l'égard de Rome – d'Artaban II à Vologèse I (ca 11–ca 79 de n. è.) et le facteurs qui la conditionnaient*. Uniwersytet Jagellónski, Rozprawy Habilitacyjne 74. Krakow.

Dąbrowa, E. (1987) 'Les premiers "otages" parthes à Rome'. *Folia Orientalia* 24: 63–71.

Dąbrowa, E. (1989a) 'Les héros de luttes politiques dans l'état parthe dans la première moitié du Ier siècle de notre ère'. *IrAnt* 24: 311–22.

Dąbrowa, E. (1989b) 'Roman Policy in Transcaucasia from Pompey to Domitian', in French and Lightfoot (1989) 67–76.

Dąbrowa, E. (2006) 'Marc Antoine, les Parthes et l'Armenie', in Traina (2006) 343–52.

Dąbrowa, E. (ed) (2009) *Orbis Parthicus: Studies in Memory of Professor Józef Wolski* (*Electrum* 15). Krakow.

Dąbrowa, E. (2017) 'Tacitus on the Parthians'. *Electrum* 24: 171–89.

Dąbrowa, E. (2021) 'Parthian-Armenian Relations from the 2nd Century BCE to the Second Half of the 1st Century CE'. *Electrum* 28: 41–57.

Daryaee, T. (ed) (2017) *King of the Seven Climes: A History of the Ancient Iranian World (3000 BC–651 CE)*. Irvine.

Debevoise, N.C. (1938) *A Political History of Parthia*. Chicago.

Dilleman, L. (1962) *Haute Mésopotamie orientale et pays adjacents, contribution a la géographie de la région du ve s. avant l'ère chrétienne au vi^e s. de cette ère*. Paris.

Dirven, L. (ed) (2013) *Hatra: Politics, Culture and Religion between Parthia and Rome*. Stuttgart.

Dodds, C.H. (1911) 'Chronology of the Eastern Campaigns of the Emperor Lucius Verus'. *Num. Chron.* 11: 209–67.

Eck, W., Santangelo, F., and Vössing, K. (eds) (2022) *Emperor, Army, and Society: Studies in Roman Imperial History for Anthony R. Birley*. Bonn.

Fabian, L. (2020) 'The Arsakid Empire', in Von Reden (2020) 205–39.

Formisano, M. and Böme, H. (eds) (2011) *War in Words: Transformations of War from Antiquity to Clausewitz*. Berlin.

Freeman, P. and Kennedy, D.L. (eds) (1986) *The Defence of the Roman and Byzantine East: Proceedings of a Colloquium Held at the University of Sheffield in April 1986*. Oxford.

French, D. and Lightfoot, C.S. (ed) (1989) *The Eastern Frontier of the Roman Empire. Proceedings of a Colloquium Held at Ankara in September 1988*. Oxford.

Frendo, D. (2003) 'Roman Expansion and the Graeco-Iranian World: Carrhae, Its Explanation and Aftermath in Plutarch'. *BAI* 17: 71–81.

Frézouls, E (1995) 'Les relations romano-parthes avant l'époque flavienne', in Frézouls and Jacquemin (1995) 479–98.

Frézouls, E. and Jacquemin, A. (eds) (1995) *Les relations internationales. Actes du Colloque de Strasbourg (15–17 juin 1993)*. Paris.

Fromentin, V., Bertrand, E., Coltelloni-Trannoy, M., Molin, M., and Urso, G. (eds) (2016) *Cassius Dion: nouvelles lectures*. Bordeaux.

Frye, R.N. (1984) *The History of Ancient Iran*. Munich.

Gabba, E. (1966) 'Sulle influenze reciproche degli ordinamenti militari dei Parti e dei Romani', in *Atti del Convegno sul tema: La Persia e il mondo greco-romano (Roma 11–14 aprile 1965)*. Rome, 51–73.

Gilliam, J.F. (1961) 'The Plague under Marcus Aurelius'. *AJPhil* 82: 225–51.

Gilmartin, K. (1973) 'Corbulo's Campaigns in the East: An Analysis of Tacitus' Account'. *Historia* 22: 583–626.

González, J. (2000a) 'Reflexiones en torno a la cronología de las campañas párticas de Trajano', in González (2000b) 203–35.

González, J. (ed) (2000b) *Trajano. Emperador de Roma*. Rome.

González, J. (2018) '*Parthia capta*: la última res gesta del emperador Trajano Parthia capta: Trajan Emperor's last res gesta'. *Veleia* 35: 13–20.

Gradoni, M.K. (2007–2009) 'The Parthian Campaigns of Septimius Severus Causes and Roles in Dynamic Legitimation'. *AJAH* 6–8.1: 19–40.

Gregoratti, L. (2011) 'A Parthian Port on the Persian Gulf: Characene and its Trade'. *Anabasis* 2: 209–29.

Gregoratti, L. (2012) 'Parthian Women in Flavius Josephus', in Hirschberger (2012) 183–92.

Gregoratti, L. (2013a) 'The Journey East of the Great King, East and West in the Parthian Kingdom'. *Parthica* 15: 43–52.

Gregoratti, L. (2013b) 'Hatra: On the West of the East', in Dirven (2013) 49–58.

Gregoratti, L. (2013c) 'The Caucasus: A Communication Space between Nomads and Sedentaries (1st BC–2nd AD)', in Magnani (2013) 477–93.

Gregoratti, L. (2015a) 'In the Land West of the Euphrates: The Parthians in the Roman Empire', in Militello and Oniz (2015) 731–35.

Gregoratti, L. (2015b) 'A Tale of Two Great Kings: Artabanus and Vologaeses', in Krasnowolska and Rusek-Kowalska (2015) 203–10.

Gregoratti, L. (2016) 'Contributions to the History of Rome in the East: Building a Province, Becoming an Empire'. *Civiltà del Mediterraneo* 27: 5–20.

Gregoratti, L. (2017a) 'Sinews of the other Empire: Parthian Great King's Rule over Vassal Kingdoms', in Teigen and Seland (2017) 95–104.

Gregoratti, L. (2017b) 'The Arsacid Empire', in Daryaee (2017) 125–53.

Gregoratti, L. (2017c) 'Corbulo versus Vologaeses: A Game of Chess for Armenia, in Looking East. Iranian Culture and History under Western Eyes'. *Electrum* 24: 107–21.

Gregoratti, L. (2018) 'Tacitus and the Great Kings', in Whiskin and Bagot (2018) 21–34.

Gregoratti, L. (2020) 'Augustus and the Parthians', in Pimentel, Lóio, Furtado, and Rodrigues (2020) 79–93.

Groebe, P. (1907) 'Der Schlachttag von Karrhae'. *Hermes* 42.2: 315–22.

Guey, J. (1937) *Essai sur la guerre partique de Trajan (114–117)*. Bucharest.

Hammond, M. (1934) 'Corbulo and Nero's Eastern Policy'. *Harv. Stud.* 45: 81–104.

Hartmann, U. (2009) 'Ein Arsakide im Heer des Septimius Severus. Überlegungen zu den Hintergründen des zweiten Partherkrieges', in Dąbrowa (2009) 249–66.

Hartmann, U. (2019) 'Die Imperiale Politik des Vologaises III. und das Bellum Parthicum des Lucius Verus'. *Anabasis* 10: 161–213.

Hartmann, U. (2022) 'Die parthische Niederlage bei Dura-Europos', in Hartmann, Stickler, and Schleicher (2022) 371–411.

Hartmann U., Stickler, T., and Schleicher, F. (eds) (2022) *Imperia sine fine? Der römisch-parthische Grenzraum als Konflikt- und Kontaktzone vom späten 1. bis zum frühen 3. Jahrhundert n. Chr. (Konferenz Jena 2019)*. Stuttgart.

Heil, M. (1997) *Die orientalische Außenpolitik des Kaisers Nero*. Munich.

Henderson, B.W. (1901) 'The Chronology of the Wars in Armenia, A.D. 51–63'. *CR* 15: 159–63, 204–13, 266–74.

Hirschberger, M. (ed) (2012) *Jüdisch-hellenistische Literatur in ihrem interkulturellen Kontext*. Frankfurt.

Jones, K.R. (2017) 'Marcus Antonius' Median War and the Dynastic Politics of the Near East', in Schlude and Rubin (2017a) 51–63.

Juntunen, K. (2020) 'The Incident at Elegeia: The Meaning of στρατόπεδον in Ioannes Xiphilinus' Epitome of Cassius Dio (S.297,14–21)'. *Arctos* 54: 167–96.

Kahrstedt, U. (1950) *Artabanos III. und seine Erben*, Dissertationes Bernenses I 2. Bern.

Karras-Klapproth, M. (1988) *Prosopographische Studien zur Geschichte des Partherreiches auf der Grundlage antiker literarischer Überlieferung*. Bonn.

Kovacs, F.L. (2016) *Armenian Coinage in the Classical Period*. Classical Numismatic Studies No. 10. Lancaster, PA.

Krasnowolska, A. and Rusek-Kowalska, R. (eds) (2015) *Studies on the Iranian World I: Before Islam*. Krakow.

Kreißig, H. and Kühnert, F. (eds) (1985) *Antike Abhängigkeitsformen in den griechischen Gebieten ohne Polisskultur und den römischen Provinzen. Actes du colloque sur l'esclavage, Iéna, 29 septembre–2 octobre 1981*. Berlin.

Kromayer J. (1896) 'Kleine Forschungen zur Geschichte des Zweiten Triumvirats. IV. Der Partherzug des Antonius'. *Hermes* 31: 70–104.

Lepper, F.A. (1948) *Trajan's Parthian War*. London.

Lerouge, C. (2007) *L'image des Parthes dans le monde greco-romain. Du debut du Ier siecle av. J.-C. jusqu'a la fin du Haut-Empire romain*. Stuttgart.

Lightfoot, C.S. (1990) 'Trajan's Parthian War and the Fourth-Century Perspective'. *JRS* 80: 116–26.

Linz, O. (2009) *Studien zur römischen Ostpolitik im Principat*. Hamburg.

Littman, R.J. and Littman, M.L. (1973) 'Galen and the Antonine Plague'. *AJPh* 94: 243–55.

Longden, R.P. (1931) 'Notes on the Parthian Campaigns of Trajan'. *JRS* 21: 1–35.

Magnani, S. (ed) (2013) *Mountain Areas as Frontiers and/or Interaction and Connectivity Spaces*. Rome.

Marciak, M. (2014a) 'The Cultural Environment of Adiabene in the Hellenistic, Parthian, and Sasanian Periods', *Parthica* 16: 111–50.

Marciak, M. (2014b) *Izates, Helena, and Monobazos of Adiabene: A Study on Literary Traditions and History*. Wiesbaden.

Mielczarek, M. (1990) 'Die parthische Panzerreiterei bei Carrhae: aus den Studien über Plutarchus, Crassus XXIV–XXVII'. *Fasciculi Archaeologiae Historicae* 4: 31–38.

Militello, P. and Oniz, H. (eds) (2015) *Proceedings of the 15th Symposium on Mediterranean Archaeology, Held at the University of Catania, March 3–5th, 2011*. Vol. 2. Oxford.

Momigliano, A. (1975) 'Corbulone e la politica romana verso i Parti', in *Quinto contributo alla storia degli studi classici e del mondo antico*, I. Rome, 649–56.

Motta, D. (2020) 'Caracalla e i Parti in Erodiano: fra disegno ecumenico e inganno'. *ὅρμος – Ricerche di Storia Antica* 12: 270–89.

Nabel, J. (2017) 'The Seleucids Imprisoned: Arsacid-Roman Hostage Submission and Its Hellenistic Precedents', in Schlude and Rubin (2017a) 25–50.

Nedergaard, E. (1988) 'The Four Sons of Phraates IV'. *ActaHyp* 1: 102–15.

Neusner, J. (1976) 'The Jews East of the Euphrates and the Roman Empire I. 1st–3rd Centuries A.D.', in Temporini and Haase (1976) 46–69.
Neusner, J. (1984) *A History of the Jews in Babylonia I: The Parthian Period.* Brown Judaic Studies. Chico, CA.
Olbrycht, M.J. (1998) *Parthia et ulteriores gentes. Die politische Bezihungen zwischen dem arsakidischen Iran und den Nomaden der eurasiatischen Steppen.* Munich.
Olbrycht, M.J. (2012) 'The Political-Military Strategy of Artabanos/Ardawān II in 34–37'. *Anabasis* 3: 215–37.
Olbrycht, M.J. (2016) 'Manpower Resources and Army Organisation in Parthia'. *AncSoc* 46: 291–338.
Oltramare, A. (1938) 'Auguste et les Parthes'. *REL* 16: 121–38.
Overtoom, N.L. (2017a) 'The Parthians' Unique Mode of Warfare: A Tradition of Parthian Militarism and the Battle of Carrhae'. *Anabasis* 8: 95–112.
Overtoom, N.L. (2017b) 'The Parthian Rival and Rome's Failure in the East: Roman Propaganda and the Stain of Crassus'. *Acta Ant. Hung.* 57: 1–22.
Patterson, L.E. (2015) 'Antony and Armenia'. *TAPA* 145: 77–105.
Patterson, L.E. (2020) 'Nero, the Reign of Tigranes VI, and the Annexation of Armenia'. *Latomus* 79: 761–78.
Pimentel, C., Lóio, A.M., Furtado, R., and Rodrigues, N.S. (eds) (2020) *Augustan Papers. New Approaches to the Age of Augustus on the Bimillennium of his Death.* Zürich.
Pucci, M. (1979a) 'Qualche osservazione sulla tradizione letteraria della rivolta ebraica al tempo di Traiano (115–117 d.C.)'. *RSA* 9: 61–67.
Pucci, M. (1979b) 'Traiano, la Mesopotamia e gli Ebrei'. *Aegyptus* 59: 168–89.
Pucci, M. (1981) *La rivolta ebraica al tempo di Traiano.* Pisa.
Regling, K. (1907) 'Crassus' Partherkrieg'. *Klio* 7: 357–94.
Sanford, E.M. (1937) 'Nero and the East'. *Harv. Stud.* 48: 75–103.
Schieber, A.S. (1979) 'Antony and Parthia'. *RSA* 9: 105–24.
Schippmann, K. (1980) *Grundzüge der parthischen Geschichte.* Darmstadt.
Schlude, J.M. (2012) 'The Parthian Response to the Campaign of Crassus Parthia'. *Latomus* 71.1: 11–23.
Schlude, J.M. (2015), 'The Early Parthian Policy of Augustus'. *Anabasis* 6: 139–56.
Schlude, J.M. (2020) *Rome, Parthia, and the Politics of Peace: The Origins of War in the Ancient Middle East.* Routledge Studies in Ancient History 12. New York.
Schlude, J.M. and Rubin, B. (*eds*) (2017a) *Arsacids, Romans and Local Elites: Cross-Cultural Interactions of the Parthian Empire.* Oxford.
Schlude, J.M. and Rubin, B. (2017b) 'Finding Common Ground: Roman Parthian Embassies in the Julio-Claudian Period', in Schlude and Rubin (2017a) 65–91.
Schur, W. (1932) *Die Orientpolitik des Kaisers Nero.* Klio Beiheft XV. Leipzig.
Schur, W. (1925) 'Untersuchungen zur Geschichte der Kriege Corbulos'. *Klio* 19: 75–96.
Schottky, M. (2019) 'Die arsakidische Nebenlinie in Armenien'. *Anabasis* 10: 96–112.

Sellwood, D. (1980^2) *An Introduction to the Coinage of Parthia*. London.
Sheldon, R.M. (2010) *Rome's Wars in Parthia: Blood in the Sand*. London.
Sillar, S. (2002) 'The Eastern Expedition of Lucius Verus (162–166 AD)'. *JAC* 17: 13–38.
Smith, F. (1916) 'Die Schlacht bei Carrhä'. *Historische Zeitschrift* 115: 237–62.
Sordi, M. (ed) (1985) *La pace nel mondo antico*. Milan.
Syme, R. (1970) 'Domitius Corbulo'. *JRS* 60: 27–39.
Tarn, W. (1932) 'Tiridates II and the Young Phraates', in *Mélanges Gustave Glotz*, I. Paris, 831–37.
Teigen, H. and Seland, E. (eds) (2017) *Sinews of Empire: Networks in the Roman Near East and Beyond*. Oxford.
Temporini, H. and Haase, W. (eds) (1976) *Aufstieg und Niedergang der römischen Welt*, II, 9,1. Berlin.
Timpe, D. (1962) 'Die Bedeutung der Schlacht bei Carrhae'. *MH* 19: 104–29.
Timpe, D. (1975) *Zur augusteischen Partherpolitik zwischen 30 und 20 v. Chr. WJA, N.F.* 1: 155–69.
Traina, G. (ed) (2006) *Studi sull'eta di Marco Antonio*. Rudiae 18. Lecce.
Traina, G. (2009) 'Note in margine alla battaglia di Carre', in Dąbrowa (2009) 235–47.
Traina, G. (2010) *La resa di Roma. 9 giugno 53 a. C., battaglia a Carre*. Bari.
Traina, G. (2011) 'Imperial Views of the Battle of Carrhae', in Formisano and Böme (2011) 211–19.
Verstandig, A. (2001) *Histoire de l'Empire Parthe* (-250–227). Brussels.
Vervaet, F.J. (1999) 'Tacitus, Domitius Corbulo and Traianus' *Bellum Parthicum*'. *Ant. Class.* 68: 289–97.
Vervaet, F.J. (1999) 'CIL IX 3426: A New Light on Corbulo's Career, with Special Reference to his Official Mandate in the East from AD 55 to AD 63'. *Latomus* 58: 574–99.
Vervaet, F.J. (2002) 'Domitius Corbulo and the Senatorial Opposition to the Reign of Nero'. *AncSoc* 32: 135–93.
Vervaet, F.J. (2003) 'Domitius Corbulo and the Rise of the Flavian Dynasty'. *Historia* 52: 436–46.
Vogt, J. (1952) 'Die Tochter des Grosskönigs und Pausanias, Alexander, Caracalla', in Weinreich (1952) 163–82.
Von Reden, S. (ed) (2020) *Handbook of Ancient Afro-Eurasian Economies*. Oldenburg.
Weinreich, O. (ed) (1952) *Satura: Früchte aus der antiken Welt*. Baden-Baden.
Wheeler, E.L. (1997) 'The Chronology of Corbulo in Armenia'. *Klio* 79: 383–97.
Whiskin, M. and Bagot, A. (eds) (2018) *Iran and the West: Cultural Perceptions from the Sasanian Empire to the Islamic Republic*. New York.
Van Wijlick, H. (2015) 'Constructing Alliances; A New Perspective on Mark Antony's Parthian Campaign of 36 BC'. *JAC* 30: 265–71.

Wheeler, E.L. 'Verus' Parthian War: Some Notes on Recent Work', in Eck, Santangelo and Vössing (2022) 147–59.

Wiesehöfer, J. (2010) 'Augustus und die Parther', in Aßkamp and Esch (2010) 187–95.

Wirth, G. (1963) 'Zur Tigrisfahrt des Kaisers Trajan'. *Philologus* 102: 288–300.

Wolski, J. (1967) 'L'aristocratie parthe et les commencements du féodalisme en Iran'. *IrAnt* 7: 133–44.

Wolski, J. (1969) 'Le rôle et l'importance des mercenaires dans l'état parthe'. *IrAnt* 5: 103–15.

Wolski, J. (1981) 'L'aristocratie foncière et l'organisation de l'armée parthe'. *Klio* 63: 105–12.

Wolski, J. (1981/1984) 'La périodisation de l'époque parthe en Iran'. *Folia Orientalia* 22: 13–21.

Wolski, J. (1985) 'Die abhängige Bevölkerung im Partherreich', in Kreißig and Kühnert (1985) 80–87.

Wolski, J. (1993) *L'Empire des Arsacides*. Acta Iranica 32. Leuven.

Yarshater, E. (ed) (1983) *The Cambridge History of Iran. Vol. 3.1–2: The Seleucid, Parthian, and Sasanid Periods*. Cambridge.

Ziegler, K.-H. (1964) *Die Beziehungen zwischen Rom und dem Partherreich: Ein Beitrag zur Geschichte des Völkerrechts*. Wiesbaden.

CHAPTER 13

Parthia, Rome, and the Horizons of Ancient Diplomacy

Jake Nabel

The concept of diplomacy as a distinct sphere of social activity is modern, not ancient, and the boundaries of the concept have been drawn in a way that reflects the power arrangements of the current age. According to Ernest Satow's standard handbook, diplomacy is 'the conduct of official relations between the governments of independent states'.[1] Though it aimed at universal applicability, this definition was a product of its time. It grew out of, and was tailored to, the Westphalian system of European nation-states that dominated world politics in the nineteenth and twentieth centuries. As Iver Neumann points out, moreover, Satow's formulation is prescriptive as much as descriptive: it tells the reader what diplomacy *should be*, not simply what it is.[2] Yet the normative aspect of the definition has been concealed by its widespread adoption among political scientists, the general public, and professional diplomats themselves. Whether in the ministry or on the street, Satow's understanding of diplomacy is the dominant one today.[3]

Ancient historians should engage with this theory of diplomacy, and they should contribute to recent efforts among political scientists to rethink it. Over the past two decades, scholars of international relations have challenged conventional definitions of diplomacy as Eurocentric, historically contingent, and unable to account for the increasing importance of non-state actors such as transnational firms, non-governmental organisations, international institutions, and diasporic ethnic or religious groups.[4] Scholars of antiquity,

1 Satow and Roberts (2018) xxix, 3. The phrase has been part of Satow's text since its first edition was published in 1917.

2 Neumann (2012) 1–2.

3 Echoes of Satow in political science literature and practitioner's manuals include Nicolson (1939) 14–16; Morgenthau (1948) 105, 419; Plischke (1973) 332–33; Watson (1982) 1; Hamilton and Langhorne (2010) 1; Berridge (2015) 1–3. The continued prevalence of the view is underlined in Constantinou and Sharp (2016) 16.

4 Jönsson and Hall (2005); Sharp (2009); Pigman (2010) 3–4; Neumann (2012); McConnell et al. (2012); Opondo (2016); Kuus (2018); Balzacq et al. (2020a); Pannier (2020) 25–27. A different but related critique of the conventional definition is Der Derian (1987).

 | DOI:10.1163/9789004710771_014

for their part, have increasingly drawn on frameworks from political science to analyse ancient interstate relations, but the theoretical literature on diplomacy remains largely untapped.[5] Ancient history can play a more prominent role in diplomacy's paradigm shift. Political scientists care about diplomacy's ancient past: they look for ways to overcome the Greco-centrism of available treatments, and they have arguably been more proactive in undertaking cross-cultural comparative histories of ancient diplomacy than have antiquity's siloed specialists.[6] As the study of diplomacy decouples from its conventional pairing with the nation-state, the political formations of the ancient world – useful precisely because antiquity was *not* an era of nations – can provide alternate models that are good to think with. As Neumann puts it, 'debates about diplomacy's future stand to gain from an awareness of diplomacy's pasts'.[7]

Parthian foreign relations with Rome are fertile ground for a theoretically oriented history of ancient diplomacy, because the subject shows both the utility and the limitations of conventional views of diplomacy as a state-bound form of communication. Some aspects of Parthian-Roman high politics can be fruitfully analysed as diplomacy in Satow's sense, with the caveat that empires and nations are different political formations. Summits, treaties, and the exchange of letters and ambassadors all accord well with conventional definitions. Other modes of interaction, however, do not, and these show the limits of a state-based approach. Into this category fall sub-state relations with Parthian and Roman dependencies; the employment of non-state agents for diplomatic missions; and the cross-border movement of dynasts outside the remit of the reigning king or emperor. The span of the Parthian-Roman relationship from

5 Aissaoui (2020); Mari and Wendt (2022); and Andres (2022) 19–28 are the rare studies to bring diplomatic theory to bear on ancient cases. Other discussions employ political science theory, but do not interrogate the concept of diplomacy. Low (2007) and Hall (2007) offer theoretically informed accounts of interstate politics in Archaic and Classical Greece. Eckstein (2006) (2008) and Overtoom (2020) draw on the neorealism of theorist Kenneth Waltz (1979) in studies of Roman and Parthian imperialism, respectively; Wang (2011) offers a similar perspective on pre- and early modern China. For a critique of the realist tradition in ancient Mediterranean history, see Lendon (2002); Hyland (2018) challenges the 'balancing' paradigm in Greco-Persian relations. Burton (2011) also treats Roman imperialism, but through the theoretical lens of constructivism rather than realism. For the ancient Near East, see Cohen (1996) 246; Cohen and Westbrook (2000); Cohen (2001); Podany (2010).

6 Jönsson and Hall (2005) 136–43; Neumann (2012) 20–33; Sofer (2013) 1–8; Balzacq et al. (2020b) 2–6. Raymond Cohen's work has spurred the incorporation of ancient Near Eastern case studies into the political science literature on diplomacy: see above, n. 5.

7 Neumann (2012) 4.

the first century BCE to the third CE offers numerous examples of both types of engagement.

The following discussion therefore assesses the role of the state in the conduct of Parthian-Roman high politics, delineating not only its involvement in diplomacy, but also its absence, circumvention, or irrelevance. Narrative political histories of Parthian-Roman interstate affairs can be found elsewhere, and there are numerous treatments of the relative prevalence of war and peace.[8] But the nature of diplomacy is assumed in those works; here, it is the issue. The projection of state-centric theorisations of diplomacy onto Parthia and Rome can homogenise and elide diverse features of their interaction, confining the reconstruction of foreign affairs to what the modern analyst expects to see.[9] A fuller accounting of these divergences will lead to a more complete view of Parthian-Roman interstate politics. The point, in the end, is not simply to write better histories of pre-modern diplomacy, but to recognise that the study of antiquity and modern theorising are interrelated endeavours.[10] Political scientists have begun to use the remote past to destabilise the theoretical assumptions of the present. Ancient historians should join them.

1 Theory and Evidence

Satow's definition assumes the existence of multiple sovereign states as a precondition for diplomacy. Satow was concerned with nations, but empires are also states, and they are moreover a type of political formation with a much older history.[11] It is therefore common to speak of diplomacy as a tool of empires, including ancient ones.[12] Indeed, diplomacy's existence in pre-modern states

8 Recent years have seen a profusion of narrative treatments of Parthian high politics: see van Kooten (2015) 508–85; Harl (2016); Schlude (2020); Fabian (2020) 209–17; Curtis and Magub (2020); Canepa (2020) 291–94; Ellerbrock (2021) 22–70. Overtoom (2020) and Olbrycht (2021) treat the early Arsacid period. Edwell (2020) begins with the tail end of Arsacid rule in Iran; Edwell (2008) covers the history of the Roman-Parthian frontier with a local focus. Shayegan (2011) covers Parthian and Sasanian ideology with a view towards engagement with Rome. Older treatments include Debevoise (1938); Ziegler (1964); Schippmann (1980); Bivar (1983); Dąbrowa (1983); Wolski (1993). On diplomacy and war, see Campbell (1993); Overtoom (2016); Schlude and Rubin (2017b); Schlude (2020).

9 Cf. Skaff (2012) 8–9, a related critique of studies of Sino-Turkic diplomacy under China's Sui and Tang dynasties.

10 Morley (2004) 6; Lawson (2018) 76.

11 Colás (2007) 3–4; Burbank and Cooper (2010) 1–3, 8–11; see also Morrison (2001) 2.

12 Sinopoli (1994) 163, 167; Black (2010) 18; Burbank and Cooper (2010) 53, 341 ('interempire diplomacy'); MacKenzie (2016).

is proof, for some, that its modern institutional apparatus – foreign ministries, permanent embassies, professionalised agents – is an incidental feature of diplomatic exchange rather than an essential one. Despite Henry Kissinger's view that 'empires have no interest in operating within an international system; they aspire to *be* the international system', there is ample evidence for diplomatic communication between two or more empires, and between empires and other kinds of polities.[13]

Satow's precondition of independence is more problematic for ancient historians, however, because this is a complex variable in imperial environments. The post-Westphalian world order regards nation-states as both sovereign and unitary, and qualified to conduct diplomacy by virtue of those attributes.[14] That logic imposes at least notional order and coherence on the modern international environment. Sovereignty within and among empires, by contrast, is 'shared out, layered, overlapping', and thus harder to parse.[15] Ancient empires could subsume various types of political formations, including tribes, city-states, and kingdoms. The incorporation of these units into imperial structures was a fitful, unsystematic, and ambiguous process that could leave the question of local autonomy unclear to contemporaries, to say nothing of modern scholars. Subject communities might even celebrate empires as the bestowers and guarantors of freedom, and such discourses have to be taken seriously even if not at face value.[16] In the context of Parthian-Roman relations, an additional complication is the contested nature of the polities along the Euphrates frontier. While some scholars have described Hatra, for example, as a city precariously balanced between the orbits of the two empires, Albert de Jong locates it firmly within the Parthian sphere.[17] If diplomacy requires sovereignty but sovereignty is ambiguous, diplomacy becomes difficult to assess.

The intractability of this problem is evident in the conviction among some ancient historians that diplomacy could take place *within* empires as part of routine communications between central authorities and their provincial subjects. 'The modern understanding of diplomacy tends to assume that it is primarily a means for conducting relations between *sovereign* states. This is not an assumption shared with the ancient world', Claude Eilers writes of the Roman

13 Kissinger (1994) 21, emphasis in the original.
14 Nathan (2002) 2–3; Jayasuriya (2005); Pigman (2010) 19–21; Neumann (2012) 41–42.
15 Quotation from Burbank and Cooper (2010) 17; see also Subrahmanyam (2001) 43; Benton (2010) 2–9; Khatchadourian (2016) 22–23.
16 Ma (2018) 279–80.
17 de Jong (2013) 148–53; cf. Anderson (2017) 142–44.

case.[18] The statement implies that the Romans (alone or together with other ancient peoples) defined diplomacy in a different way than moderns do. But they did not define it at all: they had no word for it, no professionals dedicated to its practice, and no concept of it as a distinct area of social life. It is true that the Romans could interface with both domestic subjects and sovereign foreigners through the same category of persons, namely *legati* and *presbeutai* – a significant dynamic, to be sure, and one that Rome shared with Parthia.[19] But the broader implications of the ancient practice remain unclear. Should modern theories of diplomacy accommodate intra-state communication? Can diplomacy overlap with internal administration?[20] Why is sovereignty considered essential to modern notions of diplomacy if it was incidental to the Roman iteration? The ambiguous nature of sovereignty in and among ancient empires defies tidy explanation. But grappling with this ambiguity can help ancient historians go beyond mere description of the *legatio* and the *presbeia* and contribute to interdisciplinary discussions about diplomacy on the level of theory.

Attention to diplomacy's modern theoretical vocabulary (and not just ancient terminology) is especially incumbent on students of Parthian-Roman relations, because a lack of evidence hinders reconstruction of the relevant concepts and institutions on the Arsacid side. Nearly all of the sources for foreign relations between the two empires are Roman. This lopsided evidentiary foundation does not only mean that the events of high politics are presented solely from the Roman perspective, though that problem would be substantial enough on its own. It also means that the Parthian way of thinking about diplomatic activity is obscure. There is virtually no evidence for who served as ambassadors, to whom they reported, how much agency they had, what languages they spoke, and what sort of interlocutor they saw in the Romans. The Arsacids and those close to them may have followed Hellenistic precedents, especially during their empire's formative centuries. In a rare piece of Roman documentary evidence for the reception of a Parthian envoy, for instance, a Roman governor commands that the Parthian be offered *xenia*, using the Greek

18 Eilers (2009b) 1, emphasis in the original. See also Campbell (2001) 3; Torregaray Pagola (2016) 116–17.

19 The Greek inscription known as the 'Letter of Ardawān/Artabanus II to Susa' records that a citizen of that city – a Parthian subject territory – undertook embassies to the Arsacid king (line 7: *presbeiai*). For the text, see De Rossi (2004) 121 no. 218; Merkelbach and Stauber (2005) 86 no. 407; Hackl et al. (2010) 2.486; Rougement (2012) 25 no. 3; for recent discussions of the inscription, see Gregoratti (2012) 130–32; Wiesehöfer (2015) 339; Olbrycht (2017) 7–8, 14–15.

20 Thus Eck (2009) 193.

term for diplomatic hospitality in a Latin letter.[21] But was *xenia* the operative concept for the envoy himself, especially in an era when the Greek language was falling into disuse in Parthian territory? Conclusive evidence is lacking.[22] In short, while the application of modern theory runs the risk of analysing ancient diplomacy in terms foreign to its practitioners, such anachronism can be salutary, since it does not allow Roman norms to be the baseline for the discussion.

Roman evidence dominates the study of the topic, then, but the material has major shortcomings for the issues diplomatic theorists and historians investigate. Most of the relevant passages come from historiographical texts in Greek or Latin whose purpose was to tell true tales about the past for edification, moral instruction, and pleasure. The job of the historian was *not*, however, the systematic exposition of political institutions or the literal reproduction of documents, except where these considerations were essential to the narrative.[23] Whereas historians of the Bronze Age can study epistolary correspondence among the rulers of the Eastern Mediterranean interstate system – a body of evidence that directly reflects the content, mediums, and norms of diplomatic communication at the highest level – students of Parthian-Roman relations have no such opportunity.[24] Royal letters, treaty texts, field reports, dispatches to the central government: all once existed but are now only haphazardly preserved in the impressionistic summaries of historians whose fundamental concerns were literary, not documentary.

The meager evidence for diplomatic practice is due to the vagaries of survival, but as for theory, there never was any to begin with. 'Explicit theorizing was not a Roman characteristic' even in the realm of political thought, where the Romans preferred moral and didactic exposition to the abstract thinking of the Greeks.[25] But the Greeks came no closer than the Romans to formulating a

21 *P Dura* 60 B (c. 208 CE), line 7; for discussion, see Chaumont (1987); Iovine (2023) 207–17.

22 Cf. Lenfant (2022) on *xenia* in earlier Greco-Persian relations. For the decline of the Greek language in Parthia from the first century CE onward, see Wiesehöfer (2015) 339. On the Arsacid relationship to Greek culture more generally, see Wolski (1983); Dąbrowa (2011); Olbrycht (2017); Messina and Versluys (2022).

23 On the instructional aims of Greek and Roman historians, see, e.g., Chaplin (2000); Nicolai (2007) 13–14; Turpin (2008); Stadter (2009) 459–70; Roller (2018) 17–23. Where descriptions of political institutions are included, they are often written by outsiders and for outsiders, e.g., Polybius' account of the Roman 'constitution' (Polyb. 6.2–18; cf. Pelling (2007) 244–50). On the use of documents, see Marincola (1997) 103–105; Rhodes (2007).

24 On the epistolary evidence for Bronze Age diplomacy in the Eastern Mediterranean, see Cohen and Westbrook (2000); Liverani (2001); Lafont (2001); Mynářová (2008); Podany (2010).

25 Quotation from Wiedemann (2000) 517.

concept of diplomacy that was distinct from other areas of social life. Figures like Thucydides, Aeschines, and Demosthenes were not 'thinkers of diplomacy', even if their texts have proven useful to political scientists who are.[26] Delineating the role of a *presbeutēs* or a *legatus* is not the same as delineating a concept of diplomacy. The inhabitants of the ancient Mediterranean did the former, but not the latter. As with economic history, theoretical and comparative work on ancient diplomacy requires the importation of modern categories alien to the ancient peoples under consideration.

The study of Parthian-Roman diplomacy does, however, have one major anchor in the ancient evidence, and that is the Roman recognition of Parthia as a political unit similar to itself in size, power, and form. Recent work in political science shows how diplomacy gives its practitioners a new understanding of themselves by bringing them face to face with an Other beyond their borders and beyond their experience; the job of the diplomat is 'to mediate estrangement, while retaining separateness'.[27] This dynamic was a key facet of Parthian-Roman relations. The literature on the Roman configuration of Parthia as its own cultural antithesis is vast, and deservedly so: no other foreign power took up so much space in the Roman imagination.[28] But while expressions of prejudice and xenophobia were many, so too were acknowledgements of coequality, even if some authors were 'careful not to concede full parity'.[29] The geographer Strabo encapsulates this tendency when he calls the Parthians *antipaloi*, a word that signifies an evenly matched opponent or antagonist.[30] Before Parthia, Rome had no peer in the world, no one to hold a mirror to its empire or bring its exceptionalism into question. The Arsacids challenged Rome's imperial solipsism. Once it became clear that war would not make either side go away, diplomacy was essential for finding a *modus vivendi* but also for articulating a new imperial identity in the face of a counterpart. The absence of evidence allows little insight into this process on the Parthian side,

26 Quotation from Constantinou and Sharp (2016) 14 on Aeschines and Demosthenes. Some ancient historians see indications that the Greeks theorised interstate relations: see Lendon (2006); Hall (2007) 85. But diplomacy as a discrete realm of activity makes no appearance in Greek political thought.

27 Quotation from Constantinou (2013) 142, discussing Der Derian (1987) and Sharp (2009). See also Kuus (2017) 86–87.

28 Key studies include Sonnabend (1986); Lerouge (2007); Shayegan (2011) 332–40; Schneider (2012); Wiesehöfer and Müller (2017); Alidoust (2020).

29 Quotation from Syme (1958) 530, discussing the representation of Parthia in the Roman historian Tacitus.

30 Strabo 11.9.2. For similar expressions, see Vell. Pat. 2.101.2; Plin. *HN* 5.88; Joseph. *AJ* 18.46; Tac. *Ann.* 2.56.1; 2.60.4, 12.10.2, 15.13.2; Cass. Dio 40.14.3; Hdn. 4.10.2.

but the Roman sources are ample testimony: Rome's self-conception changed through interaction with Parthia, and through diplomatic interaction not least.

Despite the profound differences between empires and nations, then, the mutual engagement of Parthia and Rome as coherent political units justifies the application of a diplomatic framework to their interactions. Within both empires, of course, myriad forms and degrees of sovereignty existed. Provincial integration into Arsacid or Roman imperial structures was not a binary matter, and diffuse notions of territoriality meant that the political landscape of the Euphrates frontier was complex, shifting, and ambiguously demarcated. Yet the Parthian-Roman relationship was not simply a kaleidoscopic jumble of greater and lesser political units colliding or coexisting in haphazard fashion. Both sides leveraged centralised state power to make war on each other, but also to represent, mediate, and negotiate in their own interests. Despite the differential forms that states took in the ancient world, then, conceptions of diplomacy as a state-bound form of communication therefore have considerable explanatory power – though also considerable limitations.

2 State-Bound Forms

Three features of Parthian-Roman interaction are reconcilable with the state-centric framework around which the modern concept of diplomacy is built. These are the direct exchange of letters and messengers between central governments that claimed sovereign power; summits at strategic and symbolic meeting sites; and treaties. The Roman sources are reliable evidence that these instruments were in regular use for the duration of the Parthian-Roman relationship. They are *not* reliable evidence for how the Parthians understood them, though in some cases extrapolation from Near Eastern sources may offer insight.

Parthian and Roman leaders exchanged letters and messengers from their first encounters to their last. The main terms for envoys were *legatus* in Latin and *presbeutēs* in Greek. The Parthian word was *frēštag* (cf. Middle Persian *frēstag*), but it is unattested before the Sasanian period.[31] It is unclear which of these languages predominated in Parthian-Roman interaction, and in what proportion. Greek may have been a lingua franca, since elites in both empires were well versed in it.[32] On the other hand, reliable evidence from Late Antiquity (though still Roman and literary) shows formal support for Greek/Middle

31 For attestations, see Gignoux (1972) 61 n. 89; Durkin-Meisterernst (2004) 85, 159–61.

32 On the Arsacids and the Greek language, see Wiesehöfer (2000); Olbrycht (2014).

Persian bilingualism through interpreters, translations, and cross-referenced documents.[33] Yet the late Sasanian period may not reflect the practice of the Arsacids, who stood closer to Iran's Hellenistic past. One source mentions that an Arsacid visitor to Rome in 66 CE required public translation, presumably into Latin. But from what source language, Parthian or Greek? Scholars do not agree.[34] The lack of consensus on such points shows the limitations of the evidence on even basic questions of Parthian-Roman communication.

Although messengers were used to conduct diplomacy, they were not quite diplomats, at least not in the modern sense. They did not reside in Ctesiphon or Rome on a permanent basis. Their work was not part of a dedicated institution. And their business was more often to execute policy than to craft it.[35] To take one example from 63 CE, Tacitus describes the arrival in Rome of Parthian *legati*, which is the usual Latin word for ambassadors or envoys. These agents spoke with the emperor Nero, but only to deliver 'the instructions of King Vologases [= Walgaš I, the Arsacid king of kings of Parthia] and a letter along the same lines'.[36] The message was thus conveyed in both oral and written form, but the Arsacid king controlled the content in both cases. Moreover, when the emperor rejected the proposed conditions, he sent the representatives back to Parthia with a counter-offer. They could not, it seems, respond to it on their own initiative.[37] Romans were sent to the Arsacid kings under similar parameters.[38] Additional information about the meetings between messengers and sovereigns might clarify the scope of the former to exercise agency, but the brief descriptions in Classical historiography fall short in this regard. As far as the sources show, messengers were conduits for Parthian-Roman diplomacy rather than its practitioners.

Even if their room to maneuver was limited, however, messengers held great importance in matters of interstate recognition, legitimation, and community. It mattered where a messenger was sent, because to receive one was to be

33 Dignas and Winter (2007) 119–51; Lee (2008) 108–11; and Morley (this volume).

34 See Suet. *Ner.* 13.2 on Tiridates' public speech at Rome. Peretz (2006) 465–66 favours Parthian as the language of Tiridates' original utterance; Millar (1988) 373 is non-committal on Parthian or Greek.

35 On the relative agency (or lack thereof) of the ancient ambassador, see Giovannini (2007) 96–97; Torregaray Pagola (2013) 243; Ager (2017) 294, 296.

36 Tac. *Ann.* 15.24–5; quotation from 15.24.1: *legati Parthorum mandata regis Vologaesis litterasque in eandem formam attulere*. For discussion, see Ziethen (1994) 182.

37 Tac. *Ann.* 15.25.3. Most evidence for Parthian embassies comes from historians like Tacitus, but see above, n. 21, for a rare documentary reference (albeit still in a Roman source).

38 See, e.g., Lucullus' dispatch of messengers to Frahād/Phraates III in 69/8 BCE: Plut. *Luc.* 30.1; App. *Mith.* 87; Cass. Dio 36.3.1–2.

recognised as a source of political authority and sovereign power. The sensitivity of rulers to such matters is well illustrated in a statement that Cassius Dio attributes to Augustus' advisor Maecenas:

> Moreover, in other respects too, it seems to me that you would be arranging things in the best way if you brought before the Senate the embassies that come from enemies and from allied kings and peoples. For it is honourable and important that the deliberative body appear to be in charge of all these [embassies], and that there appear to be many opponents to those of them who act unjustly.[39]

The opinion is offered as part of a debate about whether and to what extent Augustus should steer Rome away from a Republican form of government and towards an autocratic one (and how to opt for the latter without seeming to). But it is a powerful comment on how the reception of diplomatic messengers contributed to the maintenance of sovereign power both in actuality and in appearance. During the Republic, the Senate had control over foreign policy and was therefore the institution that heard foreign embassies. Maecenas argues that this custom must be preserved if Augustus wishes to treat the Senate correctly, even if his regime is appropriating the actual power to answer such delegations. Dio is probably using Maecenas to lament the lapse of this practice in his own time, but in earlier periods the senatorial sentiment mattered, not least when the envoys in question were Parthian.[40] After Vespasian won the emperorship in the civil war of 69 CE, for example, the Parthian envoys who came to congratulate him were forwarded on to the Senate.[41] The way that Rome received foreign embassies showed what kind of polity it was; the empire and its self-sustaining fictions took shape through interaction with the outside world, especially Parthia.

---

39 Cass. Dio 52.31.1: καὶ μέντοι καὶ τἆλλα ὧδ' ἄν μοι δοκεῖς ἄριστα διατάξαι, ἂν πρῶτον μὲν τὰς πρεσβείας τάς τε παρὰ τῶν πολεμίων καὶ τὰς παρὰ τῶν ἐνσπόνδων καὶ βασιλέων καὶ δήμων ἀφικνουμένας ἐς τὸ συνέδριον ἐσάγῃς (τά τε γὰρ ἄλλα καὶ σεμνὸν καὶ ἀξιόλογόν ἐστι τό τε τὴν βουλὴν πάντων κυρίαν δοκεῖν εἶναι, καὶ τὸ πολλοὺς τοὺς ἀντιπάλους τοῖς ἀγνωμονοῦσιν αὐτῶν φαίνεσθαὶ); cf. Cass. Dio 53.21.6.

40 For the passage as a comment on diplomatic reception in Dio's time, see Millar (1988) 348–49; Sidebottom (2007) 6–7. Talbert (1984) 423 discusses how the emperor benefitted from consultation of the Senate on diplomatic business. On Maecenas as a mouthpiece for Dio's views, see Burden-Strevens (2020) 19 n. 72, 270–74. On the 'constitutional debate' between Agrippa and Maecenas, see also Fomin (2016).

41 Tac. *Hist.* 4.51; cf. Suet. *Vesp.* 6.4.

The absence of evidence makes it impossible to know whether such a dynamic prevailed within the Arsacid Empire as well, though one statement from an early Sasanian king offers at least a suggestion that it did. After his seizure of the throne in 293 CE, the ruler Narseh erected a monument at the site of Paikuli with inscriptions in Middle Persian and Parthian. In a concluding passage, the ruler boasts that 'whoever came to our court himself or sent messengers, gifts, letters, and hostages as promises of loyalty to our court, he received fame and other kinds of things'.[42] The referents of this phrase are unspecified, but the text has just rehearsed a long list of dignitaries, some of whom were clearly part of the Sasanian Empire, and others – including 'Caesar and the Romans' – who were not.[43] If the status of the senders is ambiguous, however, it is clear that the king celebrates messengers because of the power of recognition. Narseh's accession rested on internal *and* external acknowledgement of his kingship, and the importance of Rome as the Sasanian Empire's most important interlocutor explains why the western empire merits such a prominent mention. How well these Sasanian ideas reflect Arsacid precedents must remain an open question, but the composition of a Paikuli text in Parthian – a language that remained 'a major means of communication for Sasanian royalty' in the third century CE – suggests at least some continuity in this area.[44]

A more direct venue of interaction between Parthia and Rome was the summit. Contemporary diplomatic handbooks distinguish between routine diplomacy conducted via intermediaries and summit diplomacy, a face-to-face meeting between the political leaders whom the diplomats ordinarily serve.[45] Such conferences could take place between Parthian and Roman rulers at the highest levels of power, up to and including the kings and emperors themselves. The highest profile meetings were between Gaius Caesar and

42 NPi 94 = Skjærvø (1983) 1.73, 2.130: *W MNW BNPŠE OL BBA ZY LNE YATWN 'yny' plys[tky] W d'šny W npyky W np'(k) p[t]st[w](k'n?) OL BBA ZY [LNE ŠDRWN- OLE ? hws?]lwbyhy W CBW ZY AHRN gwnky YHSNNt*. I lightly adapt Skjærvø's translation. The Middle Persian text is quoted; the Parthian version of this section is almost entirely lost. Note that the discovery of new blocks of the Paikuli inscription has led to several addenda to, and proposed emendations of, Skjærvø's text: see Cereti and Terribili (2014); Cereti et al. (2019); Cereti (2021).

43 NPi 91–93. On the relationship of the dignitaries mentioned in these sections to the Sasanian Empire, see Skjærvø (1983) 2.120; Weber and Wiesehöfer (2010) 119–21.

44 Quotation from Gyselen (2016) 61. The fragmentary state of the Middle Persian and Parthian texts makes comparison difficult, but the correspondence seems, in general, close; see Cereti and Terribili (2014) 348.

45 Berridge (2015) 184–96; Dunn and Lock-Pullan (2016).

Frahātak/Phraataces in 2 BCE;[46] Lucius Vitellius and Ardawān/Artabanus II in *c.*36/7 CE;[47] Corbulo and Tirdād/Tiridates in 63 CE;[48] Nero and Tirdād in 66 CE;[49] and perhaps Caracalla and Ardawān/Artabanus IV, though the historicity of this last episode is doubtful.[50] The participants were kings and emperors, members of the ruling dynasty, or else political figures close to the top ranks. Their meetings were bilateral, because no third party was big or powerful enough to merit a seat at the table.

While the exclusively Roman evidence permits only a skewed view of what transpired at summits between imperial leaders, it seems that such events fostered a sense of coequality and equilibrium. That is not to say that they brought about peace, necessarily, although some scholars maintain that relations were less violent in the first century CE when most summits took place.[51] Rather, personal interaction helped Parthians and Romans accept the power of the other as a permanent fact, an embedded feature of the interstate environment instead of an obstacle to be overcome. The eyewitness account of the Roman author Velleius Paterculus, for instance, describes the summit of 2 BCE with awe at the communion of such mighty dignitaries:

> This spectacle of the Roman army standing on one side and the Parthian on the other as the two most eminent heads of empires and of men met – an exceedingly brilliant and memorable sight – it was my luck to witness during the beginning of my military service when I was a tribune.[52]

---

46 Vell. Pat. 2.101; Cass. Dio 55.10a.4; cf. Plin. *HN* 6.141; Plut. *Mor.* 207e. Strabo 1.1.17 may also refer to Gaius' campaign: see Drijvers (1998) 290.

47 Joseph. *AJ* 18.96, 101–105; Suet. *Calig.* 14.3; *Vit.* 2.4; Cass. Dio 59.27.2–3. On the date, see Dąbrowa (1983) 107.

48 Tac. *Ann.* 15.27–31; Cass. Dio 62.23; discussion in Ziethen (1994) 183–85. Cass. Dio 62.22.2–3 also mentions an earlier conference on the Euphrates river between Corbulo and Monaeses.

49 The only substantive account of the so-called 'Golden Day' is Cass. Dio 63.1–8; shorter references in Suet. *Ner.* 13; Tac. *Ann.* 16.23.2–24.1; Plin. *HN* 30.17. For an effort to reconstruct this event from the Parthian perspective, see Nabel (2019c).

50 Herodian 4.10–11 has Caracalla meeting Ardawān/Artabanus in Parthia after the Arsacid king agreed to marry the Roman emperor to his daughter. But the whole episode seems to be fabricated out of whole cloth. Dio (78.1.1) writes that Caracalla was unsuccessful in his request to marry Ardawan's daughter, and nowhere else is Herodian's account corroborated. For discussion, see Timpe (1967); Zimmermann (1999) 210; Huber and Hartmann (2006) 504.

51 For this view, see, e.g., Campbell (1993) 220–34; Schlude (2020) 92–155.

52 Vell. Pat. 2.101.2–3: *Quod spectaculum stantis ex diverso hinc Romani, illinc Parthorum exercitus, cum duo inter se eminentissima imperiorum et hominum coirent capita, perquam clarum et memorabile sub initia stipendiorum meorum tribuno militum mihi visere contigit.*

To judge by Velleius' testimony, the summit was a ceremonious meeting of partners in world rule, not a grudging standoff between inimical rivals. The event he attended was staged on the Euphrates, moreover, as were other Parthian-Roman summits. These collaborative acts of demarcation made the river a symbolic border for many decades, though the reality was more complicated.[53] Even if the arrangements they formalised were already substantively resolved before the conferences took place – a question that the available evidence cannot answer – summits reconciled Parthia and Rome to the other's existence and reshaped how the interstate environment was understood on both sides of the Euphrates.

Finally, Parthia and Rome concluded treaties. The operative terms are *foedus* in Latin and *spondai* in Greek, though the Romans could also describe less formal agreements as *amicitia* ('friendship') or *societas* ('association'). While contemporaneous Arsacid sources preserve no correspondents, the Parthian word *pādixšīr* is known from later texts, and it may have been the Middle Iranian term of art.[54] The obscurity of the Parthian terminology could be attributable to the non-survival of Iranian textual sources, but it might also reveal a more substantive divergence between Parthian and Roman conceptions of what treaties were and how they operated. Oral agreements with external peoples played a role in Roman foreign relations, but many Roman *foedera* were written down, and copies were deposited in public buildings. In some cases, the texts were even publicly displayed.[55] But the Parthians shared in the predominately oral culture of pre-Islamic Iran in which religious, literary, and even political matters belonged to the realm of the spoken word, not the written one.[56] For Roman-Sasanian relations in Late Antiquity, Greco-Roman historiography offers much more ample evidence for the scribal practices behind treaties, and for the Persian side as well as the Roman: when the two sides drew up a treaty in 561/2 CE, for instance, separate translations in Greek and Middle Persian were commissioned, cross-checked, copied, sealed, and ratified.[57] The

53 On the Euphrates as a symbolic frontier, see Wheeler (2007) 235–36.

54 For *pādixšīr*, see ŠKZ 35 = Huyse (1999) 48. But note that the passage is describing Sasanian scribal protocol, not an agreement with a foreign state, and the ŠKZ's Greek equivalent of *pādixšīr* is not *spondai*; cf. Perikhanian (1983) 662 n. 4.

55 On the inscription and deposition of Roman *foedera*, see Meyer (2004) 95–97; Gladhill (2016) 1–68; but cf. Wheeler (2002) 289 on oral agreements. Cic. *Balb.* 53 speaks of a publicly displayed treaty.

56 On orality in pre-Islamic Iran and its impact on extant Parthian evidence, see Boyce (1983); Shayegan (2012) 73–108; Wiesehöfer (2020) 477–78.

57 The key primary source for the treaty of 561/2 CE is Menander Protector fr. 6.1. On official bilingualism in Roman-Sasanian relations, see the literature cited above, n. 33.

Arsacids may have employed such procedures in the negotiation of treaties with Rome, but there is no direct evidence that they did so. Roman sources contain numerous allusions to treaties, but details are few. Even the Roman understanding of their terms is obscure, to say nothing of how they may have looked through Parthian eyes.

In the absence of detailed source material, modern assumptions about treaties have sometimes filled in the gaps. The classic study is that of Karl-Heinz Ziegler, whose narrative of Parthian-Roman political history had treaties at its centre. On his reckoning, Parthia and Rome struck nine or ten formal agreements over the course of their relationship.[58] For Ziegler, these instruments laid the foundations for a system of international law between sovereign states that would develop still further when the Sasanians came to power. But his emphasis on the normative and regulatory power of treaties has struck some as overly legalistic and schematic, and the complete absence of reliable texts for these agreements allowed for ample interpretive license of their force and scope.[59] The summary descriptions in the literary sources suggest that Parthian-Roman treaties could delimit territory, provide for mutual defence, apportion suzerainty over local rulers, and restore captured people or property. But attempts to explain Parthian and Roman behaviour on the putative basis of treaties (or their lapses) have foundered, because the specifics regarding their terms, duration, and other matters do not survive. While the Parthians and Romans certainly made treaties and saw them as a component of their relations, an evidentiary gap precludes access to these instruments, and modern legalistic frameworks cannot bridge the distance.

Messengers, summits, and treaties were all modes of Parthian-Roman engagement that are commensurable with state-centric theorisations of diplomacy. Parthia and Rome were sprawling, diffuse, heterogeneous empires with a level of integration and state institutionalism that fell far short of modern norms. Moreover, the frontier between them was a variegated patchwork of dependencies, and questions of territoriality and sovereignty were ambiguous and contested. But direct and indirect communication between Parthian and Roman rulers solidified and demarcated, at least conceptually, the empires that they governed. Through these media, Parthia and Rome gained a clearer

58 Ziegler (1964) 19–32 (Sulla, a questionable case; Lucullus; Pompey), 48–50, 54–57 (Augustus), 58–64 (Tiberius/Germanicus; Vitellius), 73–77 (Nero), 114–15 (Marcus Aurelius), 132 (Severus), 134 (Macrinus). For the Republican period, see also Keaveney (1981); Keaveney (1982).

59 For this critique of Ziegler, see Wheeler (2002) 290, an 'extreme case for the non-existence of a *foedus Parthicum* before 218'.

sense of the other – and themselves – as political units that existed in counterpoise, though not necessarily in peace. Major issues of recognition, sovereignty, and legitimacy were at stake in the sending and receiving of embassies, while summits and treaties allowed the two sides to engage as singular entities rather than motley conglomerations of greater and lesser territories. If the first centuries CE were characterised by a 'division of the world' between Parthia and Rome, state-oriented diplomacy helped draw the fault lines, and it helped define the centres of gravity in the east and west.[60]

3 Non-State Forms

Other modes of interaction in Parthian-Roman high politics cannot, however, be satisfactorily explained by state-bound conceptions of diplomacy. Many scholars have explored the porousness and permeability of the Euphrates frontier in recent studies of religion, trade, and visual culture, and few today would argue that political borders severely limited trans-imperial flows in these areas of social life.[61] Because of its traditional state-centric orientation, however, diplomacy has been largely untouched by such analysis. A more capacious understanding of diplomacy's relationship to state structures can better explain key dynamics. Given the ancient empire's light footprint in comparison with modern nations, Parthian-Roman relations must be surveyed for state-adjacent modes of foreign affairs that do not adhere to contemporary ideas about diplomacy's limits.

One such mode of engagement was what political scientists call 'sub-state diplomacy', a practice that shows the stark divergence between ancient and modern notions of territorial sovereignty. For students of the modern world, sub-state diplomacy is a way to interpret how 'central governments have gradually lost their monopoly on external activity' to regional and municipal authorities as transnational networks densify through globalisation.[62] But the idea is also well-suited to the ancient world, when central governments had no discrete concept of diplomacy and therefore no drive to monopolise it. Sowing

60 The phrase 'division of the world' (*divisio orbis*) is from Justin's epitome of Pompeius Trogus (41.1.1).

61 On religion, see Boyce and Grenet (1991) 197–490; Dirven (1999); Herman (2012). Trade: Young (2001) 123–68; Graf (2018); Gregoratti (2019). Visual culture: Hauser (2014); Dirven (2016); Anderson (2017). High-resolution studies of local cities and kingdoms further illustrate the fluidity of the frontier zone: see Sommer (2005); Marciak (2014); Luther (2015); Kaizer (2017).

62 Quotation from Puybareau and Takam Talom (2020) 167.

division was a ubiquitous technique of conquest and territorial annexation when ancient empires were on the march, of course. But even in systems with a more or less stable status quo, regional dependencies of larger imperial structures could and did maintain relations with foreign states. Such channels ran parallel or even in opposition to the diplomatic networks of the empires to which they belonged.

One example involving especially far-flung communication comes from the mid-first century CE, when the Romans under Nero received an embassy from the Hyrcanians, a people who lived on the southeastern shores of the Black Sea. The political organisation of Hyrcania is not well understood, but it had been a Parthian dependency from the earliest days of the empire, and in 60 CE it was in revolt.[63] Its rebellion coincided with a Parthian-Roman conflict over Armenia, and coordination between Hyrcania and Rome was probably a matter of military expedience first and foremost. The shape of the embassy, however, went beyond the bare minimum of communication: the Hyrcanian messengers asked for an alliance not from Corbulo, the Roman commander in the Armenian theatre, but from Nero himself. Upon their return to the east, Corbulo gave them a military escort that helped see them home safely.[64] The resolution of their rebellion against Arsacid rule is unclear, as is the outcome of their embassy.

The meagre evidence mires the Hyrcanian episode in obscurity to begin with, but the confusion is compounded by the unstated application of conflicting frameworks for the study of ancient diplomacy. For Werner Schur, the embassy indicated that Hyrcania had become a sovereign and independent state; others disagree and prefer the view that it remained subordinate to the Arsacid Empire.[65] The debate is illustrative, though not only in the way that its participants intended. Its source of contention was not just the status of Hyrcania (for which there is, at present, no meaningful evidence beyond Tacitus), but rather incompatible ideas about diplomacy. For Schur, a

63 Tacitus is the sole Roman source for the embassy (*Ann.* 14.25.2; cf. 13.37.5, 15.1.1, 15.2.4), and he mentions it during his narrative of events in 60 CE. On the chronology, see Wheeler (1997) 391–92.

64 Tac. *Ann.* 14.25.2. Tacitus says that Corbulo conducted the Hyrcanians to the shores of the Red Sea and that they avoided Parthian territory on the way home (*dato praesidio ad litora maris rubri deduxit, unde vitatis Parthorum finibus patrias in sedes remeavere*), but the geography does not make sense, and a textual corruption has been suspected. For discussions of this problem, see Woodcock (1939) 114; Wellesley (1955) 136–37; Bivar (1983) 83; Potter (1991) 287 n. 40; Heil (1997) 87 n. 8; Olbrycht (1998) 181–82.

65 Schur (1923) 17–21; for criticisms of the view, see Dąbrowa (1984) 146; Schottky (1991) 120; Heil (1997) 87–89; Olbrycht (1998) 182–83.

diplomatic exchange like the one between the Hyrcanians and Nero implied sovereign power and its interstate recognition; for his opponents, diplomatic contact with the outside world was compatible with dependent status. Disagreements, then, are fuelled by conflicting theorisations of diplomacy every bit as much as divergent interpretations of the evidence. Grappling with sub-state diplomacy in the abstract might not resolve this debate, but it would at least clarify the operating assumptions of the scholars on either side.

Sub-state diplomacy emerges as a theme again in an inscription that points to a second mode of state-adjacent diplomacy: the delegation of embassies to intermediaries otherwise unconnected to the upper tiers of Parthian and Roman administration. The text is in Palmyrene Aramaic and dates to 17–19 CE or not long thereafter.[66] The stone is fragmentary and poorly preserved, but it seems clear enough that the text commemorates a Palmyrene named Alexander who undertook a mission for Germanicus Julius Caesar, the adopted son of the Roman emperor Tiberius.[67] The details of Alexander's assignment are not recorded, but his destinations apparently are: the assignment took him to Emesa, a Syrian kingdom to the west of Palmyra, and to Mesene/Characene, a Parthian dependency on the Persian Gulf in southern Mesopotamia.[68] The fragmentary text leaves much unclear, and even if it were intact, questions would probably remain. In what capacity did Germanicus send Alexander to these places? What was his business? Why him? Still, his travel to Mesene, a place far beyond Rome's eastern frontiers, has led many scholars to label the endeavour a diplomatic mission or an embassy.[69]

The text may attest, therefore, another case of sub-state diplomacy, but with the added dimension that diplomacy piggybacked on Palmyra's trans-imperial commercial networks and political influence. Alexander's mission ran parallel

66 *PAT* 2754 = Cantineau (1931) 139 no. 18 = De Rossi (2004) 87 no. 146; English translation in Matthews (1984) 164.

67 Line 3: '… previously(?), and Germanicus sent him' ([…]*h lqdmyn wšdrh grmnqs*). Germanicus spent his final years in the Roman east and died there in 19 CE. He may have visited Palmyra, but there is no unequivocal evidence that he did so, and the city is absent from Tacitus' account of his eastern command (*Ann.* 2.53–72). See further Richmond (1963) 44; Kaizer (2002) 37; Sommer (2018) 68–70; Damon and Palazzolo (2019) 202.

68 Line 4 of the inscription mentions Mesene (Aramaic *myšny*[']), apparently under the rule of one Orbazes (*'rbz*), otherwise unknown; see Schuol (2000) 322–23. Line 5 as restored by Cantineau refers to a king [*šm*]*šgrm* = Samsigeramus II, an Emesene ruler, but the *PAT* edition leaves the name unrestored. On Samsigeramus and his dynasty, see Millar (1993) 302; Ball (2000) 34–36.

69 Matthews (1984) 164; Dirven (1999) 19–20; Young (2001) 130; Smith (2013) 24; Grout (2016) 218. More cautious is Kaizer (2002) 37, who stresses that the nature of Alexander's business is not specified.

to diplomatic exchanges between Germanicus and the Arsacid king Ardawān/Artabanus II. Tacitus writes of what must have been a nearly contemporaneous exchange of messages between the rulers in 18 CE, and a cordial one at that.[70] There was no tension, it seems, in maintaining diplomatic contacts at both imperial and local levels. Moreover, by leveraging Palmyrene aid to communicate with the rulers of Mesene, Rome was not only tolerating but *utilising* a political formation whose networks penetrated deep into the Parthian west. Those networks were to no small degree commercial, as the 'caravan inscriptions' of Palmyra show; these (mostly dedicatory) texts attest a well-trafficked corridor of Palmyrene trade between Syria and southern Mesopotamia that flourished in the first two centuries CE.[71] But Palmyra was an empire-straddling *political* power no less than a commercial one. The same Palmyrenes who captained interstate trade could hold political offices in southern Mesopotamia and be no less Palmyrene for it. One served as a 'satrap of the Thilouanians [a community in modern day Bahrain] for Meherdates, king of Spasinou Charax'.[72] Other titles depend on epigraphic restorations, but another Palmyrene may have been an *archōn* in Phorath, a settlement not far from Spasinou Charax, while a third was *archōn* of Mesene.[73] In a different case, a Palmyrene who was active in the Parthian town of Vologesias was invested with *dynasteia*. The fragmentary text leaves unclear whether he wielded this power in Palmyra, Vologesias, both, or in the space between them; but the ambiguity itself reveals the diffuse and wide-ranging political authority of the caravan trade leaders.[74] Finally, yet another inscription celebrates a Palmyrene who, apparently on his own initiative, undertook an embassy to the king of Elam, another Parthian dependency.[75] While many details have fallen out of the record in these

70 Tac. *Ann.* 2.58. For the political background to the passage, see Olbrycht (2016) 622; Schlude (2020) 125–26.

71 See Gawlikowski (1994) 32–33, who delineates a corpus of thirty-four Palmyrene inscriptions that pertain to trade; Yon (2002) 263–64, table 11 adds one more. For further background on this corpus as a source for Palmyrene commerce, see Young (2001) 123–68; Grout (2016) 214–16; Andrade (2018) 4–6; Graf (2018) 483–86; Gregoratti (2020) 464.

72 *PAT* 1374 = *Inv.* 10.38 (131 CE), Greek text, lines 3–5: σατρά- | [π]ην θιλουανῶν Μεερδάτου | βασιλέως (Σ)πασίνου Χάρακος. A corresponding term is not preserved in the Aramaic text.

73 *Archōn* of Phorath: *PAT* 1412 = *Inv.* 10.112 (140 CE), Greek text, lines 2–4: [ἄρχοντα] | [Φορ]αθῶν τῆς περ[ὶ Σπασίνου] | [Χά]ρακα. The Aramaic equivalent of the title does not survive in the Aramaic, though the toponym Mesene (line 2: *m*(?)*yšn*) does appear. *Archōn* of Mesene: De Rossi (2004) 90 no. 151 (89–188 CE), line 2: τὸν ἄρχ[οντα Μ]αισ[η]νῶν.

74 *PAT* 1062 (145 CE), Greek text, section B, line 2. For discussion, see Kaizer (2002) 63–64, 150–51; Yon (2002) 110; Gregoratti (2015b) 143–47.

75 *PAT* 1414 = Inv. 10.114 (138 CE), Greek text, lines 7–9: π[ρεσβεύσαντα] αὐθαιρέτως | [πρὸς Ὀρώδην τὸν Β]ασιλέα τῆς Αἰλ[υ-] | [μήνης] (= Elymais). Cf. line 13 of the Greek, where the

lacunary inscriptions, there is enough to say that Palmyrene merchants operating in Parthian territory were not relegated to the margins to eke out a paltry trade. The leaders of the caravans were powerful trans-imperial agents who might even have wielded more political power east of the frontier than they did west of it. When Rome freerode on such networks in order to open relations with Parthian client kingdoms, it not only condoned but in fact took advantage of the interstate diplomatic networks that one of its own dependencies had cultivated, which were in some ways more robust and efficacious than anything the Roman imperial government itself could manage.

A final state-adjacent form of Parthian-Roman diplomacy came from the circulation of royal dynasts between imperial centres. In contrast to other ancient dynasties, the ruling families of Parthia and Rome did not intermarry.[76] But other sorts of peregrinations were possible. There are no unambiguous cases of Roman dynasts at the Arsacid court, though some tried to reach it, and other types of Roman elites could seek refuge there.[77] Arsacids at the courts of Roman emperors, on the other hand, are securely attested. The Arsacid residents of the Julio-Claudian period are the most numerous group,[78] but members of the Armenian branch of the Arsacid family lived (and died) in Rome

---

full name of Elymais is preserved, and also lines 7–8 of the Aramaic: [......] *bšwšn* | [......] *wrwd mlk*. For discussion, see Potts (1999) 401.

76 Caracalla's idiosyncratic plan to marry the daughter of the Parthian king Ardawān/Artabanus IV is one possible exception, but the story may be completely unhistorical; see above, n. 50. One marriage produced through diplomacy was the union of Frahād/Phraates IV and Thea Musa, on whom see Bigwood (2004); Huber and Hartmann (2006). But the only literary source for the episode is Josephus (*AJ* 18.39–43), and he calls Musa an Italian concubine (παιδίσκη); there is no indication that she was connected by kinship to the Julio-Claudian family. On intermarriage, see Mathisen (2009) for Roman Late Antiquity, and Dąbrowa (2018); Madreiter and Hartmann (2021) for Parthian practices.

77 Parthia harboured two false Neros (sources and literature in Nabel (2019a) 619 n. 83), but no real Julio-Claudians, though Nero allegedly thought of fleeing there (Suet. *Ner.* 47.2). Cassius Dio says that Macrinus dispatched his son Diadumenian to the Parthian court in the face of rebellion, though the child was intercepted and assassinated en route. Other sources make no mention of this, however (Cass. Dio 79.39–40; cf. SHA *Macr.* 10.3–6; *Diad.* 9.4; Herodian 5.4.12). Some members of the senatorial class went to the Arsacid court, or tried to, during the civil wars of the Late Republic; see Nabel (2019b) 329–30 for cases and sources. In 32 CE, one Rubrius Fabatus was arrested en route to Parthia (Tac. *Ann.* 6.14.2). The grounds for detention may have been Augustus' travel ban on senators, but Fabatus' rank is unknown; see Woodman (2017) 146–47; Damon and Palazzolo (2019) 194.

78 Sources are tabulated in Nabel (2017) 11, tables 1–2. For the background on these figures, see Dąbrowa (1987); Wiesehöfer (2010); Strothman (2012); Gregoratti (2015a). They are usually discussed under the heading of Roman hostageship, on which see Allen (2006); see Strothmann (2012) 91–92 n.36; Wheeler (2016) 193 n.157; and Nabel (2017) 46–65 for the argument that the Parthians would have understood their circulation as a matter of political fosterage.

during the Antonine period,[79] and Septimius Severus was joined on his second Parthian expedition by a brother of the reigning Arsacid king.[80] Additional glimpses survive of Parthian nobility in Roman military service, and the circulation of soldiers can be demonstrated at other social levels as well.[81] But it was the cross-border movement of royalty that brought dignitaries from the very highest levels of Parthian imperial power to the heart of Rome – and sometimes saw them return in the opposite direction, as well.

These exchanges are often analysed under the label of 'diplomacy', but if they indeed belong to that sphere, they demonstrate the inadequacy of defining this concept in state-centric terms. The unit at the centre of the circulation of Arsacids was the dynasty, not the state, and the Arsacid family is a banner example of the principle that dynasties need not be coterminous with the states that they rule. Reigning kings tend to be closely associated with state sovereignty, of course, to the extent that a king may also be called a 'sovereign'. But kings are also the managers of ruling families that must walk a fine line between intergenerational reproduction on the one hand and dynastic stability on the other: too few dynasts forebodes extinction, while too many in one place runs the risk of internecine conflict.[82] Arsacid residence in Rome offered relief from some of these pressures, but it also meant that the dynasty became internationalised; Arsacids could live outside of the Parthian state's territory and play no formal part in its administration. This development meant that conflicts for the Arsacid kingship could play out on both sides of the Euphrates, in Rome no less than in Parthia. Moreover, there is no evidence that the Arsacids in the west fulfilled any of the functions of a modern diplomat for the reigning kings in Parthia. So far as the sources reveal, they were not in contact with the king himself; they did not represent him or his interests to

79 For Armenian Arsacids at Rome during the Antonine period, see *CIG* 3.6559 = *IG* 14.1472 = *OGIS* 382, a funerary inscription set up by 'Aurelius Pacorus, king of Greater Armenia' for his brother Merithates. According to most scholars, this Pakōr/Pacorus was the Armenian Arsacid deposed in 164 CE as part of Lucius Verus' Parthian campaign (Fronto, *Ep. ad Verum imp.* 2.16 = van den Hout (1988) 126); for discussion, see Vinogradov (1992) 20; Ricci (1996) 581–83; van den Hout (1999) 302; Gnoli (2007) 71–74; de Jong (1997) 315 n. 14; Schottky (2010) 209–12; Frasson (2016) 170–71 n. 38.

80 Cass. Dio 75.9.3; discussion in Hartmann (2009).

81 For high-ranking Parthians in Roman service, see esp. *CIL* 3.8746 = *ILS* 2532 (the epitaph of one Tirdād/Tiridates, perhaps the son of a pretender to the Arsacid throne); and Tac. *Ann.* 6.37.3 on Ornospades. The Roman army had several units that were dubbed 'Parthian', though it is not clear how well the label describes the geographic origin or ethnicity of the troops that belonged to them. On these units, see Wheeler (2016) and (2017). On the persistence of such cross-border military service into the Sasanian period, see Heil (2006).

82 Duindam (2016) 88.

the Roman emperor; and they held no Parthian administrative post or office. If this was diplomacy, it was a variety that transcended the state and centred instead on the political unit of the dynasty.

The residence of Arsacids at Rome also facilitated and developed a final form of Roman sub-state diplomacy with Parthia: the embassy of the Parthian nobility. This was a group of elite Parthian clans that supported Arsacid supremacy but wielded considerable power of its own accord.[83] If the reigning Arsacid king could manipulate his dynasty by sending his family members to live at Rome, the Parthian nobles could do the same thing by bringing them back – a feat they achieved several times over the course of the first century CE. The key to their success was direct appeal to the Roman emperor, though their embassies could be received before the Senate as well.[84] Such incidents appropriated the channels of diplomacy that were otherwise the prerogative of the Arsacid king, whose grasp on Parthian sovereignty would be shaken through these collaborations among recalcitrant nobles, the Roman branch of the Arsacid family, and the Roman emperor himself. And though the resulting wars usually saw the reigning Arsacid reassert his power, such exchanges make one feature of the Parthian-Roman relationship abundantly clear: Parthia and Rome did not interact, even in the realm of diplomacy, as monolithic and unitary states. As the Arsacid family spilled over the Euphrates frontier, Parthian nobles could make their own use of diplomatic engagement to intervene – with the Roman emperor's help – in the conflicts that produced their empire's king of kings.

If Parthian-Roman diplomacy was partly the prerogative of centralised imperial authorities, then, it was also the domain of agents who variously partnered with, circumvented, or simply ignored those authorities. Local subjects, commercial chieftains, dynasts, noblemen – all skirted the formal structures of the state to bend diplomatic activity to their own ends. The composite, distributed nature of imperial sovereignty meant that neither Parthian nor Roman rulers could monopolise interstate communication, even when high politics were at stake; moreover, they may never have attempted or even thought to do

83 On the Parthian nobility, see Ziethen (1994) 178–80; Hauser (2005) 193–94; Dąbrowa (2013); Fabian (2020) 223–25; Ellerbrock (2021) 83. On the continuity of these families into the Sasanian period, see Pourshariati (2008).

84 There are four cases. For Vonones (*c.*8 CE), see *Mon. Anc.* 33; Joseph. *AJ* 18.46; Tac. *Ann.* 2.2.1. Frahād/Phraates (35 CE): Tac. *Ann.* 6.31.2; Cass. Dio 58.26.2. Tirdād/Tiridates (35/6 CE): Tac. *Ann.* 6.32.2–3; Cass. Dio 58.26.2. Mihrdād/Meherdates (49 CE): Tac. *Ann.* 11.10.4, 12.10–11 (a passage in which Parthian ambassadors address the emperor and the Senate).

so. Diplomacy was the business of the state, but also of those who operated in parallel, in tandem, or in opposition to it.

4 Conclusion

Parthian-Roman relations show the broad applicability of the Satow model, but they reveal its shortcomings as well. The prominence of embassies, summits, and treaties demonstrates that the intimate relationship between diplomacy and state structures is not a unique feature of modern Europe. Such exchanges played a formative role in defining and locating sovereignty in Rome and Parthia, and they promoted a view of an interstate system populated by comparable imperial units. But sub-state diplomacy, cross-border intermediaries, and trans-imperial dynasts show the robust impact of diplomatic exchanges beyond the purview of imperial governments. Central authorities were variously uninterested, unable, or unwilling to suppress these forms of interaction, and in some cases they may even have promoted them. The period calls for a theorisation of diplomacy that accounts for the role of the state, but also a host of sub-state actors whose foreign horizons were not constrained by their membership in larger imperial structures.

The interplay between these two factors means that Parthian-Roman relations, far from being an arid curiosity of the past, are a salutary challenge for the future. Ancient history's failure to conform to modern theorisations of diplomacy is an opportunity, not a dead end. As the predominance of the nation-state in modern analyses of diplomacy erodes, the pre-modern world can show where the practice has been before, and perhaps where it is going next. Parthia and Rome may be far removed in time from the current era, but their relationship was governed by modes of interstate engagement whose divergence from modern norms makes them valuable resources for a world in which the nation is losing its monopoly over diplomatic activity. Ancient comparanda are useful in breaking the stranglehold of modern Europe over the analysis of high politics among states, and the inherently intercultural and comparative venture of Roman-Iranian relations is especially instructive in this regard.

But if political scientists can benefit from mining the past to understand the present, historians should be prepared to think in the opposite direction. Scholars of ancient diplomacy should contribute to trans-historical theorisations of the concept, improving the study of diplomacy in the abstract as well as in the temporal contexts of their expertise. All too often, historians of antiquity tacitly adopt the state-centric model without an explicit declaration of their theoretical orientations. An undertheorised 'diplomacy' is then used to explain the past, when instead the past could be used to support a better theorisation of diplomacy. Recent work in political science has shown that these two goals are interrelated, and scholars of international relations have begun to meet the challenge by expanding their purview deeper back in time. Ancient historians should work forward to meet them.[85]

Bibliography

Ager, S.L. (2017) 'Diplomatic Communication in the Ancient Mediterranean', in Naiden and Talbert (2017) 291–310.

Aissaoui, A.I. (2020) 'Diplomacy in Ancient Times: The Figure of Udjahorresnet: An International Relations Perspective'. *Journal of Ancient Egyptian Interconnections* 26: 12–34.

Alcock, S., D'Altroy, T., Morrison, K., and Sinopoli, C. (eds) (2001) *Empires: Perspectives from Archaeology and History*. Cambridge.

Alidoust, F. (2020) *Natio Molestissima. Römerzeitliche Perserbilder von Cicero bis Ammianus Marcellinus*. Gutenberg.

Allen, J. (2006) *Hostages and Hostage-Taking in the Roman Empire*. Cambridge.

Anderson, B. (2017) 'Beyond Rome/Parthia: Intersections of Local and Imperial Traditions in the Visual Record of Hatra', in Schlude and Rubin (2017a) 137–58.

Andrade, N. (2018) *The Journey of Christianity to India in Late Antiquity: Networks and the Movement of Culture*. Cambridge.

Andres, H. (2022) Bruderzwist: Strukturen und Methoden der Diplomatie zwischen Rom und Iran von der Teilung Armeniens bis zum Fünfzigjährigen Frieden. Stuttgart.

Aßkamp, R. and Esch, T. (eds) (2010) *Imperium – Varus und seine Zeit: Beiträge zum internationalen Kolloquium des LWL-Römermuseums am 28. und 29. April 2008 in Münster*. Münster.

Ball, W. (2000) *Rome in the East: The Transformation of an Empire*. London.

Balot, R.K. (ed) (2009) *A Companion to Greek and Roman Political Thought*. Malden, MA.

Baltrusch, E. and Wilker, J. (eds) (2015) *Amici – Socii – Clientes? Abhängige Herrschaft im Imperium Romanum*. Berlin.

Balzacq, T., Charillon, F., and Ramel, F. (eds) (2020a) *Global Diplomacy: An Introduction to Theory and Practice*. Translated by William Snow. Cham.

Balzacq, T., Charillon, F., and Ramel, F. (2020b) 'Introduction', in Balzacq, Charillon, and Ramel (2020a) 1–16.

85 I thank John Hyland and Khodadad Rezakhani for their editorial guidance, and Peter Katzenstein for critical discussion and bibliography. Translations are my own except where otherwise indicated.

Bang, P.F. and Kołodziejczyk, D. (eds) (2012) *Universal Empire: A Comparative Approach to Imperial Culture and Representation in Eurasian History*. Cambridge.

Bayly, C.A., Bang, P.F., and Scheidel, W. (eds) (2020) *The Oxford World History of Empire. Volume Two: The History of Empires*. Oxford.

Benton, L. (2010) *A Search for Sovereignty: Law and Geography in European Empires, 1400–1900*. Cambridge.

Berridge, G.R. (2015) *Diplomacy: Theory and Practice*. 5th edn. London.

Biggs, T. and Blum, J. (eds) (2019) *The Epic Journey in Greek and Roman Literature*. Cambridge.

Bigwood, J.M. (2004) 'Queen Mousa, Mother and Wife (?) of King Phraatakes of Parthia: A Re-Evaluation of the Evidence'. *Mouseion, Series III* 4: 35–70.

Bivar, A.D.H. (1983) 'The Political History of Iran under the Arsacids', in Yarshater (1983) 21–99.

Black, J. (2010) *A History of Diplomacy*. London.

Börm, H. and Wiesehöfer, J. (eds) (2010) *Commutatio et Contentio: Studies in the Late Roman, Sasanian, and Early Islamic Near East in Memory of Zeev Rubin*. Düsseldorf.

Boyce, M. (1983) 'Parthian Writings and Literature', in Yarshater (1983) 1149–65.

Boyce, M. and Grenet, F. (1991) *A History of Zoroastrianism. Volume 3: Zoroastrianism under Macedonian and Roman Rule*. Leiden.

Burbank, J. and Cooper, F. (2010) *Empires in World History: Power and the Politics of Difference*. Princeton.

Burden-Strevens, C. (2020) *Cassius Dio's Speeches and the Collapse of the Roman Republic: The* Roman History, *Books 3–56*. Leiden.

Burton, P.J. (2011) *Friendship and Empire*. Cambridge.

Campbell, B. (1993) 'War and Diplomacy: Rome and Parthia, 31 BC–AD 235', in Rich and Shipley (1993) 213–40.

Campbell, B. (2001) 'Diplomacy in the Roman World (c.500 BC–AD 235)'. *Diplomacy & Statecraft* 12: 1–22.

Canepa, M.P. (2020) 'The Parthian and Sasanian Empires', in Bayly, Bang, and Scheidel (2020) 290–324.

Canevaro, M. and Gray, B. (eds) (2018) *The Hellenistic Reception of Classical Athenian Democracy and Political Thought*. Oxford.

Cantineau, J. (1931) 'Textes palmyréniens provenant de la fouille du Temple de Bel'. *Syria* 12.2: 116–42.

Carney, E. and Müller, S. (eds) (2021) *The Routledge Companion to Women and Monarchy in the Ancient Mediterranean World*. London.

Cereti, C.G. (2021) 'Narseh, Armenia, and the Paikuli Inscription'. *Electrum* 28: 69–87.

Cereti, C.G. and Terribili, G. (2014) 'The Middle Persian and Parthian Inscriptions on the Paikuli Tower. New Blocks and Preliminary Studies'. *IrAnt* 49: 347–82.

Cereti, C.G., Terribili, G., and Colliva, L. (2019) 'Il monumento di Paikuli (Kurdistan, Iraq). Ricerche e nuove attività sul campo'. *Scienze dell'Antichità* 25: 3–10.

Chaplin, J. (2000) *Livy's Exemplary History*. Oxford.

Chaumont, M.L. (1987) 'Un document méconnu concernant l'envoi d'un ambassadeur parthe vers Septime Sévère (P. Dura 60 B.)'. *Historia* 36: 422–47.

Cobb, M. (ed) (2019) *The Indian Ocean Trade in Antiquity: Political, Cultural and Economic Impacts*. London.

Cohen, R. (1996) 'On Diplomacy in the Ancient Near East: The Amarna Letters'. *Diplomacy & Statecraft* 7: 245–70.

Cohen, R. (2001) 'The Great Tradition: The Spread of Diplomacy in the Ancient World'. *Diplomacy & Statecraft* 12.1: 23–38.

Cohen, R. and Westbrook, R. (eds) (2000) *Amarna Diplomacy: The Beginnings of International Relations*. Baltimore.

Colás, A. (2007) *Empire*. Cambridge.

Constantinou, C.M. (2013) 'Between Statecraft and Humanism: Diplomacy and Its Forms of Knowledge'. *International Studies Review* 15: 141–62.

Constantinou, C.M., Kerr, P., and Sharp, P. (eds) (2016) *The SAGE Handbook of Diplomacy*. Los Angeles.

Constantinou, C.M. and Sharp, P. (2016) 'Theoretical Perspectives in Diplomacy', in Constantinou, Kerr, and Sharp (2016) 13–27.

Curtis, V.S., Pendleton, E.J., Alram, M., and Daryaee, T. (eds) (2016) *The Parthian and Early Sasanian Empires: Adaptation and Expansion*. Oxford.

Curtis, V.S. and Magub, A. (2020) *Rivalling Rome: Parthian Coins & Culture*. London.

Dąbrowa, E. (1983) *La politique de l'État parthe à l'égard de Rome-d'Artaban II à Vologèse I (ca 11–ca 79 de n.e.) et les facteurs qui la conditionnaient*. Krakow.

Dąbrowa, E. (1984) 'Vologèse Ier et l'Hyrcanie'. *IrAnt* 19: 141–47.

Dąbrowa, E. (1987) 'Les premiers "otages" parthes à Rome'. *Folia Orientalia* 24: 63–71.

Dąbrowa, E. (2011) *Studia Graeco-Parthica: Political and Cultural Relations between Greeks and Parthians*. Wiesbaden.

Dąbrowa, E. (2013) 'The Parthian Aristocracy: Its Social Position and Political Activity'. *Parthica* 15: 53–62.

Dąbrowa, E. (2018) 'Arsacid Dynastic Marriages'. *Electrum* 25: 73–83.

Dalziel, N. and MacKenzie, J.M. (eds) (2016) *The Encyclopedia of Empire*. Hoboken.

Damon, C. and Palazzolo, E. (2019) 'Defining Home, Defining Rome: Germanicus' Eastern Tour', in Biggs and Blum (2019) 194–210.

de Jong, A. (1997) *Traditions of the Magi: Zoroastrianism in Greek and Latin Literature*. Leiden.

de Jong, A. (2013) 'Hatra and the Parthian Commonwealth', in Dirven (2013) 143–60.

De Rossi, F.C. (2004) *Iscrizioni dello Estremo Oriente Greco*. Bonn.

Debevoise, N.C. (1938) *A Political History of Parthia*. Chicago.

Der Derian, J. (1987) *On Diplomacy: A Genealogy of Western Estrangement*. Oxford.
de Souza, P. and France, J. (eds) (2008) *War and Peace in Ancient and Medieval History*. Cambridge.
Dignas, B. and Winter, E. (2007) *Rome and Persia in Late Antiquity: Neighbours and Rivals*. Cambridge.
Dirven, L. (1999) *The Palmyrenes of Dura-Europos: A Study of Religious Interaction in Roman Syria*. Leiden.
Dirven, L. (ed) (2013) *Hatra: Politics, Culture and Religion between Parthia and Rome*. Stuttgart.
Dirven, L. (2016) 'The Problem with Parthian Art at Dura', in Kaizer (2016) 68–88.
Dittmann, R. (ed) (2000) *Variatio delectat: Iran und der Westen. Gedenkschrift für Peter Calmeyer*. Munich.
Drijvers, J.W. (1998) 'Strabo on Parthia and the Parthians', in Wiesehöfer (1998) 279–93.
Duindam, J. (2016) *Dynasties: A Global History of Power, 1300–1800*. Cambridge.
Dunn, D.H. and Lock-Pullan, R. (2016) 'Diplomatic Summitry', in Constantinou, Kerr, and Sharp (2016) 231–41.
Durkin-Meisterernst, D. (2004) *Dictionary of Manichaean Middle Persian and Parthian*. Turnhout.
Eck, W. (2009) 'Diplomacy as Part of the Administrative Process of the Roman Empire', in Eilers (2009a) 193–207.
Eckstein, A.M. (2006) *Mediterranean Anarchy, Interstate War, and the Rise of Rome*. Berkeley.
Eckstein, A.M. (2008) *Rome Enters the Greek East: From Anarchy to Hierarchy in the Hellenistic Mediterranean, 230–170 BC*. Malden, MA.
Edwell, P. (2008) *Between Rome and Persia: The Middle Euphrates, Mesopotamia and Palmyra under Roman Control*. London.
Edwell, P. (2020) *Rome and Persia at War: Imperial Competition and Contact, 193–363 CE*. London.
Eilers, C. (ed) (2009a) *Diplomats and Diplomacy in the Roman World*. Leiden.
Eilers, C. (2009b) 'Introduction', in Eilers (2009a) 1–13.
Ellerbrock, U. (2021) *The Parthians: The Forgotten Empire*. London.
Erdkamp, P. (ed) (2007) *A Companion to the Roman Army*. Malden, MA.
Erdkamp, P., Verboven, K., and Zuiderhoek, A. (eds) (2020) *Capital, Investment, and Innovation in the Roman World*. Oxford.
Fabian, L. (2020) 'The Arsakid Empire', in von Reden (2020) 205–39.
Fomin, A. (2016) 'Speeches in Dio Cassius', in Lange and Madsen (2016) 217–37.
Frasson, F. (2016) 'Armenia and Armenians in Asinius Quadratus' Παρθικά', in Gazzano, Pagani, and Traina (2016) 163–94.

Freeman, P., Bennett, J., Fiema, Z.T., and Hoffmann, B. (eds) (2002) *Limes XVIII: Proceedings of the XVIIIth International Congress of Roman Frontier Studies, Held in Amman, Jordan (September 2000)*. Vol. 1. Oxford.

Gawlikowski, M. (1994) 'Palmyra as a Trading Centre'. *Iraq* 56: 27–33.

Gazzano, F., Pagani, L., and Traina, G. (eds) (2016) *Greek Texts and Armenian Traditions: An Interdisciplinary Approach*. Berlin.

Gignoux, P. (1972) *Glossaire des inscriptions pehlevies et parthes*. London.

Giovannini, A. (2007) *Les relations entre États dans la Grèce antique du temps d'Homère à l'intervention romaine (ca. 700–200 av. J.-C.)*. Stuttgart.

Gladhill, B. (2016) *Rethinking Roman Alliance: A Study in Poetics and Society*. Cambridge.

Gnoli, T. (2007) *The Interplay of Roman and Iranian Titles in the Roman East*. Vienna.

Gofas, A., Hamati-Ataya, I., and Onuf, N. (eds) (2018) *The SAGE Handbook of the History, Philosophy and Sociology of International Relations*. London.

Graf, D.F. (2018) 'The Silk Road between Syria and China', in Wilson and Bowman (2018) 443–529.

Gregoratti, L. (2012) 'The Importance of the Mint of Seleucia on the Tigris for Arsacid History: Artabanus and the Greek Parthian Cities'. *Mesopotamia* 47: 129–36.

Gregoratti, L. (2015a) 'In the Land West of the Euphrates: The Parthians in the Roman Empire', in Militello and Oniz (2015) vol. 2, 731–35.

Gregoratti, L. (2015b) 'The Palmyrene Trade Lords and the Protection of the Caravans'. *Aram* 27: 139–48.

Gregoratti, L. (2019) 'Indian Ocean Trade: The Role of Parthia', in Cobb (2019) 52–72.

Gregoratti, L. (2020) 'Temples and Traders in Palmyra', in Erdkamp, Verboven, and Zuiderhoek (2020) 461–80.

Grout, J.B. (2016) 'The Role of Palmyrene Temples in Long-Distance Trade in the Roman Near East', PhD Dissertation, University of London.

Gyselen, R. (2016) 'The Parthian Language in Early Sasanian Times', in Curtis, Pendleton, Alram, and Daryaee (2016) 61–68.

Hackl, U., Jacobs, B., and Weber, D. (eds) (2010) *Quellen zur Geschichte des Partherreiches: Textsammlung mit Übersetzungen und Kommentaren*. 3 vols. Göttingen.

Hall, J.M. (2007) 'International Relations', in Sabin, van Wees, and Whitby (2007) vol. 1, 85–107.

Hamilton, K. and Langhorne, R. (2010) *The Practice of Diplomacy: Its Evolution, Theory and Administration*. 2nd edn. London.

Harl, K.W. (2016) 'Rome's Greatest Foe: Rome versus Parthia and Sassanid Persia', in Lacey (2016) 103–53.

Hartmann, U. (2009) 'Ein Arsakide im Heer des Septimius Severus : Überlegungen zu den Hintergründen des zweiten Partherkrieges'. *Electrum* 15: 249–66.

Hauser, S. (2005) 'Die ewigen Nomaden: Bemerkungen zu Herkunft, Militär, Staatsaufbau und nomadischen Traditionen der Arsakiden', in Meißner, Schmitt, and Sommer (2005) 163–208.

Hauser, S. (2014) '"Parthian Art" or "Arts in the Arsacid Empire": Hatra and Palmyra as Nodal Points for Cultural Interaction', in Jacobs (2014) 127–78.

Heil, M. (1997) *Die orientalische Außenpolitik des Kaisers Nero*. Munich.

Heil, M. (2006) 'Perser im spätrömischen Dienst', in Wiesehöfer and Huyse (2006) 143–79.

Herman, G. (2012) 'The Jews of Parthian Babylonia', in Wick and Zehnder (2012) 141–50.

Huber, I. and Hartmann, U. (2006) '"Denn ihrem Diktat vermochte der König nicht zu widersprechen ..." Die Position der Frauen am Hof der Arsakiden', in Panaino and Piras (2006) 485–517.

Huyse, P. (1999) *Die dreisprachige Inschrift Šābuhrs I. an der Ka'ba-i Zardušt* (*ŠKZ*). London.

Hyland, J. (2018) *Persian Interventions: The Achaemenid Empire, Athens, and Sparta, 450–386 BCE*. Baltimore.

Iovine, G. (2023) *Latin Military Papyri of Dura-Europos (P.Dura 55–145): A New Edition of the Texts, with Introduction and Notes*. Cambridge.

Jacobs, B. (ed) (2014) *'Parthische Kunst' – Kunst Im Partherreich: Akten Des Internationalen Kolloquiums in Basel, 9. Oktober 2010*. Duisburg.

Jayasuriya, K. (2005) 'Breaking the "Westphalian" Frame: Regulatory State, Fragmentation and Diplomacy', in Robertson and East (2005) 39–54.

Jönsson, C. and Hall, M. (2005) *Essence of Diplomacy*. London.

Kaizer, T. (2002) *The Religious Life of Palmyra: A Study of the Social Patterns of Worship in the Roman Period*. Stuttgart.

Kaizer, T. (ed) (2016) *Religion, Society and Culture at Dura-Europos*. Cambridge.

Kaizer, T. (2017) 'Empire, Community, and Culture on the Middle Euphrates: Durenes, Palmyrenes, Villagers, and Soldiers'. *BICS* 60: 63–95.

Keaveney, A. (1981) 'Roman Treaties with Parthia circa 95–circa 64 B.C.'. *AJPhil* 102: 195–212.

Keaveney, A. (1982) 'The King and the War-Lords: Romano-Parthian Relations Circa 64–53. B.C.'. *AJPhil* 103.4: 412–28.

Kemezis, A. (ed) (2015) *Urban Dreams and Realities in Antiquity*. Leiden.

Khatchadourian, L. (2016) *Imperial Matter: Ancient Persia and the Archaeology of Empires*. Oakland, CA.

Kissinger, H. (1994) *Diplomacy*. New York.

Kuus, M. (2017) 'Diplomacy and the Other – and Wagner'. *New Perspectives: Interdisciplinary Journal of Central & East European Politics and International Relations* 25: 85–90.

Kuus, M. (2018) 'Transnational Institutional Fields: Positionality and Generalization in the Study of Diplomacy'. *Political Geography* 67: 156–65.

Lacey, J. (ed) (2016) *Great Strategic Rivalries: From the Classical World to the Cold War.* Oxford.

Lafont, B. (2001) 'International Relations in the Ancient Near East: The Birth of a Complete Diplomatic System'. *Diplomacy & Statecraft* 12: 39–60.

Lawson, G. (2018) 'International Relations as a Historical Social Science', in Gofas, Hamati-Ataya, and Onuf (2018) 75–89.

Lange, C.H. and Madsen, J.M. (eds) (2016) *Cassius Dio: Greek Intellectual and Roman Politician.* Leiden.

Lee, A.D. (2008) 'Treaty-Making in Late Antiquity', in de Souza and France (2008) 107–19.

Lendon, J.E. (2002) 'Primitivism and Ancient Foreign Relations'. *CJ* 97.4: 375–84.

Lendon, J.E. (2006) 'Xenophon and the Alternative to Realist Foreign Policy: *Cyropaedia* 3.1.14–31'. *JHS* 126: 82–98.

Lenfant, D. (2022) 'The Role of *Xenia* in Diplomatic Relations between Greek Cities and the Persian Empire', in Mari and Wendt (2022) 81–93.

Lerouge, C. (2007) *L'image des Parthes dans le monde gréco-romain.* Stuttgart.

Liverani, M. (2001) *International Relations in the Ancient Near East, 1600–1100 BC.* Houndmills.

Lo Cascio, E. and Tacoma, L.E. (eds) (2016) *The Impact of Mobility and Migration in the Roman Empire: Proceedings of the Twelfth Workshop of the International Network Impact of Empire (Rome, June 17–19, 2015).* Leiden.

Low, P. (2007) *Interstate Relations in Classical Greece: Morality and Power.* Cambridge.

Luther, A. (2015) 'Das Königreich Adiabene zwischen Parthern und Römern', in Baltrusch and Wilker (2015) 275–300.

Ma, J. (2018) 'Whatever Happened to Athens? Thoughts on the Great Convergence and Beyond', in Canevaro and Gray (2018) 277–97.

MacKenzie, J.M. (2016) 'Diplomacy and Empire', in Dalziel and MacKenzie (2016).

Madreiter, I. and Hartmann, U. (2021) 'Women at the Arsakid Court', in Carney and Müller (2021) 234–45.

Marciak, M. (2014) *Izates, Helena, and Monobazos of Adiabene: A Study on Literary Traditions and History.* Wiesbaden.

Mari, F. and Wendt, C. (eds) (2022) *Shaping Good Faith: Modes of Communication in Ancient Diplomacy.* Stuttgart.

Marincola, J. (1997) *Authority and Tradition in Ancient Historiography.* Cambridge.

Marincola, J. (ed) (2007) *A Companion to Greek and Roman Historiography.* 2 vols. Malden, MA.

Mathisen, R. (2009) 'Provinciales, Gentiles, and Marriages between Romans and Barbarians in the Late Roman Empire'. *JRS* 99: 140–55.

Matthews, J.F. (1984) 'The Tax Law of Palmyra: Evidence for Economic History in a City of the Roman East'. *JRS* 74: 157–80.

McConnell, F., Moreau, T., and Dittmer, J. (2012) 'Mimicking State Diplomacy: The Legitimizing Strategies of Unofficial Diplomacies'. *Geoforum* 43: 804–14.

Meißner, B., Schmitt, O., and Sommer, M. (eds) (2005) *Krieg-Gesellschaft-Institutionen: Beiträge zu einer vergleichenden Kriegsgeschichte*. Berlin.

Merkelbach, R. and Stauber, J. (eds) (2005) *Jenseits des Euphrat: griechische Inschriften: Ein epigraphisches Lesebuch*. Munich.

Messina, V. and Versluys, M.J. (2022) 'Greeks in Asia: Understanding Hellenism in Susa and Seleucia on the Tigris', in Spier, Potts, and Cole (2022) 193–99.

Meyer, E.A. (2004) *Legitimacy and Law in the Roman World:* Tabulae *in Roman Belief and Practice*. Cambridge.

Militello, P. and Oniz, H. (eds) (2015) *Soma 2011: Proceedings of the 15th Symposium on Mediterranean Archaeology, Held at the University of Catania, March 3–5th, 2011*. 2 vols. Oxford.

Millar, F. (1988) 'Government and Diplomacy in the Roman Empire during the First Three Centuries'. *International History Review* 10: 345–77.

Millar, F. (1993) *The Roman Near East, 31 BC–AD 337*. Cambridge, MA.

Morgenthau, H.J. (1948) *Politics among Nations: The Struggle for Power and Peace*. New York.

Morley, N. (2004) *Theories, Models and Concepts in Ancient History*. London.

Morrison, K.D. (2001) 'Sources, Approaches, Definitions', in Alcock, D'Altroy, Morrison, and Sinopoli (2001) 1–9.

Mynářová, J. (2008) *Language of Amarna – Language of Diplomacy: Perspectives on the Amarna Letters*. Prague.

Nabel, J. (2017) 'The Arsacids of Rome: Royal Hostages and Roman-Parthian Relations in the First Century CE', PhD Dissertation, Cornell University.

Nabel, J. (2019a) 'Lucan's Parthians in Nero's Rome'. *CPhil* 114: 604–25.

Nabel, J. (2019b) 'Remembering Intervention: Parthia in Rome's Civil Wars'. *Historia* 68: 327–52.

Nabel, J. (2019c) 'Tiridates in the Forum, Peroz on His Knees: Religion and Reputation in Ancient Iranian Diplomacy'. *Anabasis* 10: 214–36.

Naiden, F.S. and Talbert, R.J. (eds) (2017) *Mercury's Wings: Exploring Modes of Communication in the Ancient World*. Oxford.

Nathan, J.A. (2002) *Soldiers, Statecraft, and History: Coercive Diplomacy and International Order*. Westport, CT.

Neumann, I.B. (2012) *At Home with the Diplomats: Inside a European Foreign Ministry*. Ithaca, NY.

Nicolai, R. (2007) 'The Place of History in the Ancient World', in Marincola (2007) vol. 1, 13–26.

Nicolson, H. (1939) *Diplomacy*. London.

Olbrycht, M. (1998) *Parthia et Ulteriores Gentes: Die politischen Beziehungen zwischen dem arsakidischen Iran und den Nomaden der eurasischen Steppen*. Munich.

Olbrycht, M. (2014) 'Parthians, Greek Culture, and Beyond', in Twardowska, Salamon, Sprawski, Stachura, and Turlej (2014) 129–42.

Olbrycht, M. (2016) 'Germanicus, Artabanos II of Parthia, and Zeno Artaxias in Armenia'. *Klio* 98.2: 605–33.

Olbrycht, M. (2017) 'Greeks in the Arsacid Empire', in Wiesehöfer and Müller (2017) 3–27.

Olbrycht, M. (2021) *Early Arsakid Parthia (ca. 250–165 B.C.)*. Leiden.

Opondo, S.O. (2016) 'Diplomacy and the Colonial Encounter', in Constantinou, Kerr, and Sharp (2016) 39–53.

Overtoom, N.L. (2016) 'The Rivalry of Rome and Parthia in the Sources from the Augustan Age to Late Antiquity'. *Anabasis* 7: 137–74.

Overtoom, N.L. (2020) *Reign of Arrows: The Rise of the Parthian Empire in the Hellenistic Middle East*. New York.

Panaino, A. and Piras, A. (eds) (2006) *Proceedings of the 5th Conference of the Societas Iranologica Europœa, Ravenna, 6–11 Oct 2003*. Vol. 1. Milan.

Pannier, A. (2020) 'Bilateral Relations', in Balzacq, Charillon, and Ramel (2020a) 19–33.

Pelling, C. (2007) 'The Greek Historians of Rome', in Marincola (2007) vol. 1, 244–58.

Peretz, D. (2006) 'The Roman Interpreter and His Diplomatic and Military Roles'. *Historia* 55: 451–70.

Perikhanian, A. (1983) 'Iranian Society and Law', in Yarshater (1983) 627–80.

Pigman, G.A. (2010) *Contemporary Diplomacy: Representation and Communication in a Globalized World*. Cambridge.

Plischke, E. (1973) 'Treatment of "Diplomacy" in International Relations Textbooks'. *World Affairs* 135: 328–44.

Podany, A.H. (2010) *Brotherhood of Kings: How International Relations Shaped the Ancient Near East*. Oxford.

Potter, D.S. (1991) 'The Inscriptions on the Bronze Herakles of Mesene: Vologaeses IV's War with Rome and the Date of Tacitus' *Annales*'. *ZPE* 88: 277–90.

Potts, D.T. (1999) *The Archaeology of Elam: Formation and Transformation of an Ancient Iranian State*. Cambridge.

Pourshariati, P. (2008) *Decline and Fall of the Sasanian Empire: The Sasanian-Parthian Confederacy and the Arab Conquest of Iran*. London.

Puybareau, B. and Takam Talom, R. (2020) 'Sub-State Diplomacies: Regions, Parliaments, and Local Authorities', in Balzacq, Charillon, and Ramel (2020a) 167–81.

Rhodes, P.J. (2007) 'Documents and the Greek Historians', in Marincola (2007) vol. 1, 56–66.

Ricci, C. (1996) 'Principes et reges externi (e loro schiavi e liberti) a Roma e in Italia: Testimonianze epigrafiche di età imperiale'. *Rendiconti lincei* series 9, 7.3: 561–92.
Rich, J. and Shipley, G. (eds) (1993) *War and Society in the Roman World*. London.
Richmond, I.A. (1963) 'Palmyra under the Aegis of Rome'. *JRS* 53: 43–54.
Robertson, J. and East, M.A. (eds) (2012) *Diplomacy and Developing Nations: Post-Cold War Foreign Policy-Making Structures and Processes*. London.
Roller, M.B. (2018) *Models from the Past in Roman Culture: A World of* Exempla. Cambridge.
Rougemont, G. (2012) *Inscriptions grecques d'Iran et d'Asie centrale*. London.
Rowe, C. and Schofield, M. (eds) (2000) *The Cambridge History of Greek and Roman Political Thought*. Cambridge.
Sabin, P., van Wees, H., and Whitby, M. (eds) (2007) *The Cambridge History of Greek and Roman Warfare*. 2 vols. Cambridge.
Satow, E.M. and Roberts, I. (2018) *Satow's Diplomatic Practice*. 7th edn. Oxford.
Schippmann, K. (1980) *Grundzüge der parthischen Geschichte*. Darmstadt.
Schlude, J.M. (2020) *Rome, Parthia, and the Politics of Peace: The Origins of War in the Ancient Middle East*. London.
Schlude, J.M. and Rubin, B.B. (eds) (2017a) *Arsacids, Romans, and Local Elites: Cross-Cultural Interactions of the Parthian Empire*. Oxford.
Schlude, J.M. and Rubin, B.B. (2017b) 'Finding Common Ground: Parthian-Roman Embassies in the Julio-Claudian Period', in Schlude and Rubin (2017a) 65–92.
Schneider, R.M. (2012) 'The Making of Oriental Rome: Shaping the Trojan Legend', in Bang and Kołodziejczyk (2012) 76–129.
Schottky, M. (1991) 'Parther, Meder, und Hyrkanier: Eine Untersuchung der dynastischen und geographischen Verflechtungen im Iran des 1. Jhs. n. Chr'. *AMI* 24: 61–134.
Schottky, M. (2010) 'Armenische Arsakiden zur Zeit der Antonine: Ein Beitrag zur Korrektur der armenischen Königsliste'. *Anabasis* 1: 208–24.
Schuol, M. (2000) *Die Charakene: Ein mesopotamisches Königreich in hellenistisch-parthischer Zeit*. Stuttgart.
Schur, W. (1923) *Die Orientpolitik des Kaisers Nero*. Wiesbaden.
Sharp, P. (2009) *Diplomatic Theory of International Relations*. Cambridge.
Shayegan, M.R. (2011) *Arsacids and Sasanians: Political Ideology in Post-Hellenistic and Late Antique Persia*. Cambridge.
Shayegan, M.R. (2012) *Aspects of History and Epic in Ancient Iran: From Gaumāta to Wahnām*. Washington, DC.
Sidebottom, H. (2007) 'International Relations', in Sabin, van Wees, and Whitby (2007) vol. 2, 3–29.
Sinopoli, C. (1994) 'The Archaeology of Empires'. *Annual Review of Anthropology* 23: 159–80.

Skaff, J.K. (2012) *Sui-Tang China and Its Turko-Mongol Neighbors: Culture, Power, and Connections, 580–800*. Oxford.

Skjærvø, P.O. (1983) *The Sassanian Inscription of Paikuli. Part 3.1: Restored Text and Translation; Part 3.2: Commentary*. Wiesbaden.

Smith, A.M. (2013) *Roman Palmyra: Identity, Community, and State Formation*. New York.

Sofer, S. (2013) *The Courtiers of Civilization: A Study of Diplomacy*. Albany, NY.

Sommer, M. (2005) *Roms orientalische Steppengrenze: Palmyra-Edessa-Dura-Europos-Hatra*. Stuttgart.

Sommer, M. (2018) *Palmyra: A History*. Translated by Diana Sommer-Theohari. London.

Sonnabend, H. (1986) *Fremdenbild und Politik: Vorstellungen der Römer von Ägypten und dem Partherreich in der späten Republik und frühen Kaiserzeit*. Frankfurt am Main.

Spier, J., Potts, T., and Cole, S.E. (eds) (2022) *Persia: Ancient Iran and the Classical World*. Los Angeles.

Stadter, P.A. (2009) 'Character in Politics', in Balot (2009) 456–70.

Strothmann, M. (2012) 'Feindeskinder an Sohnes statt. Parthische Königssöhne im Haus des Augustus', in Wick and Zehnder (2012) 83–102.

Steel, C. and van der Blom, H. (eds) (2013) *Community and Communication: Oratory and Politics in Republican Rome*. Oxford.

Subrahmanyam, S. (2001) 'Written on Water: Designs and Dynamics in the Portugese *Estado Da Índia*', in Alcock, D'Altroy, Morrison, and Sinopoli (2001) 42–69.

Syme, R. (1958) *Tacitus*. Oxford.

Talbert, R.J. (1984) *The Senate of Imperial Rome*. Princeton.

Timpe, D. (1967) 'Ein Heiratsplan Kaiser Caracallas'. *Hermes* 95: 470–95.

Torregaray Pagola, E. (2013) 'The Roman Ambassador's Speech: Public Diplomacy on the Diplomatic Stage', in Steel and van der Blom (2013) 229–45.

Torregaray Pagola, E. (2016) 'Diplomatic Mobility and Persuasion between Rome and the West (I–II AD)', in Lo Cascio and Tacoma (2016) 116–31.

Turpin, W. (2008) 'Tacitus, Stoic *Exempla*, and the *Praecipuum Munus Annalium*'. *Cl. Ant.* 27: 359–404.

Twardowska, K., Salamon, M., Sprawski, S., Stachura, M., and Turlej, S. (eds) (2014) *Within the Circle of Ancient Ideas and Virtues: Studies in Honour of Professor Maria Dzielska*. Krakow.

van den Hout, M.P.J. (ed) (1988) *M. Cornelii Frontonis Epistulae*. Leipzig.

van den Hout, M.P.J. (1999) *A Commentary on the Letters of M. Cornelius Fronto*. Leiden.

van Kooten, G. (2015) 'Matthew, the Parthians, and the Magi: A Contextualization of Matthew's Gospel in Roman-Parthian Relations of the First Centuries BCE and CE', in van Kooten and Barthel (2015) 496–646.

van Kooten, G. and Barthel, P. (eds) (2015) *The Star of Bethlehem and the Magi: Interdisciplinary Perspectives from Experts on the Ancient Near East, the Greco-Roman World, and Modern Astronomy*. Leiden.

Vinogradov, J.G. (1992) 'The Goddess Ge Meter Olybris: A New Epigraphic Evidence from Armenia'. *East and West* 42: 13–26.

von Reden, S. (ed) (2020) *Handbook of Ancient Afro-Eurasian Economies, Volume 1: Contexts*. Berlin.

Waltz, K.N. (1979) *Theory of International Politics*. New York.

Wang, Y. (2011) *Harmony and War: Confucian Culture and Chinese Power Politics*. New York.

Watson, A. (1982) *Diplomacy: The Dialogue between States*. London.

Weber, U. and Wiesehöfer, J. (2010) 'König Narsehs Herrschaftsverständnis', in Börm and Wiesehöfer (2010) 89–132.

Wellesley, K. (1955) 'The Date of Composition of Tacitus, *Annals* II'. *Rh. Mus.* 98: 135–49.

Wheeler, E.L. (1997) 'The Chronology of Corbulo in Armenia'. *Klio* 79: 383–97.

Wheeler, E.L. (2002) 'Roman Treaties with Parthia: Völkerrecht or Power Politics?', in Freeman, Bennett, Fiema, and Hoffmann (2002) 287–92.

Wheeler, E.L. (2007) 'The Army and the *Limes* in the East', in Erdkamp (2007) 235–66.

Wheeler, E.L. (2016) 'Parthian *Auxilia* in the Roman Army. Part I: From the Late Republic to c. 70 A.D.', in Wolff and Faure (2016) 171–222.

Wheeler, E.L. (2017) 'Parthian *Auxilia* in the Roman Army, Part II: From the Flavians to the Late Empire'. *Revue Internationale d'histoire Militaire Ancienne* 5: 103–37.

Wick, P. and Zehnder, M.P. (eds) (2012) *The Parthian Empire and Its Religions: Studies in the Dynamics of Religious Diversity*. Gutenberg.

Wiedemann, T. (2000) 'Reflections of Roman Political Thought in Latin Historical Writing', in Rowe and Schofield (2000) 517–31.

Wiesehöfer, J. (ed) (1998) *Das Partherreich und seine Zeugnisse = The Arsacid Empire – Sources and Documentation: Beiträge des internationalen Colloquiums, Eutin (27.–30. Juni 1996)*. Stuttgart.

Wiesehöfer, J. (2000) '"Denn Orodes war der griechischen Sprache und Literatur nicht Unkundig". Parther, Griechen und griechische Kultur', in Dittmann (2000) 703–21.

Wiesehöfer, J. (2010) 'Augustus und die Parther', in Aßkamp and Esch (2010) 187–95.

Wiesehöfer, J. (2015) 'Greek Poleis in the Near East and Their Parthian Overlords', in Kemezis (2015) 328–46.

Wiesehöfer, J. (2020) 'Evidence for Arsakid Economic History', in von Reden (2020) 477–96.

Wiesehöfer, J. and Huyse, P. (eds) (2006) *Ērān ud Anērān: Studien zu den Beziehungen zwischen dem Sasanidenreich und der Mittelmeerwelt*. Stuttgart.

Wiesehöfer, J. and Müller, S. (eds) (2017) *Parthika: Greek and Roman Authors' Views of the Arsacid Empire*. Wiesbaden.

Wilson, A. and Bowman, A. (eds) (2018) *Trade, Commerce, and the State in the Roman World*. Oxford.

Wolff, C. and Faure, P. (eds) (2016) *Les Auxiliaires de l'armée Romaine: Des Alliés Aux Fédérés: Actes Du Sixième Congrès de Lyon (23–25 Octobre 2014)*. Paris.

Wolski, J. (1983) 'Sur le "philhellénisme" des Arsacides'. *Gerion* 1: 145–56.

Wolski, J. (1993) *L'Empire des Arsacides*. Lovanii.

Woodcock, E.C. (1939) *Tacitus, Annals Book XIV*. London.

Woodman, A.J. (2017) *The Annals of Tacitus: Books 5 and 6*. Cambridge.

Yarshater, E. (ed) (1983) *The Cambridge History of Iran. Vol. 3.1–2: The Seleucid, Parthian, and Sasanid Periods*. Cambridge.

Yon, J.-B. (2002) *Les notables de Palmyre*. Beirut.

Young, G.K. (2001) *Rome's Eastern Trade: International Commerce and Imperial Policy, 31 BC–AD 305*. London.

Ziegler, K.-H. (1964) *Die Beziehungen zwischen Rom und dem Partherreich: Ein Beitrag zur Geschichte des Völkerrechts*. Wiesbaden.

Ziethen, G. (1994) *Gesandte vor Kaiser und Senat: Studien zum römischen Gesandtschaftswesen zwischen 30 v. Chr. und 117 n. Chr.* St. Katharinen.

Zimmermann, M. (1999) *Kaiser und Ereignis: Studien zum Geschichtswerk Herodians*. Munich.

CHAPTER 14

Early Sasanian Expansion and Military Institutions

Katarzyna Maksymiuk

The western policy of the Sasanians (224–651 CE), the dynasty that ruled in Iran for nearly 400 years,[1] is usually researched only in the context of relations with the Roman Empire, something that leads to oversimplifications, creating a flat picture of events as a conflict between two empires. The considerations below aim not only at reconstruction of military actions, which naturally remains an indispensable starting point for further research, but, above all, attempt to analyse the reasons and goals of the Iranian kings in waging successive western campaigns. The chronological scope of the work covers the first two centuries of the Sasanians, the period of the struggle to consolidate the power of the new dynasty in Iran and the struggle to signify the position of Eranshahr in West Asia.[2]

The origin of the Sasanian dynasty remains a subject of scholarly dispute.[3] This situation results from the propagandistic efforts of the Sasanian rulers to justify their right to exercise royal power.[4] All proposals remain in the realm of conjecture. A variety of source materials allow reconstruction of Sasanian activities in the west. Of course, they are written from a particular perspective and, more importantly, unevenly distributed over the period. The primary sources for the study of the early Sasanian period are the products of material

1 Christensen (1944); Schippmann (1990); Wiesehöfer (1996); Dignas and Winter (2007); Pourshariati (2008); Daryaee (2009); Jackson Bonner (2020).

2 See also Maksymiuk (2024).

3 Widengren (1971); Shayegan (2008) 115–19; Daryaee (2010); Olbrycht (2016a); Rezakhani (2017a) 39–42.

4 For traditions of Sasan as a descendant of the Achaemenid kings, see *Kār-nāmag ī Ardašīr ī Pābagān*, 1.6–7; 2.17; Dīnawarī 30 (Guirgass and Kratchkovsky (1888)). Based on an analysis of the Indo-Parthian coins of Farn-Sasan and the early issues of Ardashir, Olbrycht (2016a) has put forward a very interesting idea that Sasan may have come from the branch of the Arsacid Gondopharid line (the Indo-Parthian Dynasty), the rulers of Sakastan and Sindh; Daryaee (2010) draws attention to the genealogy contained in the Bundahišn (35.36). Juxtaposing this genealogy with the information given by Ṭabarī (814), that Sasan was the custodian of the fire temple of Anahita in Istakhr, and the content of ŠKZ (25/20/46), Daryaee proposes the idea that this was the official position of the Persian court by which Ardashir could justify his claim to the crown on both priestly and royal grounds.

 | DOI:10.1163/9789004710771_015

culture, including inscriptions, monumental reliefs of the rulers, and coins.[5] Particularly noteworthy are the royal inscriptions created in the third century, which due to the length of the texts can be treated as 'chronicles'. The text of the *Res Gestae Divi Saporis* (ŠKZ), the great trilingual (Middle Persian, Parthian, and Greek) inscription on the Ka'be-ye Zardošt at Naqš-e Rostam, describes three wars with Rome waged by Shapur I (242–272). ŠKZ also lists the members, with their titles, of the court of Pabag, Ardashir (224–242), and Shapur.[6] The second text is a bilingual (Middle Persian and Parthian) inscription from Paikuli (NPi) depicting the circumstances of coming to power of *shāhānshāh* Narseh (293–302), in 293, in the course of the dynastic war which followed the death of Wahram II.[7]

To reconstruct the history of the Sasanians we must also rely on Middle Persian sources written in the late Sasanian period. The *Kār-nāmag ī Ardašīr ī Pābagān* (the Book of the Deeds of Ardashir, son of Pabag) is a compilation from the sixth to seventh century about Ardashir I's rise to power and his foundation of the Sasanian dynasty.[8] The *Xwadāy-Nāmag* (Book of Lords/Kings)[9] might have never existed,[10] but information on the Sasanian Empire is included in the classical Arabic histories which were written in the ninth and tenth centuries, for example Ṭabarī and Dīnawarī.

Another group consists of the accounts of Armenian historiography, most notably the *History of the Armenians* of Agathangelos,[11] which describes the conversion of King Trdat (Tirdad/Tiridates) of Armenia to Christianity, and the *History of Armenia* of Moses Khorenatsʿi,[12] covering the history of the country up to the end of the native Arsacid dynasty in 428. While they are the product of Christian historiography, Armenia was inextricably linked to Iran during the period in question, and the authors were unable to 'dissociate' themselves from the Parthian origins of their kings. The greater problem in interpreting these accounts is not their religious nature, but the distorted chronology of events.[13]

Sasanian Iran appears in Roman sources, mostly chronicles of wars, on the occasion of conflicts with the empire. Unfortunately, a lack of continuity in

---

5 Cereti (1997); Mousavi and Daryaee (2012); Schindel (2013); Shavarebi (2015).
6 Kettenhofen (1982) 12–8, with further literature.
7 Humbach and Skjærvø (1983); Cereti (2021).
8 Panaino (1994). For the text see Antia (1900).
9 Shahbazi (1990) with further literature.
10 Discussion on existence of the *Xwadāy-Nāmag* in Jackson Bonner (2011) 19–28; Hoyland (2018) 1–23.
11 Thomson (1976).
12 Thomson (1978).
13 Kettenhofen (1995b); Traina (2018).

the reporting of events, as well as problems with chronology, especially in the second half of the third century, makes research into events in Iran difficult. The accounts of Cassius Dio and Herodian cover only the beginning of the third century, and the *Historia Augusta*, the only Roman history written at the beginning of the fourth century, lacks an account of the years 244–259.[14] An exceptional work in the analysis of the western expansion of the Sasanians in the fourth century, and especially in the study of the Iranian military, is that of Ammianus Marcellinus,[15] who, through his participation in Roman campaigns in the Near East, was able to draw on his own experience. Of the numerous works written in the later period, Agathias[16] and Zonaras[17] are also of some importance. Mention should finally be made of Syriac sources such as the *Chronicle of Khuzistan*,[18] the *Chronicle of Seert*[19] or the work of Michael the Syrian,[20] which, despite their Christian character and depiction of Iran, often provide information not found in other accounts.

1 Sasanian Campaigns in the Third Century

Until the mid-second century, the 'border' between Rome and Iran was largely constituted by the Upper Euphrates. Mesopotamia was considered an integral part of the Arsacid state. The situation changed dramatically with the actions of the emperor Septimius Severus (193–211), who not only captured Ctesiphon (Cass. Dio 75.9.1–12.5; Hdn. 3.9.1–12), but more importantly created the Roman provinces of Mesopotamia and Osrhoene, thereby undermining the authority of Arsacid power.[21] Of course, this event could not by itself have led to the collapse of Parthian power in Iran, but it coincided with an escalation of dynastic conflict that escalated into civil war. The seizure of the throne by Vologases VI in ca. 207, who succeeded his father, did not improve the situation in the western part of the country. The ruler was unable to respond to the revolt by the Medes and the Persians, spreading since the year 205/206. Pabag consolidated his power in Pars with the help of his sons, Shapur and Ardashir. He dethroned

14 Edwell (2010).
15 Brodka (2009) 78–88.
16 Cameron (1969).
17 Mazzarino (1971).
18 Hoyland (1997) 182–89.
19 Wood (2013).
20 Chabot (1899–1910); Duval (1970).
21 Angeli Bertinelli (1976) 39; Edwell (2008) 26–28; Marciak (2017) 388–92; Schlude (2020) 183 n. 79; Börm (2019) 105.

the king of Istakhr, Gozīhr in 211/212, while his younger son Ardashir consolidated himself in Darabgerd.[22] Vologases could not pacify the situation in Pars, because in ca. 213 his younger brother Artabanus (Ardawan) IV (216–224) challenged his right to the throne by inciting a rebellion in Media, which turned the king's attention northward. Taking advantage of the civil war,[23] the Roman emperor Caracalla (211–217) subjugated Edessa to Rome in 214. Two years later the Romans invaded Adiabene (controlled by Artabanus), where they exhumed the bones of the royals in Arbela (Cass. Dio 75.1.1–2). After the murder of Caracalla, his successor Macrinus (217–218), after an inconclusive battle near Nisibis, made peace in 218 (Hdn. 5.1.4).

Somewhat on the margins of the main events, Ardashir gradually expanded his influence. The first target of Ardashir's attack was al-Baḥrayn, most likely to gain control over the western shores of the Persian Gulf.[24] In the course of the victorious campaign Ardashir killed the local ruler Sanaṭrūq (Ṭab. 1.820).[25] The reflection of Ardashir's struggle with the Arabs can be heard in the story of 'auspicious worm' – the coalition of Haftwād (Ṭab. 1.817; *Kār-nāmag ī Ardašīr ī Pābagān* 6.1–25). Ṭabarī mentions intensive actions taken by Ardashir on the banks of the Persian Gulf (1.820).[26] It seems that in the long perspective Ardashir planned to capture all of Arabia, and subsequently the entire maritime coastline of the Persian Gulf.[27] Information about the fight with Haftwād suggests that Ardashir intended to subdue the water route not only for economic, but also for military reasons.[28] The main intention of the policy of Ardashir in this region was gaining control over the border zone of the Arabian Desert. This goal was the foundation of the close relationship between the Sasanians and Arabs from Hira.[29] According to Ṭabarī, the important tribal confederation of Tanukh inhabiting the vicinity of Hira intended to resist Ardashir. The Sasanian ruler turned directly to the clan of the Lakhmids, which had played a meaningful role in the city (1.821). These agreements seem to have resulted in the Lakhmids' later cooperation with the Sasanians (Ṭab. 1.833; NPi H18,03–H1,04).

22 On the chronology of the Pabag revolt and the interactions between Pabag and his sons Ardashir and Shapur see Daryaee (2010).
23 Olbrycht (2016a).
24 Before the defeat of Ardawan; Piacentini (1985); Kennet (2007).
25 The appearance of the Parthian name Sanatruq/Sinatrices, so common among the Arsacids, may mean that this local ruler was an Arsacid vassal; Bosworth (1983) 603–604.
26 Potts (1990) 233; Kervran (1994).
27 Potts (2008); Hojabri-Nobari *et al.* (2011); Daryaee (2016a) 41; Dmitriev (2017) 731.
28 Winter (1988) 77.
29 Fisher (2011a); Fisher and Wood (2016); Wood (2016) 789–90.

The possible connection of Ardashir's military activities in this area with his building activity, clearly demonstrated by the settling of new cities (Ṭab. 1.820),[30] may point to the beginnings of an infrastucture and autonomous source of revenue that, in a relatively short time, enabled 'the Sasanians to become more powerful than the Parthians'.[31] To accomplish this, taking control over the trade in the Gulf region was a *sine qua non* condition.

After establishing his power in the Persian Gulf, Ardashir was ready to come forward against the Arsacid dynasty, with the support of the rulers of Adiabene (Nodšēragān) and Media (Māh). The battle of Hormzdagan plain in 224 does not appear in the Roman sources, and reconstruction of the events is based on the account of Ṭabarī (1.818–19). The exact location of the battlefield is unknown, somewhere between Isfahan and Nehavand in Media. According to Ṭabarī, the direct armed clash was preceded by an exchange of correspondence between Artabanus and Ardashir, in which the exact time and place of the battle were agreed upon. Ardashir arrived before the agreed time of the battle to gain a strategic advantage for his army: 'He dug out a ditch to protect himself and his army, and took possession of a spring there' (Ṭab. 1.819). We do not have a detailed description of the course of the battle; Ṭabarī only mentions that in the first phase Shapur killed Dadhbundadh (the secretary), and Ardashir achieved the decisive victory by killing the king Artabanus with his own hands. This moment of victory has been depicted on the mounted combat relief at Fīrūzābād. After defeating the Parthian troops Ardashir declared himself *shāhānshāh* (king of kings) of Iran (Cass. Dio 80.3.2; Hdn. 6.2.1; Agath. *Hist.* 2.26.2; Ṭab. 1.819).[32]

Ardashir's subsequent military actions were determined by the necessity of removing remaining supporters of the overthrown Arsacid dynasty, and strengthening his own position in the state (Ṭab. 1.823). After defeating Artabanus, Ardashir launched an immediate attack against the Arab city of Hatra (Cass. Dio 80.3.2).[33] Several factors were responsible for this attack. The first were the

30 One of the ports on the northern side of the Persian Gulf was al-Khaṭṭ. 'He built eight cities: … in al Baḥrayn, Fasā(?) Ardašīr' (Ṭab. 1.820); 'the city of Fūrān Ardašīr, and these were in Baḥrayn' (Dīnawarī 47 (Guirgass and Kratchkovsky (1888))). The *Kār nāmag ī Ardašīr ī Pābagān* (4.8) also describes his founding of the port: 'he saw the ocean before his eyes, he offered thanksgiving to God, called that place the city of Bokht Ardašir'. The location of the city Rēv Ardašīr (Ṭab. 1.820) is still disputed, but it may have been on the coast of Pars.

31 Payne (2018) 230.

32 Bivar (1983) 96; Felix (1985) 26; Wiesehöfer (1986); Alram (2003) 22–23; Daryaee (2010); Patterson (2013); Syvänne and Maksymiuk (2018) 19–34.

33 Wiesehöfer (1982) 439–40; Dodgeon and Lieu (2002) 13; Sommer (2003) 44–46.

strong bonds of the Arab aristocracy with the Arsacid royal house.[34] The second was the strategic position of the city, whose role increased at the turn of the second and third centuries, when Rome established the province of Mesopotamia and subordinated Osrhoene. After the Sasanians seized power, a reversal of alliances took place that proved particularly dangerous for the new dynasty, as Hatra remained loyal to the court of Ctesiphon and entered the Roman defence system.[35] A third factor, of no less importance, was Ardashir's desire to control trade routes and probably the trade competition between Hira and Hatra.[36]

The strongest bastion of resistance against the new Persian dynasty was Armenia, whose kings were tied with the Arsacid rulers by blood.[37] After the battle on the plain of Hormzdagān, the sons of Artabanus found shelter in Armenia, where they were preparing for further struggle with Ardashir. In defence of the Parthian royal house stood the Arsacids of Armenia, the Parthian clan of Karen, and the countries of Media Atropatene (Āδurbāyagān), Albania (Arrān), and Iberia (Viruzān) (Cass. Dio 80.3.3; Zonar. 12.15; Agathangelos 1.19; Moses Khorenatsʿi 2.71). In order to secure his power in Iran, Ardashir had to take military action against the Arsacids of Armenia.[38] However, the first Persian campaign in Armenia, in ca. 227/228,[39] was thwarted by King Xusrō (Khosrov in Agathangelos 1.19).[40]

Ardashir's aggressive actions against Hatra and Armenia led to the conflict with the Roman Empire. The Persian army entered Mesopotamia in 230. In the first phase Ardashir besieged Nisibis, but the intrusion into Iranian territory by the Roman army, heading towards Ctesiphon, forced the ruler to move his troops south. In the winter of 232, both armies stood in proximity to the Persian capital, but there was probably no direct clash at that time (Zonar. 12.15; Hdn. 6.2.5; 6.5.1–6.6.6; Oros. 7.18.7; Eutr. 8.23).[41] The campaign of Severus Alexander in 231–233 strengthened Armenia's rapprochement with Rome on military grounds (SHA *Alex. Sev.* 61.8). Ardashir entered Mesopotamia again between 238 and 240, occupying Carrhae, Nisibis, Edessa and Hatra

34 Maksymiuk (2017a).

35 Kettenhofen (1982) 20; Hauser (2013) 130–33.

36 Young (2001) 188–90; Altaweel and Hauser (2004).

37 Dąbrowa (1983) 144–47; Olbrycht (2016a); Dąbrowa (2018) 77–80.

38 Weber 2016.

39 Shavarebi (2014) 128.

40 A problem of the reliability of the Armenian accounts is the distorted chronology. Perhaps the events described do not refer to a king named Xusrō but to Trdat II (r. 216/7–252?), but in this case the identification of Xusrō with Trdat conflicts with Moses Khorenatsʿi's identification of the duration of Xusrō's reign of forty-eight years (2.74). Cass. Dio (80.3.3), when mentioning the anti-Persian coalition, does not mention the name of the Armenian king. See Dodgeon and Lieu (2002) 298 n. 10.

41 Rösger (1978); Potter (1987).

(Zonar. 12.18).[42] There is no information about the actions of the Armenian army during this Persian invasion.

It should be pointed out that Ardashir had not entirely broken the Arsacid rule by the end of his reign. The defiance of the Arabs, the alliance of Hatra with Rome and, most of all, the survival of the Arsacids on the throne of Armenia remained clear and present dangers for the new Sasanian dynasty.

The enthronement of Shapur I coincided with the Roman army's attack on Mesopotamia (ŠKZ 4–5/3/7–8).[43] Gordian III (238–244) took immediate action in 243, occupying Carrhae and Nisibis,[44] but Shapur defeated him in the battle of Mešik/Peroz-Shapur[45] the following year (ŠKZ 5/4/9; Eutr. 9.2).[46] There is no certainty whether the emperor was killed during the battle (as suggested by the ŠKZ account) or whether he was assassinated by conspirators shortly after the battle (Amm. Marc. 23.5.17; SHA *Gord.* 30.1–9; Fest. 22; Eutr. 9.3); but his death during the war with Iran remains an undisputed fact. This event was masterfully exploited for propaganda by Shapur. On the triumphal reliefs at Darabgerd, Naqš-e Rostam VI, Bishapur I, Bishapur II and Bishapur III, Gordian is depicted lying under the hooves of Shapur's horse.[47]

More important than the propagandistic were the political effects of the victory observable on both sides of the conflict. The goals of the first Sasanians are illustrated by the provisions of the peace treaty concluded with Philip the Arab (244–249), which is confirmed not only by the written accounts, but also by the issues of commemorative coins. Coins minted by Shapur contained an elaborate legend emphasising the victorious side of the conflict.[48] Roman coins bore only a laconic reference to the making of peace.[49] Western accounts do not mention any financial regulations of the treaty, which were, presumably, the 'tribute' named in the ŠKZ (5/4/9), and actually concerned the redemption of Roman captives.[50] Significantly, Shapur did not make any territorial claims towards the Imperium Romanum. He was satisfied with the emperor's

42 Kettenhofen (1995a).

43 Kettenhofen (1983) 152.

44 Loriot (1971); Hamacher (1983); Edwell (2008) 169–72.

45 The town Mešik (Μισιχή) was refounded by Shapur I and called Peroz-Shapurto commemorate his victory over Gordian.

46 McDonald (1981); Huttner (2008) 188.

47 For an overview and discussion in Shavarebi (2015); *contra* Overlaet (2009), who identifies the Roman figure from reliefs at Darabgerd, Bishapur II, and Bishapur III as the usurper emperor Uranius Antoninus of Emesa.

48 Alram, Blet-Lemarquand, and Skjærvø (2007) 23: *ēn ān ka-š firipōs kēzar ud hrōmāy pad bāz ud bandag<īh> estād hēnd* ('This (was at) that (time) when the Caesar Philip and the Romans stood in tribute and subjection to him').

49 RIC IV Philip I 69, 71: *Pax fundata cum Persis* ('Peace founded with the Persians').

50 Maksymiuk (2016) 149.

declaration that Rome pledged not to intervene in Armenia's affairs (Evagr. *Hist. eccl.* 5.7; ŠKZ 5–6/4/10; Zonar. 12.19).[51]

The terms of the treaty, undoubtedly dictated by the Persian side, show that Shapur's main goal was to gain control of Armenia and remove the representatives of the Arsacid dynasty from the throne of that kingdom.Armenia was conquered by the Sasanian forces only in 252/253,[52] and Shapur I gave the title of *wuzurg šāh Arminān* to Hormizd-Ardashir, his eldest son.[53] Significantly, it was only Hormizd-Ardashir among the sons of Shapur listed in the ŠKZ who bore the title of *vuzurg šāh*, while others are titled only as *šāh* (ŠKZ 23/18–19/42–43).

In order to prevent a Roman intervention in support of the Arsacids in Armenia, Shapur attacked Syria and Cappadocia. According to a reconstruction of the events based on the ŠKZ, the military actions of the Persian army consisted of three individual actions, the attack on Syria in 253, the attack on Cappadocia in 253, and the attack on Dura and Circesium in 256.[54]

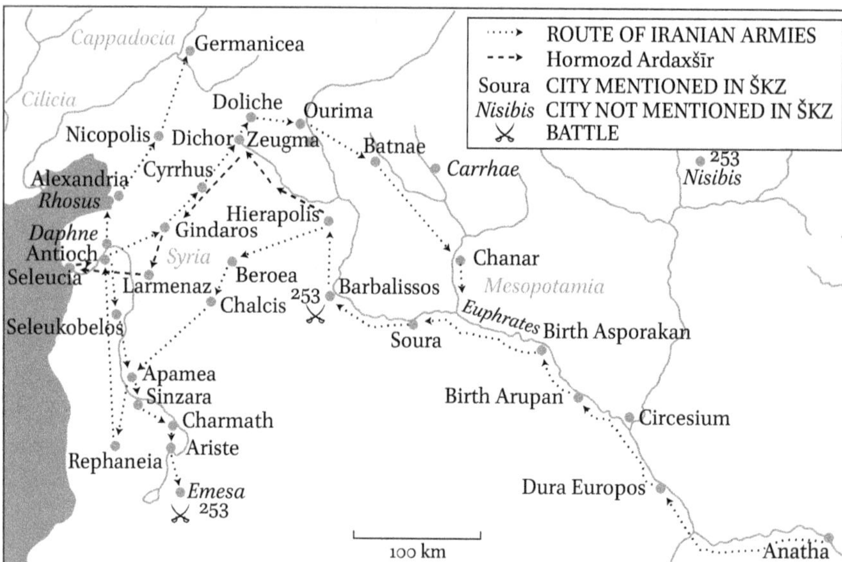

FIGURE 14.1 Military operations of Shapur I in 253–256 (according to ŠKZ)
PRODUCED BY K. MAKSYMIUK

51 Pekary (1961); Winter (1988) 97–107; Körner (2004); Dignas and Winter (2007) 119–22; Mosig-Walburg (2009) 64.

52 See Hartmann (2006) 106 n. 5.

53 ŠKZ 23/18/41: *Ohrmezd Ardašīr vuzurg šāh Arminān* / *Ohrmezd Ardašīr vuzurg šāh Arminīn* / Ὀρμισδαρταξιρ τοῦ μεγάλου βασιλέως Ἀρμενίας ('Hormizd-Ardashir, great king of Armenia'). See Daryaee (2017).

54 Alföldi (1937); Enßlin (1949) 33–35; Pekary (1962); Kettenhofen (1982) 79–83; Strobel (1993) 220–56; Mosig-Walburg (2009) 43–44.

Shapur's first expedition to Syria began in spring 253.[55] The Persian army passed through Peroz-Shapur. Then the Persians moved along the left bank of the Euphrates and crossed the river near Anatha. Shapur began a hasty forced march along the right bank of the Euphrates without delaying it by the sieges of the Roman fortresses on the Euphrates. He decided to capture only the cities that did not resist, or where the defiance was insignificant. Probably his aim was to break down the Roman armies, preempting their expected invasion of Armenia. In this way, Shapur reached Dura, the strategic point of the march. The encounter with the Roman legions took place in northern Mesopotamia, leading to the battle of Barbalissus in 253 (ŠKZ 6/5/11).[56] Then the Persian army headed north to Hierapolis, where it divided into smaller groups with the aim of plundering Syrian cities. Part of the army entered the upper Orontes area, reaching Rephanea. The second group, presumably commanded by Hormizd-Ardashir, headed towards Zeugma to cut off the Roman troops stationed in Mesopotamia. The supposed target of the 'northern' group was Antioch. The city was captured by the combined troops of Hormizd-Ardashir and Shapur (Zos. 1.27.2; ŠKZ 9/7/15; Ṭab. 1.827; SHA *Tyr. Trig.* 2.2).[57]

Following this, the Persian army divided again and set off in several directions. The 'northern' group under the command of Hormizd-Ardashir headed from Nikopolis to Cappadocia, reaching Germanicea (ŠKZ 10/8/17). These actions could have been an isolated military action protecting Armenia from a direct Roman attack from Cappadocia, after it was abandoned by the *wuzurg šāh Arminān*. The 'southern' group reached Emesa, where it was defeated (Mal.12.1–4).[58] As it was retreating towards the Euphrates, it was crushed by Palmyra's ruler Septimius Odaenathus coming from the north (Mal. 12.4–10). Unfortunately, the lack of historical data makes it impossible to reconstruct events until 256, the time of the destruction of Dura-Europos.[59] The campaign probably ended in 258.

The primary source for reconstruction of the operations of Shapur in the West is the ŠKZ inscription, which describes three wars with Rome, listing the cities that the Persian army captured. As the majority of the thirty-seven cities mentioned in the ŠKZ were captured in the course of the second campaign

55 Perhaps the campaign was preceded by a preventive attack of Hormozd in Lesser Armenia; see Satala in the ŠKZ 11/8/18.

56 Baldus (1971) 240–46.

57 Hartmann (2006). The problem of the capture of Antioch by Shapur in the mid-third century is debated, both in terms of number of times when the Persian army entered the city, and the dating of this event (252, 253, 256 or 260); see discussion with literature in Goltz and Hartmann (2008) 258–59.

58 Overlaet (2009).

59 McDonald (1986).

in 253, based on the analysis of archaeological and numismatic sources, it is reasonable to assume that Dura-Europos was not destroyed until 256. A date after 253 is also confirmed by the fact that the city was not mentioned until the end of the Persian campaign (ŠKZ 10/8/17). The temporary occupation of Dura by the Persians, of which the starting point is evidenced by the presence of Middle Persian inscriptions at the synagogue, is still under discussion.[60]

In 260, during the third expedition, the Persian army besieged the Mesopotamian cities of Carrhae and Edessa (ŠKZ 12/9/20). The direct clash of the Persian king with the emperor ended with the defeat of Valerian (253–260) in the battle of Edessa (ŠKZ 14–15/11/24–25; Eutr. 9.7;[61] Fest. 23; Lactant. *De mort. pers.* 7.13).[62] Persian troops plundered Syria, where without much resistance from the Roman troops, they retook Antioch (ŠKZ 19/14/31) and the southern coast of Asia Minor (ŠKZ 18/14/31). The second division of the Iranian army made for Tyana, from which it raided Iconium (ŠKZ 20/15/34), and then attacked Cappadocia and Lesser Armenia. In 261 the army retreated to Iran.

At the end of 261, after Septimius Odaenathus had defeated the usurper Quietus at Emesa, the emperor Gallienus (260–269) granted him the *imperium maius* in the East (Zonar. 12.24).[63] The Roman army, led by Odaenathus, entered the territories abandoned by the Persians and captured and destroyed Nisibis in 262 (Zos. 1.39.1; SHA *Gall.* 10.3), after which, without encountering resistance from the Persians, he reached Ctesiphon (Fest. 23; Zos. 1.39.2).[64]

Shapur made no major effort to block his way to Asurestan. It is likely that Odaenathus' actions were not perceived as a threat. According to Zosimos (1.39.1), the Palmyrene was to gather under his command the Roman troops which survived Valerian's defeat. However, the information that Odaenathus was accompanied by *rusticiani Syriae* (Oros. 7.22.12; Fest. 23) is significant for the combat ability of the army that moved towards Ctesiphon. The source accounts do not agree whether Odaenathus undertook only one expedition. Two campaigns are mentioned only by Zosimos (1.39.1) and Syncellus (716–717),[65] but no doubt Ctesiphon was not captured.[66]

---

60 Grenet (1988); for an overview and discussion see Baird (2012).

61 '... in ignominious slavery among the Parthians (sic!)'. The distinction between the Parthians and the Persians was not always clear to the Romans; see Chauvot (1992).

62 Stolte (1971); Luther (2006); Kettenhofen (2001); Goltz and Hartmann (2008).

63 Hartmann (2008) 351.

64 de Blois (1975); Hartmann (2001) 162–230.

65 Maksymiuk (1998).

66 The only account in which this information survives is Syncellus (716). Information about such an important event, which was a retaliation for the defeat of Valerian for the Romans, could not have been omitted in 'Western' accounts.

The weakness of the Roman Empire, clearly demonstrated by Shapur I's campaigns, led to the assumption of control over its Middle Eastern provinces by Palmyra, under the reign of Zenobia.[67] The situation only changed in favour of Rome after the enthronement of Aurelian (270–275), who conquered Palmyra in ca. 272 and effectively demolished the city (Fest. 24; SHA *Aur.* 26.1–28.5; 31.1–10).[68] The fall of Palmyra coincided with the death of Shapur I; presumably this 'human factor' was the reason for the passivity of Persian troops at the borderline at that time.[69]

The dynasty change in Iran was perceived in the Roman Empire as a threat to its position in the East (Hdn. 6.2.2; Cass. Dio 80.3.4).[70] But an analysis of the military actions taken by the Sasanians shows that they were followers of Arsacid policies, so it is not possible to describe the first Sasanian attack on northern Mesopotamia as an attempt to restore the borders of the Achaemenid period.[71] The aim was to secure the power of the new dynasty and to reclaim territories that had previously been linked to the Arsacid state. The primary concerns were Armenia and the cities of Mesopotamia, Hatra, Carrhae, Nisibis, and Edessa.

The wars with Rome waged in the lands of the empire, were 'peripheral' for the Sasanians, from a strategic point of view. This means that the Sasanians did not seek to annex Roman provinces, as is suggested by the Western sources. There is no evidence for planned administrative activities by the first Sasanian rulers in Syria and northern Mesopotamia (only the case of Dura-Europos can

67 Southern (2008).

68 Downey (1950); Kotula (1997).

69 A source that helps establish the chronology of the Sasanian origins is an inscription on a column in Bishapur (ŠB). The text contains three dates: 58 (Sasanian era) – presumably the date of the foundation of the monument, forty years after Ardashir took power, and twenty-four years after the succession of Shapur. If we assume that Shapur's death occurred in 272, his reign must have lasted about thirty years, then we obtain the year 266 as the date of the foundation of the inscription. The beginning of the Sasanian era would be in 208, the coronation of Ardashir in 226, and the beginning of Shapur's independent reign in 242. These dates are supported by the account of Agathias (*Hist.* 4.24.1), who says that Ardashir 'seized the throne of Persia ... in the fourth year of the reign of Severus Alexander, five hundred and thirty-eight years after Alexander the Great and that he reigned for fourteen years and ten months'. The only difference between Agathias' transmission and the Bishapur inscription is the specific duration of Ardashir's reign. However, the Bishapur text is more reliable. In the ŠKZ (4–5/3/7–8) there is a chronological connection between Shapur's assumption of power and the march of the Roman army to Asurestan in 242 (Gerhardt and Hartmann (2008) 1057). Thus, it seems that the relevant years of Shapur's reign are 242–272.

70 Frendo (2002); Gariboldi (2016).

71 Kettenhofen (2002a); Maksymiuk (2004) 137; Shayegan (2011) 13–29.

be discussed, which was mentioned earlier). An argument against the thesis of a desire to annex Roman provinces is the destruction of cities and the mass deportation of their inhabitants deep into Iran (ŠKZ 20–21/15–16/34–35; Zos. 1.27.2; Ṭab. 1.827).[72] This is particularly evident during the actions, described in the ŠKZ as a second clash with Rome, which were primarily aimed at securing Hormizd-Ardashir's power in Armenia, against Roman intervention. The Persian attack on Syria and Cappadocia made it impossible to send legions into Armenia, further shifting the arena of military action to Roman territory.

The victorious wars of Shapur were an undeniable success for Iran, and in a propagandistic sense, naturally, the western Roman provinces were not treated as 'peripheral' or insignificant. On the contrary, the glory of *shāhānshāh* was described in the ŠKZ, not only in Middle Persian and Parthian, but also in Greek, the language of the 'Roman East'. Emperors Gordian, Philip and Valerian are depicted on the most prominent Sasanian iconographic evidence commemorating victories over the Roman Empire – triumphal rock reliefs at Darabgerd, Naqš-e Rostam VI, Bishapur I, Bishapur II and Bishapur III.[73]

Shapur continued to pursue Ardashir's political and military objectives. He removed the Arsacids from power in Armenia, bringing his son to the throne. With his victorious campaigns against Rome, he not only secured Sasanian power in Armenia and strengthened its influence in the lands of the Caucasus,[74] but from an ideological point of view, he confirmed the legitimacy of his power. The king's victories were a sign of Ahuramazda's benevolence and a confirmation of the Sasanians' rights to the throne of Iran through the god-granted *xwarrah*.[75]

The legitimation of the Sasanian power in Iran was of particular importance in the northwestern provinces (Armenia and Iberia).[76] The 'Arsacid blood' motif and the memory of Parthian rule in Iran persisted in Armenian tradition well into Late Antiquity (Procop. *Pers.* 2.1.32).

It must be emphasised here that economic objectives also determined the military activities of the Sasanians. The first Sasanian rulers sought to control trade routes to the Far East, primarily to control the 'silk trade'.[77] The main centres of the trade in the first half of the third century were Hatra, Dura-Europos,

72 Mosig-Walburg (2010).

73 Shavarebi (2015).

74 The king of Iberia is mentioned in ŠKZ 31/25/60: *(H)amāzāsp Wiruzān šāh* / *(H)amāzāsp Wiržān šāh* / Ἀμαζάσπου τοῦ βασιλέως τῆς Ἰβηρίας. See Stickler (2021).

75 Gnoli (1999); Börm (2015).

76 Börm (2019) 106; for an overview see Hartmann (2019).

77 Payne (2018) with further literature. On the 'silk trade' in the Parthian Empire on the example of Palmyra, see Gregoratti (2015); Gregoratti (2020).

and Palmyra,[78] all associated with the Roman Empire.[79] The first military actions taken by Sasanians on the Persian Gulf interfered with Roman trade interests as the empire controlled the 'Nabatean' and 'Palmyrene' routes. Shapur I took firm control over the port of Spasinou Charax, which played an extremely important role in the Arsacid financial and administrative system.[80] This enabled his monopolisation of the trade on the Persian Gulf. The importance of this region to the Sasanian monarchy is evidenced by the fact that Shapur I granted the title of *šāh Mēšān* to his son Shapur (ŠKZ 23/18/42). The vigorous construction and settlement activity that resulted in the creation of a kind of 'urban network' under the direct authority of the king on the northern coast of the Persian Gulf resulted in the decline of earlier merchant centres (e.g., the sites of al-Dur).[81] The acquisition by Shapur of the southern Mesopotamian coastal region of Mēšān (Mesene/Charakene) forced Palmyra to seek new trade routes. It is likely that the trade agreements were included in the failed attempt at negotiations between Odaenathus and Shapur I, mentioned by Petros Patrikios (*FGrH* 4, F 10). After the destruction of Hatra (in 240) and Dura-Europos (in 256), Palmyra monopolised the trade in Syria. The trade rivalry may be the explanation for the destruction of Nisibis by Odaenathus in 262. It seems that after the destruction of Palmyra by Aurelian, the trade routes moved from the Euphrates to the Tigris.[82]

In the late third century, the restoration of influence in Armenia became a priority in the eastern policy of the Roman Empire due to the military campaigns of Shapur I, who made Armenia a *de facto* province of the Sasanian kingdom. According to Armenian sources, Trdat (Tiridates) son of King Xusrō of Armenia, of the Arsacid dynasty, was carried away as a child during the Persian invasion in 252 and grew up in Roman Cappadocia (Agathangelos 1.36; Moses Khorenatsʿi 2.76). Considering the fact that the *cognomen ex virtute*,

78 Rezakhani (2010) points out that, according to Chinese sources, it is impossible to assume existence of any organised routes west of Transoxiana, which is unsurprising, given the direct Chinese domination over the states of modern Xinjiang. The western leg of the 'Silk Road' would consist of multiplicity of local routes and in fact it would be more appropriate to talk about osmotic movement of the goods rather than a 'Road'. For this reason the term the Silk Road cannot be defined as 'route of commercial contact between China and Rome/Byzantium/Europe'.

79 Luther (2004); de Blois (2016) 36–37.

80 Potts (1988a). On the role of Characene in the Parthian Empire, see Gregoratti (2011).

81 Payne (2018) 231–32.

82 Payne (2018) 235: 'In place of the Palmyrenes, an alternative network began to form in the third century that linked the Sasanian royal foundations with the northern Mesopotamian cities of Edessa and Nisibis via the Tigris rather than the Euphrates'.

Persicus Maximus (*P. Oxy.* XIV 1631; *P. Oxy.* XIV 1713),[83] appears in the titulary of the emperor Probus (276–282) in 279, one can cautiously hypothesise that a Persian-Roman agreement was concluded at this time, but there is no way to determine its terms. In 282 Probus began preparations for a Persian campaign, during which he was assassinated (SHA *Prob.* 20.1). A renewed war with Rome[84] took place in 283 when Carus (282–283) conquered Ctesiphon (Eutr. 9.18; Fest. 24; Moses Khorenatsʿi 2.79).[85] This success was annulled by the victory of Wahram II (276–293) over Numerianus (283–284) in northern Mesopotamia.[86]

In 290, Emperor Diocletian (284–305) re-established a representative of the Arsacid dynasty, Trdat III, on the throne of Armenia (Moses Khorenatsʿi 2.82; Agathangelos 3.18).[87] In 296, *shāhānshāh* Narseh attacked Roman Armenia and Syria (Amm. Marc. 23.5.11; Moses Khorenatsʿi 2.79–82; Zonar. 12.31). The army dispatched against him, commanded by Galerius, was crushed by the Iranian forces between Carrhae and Callinicum (Oros. 7.25.9; Eutr. 9.24). The Persian army then retreated to the southwest border of Armenia, presumably to prevent Galerius from attacking Asurestan. Diocletian's reaction was firm. The emperor went to Antioch in 298 (Mal. 13.6), at the same time ordering Galerius to call up frontier troops from Dacia (Fest. 25), Illyricum, and Moesia (Eutr. 9.25), and presumably also a Gothic contingent (Jord. *Get.* 21). Once the Romans had regrouped, Diocletian, with the core of the army, moved from Syria to northern Mesopotamia, while the newly recruited army under Galerius headed for Satala, the garrison town on the Cappadocian *limes*, to launch an attack on Mesopotamia from the north. Galerius' army unexpectedly encountered Narseh's camp at Osḫa in the Basean region, not far from Satala (Aur. Vict. *Caes.* 39.34; Eutr. 9.25; Lactant. *De mort. pers.* 9.6). Galerius won a great victory, seizing the harem and the royal treasury, but Narseh himself managed to escape.[88] The defeat of the Persian army at Satala was primarily the result of border defence reforms introduced by Diocletian, which increased the strategic

83 On Probus' Persian campaign and the possible role of Trdat in these events see Kreucher (2003) 158–61.

84 Börm (2019) 107, offers an interesting idea that the most interested in escalation of the conflict with the Sasanians were the emperors whose position in *Imperium Romanum* was gravely threatened: 'Niemals wurde ein Augustus gestürzt, der zuvor römische Truppen als Feldherr siegreich gegen Parther oder Perser ins Feld geführt hatte'.

85 Halfmann (1986) 242; Winter (1988) 130–37.

86 Felix (1985) 100–102.

87 According to the inscription from Paikuli describing the events of 293 it must be assumed that Narseh as *šāh Arminān* retained the power in Persarmenia (*Armenia Maior*); Kettenhofen (1995b) 48–55; Kettenhofen (2008) 484–90; Weber and Wiesehöfer (2010) 103–105.

88 Zuckerman (1993); Leadbetter (2002); Mosig-Walburg (2009) 91–121.

mobility of the Roman army.[89] According to the peace treaty concluded in 298, Narseh withdrew from Armenia, renounced the *regiones Transtigritanae* and recognised the Roman protectorate in Iberia (Petros Patrikios, *FGrH* 4, F 14).[90] In addition to territorial changes, an economically crucial decision was taken regarding Nisibis. Diocletian enforced a clause in the treaty under which Nisibis was to be the only place of commercial exchange between the two states (Petros Patrikios, *FGrH* 4, F 14).

2 Sasanian Campaigns in the Fourth Century

When taking power in Iran, Shapur II (309–379) faced almost the same military challenges as the founder of the dynasty, Ardashir. The first step he was forced to take, starting in 325, was the pacification of the rebellious Arab tribes. Shapur initially reached Hajar, and the next campaign took place in eastern Arabia and Syria against Arabs from the al-Yamāmah region and the lands of the Bakr and Taglib. The bloody pacification of the Arab tribes is confirmed by the nickname *Dū al-Aktāf* given to the king (Ṭabari 844). What seems particularly important is that Shapur established a defensive system in southern Mesopotamia, called *War ī Tāzīgān* (the Wall of the Arabs), which appears to have been close to Hira. Control of this area was delegated to allies from the Lakhmid Arab clan.[91]

The treaty of 298 had wasted the achievements of Shapur I in northern Mesopotamia and the Caucasus region. Particularly unfavourable for the Sasanians was the restoration of the Arsacids to the throne in Armenia. The takeover of Armenia and northern Mesopotamia determined the military operations of Shapur II in the west of the empire, which pursued the political agenda of the founders of the dynasty. In the letter attributed to Shapur II by Ammianus (Amm. Marc. 17.5.6), we find a direct reference to the section of the ŠKZ in which Shapur I explains the reason for starting the second campaign against Rome (6/4/10).[92]

89 Several factors contributed to Narses' defeat, including the effect of surprise, the assistance of the local population favouring the Romans, the terrain favouring Roman infantry over the heavy Sasanian cavalry, and the custom of having a harem and treasury in the military camp, which limited the army's mobility.

90 Winter (1988) 152–215; Winter (1989); Marciak (2017) 37–39.

91 Fisher (2011b) 185; Daryaee (2016b) 81–82.

92 *ud kēsar did druxt,* (*ud*) *ō armin vināh kird* / *ud kēsar bid druxt, ō armin vinās kard* / καὶ ὁ Καῖσαρ πάλιν ἐψεύσατο καὶ εἰς τὴν Ἀρμενίαν ἀδικίαν ἐποίησεν ('and Caesar lied again, and did wrong to Armenia'); see Shayegan (2008) 121–22.

The adoption of Christianity in Armenia by Trdat during the early years of Shapur II resulted in the kingdom becoming closer to Rome, which made the situation in Iran more difficult (Euseb. *Hist. eccl.* 9.8.2–4; Sozom. *Hist. eccl.* 2.8.1; Moses Khorenatsʿi 3.5).[93] Probably already in 336, Shapur II intervened in Armenia. He succeeded in capturing King Tiran (330?–338) and the crown prince Arshak (*Epic Histories* 3.20), but agreed to the release of the royal family in 338.

In 337, the Iranian army invaded northern Mesopotamia. Although the Persians defeated the Roman troops around Singara twice,[94] despite several attempts (in 337, 346, and 350), Shapur was unable to achieve the main goal of his expedition, namely Nisibis, which was still under Roman control (Amm. Marc. 18.5.7; Fest. 27; Michael the Syrian 7.3; Zos. 3.8.2; *Chron. Pasch.* 536–539).[95]

The invasion of nomadic invaders from Central Asia forced Shapur to pay attention to the East (Amm. Marc. 14.3.1).[96] The sudden break in the military operations in Mesopotamia[97] and Shapur's agreement to enthrone Arshak II (*c.*350–364) in Armenia[98] show how serious a threat this was. Its severity is even more clearly illustrated by the necessity of the redeployment of the entire Iranian army and the personal presence of the *shāhānshāh* in north-eastern Iran.[99] Although Ammianus Marcellinus (16.9.4) refers to them by the name *Chionitae*, the clear explanation of their ethnic origin remains a matter of scholarly debate.[100] While uncertain, it is possible that this was referring to the ruling clan of the Huns or Kidarites, a part of the Chionite tribes.[101] It took nearly ten years for Shapur to neutralise the threat from the east and form an alliance with the king Grumbates (Amm. Marc. 17.5.1).

Warfare with Rome was resumed in the spring of the year 359. The Iranian army crossed the Great Zab and next the Tigris north of Singara (Amm. Marc. 18.6.19). Then the Iranian army conducted a false attack towards the Euphrates, bypassing Nisibis (Amm. Marc. 18.7.8). After twelve days' march,

93 Hewsen (1986); Chaumont (1996); Thomson (1997); Kettenhofen (2002b).
94 On the first battle of Singara, see Mosig-Walburg (1999) 330–84.
95 Lightfoot (1988); Burgess (1999) 7–17; Mosig-Walburg (2002) 329–47; Ross (2014); Dmitriev (2015); Maksymiuk (2015) 50–54.
96 Jackson Bonner (2017) 98–99; Dmitriev (2019) 49–63.
97 There was only a failed attempt to capture the town of Batnae in 353.
98 Tiran lost his life to Shapur.
99 Howard-Johnston (2010) 41–43.
100 The word the Huns used in the literature is not an ethnic name; see Kim (2013); de la Vaissière (2015); Payne (2015).
101 Payne (2015) 284; Rezakhani (2017a) 87–93; Rezakhani (2017b).

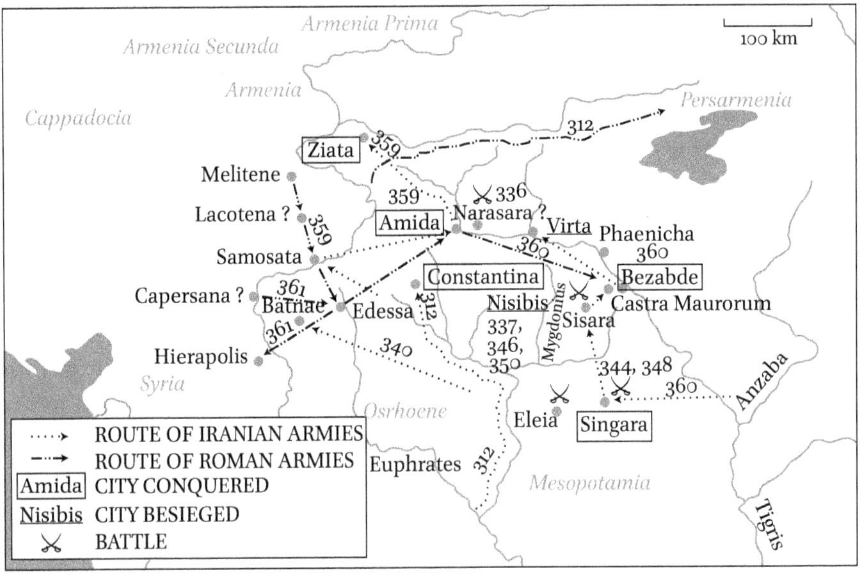

FIGURE 14.2 Military operations of Shapur II in 312–360
PRODUCED BY K. MAKSYMIUK

Shapur reached Bezabde[102] (Amm. Marc. 18.7.9). Unexpectedly for the Roman army, he did not attack Nisibis from the west, nor did he attempt the crossings of the Euphrates at Zeugma and Capersana, but attacked Amida (Amm. Marc. 18.8.1).[103] Amida was a gateway to Roman Armenia, and secured the province of Mesopotamia from attacks from the north. The city was captured after a seventy-three-day siege, on October 4, 359 (Amm. Marc. 18.9.1–4; 19.1.1–9.9).[104] In 360, Shapur captured two more Roman strongholds, Singara (Amm. Marc. 20.6.1–9) and Bezabde (Amm. Marc. 20.7.1–15). The events of 359 and 360 should be considered in terms of a single campaign.[105]

The political programme of Shapur was the cancellation of the 298 treaty and moving the frontier westward. This was impossible without taking over Nisibis. The destruction of Amida and capture of Singara and Bezabde isolated the eastern Trans-Tigritania region[106] and, more importantly, made it possible

102 About 150 km from Constantina, west of Nisibis.

103 Ursicinus expected that once Shapur had passed Nisibis, the Iranian army would attempt to cross the Euphrates. He moved west to assure the destruction of the crossings in Zeugma and Capersana.

104 Blockley (1988); Lightfoot (1989); Leński (2007a) 219–23; Seager (2010); Farrokh, Maksymiuk, and Sánchez Garcia (2018).

105 Maksymiuk (2018a).

106 The eastern Trans-Tigritania regions: Arzanena, Moxoena, Zabdicena, Rehimena and Corduena.

to attack Nisibis from the north as well, while blocking the communication routes of the Roman army. This was a crucial part of the plan to take Nisibis. Strengthening the fortifications of Bezabde also proves that Shapur's aim was not limited to simply plundering Roman cities, but extended to establishing permanent control over northeastern Mesopotamia. The fall of Amida also had serious consequences for the Roman position in the East. The Roman administrative-military machine had been thrown into confusion.[107] What is more important, the loyalty of Armenia and Iberia to the emperor was questioned (Amm. Marc. 20.11.1, 21.6.7–8).

The expedition of Julian the Apostate (361–363) in 363 proved of utmost importance to the balance of power at the border. The emperor set off from Antioch, marching across Mesopotamia. In Callinicum his army was joined by a fleet coming from Samosata. The towns on the trail along the Euphrates to Ctesiphon did not even attempt any defence. The first encounter with the Iranian forces took place as far as Ozogardana (Amm. Marc. 24.2.4–8). After the conquest of strongholds in the south, the emperor reached Asurestan. Although the Roman army managed to vanquish the Iranians commanded by Surena in the close vicinity of the walls of Ctesiphon, they were not able to capture the city itself (Amm. Marc. 24.7.1–8.7; Zos. 3.25.5–7). The campaign finished with the emperor's death in the battle of Toummara on 26 June (Amm. Marc. 25.3.1–8; Zos. 3.29.1–4).[108]

According to the subsequent peace treaty (Fest. 29; Zos. 3.31.1–2; Eutr. 10.17; Agath. *Hist.* 4.26.6–7), Jovian renounced his rights to the Trans-Tigritania, including Singara, Castra Maurorum, Nisibis, and fifteen other fortresses (Amm. Marc. 25.7.9). The treaty had a vast importance from the perspective of military infrastructure. Transfer of the fortresses of Singara and Nisibis destroyed the Roman defensive system in northeastern Mesopotamia. The treaty demanded the withdrawal of Rome's backing for Armenian rulers (Amm. Marc. 24.7.12). The acquisition of four-fifths of the South Caucasus into the Iranian sphere of influence gave it a clear strategic advantage.[109]

Shapur took further action in Iberia by removing the pro-Roman Saurmag (361–363/370–378) from the throne (Amm. Marc. 27.12.4).[110] In 364, under the

107 Jackson Bonner (2017) 101.

108 Contra Tabarī 842: 'One day, however, Luliyānūs was seated in his chamber when a stray arrow, from an invisible hand, struck him in the heart and killed him'. See Brok (1959); Austin (1972); Ridley (1973); Trümpelmann (1975); Kaegi (1981); Fornara (1991); Dąbrowa (2007); Mosig-Walburg (2009) 283–304; Brodka 2009, 76–105.

109 Blockley (1984) 34–37; Chrysos (1993); Seager (1996); Dignas and Winter (2007) 131–34; Mosig-Walburg (2009) 305–24; Howard-Johnston (2013) 872.

110 Toumanoff (1963) 460–61; Leński (2002) 165; Haas (2008).

pretext of help for the *nakharars* who were rebelling against King Arshak, Shapur invaded Armenia (Amm. Marc. 24.7.12). Ammianus Marcellinus testifies that Arshak had joined Julian's expedition against the Sasanians (Amm. Marc. 23.3.5; 24.7.8). The king was imprisoned in the Castle of Oblivion in Khuzistan, where he died.[111]

Shapur sent troops into Armenia again in 370, and after a long siege he captured the fortress of Artagerk‘, where Queen Pharandzem and Pap (371–374), the son of Arshak II, resided (Amm. Marc. 27.12.11–12). Only Pharandzem was taken prisoner, because Pap had fled to the Romans just before the end of the siege. In 371, Pap returned to the country with the assistance of the emperor Valens (364–378). At the same time, the emperor sent troops to Iberia to restore Saurmag to the throne (Amm. Marc. 27.12.16). However, he finally agreed to the division of the country between Saurmag and Aspacures (Amm. Marc. 27.12.17), which was perceived by Shapur as a violation of the treaty of 363 (Amm. Marc. 27.12.18). The next year the Roman forces, supported by Armenian contingents, decisively defeated the Iranians in a battle at Bagavan (Amm. Marc. 29.1.1–2). King Pap, suspected of the intention to make a secret deal with Shapur II, was assassinated in 374, maybe at the behest of the emperor (Amm. Marc. 30.1.18–21).[112] Then Valens installed Varazdat (374–378) on the throne, only to see him overthrown by Mushegh Mamikonean. After 378 the brother of Mushegh, Manuel Mamikonean, ruled as (*a*)*sparapet* (commander in chief) and a regent of Pap's two very young sons, Arshak and Vałarshak, to one of whom he gave his daughter in marriage (*Epic Histories* 5.37). He concluded a treaty with Shapur II which resulted in the *shāhānshāh* sending diadems to Zarmandukht and her sons (*Epic Histories* 5.38). After the death of Manuel Mamikonean, Armenia was partitioned as a result of the agreement between Theodosius I (378–395) and Shapur III.[113] In 394, the throne of Persarmenia was assumed by the Persian king's brother Vram Shapuh.[114] In 428, Wahram V (420–439) changed the status of Armenia by introducing a military administration (the marzpanate).[115] Iranian rulers introduced this administrative system in the most strategic border provinces. The *marzbān* was appointed by the *shāhānshāh*, and in the case of Armenia, the kings also appointed representatives of the local *nakharars* to this office. The marzpanate

111 Baynes (1910); Gutmann (1991) 162–91; Maksymiuk (2017b) 108.
112 Garsoian (1967); Gutmann (1991) 172; Seager 1996, 277–78; Leński (2002) 167–82; Mazza (2003); Leński (2007b).
113 Doise (1945); Blockley (1987); Greatrex (2000).
114 Daryaee (2009) 43.
115 Garsoïan (1997); Traina (2004).

period (*Marzpanakan Hayastan*) in the history of Armenia lasted until the Arab conquests in the seventh century.

The frontier strategies of the Sasanian rulers, which included the mass deportations of the borderland population, must also be noted here.[116] These relocations mainly affected the rebellious Arabian tribes (Ṭabari 839) and the Armenian subjects (*Epic Histories* 4.24; 4.54–4.55; 5.7). Shapur II strove not only to pacify the population, but also to change the ethnic composition of northern Mesopotamia. This goal was confirmed by the fact that the inhabitants of Nisibis were relocated to the former Roman territory (Amm. Marc. 24.7.11, 25.9.1–6; Ṭab. 1.843), and on Shapur's orders, Iranians were settled in Nisibis (Ṭab. 1.843).

At the end of the fourth century, the Sasanians managed to implement their strategic plans on the western borders. Not only did they take control of northern Mesopotamia, but this ultimately led to the removal of the Arsacid dynasty in Armenia. It seems that the division of Armenia between Iran and Rome removed the constant impulse of military actions, and Iran's western frontier remained stable for another century.

3 Sasanian Military Organisation

Undoubtedly, one of the factors in the Sasanians' successes in the third and fourth centuries was the army. Efficient military organisation was the crucial factor in waging the campaigns discussed above. It is notable that the successful performance in combat with varied enemies in different geographical settings could not and did not depend solely on quality of armament or the troops' perfected fighting skills. These factors, of great importance to the Sasanian army, are insufficient in themselves to result in numerous victories against varied types of armies commanded in varied ways, originating from different traditions. Early Sasanian warriors faced not only their kin in civil strife against the Arsacids, but also the Roman military machine, Kushan imperial forces, and nomadic hordes. Preserving fighting efficiency in such varied combat environments was the result of personal and material excellence, but most of all a chain of command which allowed the execution of tasks within a strategic framework. The shape of this highly efficient chain of command can be reconstructed only through detailed examination of the preserved military titles and the interrelations between them. Due to the shortage

116 Kettenhofen (1996).

of direct sources, the command structure needs to be reconstructed based on the surviving clues of which the military titles are of greatest importance. It would be no exaggeration to state that without the detailed study of the titles and the relations between them, there is no other way to understand Sasanian operational practice.

Despite the numerous accounts describing the wars waged by the Sasanian kings on the western frontiers of the empire, it is not easy to map the actual organisation of the army during this period. An extremely valuable source is the ŠKZ, which lists hierarchically the members of the court of the early Sasanians, including those bearing military titles. Sadly, it does not specify their powers. The remaining accounts of the 'Iranian circle' were written at a great distance from the considered period; hence we are not sure how the information about the organisation of the army contained for example in the *Kār-nāmag ī Ardašīr ī Pābagān* or the work of Ṭabarī relates to the reality of the early Sasanian period.

Another problem is that information about the Iranian army largely derives from 'foreign' accounts. Of course, one should appreciate the value of Ammianus Marcellinus, who, being a soldier himself, had a better understanding of certain mechanisms than 'civilian' historians. It should be remembered, however, that his knowledge of the functioning of the Iranian state was limited, and that is why he made mistakes when describing the Persian army, using a kind of 'tracing paper' search for analogies between the organisation of the Roman and Iranian armies. A description of the army left by Heliodorus in the *Aethiopica* might appear to be reliable, but we can state with certainty that it is just a fixed literary topos aiming at a sense of aesthetic exotica. The problems of nomenclature must be re-emphasised here. It is obvious that we obtain our data from the Roman, Armenian, Greek, Syriac, or classical Arabic sources. The authors of these works, using their native languages, were not familiar with Middle Persian terminology, so they presented the nomenclature of military titles in a distorted form or in tight relation to the habits of their own times. This is particularly evident in Greek and Arabic sources, where the titles of significant figures in Sasanian history are sometimes believed to be their proper names.

Sasanian military organisation was based on the system of their Arsacid predecessors.[117] The *wašt*, commanded by the *wašt sālār*, was a small detachment of troops. A larger unit of possibly one thousand troops was the *drafš* with the *drafš sālār* in command. The largest known unit was the *gund* of possibly 12,000 warriors, led by the *gund sālār* (*Kār-nāmag ī Ardašīr ī*

117 Widengren (1976) 281. Presumably, the internal structure of the empire initially remained essentially unchanged; see Börm (2017) 552.

Pābagān 5.2; 6.1; 8.1). But there is no certainty on the use of a decimal system by the Sasanian army.[118]

A detailed analysis of the composition of the Arsacid and Sasanian armies by Marek Olbrycht gives an approximate figure of around 120,000–150,000 men, taking into account all possible provincial recruits, conscripts, and auxiliaries combined with a 'professional' core.[119] The *Nāma-ye Tansar* (12)[120] describes two major groups of warriors in the Sasanian military, the cavalry and the infantry. The cavalry (the *aswārān*), consisting primarily of *bandagān*[121] and recruited from Iranian aristocracy or *āzādān*,[122] was divided into heavy lancers and mounted archers (Amm. Marc. 24.2.5; 25.12–18). These cavalry units were the backbone of both Arsacid and Sasanian armies,[123] and remained the main battle force and core of the Sasanian army throughout the empire's existence (Amm. Marc. 23.6.83).[124]

A mural painting from Dura-Europos displays a fully armoured cavalryman (man and horse) of the Iranian type; its dating is debated among scholars (late second to early third century).[125] A detailed description of armament of the heavy cavalry of Shapur II was given by Heliodorus (*Aeth.* 9.15.1; 9.15.5)[126] and Ammianus (25.1.12–13). As in the preceding Parthian dynasty, horse archery was one of the *spāh*'s vital military assets from the outset of the Sasanian dynasty.[127] Herodian (6.5.5–10) attributes the destruction of Severus Alexander' invasion forces at the Battle of Ctesiphon in 233 to the effective horse archery forces of Ardashir I. The light cavalry of the Sasanian army was recruited among the allied troops fighting under the command of their own chieftains, such as the Albanians, Sakas, Gelani, or Chionites in the army of Shapur II in 359 (Amm. Marc. 19.2.3).

The sources rarely mention infantry (the *paygān*) when describing the army of the Sasanians' Arsacid predecessors, resulting in the common opinion that they did not include infantry as a battlefield support for their cavalry.[128]

118 Farrokh, Karamian, and Maksymiuk (2018) 13–14.
119 Olbrycht (2016b) 292–96.
120 Boyce (1968).
121 Shahbazi (1986).
122 De Blois (1985) 5–7.
123 Olbrycht (2001); Hauser (2006); Potts (2007).
124 Bivar (1972) 289.
125 Shahbazi (1986); Mielczarek (1993).
126 On the low reliability of Heliodorus' description of Sasanian heavy cavalry, see Skupniewicz and Maksymiuk (2018).
127 Nikonorov (2005).
128 Mielczarek (1993) 31; contra Hauser (2006) 311.

The Classical sources (e.g., Procop. *Pers.* 1.14.25) have often provided negative assessments of Sasanian infantry combat performance, but on top of repeating a topos of Iranian infantry being a blind and worthless crowd, Ammianus Marcellinus (26.12.49) mentions units which he compares to the gladiators. Touraj Daryaee describes the *paygān* as being lightly armed with only spears, and for battlefield protection, having a shield but no armour.[129] An indication of the *paygān*'s professional status is provided by their registration on the state's rolls, allowing them to be paid (like the *aswārān* cavalry) for their military services to the empire.[130]

The sources emphasise Arsacid inability to take cities, which seems overstated in light of historical facts like the conquest of Seleucid Mesopotamia. Plutarch (*Crass.* 21) mentioned that Surena (a member of the Suren clan) distinguished himself in storming the walls of Seleucia. Ammianus Marcellinus (19.5.6) provides a glimpse of Sasanian ballistic weapons during Shapur II's siege of Amida in 359. A notable example of its weaponisation of rivers occurred during Shapur's first siege of Nisibis in 337/338 (Theod. *Hist. eccl.* 1.11–12). The use of skilled engineering units can be observed in the example of operations during the siege of Nisibis in 350.[131]

The holding of all of the higher-ranking military commands was a privilege of the first three categories of society: the *šahrdārān* (kings, royal nobility), the *wāspuhragān* (members of the Sāsānian family), and the *wuzurgān* (the most important noble families, first of all of Parthian descent), while the lesser positions were reserved for the *āzādān*. Ṭabarī ranks the *spāh*'s three most important military titles as the *(h)argbed*, the *artēštārān sālār*, and the *spāhbed* (1.869).[132] The *(h)argbed* is translated as a commander of a castle or fortress.[133] The eunuch Tīr-dād was the *(h)argbed* of the fortress of Darabgerd.[134] However, the ŠKZ mentions the specific word *dizbed* meaning 'commander of a fortress'

129 Daryaee (2009) 46.

130 Howard-Johnston (1995) 219.

131 Skupniewicz and Maksymiuk (2018).

132 'The third was called *Kārd*[*ār*], supreme commander of the army, the name of his rank in Persian being *Rathāshtārān Sālār*; this is a rank higher than that of *al-Iṣbahbadh* and is near to that of *al-Arjabadh*'.

133 Chaumont (1986) 400–401.

134 Ṭab. 1.815: 'This last had a eunuch called Tīrā, whom he had appointed castellan (*arjabadh*) of Dārābjird. When Ardashir reached the age of seven years, his father took him to Juzhir, whose seat was at al-Bayḍā', made his son stand before Juzhir, and asked the latter if he would attach the boy to Tīrā as a foster child of his and as the future castellan in Tīrā's stead'.

(32/26/63);[135] therefore the problem is to determine the military functions of the *(h)argbed*.[136] An inscription from Paikuli confirms the high rank of this office in the early Sasanian period (NPi B11,01–B16,02: *Šābuhr ī hargbed*). If one accepts the identification of the *(h)argbed* Šābuhr mentioned in the NPi with the *archapetes* Barsaborsos (the *(h)argbed* Burz-Šābuhr), one of the dignitaries who received the Roman envoys to Narseh's court to settle the peace treaty in 298, the term should be associated with diplomatic authority.[137]

Early Sasanian inscriptions include the title *hazāruft/hazārbed/chiliarch*.[138] In the ŠKZ the *hazāruft* is mentioned immediately after the members of the royal house and the *bidaxš*.[139] There is no clear confirmation in the sources that the *hazāruft* would be the commander of the guards-regiment and responsible for the safety of the king.[140] However, the bearer of the title would undoubtedly participate in diplomatic activities, as evidenced by the negotiations conducted by the *hazāruft* Affarbān in 298.[141] More importantly, they are never identified as *spāhbed*.[142]

The term *spāhbed* ('chief of an army', 'general')[143] is confirmed in the ŠKZ and NPi, where a *spāhbed* by the name of Raxš is mentioned.[144] The title *artēštārān sālār*, which only appears in sources for the fifth century, should seemingly be identified with the title of the *spāhbed*. Without a doubt, this title defined the overall military leader but not the cavalry commander. This thesis could be tentatively supported by the remarks of a minor Pahlavi text, the *Abar Wizārišn ī Čatrang ud Nihišn ī Nēw-Ardaxšīr* ('On the Explanation of Chess

135 *tīrmihr ī šahrgird dizbed* / *tīrmihr šahrgard dizbad* / Τιρμερ Σαρακάρτων καστελλοφύλακος ('Tīrmihr, commander of the fortress of Šahrgird').

136 See Wiesehöfer (1996) 188, identifying this as 'supreme tax collector'.

137 Petros Patrikios, *FGrH* 4, F 14: 'Narseh sent everyone else away inside the royal palace and contented himself with Affarban, Archapetes, and Barsaborses being present, one of whom was praetorian prefect, and the other had command of written language. He then allowed Probus to present a detailed report of his legation'. See Shayegan (2003).

138 ŠKZ 29/23–24/56–57; ŠKZ 31/25/62: *pābag ī hazāruft* / *pābag hazāruft* / Παβάκ ἁζαροπτ; NPi B11,01–B16,02: *Ardaxšahr ī hazārbed*. See Khurshudian (2015) 21–68, 101–19.

139 In the opinion of Ghafouri (2015), it was the *bidaxš* who exercised the power as an actual regent with full military and political competences in the case of the Sasanian king's absence resulting from sudden death or commanding distant campaigns.

140 This only appears in Petros Patrikios, *FGrH* 4, Fr. 14: ὕπαρχος πραιτωρίων as *praefectus praetorio*; see Frye (1983) 249–50.

141 Maksymiuk (2018b) 593–94.

142 Gyselen (2004).

143 Khurshudian (2015) 195–218.

144 ŠKZ 30/24/58: *raxš ī spāhbed* / *raxš spāhbed* / Ῥοξ σπαπίτου; NPi B11,01–B16,02: *Raxš ī spāhbed*.

and Backgammon'), probably belonging to the late Sasanian period,[145] which describes how the game of chess was introduced to Persia from India and how Persian sages invented the game of backgammon. According to this text, chess is a kind of battle game and the chessboard is like a battlefield. In this 'manual' the chess figure (currently called the queen) appears as the *artēštārān sālār*, or chief of the warriors, and the *aswārān sālār* (horse or knight at present) as the chief of horsemen.[146] The *aswārān sālār* would correspond in such a case to the title of *aspbed* held in the ŠKZ by Pērōz, who occupies the third rank among the great dignitaries.[147] However, there is no doubt that with time the *spāhbed* title came to denote the highest military rank.[148] Its competences expanded to diplomatic missions as well as commanding the army, as confirmed by the actions of a member of the Suren clan during the reign of Shapur II. According to Classical sources, Surena commanded the Persian army against Julian (Amm. Marc. 24.3.1; Zos. 3.19.1–2), and what is of greater importance, he carried out the negotiations which resulted in the peace treaty of 363 (Amm. Marc. 25.7.5; Zos. 3.31.1). He was also active in the negotiations with the Emperor Valens (Amm. Marc. 30.2.5). Ammianus Marcellinus describes this very high dignitary's status (Amm. Marc. 30.2.5: *Surena potestatis secundae post regem*, 'the second person after the king'). The elevation of the *spāhbed* in comparison with the *aspbed* in the span of over 200 years may have resulted from the reorganisation of the Sasanian army. So, over time the *aspbed*, who is listed among the court dignitaries in the ŠKZ in a much higher position than the *spāhbed*, was eventually subordinated to the other office.

It seems that the command system of the Sasanian army as described in Ṭabarī was fully formed by the end of the fourth century. The army was headed by the *spāhbed* or *artēštārān sālār*, to whom the commanders of individual formations reported, especially the *aspbed* or *aswārān sālār* (the commander of the entire cavalry corps)[149] and the *paygān sālār* (the commander of the infantry).[150] The cavalry commander had a higher rank than the infantry commander. The military formation directly subordinated to the king was the 'royal guard' (Amm. Marc. 19.1.5). At the head of this formation was the *puštigbān-sālār* (commander of the royal guard),[151] which was one of the most prestigious

145 Panaino (1999) 83–91.

146 Daryaee (2002) 300; Maciuszak (2003) 95–97.

147 ŠKZ 31/25/62: *pērōz ī aspped* / *paryōz asppad* / Πηρώζου ἀσπιπίδου 'commander of the cavalry'. However, it was not mentioned in NPi; Chaumont (1987).

148 Christensen (1944) 524–25.

149 Tafażżoli (2000) 13; Farrokh, Karamian, and Maksymiuk (2018) 29–30.

150 Tafażżoli (2000) 14; Foss (2002) 170.

151 Foss (2002) 170.

positions in the court.[152] According to Ṭabarī, it was established by Shapur II during the first campaign against the Arab tribes, but the determination of the number of soldiers as 1,000 may suggest its earlier origins.[153]

Representatives of the regional administration were the *marzbān*[154] and the *pāygōsbān*.[155] However, the relation between the *marzbān* (guardian of the borders) and the *spāhbed* cannot be firmly determined from the sources. The title of the *marzbān* does not appear in the ŠKZ and in the inscription from Paikuli,[156] but is confirmed by the Parthian inscriptions from Old Nisa.[157] This title can be found in literary texts, which were probably written in the fifth or sixth century.[158] The function of the *šahrab* is mentioned several times in the ŠKZ,[159] but there is no confirmation of the military connotations of this office in the early Sasanian period. Although Wiesehöfer defines the *pāygōsbān* as the military commander of a province,[160] the exact relationship between the *šahrab*, the *marzbān*, and the *pāygōsbān* cannot, in fact, be determined. The additional confusion in the definition of military competences is introduced by the 'Western' sources. According to Ammianus Marcellinus in the fourth century, the military governor of a province was named *vitax* (*bidaxš*), which he identifies with the *magistri equitum*.[161] Based on the ŠKZ (29/23/56; 31/25/61; 33/27/64) and NPi (B11,01–B16,02), we can only conclude that the *bidaxš* was the highest dignitary mentioned after the members of the royal house, but he was not a cavalry commander, as Ammianus Marcellinus suggests. One can cautiously assume that the *marzbān* was administrator of the border provinces with combined civilian and military competences. In such a case, the

152 Tafażżoli (2000) 12.

153 Ṭab. 1.839: 'He selected one thousand cavalrymen from among the stoutest and most heroic of the troops'.

154 Gignoux (1984); Khurshudian (2015) 76–95.

155 Khurshudian (2015) 69–76.

156 The inscription from Eqlid mentioning the *marzbān* of Bishapur dates back only to the fifth century; Gropp (1969).

157 Khurshudian (2015) 80.

158 *Kār-nāmag ī Ardašīr ī Pābagān* 1.3: *pābag marzobān ud šahryār ī pārs būd* ('Pābag was the *marzbān* and the ruler of Pars'); *Acta Mar Ḳardaghi* 5: Mār Kardag as the *marzbān* of Nisibis (Abbeloos (1890)); *Epic Histories* 5.38: Suren the *marzpan* [*Մարզպան*] of Armenia.

159 32/26/63: *ardašīr ī gōmān šahrab* / *ardašīr gōmān xšahrab* / Ἀρταξας Γωμαν δατράπου ('Ardašīr, satrap of Gōmān').

160 Wiesehöfer (1996) 198.

161 Amm. Marc. 23.6.14: *Sunt autem in omni Perside hae regiones maximae, quas Vitaxae id est magistri equitum curant, et reges et satrapae* ('Now there are in all Persia these greater provinces, ruled by *vitaxae*, or commanders of cavalry, by kings, and by satraps').

marzbāns in the sense of 'guardian of the borders' should include both the *kanārang* in the Abaršahr[162] and the *bidaxš* in Armenia and Iberia.[163]

4 Conclusion

Sasanian policies in the west of the empire are of fundamental importance for understanding the history of the Middle East in Late Antiquity. It cannot be denied that the most powerful opponent of Iran in the third and fourth centuries was the Roman Empire. At the same time, the statements of Cassius Dio and Herodian about the alleged desire of the Sasanians to restore the borders of the Achaemenid period, and consequently their desire to drive the Romans out of Asia, must be rejected. The main goal of the early Sasanian rulers was the removal of the opposition associated with the Arsacid dynasty and the legitimisation of the power of the new dynasty. Both of these goals can be noticed in the actions of Ardashir I and his successor Shapur I. The attack on Hatra (captured only a dozen years after defeating Artabanus) and, above all, the military actions against Armenia, ruled by a side line of Arsacids, brought them into conflict with the Roman Empire, although in principle they should be regarded as 'internal affairs' of Iran.[164]

Of course, the military clash with Rome for control over Northern Mesopotamia and the restoration of the frontier on the Euphrates was inevitable. In this case, the Sasanians were pursuing a programme of integrating lands previously subordinated to the Arsacids on the one hand, and legitimising their power through military successes against an external enemy on the other. Shapur perfectly understood the propaganda significance of defeating the Roman Empire. He produced a series of monumental royal reliefs depicting the defeated emperors Gordian and Valerian and Philip asking for peace. The ŠKZ inscription commemorating these events was written not only in Middle Persian and Parthian, but also in Greek. After the undoubted success of capturing Hatra and installing Shapur's son Hormizd-Ardashir on the throne of Armenia, the *shāhānshāh* made a spectacular mark of his position in the Middle East by capturing Antioch and deporting its inhabitants to the newly founded city of Weh Antiok Shapur (Better Antioch of Shapur) in Khuzistan.

162 Khurshudian (2015) 95–100.

163 Herman (2010) 58–82; Rapp (2014) 19.

164 Daryaee (2021) does 'not think in any point in time Armenia was considered to be as part of Ērānšahr and was always thought to be non-Iranian'.

The implementation of the western programme of the first two Sasanians collapsed after Shapur's death, when an internal conflict broke out within the dynasty, from which the last surviving son of Shapur, Narseh, emerged victorious. First Diocletian restored the Arsacid dynasty to power in Armenia, and after a humiliating defeat at the battle with Galerius at Satala, Narseh, by the terms of peace set in 298, gave up the *regiones Transtigritanae* and ceded Nisibis, the key stronghold in northern Mesopotamia. The Romans gained far-reaching influence in Mesopotamia, Armenia, and Iberia.

Iran did not return to active actions in the West until forty years later. Shapur II began a long-lasting war aimed at regaining northern Mesopotamia and Armenia. Shapur achieved his goal after the disastrous Persian expedition of Julian the Apostate, when in 363 Jovian made an 'ignominious peace' indemnifying his strongholds in northeast Mesopotamia and surrendering Roman influence in Armenia and Iberia. After Shapur's death, an Iranian-Roman agreement divided Armenia, and from 428 part of the kingdom became an Iranian province, marking the end of Arsacid rule in Armenia. The revision of the treaty of 298 and the partition of Armenia ended a period in Iran's Western policy described by Jan Willem Drijvers as a period of confrontation.[165]

The Sasanian successes throughout this period of warfare depended on the organisation and command of a highly effective military system. It seems that the best summary of the above considerations about the Sasanian army will be the opinion given by the Roman historian and soldier Ammianus Marcellinus, who personally witnessed its combat value in the fourth century: 'Through military training and discipline, through continuous exercises in warfare and military manoeuvres, they cause fear even in large armies' (23.6.83).[166]

Bibliography

Abbeloos, J.B. (1890) *Acta Mar Ḳardaghi Assyriae praefecti qui sub Sapore II martyr occubuit syriace juxta manuscriptum Amidense una cum versione Latina*. Brussels.

Alföldi, A. (1937) 'Die Hauptereignisse der Jahre 253–261 n. Chr. im Orient im Spiegel der Münzprägung'. *Berytus* 4: 41–68.

Alram, M. (2003) 'Numismatics and History – An Outline. a) Ardashir I', in Alram and Gyselen (2003) 21–31.

Alram, M., Blet-Lemarquand, M., and Skjærvø, P.O. (2007) 'Shapur, King of Kings of Iranians and Non-Iranians', in Gyselen (2007) 11–40.

165 Drijvers (2009).

166 Drijvers (2006) 54.

Alram, M. and Gyselen R. (eds) (2003) *Sylloge nummorum sasanidarum: Paris-Berlin-Wien. Band I, Ardashir I. – Shapur I.* Vienna.

Altaweel, M. and Hauser, S. (2004) 'Travelling via Hatra: Trade Routes in the Eastern Jazira According to Evidence from Ancient Sources and Modern Satellite Imagery'. *Baghdader Mitteilungen* 35: 57–84.

Amin, D., Kasheff, M., and Shahbazi, A.S. (eds) (1990) *Iranica Varia: Papers in honor of Professor Ehsan Yarshater*. Leiden.

Angeli Bertinelli, M.G. (1976) 'I Romani oltre l'Eufrate nel II secolo d. C. (le province di Assiria, di Mesopotamia e di Osroene)', in Temporini and Haase (1976) 3–45.

Antia, E.K. (1900) *Kârnâmak-i Artakhshîr Pâpakân: The Original Pahlavi text, with Transliteration in Awesta Characters, Translation into English and Gujârati and Selections from the Shâhnâmeh*. Bombay.

Areshian, G.E., Avetisyan, P., and Grekyan, Y.H. (eds) (2017) *Bridging Times and Spaces: Papers in Ancient Near Eastern, Mediterranean and Armenian Studies: Honouring Gregory E. Areshian on the Occasion of his Sixty-fifth Birthday*. Oxford.

Austin, N.H. (1972) 'Julian at Ctesiphon: A Fresh Look at Ammianus' Account'. *Athenaeum* 50: 301–309.

Baird, J. (2012) 'Dura Deserta: The Death and Afterlife of Dura-Europos', in Christie and Augenti (2012) 307–29.

Bakhos, C. and Shayegan, M.R. (eds) (2010) *The Talmud in Its Iranian Context*. Tübingen.

Baldus, H.R. (1971) *Uranius Antoninus: Munzprägung und Geschichte*. Bonn.

Baynes, N. (1910) 'Rome and Armenia in the Fourth Century'. *English Historical Review* 25: 625–43.

Bivar, A.D.H. (1972), 'Cavalry Equipment and Tactics on the Euphrates Frontier'. *DOP* 26, 273–91.

Bivar, A.D.H. (1983) 'The Political History of Iran under the Arsacids', in Yarshater (1983) 21–99.

Blockley, R.C. (1984) 'The Romano-Persian peace treaties of ad 299 and 363'. *Florilegium* 6: 28–49.

Blockley, R.C. (1987) 'The Division of Armenia between the Romans and the Persians at the End of the Fourth century A.D.'. *Historia* 36: 222–34.

Blockley, R.C. (1988) 'Ammianus Marcellinus on the Persian Invasion of A.D. 359'. *Phoenix* 42: 244–60.

Börm, H. (2015) 'Dynastie und Charisma im Sasanidenreich', in Boschung and Hammerstaedt (2015) 253–80.

Börm, H. (2017) 'Kontinuität im Wandel: Begründungsmuster und Handlungsspielräume der iranischen Monarchie in arsakidischer und sasanidischer Zeit', in Rebenich (2017) 545–64.

Börm, H. (2019) 'Die Grenzen des Großkönigs? Überlegungen zur arsakidisch-sasanidischen Politik gegenüber Rom', in Schleicher, Stickler, and Hartmann (2019) 99–123.

Börm, H. and Wiesehöfer, J. (eds) (2010) *Commutatio et contentio: Studies in the Late Roman, Sasanian, and Early Islamic Near East: In Memory of Zeev Rubin*. Düsseldorf.

Boschung, D. and Hammerstaedt, J. (eds) (2015) *Das Charisma des Herrschers*. Paderborn.

Bosworth, C.E. (1983) 'Iran and the Arabs before Islam', in Yarshater (1983) 603–604.

Boyce, M. (1968) *The Letter of Tansar*. Rome.

Brodka, D. (2009) *Ammianus Marcellinus: Studien zum Geschichtsdenken im vierten Jahrhundert n. Chr*. Kraków.

Brok, M. (1959) *De perzische expeditië van Keizer Julianus volgens Ammianus Marcellinus*. Groningen.

Burgess, R.W. (1999) *Studies in Eusebian and Post-Eusebian Chronography*. Stuttgart.

Cameron, A. (1969) 'Agathias on the Sassanians'. *DOP* 23/24: 67–183.

Cameron, A. and Lawrence, C. (eds) (1995) *The Byzantine and Early Islamic Near East III: States, Resources and Armies*. Princeton.

Cereti, C. (1997) 'Primary Sources for the History of Inner and Outer Iran in the Sasanian Period'. *Archivum Eurasiae Medii Aevi* 9: 17–71.

Cereti, C. (2021) 'Narseh, Armenia, and the Paikuli Inscription'. *Electrum* 28: 69–87.

Chaumont, M.L. (1987) 'Aspbed'. *Enc. Ir.* 2.8: 791–92.

Chaumont, M.L. (1996) 'Une visite du roi d'Arménie. Tiridate III à l'empereur Constantin à Rome?' *Byzantina-Sorbonensia* 12: 55–66.

Chauvot, A. (1992) 'Parthes et Perses dans les sources du IVe siècle', in Christol (1992) 115–25.

Christensen, A. (1944), *L'Iran sous les Sassanides*. Copenhagen.

Christie, N. and Augenti, A. (eds) (2012) *Vrbes Extinctae: Archaeologies of Abandoned Classical Towns*. Farnham.

Christol, M. (ed) (1992) *Institutions, société et vie politique dans l'empire Romain au IVe siècle AP. J.-C.* Rome.

Chrysos, E. (1993) 'Räumung und Aufgabe von Reichsterritorien. Der Vertrag von 363'. *BJ* 193: 165–202.

Criscuolo, U. (ed) (2003) *Da Costantino a Teodosio il Grande: cultura, società, diritto: atti del convegno internazionale, Napoli 26–28 aprile 2001*. Naples.

Curtis, V.S., Pendleton, E. and Alram, M. (eds) (2016) *The Parthian and Early Sasanian Empires: Adaptation and Expansion*. Oxford.

Dąbrowa, E. (1983) *La politique de l'état Parthe à l'égard de Rome – d'Artaban II à Vologèse I (ca 11–ca 79 de n.e.) et les facteurs qui la conditionnaient*. Kraków.

Dąbrowa, E. (2007) 'Naval Operations during Persian Expedition of Emperor Julian (363 AD)', in Lewin and Pellegrini (2007) 237–42.

Dąbrowa, E. (2018) 'Arsacid Dynastic Marriages'. *Electrum* 25: 73–83.

Daryaee T. (2002) 'Mind, Body, and the Cosmos: Chess and Backgammon in Ancient Persia'. *Iran. Stud.* 35.4: 281–312.

Daryaee, T. (2009) *Sasanian Persia: The Rise and Fall of an Empire*. London.

Daryaee, T. (2010) 'Ardaxšīr and the Sasanians' Rise to Power'. *Anabasis* 1: 236–55.

Daryaee, T. (2016a) 'The Sasanian "Mare Nostrum": The Persian Gulf'. *International Journal of the Society of Iranian Archaeologists* 2.3: 40–46.

Daryaee, T. (2016b) 'If these Walls Could Speak: The Barrier of Alexander, Wall of Darband and Other Defensive Moats', in Pellò (2016) 79–88.

Daryaee, T. (2017) 'A Note on the "Great King of Armenia"', in Areshian, Avetisyan, and Grekyan (2017) 85–87.

Daryaee, T. (2021) 'Armenia and Iran: The Birth of Two Nations in Late Antiquity'. *Electrum* 28: 59–67.

Daryaee, T. (ed) (2018) *Sasanian Iran in the Context of Late Antiquity*. Leiden.

de Blois, F. (1985) '"Freemen" and "Nobles" in Iranian and Semitic Languages'. *Journal of the Royal Asiatic Society of Great Britain and Ireland* 1: 5–15.

de Blois, L. (1975) 'Odenatus and the Roman-Persian War of 252–264 A.D.'. *Talanta* 6: 7–23.

de Blois, L. (2016) 'Rome and Persia in the Middle of the Third Century AD (230–266)', in Slootjes and Peachin (2016) 33–44.

de la Vaissière, E. (2015) 'The Steppe World and the Rise of the Huns', in Maas (2015) 178–82.

den Boeft, J. (ed) (2007) *Ammianus after Julian: The Reign of Valentinian and Valens in Books 26–31 of the Res Gestae*. Leiden.

Chabot, J.B. (1899–1910) *Chronique de Michel le Syrien*. Paris.

Dignas, B. and Winter, E. (2007) *Rome and Persia in Late Antiquity: Neighbours and Rivals*. Cambridge.

Dirven, L. (ed) (2013) *Hatra: Politics, Culture and Religion between Parthia and Rome*. Stuttgart.

Dmitriev, V. (2015) '«Ночное» сражение под Сингарой (340-е гг. н. э.)', in Nikonorov and Alyokshin (2015) 228–59.

Dmitriev, V. (2017) 'The Sasanian Navy Revisited: An Unwritten Chapter in Iran's Military History'. *The International Journal of Maritime History* 29.4: 727–37.

Dmitriev, V. (2019) 'Восточные кампании Шапура II'. Восток. *Афро-Азиатские общества: история и современность* 2: 49–63.

Dodgeon, M. and Lieu, S.N.C. (2002), *The Roman Eastern Frontier and the Persian Wars A.D. 226–363: A Documentary History*. London.

Doise, J. (1945) 'Le partage de l'Arménie sous Théodose 1er'. *REA* 7: 274–77.

Downey, G. (1950) 'Aurelian's Victory over Zenobia at Immae A.D. 272'. *TAPA* 81: 57–68.

Drijvers, J.W. (2006) 'Ammianus Marcellinus' Image of Sasanian Society', in Wiesehöfer and Huyse (2006) 45–69.

Drijvers, J.W. (2009) 'Rome and the Sasanid Empire: Confrontation and Coexistence', in Rousseau (2009) 441–54.

Duval, R. (1970) *La littérature syriaque*. Paris and Amsterdam.

Duyvesteyn, I. and Heuser, B. (eds) (2024) *The Cambridge History of Strategy, Volume 1: From Antiquity to the American War of Independence*. Cambridge.

Edwell, P. (2008) *Between Rome and Persia: The Middle Euphrates, Mesopotamia and Palmyra under Roman Control*. New York.

Edwell, P. (2010) 'The Sources for Rome's Wars with Shapur I: Eurocentric and Eastern Perspectives'. *Ancient West and East* 9: 155–79.

Enßlin, W. (1949) *Zu den Kriegen des Sassaniden Schapur* I. Munich.

Erdkamp, P., Verboven, K., and Zuiderhoek, A. (eds) (2020) *Capital, Investment, and Innovation in the Roman World*. Oxford.

Farrokh, K. Maksymiuk, K., and Sánchez Garcia, J. *The Siege of Amida (359 CE)*. Siedlce (2018).

Farrokh, K., Karamian, G., and Maksymiuk, K. (2018) *A Synopsis of Sasanian Military Organization and Combat Units*. Siedlce.

Felix, W. (1985) *Antike literarische Quellen zur Außenpolitik des Sasanidenstaates, I: 224–309*. Vienna.

Fisher, G. (2011a) 'Kingdoms or Dynasties? Arabs, History, and Identity before Islam'. *JLA* 4: 245–67.

Fisher, G. (2011b). *Between Empires: Arabs, Romans, and Sasanians in Late Antiquity*. Oxford.

Fisher, G. and Wood, P. (2016) 'Writing the History of the "Naṣrid" Dynasty at al-Ḥīrah: The Pre-Islamic Perspective'. *Iran. Stud.* 49: 247–90.

Fornara, C. (1991) 'Julian's Persian Expedition in Ammianus and Zosimus'. *JHS* 111: 1–15.

Foss, C. (2002) 'The Sellarioi and Other Officers of Persian Egypt'. *ZPE* 137: 169–72.

Freeman, P. (ed) (2002) *Limes XVIII: Proceedings of the XVIIIth International Congress of Roman Frontier Studies Held in Amman, Jordan (September 2000)*. Oxford.

Frei-Stolba, R. and Herzig, H.E. (eds) (2004) *Siedlung und Verkehr im Römischen Reich: Römerstraßen zwischen Herrschaftssicherung und Landschaftsprägung*. Bern.

French, D. and Lightfoot, C.S. (eds) (1989) *The Eastern Frontier of the Roman Empire: Proceedings of a Colloquium Held at Ankara in September 1988*. Oxford.

Frendo, D. (2002) 'Cassius Dio and Herodian on the First Sasanian Offensive against the Eastern Provinces of the Roman Empire (229–232)'. *Bulletin of the Asia Institute* 16: 25–36.

Frye, R.N. (1983) 'Achaemenid Echoes in Sasanian Times'. *AMI, Ergänzungsband* 10: 247–52.

Gariboldi, A. (2016) 'The Birth of the Sasanian Monarchy in Western Sources', in Curtis, Pendleton, and Alram (2016) 47–52.

Garsoïan, N.G. (1967) 'Politique ou orthodoxie? L'Arménie au quatriéme siécle'. *REArm* 4: 297–320.

Garsoïan, N.G. (1997) 'The Marzpanate (428–652)', in Hovannisian (1997) 95–115.

Gatier, P.-L., Helly, B., and Rey-Coquais J.-P. (eds) (1988) *Géographie historique au Proche-Orient (Syrie, Phénicie, Arabie, grecques, romaines, byzantines): Actes de la Table Ronde de Valbonne, 16–18 septembre 1985*. Paris.

Gerhardt, T. and Hartmann, U. (2008) 'Fasti', in Johne (2008) 1055–1198.

Ghafouri, F. (2015) 'Sasanian Bidaxšs in Shāhnāmeh and Historical Sources'. *Nāme-ye Farhangestān* 14.4: 168–210. [in Persian]

Gignoux, P. (1984) 'L'organisation administrative sasanide: le cas du marzbān'. *Jerusalem Studies in Arabic and Islam* 4: 1–29.

Gnoli, G. (1999) 'Farr(ah)'. *Enc. Ir.* 9.3: 312–19.

Goltz, A. and Hartmann, U. (2008) 'Valerianus und Gallienus', in Johne (2008) 223–95.

Greatrex, G. (2000) 'The Background and Aftermath of the Partition of Armenia in AD 387'. *AHB* 14: 35–48.

Gregoratti, L. (2011) 'A Parthian Port on the Persian Gulf: Characene and Its Trade'. *Anabasis* 2: 209–29.

Gregoratti, L. (2015) 'The Palmyrene Trade and the Protection of the Caravans'. *ARAM* 27 1–2: 139–48.

Gregoratti, L. (2020) 'Temples and Traders in Palmyra', in Erdkamp, Verboven, and Zuiderhoek (2020) 461–80.

Grenet, F. (1988) 'Les Sassanides a Doura-Europos (253 ap. J.-C.). Réexamen du matériel épigraphique iranien du site', in Gatier, Helly, and Rey-Coquais (1988) 133–58.

Gropp, G. (1969) 'Einige neuentdeckte Inschriften aus sasanidischer Zeit' in Hinz, Borger, and Gropp (1969) 237–60.

Guirgass, V. and Kratchkovsky, I. (eds) (1888) *Dinawari, Kitāb al-Akhbār al-Tiwāl*. Leiden.

Gutmann, B. (1991) *Studien zur römischen Außenpolitik in der Spätantike (264–395 n. Chr.)*. Bonn.

Gyselen, R. (2004) 'Spāhbed'. *Enc. Ir.* online edition, https://iranicaonline.org/articles/spahbed (accessed on 23 May 2016).

Gyselen, R. (ed) (2007) *Des Indo-Grecs aux Sassanides: données pour l'histoire et la géographie historique*. Bures-sur-Yvette.

Haas, C. (2008) 'Mountain Constantines: The Christianization of Aksum and Iberia'. *JLA* 1: 101–26.

Halfmann, H. (1986) *Itinera principium. Geschichte und Typologie der Kaisereisen im Römischen Reich*. Stuttgart.

Hamacher, B. (1983) 'Der Heerzug Gordians III. durch Kleinasien 242'. *Münstersche Numismatische Zeitung* 13: 19–32.

Hartmann, U. (2001) *Das palmyrenische Teilreich*. Stuttgart.

Hartmann, U. (2006) 'Mareades-ein sasanidischer Quisling?', in Wiesehöfer and Huyse (2006) 105–42.

Hartmann, U. (2008) 'Das palmyrenische Teilreich', in Johne (2008) 343–78.

Hartmann, U. (2019) 'Iberien in der Historia Augusta' in Schleicher, Stickler, and Hartmann (2019) 25–68.

Hauser, S. (2006) 'Was There No Paid Standing Army? A Fresh Look on Military and Political Institutions in the Arsacid Empire', in Mode and Tubach (2006) 295–319.

Hauser, S. (2013) 'Where is the Man of Hadr, Who Once Built It and Taxed the Land by the Tigris and Chaboras? On the Significance of the Final Siege of Hatra', in Dirven (2013) 119–42.

Herman, G. (2010) 'Persia in Light of the Babylonian Talmud: Echoes of Contemporary Society and Politics: hargbed and bidaxš', in Bakhos and Shayegan (2010) 58–82.

Hewsen, R.H. (1986) 'In Search of Tiridates the Great'. *Journal of the Society for Armenian Studies* 2: 11–49.

Hinz, W., Borger, R., and Gropp G. (eds) (1969) *Altiranische Funde und Forschungen*. Berlin.

Hojabri-Nobari, A., Mousavi Kouhpar, S.M., Vahdatinasab, H., and Khosrowzadeh, A. (2011) 'Trade and Cultural Contacts between Northern and Southern Persian Gulf during Parthians and Sasanians: A Study Based on Pottery from Qeshm Island'. *International Journal of Humanities* 18: 89–115.

Hovannisian, R.G. (ed) (1997) *The Armenian People from Ancient to Modern Times*. New York.

Howard-Johnston, J. (1995) 'The Two Great Powers in Late Antiquity: A Comparison' in Cameron and Lawrence (1995) 157–226.

Howard-Johnston, J. (2010) 'The Sasanian Strategic Dilemma', in Börm and Wiesehöfer (2010) 37–69.

Howard-Johnston, J. (2013), 'Military Infrastructure in the Roman Provinces North and South of the Armenian Taurus in Late Antiquity', in Sarantis and Christie (2013) 853–91.

Hoyland, R.G. (1997) *Seeing Islam as Others Saw It: A Survey and Evalutaion of Christian, Jewish and Zoroastrian Writings on Early Islam*. Princeton.

Hoyland, R.G. (2018) *The 'History of the Kings of the Persians' in Three Arabic Chronicles: The Transmission of the Iranian Past from Late Antiquity to Early Islam*. Liverpool.

Huttner, U. (2008) 'Von Maximinus Thrax bis Aemilianus', in Johne (2008) 161–221.

Jackson Bonner, M.R. (2011) *Three Neglected Sources of Sasanian History in the Reign of Khusraw Anushirvan*. Paris.

Jackson Bonner, M.R. (2017) 'A Brief Military History of the Later Reign of Šāpur II'. *Historia i Świat* 6: 97–105.

Jackson Bonner, M.R. (2020) *The Last Empire of Iran*. Piscataway.

Johne, K.P. (eds) (2008) *Die Zeit der Soldatenkaiser: Krise und Transformation des Römischen Reiches im 3. Jahrhundert n. Chr. (235–284)*. Berlin.

Kaegi, W.E. (1981) 'Constantine's and Julian's Strategies of Strategic Surprise against the Persians'. *Athenaeum* 69: 209–13.

Kennet, D. (2007) 'The Decline of Eastern Arabia in the Sasanian Period'. *Arabian Archaeology and Epigraphy* 18: 86–122.

Kervran, M. (1994) 'Forteresses, entrepôts et commerce: une histoire à suivre depuis les rois sassanides jusqu' aux princes d'Ormuz'. *Res Orientales* 6: 325–50.

Kettenhofen, E. (1982) *Die römisch-persischen Kriege des 3. Jahrhunderts. n. Chr. Nach der Inschrift Sāhpuhrs I. an der Ka'be-ye Zartošt (ŠKZ).* Wiesbaden.

Kettenhofen, E. (1983) 'The Persian Campaign of Gordian III and the Inscription of Sapuhr I at the Ka'be-ye Zartošt', in Mitchel (1983) 151–71.

Kettenhofen, E. (1995a) 'Die Eröberung von Nisibis und Karrhai durch die Sasaniden in der Zeit Kaiser Maximins (235/236 n.Ch.)'. *IrAnt* 30: 159–77.

Kettenhofen, E. (1995b) *Tirdād und die Inschrift von Paikuli: Kritik der Quellen zur Geschichte Armeniens im späten 3. und frühen 4. Jh. n. Chr.* Wiesbaden.

Kettenhofen, E. (1996) 'Deportations II: In the Parthian and Sasanian period'. *Enc. Ir.* 7.3: 297–308.

Kettenhofen, E. (2001) 'Das Jahr 7 Kaiser Valerians'. *Nāme-ye Irān-e Bāstān* 1: 17–22.

Kettenhofen, E. (2002a) 'Die Einforderung der achaimenidischen Territorien durch die Sasaniden – eine Bilanz', in Khalifeh-Soltani and Kurz (2002) 49–75.

Kettenhofen, E. (2002b) 'Die Anfänge des Christentums in Armenien'. *Handēs Amsōreay* 116: 45–104.

Kettenhofen, E. (2008), 'Die kaukasischen Reiche' in Johne (2008) 475–500.

Khalifeh-Soltani, I. and Kurz, S. (eds) (2002) *Yādnāme-ye Iradj Khalifeh-Soltani: Festschrift Iradj Khalifeh-Soltani zum 65. Geburtstag.* Aachen.

Khurshudian, E. (2015) *Государственные институты парфянского и сасанидского Ирана.* Almaty.

Kim, H.J. (2013) *The Huns, Rome, and the Birth of Europe.* Cambridge.

Körner, C. (2004) 'Der Perserfriede von 244 n. Chr. und Meilensteinfunde aus Kappadokien', in Frei-Stolba and Herzig (2004) 317–29.

Kotula, T. (1997) *Aurèlien et Zénobie. L'Unité ou la division del'Empire.* Wrocław.

Kraus, C.S., Marincola, J., and Pelling, C. (eds) (2010) *Ancient Historiography and its Contexts: Studies in Honour of A.J. Woodman.* Oxford.

Kreucher, G. (2003) *Der Kaiser Marcus Aurelius Probus und seine Zeit.* Stuttgart.

Leadbetter, B. (2002) 'Galerius and the Eastern Frontier', in Freeman (2002) 85–89.

Leński, N. (2002) *Failure of Empire: Valens and the Roman State in the Fourth Century* A.D. Berkeley.

Leński, N. (2007a) 'Two Sieges of Amida (AD 359 and 502–503) and the Experience of Combat in the Late Roman Near East'. *British Archaeological Reports International Series* 1717: 219–36.

Leński, N. (2007b) 'The Chronology of Valens' Dealings with Persia and Armenia, 364–378 CE', in den Boeft (2007) 95–127.

Lewin, A. and Pellegrini, P. (eds) (2007) *The Late Roman Army in the Near East from Diocletian to the Arab Conquest: proceedings of a colloquium held at Potenza, Acerenza and Matera, Italy (May 2005)*. Oxford.

Lightfoot, C.S. (1988) 'Facts and Fiction – The Third Siege of Nisibis'. *Historia* 37: 105–25.

Lightfoot, C.S. (1989) 'Sapor before the Walls of Amida', in French and Lightfoot (1989) 285–94.

Loriot, X. (1971) 'Itinera Gordiani Augusti, I, Un voyage de Gordien III à Antioche en 239 après J.C.?'. *Bulletin de la Societè Française de Numismatique* 26.2: 18–21.

Luther, A. (2004) 'Dura-Europos zwischen Palmyra und den Parthern: Der politische Status der Region am Mittleren', in Rollinger, Ulf, and Schnegg (2004) 327–51.

Luther, A. (2006) 'Roms mesopotamische Provinzen nach der Gefangennahme Valerians (260)', in Wiesehöfer and Huyse (2006) 203–19.

Maas, M. (ed) (2015) *The Cambridge Companion to the Age of Attila*. Cambridge.

Maciuszak, K. (2003) 'Persian Checkmate – "The King Is Oppressed". On the Origin of the Chessmen's Names'. *Studia Etymologica Cracoviensia* 8: 91–101.

Maksymiuk, K. (1998) 'Odenat – Rzymianin czy Palmyreńczyk'. *Eos* 85: 143–54.

Maksymiuk, K. (2004) 'Some Remarks on Political Relations between Sassanian Persia and Rome during the Principate'. *Eos* 91: 131–39.

Maksymiuk, K. (2015) *Geography of Roman-Iranian Wars: Military Operations of Rome and Sasanian Iran*. Siedlce.

Maksymiuk, K. (2016) 'Die finanziellen Abrechnungen in den persisch-römischen Kriegen in den Zeiten der Sasaniden'. *Historia i Świat* 5: 149–57.

Maksymiuk, K. (2017a) 'The Capture of Hatra in Light of Military and Political Activities of Ardašīr I'. *Historia i Świat* 6: 89–95.

Maksymiuk, K. (2017b) 'The Sasanian Relief at Salmās – New Proposal', in Maksymiuk and Karamian (2017) 97–112.

Maksymiuk, K. (2018a) 'Strategic Aims of Šāpur II During the Campaign in Northern Mesopotamia (359–360)'. *Historia i Świat* 7: 87–97.

Maksymiuk, K. (2018b) 'The Two Eyes of the Earth: The Problem of Respect in Sasanid-Roman Relations'. *GRBS* 58.4: 591–606.

Maksymiuk, K. (2024) 'The Sassanian Empire's Strategies', in Duyvesteyn and Heuser (2024) 191–203.

Maksymiuk, K. and Karamian, G. (eds) (2017) *Crowns, Hats, Turbans and Helmets: The headgear in Iranian history. Volume I: Pre-Islamic Period*. Siedlce.

Marciak, M. (2017) *Sophene, Gordyene, and Adiabene: Three Regna Minora of Northern Mesopotamia between East and West*. Leiden.

Maroth, M. (1979) 'Le siège de Nisibe en 350 ap. J.-Chr. d'après des sources syriennes'. *Les Annales Archeologiques Arabes Syriennes* 27: 239–43.

Masson, V.M. (ed) (2001) *The Role of Ahalteke Horse in the Formation of World Horse-Breeding: Materials for the International Conference*. Ashgabat.

Mazza, M. (2003) 'Bisanzio e Persia nella tarda antichità: note su guerra e diplomazia nella seconda metà del IV secolo d.C.', in Criscuolo (2003) 405–40.

Mazzarino, S. (1971) 'Anonymus post Dionem e la "topica" delle guerre romano-persiane 242/3 d.C.–283/(4?) d.C.'. *Accademia Nazionale dei Lincei Quaderno* 160: 655–78.

McDonald, D.J. (1981) 'The death of Gordian III – another tradition'. *Historia* 30: 502–508.

McDonald, D.J. (1986) 'Dating the Fall of Dura-Europos'. *Historia* 35: 45–68.

Mielczarek, M. (1993), *Cataphracti and Clibanarii: Studies on the Heavy Armoured Cavalry of the Ancient World*. Łódź.

Mitchel, St. (ed) (1983) *Armies and Frontiers in Roman and Byzantine Anatolia: Proceedings of a Colloquium Held at University College, Swansea, in April 1981*. Oxford.

Mode, M. and Tubach, J. (eds) (2006) *Arms and Armour as Indicators of Cultural Transfer: The Steppes and the Ancient World from Hellenistic Times to the Early Middle Ages*. Wiesbaden.

Mosig-Walburg, K. (1999) 'Zur Schlacht bei Singara'. *Historia* 48: 330–84.

Mosig-Walburg, K. (2002) 'Zur Westpolitik Shapurs II'. *StIr* 25: 329–47.

Mosig-Walburg, K. (2009) *Römer und Perser: vom 3. Jahrhundert bis zum Jahr 363 n. Chr.* Gutenberg.

Mosig-Walburg, K. (2010) 'Deportationen römischer Christen in das Sasanidenreich durch Shapur I. Eine Neubewertung'. *Klio* 92: 117–56.

Mousavi, A. and Daryaee, T. (2012) 'The Sasanian Empire: An Archaeological Survey, c. 220–640 AD', in Potts (2012) 1076–94.

Nikitin, A.B. (1994) 'Die Münzen des letzten indo-parthischen Königs von Sīstān. Ein Abschied von "Ardamitra"'. *Numismatische Zeitschrift* 102: 167–70.

Nikonorov, V.P. (2005) 'К вопросу о парфянском наследии в сасанидско Иране: военное дело', in Nikonorov (2005) 141–79.

Nikonorov, V.P. (ed) (2005) *Центральная Азия от Ахеменидов до Тимуридов: археология, история, этнология, культура. Материалы международной научной конференции, посвященной 100-летию со дня рождения Александра Марковича Беленицкого (Санкт-Петербург, 2–5 ноября 2004 года)*. St. Petersburg.

Nikonorov, V.P. and Alyokshin, V.A. (eds) (2015) *Академическое востоковедение в России и странах ближнего зарубежья (2007–2015)*. St. Petersburg.

Olbrycht, M. (2001) 'The Origins of the Arsacid Parthian Cavalry: Some Remarks', in Masson (2001) 108–11.

Olbrycht, M. (2016a) 'Dynastic Connections in the Arsacid Empire and the Origins of the House of Sāsān', in Curtis, Pendleton, and Alram (2016) 23–35.

Olbrycht, M. (2016b) 'Manpower Resources and Army Organisation in the Arsakid Empire'. *Anc. Soc.* 46: 291–338.

Overlaet, B. (2009) 'A Roman Emperor at Bishapur and Darabgird: Uranius Antoninus and the Black Stone of Emesa'. *IrAnt* 44: 461–530.

Panaino, A. (1994) 'The Two Astrological Reports of the Kārnāmag ī Ardaxšīr ī Pābagān (III, 4–7; IV,6–7)'. *Die Sprache, Zeitschrift für Sprachwissenschaft* 36: 181–98.

Panaino, A. (1999) *La novella degli scacchi e delle tavola reale*. Milan.

Patterson, L.E. (2013) 'Caracalla's Armenia'. *Syll. Class.* 2: 27–61.

Payne, R. (2015) 'The Reinvention of Iran: The Sasanian Empire and the Huns', in Maas (2015) 282–99.

Payne, R. (2018) 'The Silk Road and the Iranian Political Economy in Late Antiquity: Iran, the Silk Road, and the Problem of Aristocratic Empire'. *Bulletin of the School of Oriental and African Studies* 81.2: 227–50.

Pekary, T. (1961) 'Le "tribute" aux Perses et les finances de Philippe l'Arabe'. *Syria* 38: 275–83.

Pekary, T. (1962) 'Bemerkungen zur Chronologie des Jahrzehntes 250–260'. *Historia* 11: 123–28.

Pellò, S. (ed) (2016) *Borders: Itineraries on the Edges of Iran*. Venice.

Piacentini, V.F. (1985) 'Ardashir I Papakan and the Wars against the Arabs: Working Hypothesis on the Sasanian hold of the Gulf'. *Proceedings of the Seminar for Arabian Studies* 15: 57–78.

Potter, D. (1987) 'Alexander Severus and Ardashir'. *Mesopotamia* 12: 147–57.

Potts, D. (1988a) 'Arabia and the Kingdom of Characene', in Potts (1988) 137–67.

Potts, D. (ed) (1988b) *Araby the Blest: Studies in Arabian Archaeology*. Copenhagen.

Potts, D. (1990) *The Arabian Gulf in Antiquity*. Oxford.

Potts, D. (2007) 'Cataphractus and kamāndār: Some Thoughts on the Dynamic Evolution of Heavy Cavalry and Mounted Archers in Iran and Central Asia'. *BAI* 21: 149–58.

Potts, D. (2008) 'The Sasanian Relationship with South Arabia: Literary, Epigraphic and Oral Historical Perspectives'. *StIr* 37: 197–213.

Potts, D. (ed) (2012) *A Companion to the Archaeology of the Ancient Near East*. Malden, MA.

Potts, D. (ed) (2013) *The Oxford Handbook of Ancient Iran*. Oxford.

Pourshariati, P. (2008) *Decline and Fall of the Sasanian Empire: The Sasanian-Parthian Confederacy and the Arab Conquest of Iran*. London.

Rapp, St. H., Jr. (2014), *The Sasanian World through Georgian Eyes: Caucasia and the Iranian Commonwealth in Late Antique Georgian Literature*. London.

Rebenich, S. (ed) (2017) *Monarchische Herrschaft im Altertum*. Berlin.

Rezakhani, K. (2010) 'The Road that Never Was: The Silk Road and Trans-Eurasian Exchange'. *Comparative Studies of South Asia, Africa and the Middle East* 30: 420–33.

Rezakhani, K. (2017a) *Reorienting the Sasanians: Eastern Iran in Late Antiquity*. Edinburgh.

Rezakhani, K. (2017b) 'Saansaan Pirosen: Ammianus Marcellinus and the Kidarites'. *Dabir* 3: 44–50.
Ridley, R.T. (1973) 'Three Notes on Julian's Expedition (363)'. *Historia* 22: 317–30.
Rollinger, R., Ulf, C., and Schnegg, K. (eds) (2004) *Commerce and Monetary Systems in the Ancient World. Means of Transmission and Cultural Interaction*. Stuttgart.
Rösger A. and von Haehling, R. (eds) (1978) *Studien zum Herrscherbegriff der Historia Augusta und zum antiken Erziehungswesen*. Frankfurt am Main.
Rösger, A. (1978) 'Die Darstellung des Perserfeldzugs des Severus Alexander in der Historia Augusta', in Rösger and von Haehling (1978) 167–74.
Ross, A.J. (2014) 'Constantius and the Sieges of Amida and Nisibis: Ammianus' Relationship with Julian's Panegyrics'. *AC* 57: 127–54.
Rousseau, P. (ed) (2009) *A Companion to Late Antiquity*. Chichester.
Sarantis, A. and Christie, N. (eds) (2013) *War and Warfare in Late Antiquity: Current Perspectives*. Leiden.
Schindel, N. (2013) 'Sasanian Coinage', in Potts (2013) 814–39.
Schippmann, K. (1990), *Grundzüge der Geschichte des sasanidischen Reiches*. Darmstadt.
Schleicher, F., Stickler, T., and Hartmann, U. (eds) (2019) *Iberien zwischen Rom und Iran: Beiträge zur Geschichte und Kultur Transkaukasiens in der Antike*. Stuttgart.
Schlude, J.M. (2020), *Rome, Parthia, and the Politics of Peace: The Origins of War in the Ancient Middle East*. London.
Seager, R. (1996) 'Ammianus and the Status of Armenia in the Peace of 363'. *Chiron* 26: 275–84.
Seager, R. (2010) 'Rome and Persia 357–9: the Role of Tamsapor', in Kraus, Marincola, and Pelling (2010) 155–62.
Shahbazi, A.S. (1986) 'Army i: Pre-Islamic Iran'. *Enc. Ir.* 2.5: 489–99.
Shahbazi, A.S. (1990) 'On the Xwadāy-nāmag', in Amin, Kasheff, and Shahbazi (1990) 208–29.
Shavarebi, E. (2014) 'A Reinterpretation of the Sasanian Relief at Salmās'. *Iran and the Caucasus* 18: 115 33.
Shavarebi, E. (2015) 'Roman "Soldatenkaiser" on the Triumphal Rock Reliefs of Shāpūr I: A Reassessment'. *Historia i Świat* 4: 47–63.
Shayegan, M.R. (2003) 'Hazārbed'. *Enc. Ir.* 12.1: 93–95.
Shayegan, M.R. (2008) 'On the Rationale behind the Roman Wars of Šābuhr II the Great', *Bulletin of the Asia Institute* 18: 111–33.
Shayegan, M.R. (2011), *Arsacids and Sasanians: Political Ideology in Post-Hellenistic and Late Antique Persia*. Cambridge.
Skupniewicz, P. and Maksymiuk, K. (2018) 'Persian Riders in the Aethiopica of Heliodorus. A Reliable Source?'. *Historia i Świat* 7: 99–110.
Slootjes, D. and Peachin, M. (ed) (2016) *Rome and the Worlds beyond its Frontiers*. Leiden.

Sommer, M. (2003) *Hatra: Geschichte und Kultur einer Karawanenstadt im römisch-parthischen Mesopotamien*. Mainz.

Southern, P. (2008) *Empress Zenobia: Palmyra's Rebel Queen*. London.

Stickler, T. (2021) 'Armenien und Iberien zwischen Rom und Iran: wechselseitige Bezüge, parallele Entwicklungen'. *Electrum* 28: 189–206.

Stolte, B. (1971) 'The Roman Emperor Valerian and Sapor I, King of Persia'. *RSO* 1: 157–62.

Strobel, K. (1993) *Das Imperium Romanum im "3. Jahrhundert": Modell einer historischen Krise?: zur Frage mentaler Strukturen breiterer Bevölkerungsschichten in der Zeit von Marc Aurel bis zum Ausgang des 3. Jh. n. Chr.* Stuttgart.

Syvänne, I. and Maksymiuk, K. (2018) *The Military History of the Third Century Iran*. Siedlce.

Tafażżoli, A. (2000), *Sasanian Society: I. Warriors II. Scribes III. Dehqāns*. New York.

Temporini, H. and Haase, W. (eds) (1976) *Aufstieg und Niedergang der römischen Welt: Geschichte und Kultur Roms im Spiegel der neueren Forschung/2, Principat. Bd. 9, Politische Geschichte: (Provinzen und Randvölker: Mesopotamien, Armenien, Iran, Südarabien, Rom und der Ferne Osten). Halbband 1*. Berlin.

Thomson, R.W. (1976) *Agathangelos' History of the Armenians*. Albany, NY.

Thomson, R.W. (1978) *Moses Khorenats'i: History of the Armenians*. Cambridge, MA.

Thomson, R.W. (1997) 'Constantine and Trdat in Armenian Tradition'. *AOrientHung* 50: 277–89.

Toumanoff, C. (1963) *Studies in Christian Caucasian History*. Georgetown.

Traina, G. (2004) 'La fine del regno d'Armenia'. *La Persia e Bisanzio, Atti dei Convegni dei Lincei, Roma 14–18 ottobre 2002, Accademia Nazionale dei Lincei 201*: 353–72.

Traina, G. (2018) 'The Rise of the Sasanians: The Armenian Evidence', in Daryaee (2018) 107–19.

Trümpelmann, L. (1975) 'Triumph über Julian Apostata'. *JNG* 25: 107–11.

Weber, U. (2016) 'Narseh'. *Enc. Ir.*, online edition, http://www.iranicaonline.org/articles/narseh-sasanian-king (accessed on 23 May 2016).

Weber, U. und Wiesehöfer, J. (2010), 'König Narsehs Herrschaftsverständnis', in Börm and Wiesehöfer (2010) 89–132.

Widengren, G. (1971) 'The Establishment of the Sasanian Dynasty in the Light of New Evidence'. *Accademia Nazionale dei Lincei Quaderno* 160: 711–82.

Widengren, G. (1976) 'Iran, der grosse Gegner Roms. Königsgewalt, Feudalismus, Militärwesen', in Temporini and Haase (1976) 219–306.

Wiesehöfer, J. (1982) 'Die Anfänge sassanidischer Westpolitik und der Untergang Hatras'. *Klio* 64: 437–47.

Wiesehöfer, J. (1986) 'Ardašīr I. i. History'. *Enc. Ir.* 1.4: 371–76.

Wiesehöfer, J. (1996) *Ancient Persia: From 550 BC to 650 AD*. London.

Wiesehöfer, J. and Huyse, P. (eds) (2006) *Erān ud Anērān: Studien zu den Beziehungen zwischen dem Sasanidenreich und der Mittelmeerwelt. Beiträge des internationalen Colloquiums in Eutin, 8.–9. Juni 2000*. Stuttgart.

Winter, E. (1988) *Die sasanidisch-römischen Friedensverträge des 3. Jahrhunderts n. Chr. Ein Beitrag zum Verständnis der aussenpolitischen Beziehungen zwischen den beiden Grossmächten*. Frankfurt am Main.

Winter, E. (1989) 'On the Regulation of the Eastern Frontier of the Roman Empire in 298 AD', in French and Lightfoot (1989) 555–71.

Wood, P. (2013) *The Chronicle of Seert: Christian Historical Imagination in Late Antique Iraq*. Oxford.

Wood, P. (2016) 'Al-Ḥīra and Its Histories'. *JAOS* 136: 785–99.

Yarshater, E. (ed) (1983) *The Cambridge History of Iran. Vol. 3: The Seleucid, Parthian and Sasanian Periods. Part 1*. Cambridge.

Young, G.K. (2001) *Rome's Eastern Trade: International Commerce and Imperial Policy 31 BC–AD 305*. London.

Zuckerman, C. (1993) 'Les campagnes des tétrarques, 296–298. Notes de chronologie'. *Antiquité Tardive* 2: 65–70.

CHAPTER 15

Cavalry Arms and Armour from the Achaemenids to the Sasanians: Iconography and Methods of Reconstruction

Patryk Skupniewicz and Katarzyna Maksymiuk

The development of arms and armour in the pre-Islamic Iranian empires occurred against a backdrop of twelve centuries of cultural change. Nevertheless, one of the aspects of Iranian identity in the *longue durée* is a surprising consistency in the forms of arms and armour, resulting either from continuity in preference of some weapons over others, or revival of old forms in new contexts. Naturally, this continuity and periodic revivals did not exclude the abandoning of outdated and adoption of new forms. Sadly, there are only a scanty number of preserved examples of Iranian weapons from the period under discussion. Any study of ancient Iranian arms and armour must rely heavily on literary sources and iconography, which often prove discrepant. As a result, a traditional survey of military equipment cannot be carried out without an examination of the formal aspects of both art and literature.

Xin Wu has defined a dichotomy between 'art historians' and 'textual historians' in methodologies of approaching ancient Iranian art.[1] It goes without saying that this observation generally applies to the researchers of the other epochs of ancient Iranian history and perhaps ancient history *en masse*. For later periods, the 'textual' method should probably include not only identifications of historical events but also identification of historical or mythical personages. Also, somehow, the diminishing phrase '*merely motifs*' reveals the inclination of the author, who seems to perceive motifs as inferior to textual or historical events. In fact, it is difficult to agree with 'textual' interpretations which fail to explain the reasons for certain iconographic developments,[2] which remain faithful to the fixed formal paradigms. The important observation of Tonio Hölscher on Greco-Roman iconography would apply here with all its consequences: 'War in art is not war but art'.[3] Thus art is a subject of

1 Wu (2014) 223–29.

2 Wu (2014) 228–29, 232–34; Sevinč, Korpe, Tombul, Rose, Strahan, Kiesewetter, and Wallrodt (2001).

3 Hölscher (2003) 2.

 | DOI:10.1163/9789004710771_016

research in itself, and can only be treated as a historical source for reconstruction of a course of events very carefully and with strict methodological boundaries.

The very idea of polarisation between textual and visual methodologies does not seem fully correct either, as it should be complemented with the study of art as a source for reconstruction of the elements of material culture. Thus, we need to perceive this study rather as tripartite, involving research on the artwork itself, the related events, and the material culture of the time. This aspect of research becomes especially important in the absence of relevant archaeological material and textual information. Its main weakness is shared with the textual historians' approach – it attempts to 'squeeze' from art some information from beyond the visual sphere, and is therefore prone to being affected by visual traditions, aesthetic preferences, short-lived fashions, and the decorum of the time. The way a personage wanted to be depicted might depend to a lesser extent on the actual look of the time and to a greater extent on visual tradition, attempts to imitate ancestors, or mythological/religious patterns. However, this approach attempts to see 'war' in 'art' in aspects other than textual-historical methodology. Visual arts present as many challenges as fictional literature in terms of providing firm sources of information about material culture; both are usually focused on elites, and biased by fixed 'phrases' and the conveyance of ideological contents.[4] The aesthetic factor must not be forgotten – neither art nor narrative literature normally depicts 'the world outside the window', nor is either a documentary relation of historical events.

When associating depictions with actual objects, it is important to remember the tendency of Near Eastern art to put symbolic meanings into realistic material environments. The careful and detailed depiction of the objects does not necessarily reflect their actual use at the time when they were depicted. The study of depicted elements of material culture must not ignore fixed compositional formulae and motifs, as well as the geographical distribution of the objects.

A critical approach to the contexts of the visual sources should result not only in a more reliable reconstruction of the panoplies of Persian warriors, but more interestingly, in a greater understanding of the weapons' cultural origins and significance, and the relationship between artistic depictions and actual weapons and combat techniques. While a comprehensive survey is not possible here, this chapter will provide a case study in such approaches by focusing on depictions of cavalry, a vital component in ancient Iranian armies.

4 Skupniewicz and Maksymiuk (2018).

1 The Role of Cavalry in Ancient Iran

Already significant in the Achaemenid armed forces, cavalry became the backbone of the Iranian army of the Arsacid period and remained its core throughout the Sasanian era.[5] Although infantry never disappeared from Iranian forces, its status in most periods was greatly inferior to the horsemen. Except for in pastoral-nomadic societies where horse riding was a basic and common skill and horses were a condition of day-to-day survival, possession of horses always involved huge expenditure which provided no (or negligible) economic return in comparison with oxen, donkeys, or even mules. The importance of horses lay in warfare, communication, and leisure, and in sedentary societies the ability to afford horses was a mark of high social standing. In pre-modern societies the hard work was done by oxen, which, despite being radically slower than horses, could pull greater loads. Donkeys and camels were less demanding, slower but able to carry a heavy weight over distances and terrain difficult for horses. The European privileged classes, originating from former military aristocracies, often bore names deriving from horse-riding (*hippeis*, *equites*, *chevaliers*, or *Rittern*). This process is less clear in Iran, where terms for the gentry referred to freedom, being born of 'free' social standing (*āzādān*), grandeur (*wuzurgān*), and administrative obligations or scale of land ownership (*dehkanān*).[6] This might result from the existence of elite infantry forces in the Achaemenid period, and the almost constant contact with steppe nomads for whom the ability to ride a horse was not a matter of social differentiation. Nevertheless, horse troops were an important part of Iranian military forces throughout the Achaemenid period and even predate the imperial phase, as evidenced by the seal of Cyrus I (grandfather of Cyrus the Great and contemporary of Ashurbanipal).[7] Despite the presence of elite infantry units, Achaemenid cavalry was a prestigious force accompanying the commanders and kings in battles. Riders in lavish robes became a common iconographic motif even without a direct military or hunting context.[8]

5 Nicolle (1996) 6–9; Nikonorov (2005) 146–50, 153–55.
6 Farrokh, Karamian, and Maksymiuk (2018) 18–20, 28.
7 Harper (2006) 14, 46, fig. 16.
8 Tuplin (2010) 104–20; Gabrielli (2006) 68–83; Manning (2021) 183–87. The primary importance of iconography as a source for the study of Achaemenid cavalry is emphasised in Tuplin's seminal article, which lists pictorial before textual sources; interpretation of iconography also plays an important part in other studies of Achaemenid armies including Bittner (1985); Head (1992); Sekunda (1992); and Manning (2021).

The situation became even more radical in the course of the Arsacid dynasty,[9] when nobility was clearly and directly identified with horse-riding. This became one of the long-lasting features of Iranian elite identity. Justin's epitome of Pompeius Trogus (41.3) reports that

> They ride on horseback on all occasions; on horses they go to war, and to feasts; on horses they discharge public and private duties; on horses they go abroad, meet together, traffic, and converse. Indeed, the difference between slaves and freemen is that slaves go on foot, but freemen only on horseback.

This does not seem very different from an account by Sir John Chardin, a seventeenth-century traveller to Persia, who states: 'For the Men, they ride on Horseback, but never walk'.[10]

By the early Achaemenid period, the Iranian cavalry had already adopted the steppe-nomadic skill of mounted archery, which differed from Mesopotamian methods in the employment of *gorytoi*, integrated quivers and arrow-cases hung at the left hip, instead of quivers slung behind the right shoulder.[11] Another process, already initiated in Achaemenid times, was the gradual increase in the amount of armour worn by the warriors and mounts, gradually leading to a distinction between specialised units of lightly armoured archers and heavily armoured lancers.[12] It must, however, be remembered that horse archers were also described as wearing armour, and heavy riders are usually shown carrying archery equipment. It is believed that the armoured troops consisted of men of higher status,[13] but this opinion might be a direct projection of European medieval realities and the rational conclusion that costly armour determined social standing. It is certainly true that items of armour were of great cost, and that armoured riders could not originate from the lower echelons of society (providing that they owned their equipment). However, it cannot be excluded that tactical principles governed assignment to the archers or the lancers. It seems that the lancers were tactically less mobile and played a role which can be compared to Byzantine *defensores*, which decreased the chance of gaining military glory (which usually depends on the number of kills, more difficult to

9 Hauser (2006) 301; Nicolle (1996) 6–9; Nikonorov (2005) 142–44.
10 Chardin (1988) 193.
11 Skupniewicz (2021a).
12 Mielczarek (1993) 46–67; Hauser (2006) 299–301, 303; Olbrycht (2010) 67–73; Potts (2007); Skupniewicz (2006); Charles (2015); Olbrycht (2015) 369–74.
13 Wolski (2000) 98–99.

achieve in a defensive situation), usually greatly desired by young men in warrior societies, regardless of social standing. Ammianus Marcellinus (25.1.12–13) also reports that both the archers and lancers were armoured, while Maurice's *Strategikon* (11.1.113–114) describes the Persians as armoured mounted bowmen, without even mentioning the lancers. Finally, both Arsacid and Sasanian iconography shows a certain reluctance to depict men in armour, and prefers to represent kings and nobles unarmoured, which would not be the case if armour was intrinsically linked with high social standing.

The society of Ancient Iran produced an elitist culture where artistic resources were rarely 'wasted' on depicting commoners. At the same time the pictorial principles were quite rigid. The most obvious reason for such aesthetics was the conscious choice of visual language to be employed in state-sponsored art, imitated to an extent by the people in power. Naturally, the principles differed in the Achaemenid, Parthian, and Sasanian eras, but the dominant forms of artistic expression depended on approval from the court.[14] Artisans of lower status could not be free from the paradigm simply because the commissioners and consumers also expected the works within particular frames. One might conclude from the repetitive motifs fluctuating around unchanging formulae, technical conservatism, and class distinction between artisans and aristocrats that the noble commissioners did not enjoy being surprised by the commoners whose life depended on their work. Thus, the assumption can be made that depictions of commoners in Iranian art are unlikely, and if they do exist, they are shown serving their masters. After the Achaemenid era, when elite infantry was still an important part of the Iranian army, , it is uncommon to find depictions of infantry soldiers since the role of infantry was humble, both tactically and socially, in the Arsacid and Sasanian periods,. What clearly defines cavalry warriors in art, removing any doubt about whether the depiction represents a dismounted rider or elite infantryman, is surely showing them mounted. The sections that follow will survey specific aspects of the ancient Iranian depictions of mounted warriors, beginning with the representation of several key items of cavalry equipment, including body armour, headgear, leg protection, and offensive weaponry. In each case, it will summarise the available evidence and analyse points of continuity and development from the Achaemenid through Sasanian periods, before turning in the final section to the iconography of the cavalry battle.

14 For basic overviews of the Iranian arts in these eras, see Garrison (2013); Canepa (2013); Porada (1962); Ghirshman (1964); Ghirshman (1962), Schlumbacher (1983); Shepherd (1983); Boardman (2000).

2 The Panoply of Iranian Cavalry in the Light of Iconography

2.1 *Torso Armour and Limb Defences*

The dominant type of body armour shown on objects of Achaemenid art consists of a stiff corselet with large shoulder pieces and a high collar protecting the back of the neck, usually reaching the top of the head. The lower part of the torso was protected by *pteryges*, usually rather short and covering the hips; however, a supposedly longer version can be found on a slab from Bin Tepe or Sardis (BM inv. 1889,1021.1) and a relief from Aktepe,[15] which might be another form of protective skirt. Although the Sardis relief might belong to the pre-Achaemenid period and represent a local Lydian tradition, the similarity of later, firmly Achaemenid examples is so striking that, in terms of arms and armour development, they belong to a single evolutionary line. A very short form of *pteryges* with semicircular or rectangular lappets placed at the lower edge of the cuirass can be seen on the Manisa stele (inv. 3389) and a pectoral in the Miho Museum (inv. H-26 W-25.7).[16] Depictions on the Çan sarcophagus[17] and the Alexander Mosaic[18] show corselets of a red or reddish colour, which suggests that the armour was made from dyed leather or linen. The pectoral in the Miho Museum shows the cuirass in light blue, which might be a colour code for iron or another variation of dyed organic material. The tight corselets of the Persian riders on the Tatarli beam[19] are painted red, blue, and white, which proves that different colours could be used at the same time, and therefore are more likely to refer to social coding than material. They might also have a purely aesthetic function. Assuming that red was a colour of warriors,[20] associated with Verethraghna, it may have marked the warriors around the great king and the hero on the Çan sarcophagus as members of the highest social stratum, whereas the riders on the Tatarli beam could belong to different ranks; but the question would remain why the armoured rider on the Miho Museum pectoral does not share the same colour, even though he is a central figure on a golden pendant and undoubtedly a personage of the highest social and military standing.

Linen armour had gained great popularity already before the Persian-Greek wars and was used in the entire Mediterranean area.[21] It provided reasonable

15 Baughan (2008).
16 Bernard and Inagaki (2000); Casabonne and Gabrielli (2006).
17 Sevinč et al. (2001); Ma (2008).
18 Cohen (1997); Collins (2012).
19 Summerer (2007).
20 Tafażżoli (2000) 1.
21 Aldrete, Aldrete, and Bartell (2013).

protection against arrows, which must have been a crucial factor for Achaemenid riders, who usually operated in environments full of missiles. It was inferior to metallic armour and frequently reinforced with scales, at least in the most vulnerable parts. The cuirass from Vergina[22] proves that metallic armour followed the shape normally associated with linen armour. The horizontal lines decorated with meanders on the Bozokir frieze[23] clearly repeat the lines across the Vergina armour and are commonly found in Greek art. This would support Sekunda's supposition that the personage on the Bozokir relief might indeed represent a cavalryman.[24]

The line across the front of the torso of the personage on the stele in the Manisa Museum suggests an opening, which would by no means be required in the case of linen construction. Naturally some kind of opening is needed when arming, but choosing an opening at the centre of the chest and abdomen, where the vital organs are located, must have resulted from technical or technological considerations. The smaller sheets to create the box around the man's torso must have been easier to make, and a central opening might have allowed arming and disarming of the warrior with little or no assistance. Neither the armour from Vergina nor the cuirass depicted on the wall of the tomb of Lyson and Kallikles[25] has a central opening, but they consist of several sheets covering the front, back, and flanks. The cuirasses of Persian elite infantry soldiers depicted on the seals strongly suggest that the corselet was built as a 'box' with pronounced flank plates. None of them shows a central opening in the front, which may result from the profile view of the torsos of the combatants. It must be pointed out that the cuirass with front opening, associated with Hellenistic patterns, appears in a late Parthian relief from Tang-e Sarvak[26] and later gains popularity in Gandhara and Central Asia, finally reaching Korea. Despite its imperfections, the relief from Manisa may provide a link between Greek and Achaemenid armament and later Iranian tradition. The tight armoured 'vest' closing in the front with plates moving on hinges is also attested in the art of Gandhara, from where, seemingly, it was adopted in the iconography of East Turkestan, as evidenced by the murals from Kizil.[27] The third- or fourth-century cuirasses from Korea and Japan attest the adoption of this fashion in East Asia. There the plates were not solid but consisted of

22 Connolly (1984) 81.

23 Relief from Bozkir, southern Turkey (now lost). Image in Sekunda (1992) 25.

24 Sekunda (1992).

25 Miller (1993).

26 Kawami (2013) 756–61; Kawami (1987) 106–108, 201–204. Pl. 49, figs. 19–20; Nicolle (1996) 8; Skupniewicz (2016) 63; Skupniewicz (2021b).

27 Ghose (2008).

riveted elements constituting stiff sheets. When looking closely at the cuirass of the personage depicted on the capital currently held in Taq-e Bostan, the same pattern can be observed.[28] This may lead to the conclusion that either the construction was re-adopted in Iran after reaching as far as Japan, or was never abandoned, although the gap between Tang-e Sarvak[29] and the capital leaves room for speculation. Importantly, Tang-e Sarvak predates Far Eastern examples, while the Taq-e Bostan capital is from some two centuries later. It seems important to note that the Korean suits of armour are accompanied by lamellar helmets, the form of head protection depicted on the Taq-e Bostan capital. Similarly, the armour shown in the battle scene on the famous Orlat plate[30] may schematically represent the same type of construction. The pronounced front opening is characteristic for both Tan-e Sarvak and Korean/Japanese cuirasses. The horizontal segments which fill the torso parts differ from the limb coverings, and may correspond to the construction of the East Asian examples, even though no rivets are clearly shown. Naturally, the upper parts of the Tang-e Sarvak plates, just like the Korean cuirasses, leave the shoulders open, leaving their protection to other, more flexible, means. This must have allowed archery and wielding of a long lance with both hands. A similar, though not identical, principle can be found on the famous 'clibanarius graffito' from Dura-Europos (and the far less famous graffito of the heavily armoured warrior fighting on foot), where the belly is covered with large rectangular plates while shoulders and arms are protected by smaller plates and/or scales or mail. Large pieces of metal affixed to another must have produced quite stiff protection, preventing penetration of the vital organs, while the soft cover over the shoulders and arms provided necessary flexibility. The differentiation of large and small scales or mail was pointed out by Kubik in his study.[31] It does not contradict the ideas of Wójcikowski, who noted the 'hybrid' nature of the armour but did not draw tactical conclusions.[32] Although the shoulders are important targets in sword combat, they are far less important in lance or missile combat (modern bullet-proof vests cover comparable areas, allowing for greater shoulder movement). The apparent reappearance of a front-closed vest, with large plates on the hinges, in the Safavid and Qajar *chahr-aina* might be an imitation of older patterns or a response to similar threats on the battlefield. Another analogy might be the development of cord-and-plaque armour in post-Han

28 Skupniewicz (2022).
29 Kawami (2013) 756–61; Kawami (1987) 106–108, 201–204. Pl. 49, figs. 19–20.
30 Nikonorov (1997) 75, 43.
31 Kubik (2020).
32 Wójcikowski (2013).

China, consisting of protective plastrons on the front and back of the torso. The principle seems to remain the same, but construction differences appear too big to allow for the idea of Iranian influence.

The rider on the Payava tomb (BM 1848,1020.142), possibly Autophradates, is shown wearing a muscle cuirass just like Payava and his father, depicted standing on foot in another part of the monument. It is probably the first example of the special significance of that type of armour in Iranian art, a feature recognisable in Sasanian era. What seems worth noting is the visual or iconographic correspondence, or perhaps even interchangeability, of plain cuirasses with shoulder lappets usually identified as forms of *linothōrakes* and muscle cuirasses. It must have referred to their elite nature. Muscle cuirasses appear frequently in iconography of the land 'between the empires', Dura-Europos, Hatra, and Palmyra, where they are often mixed with elements of Iranian or local dress. In Parthian iconography they are depicted in the art of Old Nisa, which raises the question whether they are iconographic motifs or actual pieces of armament. In the Sasanian era, muscle cuirasses appear on the coins of the Kushanshahs.[33] The royal rock reliefs, when showing the kings in combat, depict them in plain torso covers, which most likely represent cuirasses. However, due to a repeatedly appearing vertical line across these cuirasses, it is impossible to exclude continuation of the tradition of cuirasses with frontal opening seen on the Manisa Museum stele, the Tang-e Sarvak relief, and the Taq-e Bostan capital. The semantic proximity between plain cuirasses and muscle cuirasses, being part of the Roman iconography of power evidenced by the imperial *statua loricata* type, and later adopted in Iranian iconographies for religious purposes, supports the idea of Sasanian kings being depicted in that form of high-status armament. It should be borne in mind that the Sasanian royal reliefs were coloured, and thus it cannot be excluded that the musculature was painted as well. Although heroic nudity did not correspond with Iranian imagery, the impact of the Hellenistic cultures surrounding Arsacid Iran both from west and east must have affected the iconography.

Starting with Herodotus (1.215), the written sources emphasise employment of scale armour by Iranian elite warriors. These either did not carry sufficient semantic impact, being too common to make an effort of showing them, or might reveal local trends within the Achaemenid imperial war-machine and would not appear in the regions where the warriors were depicted in armour. Therefore, this could be an actual armament fashion or an iconographic tendency. Scale, lamellar, and smaller-plate armour are better represented in the post-Achaemenid iconography, but at the fringes of the empire. This fact

33 Skupniewicz (2015).

might be explained by a relative reluctance to depict combat in armour, or even combat between humans, in Arsacid and Sasanian art. An example of scale was shown on the rider on the Parthian tile from Babylonia, now in the British Museum (inv. 91908). Scale armour, deriving from Iranian art, was also shown on the Himyarite relief (BM inv. 2014,6002.1), but this object should be treated as an analogy rather than direct evidence. Together with the account of Trajan's Column, it allows the reconstruction of scale 'overalls' protecting torso and limbs with a layer of metallic scales. The construction may have derived from the Scythians. Scale elements were shown on reliefs in Tang-e Sarvak (scale collar), Firuzabad, and Naqš-e Rostam (scale skirts and coifs); in the former case the scale elements characterised the defeated enemies, but in the latter the same scale elements are worn by victorious kings as well as their enemies, and they have lost their identification value. Apparently, the popularity of the construction did not allow its limitation to enemies alone. At first glance, the torsos of the warriors shown on the late Sasanian *spāhbēdan* seals are also covered by scales, but their small size does not allow us to say with certainty that they represent carefully elaborated rings of mail. The terracotta tile from the Metropolitan Museum (inv. 40.58), depicting a mounted archer in scale armour, was heavily reconstructed and cannot be treated as a reliable source. Fragments of a scale pattern around the neck of the personage shown on post-Sasanian Chal Tarkhan stuccos (BM inv. 140660), similar to that worn by the rider on Tang-e Sarvak, allows us to suppose that the scale collar had some semantic content. It must be noted that similarly sized rows of scales were found in Pasargadae and are located in the Metropolitan Museum of Art (inv. 1978.93.15).

Lamellar body protection is depicted on the reliefs from Hatra and Palmyra where it is shaped in the form of a tight corselet, not dissimilar from the examples on Assyrian reliefs. It is hard to believe in such a *longue durée* and it seems more likely that they followed the example of Roman cuirasses in locally available technology. The rider on the Tang-e Sarvak frieze has his legs covered by lamellar leggings, which find a direct analogy in Korean art, emphasising the idea of transfer of military technology relating to heavy cavalry between Iran and East Asia. The barding of the mount in Taq-e Bostan's great aivan is fully lamellar. The great popularity of lamellar armour throughout the ages results from its relative technological simplicity and decent protection which does not compromise mobility.

Large, square plates were found in Old Nisa, where they could be part of looted foreign armour.[34] They seemingly correspond with long protective

34 Pilipko (2001).

coats known from Indo-Saka iconography and body-armour in the Khalchayan friezes, but such construction is not clearly confirmed in Arsacid or Sasanian Iran.[35] The graffiti from Dura-Europos include large plates, but they are rectangular, and the entire construction is not reminiscent of those from Iran's eastern neighbours. The popularity of such armour was rather short-lasting and local.

Mail was most likely a Roman import in the Iranian war kit, however keenly adopted. Mail sleeves, shoes, and skirts are shown on the Firuzabad frieze.[36] The latter element appears on the capital in Taq-e Bostan, together with a mail coif and additional neck protection hanging from the helmet. A rider in a long hauberk was shown on the iconic depiction from the great aivan at Taq-e Bostan. Sasanian monumental art was very careful in showing the rings of the mail fragments, in contrast with Roman iconography where the rings were stylised by drilled holes or painted on flat surfaces.[37] When it comes to minor or simplified forms of art like graffiti or seals, it is difficult to differentiate the mail from scales. Functionally, however, both forms might have had a similar function of offering flexible forms of protection; this idea is strongly supported by the Firuzabad frieze, where the same elements were shown as chain mail or scale, depending on the side which the personages represent. The relation of both protective forms is further enhanced by archaeologically confirmed Roman mail covered with scales, which may correspond with several-layered mails from Islamic literary depictions of late Sasanian armament.

It must be noted that the sleeves of Sasanian warriors are mail, while Arsacid warriors wear segmented arm protection known since Hellenistic times, and possibly used already by Achaemenid heavy riders, as suggested by Xenophon's prescription for Athenian armoured cavalry (*Eq. mag.* 12.5). Intriguingly, the asymmetrical application of the *cheira* was later found in gladiatorial kit. Segmented limb defences are depicted on Indo-Saka coins, the Khalchayan frieze, all combat scenes in Sasanian royal reliefs, and also on Kushano-Sasanian coins and Sogdian religious iconography. It seems that their flexibility was a factor which assured great popularity over time among warriors whose combat technique involved active employment of both hands, whether for archery or wielding long lances. This argues that the sleeves or leggings of that type could consist only partly, if at all, of tubular rings. It is possible that initially the Achaemenid construction was stiffer and was only

35 Pugačenkova (1971); Gorelik (1982a); Olbrycht (2015); Skupniewicz and Lichota (2017).
36 von Gall (1990); Nikonorov (2020).
37 Bivar (1972).

gradually perfected to flexible forms. The segmented sleeves of the warriors on the Orlat plate do not restrict movement but evidently are not made of rings.

2.2 *Headgear*

There are several types of headgear worn by armoured riders in Achaemenid art, but the most numerically dominant is the headdress covering the back of the head and appearing flat on top. It some cases it clearly represents a tiara or *kyrbasia*, a typical headgear of Achaemenid and Hellenistic Persians. The head covering of the armoured rider on the Çan sarcophagus provides an interesting example of the helmet with a flat top and covered sides and back of the head. This sort of head protection does not have any analogy within the standard typology of ancient helmets. The only example of surviving war kit clearly related to that helmet is a bronze example excavated in the sanctuary of Poseidon in Isthmia.[38]

The group of so-called Kuban-type helmets made of cast bronze is shaped with a conical crown and sides falling lower alongside the face. In many examples there is a characteristic protrusion over the nose, present in Isthmian examples as well. Persian soldiers wearing Kuban-type helmets are present in Achaemenid and Greek iconography, where they are usually associated with warriors on foot. This led Mihail Gorelik[39] to the premature conclusion that the type was reserved for infantry while the preferred cavalry helmet was a tall conical one with a spike finial on top. His argument on the one hand assumed that all warriors depicted on foot are in fact infantry and cannot be dismounted riders, and on the other hand ignored the fact that the majority of Kuban-type helmets were excavated in the *milieu* of nomadic-steppe riders. There might be a genetic relation between Kuban-type helmets and the Geto-Dacian examples, but it has not been researched yet. Except for an example from Isthmia, none of these helmets had the flattened top, so only the former can serve as an analogy or source for the headgear depicted on the Çan sarcophagus. It is also possible that the Kuban or related Geto-Dacian types were actually helmets with an additional crown or covering, or even in some cases a tiara, which was a headdress marking the high status of its wearer, worn over a tight-fitting helmet. What makes this option less credible, except for the Isthmia helmet, is the lack of any decorations of flutes which would have appeared on the crown.

It is reasonable to assume that the seal depictions which share the iconographic type with the combat scene on the Çan sarcophagus depict a similar type of attire. Therefore, even though the small size of the pictures on the seals

38 Woźniak (2010) 36–38.

39 Gorelik (1982b).

or the portrayal of the armoured archer on the Miho Museum pectoral do not allow precise identification of the depicted type of the headgear, a clear relation between the visual formula and remaining parts of the war kit allows association with flat-topped helmets rather than tiaras. In case of the pectoral in the Miho Museum, another option presents itself: two infantry attendants of the main personage wear Kuban-type helmets, so the possibility that the cavalryman also wears the same type of headgear cannot be excluded. The existence of actual helmets with flat top and cheek pieces enhances this supposition. It is possible that the flat-topped helmets imitated or followed the visual pattern of tiaras with flopped-over upper parts. Clearly tiaras were worn by armoured riders, as evidenced by the Alexander Mosaic, but they are also worn by the riders on the Tatarli beam and the Oxus Treasure plaque (BM inv. 123949), where one may see textile or felt armour in the shape of a short tunic. Another difficulty with that interpretation is that the unarmoured mounted hunters shown on the seals wear the tiaras depicted in a very similar way to some armoured riders from sigillography.[40]

Incidentally, it might be noted that the development of medieval European flat-topped pot helmets, which later led to the creation of enclosed helmets which in turn resulted in the formation of the great helmets, occurred in approximately the same area as the adoption of 'Phrygian cap' shapes in European helmets, and might have been inspired by access to ancient iconography.

Another type of headgear worn commonly by Persian armoured riders was the *pilos*-type helmet, both with and without plumage. This type of head protection is attested on friezes from Bozokir and the tomb of Payava. The main rider on the Payava tomb battle scene has his face and lower head damaged, but the height of the helmet and analogy with other scenes from the same tomb strongly suggest a *pilos*-type. On the other seal the rider fleeing from the victorious Persian wears the same type of helmet, also decorated with a plume.[41]

The riders on the relief from Bin Tepe or Sardis seem to wear close-fitted headgear. They do not seem to be shown bare-headed, as no traces of the coiffure are shown, despite Xenophon's record of going to battle without helmets as a mark of courage (*An.* 1.8.17–18). Only Kuban-type helmets might correspond with this depiction, especially as the elite infantry soldiers shown on sigillographic material and the pectoral in the Miho Museum are equally rounded and seemingly close to the skull. The same sigillographic material

40 Boardman (2001) 305–309, 312–16, 331 (fig. 843), 334 (fig. 863), 337 (figs. 885, 886, 888, 889), 339 (fig. 905), 342 (figs. 924–929).

41 Nikonorov (2020) 177–78, 224–25.

proves that body armour types were interchangeable between elite infantry and elite cavalry.

The helmet of the rider shown on the stele in the Manisa Museum has an egg-shaped body with what looks like large triangular cheek pieces and a neck guard. The body is topped with the small comb at the back. As mentioned earlier, it is difficult to definitely state whether the face represents the warrior's actual face or a protective mask. This headgear is exceptional among ancient helmets. The small comb at the back is reminiscent of the headgear depicted on the seal of Cyrus.[42] It is difficult to place it in any typology, but it is possible to link the large triangular cheek-pieces with some Assyrian examples. The small comb on top of the helmet may also relate to Assyrian or Urartian helmets. The general shape and the shape of the comb itself seem different, but there might be a genetic link between them. A practical function of the comb seems unlikely, and it seems that it served as an object of function distinction.

Tall conical helmets must have survived the fall of the Achaemenids, as this shape proved very popular in post-Sasanian Sogdiana and is attested by examples from the British Museum found in Nineveh (inv. 22495) and from the Khalili Collection (inv. W 1415). Although there are no direct and undisputed examples that bridge the gap between Achaemenid and late Sasanian specimens, the headdress of the personages on the Parthian reliefs at Tang-e Sarvak and Hung-e Kamalwand[43] are very reminiscent of helmets of that kind; but no other traces of armament have been preserved. The rider depicted on the Hung-e Kamalwand relief carries a lance and can be identified as a warrior, but the personage on the Tang-e Sarvak relief carries a *cornucopia*, which makes identification more difficult. Kubik has remarked that the helmets in Iranian art appear independently from armour, but armoured persons appear also in non-combat situations, as evidenced clearly by the capital at Taq-e Bostan.

The 'morion-type' derivates of the Hellenistic headgear of Athena-type goddesses and the Hellenistic helmets of the two heads from Old Nisa seem to represent artistic convention rather any actual type of headgear.[44] Thus it is difficult to reconstruct the development of helmets throughout the Arsacid period. The Babylonian tile, mentioned earlier, shows what may be a flattened version of earlier Kuban type helmets, and may have been transferred to the low conical helmets of the Parthians depicted on the Firuzabad frieze.

The tall helmets imitating tiaras became popular in the early Sasanian era, standing in for the 'Sasanian' helmet, and continued in use until the very end

42 Harper (2006) 14, 46, fig. 16.
43 Kawami (1987) 201–204.
44 Skupniewicz and Maksymiuk (2019).

of the era, while the last century of the dynasty saw the appearance of lamellar helmets, constructed either from two halves or as *Bandenhelme*, which outlived the period. The later Sasanians also adopted lamellar helmets, consisting of narrow strips of metal affixed together, reinforced by the horizontal band running around the skull. This type was borrowed from the East, either as an immediate result of conquering Kushanshahr or due to later contacts with Hephthalites or, even later, with Turks. The popularity of this construction among the Steppe Peoples and in East Asia was extremely long-lasting, surviving antiquity and continuing until the nineteenth/early twentieth century, which makes the actual time of adoption difficult to determine. Hephthalite influence seems, however, most likely, given the similarities between Iranian art and Central Asian iconography pre-dating Turkic expansion.

2.3 *Leg Covers*

There are two types of leg coverings of armoured riders represented in Achaemenid iconography. The first are *parapleuridia*, a part of cavalry defensive equipment mentioned by Xenophon (*An.* 1.8.6; *Cyr.* 6.4.1) and Arrian (*Tact.* 4.1), and identified with the large elements covering riders' legs by Bernard,[45] Head,[46] and Sekunda.[47] It seems unlikely that such big 'wings' could have been worn by the rider, and therefore they should be affixed to the saddle in a way similar to the *guardamonte* used by Argentinian *gauchos*. This analogy seems even more convincing when we consider that no scales or other form of metallic reinforcements are shown in any depiction of armoured riders. A terracotta slab from Khumbuz Tepe[48] is sometimes interpreted as a depiction of *parapleuridia* made of plates or rectangular scales, but the small fragment does not allow us to assess whether it shows the discussed element or a type of armoured long coat which survived later among Indo-Scythians or the nomadic invaders of Bactria. The fact that all depictions of Achaemenid *parapleuridia* show them as plain, and that long Saka armoured coats were reinforced with plates, supports the idea that Khumbuz Tepe represents the latter. The *parapleuridia* shown on assembled examples, just like *guardamonte*, are plain and most likely made of the same material, thick leather. Although they are occasionally reconstructed as huge metallic sheets or made of scales, such assumptions do not find confirmation in iconographic material. Large, plain sheets of metal might not seem impossible, but their weight would be immense and

45 Bernard (1964) 198–201.
46 Head (1992) 36, 70, Pl. 6.
47 Sekunda (1992) 25.
48 Mambetullaev (1977).

their value rather limited, especially when no peytral was used to protect the steed's chest. Although Xenophon (*Cyr.* 6.4.1) mentions armoured frontlets, they are not depicted in Achaemenid art. Also, attaching heavy sheets of metal to the horse's flanks with just one end would cause technical problems, like the flopping of the entire part in motion. The very existence of *parapleuridia* suggests that the saddles of the Achaemenid era were more developed than mere saddle-blankets. This is also suggested by the existence of breast straps and girths, which would hardly be necessary if there was not even a soft cushion or saddle. Even *parapleuridia* of hard leather would require a saddle of some kind. This suggest that the thighs and knees required the most protection and the feet, ankles, and lower shins were either of lesser importance, or too difficult to target. Perhaps shorter *parapleuridia* facilitated easier mounting.[49]

The word *parapleuridia* disappears early in the fourth century, which allowed Shahbazi[50] to conclude that the item was also withdrawn from use. It is difficult to accept this statement, as models of arms and armour were only gradually abandoned in pre-modern eras and discrepancies may naturally occur between literature, archaeology, and art. The element was allegedly replaced by the armoured leggings which might have allowed greater comfort in mounting and dismounting. The fact is, however, that armoured leggings are not frequently found in iconography. Scale leggings are often found in the tombs of Pontic Scythians, so the element used by the painters originated from the north rather than the east. The relief from the Pergamum metope shows the armour of either charioteers or heavy cavalry (the latter being more likely), which consists of a cuirass with lappets and *pteryges* (the set repeatedly occurring on Achaemenid depictions of armoured riders), laminated sleeves (some of the warriors shown on seals may wear that type of arm protection, too), and a helmet with a mask; the horse armour there is made up of a plumed chamfron and wide rectangular element of two rows of straps. A Hellenistic, probably Seleucid, figurine in the Louvre (inv. AO 25538) shows heavily armoured riders wearing muscle cuirasses with segmented sleeves and *pteryges* and hemispherical helmets. One of them has his feet covered with segmented leggings above the ankle, while the other's legs are badly damaged. Payava, in the scene with his father, does not wear any leg protection, and on his tomb's battle relief, the victorious rider, armed in the same way, is shown on a horse

49 A quite similar element was developed in late mediaeval and Renaissance jousting armour, which normally offered the additional protection of iron sheets for the fronts of the thighs and sometimes covered the lower legs with a hanging textile. This was not used in battle attire and was purely sporting equipment.

50 Shahbazi (2012).

with *parapleuridia*. In none of the three Hellenistic examples is real protection of the legs depicted, while one of the figurines shows the feet covered with what appears to be a form of armoured boots. Assuming a dose of uniformity between these examples, it would seem that Seleucid cataphracts had their legs uncovered from thigh to ankle. The wide peytral depicted on the metope from Pergamum might provide an explanation, if wide enough to cover the mount's chest and spread over the rider's legs. The Himyarite slab from the British Museum (inv. 2014,6002.1) shows a heavy rider whose legs are hidden under the front part of the mount's thick and heavy caparison. Therefore, even if the word *parapleuridia* went out of use, it is possible that the idea of protecting the legs of the riders with plastrons attached to the saddle and combined with a peytral survived into later antiquity.

Another type of leg covering of the armoured riders shown in Achaemenid art is the loose trousers and tight leggings reaching from the ankle to above the knee and cut into a sharp triangle, which suggests that they were affixed by a strap to a belt under the tunic. It is possible that such plain chaps were integrated with the footwear and perhaps created a kind of sock. This kind of riders' dress can be clearly seen on the stele from Manisa, the Çan sarcophagus, the pectoral in the Miho Museum, and the seal in the Metropolitan Museum of Art. The upper triangular extension might at first sight suggest a form of greaves, but the same element is worn by the rider depicted in full sculpture on the silver rhyton from Erebuni.[51] The so-called Median warriors in the Persepolis Apadana reliefs probably wear the same type of chaps. Therefore, the element should be considered a piece of the riding costume rather than armour. Naturally, thick chaps provided some protection in combat, but incomparable to metallic greaves, metallic laminated or scale leggings, or *parapleuridia*. Perhaps it is an over-reaching conclusion, but as the ankles of the riders with *parapleuridia* are tightly clad and no leggings are visible, they might be wearing the chaps as a standard piece of equine kit. It is clear that the warriors on the Hermitage seal and the Tatarli beam are not wearing even that kind of leg guard. It is also worth noting that the shin chap or a high boot attached by a strap to the belt remained in use in Iran into the Seljuk era, also in military contexts. The loose version of this type of covering is well attested in the famous Shapur cameo (Bibliothèque Nationale de France, inv. camée. 360). The legs of the armoured riders in Parthian and Sasanian times were usually depicted as covered with defensive leggings, lamellar as shown on the Tang-e Sarvak frieze or segmented as can be seen on the Firuzabad and Naqš-e Rostam friezes.

51 Treister (2015).

2.4 *Offensive Weapons*

The most common offensive weapon in the discussed examples is a cavalry lance, of a length allowing the rider to comfortably reach an infantry opponent. It is difficult to identify all examples with the short *palton* described by Xenophon (*Hell.* 3.4.14) as the main weapon of Persian cavalry. The lance wielded on the Çan sarcophagus is of considerable length, but the other depictions fit perfectly with a weapon fit both for hurling and hand-to-hand combat. The radical change in the principles of lance wielding was brought about by Alexander's invasion.[52] Darius III, seeing the advantage of longer Macedonian lances, ordered his troops to extend the length of the Persian weapons as well (Diod. 17.55.1). It seems, however, that the very idea of the long lances might have been borrowed by the Macedonians from the Scythians.[53] Warriors armed with long lances, usually held two-handed at hip level, dominated the scenes of mounted combat in the Arsacid and Sasanian eras. The shorter-shafted weapons remained in the iconography of heroes fighting beasts, represented on toreutics. Because the proportions of Parthian and Sasanian art are customarily not reliable in assessment of the actual sizes, it is difficult to state whether the longest lances shown in hunting scenes are of the same type as those used in battle, or represent a slightly shorter variety. The scenes depicting hunting on foot include shorter spears, while javelins are hurled by both riders and foot-hunters. The latter weapon is indirectly confirmed in hands of heavy cavalry, as the emperor Julian was mortally wounded by an apparently Persian *hasta equitata* during his combat with Iranian heavy cavalry. In Ammianus' account (25.3.6), this is the only instance when a lance different from a *contus* is associated with Sasanian cavalry.[54] Even if the weapon was hurled by a Roman soldier, it could not have been very different from the Iranian equivalents in order to be identified as such.

The two-handed grip on the lance prevented the employment of a shield. In terms of weapon-wielding this was a great advantage, while the compromised protection was counteracted by strengthened armour. Nevertheless, shields did not go out of use completely. They were used by heavy riders as protection against arrows and in close combat with swords or maces. The combination of shorter lances and bucklers was typical for Arabic warriors, as attested by art from Palmyra, Hatra, and Dura-Europos, but the same combination is shown in the hands of the heavily armoured rider of the Iranian style on the Himyarite relief. The combination of buckler and spear in the hands of the rider, whether armoured or not, is a characteristic of armament predominant

52 Olbrycht (2015) 369–70; Bugh (2020).
53 On Scythian long lances, see Chernenko (1984).
54 Nikonorov (2005) 158–59.

in the pre-Islamic Arab Peninsula and Eastern Syria, but was seemingly adopted by the Sasanians at a late stage, as evidenced by the monumental rider in the great aivan of Taq-e Bostan. This, however, may be an adoption under Byzantine influence, as the shield carried by the rider is significantly bigger than Arabic bucklers, though not as large as the shield prescribed by the Byzantine manuals or depicted in artwork.

There is only one depiction of a sword in portrayals of Achaemenid armoured riders, on the Çan sarcophagus. It is not shown in combat but left securely sheathed, and instead of a *makhaira*, prescribed by Xenophon for cavalrymen (*Eq. mag*. 12.11), it is a typical Persian *akinakes*, also worn by the riders on the Erebuni rhyton and several hunting seals. It seems that this was a weapon of prestige and special significance, perhaps even an insignia of high social and military status.

The swords in the Arsacid era grew in length, probably under influence from China, where long-bladed swords appeared in pre-Qing times and gained popularity during the Han dynasty. The late Arsacid and Sasanian swords reveal similarity to the Chinese specimens, not only in the system of suspension by scabbard slide[55] and in sheer length, but also in the employment of mushroom shaped pommels. The innovation must have been transferred through the Kushan state. The adoption of long-bladed swords was an obvious response to the requirements of cavalry combat, where reach was of utmost importance, but also required more advanced metallurgical techniques, as a longer, narrower blade was easier to break. The swords depicted in early Arsacid coinage happen to be short and wide, reminiscent of modern Iranian kindjals (*qameé*), and seemingly not unrelated to Roman *gladii*. The relation seems superficial and functional rather than genetic (similar usage generates similar tools), but technological transfer cannot be excluded, as a combination of Roman and Iranian armament is attested in Palmyra. However, the length of the swords extended in the iconography of the common era. The swords of the Sasanian dynasty make a functionally unified group, differing in the shape of guards, pommels, and sheath decoration.[56] Masia believed that the small pommels enforced a certain technique of wielding the sword with the forefinger hooked by the cross-guard.[57] This idea does not seem fully reliable. Blades of such length would require large, metallic pommels to balance, while the Sasanian depictions show them as relatively small. Also, preserved pommels are not metallic but of semiprecious metals, while the preserved Arsacid and

55 Trousdale (1975).
56 Skupniewicz (2019).
57 Masia (2000).

Sasanian blades have neither metallic nor stone pommels, suggesting that there were either no pommels at all or ones made of organic materials. Also, the habit of hooking the cross-guard with the index finger was historically attested for different kinds of swords, often light ones. The idea of a very balanced weapon is required only in a particular paradigm of fighting, where control over movement is preferred over the power of momentum.[58] Momentum was a particularly important element for a horseman, whose advantage was velocity and who, while galloping, was less concerned about recovery after the cut but placed a crucial importance on the immediate power of a single blow. The pommels might not necessarily balance the blades, but would prevent the hilt from slipping away without compromising the momentum. Hooking the cross-guard with the index finger would work similarly, and the special place to lock the finger on the handle of late or post-Sasanian swords, which have a minimalist pommel integrated into the hilt, served the same purpose. The suspension system of Parthian and Sasanian long-bladed swords was the scabbard slide, an element adopted together with the type of blade from East Asia. The swords identified as Sasanian used P-shaped straps as a suspension system.[59]

On the pectoral in the Miho Museum and on the beam from the Tatarli tumulus, the riders were shown as archers, which is the role usually ascribed to light, unarmoured cavalry. The riders on the Tatarli beam may represent unarmoured warriors in strangely short tunics, but it is more likely that they are wearing light, soft-padded forms of armour. They all use double recurved composite bows of a 'Scythian' type suitable for shooting from horseback. This shows again that archery equipment was employed by both armoured and unarmoured riders already in Achaemenid period.

A majority of depictions of heavily armoured Parthian and Sasanian riders include archery equipment, even if no piece of iconography definitively presents a rider wearing armour while shooting. The relief in the Metropolitan Museum was mentioned earlier, but there are doubts about the amount of modern reconstruction involved. The graffito from Dura-Europos is rather crude and indeed is likely to depict an armoured rider, but the elements are unclear. Nevertheless, there is little doubt that Arsacid and Sasanian 'cataphracts' became archers when needed. Parthians preferred a developed form of the former *gorytos* which combined the bow-case with two narrow tubes for arrows. With the Sasanian dynasty came a fashion for large tubular quivers, while the bows were carried most likely unstrung. The iconography of hunting scenes suggests that the idle strung bows were carried either in the hand, on

58 Skupniewicz (2019).

59 Skupniewicz (2019).

the shoulder, or on the neck. In the late sixth/early seventh century, hour-glass quivers were adopted from the Turks.

Despite changes in technological details, the armament of Iranian armoured riders changed in an evolutionary rather than a revolutionary manner. The armoured riders known from Achaemenid iconography differ from their Parthian and Sasanian successors in types of archery equipment, length and types of swords, sizes of lances, and the manufacturing technique of the protective corselets, but the paradigm seemingly remained intact. Rapid movement and the ability to use bows and lances remained features of the Iranian art of war. This cannot be explained by mere 'conservatism' but by unchanged strategic challenges to which the army had to respond. Stretched between the great steppe and the Mediterranean world, Iran required highly mobile troops ready to face large nomadic hordes and the drilled infantry forces of Greece, Macedon, and Rome.

3 The Iconography of Mounted Combat

The compositional formulae of mounted combat reveal surprising longevity throughout Iranian antiquity, and are related in many cases to Classical Greek and Hellenistic art; thus, the resemblance between the visual structures of Roman cavalry tombstones and Iranian iconography seems justified. Naturally, it is possible to trace the lines of development, and there are changes enforced by tactical evolution, but they seem to remain within a relatively small number of repetitive motifs or designs, adopted to the current tastes. The following formulae can be distinguished:

3.1 *Rider Confronting an Enemy Standing on Foot*

This formula includes scenes where the rider confronts the infantryman running towards him. Usually, the footman is shown larger than natural to match the *isokephalia*. Examples showing the rider using an overhead thrust, or hurling the weapon, can be found in Achaemenid Anatolian seals,[60] the Thracian painting in the Alexandrova tumulus,[61] the Scythian golden plaque from the Geramesova kurgan,[62] and Etruscan urns.[63] Examples of the underhand thrust,

60 Tuplin (2020) 423–29, 450–52, figs. 42–50; Boardman (1970) 303–57; Vassileva (2010) 37–46; Skupniewicz (2018) 324–25; Skupniewicz (2020).

61 Vassileva (2010) 39–44.

62 Piotrovsky and Galanina (1986) 92–93, pl. 136.

63 Pirson (2014) 252–74.

suggesting the employment of a longer-shafted weapon, are found in the Kinch mural,[64] the bullae from Old Nisa,[65] and a now-lost fragment of metallic decoration from the same site.[66] The same design was used in Sasanian hunting scenes which will be briefly discussed below. Although Nikonorov's classification of the schemes employed in Sasanian battle reliefs applies to cavalry combat, 'Rider confronting an enemy standing on foot' corresponds to his 'confrontation scheme' and was probably its origin.[67]

3.2 *Rider Confronting a Standing Foot Enemy with Dead Body Stretched Horizontally Below*

This formula is dominant in Sasanian silver, with its several varieties.[68] However, it clearly derives from much earlier battle scenes, as evidenced by the seal of Cyrus of Anshan (Cyrus the Great's grandfather),[69] the relief on the Payava tomb, and Anatolian seals.[70]

3.3 *Rider Attacking the Victim Shown Obliquely*

This model gained great popularity, as it allows clear, definite, and immediate determination of the victor, a feature seemingly more important in Iranian aesthetics than dramatic ambiguity. In his classification of the iconographic models used in Sasanian battle reliefs, Nikonorov classified this moment as the 'culmination scheme'.[71] The diagonal direction of the attack was shown on the neck and suspension shield of the sheath from Oxus.[72] In Achaemenid art, it was applied to hunting scenes as well as battles, of which the scenes from the Çan sarcophagus, a seal in the Metropolitan Museum of Art,[73] and the 'Alexander sarcophagus' from Sidon are the most renowned. Presenting the victim kneeling on one leg and with outstretched arms, the second probably reflects the rhetoric of the visual language of the time, but it simultaneously tilted the silhouette and marginalised it within the pictorial field. Later this

64 Chaniotis (2005) 196, il. 10.1.
65 Gaibov and Košelenko (2008).
66 Pilipko (2001) 322.
67 Nikonorov (2020) 226.
68 Skupniewicz (2020).
69 Harper (2006) 14–18, 46, fig. 16.
70 Tuplin (2020) 420–22, 450, fig. 39–41. Outside of Iranian contexts the scene can be found in Assyrian art, where the formulae are included in narrative schemes and do not fully correspond with their application in Iranian iconography. The formula is also represented in Greek and Thracian art.
71 Nikonorov (2020) 220, 226.
72 Stronach (1998); Boardman (2006) 116–19.
73 Inv. No. MMA 41.160.653; Tuplin (2020) 427, 451, fig. 47.

posture was supplemented with the figure of the dead horse, which allowed identification of the victim as a cavalryman. In the authors' opinion, the persons depicted over fallen horses in Sasanian rock reliefs, identified by von Gall as 'fleeing' and later followed by Skupniewicz, are in fact a development of this formula.[74]

3.4 *Confrontation between Riders*

Direct confrontation between two riders, or the 'confrontation scheme' in Nikonorov's terminology,[75] seemingly did not gain much popularity in Iran. In ancient Iranian aesthetics the heraldic layout was found quite attractive, and its popularity in the West might have influenced a conditional adoption where duelling riders clash and one is already hit.[76] In narrative terms, this refers to Nikonorov's 'culmination scheme', but remained visually close to the 'confrontation scheme'.

3.5 *Rider Striking Vertically Downward*

The galloping rider thrusting at a target stretched out below is a motif found in Sasanian toreutics which did not gain wider popularity in other media.[77] Most likely, it adapted the idea of a standing monarch on foot, thrusting at a fallen enemy below him, to a mounted scene. It may have also been influenced by Romano-Byzantine aesthetics.

3.6 *Pursuit of a Fleeing Mounted Enemy*

The motif of chasing an opponent who attempts to escape his doom is likely of Assyrian origin and is attested in Achaemenid art.[78] Although some examples of Sasanian reliefs echo the formula, they seem to combine features of the models 'Rider attacking the victim shown obliquely' and 'Pursuit of a fleeing mounted enemy' rather than adapting the old model to a new reality, as proposed by von Gall and sustained by Skupniewicz but recently contested by Nikonorov.[79]

In Sasanian art the formulae developed to depict combat between warriors were applied mainly to scenes of heroic combat with dangerous beasts, with only few examples depicting warfare. The actual combat scenes are almost

74 Von Gall (1990); Skupniewicz (2015) contra Nikonorov (2020).

75 Nikonorov (2020) 221–25.

76 Skupniewicz (2015).

77 Skupniewicz (2020).

78 Bernard and Inagaki (2000); Casabonne and Gabrielli (2006); Tuplin (2020) 417–19, 449–50.

79 Von Gall (2008) 149; Skupniewicz (2015); Nikonorov (2020).

exclusively reserved for royal rock reliefs, while hunting depictions are found mainly in toreutics and sigillography.[80] Application of the formula to hunting scenes occurred quite early, which suggests a mutual exchangeability of the topic embedded in the formula itself. Here, examples are known from the golden sheath from the Oxus treasure and the Parthian tile found in Babylonia in the British Museum. The Palmyran mosaic associated with Odaenathus is the example chronologically closest to Sasanian art, which already adapts the full formula to scenes of men fighting beasts.[81] Clearly the hunting scenes developed from the combat pattern, as they position four-legged animals in an unnatural, vertical manner.[82] The late or post-Sasanian relief discovered in Taq-e Bostan, published by Moradi and Compareti, is evidence that application of the formulae to battle scenes was either revived in the later period, or had continued in non-extant media.[83]

One of the reasons for the shift from combat to hunt could be the intrinsic connection of the formulae with the rider's triumph over infantrymen. It seems likely that in late Arsacid and Sasanian reality, infantry's social inferiority diminished the meaning of the depictions of triumph over it. Sasanian kings were warriors far too great to be bothered with killing commoners on foot, and their glory came from defeating equal opponents. The employment of very long lances held with both hands at hip level also affected the layouts. The formulae were already modified to the needs of mounted combat, but at a cost – the scenes lost their compact nature and could not be framed within a circle or square, which made them unsuitable for decoration of plates or stamp seals. The reappearance of the modified formula in the late/post-Sasanian environment may come from the reintroduction of elite infantry troops such as the Daylamites in Iranian forces, or a general increase of infantry in the late period. The desperate measures, including enlisting every man, taken before the battle of Nihavand, which echo the apocalyptic picture of King Vištaspa's struggle depicted in *Āydāgar e Zāreran*, make it clear that the army could no longer consist mainly of cavalry. Similarly, the Arab foe used large infantry forces, so the readoption of the old formula may have reflected the actual situation. Also, centuries of using the model to depict the hunt might additionally dehumanise the opponents of the victorious horsemen. As said above, it is

80 Harper (2006); Harper and Meyers (1980); Skupniewicz (2020).

81 Gawlikowski (2005); Skupniewicz (2020).

82 Further examples can be found on depictions of the lion hunt from a chariot, both on Assyrian reliefs and Achaemenid seals, but including chariotry in the current consideration seems a digression too far.

83 Moradi and Compareti (2019); Skupniewicz (2020).

possible that the combat scenes which included riders defeating infantry were not really abandoned but found no representation in examples discovered so far. When Gaibov and Košelenko published their study on the depiction of the rider attacking a foot soldier on the seals found in Old Nisa, they correctly emphasised the novelty of the topic in Parthian art without placing the formula in the evolutionary chain between the Achaemenids and the Sasanians.

4 Conclusion

The Iranian artistic representations of cavalry arms and armour subordinated the actual realities of battle to the needs of visual language. The selection of particular items in the artwork of different dynasties was governed by their semantic content within the frames of compositional formulae. The message had to fit into pictorial formats, which limited the choice of available layouts. Just as descriptions of arms and armour in literature are ruled by literary *topoi*, as the authors demonstrated in the case of Heliodorus' *Aethiopica*,[84] the visual material has its own principles which must be taken into consideration when attempting to reconstruct the actual arms and armour of a particular time. The most reliable sources for reconstructing actual armament would be archaeological finds. Otherwise, research must emphasise that iconography is in the first place art and descriptions are in the first place literature; it is their secondary role, somewhat accidental, to document the events and practices of their time. The preference for depicting high-status personages wearing cuirasses instead of other forms of armour had a cultural aspect. Similarly, the Achaemenid preference for showing the noblemen carrying *akinakai* over single-edged swords seems connected to their function as status markers. The length of lances in art and their use in combat were modified with respect to the actual fashion of the era, but the depiction of weapons being broken had symbolic dimensions.[85] In depictions of cavalry and other elements of ancient warfare, assessment of iconography's documentary value can become reliable only after aesthetic analysis.

84 Skupniewicz and Maksymiuk (2018).
85 Skupniewicz (2018) 330.

Bibliography

Aldrete, G.S., Bartell, S.M., and Aldrete, A. (2013) *Reconstructing Ancient Linen Body Armor: Unraveling the Linothorax Mystery*. Baltimore.

Balakhvantsev, A.S. and Makkaveev, N.A. (eds) (2019) *Эпоха империй. Восточный Иран от Ахеменидов до Сасанидов: история, археология, культура. Материалы международной научной конференции, посвященной памяти Бориса Анатольевича Литвинского*. Moscow.

Baughan, E. (2008) 'Persian Riders in Lydia? The Painted Frieze of the Aktepe Tomb Kline'. *Proceedings of the XVII International Congress of Classical Archaeology*: 24–36.

Bemmann, J. and Schmauder, M. (eds) (2015) *Complexity of Interaction along the Eurasian Steppe Zone in the First Millennium CE*. Bonn.

Bernard, P. (1964) 'Une pièce d'armure perse sur un monument lycien'. *Syria* 41.3–4: 195–212.

Bernard, P. and Inagaki, H. (2000) 'Un torque achéménide avec une inscription grecque au musée Miho (Japon)'. *CRAI* 144: 1371–1437.

Bittner, S. (1985) *Tracht und Bewaffnung persischen Heers zur Zeit der Achaimeniden*. Munich.

Bivar, A.D.H. (1972) 'Cavalry Equipment and Tactics on the Euphrates Frontier'. *DOP* 26: 273–91.

Boardman, J. (1970) *Greek Gems and Finger Rings: Early Bronze Age to Late Classical*. London.

Boardman, J. (2000) *Persia and the West: An Archaeological Investigation of the Genesis of Achaemenid Art*. London.

Boardman, J. (2001) *Greek Gems and Finger Rings: Early Bronze Age to Late Classical*. 2nd edn. London.

Bongard-Levin, G. (ed) (1982) *Древняя Индия: Историко-культурные связи*. Moscow.

Brice, L.L. (ed) (2020) *New Approaches to Greek and Roman Warfare*. Hoboken, NJ.

Bugh, G.R. (2020) 'Greek Cavalry in the Hellenistic World: Review and Reappraisal', in Brice (2020) 65–80.

Canepa, M.P. (2013) 'Sasanian Rock Reliefs', in Potts (2013) 856–77.

Casabonne, O. and Gabrielli, M. (2006) 'Brèves remarques sur un torque achéménide au musée Miho (Japan)'. *Colloquium Anatolicum* 5: 85–90.

Chaniotis, A. (2005) *War in the Hellenistic World: A Social and Cultural History*. Oxford.

Chardin, J. (1988) *Travels in Persia 1673–1677*. New York.

Charles, M.B. (2015) 'Achaemenid Elite Cavalry: From Xerxes to Darius III'. *CQ* 65: 14–34.

Chernenko, E.V. (1984) *Длинные копья скифов, Древности Евразии в скифо-сарматское Время*. Moscow.

Cohen, A. (1997) *The Alexander Mosaic: Stories of Victory and Defeat*. Cambridge.

Collins, A. (2012) 'The Royal Costume and Insignia of Alexander the Great'. *AJPhil* 133: 371–402.

Connolly, P. (1984) 'Hellenistic Warfare', in Ling (1984) 81–90.

Dusinberre, E. and Garrison, M. (eds) (2020) *The Art of Empire in Achaemenid Persia. Festschrift in Honor of Margaret Cool Root*. Leiden.

Erofeeva, I.V. (ed) (2010) *Роль номадов евразийских степей в развитии мирового военногоискусства. Научные чтения памяти Н.Э. Масанова*. Almaty.

Farrokh, K., Karamian, G., and Maksymiuk, K. (2018) *A Synopsis of Sasanian Military Organization and Combat Units*. Siedlce.

Gabrielli, M. (2006) *Le cheval dans l'Empire Achéménide*. Istanbul.

Gaibov, V.A. and Košelenko, G.A. (2008) 'A Horseman Charging a Foot Soldier: A New Subject in Parthian Glyptic Art'. *Parthica* 10: 99–107.

Garrison, M.B. (2013) 'Royal Achaemenid Iconography', in Potts (2013) 566–95.

Gawlikowski, M. (2005) 'L'apothéose d'Odeinat sur un mosaïque récemment découverte à Palmyre'. *CRAI* 149: 1293–1304.

Ghirshman, R. (1962) *Persian Art. 249 B.C.–A.D. 651: The Parthian and Sassanian Dynasties*. New York.

Ghirshman, R. (1964) *The Arts of Ancient Iran: From Its Origins to the Time of Alexander the Great*. New York.

Ghose, R. (2008) 'The Kizil Caves: Date, Art, and Iconography', in Ghose (2008) 40–53.

Ghose, R. (ed) (2008) *Kizil on the Silk Road: Crossroads of Commerce & Meeting of Minds*, Mumbai.

Gorelik, M.V. (1982a) 'Кушанский доспех: [О панцирном защит. вооружении воинов Кушан. империи]', in Bongard-Levin (1982) 82–112.

Gorelik, M.V. (1982b) 'Защитное вооружение персов и мидян ахеменидского времени'. *Vestnik Drevneī Istorii* 3 (161): 90–106.

Harper, P.O. and Meyers, P. (1981) *Silver Vessels of the Sasanian Period. Vol. 1: Royal Imagery*. New York.

Harper, P.O. (2006) *In Search of a Cultural Identity: Monuments and Artifacts of the Sasanian Near East, 3rd to 7th Century A.D.* New York.

Hauser, S. (2006) 'Was There No Paid Standing Army? A Fresh Look on Military and Political Institutions in the Arsacid Empire', in Mode and Tubach (2006) 295–319.

Head, D. (1992) *The Achaemenid Persian Army*. Stockport.

Hölscher, T. (2003) 'Images of War in Greece and Rome: Between Military Practice, Public Memory, and Cultural Symbolism'. *JRS* 93: 1–17.

Kawami, T.S. (1987) *Monumental Art of the Parthian Period in Iran*. Leiden.

Kawami, T.S. (2013) 'Parthian and Elymaean rock reliefs', in Potts (2013) 751–65.

Kennet, D. and Luft P. (eds) (2008) *Current Research in Sasanian Archaeology, Art and History. Proceedings of a Conference held at Durham University, November 3rd and 4th, 2001 Organized by the Centre for Iranian Studies, IMEIS and the Department of Archaeology of Durham University*. Oxford.

Kubik, A. (2020) 'A New View on the Possible Reconstruction of the "Famous Clibanarius" Graffito from Dura Europos', in Niknami and Hozhabri (2020) 309–18.

Ling, R. (ed) (1984) *The Cambridge Ancient History: Plates to Vol. VII, Part 1 (new edition)*. Cambridge.

Ma, J. (2008) 'Mysians on the Çan Sarcophagus? Ethnicity and Domination in Achaimenid Military Art'. *Historia* 57: 243–54.

Maksymiuk, K. and Karamian, G. (eds) (2017) *Crowns, Hats, Turbans and Helmets: The Headgear in Iranian History. Volume I: Pre-Islamic Period*. Siedlce.

Mambetullaev, M. (1977) 'Рельефное изображение всадника на керамической фляге из Хамбуз-Тепе'. *Sovetskaia Arkheologiia* 3: 278–81.

Manning, S. (2021) *Armed Force in the Teispid-Achaemenid Empire: Past Approaches, Future Prospects*. Stuttgart.

Masia, K. (2000) 'The Evolution of Swords and Daggers in the Sasanian Empire'. *IrAnt* 35: 185–290.

Mielczarek, M. (1993) *Cataphracti and Clibanarii: Studies on the Heavy Armoured Cavalry of the Ancient World*. Łódź.

Miller, S.G. (1993) *The Tomb of Lyson and Kallikles: A Painted Macedonian Tomb*. Mainz.

Minta-Tworzowska, D. (ed) (2018) *Estetyka w archeologii. Obrazowanie w pradziejach i starożytności*. Gdańsk.

Mode, M. and Tubach, J. (eds) (2006) *Arms and Armour as Indicators of Cultural Transfer: The Steppes and the Ancient World from Hellenistic Times to the Early Middle Ages*. Wiesbaden.

Moradi, Y. and Compareti, M. (2019) 'A Sasanian Figured Relief Plaque from Taq-e Bostan'. *Parthica* 21: 179–88.

Nicolle, D. (1996) Sassanian Armies: The Iranian Empire, Early 3rd to Mid 7th Centuries AD. Stockport.

Niknami, K.A. and Hozhabri, A. (eds) (2020) *Archaeology of Iran in the Historical Period*. Cham.

Nikonorov, V.P. (1997) *The Armies of Bactria 700 BC–450 AD*. Stockport.

Nikonorov, V.P. (2005) 'К вопросу о парфянском наследии в сасанидско Иране: военное дело', in Nikonorov (2005) 141–79.

Nikonorov, V.P. (2020) 'Сасанидские боевые рельефы и происхождение темы конной дуэли на пиках в прокламативном искусстве доисламского Ирана'. *Arkheologicheskiye Vesti* 29: 215–38.

Nikonorov, V.P. (ed) (2005) *Центральная Азия от Ахеменидов до Тимуридов: археология, история, этнология, культура. Материалы международной научной конференции, посвященной 100-летию со дня рождения Александра Марковича Беленицкого (Санкт-Петербург, 2–5 ноября 2004 года)*. St. Petersburg.

Olbrycht, M. (2010) 'К вопросу о происхождении конницы катафрактов в Иране и Средней Азии', in Erofeeva (2001) 66–85.

Olbrycht, M. (2015) 'Arsacid Iran and the Nomads of Central Asia – Ways of Cultural Transfer', in Bemmann and Schmauder (2015) 333–90.

Pilipko, V.N. (2001) *Старая Ниса. Основные итоги археологического изучения в советский период*. Moscow.

Piotrovsky, B. and Galanina, L. (1986) *Scythian Art: The Legacy of the Scythian World: Mid-7th to 3rd Century B.C.* Leningrad.

Pirson, F. (2014) *Ansichten des Krieges: Kampfreliefs klassischer und hellenistischer Zeit im Kulturvergleich*. Wiesbaden.

Porada, E. (1962) *Alt-Iran. Kunst der Welt*. Serie 3. Zurich.

Potts, D.T. (2007) '*Cataphractus* and *kamandār*: Some Thoughts on the Dynamic Evolution of Heavy Cavalry and Mounted Archers in Iran and Central Asia'. *Bulletin of the Asia Institute* 21: 149–58.

Potts, D.T. (ed) (2013) *The Oxford Handbook of Ancient Iran*. Oxford.

Pugačenkova, G.A. (1971) *Скульптура Халчаяна*. Moscow.

Schlumbacher, D. (1983) 'Parthian Art', in Yarshater (1983) 1027–54.

Sekunda, N.V. (1992) *The Persian Army 560–330 BC*. Oxford.

Sevinč, N., Korpe, R., Tombul, M., Rose, C.B., Strahan, D., Kiesewetter, H., and Wallrodt, J. (2001) 'A New Painted Graeco-Persian Sarcophagus From Çan'. *Studia Troica* 11: 383–420.

Shahbazi, A.S. 'Army i: Pre-Islamic Iran'. *Encyclopædia Iranica*, II/5, pp. 489–499, available online at http://www.iranicaonline.org/articles/army-i (accessed on 30 December 2012).

Shepherd, D. (1983) 'Sasanian Art', in Yarshater (1983) 1055–1112.

Skupniewicz, P. (2006) 'O ciężkozbrojnej jeździe Sasanidów'. *Acta Universitatis Nicolai Copernici, Archeologia* 30: 151–74.

Skupniewicz, P. (2015) 'The Iconographic Function of Armor in Sasanian Art'. *Rivista degli studi orientali* 88.1/4: 251–81.

Skupniewicz, P. (2016) 'The Himyarite "Knight" and Partho-Sasanian Art'. *Historia i Świat* 5: 57–75.

Skupniewicz, P. (2018) 'Hellenistyczne estetyczne modele ujmowania walki konnej a Mozaika Aleksandra w kontekście eschatologicznym. Studium formalne', in Minta-Tworzowska (2018) 317–42.

Skupniewicz, P. (2019) 'On the Shortfalls of Iconography in the Reconstruction of Swordsmanship: The Sasanian Case', in Balakhvantsev and Makkaveev (2019) 195–212.

Skupniewicz, P. (2020) 'Scene of Fighting Tigers on a Sasanian Plate from Mes 'Aynak: Notes on the Composition'. *Acta Archaeologica Lodziensia* 66: 65–84.

Skupniewicz, P. (2021a) 'The Bow as an Insignia of Power in the Art of Ancient Iran'. *Historia i Świat* 10: 153–70.

Skupniewicz, P. (2021b) 'Notes on the Combat Scene on Tang-e Sarvak III Rock Relief'. *Acta Archaeologica Lodziensia* 67: 117–45.

Skupniewicz, P. (2022) 'The Panoply of the Armoured Personage from the Capital from Bisotun: Remarks on the Spread of the Military Technology along the Silk Road'. *Acta Archaeologica Lodziensia* 68: 165–87.

Skupniewicz, P. and Lichota, M. (2017) 'Diadem on the Head from Khalchayan Battle Scene and Possible Reconstruction of the Composition', in Maksymiuk and Karamian (2017) 69–95.

Skupniewicz, P. and Maksymiuk, K. (2018) 'Persian Riders in the Aethiopica of Heliodorus: A Reliable Source?'. *Historia i Świat* 7: 99–110.

Skupniewicz, P. and Maksymiuk, K. (2019) 'Gordāfarid, Penthesilea and Athena: The Identification of a Greek Motif in Ferdowsī's Šāh-nāma and Its Possible Association with Hellenistic Art in the East'. *MHR* 34: 123–43.

Stronach, D. (1998) 'On the Date of the Oxus Gold Scabbard and Other Achaemenid Matters'. *BAI* 12: 231–48.

Summerer, L. (2007) 'Picturing Persian Victory: The Painted Battle Scene on the Munich Wood', in Ivantchik and Licheli (2007) 3–30.

Tafażżoli, A. (2000) *Sasanian Society: I. Warriors II. Scribes III. Dehqāns*. New York.

Treister, M.Y. (2015) 'A Hoard of Silver Rhyta of the Achaemenid Circle from Erebuni'. *Ancient Civilizations from Scythia to Siberia* 21: 23–119.

Trousdale, W. (1975) *The Long Sword and Scabbard Slide in Asia*. Washington.

Trundle, M. and Fagan, G. (eds) (2010) *New Perspectives on Ancient Warfare*. Leiden.

Tuplin, C. (2010) 'All the King's Horse: In Search of Achaemenind Persian Cavalry', in Trundle and Fagan (2010) 100–82.

Tuplin, C. (2020) 'Sigillography and Soldiers: Cataloguing Military Activity on Achaemenid Period Seals', in Dusinberre and Garrison (2020) 329–459.

Vassileva, M. (2010) 'Achaemenid Interfaces: Thracian and Anatolian Representations of Elite Status'. *Bollettino di Archeologia* vol. speciale 1: 39–44.

von Gall, H. (1990) *Das Reiterkampfbild in der iranischen und iranisch beeinflußten Kunst parthischer und sasanidischer Zeit*, Berlin.

von Gall, H. (2008) 'New Perspectives on Sasanian Rock Reliefs', in Kennet and Luft (2008) 149–61.

Wójcikowski, R.S. (2013) 'The Graffito from Dura-Europos: Hybrid Armor in Parthian-Sasanian Iran'. *Anabasis* 4: 233–48.

Wolski, J. (2000) 'Servitia, ich funkcja i znaczenie w społeczeństwie irańskim okresu Arsacydów'. *Electrum* 4: 95–100.

Woźniak, M.A. (2010) *Armie starożytnej Persji. Od powstania państwa Achemenidów do upadku imperium sasanidzkiego*. Zabrze.

Wu, X. (2014) 'Warfare, History, and Elite Ideology of the Achaemenid Persian Empire'. *IrAnt* 49: 209–99.

Yarshater, E. (ed) (1983) *The Cambridge History of Iran. Vol. 3: The Seleucid, Parthian and Sasanian Periods. Part 2*. Cambridge.

CHAPTER 16

Peroz and the Hephthalites … and the Kidarites, Alkhans, and Sasanians: Warfare, Diplomacy, and the Complexity of the Sasanian East in the Fifth Century

Robert Haug

As it is remembered in the available sources, the reign of the Sasanian *shāhānshāh* Peroz (r. 459–484) was defined from beginning to end by conflicts in the east. Peroz achieved the imperial crown with the support of eastern allies, typically identified as Hephthalites. He was later killed in battle against the same Hephthalites. In between, Peroz led several campaigns to the east against both Kidarites and Hephthalites. The accounts of Peroz's reign present this as a fairly straightforward competition between an imperial Iran and a uniform and unified eastern neighbour, but this is overshadowing a great amount of complexity in the east – a region that included Khurasan, Tukharistan, Sogdiana, and Gandhara and witnessed rapid change over the course of the fifth century. At the same time, the introduction of eastern powers into the political dynamics of the Sasanian Empire continued to influence the politics of the empire up until its final moments when the last Sasanian emperor Yazdgird III (r. 632–651) fell afoul of a conspiracy between the *marzbān* of Marw, Mahuye, and the Hunnic ruler of Badhghis, the Nezak Ṭarkhan.[1] This article attempts to untangle the history of East Iran before, during, and immediately after the reign of Peroz, making use of literary sources along with numismatics, archaeology, and documentary evidence to examine the Sasanians and their neighbours in a period of transition and change during the fifth and early sixth centuries.[2] It will take into account newer understandings of the political complexity of Tukharistan, the Hindu Kush, and Gandhara to re-evaluate Peroz's reign, his military and diplomatic relationships with his Hunnic neighbours,

1 See Haug (2019) 86–87.

2 After being long neglected by historians of Iran, the Sasanian East – including topics of both warfare and diplomacy along the eastern frontier – has been the focus of several recent studies including Potts (2018), Rezakhani (2017a and 2017b), Sims-Williams (2009), and Schindel (2006).

 | DOI:10.1163/9789004710771_017

and the aftermath of his catastrophic wars within the frame of competition between various Hunnic dynasties and the Sasanian Empire itself. While our literary sources tend to simplify the events in the east down to a competition between Sasanian Iran and unified, monolithic Hunnic powers such as the Hephthalites or the earlier Kidarites over Tukharistan,[3] finer analysis and the incorporation of documentary and numismatic sources point towards a much more complex negotiation between local and imperial powers on both sides of the frontier.

1 The Kidarites before Peroz

In the mid-fourth century, a group of people known collectively as the Iranian Huns migrated south and west from the steppes of Inner Asia into Tukharistan and from there expanded into Sogdiana and Gandhara. By the end of the century a series of Hunnic dynasties had seized control of the Sasanians' eastern frontier, pushing Sasanian authority south of the Amu Darya and west of the Murghab River. These Huns were not clearly defined by our available sources and modern historians and numismatists have divided them into several related groups, including the Kidarites, Alkhans, Nezaks, and Hephthalites, largely based on coin inscriptions. It is not clear if the Huns arrived in a single migration or if each new political dynasty that appeared across the fourth and fifth centuries represented a different 'wave' of migration.[4] A further possibility is that some of these peoples already inhabited the region and rose to political prominence following the breakdown of Kushano-Sasanian authority.[5] Attempts to reconstruct the history of any one of these groups have contended with the challenges of a dearth of written sources, which has forced a reliance on the combined work of numismatists, linguists, epigraphers, archaeologists, Iranologists, Sinologists, historians, and art historians. Still, we have not been able to recover even basic political histories for these dynasties. The most important sources left by the Hunnic dynasties themselves were coins, and

3 In this essay, a maximal definition of Tukharistan will be in use, including the lands east of the Murghāb River between the Amu Darya and the Hindu Kush. This region still closely approximates ancient Bactria and remains centred on the city of Balkh.

4 Grenet (2002, 205–207) summarises the argument for multiple waves, while de la Vaissière (2003, 122) explains the single wave theory.

5 The Kushano-Sasanians were a dynasty, most likely a cadet branch of the Sasanian royal family, that ruled Tukharistan and parts of eastern Khurasan or the Kushanshahr from the mid-third to mid-fourth centuries as vassals of the Sasanian Empire.

it was through the process of collecting and cataloguing these coins that we came to define the categories into which we divide the Iranian Huns today.[6]

The Huns first made their mark on Sasanian history during the long reign of Shapur II (r. 309–379), who led multiple campaigns to the east from 350 until 376. By the end of Shapur's reign, we can identify the Kidarites, a Hunnic dynasty named for a king Kidara, who appeared in coin inscriptions as early as the 370s.[7] Numismatic and archaeological evidence presents the Kidarites as the imperial unifiers of the peoples of Tukharistan, Sogdiana, and Gandhara including both indigenous populations and the Hunnic newcomers, comparable to their Xianbei contemporaries in northern China, the Northern Wei (r. 386–535).[8] For example, a clay bulla found in northwestern Pakistan from the end of the Kidarite period identifies a Kidarite ruler as 'king of the Huns, the great Kushan-shah, and the *afshiyan* of Samarkand', thereby claiming three distinct legacies and sources of authority tied to the Huns as a collective body, the historical imperial legacy of the Kushan Empire (r. 30–375), and the local authority of the Sogdian kingdoms.[9] Contradicting the popular notions of Huns as a destructive force, archaeological evidence from the period of Kidarite rule includes numerous large-scale construction projects, often with a military angle. For example, the Kidarites and Sasanians built a parallel line of fortified 'city fortresses' along their frontier, especially in Sogdiana.[10] The height of the citadel of Samarkand increased during the Kidarite period.[11] A great defensive wall was built around Bukhara with a circumference of over 250 km.[12] Paykand, remembered in the Arabic tradition as 'the city of merchants', was founded at this time as well, most likely for defensive purposes as it secured the Bukhara Oasis from the Amu Darya crossing at Āmul (Türkmenabat).[13] As Étienne

6 Much of this work began with Göbl (1967) but new discoveries and advances in understanding the Bactrian language have required revisiting the project with Vonderovec (2014) and Pfisterer (2014) deserving special attention. Seals have similarly broadened our picture of events on the ground thanks to the work of Judith Lerner and Nicholas Sims-Williams (2011).

7 On the debates around dating the rise of the Kidarites and tensions between written and numismatic evidence, see Cribb (2010) 95–96.

8 Grenet (2002) 209.

9 Rahman, Grenet, and Sims-Williams (2006); Lerner and Sims-Williams (2011) 181–82.

10 Rante and Collinet (2013) 204. For an overview of projects in Sogdiana, see Rezakhani (2017b) 150. We are lacking similar archaeological evidence from Tukharistan, but the available surveys tend to paint the opposite picture. See Rezakhani (2017b) 99; de la Vaissière (2005) 101–102.

11 Grenet (1996) 374.

12 De la Vaissière (2005) 105.

13 Rante and Raimkulov (2013) 253–55.

de la Vaissière has explained, the period of Hunnic control in Sogdiana – which he dates from 350 until the second half of the fifth century – was a period of 'rapid development' thanks to 'significant contributions of population and to certain political stability over the course of three generations'.[14]

Although we have scattered evidence for regular Sasanian campaigns in the east from the reign of Shapur II on up, it is only in the reign of Wahram V Gur (r. 420–38) that our literary sources take a strong interest in eastern warfare. During Wahram's reign the Sasanians had to repel an invasion by the Kidarites – though attributed anachronistically to the Turks in the later Islamic sources – which may have reached as far west as Rayy.[15] The reports of this invasion and Wahram's response contain certain literary inventions – he initially ignored the Huns and instead spent his time with idle pleasures while his people negotiated a peace with the 'khāqān' – but they generally provide an account that can be confirmed through other sources. In the literary sources, the people of Iran, fearing a full invasion, offered the 'khāqān' tribute and control over Marw.[16] A stylistic shift in coins from Herat and Marw during Wahram's reign signals a temporary loss of those mints.[17] When Wahram did engage the Kidarites militarily, he won a great victory during which the 'khāqān' was killed (by Wahram's own hand in some accounts), he captured the 'khāqān's' crown and diadem, enslaved many of the Huns' wives and children including the Kidarite queen, and sent forces across the Amu Darya to attack the Kidarite domains directly. As part of a peace agreement, Wahram fixed a boundary between the Sasanians and Kidarites by building a tower near Kushmayhan, south of the Amu Darya between Marw and Āmul, and appointing a *marzbān* or march warden to govern the frontier. This victory was celebrated across the empire and much of the loot was donated to the great fire temple of Ādur-i

14 De la Vaissière (2005) 105. Similar revisions are being made about the impact of the Huns in Gandhara as well, where scholars 'have often attributed a violent destruction of the Gandharan cultural environment' to the Hūṇas, a premise that is diminishingly defensible. Filigenzi (2010) 390.

15 The sources for Wahrām's war with the Kidarites include Ibn Qutayba (Okacha (1960)) 661; Dīnawarī (Guirgass and Kratchkovsky (1888)) 57–59; Ṭab. 1.863–65; Mas'ūdī (Meynard and Courteille (1861)) 2.190–91; Bal'amī (Zotenberg (1867)) 2.119–21; Ferdowsi (Warner and Warner (1905–1925)) 7.84–92; Tha'ālibī (Zotenberg (1900)) 558–60; Gardīzī (Ḥabībī (1984)) 76–77.

16 Of the literary sources cited above, Dīnawarī states that the 'khāqān' was in Marw when Wahrām arrived in Khurasan. Ṭabarī, Bal'amī, Ferdowsi, Tha'ālibī, and Gardīzī each include some version of the people of Khurasan offering tribute to the 'khāqān' (or erroneously the Khān of China in Ferdowsi's account) and the Turks seizing control of Marw and its environs.

17 Schindel (2006) 678–79.

Gushnasp in Azerbaijan. Beyond the literary accounts, Wahram's victory was also celebrated in the stucco panels found at the fire temple of Bandiyan near Darreh Gaz, southeast of modern Ashgabat.[18] It was likely in the aftermath of this campaign that Wahram founded the city of Marw al-Rūd as a garrison along the Murghab River, on the frontier between Khurasan and Tukharistan.[19]

The momentum of Wahram Gur's victory carried into the reign of his son and successor, Yazdgird II (r. 438–457). Yazdgird resided in the east for seven years (442–449), sometimes at Nishapur and others at the newly (re)built city of Shahr-i Qumis (ancient Hecatompylos).[20] During this time, Yazdgird ordered nobles from across the empire to come and fight in the east in two-year cycles.[21] In 450, Yazdgird led a campaign into Tukharistan that resulted in the capture of Talaqan from the Kidarites.[22] There are two regions of Tukharistan known as Talaqan, one west of Balkh and the second a town east of Qunduz. We may assume that Yazdgird captured western Talaqan because it was closer to the borders of the Sasanian Empire itself and therefore easier to add to the empire. A report by the fifth-century Armenian chronicler Łazar P'arpec'i of a victory by Yazdgird over the Kidarites at Marw al-Rūd suggests western Talaqan.[23] On the other hand, archaeological surveys around Qunduz show evidence of widespread destruction and burning in the early fifth century.[24] Excavations at the fortified village of Chaqalaq Tepe, 11 km south-southeast of Qunduz, have uncovered three separate layers of burning from the late fourth through early fifth century as well as the construction of a double rampart between them.[25] Even though we cannot tie the campaigns of Yazdgird specifically to any one of these layers of destruction, it is a good indicator that warfare had reached eastern Tukharistan in the fifth century. Yazdgird's campaign of 450 was cut short by a rebellion in Armenia. Many of the nobles and clergy imprisoned following this uprising were relocated to Nishapur, where they remained until 453 or 454 when Yazdgird returned to the east to campaign against the Kidarites again.

18 For an overview of the excavation of the fire temple at Bandiyān, see Rahbar (2008).

19 Daryaee (2002) 18, 37.

20 The text of *Shahristānīha-i Īrānshahr* says it was Yazdgird I who built the city during his campaigns against Čōl, which appears to be the name of a Hunnic ruler, but this appears to be a mistake and Yazdgird II is meant. Daryaee (2002) 18, 40.

21 Ełiše (Thomson (1982)) 63–66.

22 Ełiše (Thomson (1982)) 72. The translation gives 'the land of the T'etals' or Hephthalites as the location of Yazdgird's attack, but this would be too early for the Hephthalites to be a major presence in Tukharistan. In a note, Robert Thomson writes that the manuscript gives Italakan or Ṭālaqān.

23 P'arpec'i (Thomson (1991)) 124.

24 Lyonnet (1997) 283; de la Vaissière (2005) 101–102.

25 Higuchi and Kuwayama (1970) 26.

This time the Sasanians faced a humiliating defeat. In the aftermath, Yazdgird executed the Armenian prisoners and the Armenian sources on which we rely for this campaign view the campaign primarily as a prelude to a martyrdom narrative and therefore provide us with very little detail.[26] The Kidarites were able to force Yazdgird to pay them a tribute – which may have been the practice beginning with Wahram Gur – but in 457 he stopped and instead led another campaign into Tukharistan during which he died.

In the mid-fifth century, the Kidarites controlled the eastern frontier of the Sasanian Empire, unifying control over Tukharistan and Sogdiana while also exerting control south of the Hindu Kush in Gandhara. Conflict between the Kidarites and Sasanians went back and forth with neither side able to maintain the upper hand for long. From the Sasanian perspective, this may be the result of distractions in the west including political uncertainties after the sudden death (probable assassination) of Yazdgird I, wars with Rome under both Wahram Gur and Yazdgird II, and revolts in Armenia. The Sasanians relied upon mobile campaign armies that travelled east for the purposes of fighting campaigns against the Kidarites. Therefore, events in the west could easily tie up the forces needed for sustained eastern campaigning. Meanwhile, the Kidarites provided a certain level of stability in the east that allowed them to flourish, at least for a while. This situation changed rapidly in the mid-fifth century as the Kidarites were forced out of their domains by a new 'Hunnic' power, the Hephthalites. Although we often describe this transition as a simple replacement of the Kidarites by the Hephthalites with or without Sasanian assistance, the political situation in the east was not so straight forward. From the beginning, the Kidarites co-existed and fought with another less understood Hunnic dynasty, the Alkhans, who continued to play a role in the Hunnic east well into the Hephthalite period.

2 The Alkhans before Peroz

The Alkhans were a Hunnic dynasty identifiable primarily from their coinage, which included the Bactrian inscription *Ālkhān* and a consistent tamga – both understood as dynastic symbols rather than the name of an individual – and, overtime, replaced the Sasanian-style busts with that of a beardless, mustachioed figure with an elongated skull and prominent nose. The earliest Alkhan

26 Ełišē (Thomson (1982)) 192–94; P'arpec'i (Thomson (1991)) 133. These events should be seen within a larger pattern of repression against Christians under Yazdgird II, including the removal of Christians from the military.

coins were based on dies of Shapur II and Shapur III (r. 383–388). During Shapur II's campaigns in the east, the imperial mint was moved to Kabul, likely to pay either the Sasanian soldiers or provide tribute to the Huns.[27] It is assumed that, at some point during Shapur III's reign, the Alkhans had conquered and/or looted Kabul, capturing the dies in the process.[28] This would place the rise of the Alkhans parallel to that of the Kidarites, with the Alkhans establishing themselves along the southern face of the Hindu Kush, ruling Kabulistan/Kapisa, Zabulistan, and Gandhara, and also expanding into northern India and parts of eastern Tukharistan. In many of these places, the Alkhans would have been competing with the Kidarites for authority.

The Alkhans have traditionally been conflated with the Hephthalites, often conceptualised as the southern branch of the Hephthalites. Recently the Alkhans have been recognised as their own, independent though related authority.[29] Similarly, we have evidence of blended Alkhan-Kidarite rule, especially in Gandhara. Coins struck in the southern Hindu Kush show a mixture of Kidarite and Alkhan symbolism before the emergence of pure Alkhan issues, highlighting a process of differentiation between the two.[30] A pair of silver bowls dated to the 460s–470s – around the time of the collapse of the Kidarite kingdom – display both Alkhan and Kidarite symbols and figures with clear Alkhan and Kidarite features. These were likely produced in Gandhara. One was found in the Swat Valley of Pakistan, but the other was excavated at Chilek, 30 km north of Samarkand, demonstrating the interconnected nature of the Hunnic east from Gandhara to Sogdiana.[31] The lack of many written references to the Alkhans, less than even the paltry sources for the Kidarites or Hephthalites, has made reconstructing their history incredibly difficult.

The earliest Alkhan coins imply that they had established their authority around Kabul by the end of the fourth century. Their coins remained anonymous (beyond the Alkhan inscription) until the 430s when the name Khingila began to appear on coins in both Bactrian and Brahmi.[32] It is important to note that the use of two scripts implies that Alkhan coins were intended to circulate on both sides of the Hindu Kush. Beyond the different scripts, the

27 Schindel (2003) 1:236, 253–55.

28 Schindel (2006, 677) sees this as a catalyst for decentralising Sasanian minting which then led to the inclusion of normal mint marks on coins.

29 Rezakhani (2017b) 105.

30 For an exploration of this transition through the focused study of a single coin type, see Rezakhani (2015).

31 Grenet (2002) 211–12.

32 Vondrovec (2014) 184.

coins of Khingila appear in a wide variety of styles, which may indicate that they were struck at many different mints, pointing to a large and diverse domain for the Alkhans.[33] This variety of coin types could also indicate a long reign with Khingila's name appearing on the Schøyen Copper Scroll, dated to 492–93.[34] It is assumed that Khingila led the Alkhan conquest of Gandhara from the Kidarites.[35] The one reference to Khingila in our literary sources certainly associates him with Gandhara. In Ferdowsi's *Shahnameh,* Khingila appears as Shengil, the king of India whose court Wahram Gur visited in disguise and whose daughter he married.[36] Despite this clear association between the Alkhans and Gandhara, evidence including coin issues continued to affirm Kidarite authority in Gandhara as late as 477, and it is difficult to disentangle the two Hunnic powers.[37] Similarly, a series of campaigns against the Gupta empire of India in the mid-fifth century are attributed to the Kidarites despite evidence of Alkhan authority in the form of inscriptions in the name of the Alkhan Toramana – believed to be Khingila's son although this identification is currently in dispute – in central India.[38] Regardless, it is clear that the political situation in the east was much more complicated than our sources acknowledge, especially our literary sources. Next, we may re-read the reign of Peroz in light of this complexity.

3 Peroz and the Hephthalites, Part 1

In 457, Yazdgird II died of natural causes while campaigning against the Kidarites in the east. According to the later Islamic literary sources, his son Hormizd III (r. 457–459) succeeded him, but Hormizd's claim to the imperial crown was challenged by his brother Peroz.[39] Even though there are several contradictions in these sources when it comes to specific details and a notable lack of references to the war between the sons of Yazdgird in the more

33 Vondrovec (2014) 183.
34 Though it has been argued that not all the rulers named on this dedication scroll were alive at the time of its production; see da la Vaissière (2007) 128.
35 Vondrovec (2014) 184.
36 Ferdowsi (Warner and Warner (1905–1925)) 7.109–48.
37 Grenet (2002) 211.
38 For details on these inscriptions, see Rezakhani (2017b) 110–11. Indian sources refer to the invaders as Hūnā.
39 The sources for Peroz's conflict with Hormizd include Ibn Qutayba (Okacha (1960)) 661; Dīn. (Guirgass and I. Kratchkovsky (1888)) 60; Yaʿqūbī (Muhannā (2010)) 1.163; Ṭab. 1.871–72; Balʿamī (Zotenberg (1867)) 2.127–28; Ferdowsi (Warner and Warner (1905–1925)) 7.157–58.

contemporary Armenian sources,[40] the general narrative can be laid out as follows. Peroz fled to the east and eventually reached the domains of the Hephthalites, described by Dīnawarī as including Tukharistan, Chaghaniyan, Kabulistan, and the lands behind the great river of Balkh (meaning the Amu Darya), to which Balʿamī added Gharchistan and Badakhshan.[41] Peroz's decision may have been influenced by his position as governor of Sistan,[42] keeping him away from the centres of power but close to the frontier. Peroz enlisted the aid of a Hephthalite king named Akhshunwar[43] – or Khushnawaz in Ferdowsi – requesting military aid to claim the crown for himself. Akhshunwar asked Peroz to prove that his claim was better than Hormizd's. In most sources, Peroz's seniority was the only evidence necessary to prove he was the rightful heir. Ṭabarī instead had the Hephthalite king demand evidence that Hormizd was a tyrant before lending his support. This passage includes a moralising statement about God's disfavour for tyrants and the people who live under them. In the end, Peroz proved his worth by offering Akhshunwar control over Talaqan.[44] Balʿamī reported that Akhshunwar refused to allow Peroz to leave and make war on his brother until Hormizd proved himself to be a tyrant through his actions and the people of Iran sought out Peroz, coming to Talaqan where Peroz had been established in exile.[45] After two years, Akhshunwar granted Peroz the assistance of a reported thirty thousand men with whom he invaded Iran, defeated and killed his brother Hormizd, and seized the crown for himself.[46] Thus Peroz became *shāhānshāh* through the military support of the Sasanians' eastern neighbours, the Hephthalites.

This account appears straightforward until we consider the history of the Hephthalites themselves. The war between Hormizd III and Peroz is the first time our sources mention the Hephthalites. By all accounts, the Hephthalites

40 Furthermore, the lack of coins struck in the name of Hormizd III has led some scholars to argue that Peroz succeeded his father immediately; see Schindel (2006) 680. Earlier, Arthur Christensen had proposed that the story found in the Islamic sources was a literary doubling of events from the reign of Peroz's son Kavad who twice fled to the Hephthalites and enlisted their aid in achieving the crown – once to secure his throne for the first time and again after he was forced out of office in 496. See Christensen (1944) 289 n. 5.

41 Dīn. (Guirgass and I. Kratchkovsky (1888)) 60; Balʿamī (Zotenberg (1867)) 2.127. Ferdowsi makes the king a prince of Chaghaniyan specifically (Warner and Warner (1905–1925)) 7.157.

42 Balʿamī (Zotenberg (1867)) 2.127.

43 It is important to note that Akhshūnwār was a royal title rather than a personal name; see Brunner (1984).

44 Ṭab. 1.872.

45 Balʿamī (Zotenberg (1867)) 2.127–28.

46 Only in Ferdowsi's account (1905–25, 7:158) does Peroz spare Hormizd's life.

had yet to reach the frontiers of the Sasanian Empire by 457 and the territories described by Dīnawarī and Balʿamī were mostly under the control of the Kidarite Huns, against whom both Yazdgird II and Peroz led campaigns. Kidarite power in Tukharistan was eventually broken in 467, resulting in a political void filled by the Hephthalites, but Peroz's first campaign against the Hephthalites would not occur until 474. In the meantime, we have documentary and numismatic evidence for a Sasanian presence in Tukharistan during the first half of Peroz's reign. Still, the literary sources for Peroz's rise to power are explicit, using the Arabic term *al-Hayāṭila* rather than a generalising or anachronising term like Turk for his benefactors, while also clearly identifying Tukharistan as the site of his temporary exile. Something is clearly amiss. How does the straightforward narrative provided in the literary texts obstruct the complexity of the politics of East Iran in the fifth century? To address this, we will examine the groups who did inhabit the eastern frontiers of the Sasanian Empire in the mid-fifth century, including the Kidarites and Alkhan Huns. After focusing on their interactions with Peroz, we can attempt to insert the Hephthalites into the narrative.

4 Peroz and the Kidarites

Looking at the eastern frontier of the Sasanian Empire at the outset and over the early years of Peroz's reign, it would seem unlikely that he could have sought aid from the Hephthalites, or at least not from the same Hephthalites who would become his major rivals by the end of his reign. Instead, the Kidarites dominated Tukharistan and Sogdiana. The Alkhans had seized authority in Gandhara and elsewhere south of the Hindu Kush and potentially challenged Kidarite rule in eastern Tukharistan. Peroz's early campaigns in the east were likewise clearly against the Kidarites, and it was only after the Kidarites were defeated that the Hephthalites could come to fill the political void. A focus on these campaigns against the Kidarites – which were themselves continuations of Yazdgird II's campaigns – will help situate East Iran at the time of Peroz's rise to power and identify his potential eastern benefactors.

According to the contemporary Byzantine diplomat and historian Priscus, Yazdgird II's final campaign against the Kidarites carried on into the reign of Peroz, who tried unsuccessfully to negotiate an end to the conflict through treachery as 'the Persians were being worn down by the fighting'.[47] Peroz, in

47 Priscus *Exc. de Leg. Rom.* 12. Łazar reports that in his first year Peroz released a group of Armenian nobles who had been imprisoned by his father and stationed them in Herat to

Priscus's account, attempted to trick the Kidarite king Kunchas[48] into a diplomatic marriage to a woman pretending to be the *shāhānshāh*'s sister, but the woman confessed to the king. Kunchas got his revenge against a group of Sasanian military advisors sent by Peroz to aid the Kidarites in their conflicts with their neighbours. Although this passage certainly contains outlandish dramatic elements, it may also refer to a larger truth. By the mid-fifth century, East Iran was a contested space and the Kidarites were facing challenges from multiple directions, most notably from the Sasanians and the emergent Hephthalites, but also the Alkhans. In other words, the Kidarites' imperial coalition was falling apart, and the passage may reflect the ongoing negotiations and competitions among these multiple parties. Peroz turned to Akhshunwar and the 'Hephthalites' for help gaining the crown and then later provided military aid in the form of advisors to Kunchas and the Kidarites in their wars against the Hephthalites. Such negotiations expanded all the way to Byzantium as Peroz had requested aid from his Roman rivals to the west to defend against the Kidarites, arguing that the Sasanians could act as a bulwark against Hunnic advances at a time when the memory of Hunnic attacks were still fresh and a formidable threat across the Roman world.[49]

Fighting dragged on and around 464, when the Byzantine envoy Constantius travelled to Iran to discuss the requests for Roman aid, he found Peroz 'not in the cities but on the borders between his people and the Kidarite Huns', and met him at his camp at a site called Gorgo.[50] The location of Gorgo is disputed. The site appears only in Greek sources, both Priscus and the Byzantine historian Procopius (d. after 565) in his discussion of Peroz's later campaigns against the Hephthalites. Priscus gave no indication of Gorgo's location other than to report it was on the border with the Kidarites, which included both Sogdiana and Tukharistan. Procopius placed Gorgo 'on the extreme Persian frontier', which implies, from the perspective of Constantinople, the frontier with Tukharistan.[51] In an earlier passage, he also described Gorgo as the city of the Hephthalites 'located over against the Persian frontier', which 'is consequently the centre of frequent contests concerning boundary lines between the two

perform military service; see P'arpec'i (Thomson (1991)) 159. Perhaps this was a sign of the Sasanians' desperation and need for fighting men.

48 Kunchas may be a Greek form of Xūn-qan or 'qan of the Huns'. Grenet (2002) 209. Such a title would again highlight the image of the Kidarites as imperial unifiers of the Iranian Huns.

49 Priscus *Exc. de Leg. Gent.* 15, 19; Jos. Styl. 9–10. Pseudo-Joshua specifically references the Hunnic invasions of Asia Minor in 395.

50 Priscus *Exc. de Leg. Rom.* 12.

51 Procop. *Pers.* 1.4.10.

peoples', again implying an eastern location.[52] There is a possibility that these two Byzantine sources are not describing the same location because Priscus stakes it as Peroz's camp whereas Procopius identifies Gorgo as the city of the Hephthalites. Given its location and contested nature, Gorgo may be Gozgan in western Tukharistan or a reference to the Bactrian title *Garigo Shauo* which appeared on coins struck later by the kings of Gozgan.

Based on the name, others have made the less likely connection between Gorgo and the region of Gorgan (Hyrcania) on the southeastern coast of the Caspian Sea and the modern city of Gonbad-e Kavus which was then the administrative centre of Gorgan specifically.[53] Along these lines, Łazar P'arpec'i identified Vrkan, an older form of Gorgan, as Peroz's camp during a later campaign against the Hephthalites.[54] This identification puts Gorgo about 200 km northeast of Shahr-i Qumis, which was built by Yazdgird II as his base of operation for campaigns in the east, making it plausible that his son would designate Gonbad-e Kavus, further north and east but still very far west from the frontier with the Kidarites, his centre of operations.[55] Gorgan was also the site of the largest single military construction project of the Sasanian era, the Great Wall of Gorgan, a 195 km wall accompanied by thirty-six forts stretching from the Caspian Sea to the Alburz Mountains just north of the Gorgan River. Various dates and historical contexts have been suggested for its construction – from the time of Alexander the Great to Khosrow I Anushirwan (r. 531–579) – but recent archaeological excavations dated the wall to the mid-fifth century, in line with Yazdgird and Peroz's campaigns in the east.[56] The presence of the wall, the forts, and the estimated 22,500 men it would take to garrison them[57] all indicate a major investment in this Gorgan frontier. Gonbad-e Kavus sits immediately south of the wall, making it a strong contender for an imperial camp during construction of the wall. There are also reports that Peroz founded

52 Procop. *Pers.* 1.3.2.

53 Bivar (2002). Elsewhere, Bivar (2003) also argues that the location could be Gorgānj in Khwarazm, which would represent quite a great advance by the Sasanians.

54 P'arpec'i (Thomson (1991)) 214.

55 In Łazar's account, Vrkan/Gorgan was also Yazdgird II's base of operations in the east: P'arpec'i (Thomson (1991)) 133, 155. It should also be noted that this would place Peroz's base over 400 km west of Nishapur which Yazdgird had also used as a staging ground for his eastern campaigns. It would also place Peroz's base over 750 km west of Marw, the provincial capital of Khurasan. Nishapur itself was the most western major city of Khurasan and the focus on camps further west may be an indication that much of Khurasan was not securely in Sasanian hands in the fifth century.

56 Sauer, Rekzavandi, Wilkinson, and Nokandeh (2013) 594–95; Nokandeh et al. (2006) 158–62.

57 Sauer, Rekzavandi, Wilkinson, and Nokandeh (2013) 230–34.

a city in Gorgan. According to Ṭabarī, Peroz built 'another town between Jurjān and the Gate of Ṣūl [the Great Wall of Gorgan] and called it Rūshan Fayrūz'.[58] Although details of this specific report are questioned, it adds to the general association of Peroz with construction in Gorgan and around the wall.

It took Peroz at least eight years to finally defeat the Kidarites. According to Priscus, around 467 a Sasanian embassy reported to the Byzantines that 'they had crushed the Kidarite Huns and had taken their city of Balaam', which some have associated with Balkh. This was a great victory and Priscus accused the Iranian embassy of reporting it 'in a barbaric fashion', specifying that they 'boasted about it, since they wished to advertise the very large force which they had present'.[59] Pseudo-Joshua the Stylite likewise recorded that Peroz 'subdued the Huns, seized many places within their territory, and added them to his kingdom'.[60] Peroz struck gold coins in Balkh after 467, demonstrating at least temporary Sasanian control in Tukharistan.[61] The details of this campaign are vague. Given that the Hephthalites were beneficiaries of the Kidarite defeat and able to take control of both Tukharistan and Sogdiana in the aftermath, it can be argued that either the Hephthalites joined forces with Peroz or that Peroz took credit for the Hephthalites' victories.

The defeat of the Kidarites was a long and destructive process, especially in Tukharistan. While the period of Hunnic rule in Sogdiana was an era of growth and prosperity marked by the foundation of new cities and the expansion of existing settlements, from the available archaeological surveys of Tukharistan, we see quite the opposite situation.[62] There are practically no occupational layers in the plain of Balkh dating to the fifth century.[63] Following the Hunnic invasions, irrigation systems were partially abandoned in the Wakhsh Valley north of the Amu Darya.[64] Many settlements in western Tukharistan were abandoned, including Dil'berjin Tepe and Emshi Tepe.[65] As mentioned above, layers of burning and destruction were uncovered in areas around Qunduz and eastern Tukharistan. By the seventh century, the Chinese Buddhist pilgrim

---

58 Ṭab. 1.874 (translation from Bosworth (1999) 112–13); Thaʿālibī (Zotenberg (1900)) 578.

59 Priscus *Exc. de Leg. Gent.* 22.

60 Jos. Styl. 10.

61 Interestingly, these coins include Bactrian inscriptions, including Peroz's name and title, showing the *shāhānshāh* borrowing some elements of Kidarite coinage in Tukharistan. Alram (2014) 271; Vodrovec (2014) 142.

62 Rezakhani (2017b) 99; de la Vaissière (2005) 101–102.

63 Gardin (1957) 95.

64 Litvinskii and Solov'ev (1985) 135–36.

65 De la Vaissière (2005) 102.

Xuanzang noted that Balkh was thinly populated.[66] It may be the case that the population centres shifted from the large cities found in the plains of northern Tukharistan to the mountains and valleys of the south,[67] but the archaeological evidence suggests that the fifth century saw widespread destruction and displacement in Tukharistan.

The collapse of the Kidarite kingdom was a long process and perhaps more complicated than our literary sources depict. The relationship between the Kidarites and Alkhans in Gandhara and the continued presence of the Kidarites south of the Hindu Kush until at least 477, discussed earlier, shows how the shifting of dynasties could occur through a gradual process. The simplified narrative of the Hephthalites replacing the Kidarites in the mid-fifth century reflects neither the dynamics between the Kidarites, Hephthalites, Alkhans, and Sasanians nor the full complexity of Tukharistan and the kingdoms in and south of the Hindu Kush. Thinking in terms of a unified Sasanian Iran and an equally unified Hunnic state under the Hephthalites would imply far more centralised authority than our sources show.[68] In order to explore these dynamics further, let us turn towards some documentary evidence for East Iran during the reign of Peroz that more clearly centres the role of the Alkhans in this period of transition.

5 Peroz and the Alkhans

While the Alkhans do not appear in our literary sources directly – with the exception of Khingila/Shengil in the *Shahnameh* – documentary evidence ties at least one Alkhan ruler with Peroz. Among the Bactrian Documents,[69] two documents (ea and ed) were written by a man named Meyam who bore the title 'the king of the people of Kadag, the governor of the famous (and) prosperous king of kings Peroz'.[70] These documents are dated to the years 461 and either 465 or 475 respectively, during Peroz's war with the Kidarites.

66 Beal (1884) 44.

67 Rezakhani (2010) chapters 5 and 6.

68 Rezakhani (2019) 124. Nevertheless, it is important that we remain cognizant of the real cultural, economic, and historical connections that united various Hunnic groups from northern India to Transoxiana.

69 The Bactrian Documents are a collection of just over 150 legal, economic, and religious documents from Tukharistan written in the Bactrian language dating from the fourth through eighth centuries. These documents have been edited, translated, and published by Nicholas Sims-Williams (2001, 2007).

70 Sims-Williams (2007) 108–109, 114–15.

Kadagistan was a region in the highland valleys of the Qunduz River which begins in the Hindu Kush mountains southwest of Bamiyan (where it is known as the Bamiyan River), flowing east and then north out of the mountains and through Qunduz before feeding into the Amu Darya.[71] From these documents, we understand that during the first decade of Peroz's reign, a local lord in the highlands of eastern Tukharistan recognised Sasanian authority, ruling as both king of a 'secondary state' in Kadagistan and a Sasanian governor.[72] In the context of Peroz's war with the Kidarites, we may also see Meyam as a rival for control over Tukharistan, using a connection to the Sasanian emperor to help solidify his claims in a contested space as he reached out from the highlands.

We also know Meyam from a series of both gold and silver coins that were struck in his name. These coins present Meyam in a transitional setting. The gold dinars are modelled on Kidarite issues and name Meyam as 'Lord Mehama, King' in Bactrian.[73] In contrast, the silver drachms fit clearly into the Alkhan series and feature a bust with an elongated and beardless head with a large nose and large, angled eyes. Meyam's silver issues appear with inscriptions in both Brahmi and Bactrian that are otherwise so similar it is assumed they were produced in the same mint with dies carved by the same hand but for circulation in different regions.[74] The mixture of Kidarite and Alkhan styles place Meyam in a transitional setting politically between Kidarite and Alkhan rule, and the use of both Brahmi and Bactrian situates him between Gandhara and Tukharistan and their different cultural settings. Two seals naming Meyam in Bactrian are also known.[75] Only one other Alkhan ruler, Adomano 'King of the East' (*miirosano shao*), issued both gold dinars and silver drachms. Like Meyam's issues, the gold dinars follow a Kidarite style, though effort was made to give the pictured ruler Alkhan features while the silver dirhams are Alkhan in style.[76] Inscriptions are found in both Brahmi and Bactrian on the coins of Adomano as well.

71 Inaba (2010) 447.

72 Rezakhani (2019) 129. The title being translated as governor here is *kadag-bid* which literally means 'lord of the house'. It is unique to Tukharistan, first appearing in a Bactrian Document dated to 380 with its last use dated to the late eighth century with *kadag-bid*s serving under Sasanian, Hephthalite, and Turkish overlords. Sims-Williams (2008) 92–93, 98–99.

73 Vondrovec (2014) 142.

74 Vondrovec (2014) 187–88; Vondrovec (2008) 29.

75 Lerner and Sims-Williams (2011) 82–83.

76 Vondrovce (2014) 144–5, 203–4; Alram (2014) 276.

The Schøyen Copper Scroll is an intriguing and important written source for the history of the Alkhans. It ties Meyam more closely to the Alkhans as well as eastern Tukharistan. The source is a copper scroll inscription dedicating a Buddhist stupa in a place the text calls Tālagānika at the end of the fifth century, tentatively dated to 492–93.[77] The scroll itself names several donors, four of whom are known from coin inscriptions in the Alkhan series and the Bactrian Documents: the aforementioned Khingila, Toramana, and Meyam (appearing in Brahmi as Mehama), along with Javukha. The text of the scroll specifies that the dedication took place during the reign of Meyam and, therefore, presumably within his domains as well.[78] His relationship to the other named donors is unclear and it is uncertain if this list represents a hierarchy or genealogy of Alkhan rulers. The location of the monastery and the domains of Meyam have been a point of controversy because *Tālagānika* appears to be the city of Talaqan – in this case, the eastern Talaqan, east of Qunduz, which would place it outside the domains of the Alkhans as they are traditionally understood.[79] But it would situate the monastery in close proximity to Kadagistān, immediately to its northeast.

The last donor mentioned on the Schøyen Copper Scroll, Javukha, is also known for coin issues which share several similarities with the coins of Meyam. These coins are part of a group known as the 'Eastern' Alkhan that show both Gandharan and Ṭukhārian influences. They include issues by several different rulers including the aforementioned Adomano and are named 'Eastern' due to Bactrian inscriptions reading *miirosano shao* or 'King of the East' on many examples. Like the coins of Meyam, the 'Eastern' Alkhan coins appear with both Bactrian and Brahmi inscriptions. In the coins of Javukha, his name appears as it does on the Schøyen Copper Scroll on Brahmi coins but as Zabokho, the Bactrian equivalent, on Bactrian issues.[80] We do not know where these coins were struck, but a location in eastern Tukharistan seems likely, especially considering Javukha's connection to the Schøyen Copper Scroll. In the case of Javukha/Zabokho, he struck such a variety of coins that is assumed they were intended to circulate in different cultural spheres and

77 Melzer (2006).

78 It is important to note that it is unlikely that all four of these donors were contemporaries and Mēyam is the only donor listed here we can confidently say was alive at the time of the dedication; see De la Vaissière (2007) 128.

79 Rezakhani (2019) 126. Étienne de la Vaissière (2007) has suggested that the location should instead be the town of Talagang in the Salt Range of Gandhara.

80 A single issue of Zabokho includes both Bactrian and Brahmi legends on the same coin. Here, an otherwise Bactrian coin includes a Brahmi inscription on the reverse which appears to be a personal name, (*ja*)*yatadhama gotaya*; see Vondrovec (2014) 204.

therefore employed language and iconography that would have been familiar in both Gandhara and Tukharistan, much like the coins of Meyam.[81] In one example, some coin issues of Javukha feature a horseman that 'was obviously adopted from Gupta dinars and transferred to drachms' whether to circulate within the Gupta sphere of cultural influence or as a claim on Guptan legitimacy is unclear.[82]

For purposes of discussing Peroz's rise to power and his campaigns against the Kidarites, the Schøyen Copper Scroll is important for connecting Meyam 'the king of the people of Kadag, the governor of the famous (and) prosperous king of kings Peroz' of the Bactrian Documents to Talaqan. As mentioned above, in both Ṭabarī and Balʿamī – who provide divergent accounts on this point – Talaqan played an important role in Peroz's negotiations with the so-called Hephthalite king Akhshunwar. In Ṭabarī's account, Talaqan was offered to Akhshunwar as payment for his support in the war against Hormizd III. It was assumed that this would be the Talaqan that had been won by Yazdgird II and that would more likely be western Talaqan, but as noted above, archaeological evidence shows repeated instances of burning and destruction indicative of warfare around Qunduz during the fifth century, evidence that Yazdgird's campaigns could have reached eastern Tukharistan, though this could also signal warfare between the Kidarites, Alkhans, and Hephthalites. The fifth-century Armenian historian Ełišē had described Yazdgird chasing the Kidarites into their 'impregnable desert' where they 'lived in hiding with all [their] troops' while Yazdgird 'assailed [their] provinces, regions and lands, captured many fortresses and cities, amassed captives, booty, and plunder'.[83] Peroz's gold dinar struck in Balkh around 467 demonstrates that shortly after Meyam appeared in the Bactrian Documents as a vassal governor of the *shāhānshāh* the Sasanians did have control of the central city of Tukharistan. Match this with the changing demographic and economic patterns of Tukharistan in which populations shifted from the cities of the plains to the highlands, and we can piece together a picture of the exiled prince Peroz claiming authority over a war-ravaged and depopulated Talaqan. Balʿamī made this connection more explicit, identifying Talaqan as Peroz's home in exile, which he governed in the name of Akhshunwar until it was time to seize the Sasanian crown. The construction of a Buddhist temple in Talaqan in the late fifth century by a local king who was extending his authority from the highlands into the plains – the event commemorated by the Schøyen Copper Scroll – may also be seen as an attempt at

81 Vondrovec (2014), 206.
82 Vondrovec (2014) 203, 205.
83 Ełišē (Thomson (1982)) 72.

urban recovery following a century of war and destruction. Pieced together, it becomes possible that the Akhshunwar who aided Peroz was not a Hephthalite king – a proposition questioned already by Arthur Christensen[84] – but rather Meyam or another related Alkhan or *miirosano shao*, or that Meyam was part of a larger coalition that supported Peroz's bid for the throne.[85] Indeed, when Ṭabarī summarises Peroz's war against Hormizd at the beginning of his description of Peroz's reign, it was no longer a Hephthalite army that brought him to power but 'the resources of Khurasan' and 'the men of Tukharistan and the regions neighboring on it'.[86] Considering that Peroz was sometimes named the governor of Sistan under his father, such a coalition may include parts of the Sasanian army that had served under Prince Peroz in the east (in other words, the resources of Khurasan).

Above, it was suggested based on seal inscriptions that we should think of the Kidarites not as a homogenous dynastic state, but as the imperial unifiers of the various peoples of Tukharistan, Sogdiana, and Gandhara. The literary sources represent them as a monolith, the Kidarite Huns, but the numismatic and documentary evidence give a more diverse picture of the Kidarites ruling over a variety of peoples and deploying a variety of identities to support that rule. Especially in the highlands of the Hindu Kush, this meant that the Kidarite domains were home to many small and autonomous kingdoms such as Rob – a kingdom south of Tukharistan connected to many of the Bactrian Documents – and Kadag with their own lords who could join or leave the larger confederations as conditions demanded. The Alkhans were another such confederation that clearly defined themselves as something other than the Kidarites – through distinctive imagery and inscriptions on their coins, for example. Evidence from the Schøyen Copper Scroll and numismatics indicate that the Alkhan realms were sometimes ruled using an association of up to four regional rulers and that these individual lords represented, protected, and patronised different regional religious and cultural traditions.[87] At times,

84 Christensen (1944) 289 n. 5.

85 The idea of Meyam as part of a pro-Peroz coalition was previously suggested by Rezakhani (2019) 129 n. 47. Nicholas Sims-Williams has likewise hinted at the involvement of the local lords of the Hindu Kush in Peroz's war with Hormizd, suggesting that a man named Kirdīr-Warahrān the *khāhr* of Rōb, not only engaged in the conflict but switched loyalties in the midst of it. In document ck, Kirdīr-Warahran is referred to as 'true to Peroz' while in documents cl and cm he is 'glorious through Ohrmuzd'. Sims-Williams (2007) 88–91; Sims-Williams (2008) 93–94. Kirdīr-Warahrān is also known from two seals similar to those of Mēyam. Lerner and Sims-Williams (2011) 81–82.

86 Ṭab. 1.872 (translation from Bosworth (1999) 109).

87 Alram (2014) 275–76.

these confederations could work together, as the signs of blended co-rule in Gandhara showed. In other moments, they competed, as in the Alkhan conquest of Gandhara. A local lord situated between the two, like Meyam, could negotiate their position. In such a dynamic, a figure like Peroz could represent a third option. We can imagine that as Kidarite fortunes fell in the mid-fifth century to both Sasanian and Hephthalite advances, Meyam looked to replace the Kidarites with a Sasanian emperor willing to recognise his position in the valleys of the Qunduz River and perhaps even offer to extend his authority into the plains. The dissolution of existing alliances in Tukharistan, Sogdiana, and Gandhara can help us understand Yazdgird II and Peroz's successes against the Kidarites, but also the latter's ultimate failure against the Hephthalites, who had rebuilt the imperial ties across the region in a manner that allowed them to present a unified resistance to the Sasanians.

6 Peroz and the Hephthalites, Part 2

Following the defeat of the Kidarites, Peroz was able to, temporarily, assert Sasanian authority in Tukharistan. As mentioned above, gold dinars were briefly struck in his name at Balkh and some of the local lords such as Meyam of Kadagistan recognised his authority. At least seven years elapsed between his victory over the Kidarites in 467 and Peroz's first recorded campaign against the Hephthalites. According to Procopius, this campaign occurred during the reign of the Byzantine Emperor Zeno (r. 474–475, 476–491), who sent an ambassador named Eusebius to Peroz, who then accompanied the emperor on his campaign.[88] Procopius' account of this campaign is unique. Peroz gathered his army at Gorgo and then marched on the land of the Hephthalites. When the Sasanians approached a Hephthalite force, the Hephthalites turned and fled, luring Peroz into a valley where his forces were ambushed. The Hephthalite king forced Peroz to kneel before him and pledge that he 'would never again take the field against' the Hephthalites.[89] Ṭabarī included a more dramatic account of what was likely the same campaign. In this version, when news reached the Hephthalite king Akhshunwar that Peroz was on the march, he sent a spy to infiltrate the Sasanian camp – after cutting off his hands and feet so that it looked like he was the victim of a cruel punishment and therefore believably disgruntled with the Hephthalites and seeking revenge. The Hephthalite spy fed Peroz false information and led him and his army through endless deserts

88 Procop. *Pers.* 1.3.8.

89 Procop. *Pers.* 1.3.17.

until the bulk of the Sasanian army died of thirst, including the maimed spy himself. Peroz then surrendered to Akhshunwar. The two established a boundary between them and promised to never cross it.[90] This border was associated with the tower built by Wahram Gur. Reports of this campaign include no references to fighting. Rather, Peroz was defeated by his lack of preparation and, perhaps, an expectation that he would receive local support in Tukharistan, symbolised in Ṭabarī's account by the man with no hands or feet.

Pseudo-Joshua the Stylite provided a different version of this first campaign and perhaps added a second, though it may be a detail-enriched repetition of the first. This account presents warfare in the east during Peroz's reign as a continuous process against a singular foe in which Peroz 'subdued the Huns, seized many places within their territory, and added them to his kingdom before he was eventually taken prisoner'.[91] He then made a treaty with the Huns, which we can imagine is the same treaty of the previous accounts with promises to not invade the Hephthalite domains again. Peroz was also forced to pay ransom, which he did with financial support from the Byzantine Emperor Zeno – again setting Pseudo-Joshua's narrative after 474. When Peroz returned home, he decided to break the treaty and invade the Hephthalite domains. He was again defeated and captured but this time his opponent asked for a ransom of thirty mule loads of silver – an estimated 4,500 kg of silver or over one million drachms[92] – but he could only raise twenty and was forced to leave his son Kavad behind to guarantee the payment. To raise the remaining funds, Peroz instituted a poll tax.[93] The ransom payments are believable because of the large number of coins from Peroz's reign found with Hephthalite countermarks. These coins date to the period immediately after Peroz's defeat, based on the design of his crown, which changed at that time. These issues of Peroz featuring his third crown became the model for the first Hephthalite coins.[94] Payments of Sasanian silver continued to reach the Hephthalites up to the reign of Khosrow I Anushirwan and the proliferation of the coins in finds from Tukharistan and Sogdiana to China

90 Ṭab. 1.874–76. Balʿamī (Zotenberg (1867), 2.132–37) includes an expanded version of the same story that, interestingly, includes a moment when Akhshūnwār, named Kushnawāz, chastises Peroz for repaying his aid in winning the crown from Hormizd III with such treachery.

91 Jos. Styl. 10.

92 Schindel (2006) 682.

93 Jos. Styl. 11.

94 Vondrovec (2014) 399. It is worth noting that Hephthalite coins continued to imitate Sasanian issues throughout their rule.

implies that this wealth underpinned the expansion of the Sogdian trade networks in the sixth century.[95] It may also have been the necessary tool to reunite the Kidarite domains under a new Hephthalite coalition.

During the period between the fall of the Kidarites and Peroz's first campaign, the Hephthalites were expanding their authority in Tukharistan. Peroz gathered his forces at Gorgo, possibly Gozgan, which was already an established camp by 474. Balʿamī described Peroz engaging the Hephthalites in western Tukharistan, saying that when Peroz had arrived at Balkh, only the desert of Marw stood between him and his opponent.[96] The stratagem by which the Hephthalites drew Peroz through endless desert and mountain valleys supports the impression that the Sasanians could perhaps exert authority in a city like Balkh, where Peroz had been able to strike coins, but that this was an island in a sea of Hunnic power. It is important to note that the peace treaty that concluded the campaign defined the boundary between the two powers, encouraging the impression that this boundary had been previously vague or ill-defined and that Hephthalite authority was still being shaped. For example, the earliest dated direct reference to Hephthalite authority in the Bactrian Documents appears only in 483 when the lord of Rob was charged with paying a Hephthalite tax, which is nearly a decade after the first Hephthalite victory over Peroz and only a year before the *shāhānshāh*'s death.[97] By the time of Peroz's next and final campaign, it appears that the Hephthalites had claimed authority over all Kidarite domains north of the Hindu Kush and across the Amu Darya into Sogdiana with the boundary between the Hephthalites and Sasanians defined by Wahram Gur's tower, built a half century earlier near Kushmayhan between Marw and the river.

Peroz led a final, cataclysmic campaign against the Hephthalites in 484. In all accounts, this campaign is presented as the folly of Peroz's haughtiness. His advisors regularly reminded him that he had been defeated before and that he had agreed not to cross the boundary that delimited the domains of the Sasanians from those of the Hephthalites, but Peroz ignored this advice and marched across the boundary at the head of a large army. When Akhshunwar became aware of the Sasanians' approach, he joined the chorus of people reminding Peroz of their agreement, reportedly displaying a written and sealed copy of the treaty. He also had a great ditch dug before his camp and concealed it with reeds, sticks, shrubbery, and dirt so that when Peroz charged the Hephthalite position – pursuing a group of retreating horsemen who knew

95 De la Vaissière (2005) 110–12.
96 Balʿamī (Zotenberg (1867)) 2.132.
97 Vondrovec (2014) 144; Sims-Williams (2001) 46–47.

where a secret path had been left across the ditch – the army plunged into the ditch and the Hephthalites were able to fall upon and kill the Sasanian forces, including the ill-fated Peroz.[98]

While our sources are rather light on details when it comes to military tactics, this final battle highlights some aspects about the conduct of warfare between the Sasanians and the Hephthalites. In each of Peroz's campaigns, the Hephthalites are presented as avoiding direct confrontation and instead deploying stratagems that create ambushes, relying on light and fast cavalry that can lure the enemy into a trap through feigned retreats. The Sasanians likewise appear to be relying heavily on cavalry that can pursue a retreating enemy. After all, a block of marching infantry would not fall into a ditch in such great numbers as fast charging cavalry. The Sasanian cavalry of Peroz's time included skilled mounted archers who could quickly swap their bows for swords as they closed for the charge. Such impetuous pursuit may have been a part of Sasanian strategy in the east – reports of Yazdgird II chasing the Kidarites into the desert, for example, may point towards a similar strategy – and Akhshunwar was prepared to take advantage. Ṭabarī and Balʿamī mention elephants in Peroz's army as well. In these reports, the elephants were used to drag Wahram Gur's tower and, thereby, the boundary between the Sasanians and the Hephthalites, so that Peroz could advance against Akhshunwar without violating their treaty.[99]

Despite agreement about the story in the sources,[100] there is uncertainty about where Peroz met his demise. Łazar specifies that Peroz gathered his troops in preparation for this battle at Vkran or Gorgan[101] and Procopius has the army amass at Gorgo, which the Hephthalite king described as the 'extreme Persian frontier'.[102] Many of the Islamic sources discuss the border tower built by Wahram Gur, focusing on Peroz's act of trespassing this symbol

98 The sources for Peroz's final battle against the Hephthalites and his death include P'arpec'i (Thomson (1991)) 214–15; Jos. Styl. 11; Procop. *Pers.* 1.4.1–14, 32–35; *Seert Chron.* 107–108; Ibn Qutayba (Okacha (1960)) 662; Dīn. (Guirgass and I. Kratchkovsky (1888)) 61–62; Yaʿqūbī (Muhannā (2010)) 1.163; Ṭab. 1.876–79; Masʿūdī (Meynard and Courteille (1861)) 2.195; Balʿamī (Zotenberg (1867)) 2.137–42; Ferdowsi(Warner and Warner (1905–1925)) 7.164–69; Thaʿālibī (Zotenberg (1900)) 579–82.

99 Ṭab. 1.878–79; Balʿamī (Zotenberg (1867)) 2.138–39.

100 Only Pseudo-Joshua the Stylite provides a version of Peroz's death that does not include the ditch. Instead, the Hephthalites assaulted the Sasanians and routed them. Peroz disappeared in the chaos and no one was sure what happened to him, with Pseudo-Joshua musing that he may have starved to death in some mountain valley or been eaten by forest animals (Jos. Styl. 11).

101 P'arpec'i (1991) 214.

102 Procop. *Pers.* 1.4.10.

of the boundary between the Sasanians and the Hephthalites. This tower was built near Kushmayhan, south of the Amu Darya between Marw and Āmul, which points to an invasion of Sogdiana rather than Tukharistan. Ferdowsi made this connection explicit by inserting an inscription on Wahram Gur's tower warning people to not cross the river. Peroz ignored this and crossed the Amu Darya – Ferdowsi gave the Arabic name for the river, Jayḥūn.[103] Despite telling the story of Wahram Gur's tower, Balʿamī specifically said the battle happened on the frontier of Balkh and Tukharistan.[104] Masʿūdī described the battle taking place near Marw al-Rūd – another frontier outpost associated with Wahram Gur – implying that Peroz was invading Tukharistan, but then associated Akhshunwar with Sogdiana instead.[105] Thaʿālibī explicitly stated that Peroz was campaigning in Tukharistan.[106]

The effects of Peroz's defeat were long felt. Not only was the *shāhānshāh* killed but so were many of his sons and brothers. In fact, Procopius stated that Peroz lost thirty sons in the battle, all but Kavad, while Ṭabarī specified that four sons and brothers, 'all of whom bore the title of king', were killed.[107] A daughter named Perozdukht was captured and made a concubine of Akhshunwar. She gave birth to a daughter who was later married to her uncle Kavad.[108] The Hephthalites appear to have captured the royal mint that travelled with the emperor. This included dies for the mint of Asuristan (Assyria) as well as mint workers, including engravers, and the Hephthalites began minting their own coins on the Sasanian model.[109] In the aftermath of the battle, the Hephthalites engaged in another rapid wave of expansion, this time well into Sasanian territory, which Ṭabarī described as the conquest of the whole of Khurasan.[110] In the words of Pseudo-Joshua the Stylite, when Peroz's brother Walash (r. 484–488) took the throne, he 'found the Persian treasury empty and the land ravaged by the Huns'.[111] Numismatic and archaeological evidence,

103 The poet continues to reference the Jayḥūn as the boundary between the Sasanians and Hephthalites throughout this narrative. Ferdowsi also reported that Akhshūnwār, named Kushnawāz, gathered his troops in Samarkand (Warner and Warner (1905–25) 7.164–65).

104 Balʿamī (Zotenberg (1867)) 2.139.

105 Masʿūdī (Meynard and Courteille (1861)) 2.195.

106 Thaʿālibī (Zotenberg (1900)) 579.

107 Procop. *Pers.* 1.4.2, 14, 34; Ṭab. 1.873 (translation from Bosworth (1999) 110).

108 Jos. Styl. 24.

109 Alram (2014) 278.

110 Ṭab. 1.873.

111 Pseudo-Joshua went on to muse on the nature of war: 'Doubtless it will not escape you in your wisdom how much loss and expense is incurred by kings in war, even in victory, let alone defeat' (Jos. Styl. 18).

including gaps in coin production at Abarshahr (Nishapur), Herat, and Marw and the foundation of two Buddhist stupas at Marw, point towards a period of Hephthalite occupation of Sasanian territory as far west as Nishapur from the death of Peroz until the last years of the reign of Kavad (r. 488–496, 498–531).[112] During Walash's rule, the mint of Gorgan suddenly became the third most productive in the empire, highlighting its increased status as the new frontier against the Hephthalites.[113] It is unclear if these coins were intended to pay the Sasanian troops or serve as tribute to the Hephthalites. The aftermath of Peroz's defeat and Akhshunwar's victory had long-term ramifications for both Sasanian Iran and Hephthalite Tukharistan.

The Hephthalite victory over Peroz also impacted Sasanian military tactics and administration. According to Procopius, following Peroz's defeat, the pursuit of fleeing opponents was forbidden while in enemy territory.[114] While this policy limited (but did not fully prevent) future follies born from impulsive cavalry charges, it also forced the Sasanian army to fight in a slower, more methodical way that relied more heavily on drawn-out archery battles before engaging the enemy.[115] The loss to the Hephthalites also exposed the dangers of a centralised military and likely contributed to the quadripartite division of the empire under Kavad and Khosrow Anushirwan, with the east falling under the authority of the *spāhbed* of Khurasan.[116]

Even though we may question the place of the Hephthalites in Tukharistan and their ability to assist Peroz at the beginning of his reign, by the end it is apparent that they had been able to rebuild the confederacy of the Kidarites and present a military challenge to the Sasanians. Peroz clearly aided in the rise of the Hephthalites by first eliminating their Kidarite rivals and then by presenting them with a catalyst for bringing together the peoples of Tukharistan to resist Sasanian advances. The early Hephthalite victories further gave them access to wealth in the form of ransoms and tributes that could be used to secure local loyalties. Their final victory in the face of Peroz's ill-advised aggression would similarly have propagandistic value and it allowed the Hephthalites to assemble a force capable of seizing and potentially holding Khurasan, perhaps for decades. At the same time, their victory caused disruptions within

112 Schindel (2006) 683; Grenet (2002) 212–13.

113 Schindel (2006) 683.

114 Procop. *Pers.* 1.4.33.

115 Syvänne (2021) 114–15.

116 Gyselen (2001); Rezakhani (2017a) 209.

the Sasanian realm that gave the Hephthalites the opportunity to influence politics in Iran for a generation.

7 Kavad, the Hephthalites, and the Karenids

In many of the later, Islamic literary sources, Peroz's defeat was avenged quickly by a Parthian nobleman of the Karenid clan and governor of Sijistān (Sistan) named Sukhra (d. 493). These accounts are fantastical, focusing on Sukhra's prowess as he defeated Akhshunwar, rescued many prisoners including the Mobeds who had travelled with Peroz, and regained the lost treasury.[117] The continued Hephthalite presence in Khurasan long into Kavad's reign means such a stunning reversal did not occur.[118] Yet the rise of Sukhra and the Karenid clan in the aftermath of Peroz's defeat – perhaps, as a pro-Hephthalite party allied with the Sasanians' eastern rivals[119] – was an important result of the war. During the reign of Walash, Sukhra became the power behind the throne. This status solidified when Walash's nephew Kavad successfully challenged his uncle for the throne.[120]

When the crown went to his rival, Kavad fled east to the Hephthalite court much like his father, though in this case we can be more confident that it was in fact the Hephthalites he reached. During his flight east, Kavad is noted as building ties with the Karenids. When he reached Nishapur, he met with Sukhra's son Zarmihr, who secured the daughter of his major-domo, soon to be the mother of Khosrow Anushirwan, for the prince.[121] In Balʿamī's account, Zarmihr was accompanying Kavad east as part of his entourage. The connection is so close that in Ferdowsi's account of Kavad's ascent to the throne, Sukhra rescued Kavad during his reprisal attack on the Hephthalites, returned him to Iran,

117 For a more detailed overview including the allusions to the Iranian epic tradition, see Rezakhani (2017b) 128–30.

118 This may have lasted even longer in some ways. For example, the *marzbān* of Marw during the Arab conquests of the seventh century, Māhuyē, was possibly Hephthalite himself despite being a Sasanian military governor. Haug (2019) 87.

119 Bonner (2015) 83–84.

120 The title used by Walash, as found on his coins, was *hukay* or 'Good Prince' Walash, perhaps a title meant to contrast with the not good prince Kavad, who relied on the support of the Sasanians' enemies, the Hephthalites. Rezakhani (2017b) 130–31.

121 Ṭab. 1.883–84; Balʿamī (Zotenberg (1867)) 2.145–46. In some accounts, Kavad met the mother of Anūshīrwān during his second flight east, described below. This may be more likely as Kavad was only ten years old when his father was killed and therefore most likely unable to father an heir. Sūkhrā and Zarmihr may be the same person, erroneously divided by our later sources.

and immediately placed him on the throne.[122] In this version, the Karenids by themselves raised Kavad to power without the aid of the Hephthalites. Kavad stayed at the Hephthalite court for four years, only returning when the Iranian nobility found it worthwhile to remove, blind, and kill Walash in favour of the exiled prince. In a few instances, this choice was motivated by Kavad beginning to march west with his new allies. This situation did not last long and in 496 tensions among the leading Parthian families led to a coup that replaced Kavad with his brother Zamasp (r. 496–498).[123] In response, Zarmihr helped Kavad escape his prison and flee again to the Hephthalites, where he lived for another three years before retaking the throne with the aid of thirty thousand Hephthalite troops, a force sizeable enough to convince the nobility to change their allegiance back to Kavad.[124] According to Dīnawarī and Ferdowsi, Kavad exchanged control over Chaghaniyan for this military support, which then opened Sogdiana up to Hephthalite authority. Only in 509 did Hephthalite ambassadors replace Sogdian ambassadors from Samarkand at the Chinese court.[125] Kavad also promised the Hephthalites monetary tribute, evidenced by the large number of coins from his second reign with Hephthalite countermarks.[126] The connection between Kavad and the Hephthalites continued even after his reinstatement. Hephthalite troops fought alongside the Sasanian Empire against the Byzantines in eastern Anatolia during the Anastasian War (502–506), a war fought in part to secure tribute due to the Hephthalites. Pseudo-Joshua implies that the Hephthalite mercenaries were primarily cavalry.[127] From Procopius, we learn that these Hephthalite auxiliaries were deployed in the vanguard.[128]

As the Hephthalites were helping Kavad and the Karenid Parthian family achieve and maintain power in Iran, the turn of the sixth century was also a time of expanded Hephthalite authority in Tukharistan. Through the Bactrian Documents, coins, and seals, we see the Hephthalite *Yabghu* increasing his

---

122 Ferdowsi (Warner and Warner (1905–1925)) 7.179–82.

123 Typically, the removal of Kavad from office is tied to the heretical Mazdakite movement, but this was likely cover for in-fighting among the nobility. For an overview of the relationship between Kavad and Mazdak and the problems with our sources for the Mazdakite movement, see Crone (1991).

124 The sources for Kavad's flight to the Hephthalites include Jos. Styl. 24; Procop. *Pers.* 1.6.10; Dīn. (Guirgass and I. Kratchkovsky (1888)) 67–68; Ṭab. 1.887; Balʿamī (Zotenberg (1867)) 2.151; Ferdowsi (Warner and Warner (1905–1925)) 7.197–200; Thaʿālibī (Zotenberg (1900)) 590–94.

125 De la Vaissière (2005) 110–11.

126 Göbl (1967) 1:193–94.

127 Jos. Styl. 48, 51, 55, 57, 62.

128 Procop. *Pers.* 1.8.13.

power over a growing domain. The Hephthalites clearly filled the role previously held by both the Kidarites and the Sasanians as overlords of the smaller kingdoms of Tukharistan. For example, a man named Kilman, presumably an heir or successor to Meyam, described himself in his correspondence as 'the king of the people of [Ka]dag, the governor of the famous (and) prosperous *yab[ghu]* of Hephthal' swapping out the 'famous and prosperous Peroz' of Meyam's letters for a Hephthalite overlord.[129] Meyam's appearance among Alkhan lords on the Schøyen Copper Scroll and the switch from Kidarite-style to Alkhan-style coinage may be artefacts of local resistance to Hephthalite expansion into eastern Tukharistan as Meyam's former Sasanian patrons became more intertwined with the Hephthalites.[130] At some point, the Hephthalites even took these local titles over for themselves. One Sārt Khudēbandān claimed the titles 'the glorious *Yabghu* of Hephthal, the ruler of Rob, the scribe of the Hephthalite lords, the judge of Tukharistan (and) Gharchistan, the noble lord', implying that the *yabghu* himself was now lord of Rob as well.[131] As noted above, evidence of Hephthalite expansion into Sogdiana similarly became clear in the first decade of the sixth century. As a result, Kavad's reign is the first moment that can be defined as an imperial Hephthalite period.[132] As such, the Hephthalites may be seen as a better documented and therefore 'amplified version of the Kidarites' whose power derived from uniting the various groups of East Iran.[133] Interestingly, they did so with a more distinct self-identification. Whereas the Kidarites continued to refer to themselves on coins and seals as Kushanshah, relying on the inherited imperial authority of their Kushan predecessors, the Hephthalites – in a move similar to the Alkhans – dropped this title entirely and replaced it, on coinage at least, with an ethnic signifier, the word Hephthalite inscribed in Bactrian. Unlike the Alkhans, the Hephthalites continued to model their coinage on Sasanian issues.

Although our literary sources present Peroz's reign as a time of rivalry with a unified and centralised Hephthalite enemy, it was not until the second reign of his son Kavad that Hephthalite authority was fully secured in Tukharistan and then expanded into Sogdiana. The Hephthalites were portrayed as powerbrokers in the Sasanian realm at the beginning of Peroz's reign, but it is only

129 Sims-Williams (2007) 124–25.

130 Rezakhani (2019) 130.

131 Sims-Williams (2007) 126–27.

132 Frantz Grenet has dated the imperial Hephthalite period to 457–560, but it is clearly not until after 492–493, the date of the Schøyen Copper Scroll, that Hephthalite authority extended to eastern Tukharistan and the small kingdoms of the Hindu Kush. Grenet (2002) 209.

133 Grenet (2002) 210.

after his death and through the actions of the Karenid Parthian family that Hephthalite influence came to the fore of Sasanian politics. The transition from Kidarite to Hephthalite authority was a drawn-out process while the Hephthalites secured the support of or dominance over local lords, but it was a process that also left them in a position to exert their authority further into the Sasanian domain than their predecessors had and in a more sustained fashion.

8 Conclusions: the Coming of the Turks

By 560, Hephthalite imperial authority had been swept aside by the rise of the Turks who, according to our literary sources, conquered Sogdiana while the Sasanians under Khosrow I Anushirwan took Tukharistan – which was soon enough also incorporated into the domains of the Türk Khāqānate.[134] This did not mean the total destruction of the Hephthalites or their political authority; Hephthalite dynasties such as the Nezaks ruled parts of highland Tukharistan into the eighth century.[135] Just as the Kidarites and Hephthalites had relied on local vassals to rule many parts of their domain, the Turks too ruled as a confederacy. On the ground, the Turks had a very light footprint in Sogdiana and even less so in Tukharistan, maintaining only a large enough presence to prevent rebellions and to defend the territories under their authority when faced with a direct threat;[136] they relied on local institutions to manage their domains and ensure that tribute and taxes continued to flow.[137] By the time the Turks had seized control, the local agents in large parts of Tukharistan were now the Hephthalites who had been the foreign invaders a century earlier. This dynamic between the Turks and Hephthalites underscores that although we do not know many of the details about the relationship between political and demographic changes in Late Antique Tukharistan, we do know that political and social changes were never total. Even within this structure, authority was continuously contested and the documentary and numismatic evidence points towards pockets of Sasanian authority within Tukharistan across the sixth century.[138] One Bactrian Document dated roughly to the 560s or later includes a reference to a local lord in Tukharistan making regular payments

134 Sasanian involvement in these campaigns can be debated. Bonner (2015) 111.

135 On the conflicts between the Nezak Tarkhan and the Umayyad governor of Khurasan, Qutayba b. Muslim (r. 705–715), see Haug (2019) 119–21.

136 Skaff (2002).

137 Payne (2016) 23–24.

138 On the numismatic evidence, see Gyselen (2003) 162–67.

to both the Sasanians and the Hephthalites, demonstrating an active struggle between two authorities with the local population caught in between.[139]

As we can see across this survey of several centuries of Sasanian interaction with their eastern neighbours – both military and diplomatic – East Iran was ruled through coalitions of imperial and local authorities during Late Antiquity. Within this dynamic, the Sasanians themselves were but one part. At times they could find themselves at the centre of power, ruling parts of Tukharistan directly or acting as overlords to local vassals. At other times, the Sasanians could also find themselves under the sway of other confederacies as the repeated turn towards the Hephthalites and other eastern dynasties for military and political support and the long periods of Hephthalite presence in Khurasan indicate. Our available literary sources are much more comfortable writing narratives of monolithic imperial rivals – the types of histories that make sense in the west, where the Sasanians faced the Byzantine Empire – and, as a result, they look to portray a dichotomy between a unified Sasanian Empire versus unified Hunnic dynasties, whether Kidarite or Hephthalite. Numismatic and documentary evidence complicates this depiction and illuminates how both sides operated as coalitions of peoples that could shift and evolve over time within a contested borderland. Such complexities were most abstruse in times of transition, such as the difficult fifth century that saw the shift from Kidarite to Hephthalite rule, while the Sasanians faced contests of leadership between Hormizd III and Peroz and then Kavad, Walash, and Zamasp that involved the input and support of eastern rivals.

Bibliography

Accademia nazionale dei Lincei and Istituto italiano per il Medio ed Estremo Oriente (eds) (1996) *La Persia e l'Asia centrale da Alessandro al X secolo*. Rome.

Alram, M. (2014) 'From the Sasanians to the Huns: New Numismatic Evidence from the Hindu Kush'. *Num. Chron.* 174: 261–91.

Alram, M., Klimburg-Salter, D., Inaba, M., and Pfisterer, M. (eds) (2010) *Coins, Art, and Chronology II: The First Millennium C.E. in the Indo-Iranian Borderlands*. Vienna.

Beal, S. (1884) *Si-Yu-Ki: Buddhist Records of the Western World*. London.

Belardi, W. (ed) (1996) La Persia e l'Asia Centrale da Alessandro al X Secolo. Rome.

Bivar, A.D.H. (2002) 'Gorgān v. Pre-Islamic history'. *Encyclopaedia Iranica Online*.

Bivar, A.D.H. (2003) 'Hephthalites'. *Encyclopaedia Iranica Online*.

Blockley, R. (1983) *The Fragmentary Classicising Historians of the Later Roman Empire: Eunapius, Olympiodorus, Priscus and Malchus*. Liverpool.

139 Sims-Williams (2001) 164; Sims-Williams (2008) 96.

Bonner, M.J. (2015) *Al-Dīnawarī's* Kitāb al-aḫbār al-ṭīwāl: *An Historiographical Study of Sasanian Iran*. Leuven.

Bosworth, C.E. (1999) *The History of Ṭabarī: Vol. 5: The Sāsānids, the Byzantines, the Lakhmids, and Yemen*. Albany, NY.

Braarvig, J. (ed) (2006) *Manuscripts in the Schøyen Collection: Buddhist Manuscripts*. Vol. 3. Oslo.

Brunner, C.J. (1984) 'Akšonvār'. *Encyclopaedia Iranica Online*.

Christensen, A. (1944) *L'Iran sous les Sassanides*. Copenhagen.

Cribb, J. (2010) 'The Kidarites: The Numismatic Evidence. With an Analytical Appendix by A. Oddy', in Alram, Klimburg-Salter, Inaba, and Pfisterer (2010) 91–146.

Crone, P. (1991) 'Kavād's Heresy and Mazdak's Revolt'. *Iran* 29: 21–42.

Curtis, V.S. and Stewart, S. (eds) (2008) *The Sasanian Era*. London.

Daryaee, T. (2002) *Šahrestānīhā ī Ērānšahr: A Middle Persian Text on Late Antique Geography, Epic, and History with English and Persian Translations and Commentary*. Costa Mesa, California.

De la Vaissière, É. (2003) 'Is There a Nationality of the Hephthalites?' *Bulletin of the Asia Institute* 17: 119–32.

De la Vaissière, É. (2005) *Sogdian Traders: A History*. Leiden.

De la Vaissiêre, É. (2007) 'A Note on the Schøyen Copper Scroll: Bactrian or Indian?' *Bulletin of the Asia Institute* 21: 127–30.

Di Cosmo, N. and Maas, M. (eds) (2018) *Empires and Exchanges in Eurasian Late Antiquity: Rome, China, Iran, and the Steppe, ca. 250–750*. Cambridge.

Filigenzi, A. (2010) 'Post-Gandharan/non-Gandharan: An Archaeological Inquiry into a Still Nameless Period', in Alram, Klimburg-Salter, Inaba, and Pfisterer (2010) 389–406.

Gardin, J.C. (1957) *Céramiques de Bactres*. Paris.

Göbl, R. (1967) *Dokumente zur geschichte der iranischen Hunnen in Baktrien und Indien*. Wiesbaden.

Grenet, F. (1996) 'Crise et sortie de crise en Bactriane-Sogdiane aux IVe–Ve siècles', in Belardi (1996) 367–90.

Grenet, F. (2002) 'Regional Interaction in Central Asia and Northwest India in the Kidarite and Hephthalite Periods', in Sims-Williams (2002) 203–24.

Guirgass, V. and Kratchkovsky, I. (eds) (1888) *Dīnawarī, Kitāb al-akhbār al-ṭiwāl*. Leiden.

Gyselen, R. (2001) *The Four Generals of the Sasanian Empire: Some Sigillographic Evidence*. Rome.

Gyselen, R. (2003) 'La reconquête de l'est iranien par l'empire sassanide au VI^ie^ s. d'après les sources "iraniennes"'. *Arts asiatiques* 58: 162–67.

Ḥabībī, 'A. (ed) (1984) *Ta'rīkh-i Gardīzī*. Tehran.

Haug, R. (2019) *The Eastern Frontier: Limits of Empire in Late Antique and Early Medieval Central Asia*. London.

Higuchi, T. and Kuwayama, S. (1970) *Chaqalaq Tepe: Fortified village in North Afghanistan excavated in 1964–1967*. Kyoto.
Inaba, M. (2010) 'From Kesar the Kābulšāh and Central Asia', in Alram, Klimburg Salter, Inaba, and Pfisterer (2010) 443–55.
Inalcik, H. (ed) (2002) *The Turks*. Ankara.
Kennet, D. and Luft, P. (eds) (2008) *Current Research in Sasanian Archaeology, Art and History*. Oxford.
Lerner, J. and Sims-Williams, N. (2011) *Seals, Sealings and Tokens from Bactria to Gandhara (4th to 8th Century CE)*. Vienna.
Litvinskii, B.A. and Solov'ev, V.V. (1985) *Srednevekovaia kul'tura Tokharistana*. Moscow.
Lyonnet, B. (1997) *Céramique et peuplement du Chalcolithique à la conquête arabe. Propections archéologiques en Bactriane Orientale (1974–1978), volume 2*. Paris.
Melzer, G. (2006) 'A Copper Scroll Inscription from the Time of the Alchon Huns. In Collaboration with Lore Sander', in J. Braarvig (2006) 251–314.
de Meynard, C.B. and de Courteille, A.P. (eds) (1861) *Masʿūdī. Les prairies d'or*. Paris.
Muhannā, 'A. (ed) (2010) *Tārīkh al-Yaʿqūbī*. Beirut.
Nokandeh, J., Sauer, E.W., Rekavandi, H.O., Wilkinson, T.J., Abbasi, G.A., Schwenninger, J., Mahmoudi, M., Parker, D., Fattahi, M., Usher-Wilsom, L.S., Ershadi, M., Ratcliffe, J., and Gale, R. (2006) 'Linear Barrier of Northern Iran: The Great Wall of Gorgan and the Wall of Tammishe'. *Iran* 44: 121–73.
Okacha, S. (ed) (1960) *Ibn Qutayba, Kitāb al-maʿārif*. Cairo.
Panaino, A. and Piras, A. (eds) (2006) *Proceedings of the 5th Conference of the Societas Iranologica Europaea. Volume 1: Ancient and Middle Iranian Studies*. Milan.
Payne, R. (2016) 'The Making of Turan: The Fall and Transformation of the Iranian East in Late Antiquity'. *JLA* 9: 4–41.
Pfisterer, M. (2014) *Hunnen in Indien: Die Münzen der Kidariten und Alchan aus dem Bernischen Historischen Museum und der Sammlung*. Vienna.
Potts, D. (2018) 'Sasanian Iran and Its Northeastern Frontier: Offense, Defense, and Diplomatic Entente', in Di Cosmo and Maas (2018) 287–301.
Rahbar, M. (2008) 'The Discovery of a Sasanian Period Fire Temple at Bandiyān, Dargaz', in Kennet and Luft (2008) 15–40.
Rahman, A., Grenet, F., and Sims-Williams, N. (2006) 'A Hunnish Kushan-shah'. *Journal of Inner Asian Art and Archaeology* 1: 125–31.
Rante, R. and Collinet, A. (2013) *Nishapur Revisited: Stratigraphy and Ceramics of the Qohandez*. Oxford.
Rante, R. and Raimkulov, A, (2013) 'Les fouilles de Paykend: nouveaux éléments'. *Cahiers d'Asie centrale* 21/22: 237–58.

Rezakhani, K. (2010) 'Empires and Microsystems: Late Antique Regional Economy in Central and West Asia, 500–750', PhD Dissertation, University of California, Los Angeles.

Rezakhani, K. (2015) 'A Note on the Alkhan Coin Type 39 and Its Legend'. *DABIR* 1: 24–27.

Rezakhani, K. (2017a) 'From the Kushans to the Western Turks', in Daryaee (2017) 199–225.

Rezakhani, K. (2017b) *ReOrienting the Sasanians: East Iran in Late Antiquity*. Edinburgh.

Rezakhani, K. (2019) 'Miirosan to Khurasan: Huns, Alkhans and the Creation of East Iran'. *Vicino Oriente* 23: 121–38.

Sauer, E.W., Rekavandi, H.O., Wilkinson, T.J., and Nokandeh, J. (2013) *Persia's Imperial Power in Late Antiquity: The Great Wall of Gorgān and Frontier Landscapes of Sasanian Iran*. Oxford.

Schindel, N. (2003) *Sylloge Nummorum Sasanidarum III. Shapur II. – Kawad I./2. Regierung*. Vienna.

Schindel, N. (2006) 'The Sasanian Eastern Wars in the 5th Century: The Numismatic Evidence', in Panaino and Piras (2006) 675–89.

Sims-Williams, N. (2001) *Bactrian Documents from Northern Afghanistan, I*. Oxford.

Sims-Williams, N. (2002) *Indo-Iranian Languages and Peoples: Proceedings of the British Academy* 116. Oxford.

Sims-Williams, N. (2007) *Bactrian Documents from Northern Afghanistan, II*. Oxford.

Sims-Williams, N. (2008) 'The Sasanians in the East: A Bactrian Archive from Northern Afghanistan', in Curtis and Stewart (2008) 88–102.

Skaff, J.K. (2002) 'Western Turk Rule of Turkestan's Oases in the Sixth through Eighth Centuries', in Inalcik (2002) 364–72.

Syvänne, I. (2021) 'The Three Hephthalite Wars of Peroz 474/5–484'. *Historia i Świat* 10: 95–116.

Thomson, R.W. (ed) (1982) *Ełiše: History of Vardan and the Armenian War*. Cambridge, MA.

Thomson, R.W. (ed) (1991) *The History of Łazar P'arpec'i*. Atlanta, GA.

Vondrovec, K. (2014) *Coinage of the Iranian Huns and Their Successors from Bactria to Gandhara: 4th to 8th Century CE*. Vienna.

Vondrovec, K. (2008) 'Numismatic Evidence of the Alchon Huns Reconsidered'. *Beiträge zur Ur- und Frühgeschichte Mitteleuropas* 50: 25–56.

Warner, A.G. and Warner, E. (ed) (1905–1925) *The Shāhnāma of Firdausī*. London.

Zotenberg, H. (ed) (1867) *Chronique de Abou-Djafar-Mohammed-ben Djarir-ben-Yezid Tabari, traduite sur la version persane d'Abou 'Ali Mohammed be Bel'ami*. Paris.

Zotenberg, H. (ed) (1900) *Tha'ālibī, Ghurar akhbār mulūk al-furs wa sīyarihim*. Paris.

CHAPTER 17

Persian-Roman Diplomacy in the Fifth Century

Craig Morley

> [Yazdgird I] established and always kept a profound peace with the Romans.
>
> Procop. *Pers.* 1.2.9

•••

> Unceasing embassies were sent to and fro between the sovereigns of Persia and the Roman Empire, for which there were continual occasions.
>
> SOCRATES, *Hist. eccl.* 7.8.2

In the anarchic ancient world, interstate relations were characterised by competition, usually in the form of warfare. The other chapters in this volume confirm this was certainly the case for Iran and its western neighbours. In such a competitive world, unique circumstances were needed for any long-term peace to exist between rival states. As the quotes above show, these unique circumstances existed in the Sasanians' fifth-century relationship with the Roman Empire. Unlike the third, fourth, sixth, and seventh centuries, which were all marked by long and frequent wars, the fifth century was remarkably peaceful, as only two quickly resolved conflicts erupted in 421 and 441.[1]

This chapter is interested in *how* peace was established and maintained, rather than *why* the empires needed peace in the fifth century, on which there is largely consensus. A two-stage approach will be used to achieve this. First, a narrative account of the treaties signed between the empires in the late fourth and fifth centuries will show the slow, adaptive, and iterative process that was the fifth-century peace. This will be followed by an examination of some of the

1 Luther (2014) has argued that there was perhaps a third, shorter conflict in 416/17.

 | DOI:10.1163/9789004710771_018

key diplomatic protocols which developed in this period. It will be argued that these protocols allowed diplomacy to act as a viable alternative to warfare in creating a mutually acceptable balance of power.

1 Diplomacy and the Fifth Century

Modern diplomacy is generally understood as direct dialogue or negotiation between different states, principally with the aim of maintaining stable international relations.[2] The *Oxford Dictionary* defines diplomacy as the 'management of international relations by negotiation'. This modern view is centred on the presumption that international relations are largely controlled and regulated by international law, with entrenched protocols and traditions, all of which are designed to help stimulate peace-making and peace-maintenance. The ancient world, in contrast, was much more anarchic; warfare played a more consistent and determinant role in interstate relations than diplomacy. Ancient diplomacy was less a dialogue between states and more the direct dictation of one state to another, usually at the conclusion of a military campaign. Military victory, not negotiation, was the predominant means of managing interstate relations.[3]

Nevertheless, interstate diplomacy and relations are a product of their time and develop in relation to new circumstances.[4] In the third and fourth centuries this meant the balance of power was decided by warfare, with each empire constantly striving to strengthen their position over the other.[5] Consequently, diplomacy did not develop much beyond a role in confirming military victories. The realities of the fifth century changed this, however. The driving force behind this change was the rise of new threats on other frontiers and

2 Watson (2005) xv; Eilers (2009) 1.

3 Polybius (1.37.7) and Vegetius (*De Re Mil.* 3) show the traditional Roman view that peace was only achieved through war. Eckstein (2006) made a convincing case for this kind of political realism in explaining the Roman Republic's rise to dominance in the ancient Mediterranean. Likewise, Woolf (1993) argues that Roman diplomacy was interwoven so deeply into military conflict that it was little more than just another aspect of war. Although there is less evidence from the Sasanian perspective, Eckstein's comment (2006, 13) that states live in connected systems, not isolation, would imply that by necessity, they too would have likely viewed diplomacy in the same way. Certainly, the language of fraternal Sasanian-Roman co-existence that developed in the late fourth and fifth centuries (Canepa, 2009) is not evident in Sasanian relations with other neighbours.

4 Watson (2005) xviii. See Nabel, in this volume.

5 Sauer *et al.* (2017, 248–52) summaries how military conflict decided the balance of power in this period. See Maksymiuk, in this volume.

concurrent changes in Roman emperorship.[6] The shared frontier in Armenia and the Caucasus, from where Hunnic raids threatened both empires, was particularly important in shaping the need and development of imperial diplomacy.[7] Therefore, the first, and arguably most important, thing to recognise when discussing fifth-century peace is that it was the result of self-interest and shared strategic concerns, not a sudden move towards pacifism.[8]

Despite peace increasingly becoming a necessity for both sides towards the end of the fourth century, the militaristic nature of imperial relations up to this point meant diplomacy did not have the sophistication, flexibility, or symbolic importance to act as an alternative to war in managing imperial relations. As shall be seen below, it was only through a slow, adaptive and iterative process of diplomatic action and reaction that the empires were able to establish diplomatic protocols that were strong enough to mitigate against competing imperial and religious ideologies, cultural competition, economic interests and strategic concerns.[9]

2 The Road to Peace

To understand the development of the fifth-century peace, we must analyse the treaties negotiated in this period: namely, 363, 387, 400, 408/9, 422, and 442. It should be noted that there are ongoing debates around the dating of certain treaties and clauses. These debates will not be the focus of this narrative, however. The chief concern here is the overall development of these diplomatic agreements. A narrative account will, therefore, be used to show how successive treaties built upon one another to address challenges that would usually have led to and/or been resolved through warfare.

6 On the rise of new threats on different frontiers see Drijvers (2008); Greatrex (2000); Morley (2015); Rubin (1986); Rubin (2000); Wiesehöfer (2007) 206. As Ball (2017) and Börm (2019) point out, the eastern frontier was arguably always more important to the Sasanians than their western frontier. Importantly, as noted by Börm (2016, 2019, 2021), emperors in the fifth century were more palace-bound than their predecessors and correspondingly less willing to risk military commanders winning prestige through victories over the Sasanians.

7 Greatrex (1999, 65) describes the effect the Hunnic raid of 395, which took place through the Caucasus, had on the mentality and thinking of both empires. He also highlights that a similar Hunnic raid of some sort occurred in 397 which would have likely reinforced this notion.

8 In this regard, Campbell (2001, 3) states that 'diplomacy takes place when those in control decide that it is in their material self-interest'.

9 Blockley (1992) 151, 166; Drijvers (2008) 450; Whitby (2008) 123.

The treaty of 363, signed between Shapur II and Jovian, began the process towards fifth-century peace. This treaty has traditionally been viewed in parallel with the 299 treaty, as a representation of the militaristic, competitive nature of imperial relations.[10] In 299, the Romans negotiated from a dominant position following Galerius' military victory over Narseh to create a balance of power favourable to themselves. In turn, Shapur's victory over Julian in 363 allowed the Sasanians to reverse the outcomes of 299 to their own benefit, namely forcing the Romans to surrender Nisibis and Singara.[11] The argument goes, therefore, that the treaties of 299 and 363 were two sides of the same bellicose coin. However, there are important, subtle differences between the two treaties which mean the treaty of 363 should not be seen in parallel to that of 299, but rather as the beginning of a shift in imperial priorities and imperial diplomacy itself.

Despite Sasanian hatred of the 299 treaty and the corresponding internal pressure from the noble *wuzurgān*, whose own lands would have been better protected by an extended frontier, to fully reverse that settlement to the advantage of the Sasanian Empire,[12] Shapur was seemingly determined this new treaty would last, to allow him to concentrate on the empire's other frontiers without worrying about Roman retaliation. Thus, compared to previous negotiations, Ṭabarī notes the conciliatory, pragmatic approach Shapur took with the new emperor, Jovian.[13] Although Ṭabarī's description was likely an attempt to buttress Shapur's reputation as a magnanimous ruler to some degree, the shah's concessions do indicate an acceptance of the need to gain Roman acquiescence with the new balance of power if it was to endure long enough for him to take advantage of it on other frontiers. Therefore, to win Roman acceptance, Shapur first allowed the Roman populations of Nisibis and Singara to evacuate before the Sasanian takeover.[14] Second, and more importantly, rather than recover all the lands lost in 299 and more, which he had the power to do, Shapur reclaimed only the five most strategically important *regiones Transtigritanae*,

10 Blockley (1984).

11 On the 299 treaty, see Amm. Marc. 25.7.9; Fest. 14.25.; Pet. Patric. fr. 13–14. For more detailed discussion on the territorial settlement of the 299 treaty, see Luttwak (1979) 154–55; Blockley (1984); Dignas and Winter (2007) 122–31. On the 363 treaty, see *Cod. Iust.* 4.63.4; Amm. Marc. 25.7.9–14.

12 Amm. Marc. 17.5.2–4. Sasanian hatred of the 299 treaty is noted by Barceló (1981) 74; Klein (1977) 185; Wirth (1980) 336–37. However, this view is countered by Dignas and Winter (2007, 130) who argue that given the decisive nature of the Roman victory, Diocletian's demands were relatively modest and restrained.

13 Ṭab. 843.

14 Mal. 13.27; Amm. Marc. 25.9.

alongside Nisibis and Singara, while allowing the Romans to retain all territory beyond the Nymphius River, primarily Sophene and Ingilene.[15] In allowing the Romans to hold onto these territories, Shapur created a new balance of power in Mesopotamia that was acceptable to both empires. Roman satisfaction is evident in the fact that although more bellicose Roman historians regarded the treaty as a humiliation, the majority took a balanced view and regarded it as largely acceptable.[16] Perhaps even more tellingly, although each empire tried to make economic and political gains at the other's expense following 363, neither tried to alter the Mesopotamian territorial settlement.[17] Shapur's wish for a durable agreement is also evident in two vital aspects of the 363 treaty: the written recording of the agreement and the inclusion of a thirty-year time period for which the treaty was to hold.[18] These two clauses were important in the development of imperial diplomacy and its transition away from the short-term, military-adjunct diplomacy of the third and fourth centuries to the more determinant diplomacy of the fifth century. The 363 treaty can therefore be viewed as the first example of either empire taking a more conciliatory approach to imperial diplomacy in the hopes it would maintain a balance of power that would allow them to concentrate on other frontiers.

Despite these developments, the treaty still left other *casus belli* unaddressed, the most potentially damaging of which was Armenia. The Armenian clause was ambiguous and the division of imperial authority in the region was ill-defined, causing both empires to perceive the resolution differently. The Sasanians believed the treaty awarded them a free hand to conquer Armenia as and when they were able. In contrast, the Romans believed they had only agreed to withdraw their troops and not to support the Armenian king, Arsak, militarily; not that they had granted the Sasanians full suzerainty over the

15 Zon. 13.14.4–6; Zos. 3.31; Amm. Marc. 25.7.9–11. Blockley (1992) 240.

16 Libanius (*Or.* 1.134, 18.277–8), Eutropius (10.17), Eunapius (fr. 29.1.3–6,10–15), Festus (29) and Agathias (4.26.6–7) all regard the treaty in a negative light. Although Ammianus Marcellinus (25.7.9–14) saw this treaty as a reversal of Diocletian's gains, he takes a less hostile view than most other Roman writers. Other Romans who took a similar attitude include Cedrenus (1.539.16–21), Joshua the Stylite (242.28), Orosius (7.31.1–2) Rufinus (*Hist. eccl.* 11.1), Socrates (*Hist. eccl.* 3.22.5–7), Sozomen (*Hist. eccl.* 6.3.2), Zonaras (13.14.4–6) and Zosimus (3.31.1–2). While those who saw the treaty in a positive light include Gregory of Nazianzus (*Or.* 5.15), Jerome (*Chron.* 364), Malalas (13.27), Philostorgius (*Hist. eccl.* 8.1), Themistius (*Or.* 5.66a) and Theodoret (*Hist. eccl.* 4.2.2–3), it must be noted that the Christian ecclesiastical historians amongst these likely took a positive view as a means to criticise the disastrous circumstances in which Julian had left the Roman army.

17 Lee (1993) 22. This is shown by the fact that neither attempted to alter it again for over one hundred years.

18 Amm. Marc. 25.7.14. 'Thus a peace of thirty years was concluded'.

region.[19] This confusion soon led to conflict in Armenia during the reigns of Shapur II and Valens.[20] Importantly, it has been argued that the lack of conflict in Mesopotamia during this conflict suggests there was a tacit agreement to limit fighting in the 370s to Armenia and Iberia.[21] If so, this would further indicate both sides were still satisfied with the settlement in Mesopotamia. Another limitation of the 363 treaty was trade and commerce. As in 299, Nisibis, now controlled by the Sasanians, was left as the sole official trading-centre between the two empires, which meant the Sasanians had a monopoly on custom dues. This was a blow to the Romans and soon became a grievance. Consequently, although this treaty was important in the development of imperial diplomacy, it was not enough to maintain a mutually acceptable balance of power that would be central to fifth-century peace.[22]

The next major treaty was signed in 387 between Shapur III and Theodosius I, and helped resolve the situation in Armenia.[23] Previous treaties had attempted to establish militaristic and territorial balances of power in Armenia, such as the complete surrender of all authority by one empire, paving the way for its complete domination by the other, or Roman suzerainty with a Roman nominee on throne or vice-versa with Sasanian suzerainty. Such rigid attempts were unsuitable in Armenia, which neither empire was willing to abandon completely and which was characterised by a near constant state of flux and instability.[24]

Late antique Armenia was dominated by fiercely independent and competitive noble clans, *nakharars*. The *nakharars* were experts at gaming imperial competition for their own benefit by frequently switching allegiances and ignoring territorial divisions between the empires.[25] This all contributed to

19 Blockley (1984, 36, 1987, 225) and Greatrex (2000, 36) comment on the ambiguities caused by the Armenian clause.

20 For Roman-Sasanian conflict over Armenia in the period 363–387 see Ammianus Marcellinus (Books 27, 29, and 30) and the *Epic Histories* (5.1, 34, 38).

21 Blockley (1992) 139.

22 Greatrex (2000, 44) claims that the only reason no major conflict erupted soon after was due to the 'intervention of events elsewhere'.

23 Andres (2022) provides an updated and detailed account of imperial diplomacy in this period, which was published after the completion of this chapter.

24 Armenia had long been a source of conflict between the two empires and was regularly used as a justification for war (Amm. Marc. 17.5.6, 26.4.5–6; *ŠKZ* §4.5). On the strategic and economic importance of Armenia, see Russell (2005) 25; Lang (1983) 505. Greenwood (2017, 200) illustrates the inherent internal instability in Armenia.

25 *Nakharars* ruled their vast territories as virtual kings; each had judicial and administrative control over their lands and people: see Dignas and Winter (2007) 159; Lang (1983) 510; Lenski (2002) 154. On the *nakharars*' ability to pursue their own agenda despite imperial

a fluctuating balance of power that made any long-term stability in Armenia difficult. Therefore, it was the ambitious *nakharars*, not the land, which the empires needed to control and partition if a stable balance of power was to be created. Despite several years of inconclusive fighting in Armenia and a growing realisation that the region was a continual source of conflict,[26] this was no easy task.

The difficulty of reaching a diplomatic resolution in Armenia was evident in the fact that negotiations rumbled on for six years before the treaty of 387 was finally agreed.[27] As with Shapur II's desire to focus on other frontiers in 363, it was the actions of outside forces that forced both empires to make genuine concessions to reach a mutually-satisfactory settlement. This outside force was Manuel Mamikonean, a powerful *nakharar*, who before his death in 385 managed to temporarily unite enough noble clans to begin pushing back Sasanian and Roman forces in Armenia.[28] This setback, which threatened the position of both empires, seemingly led both to reconsider their tactics, and the fallout of Manuel Mamikonean's death in 385 presented the perfect opportunity for a new diplomatic approach. Armenia and the *nakharars* were once more violently divided between pro-Roman and pro-Sasanian factions, each led by a different Arsacid king, Arsak IV and Khosrow III, respectively.[29] Internal instability in both empires, however, meant that neither had the appetite to support Arsak IV or Khosrow III.[30] Therefore, rather than be drawn into another conflict, the empires agreed to divide influence in Armenia between themselves according to the existing division amongst the *nakharars*, represented by the two Arsacid kings.[31] This partition was ratified in the treaty of 387, which also included a clause that forbade either empire from accepting the allegiance of

pressure, see Morley (2015) 169–78; Preiser-Kapeller (2010). For individual instances of *nakharar* side-switching see *Epic Histories* (4.23:147) and Sebeos (16.88).

26 Amm. Marc. 30.2.2.

27 Oros. 7.34.8; Them. *Or.* 16.212d–213a; *Pan. Lat.* 2 (12) 22.4–5.

28 *Epic Histories* 5.37–44.

29 Greatrex (2000) 43.

30 Following Shapur II's death, the Sasanian Empire was wracked by internal unrest. Ardashir II, Shapur III, and Wahram IV all had short, turbulent reigns and were forced to concentrate on their internal position, while Theodosius himself had to confront the usurper Magnus Maximus in the west. Blockley (1992, 44) and Greatrex (2000, 41) believe that the willingness for accommodation was absent during the reigns of the earlier rulers Shapur II and Valens.

31 *Epic Histories* 6.1; Procop. *Aed.* 3.1.12–15; Procopius made a chronological mistake as he credits this to the reign of Theodosius II (408–450).

a clan that belonged in other's sphere of influence.[32] This clause negated the *nakharars*' ability to destabilise imperial relations by preventing them from switching sides. Both empires' adherence to this clause, even when presented with opportunities to detach *nakharars* from their rival's influence, underlines how necessary peace was becoming.[33]

This diplomatic solution, which required a good deal of trust, stabilised the balance of power on the northern frontier. The treaty of 387, therefore, was another important step towards peace.[34] It is also important to note that, as this treaty was signed in the thirty-year time limit of the 363 treaty, there was no discussion about the frontier in Mesopotamia, again underlining the iterative nature of these treaties.[35] However, despite the progress made, 387 did not resolve all tensions between the two empires; grievances about the southern frontier in Arabia and trade restrictions still remained unanswered.

As important as the treaties of 363 and 387 were in establishing the foundations for fifth-century peace, they still followed the usual pattern of Sasanian-Roman diplomacy: they were signed to end ongoing conflicts. The Hunnic raid of 395, and to a lesser extent, those that followed in 397 and 398, proved a catalyst in disrupting diplomacy's traditional position as an adjunct of war. These attacks through the Caucasus effected both empires[36] and pushed them towards renewed negotiations that led to the treaties of 400 and 408/9. Importantly, for the first time in Sasanian-Roman relations, both treaties were concluded while the empires were at peace. As well as reconfirming adherence to the agreements signed in 363 and 387,[37] these treaties sought to strengthen the existing status quo.

The first of these two treaties was signed by Arcadius and Yazdgird. Both quickly realised the threat posed by the Huns and worked to negotiate a new treaty that would allow them to focus on this new danger. The expiration of the

32 This clause was later reconfirmed in the treaty of 562, as Menander Protector (fr. 6.1) informs us that the imperial powers agreed that 'those [*nakharars*] who in time of peace defected, or rather fled, from one to the other shall not be received, but every means shall be used to place them, even against their will, in their hands of those from whom they have fled'.

33 Lazar P'arpets'i 33.36; Jos. Styl. 249.15–23.

34 Greatrex (2000) 40; Isaac (1990) 10; Rubin (1986) 678.

35 Blockley (1992, 139) also argues that the Armenian solution of 387 was simply inserted alongside the main treaty clauses established in 363.

36 Priscus, fr. 11; Liber Caliphorum (*Chron.* 3.4) with Greatrex and Lieu (2002) 17–19.

37 Greatrex (2000, 44) agrees with this assessment.

363 treaty provided a timely opportunity to negotiate a stronger settlement.[38] Therefore, in 400, Arcadius sent an embassy to Yazdgird, led by Anthemius.[39] The bishop, Marutha, who was familiar with the Sasanian court and well-known to Yazdgird after he had reportedly relieved the shah of a chronic headache, also accompanied Anthemius to the Sasanian Empire.[40] The decision to send Marutha may, therefore, be seen as attempt to add a veneer of trust and reliability to negotiations. Continued satisfaction with the status quo in Armenia and Mesopotamia that was negotiated in 363 and 387 is evident in the fact that Roman sources only document one clause of the treaty of 400: the joint defence of the Caucasian frontier, thus reinforcing the notion this treaty was signed in direct response to the increased Hunnic threat. Although at this stage no firm agreement was made, negotiations over shared frontier defence became a recurring theme of fifth-century diplomacy.[41]

Theodosius II's ascension in 401 presented another opportunity to strengthen the burgeoning détente. In 408/9, Anthemius, now one of the leading figures in the Roman Empire, was again dispatched to the Sasanian Empire.[42] The

---

38 Although Claudian (*In Eutropium* 2.474–84) informs us that at the start of his reign, and before the signing of the treaty, Yazdgird threatened war and began planning for skirmishes against the Roman Empire, given his earlier actions of first raising items of concern and potential dispute with the Romans and also his later peaceful policy towards the west this may be seen as a manoeuvre by the shah to force Arcadius to realise the need for a new treaty.

39 There is much debate about whether the initial treaty between Arcadius and Yazdgird was signed in 400 or earlier. The argument for 400 makes use of Theodoret's note (*Hist. eccl.* 8) that while travelling in the Sasanian Empire in 400 he passed Anthemius, who was travelling back from the Sasanian court. Unfortunately, however, Theodoret does not provide precise details that would make this certain. Indeed, it is only from this passage in Theodoret that Anthemius' embassy in 400 is known. Yet, that Marutha, who played an important role in these negotiations, was known to be in the Sasanian Empire in 400 also adds weight to the treaty taking place in the same year (Blockley (1992) 48). Furthermore, it could also be argued that Arcadius would have been too distracted by the threat of Gainas to organise a treaty with the Sasanian Empire before 400.

40 Socrates, *Hist. eccl.* 7.8.1–3.

41 As Blockley (1992, 51) makes clear, there is no direct evidence of this. However, he makes two pertinent points that are important and deserve consideration here. First, if John Lydus (*De Mag.* 3.52–3) is to be believed, negotiations about the joint defence of the Caucasian Gates had been going on since Theodosius I's reign and, therefore, since no agreement had been made then it is likely that negotiations also continued during the reign of Arcadius. Secondly, see the passage of Cedrenus (1.586) where he states that in the course of these same negotiations, Arcadius sent one thousand pounds of silver to the shah clearly as a payment of some sort, and one that Yazdgird could well have put to use in the defence of the Caucasus.

42 *Cod. Iust.* 4.63.4; Socrates, *Hist. eccl.* 7.8; Sozom. *Hist. eccl.* 9.4.

negotiations that followed strengthened the existing balance of power in two ways. First, they addressed the imbalance in trade restrictions that had existed since 363. This was achieved by widening the circle of official trade centres to include Callinicum, in Roman territory, and Artaxata, in Armenia, as well as the original centre in Nisibis, in Sasanian territory.[43] The official reason given in the treaty for this was to prevent trade being used for espionage.[44] It has, therefore, been argued that this clause shows imperial cooperation in controlling cross-border travel. However, controlling cross-border travel in the ancient world, especially across a frontier like that between the Roman and Sasanian empires, was an impossible task.[45] It would also presumably be easier to control travel and prevent espionage if trade continued to be limited to a single city. Therefore, it is tempting to see the official rationale of this clause as a face-saving effort for Yazdgird to avoid internal criticism that he had given up hard-won advantages. The argument that this decision was concerned with economic considerations rather than espionage is further supported by the importance the initial Roman surrender of Nisibis in 363 had for the Sasanians in ending the Roman monopoly on custom dues.[46] The Sasanians' agreement to now end their own monopoly over custom dues underlines their willingness to compromise to achieve much-needed peace.[47] It also shows the growing appetite and ability to use diplomacy to resolve grievances that may have previously led to war.

43 *Cod. Iust.* 4.63.4. The dating for this clause, like the treaty itself, has been debated, with claims (Lee (1993) 62–63) that it was simply a reiteration of an early treaty, most likely 363 or 387. However, whatever the date at which this particular clause was first signed, the argument that it was done so earlier nevertheless still fits into the overall theory presented here, that all treaties signed in the fifth century were largely reiterations of earlier treaties.

44 Greatrex and Lieu (2002) 33.

45 On the porousness of the Roman-Sasanian frontier see Morley (2016); Garsoïan (1983).

46 Dignas and Winter (2007) 133; Blockley (1984) 36. For the importance of trade agreements in Roman-Sasanian peace treaties see Winter (1987).

47 Lee (1993, 63) highlights that the 408/9 treaty states this was done to prevent trade being used as espionage and Greatrex and Lieu (2002, 33) likewise believe this is evidence of cooperation in controlling cross-border travel. However, presumably it would be easier to prevent espionage if trade continued to be limited to the single city of Nisibis rather than widening the net to include Callinicum and Artaxata as well. Dignas and Winter (2007, 133) and Blockley (1984, 36) agree that the original clause handing over Nisibis to the Sasanian Empire in 363 was related to the Sasanians' desire to break the previous Roman monopoly on trans-border trade. Therefore, this would also suggest that any further clause on this aspect was related to the nature of trans-border trade, not espionage.

Secondly, this treaty built on the precedent set in 363 by including a hundred-year time duration.[48] Although ambitious, and perhaps more symbolic than realistic, the desire for such a long-lasting peace further underlines how vital peace had become for both empires in the fifth century.[49] The growing need for a lasting détente is perhaps further evident in the much-debated adoption of Theodosius II by Yazdgird I.[50] This adoption story is important whether it is true or not. If not real, it can be argued that Procopius used it as a literacy device to highlight the changed nature of imperial relations in this period; if true, it fits with other Sasanian concessions and attempts to stabilise their western frontier.

The next two treaties were signed in more traditional circumstances, following two short conflicts. The first, signed in 422, ended the short-lived conflict between Theodosius II and Yazdgird I, and his successor, Wahram V, which erupted due to the treatment of Christians in the Sasanian Empire and, perhaps more importantly, side-switching amongst Arabian allies.[51] Ironically, the outbreak of this conflict reinforces the nature of fifth-century peace as a slow, iterative process that encountered many obstacles. Despite some initial fighting in Mesopotamia and Armenia, both sides' attention was soon forced elsewhere, as Hunnic raiders took advantage of weakened imperial frontiers to attack both empires. Consequently, in 422 a general peace was signed that

48 Sozom. *Hist ecc.* 9.4.

49 From the Roman perspective this treaty should be viewed in the context of Anthemius' wider strategic reforms to defend the Danube against the Huns (*Cod. Theod.* 7.17.1, 7.16.2; Kelly (2008) 58; Morley (2015) 65). His desire to reconfirm peace with the Sasanian Empire was necessary to focus Constantinople's military resources on defending the northern frontier.

50 The dating here has also been debated. Greatrex and Bardill (1996) argue for an early date. They present three factors they believe point to an early date for the adoption. First, that Malalas (349) states Arcadius' death was sudden, and that, therefore, Procopius' claim (*Pers.* 1.2.1–10) that the agreement was reached on the emperor's deathbed cannot have been true. Second, that Antiochus, the Sasanian emissary sent to negotiate the deal arrived at Constantinople several years before Arcadius' death (Priscus fr. 7). Third, that Antiochus was dispatched to Constantinople on Arcadius' request, and that by the time of the emperor's death he had become a man of influence in the Roman court. Therefore, they argue that the Sasanian emissary must surely have been in Constantinople long before 408 if he was an influential man by the time of Arcadius' death. As such, Greatrex and Bardill (1996, 173) suggest that this agreement took place in 402, rather than 408/9. These claims for an early date for the agreement over Yazdgird's adoption of Theodosius have therefore also been used to argue that the treaty of 408/9 must have taken place at a different time. However, Procopius (*Pers.* 1.2.1–10), who first wrote about the adoption, and Theophanes (AM 5900) both say the agreement did take place in 408/9.

51 On the course of this conflict see Socrates, *Hist. eccl.* 7.18; Mal. 14.23; Theod. *Hist. eccl.* 5.37.6–10. Rubin (1986) 681; Rubin (2008) 641.

reconfirmed the *status quo ante bellum*. The need for a quicker resolution, from the Roman perspective, at least, is evident in the granting of unprecedented plenipotentiary powers to Helion, the Roman ambassador, which empowered him to end the conflict as quickly as possible.[52] Nevertheless, the outbreak of this war had shown that, despite the relative success of earlier treaties, peace was still vulnerable to unresolved *casus belli*, in this instance, the Arabian frontier.

The political ambiguity and geographic diffusion of Arab tribes across the Near East made it difficult for the empires to account for the Arabs in peace treaties, especially those that had previously focused on territorial divisions.[53] Such territorial divisions were wholly incompatible on the Arabian frontier were both empires exercised authority through the indirect control of allied tribes.[54] Furthermore, just as their Armenian counterparts did on the northern frontier, Arab tribal leaders exploited imperial competition for their own benefit, by switching sides when profitable opportunities arose.[55] This situation ensured the balance of power in Arabia could change quickly and unexpectedly, which reduced the likelihood of any long-term stability on the southern frontier. This diplomatic difficulty was evident in the absence of any mention of Arab tribes in earlier treaties. Fortunately, however, the 387 treaty, which helped resolve Armenian side-switching, provided a framework for the empires to reach a mutually acceptable compromise on the southern frontier.

Following Arab tribes' role in the outbreak of hostilities in 421 the empires were determined to rectify their previous omissions of Arabs in the 422 treaty. They had begun to recognise the tribes' overall importance in their relationship and that the much-needed détente could not be secured without addressing the Arabs' position. Alongside reconfirming the wider status quo,[56] the treaty included a new clause to overcome the Arab tribes' ability to exploit

52 Socrates, *Hist. eccl.* 7.20.

53 This ambiguity was caused by the many divisions within Arab society and the region, which was in a constant state of flux (Trimingham (1979) 3; Moorhead (2001) 195). Indeed, this ambiguity spawned the many different Roman terms used for the Arabs such as *Sarakenoi*, *Skenitai*, *Scenitai*, *Scenitas*, *Arabes* and *Thamudenoi* (Ptol. *Geog.* 6.7.21, 6.7.23, 6.7.4; SHA *Aurel.* 11.3; Fest. 3.16).

54 Toral-Niehoff (2013) 119. Strabo's account of the role the Arab tribes played in Roman-Parthian relations (*Geog.* 236–7) is similar to their position in Sasanian-Roman relations.

55 For example, the Nasrids wore a crown (*tāj*) that was bestowed on them by the Sasanian shahs (Ṭab. 1.821, 858).

56 *Cod. Iust.* 4.63.6. The Romans also seemingly agreed to continue to make some form of payment to the defence of the Caucasus. Ṭabarī (1.868, 872) states that yearly payments were made, whereas al-Thaʿalibi (p. 560) claims that the Romans purchased peace with a payment of two million dinars and other gifts.

imperial relations by ensuring that 'neither side would accept the Saracen [Arab] allies of the other'.[57] It prohibited either empire from accepting the loyalty of a tribe that was already allied to the other.[58] This agreement prevented the ambition and self-interests of Arab leaders from instigating or intensifying imperial conflicts, as they had done in 421. Significantly, as the flow of Arab side-switching was usually in Constantinople's favour and Arabia was more strategically important to the Sasanians, this clause shows it was not only the Sasanians who were willing to make concessions to secure the détente.[59]

A second conflict in 441 led to the signing of a new treaty in 442.[60] The causes of this war were Roman fortification of frontier towns and lack of contributions to the defence of the Caucasus.[61] For the Sasanians, the fortification of frontier towns had the potential to alter the balance of power that had been so carefully negotiated over the last eighty years. Together with Roman refusal to contribute to the defence of the Caucasus, this likely caused concern about their rival's long-term intentions. Therefore, when the Romans were distracted elsewhere,[62] Yazdgird II's brief show of force in 441 reminded them of the need for peace, and, threatened on both frontiers, the Romans quickly realised the danger they were in and came back to the negotiating table. The treaty of 442 once again reconfirmed the existing *status quo ante bellum*. The Romans also made a vague agreement about re-continuing their contributions to frontier defence in the Caucasus.[63] Procopius also tells us that the construction of any

57 Malchus fr. 1.4–7.

58 Shahid (1989) 36–37.

59 This counters Rubin's argument (1986, 677) that the Sasanians were more interested in working towards peace in the fifth century. For the strategic importance of Arabia to the Sasanian Empire see Morley (2017).

60 Greatrex (1993) and Croke (1984) have dated this war to 440, rather than 441 as is suggested here.

61 Jos. Styl. 9–10; Mal. 18.449. Luther (2014) 185; Rubin (1986) 683–84. Rubin argues that Roman failure, or even refusal, to pay contributions shows they did not value peace as much as the Sasanians. That they were willing to throw it away in order to save money on what was, relatively speaking, an expenditure of little financial consequence. However, such payments, or subsidies, were never a fully integrated clause of any treaty between the empires. Therefore, by refusing to pay, the Romans were not breaking any treaty. For more on the development of this war see Theodoret (*Hist. eccl.* 5.37.5–6), Moses of Chorene (3.67) and Elishé (61–62 [7]). There has been much debate on whether Theodoret was writing about this war or the earlier conflict of 421: Lee (1987) argues for 440 and Croke (1984) argues for 421.

62 Theod. *Hist. eccl.* 5.37. The other threats that occupied the Romans were the Huns (Marcellinus, *Chron.* 441.1) and the Vandals, against whom the Romans were preparing a major expedition.

63 Jos. Styl. 9–10; Mal. 18.449.

new fortifications along the frontier was prohibited.[64] This agreement was another major step in guaranteeing the continuation of the status quo on the Mesopotamian frontier.

Alongside the major negotiations that ratified the above treaties, and which took place 'between the sovereigns of Persia and the Roman Empire',[65] local frontier diplomacy was also important in maintaining the balance of power. For example, in 485 when Sasanian-allied Arab raids raised tensions, Shah Walash, eager to avoid being dragged into a wider conflict, ordered local dignitaries, including a Christian bishop, the regional military commander, and the Arab client leader himself to meet with the local Roman *dux* to resolve the dispute.[66] The ability of these local notables to come to an acceptable resolution highlights the effectiveness of local, as well as central diplomacy in the fifth century.[67] This incident, and others,[68] show that despite the relative success of the major peace treaties, they could not guarantee peace on their own. Effective local diplomacy was needed to help avoid outbreaks of local conflicts, which always had the potential to flare up into wider military entanglements.

As stated earlier, although there are ongoing debates surrounding dates of treaties and individual clauses, analysing the treaties of 363, 387, 400, 408/9, 422, and 442 together helps to see the broad, long-term development of the fifth-century peace. It shows how an often uneasy, but necessary peace was maintained through a slow, iterative, and adaptive process involving six separate, but interconnected treaties which built upon one another to create a, mostly, mutually acceptable balance of power. Indeed, the success of the compromises reached in the earlier treaties at the end of the fourth and beginning

64 *Pers.* 1.2.15. However, Procopius's narrative is unclear here. He confuses the wars and treaties of 422 and 442. Yet even if this was agreed in 422 rather than 442 it still nevertheless reveals the same desire to stabilise relations by seeking to further cement the status quo. Dignas and Winter (2007) 137 state that this clause was agreed in 442, not 422.

65 Socrates, *Hist. eccl.* 7.8.2.

66 *Syn. Or.* 532–4. Nechaeva (2014, 30) remarks that Roman *duces limitanei* could have held considerable influence over diplomatic processes and even strategic directions.

67 The 485 incident also reinforces the strategic priority that peaceful relations with Constantinople had for successive Sasanian shahs. Although some scholars (Blockley (1992) 135; Scholten (1998) 467) argued that direct central imperial involvement was largely absent from frontier diplomacy and that local dignitaries had to act on their own initiative, Walash's order to resolve this issue shows that local notables were still expected to advance the shahs' stated aims: in this case, preserving peace with Constantinople. For the authority of the Roman emperors over their foreign policies and frontiers see Millar (1982); Sidebottom (2007) 8–11.

68 Procop. *Pers.* 1.16.6; Theoph. Sim. 3.9.2.

of the fifth centuries is evident in the fact that the later treaties of 422 and 442 focused predominantly on reconfirming the *status quo ante bellum*.

3 The Nature of Fifth-Century Diplomacy

Diplomacy in the third and early fourth century lacked the sophistication or flexibility to act as a viable alternative to war in managing imperial relations. As seen above, however, the need for peace in the fifth century led to new diplomatic developments that made peace more possible. This section discusses why developments such as writing and storing peace treaties, the inclusion of specific time-durations, and the ritualisation of diplomacy were important in establishing and maintaining peace.

3.1 *Writing and Storing Treaties*

The first explicit example of a Roman-Sasanian treaty recorded in writing was that of 363, and later evidence shows that the treaties of 506, 532, and 562 were also written down.[69] It has been suggested that a major difference between the Romans' treaties with their northern neighbours and the Sasanians was the written recording of treaties.[70] Importantly, however, there is no evidence that any earlier Sasanian-Roman treaties were recorded in writing, which suggests that the 363 treaty was innovative in this regard. As suggested above, the decision to record the 363 treaty may have been linked with the need for a more durable détente in this period. Correspondingly, there is no explicit evidence that the treaties of 387, 400, 408/9, 422 and 442 were written down in turn. Although given the very detailed description of the jointly agreed law of 408, likely a clause of the treaty signed in the same year, which attempted to control cross-frontier travel,[71] an argument can be made that written treaties likely became a diplomatic norm after 363. Regardless of whether the subsequent fifth-century treaties were written down or not, the existence of a written record of the clauses agreed in 363, which helped form the basis of the fifth century's iterative diplomacy, was important.

Menander's description of the 562 treaty gives us an insight into how treaties were recorded.[72] Although there may have been some differences between

69 Zos. 3.31.2; Jos. Styl. 98; Men. Prot. fr. 6.1. Lee (2008) 110.

70 Lee (2008) 110–16.

71 *Cod. Iust.* 4.63.4.

72 Men. Prot. fr. 6.1. For more information on the nature and reliability of Menander's work see Blockley (1985); Baldwin (1978).

the protocols described and their fifth-century counterparts, Menander makes clear that writing down the individual clauses of treaties, making multiple copies, recording treaties in the language of both empires, securing them with official seals, and using interpreters and high ranking officials were all important in the ratification of imperial peace treaties.[73] The creation of a corps of bilingual interpreters in the Roman Empire was central to this, while the fact that the Sasanians also recorded treaties in the same way suggests a corresponding corps existed in the Sasanian Empire.[74]

The written recording of treaties had two benefits. First, agreements likely became less ambiguous and open to interpretation. Second, treaties could be stored for future consultation. As seen earlier, imperial peace treaties, especially during the fifth century, contained a vast array of complicated and, at times, highly specific clauses, including time limits, religious toleration, and economic concerns. Therefore, preserving treaties through multiple written copies, in the languages of both empires, was essential if they were to be successful and not misinterpreted as, ironically, the Armenian clause in the 363 treaty was. Indeed, Menander's comment that treaties were kept for future 'reference' suggests that these written records were consulted by officials to some degree. As has been pointed out elsewhere, although there is no direct evidence that the Romans or Sasanians kept written treaties in any sort of archive, it is difficult to imagine they did not.[75] Indeed, there seems no point in writing treaties down if they were not also stored somewhere. Roman archiving of expense details incurred by Sasanian embassies when travelling in the Roman Empire in the *scrinium barbarorum*[76] further supports the notion that they would likely archive more important treaty documents in a similar way.

The above narrative, which highlights the iterative nature of the fifth-century peace, underlines the importance of the written recording of treaties. It would have been difficult for new rulers and their advisers to build upon earlier negotiations if they did not have a clear record of what had been agreed in previous treaties. Thus, the recording and archiving of peace treaties was an important step in creating more stable peace-agreements.

73 For more on the importance of this treaty in our overall understanding of Sasanian-Roman treaties in general see Lee (2008).

74 Priscus (*Or.* 9.46) describes these interpreters as the *interpretes diversarum gentium*. The *Notitia Dignitatum* also mentions the existence of interpreters in the western empire (*Occ.* 11.52).

75 Lee (1993) 36; Nechaeva (2014) 32.

76 Const. Porph. *De. Caer.* 89–90.

3.2 *Time Duration in Treaties*

Time durations were another important symbolic and practical mechanism in the iterative nature of fifth-century peace. Our sources indicate that treaties were made between 'the sovereigns of Persia and the Roman Empire', in power at the time, not between the empires themselves.[77] This arguably suggests that the duration of treaties was linked to the lifetime of the two signatories, and once one died the treaty could be considered void by their successor, who may have wished to pursue a different agenda.[78] This was not very conducive to the creation of long-term peace, as treaties could quickly and unexpectedly become worthless (which is not to say that treaties with set time-periods always lasted the distance, however). Towards the end of the fourth century, when both empires needed a more stable relationship, they attempted to overcome this problem by introducing specific time limits for the duration of treaties. The first inclusion of a time limit was in 363, in which a thirty-year duration was agreed. This was followed by one-hundred-year time duration for the 408/9 treaty, the length of which suggests both empires were interested in, and actively seeking, long-term peace. Clearly neither signatory would expect to still be alive at the end of this treaty. The implementation of time limits allowed the treaties of 363 and 387 to be amended and strengthened with the introduction of new clauses, without having to renegotiate the whole treaty upon the death of one of the original signatories. Indeed, we have already seen that this happened in 400, 408/9, 422 and 442. In this way, time limits gave fifth-century treaties more continuity and longevity than their earlier counterparts.

The fact that no 'eternal' peace treaty was signed in the fifth century underscores the notion that peace was a necessity for both empires at this time, not an overall shift in wider relations. It has also been argued elsewhere that the Romans were traditionally reluctant to sign longer term peace treaties, which the Sasanians preferred, as this usually meant agreeing to pay subsidies for a corresponding amount of time.[79] Therefore, Roman willingness to agree longer treaties in the fifth century underlines their need for peace in this period.

---

77 Socrates, *Hist. eccl.* 7.8.2. For example, Shapur presents the 244 treaty as an agreement between himself and Philip (*SKZ* §8) and the 298 peace treaty is negotiated and described in terms of the relationship between Narseh and Galerius (Pet. Patric. fr. 13–14). The much later 628 treaty is similarly described as between Heraclius and Kavād II (*Chron. Pasch.* 628). The Eternal Peace signed by Justinian and Khosrow was an exception in this regard (Mal. 477.15–16).

78 Blockley (1992) 161.

79 Nechaeva (2014) 108.

After the Hunnic, Vandal, and Hephthalite threats had been defeated or receded, the empires returned their military attention to one another and time durations became less important. Tellingly, the first peace treaty signed in the sixth treaty, in 506, included only a seven-year time limit, much less than those agreed in the fifth century.[80] This suggests both empires were now less interested in the long-term maintenance of the status quo. Even when longer peace treaties were negotiated, they proved much more fragile than their fifth-century counterparts. In 532, Justinian and Khosrow signed their Eternal Peace, which was intended to last 'for the duration of both their [the Roman and Sasanian empires'] lives'.[81] In the end, however, the treaty lasted only eight years before Khosrow seized the opportunity presented by the decline of Roman defences to attack Justinian in 540 and start a war that raged for twenty-two years. Justinian's willingness to neglect his eastern defences and allow them to fall into disrepair in the first eight years of the treaty[82] may ironically underline the success that time limits had in the fifth century. Justinian may have believed he could rely on Khosrow to maintain peace for the duration of the treaty, just as his predecessors had largely done throughout the fifth century. However, without the simultaneous threat on their other frontiers, which had existed in the fifth century, it proved harder to ignore opportunities presented by the rival's vulnerability. This was again reinforced by the treaty of 562, which was intended to last for fifty years, but only endured for ten years before another long war broke out.[83] The negotiations for the treaty of 562 also seem to show there was more debate about the length of peace treaties in negotiations compared to those in the fifth century.[84] This is perhaps indicative of the wider unravelling of the shared fifth-century strategic priorities and shows that peace was no longer a necessity for both empires.

Time limits played an important role in the development of the temporary and ad hoc nature of diplomacy in the third and fourth century into the more adaptive and iterative diplomacy of the fifth century. However, it is also clear that time limits themselves did not guarantee peace, even in their own lifespan. The sixth century underlines that no matter how much imperial diplomacy developed, peace was always dependent on wider strategic concerns and the ambitions of individual rulers.

80 Procop. *Pers.* 1.9.24.

81 Mal. 477.15–16. For more information on the Eternal Peace see Greatrex (1998) 213–21 and Procop. *Pers.* 1.12–22. See Börm, in this volume.

82 Greatrex (1998) 218–21.

83 Men. Prot. fr. 6.1.

84 Men. Prot. fr. 6.1. Dignas and Winter (2007) 138 believe this was due to a clash between Sasanian eagerness and Roman reluctance for the long-term payment of subsidies.

3.3 *Ritualisation of Imperial Diplomacy*

The increased number of treaties signed in the fifth century, two of which were signed during times of peace rather than war, created the conditions for changes in diplomatic rituals that made imperial diplomacy and negotiations more peaceful.[85] Although there were exceptions, negotiations in the third and fourth century predominantly took place on the battlefield, in a very ad hoc and irregular manner.[86] It seems fair to assume that negotiations which take place on the battlefield are more likely to be rushed, incomplete, and not very conducive to lasting peace. Indeed, it was the negotiations between Jovian and Shapur II in the battle-charred lands of Mesopotamia that led to the ambiguous Armenian clause in 363.[87] In the fifth century, negotiations moved away from the battlefield and were more frequently conducted either at the frontier or at the Sasanian or Roman court. As we have already seen, the high point of peaceful negotiations took place in 400 and 408/9, in the Sasanian court, independent of any military action. This change in location was a catalyst in the formalisation and professionalisation of imperial diplomacy.

The tenth-century Byzantine Constantinus Porphyrogenitus' *De ceremoniis aulae byzantinae* and Ferdowsi's eleventh-century *Shahnameh* (Book of Kings) are useful in analysing the impact the fifth century had on the formalisation of diplomatic contact.[88] The *Shahnameh* describes a third-century Roman embassy dispatched to treat with Ardashir, while Constantinus Porphyrogenitus recounts a sixth-century Sasanian diplomatic mission to Constantinople led by Yazdgushnasp. By comparing the two accounts, we can begin to identify the protocols that likely developed in the intervening period, although we must be cautious in too definitively linking any differences related in the two sources to the fifth-century peace itself. Likewise, caution is also needed when using these two sources themselves. Neither was contemporary with the events it describes and the historicity of the *Shahnameh* is undermined by its poetic composition, which meant that literary techniques often took precedence over veracity.[89] Therefore, where historical accuracy can be gleaned, it must come from corroboration with other sources. Fortunately, Constantinus Porphyrogenitus' work, which was compiled from Roman archival records and used

85 Canepa (2009) 123 identifies the relationship between Arcadius and Yazdgird as an important period in the development of diplomatic interactions that prevailed in the sixth century.

86 Canepa (2009) 123; Garsoïan (1983) 574.

87 Amm. Marc. 25.7.

88 See *Shahnameh* 218; *De Caer.* 89–90.

89 For the poetic nature of the *Shahnameh* see Davidson (1994).

to give practical advice to imperial agents,[90] relates similar protocols as those mentioned by Ferdowsi, and, therefore, supports the *Shahnama*'s veracity in this regard at least. It is unlikely that this comparison will allow us to see all the new rituals that emerged from the third to the sixth centuries. However, it is likely, and has also been argued elsewhere, that the sustained period of peace in the fifth century, with its increased diplomatic contacts, was the stimulus for many changes in the protocols of imperial diplomacy.[91]

One of the first things that stands out when comparing these accounts is the number of officials involved in diplomacy, particularly in the sending and receiving of embassies. Ferdowsi mentions only two officials in the third century, the *marzbān*, who was responsible for preparing accommodations for Roman envoys, and the *kardār*, who liaised with the shah and organised an escort for the Romans.[92] In contrast, Constantinus Porphyrogenitus mentions eleven officials involved in the sixth century: the *magister officiorum*, *praefectus orbis*, *silentarius*, *comes privatorum*, *sacellarius*, *optio*, *magistrianus*, *ducici*, *labaresioi*, 'men from the workshop', and 'men from the arsenal'. These officials had individual, and often highly specific, responsibilities.[93] For example, the *comes privatorum* bore 'the expense of the bed linen' that was to be provided to the Sasanian envoy. This indicates a rise in the specialisation of imperial diplomacy; as diplomatic protocols increased, the number of officials involved in ensuring these expectations were upheld also increased. It could, however, be claimed that Ferdowsi may simply not have wanted to undermine the poetic flow of his work by including too many officials' titles in this passage or that the Sasanians may simply have not involved as many officials as the Romans. However, given the potential damage perceived insults could cause in negotiations and the ever-present competition for imperial prestige, it is unlikely that the shahs would wish to be outdone by their Roman rival's hospitality. Therefore, we can imagine this specialisation equally took place in the Sasanian Empire. An increase in the number of personnel involved in imperial diplomacy certainly suggests a corresponding increase in its importance in imperial relations. As diplomacy became an alternative to war in maintaining the balance of power in the fifth-century, emperor and shah alike were willing to expend more resources to ensure its success.

90 For more on his work see Dignas and Winter (2007) 247–48; Toynbee (1973); Sevcenko (1992). Porphyrogenitus' reliability is further strengthened by the likelihood his information came from Peter the Patrician (Dignas and Winter (2007) 248).

91 Canepa (2009) 123.

92 *Shahnameh* 218.

93 *De Caer.* 89–90.

The two accounts also describe many similar protocols surrounding the sending and receiving of diplomatic embassies: providing an escort for the foreign embassy, covering its expenses, providing accommodation and the involvement of high-ranking officials. However, the language used in the later *De ceremoniis aulae byzantinae* suggests an increase in diplomatic ritualisation. The word 'necessary' features eight times, while 'customary' also makes regular appearances when recording diplomatic expectations. Almost every aspect of Yazdgushnasp's journey in Roman territory was carefully regulated by traditions and customs that cannot be identified in the earlier third century. There were strict regulations and rituals: from the number of horses and pack animals Sasanian envoys were provided with (five horses and thirty pack animals) to a customary time limit of 103 days to complete their journey from the frontier to Constantinople. This is not to suggest that embassies and envoys were treated harshly, however. They were afforded many courtesies by the host empire: all their expenses were paid, just as they were in the third century, they were given gifts, and even a choice of travelling options when crossing from Helenopolis to Nicomedia, whether 'on foot' or 'by light vessels'. Although some of these rituals were likely used to monitor envoys and prevent espionage, they were also important in guaranteeing envoys reached their destinations safely. It was in both empires' interest to ensure negotiations went smoothly when they shared a need for peace, and these protocols helped to achieved this.

Modern diplomacy would regard these rituals as costly signalling: the act of 'making small but significant gestures that serve to prove one is trustworthy'.[94] The nature of diplomacy in the third and fourth centuries did not do much to build trust between the empires, and distrust was almost inherent in negotiations.[95] In the fifth century, however, trust, or at least reliability, became a precious commodity. Therefore, adhering to more costly and time-consuming diplomatic protocols may have helped to signal reliability, if not complete trustworthiness.[96] Fulfilling these rituals may have built up a layer of trust that was also used to show wider reliability in relation to not upsetting the fragile balance of power that maintained peace. The symbolic language of imperial interdependence in diplomatic discourse in the late fourth and fifth

94 Kydd (2005) 5.

95 Menander Protector (6.1.105–108) and Procopius (*Pers.* 1.14.4) reveal the inherent distrust in imperial negotiations.

96 Whitby (2008, 123) argues that good faith and trustworthiness were especially important in Rome's diplomacy on the eastern frontier. Lee (1991, 2009) also stresses that the lack of Roman 'dirty tricks' and differences in the use of hostages in Roman-Sasanian diplomacy were a further sign of the building of trust between the two empires.

centuries should be regarded as another element of this.[97] Failure to comply with these rituals could cause offence but also, on occasion, result in the failure of an embassy as a whole.[98] Thus, the expanded number of officials involved in supervising these rituals seems an attempt to add robustness to the ritual production-line and mitigate the risk of a single point of failure that could potentially derail negotiations and even peace as whole.

Both Ferdowsi and Constantinus Porphyrogenitus also inadvertently reveal another truism of Sasanian-Roman diplomacy: emperors and shahs never met face-to-face. This did not change in the fifth century, even as negotiations increasingly took place in the two respective courts.[99] Consequently, as diplomacy became increasingly important in maintaining the balance of power and treaties became much more complex, both rulers needed officials they could trust to negotiate on their behalf and supervise the vast array of officials that were now involved in diplomacy. Thus, during the fifth century, two of the highest-ranking officials in both empires, the shah's personal chamberlain and the *magister officiorum*, led diplomatic embassies and negotiations.[100] The often urgent need to re-assert peace meant that these officials required greater ability to adapt and react in negotiations without having to refer everything back to the sovereigns.[101] It was this need that led to plenipotentiary powers being granted to the *magister officiorum* Helion for the first time in 422, although a sixth-century example, the Sasanian diplomat, Mebodes, was also seemingly granted plenipotentiary powers by Khosrow to negotiate with the Romans, which may indicate that earlier diplomats were granted similar powers

97 Fest. *Brev.* 4.44; *Syn. Or.* 37; Theoph. Sim. 4.2.2–3. Both Maksymiuk (2018) and Garsoïan (1983, 591) make the case that this type of language did not show true mutual respect, but rather hid animosity beneath diplomatic niceties. Importantly, however, this does not refute the overall premise here, that peace, and therefore diplomatic trust, was mutually *necessary* in the fifth century, not that both empires suddenly felt inherently peaceful towards one another.

98 Amm. Marc. 17.5.1–2; Procop. *Pers.* 1.2.15; Theoph. Sim. 3.9.

99 Even when Jovian and Shapur II were on the same battlefield after Julian's death in 363 they did not negotiate in person (Amm. Marc. 25.7.5–8). Ṭabarī's claim (1.843) that Shapur II and Jovian did meet face-to-face, that they 'both fell on the ground before each other' and that Shapur 'embraced [Jovian] for what he had done' before negotiating the 363 peace treaty seems unlikely.

100 Lee (1993) 41; Lee (2008) 108–109; Nechaeva (2014) 26–29. Priscus gives accounts of *magisters* advising Roman emperors on issues beyond their frontiers. On Theodosius II and Martialis, see Priscus (fr. 2.1/59–66). On Marcian and Euphemius, see Priscus (fr. 33.2). Jones (1964) 369 described the *magister* as 'a sort of minister of foreign affairs'.

101 Ammianus Marcellinus (30.2.4) suggests fourth-century negotiators had to follow strict instructions with little room for manoeuvre.

in the fifth century.[102] It should be noted, however, that these new powers did not extend to full freedom of action and final agreements still needed to be ratified by the exchange of letters between shah and emperor.[103] The increased authority and centrality of the *magister officiorum* and the shah's personal chamberlain in negotiations gave imperial diplomacy a level of organisation and status it had previously lacked. The rising importance, and need, of such officials also suggests diplomacy had now developed into an important arm of imperial relations in its own right.

We should not overstate the professionalisation that took place, however. On the Roman side, below the *magister officiorum* the selection criteria for participating in an embassy varied widely and were dependent on status and influence as much as diplomatic training.[104] Nevertheless, it seems evident that the need for and existence of peace provided the catalyst for the increased ritualisation and professionalisation of imperial diplomacy that made peace possible.

The innovations and developments described above gave imperial diplomacy the flexibility, sustainability, and ritual status to act as a viable alternative to war in managing imperial relations in the right circumstances. Imperial diplomacy was now capable of establishing and maintaining a stable balance of power, *if* both empires had a simultaneous need for peace. Treaties were more durable and resilient against the inherent tensions and changes that could quickly take place in Roman-Sasanian relations and within the two themselves. The shared need for peace in the fifth century allowed treaties to be negotiated on more 'equal terms' than the treaties in previous centuries which existed merely to reinforce the short-term military dominance of one side over the other.[105] This created a negotiating environment in which both sides were willing to compromise when needed. The changes in diplomacy brought about by peace made negotiations and diplomacy itself more peaceful. In other words, peace begot peace in the fifth century.

102 Men. Prot. fr. 20.1.

103 Men. Prot. fr. 6.1.

104 Blockley (1992) 136; Lee (1993) 46; Scholten (1998) 454. However, there were examples of envoys being chosen for the skill and training in oratory and rhetoric (Amm. Marc. 27.5.15; John Lyd. *De. Mag.* 3.53).

105 For the use of this term see Men. Prot. fr. 20.2.33; Theoph. Sim. 3.17.2.

4 Conclusion

The nature and development of the fifth-century peace was a product of the anarchic late antique world in which the Sasanian Empire existed. More pressing wars elsewhere were the only thing that could prevent war between two competing states, and even then, peace was difficult to establish and maintain. There was no one single moment when Sasanian-Roman peace was decided upon or established; rather, the uneasy, but necessary, fifth-century détente was the result of a slow, iterative, and adaptive process that needed to constantly reaffirm and build upon itself. All six treaties which played a role in peace were directly influenced by and interconnected with each other. The need and existence of peace itself created the conditions which allowed imperial diplomacy to develop, which likely would not have been possible in other periods of Sasanian-Roman relations. It was only when diplomacy was needed as an alternative to war in creating and maintaining the balance of power that both empires were willing to endow it with the prestige and power to do so. Nevertheless, despite the development of imperial diplomacy in this period, its ability to maintain peace was always dependent on the existence of war elsewhere, which in the fifth century was largely the Hunnic threat on the Sasanians' eastern frontiers.

Bibliography

Andres, H. (2022) *Bruderzwist. Strukturen und Methoden der Diplomate zwischen Rom und Iran von der Teilung Armeniens bis zum fünfzigjährigen Frieden*. Stuttgart.

Baldwin, B. (1995) 'Menander Protector'. *DOP* 32: 99–125.

Ball, W. (2017) 'The Sasanian Empire and the East: A Summary of the Evidence and Its Implications for Rome', in Sauer (2017) 151–78.

Barceló, P.A. (1981) *Roms auswärtige Beziehungen unter der Constantinischen Dynastie (306–363)*. Regensburg.

Berthelot, H., Boiché, A., Caltot, P.-A., Diarra, M., Réveilhac, F., and Romieux-Brun, E. (eds) (2016) *Vivre et penser les frontiers dans le monde méditerranéen antique*. Bordeaux.

Binder, C., Börm, H., and Luther, A. (eds) (2016) *Diwan: Studies in the History and Culture of the Ancient Near East and the Eastern Mediterranean. Festschrift für Josef Wiesehöfer zum 65. Geburtstag*. Duisburg.

Blockley, R.C. (1984) 'The Romano-Persian Peace Treaties of A.D. 299 and 363'. *Florilegium* 6: 28–49.

Blockley, R.C. (1985) 'Subsidies and Diplomacy: Rome and Persia in Late Antiquity'. *Phoenix* 39: 62–74.

Blockley, R.C. (1992) *East Roman Foreign Policy: Formation and Conduct from Diocletian to Anastasius*. Leeds.

Börm, H. (2016) 'A Threat or a Blessing: The Sasanians and the Roman Empire', in Binder, Börm, and Luther (2016) 624–33.

Börm, H. (2019) 'Die Grenzen des Großkönigs. Überlegungen zur arsakidisch-sasanidischen Politik gegenüber Rom', in Schleicher, Stickler, and Hartmann (2019) 99–123.

Cameron, A., Ward-Perkins, B., and Whitby, M. (eds) (2000) *Cambridge Ancient History. 2nd ed. Vol. 14. Late Antiquity: Empire and Successors, AD 425–600*. Cambridge.

Campbell, B. (2001) 'Diplomacy in the Roman World (c.500 BC–AD 235)'. *Diplomacy and Statecraft* 12: 1–22.

Canepa, M.P. (2009) *The Two Eyes of the Earth: Art and Ritual of Kingship between Rome and Iran*. London.

Chaumont, M.L. (1986) 'Armenia and Iran ii: The pre-Islamic Period' *Encyclopaedia Iranica*, Online Edition, available at: http://www.iranicaonline.org/articles/armenia-ii.

Croke, B. (1984) 'Dating Theodoret's Church History and Commentary on the Psalms'. *Byzantion* 54: 59–74.

Davidson, O. (2013) *Poet and Hero in the Persian Book of Kings*. Harvard.

DeSouza, P. and France, J. (eds) (2008) *War and Peace in Ancient and Medieval History*. Cambridge.

Dignas, B. and Winter, E. (2007) *Rome and Persia in Late Antiquity: Neighbours and Rivals*. Cambridge.

Drijvers, J.W. (2008) 'Rome and the Sasanid Empire: Confrontation and Coexistence', in Rousseau (2008) 441–54.

Eckstein, A.M. (2006) *Mediterranean Anarchy, Interstate War and the Rise of Rome*. London.

Eilers, C. (2009) *Diplomats and Diplomacy in the Roman World*. Leiden.

Freeman, P. and Kennedy, D.L. (eds) (1986) *The Defence of the Roman and Byzantine East. Proceedings of a Colloquium Held at the university of Sheffield in April 1986*. Oxford.

Garsoïan, N. (1983) 'Byzantium and the Sasanians', in Yarshater (1983) 568–92.

Greatrex, G. (1993) 'The Two Fifth Century Wars Between Rome and Persia'. *Florilegium* 12: 1–14.

Greatrex, G. (1998) *Rome and Persia at War, 502–532*. Cambridge.

Greatrex, G. (1999) 'The Hunnic Invasion of the East of 395 and the Fortress of Ziatha'. *Byzantion* 1: 65–75.

Greatrex, G. (2000) 'Background and Aftermath of the Partition of Armenia in A.D. 387'. *AHB* 14: 35–48.

Greatrex, G. and Bardill, J. (1996) 'Antiochus the *Praepositus*: A Persian Eunuch at the Court of Theodosius II'. *DOP* 50: 171–97.

Greatrex, G. and Lieu, S.N.C. (2002) *The Roman Eastern Frontier and the Persian Wars: Part II 363–630. A Narrative Sourcebook*. London.

Greenwood, T. (2017) 'A Contested Jurisdiction: Armenia in Late Antiquity', in Sauer (2017) 199–220.

Herzig, E. and Kurkchiyan, M. (eds) (2005) *The Armenians: Past and Present in the Making of National Identity*. London.

Isaac, B. (1990) *The Limits of Empire: The Roman Army in the East*. Oxford.

Jones, A.H.M. (1964) *The Later Roman Empire: A Social, Economic and Administrative Survey: Volume I*. Oxford.

Kelly, C. (2008) *Attila the Hun: Barbarian Terror and the Fall of the Roman Empire*. London.

Klein, R. (1977) *Constantius II. Und die christliche Kirche*. Darmstadt.

Kydd, A.H. (2005) *Trust and Mistrust in International Relations*. Oxford.

Lang, D.M. (1983) 'Iran, Armenia and Georgia', in Yarshater (1983) 505–36.

Lee, A.D. (1987) 'Dating a Fifth Century War in Theodoret'. *Byzantion* 57: 188–91.

Lee, A.D. (1991) 'The Role of Hostages in Roman Diplomacy with Sasanian Persia'. *Historia* 40: 366–74.

Lee, A.D. (1993) *Information and Frontiers: Roman Foreign Relations in Late Antiquity*. Cambridge.

Lee, A.D. (2008) 'Treaty-Making in Late Antiquity', in de Souza and France (2008) 107–19.

Lee, A.D. (2009) 'Abduction and Assassination: The Clandestine Face of Roman Diplomacy in Late Antiquity'. *International History Review* 31.1: 1–23.

Lenski, N. (2002) *The Failure of Empire: Valens in the Fourth Century A.D.* London.

Luther, A. (2014) 'Ein "übersehener" römish-persischer Krieg um 416/17?'. *Gymnasium* 121: 183–93.

Luttwak, E. (1979) *The Grand Strategy of the Roman Empire*. Baltimore.

Maksymiuk, K. (2018) 'The Two Eyes of the Earth: The Problem of Mutual Respect in Sasanid-Roman Relations'. *Greek, Roman and Byzantine Studies* 58: 591–606.

Millar, F. (1982) 'Emperors, Frontiers and Foreign Relations, 31 BC to AD 378'. *Britannia* 13: 1–23.

Moorhead, J. (2001) *The Roman Empire Divided 400–700*. London.

Morley, C. (2015) 'Rome and the Sasanian Empire in the Fifth Century A.D.: A Necessary Peace', PhD Thesis, University of Liverpool.

Morley, C. (2016) 'Interaction and Contact between the Roman and Sasanian Empires in Late Antiquity', in Berthelot, Boiché, Caltot, Diarra, Réveilhac, and Romieux-Brun (2016) 117–26.

Morley, C. (2017) 'The Arabian Frontier: A Keystone of the Sasanian Empire', in Sauer (2017) 268–83.

Nechaeva, E. (2014) *Embassies – Negotiations – Gifts: Systems of East Roman Diplomacy in Late Antiquity*. Stuttgart.

Preiser-Kapeller, J. (2001) '*Erdumn, Uxt Carayut'iwn*: Armenian Aristocrats as Diplomatic Partners of the Eastern Roman Emperors 387–884/5 A.D.'. *Armenian Review* 52: 139–217.

Rich, J. and Shipley, G. (eds) (1993) *War and Society in the Roman World*. London.

Rousseau, P. (ed) (2008) *A Companion to Late Antiquity*. Malden.

Rubin, Z. (1986) 'Diplomacy and War in the Relations between Byzantium and the Sassanids in the Fifth Century', in Freeman and Kennedy (1986) 677–98.

Rubin, Z. (2000) 'The Sasanid Monarchy', in Cameron, Ward-Perkins, and Whitby (2000) 638–61.

Russell, J.R. (2005) 'Early Armenian Society', in Herzig and Kurkchiyan (2005) 23–40.

Sabin, P., van Wees, H., and Whitby, M. (eds) (2007) *The Cambridge History of Greek and Roman Warfare. Volume 2: Rome from the Late Republic to the Late Empire*. Cambridge.

Sauer, E. (ed) (2017) *Sasanian Persia: Between Rome and the Steppes of Eurasia*. Edinburgh.

Sauer, E., Nokandeh, J., Pitskhelauri, K., and Rekavandi, H.O. (2017) 'Innovation and Stagnation: Military Infrastructure and the Shifting Balance of Power between Rome and Persia', in Sauer (2017) 241–67.

Sevcenko, I. (1992) 'Re-reading Constantine Porphyrogenitus', in Shepard and Franklin (1992) 167–96.

Schleicher, F., Stickler, T., and Hartmann, U. (eds) (2019) *Iberien zwischen Rom und Iran*. Stuttgart.

Scholten, H. (1998) 'Römische Diplomatie im 4.Jhd. N. Chr: Eine Doppelstrategie Des Praefectus Praetorio Orientis Musonianus?'. *Historia* 47: 454–67.

Shahid, I. (1989) *Byzantium and the Arabs in the Fifth Century*. Washington, DC.

Shepard, J. and Franklin, S. (eds) (1992) *Byzantine Diplomacy*. Aldershot.

Sidebottom, H. (2007) 'International Relations', in Sabin, van Wees, and Whitby (2007) 3–29.

Spawforth, A. (ed) (2007) *The Court and Court Society in Ancient Monarchies*. Cambridge.

Toral-Niehoff, I. (2013) 'Late Antique Iran and the Arabs: The Case of al-Hira'. *JPS* 6: 115–26.

Toynbee, A. (1973) *Constantine Porphyrogentius and His World*. Oxford.

Trimingham, J.S. (1979) *Christianity Among the Arabs in Pre-Islamic Times*. London.

Watson, A. (2005) *Diplomacy: The Dialogue Between States*. London.

Whitby, M. (2008) 'Byzantine Diplomacy: Good Faith, Trust and Co-operation in International Relations in Late Antiquity', in de Souza and France (2008) 120–40.

Wiesehöfer, J. (2007) 'King, Court and Royal Representation in the Sasanian Empire', in Spawforth (2007) 58–81.

Winter, E. (1987) 'Handel und Wirtschaft in sassanidisch – (ost)römischen Verträgen und Abkommen'. *Münstersche Beiträge zur antiken Handelsgeschichte* 6.2: 46–74.

Wirth, G. (1980) 'Rom, Parther und Sassaniden: Erwägungen zu den Hintergrunden eines historischen Wechselverhaltnisses'. *Anc. Soc.* 11/12: 305–47.

Woolf, G. (1993) 'Roman Peace', in Rich and Shipley (1993) 171–94.

Yarshater, E. (ed) (1983) *The Cambridge History of Iran Volume 3(2): The Seleucid, Parthian, and Sasanid Periods*. Cambridge.

CHAPTER 18

Kavad I, Khosrow I, and the Wars with the Roman Empire

Henning Börm

In the autumn of 502 CE, the *shāhānshāh* Kavad I (488 to 531),[1] ruler of the Sasanian Empire,[2] invaded Roman territory. Despite the fact that there had previously been growing tensions between the two great powers of Late Antiquity,[3] this attack came as a surprise, because in the 125 years since 377, when the emperor Valens had concluded a treaty (*foedus*) with the Sasanians, armed conflicts between the Romans and Iranians had occurred only very rarely. The only time intense fighting took place had been around 420;[4] another brief war had ended in 441 after only a few weeks without any major battles,[5] while in the sixty years before 502 undisturbed peace had prevailed between the Sasanians and the Roman Empire. Tensions and conflicts were resolved diplomatically;[6] officials and generals of both empires cooperated at the local level,[7] the rulers called each other 'brother',[8] and envoys were exchanged regularly.[9] The cultural and economic exchange between Iran and Rome also seems to have intensified noticeably during this long period of peace.[10] Against this background, the question of how this period of peace, from which both empires had profited, ended, is of great relevance, because one could argue

1 Cf. Schindel (2013).

2 On the Sasanian Empire (224 to 651 CE) see Rubin (2008); Daryaee (2009); Wiesehöfer (2010); Börm (2017); Rezakhani (2017); Bonner (2020). Christensen (1944) is outdated but still valuable.

3 Cf. Howard-Johnston (1995). See also Shahbazi (1990) and Börm (forthcoming).

4 Socrates, *Hist. eccl.* 7.20f. Cf. Greatrex (1993). Fighting probably occurred as early as 414, but it hardly left any traces in the sources; cf. Luther (2014).

5 Marc. *Com. ad ann.* 441. Cf. Greatrex and Lieu (2002) 41; Bonner (2020) 111–14.

6 Cf. Andres 2022.

7 Barsauma ep. 2.526–27, translated in Greatrex and Lieu (2002) 48–50.

8 Euseb. *Vit. Const.* 4.11.1; Amm. Marc. 17.5.3; Procop. *Pers.* 1.16.1; Mal. 18.44; Men. Prot. fr. 6.1. However, Maksymiuk (2018) doubts that such formulations were more than polite phrases. On Roman views of the Sasanians, see McDonough (2011b); on *Rūm* as an enemy of *Ērān*, see Wiesehöfer (2005).

9 Cf. Diebler (1995).

10 Cf. Hartmann (2007); Drijvers (2009) 449–52; Greatrex (2014).

 | DOI:10.1163/9789004710771_019

that the endless wars that the Sasanians and Romans waged against each other between 502 and 630 significantly contributed to the fall of pre-Islamic Iran and the end of the Ancient World.

1 Iranians and Romans in Late Antiquity

The Parthian Empire had been the neighbour of the Romans since the first century BCE. However, our picture of the relationship between the Arsacids and the Mediterranean world is heavily distorted, because neither Greek and Roman authors nor later Near Eastern sources were interested in offering a neutral account of the events. It is, however, certain that there were repeated wars,[11] even if one should not overlook the fact that most of the time the contacts were generally peaceful.

However, between the final phase of the Arsacid rule in the early third century CE and the middle of the fourth century, the intensity of the military conflict increased for about 160 years. Northern Mesopotamia probably was under Roman influence since 166 CE, but at first the emperors avoided provoking the Arsacids by turning these areas into a province. So, when the emperor Septimius Severus, after having prevailed in a bloody civil war, decided to break with this policy of his predecessors and to annex a territory to the west of the Euphrates as the province of *Mesopotamia* after a short campaign in order to become *Parthicus Maximus* in around 200,[12] this was therefore the real turning point in the relationship between Iran and Rome. Already the contemporary Roman historian Cassius Dio remarked that the emperor had in no way made Syria safer, but rather created the occasion for endless wars.[13] In fact, Artabanus IV was already working hard to force the Romans to retreat from Mesopotamia, but although he was able to defeat Macrinus in 218, he failed to enforce these demands.[14] For this reason, the Sasanians inherited the problem of restoring the Euphrates border from their Arsacid predecessors when they took over the empire in 224.[15]

When it comes to Roman-Iranian wars, most historians think of the campaigns of those years, especially those associated with the Sasanian kings Shapur I (240 to 273), Narseh (293 to 302) and Shapur II (309 to 379). Between

11 Cf. Edwell (2020) 25–56.
12 Cass. Dio 76.9.1–4. Cf. Gradoni (2013) 7–13.
13 Cass. Dio 75.3.2–3.
14 Cf. Edwell (2020) 48–50.
15 Cf. Börm (2019b) 105–106.

224 and 377, rulers of both empires seem to have regularly had an interest in escalating conflicts. There is, however, some evidence that the decision for or against full-scale war was in most cases made by the emperors, even if the Persian kings by no means only acted defensively.[16] Between 244 and 260, Shapur I defeated three Roman emperors;[17] in 298 Narseh suffered a heavy defeat and had to cede large areas to the Romans,[18] but in 363 Shapur II was able to revise this result.[19] All in all, both empires suffered great losses, only to end up empty-handed. Neither side had won.

This chapter, however, will deal with the second long phase of intense fighting between Romans and Iranians, dominating the entire sixth and early seventh centuries. There were several reasons for the long period of mostly peaceful coexistence that started in 377 which was so very different from the violent clashes the third and fourth centuries had witnessed.[20] For one thing, around the year 400 the Roman emperors had ceased to personally lead their armies into battle. Instead, they remained permanently in Constantinople, which by then had become the main residence of the eastern half of the Roman Empire.[21] This change had far-reaching ramifications for Roman relations with Iran: the emperor, increasingly tied to his capital, was no longer able to move east as a *novus Alexander* in an attempt to stabilise his precarious position by gaining military glory,[22] while at the same time it was very risky for any ruler to give command of a Persian war to a general, because in the event of victory this man would become a danger to the emperor. So, precisely because, from a Roman perspective, no war was associated with as much prestige as an *expeditio Persica*, the *principes clausi* after Theodosius I (379 to 395) lost interest in a military escalation.[23] Therefore, since 441 at the latest,[24] they appear to have been prepared to pay regular tributes to the Sasanians in order to purchase peace, something that the Persian kings interpreted as a symbol of imperial inferiority.[25] Certainly not every war between Rome and Iran in

16 Cf. Börm (2016) 619–23.
17 Cf. Edwell (2020) 86–110.
18 Cf. Dignas and Winter (2007) 122–30.
19 Cf. Chrysos (1993); Bonner (2020) 86–89.
20 The best modern study of the wars between 233 and 363 is Mosig-Walburg (2009). See also Dignas and Winter (2007) 18–34, 70–94; Lee (2013); Börm (2016) 617–24.
21 Cf. Pfeilschifter (2013); Meier (2017); Maier (2019).
22 On the *imitatio Alexandri* cf. Kühnen (2008).
23 Cf. Börm (2016) 624–33.
24 Cf. Blockley (1992) 58.
25 Cf. Börm (2008). In my opinion, the fact that the Sasanians usually demanded gold while the Iranian monetary system was actually based on silver suggests a predominantly symbolic meaning of these payments. However, Michael Jackson Bonner argues that the

the years before 395 had been caused by emperors striving for glory;[26] but after Theodosius I the Roman rulers were clearly much more interested in a peaceful settlement of conflicts than before.[27] Moreover, they were distracted by events in the Western Roman Empire, where civil wars led to the collapse of the central government and the establishment of autonomous post-Roman *regna*.[28]

One might ask why the Sasanians did not take advantage of this new hesitancy of the emperors during the fifth century by pursuing a more aggressive policy towards the Roman Empire; but although the original goal of driving the Romans out of Northern Mesopotamia had not been achieved, the frontier established in 363 was perfectly acceptable for the Iranian kings. In that year, the Second Peace of Nisibis created a strategic situation in Mesopotamia that made it virtually impossible for the Romans to march quickly to Ctesiphon, sack the city, and then retreat.[29] This strategy had been used more or less successfully by emperors such as Lucius Verus, Septimius Severus, Carus, or Julian; but now the Sasanians controlled the city of Nisibis, and this strong fortress with its large garrison ensured that rapid Roman advances to the south were effectively prevented.[30] And when Armenia was divided in 387 and four fifths of the country, henceforth called Persarmenia, came under Sasanian control, this also created a strategic situation in the Caucasus region with which both Romans and Iranians were satisfied.[31] From then on, there was basically no longer a strategic need to wage a new, risky war; instead, the benefits of a peaceful cooperation between the great powers, 'the two eyes of the earth',[32] became increasingly clear.[33]

Sasanians needed gold for the silk trade with India; cf. Bonner (2020) 209–12. See also Howard-Johnston (2016) 603–608. I follow Schindel (2006) 123: 'Persian merchants used silver, and not gold coins, for foreign trade'.

26 Greatrex (2007) offers a very good overview of Roman strategy in the East.

27 Cf. Edwell (2013, 848–50), who considers the break of the Persian Church with Rome as a reason for more peaceful conditions in addition to the supposed threat the Sasanians faced from the Huns.

28 The fact that 'Germanic' soldiers (*foederati*) were increasingly involved in these internal wars is the reason why 'barbarian invasions' are often blamed for the fall of the Western Roman Empire. In my view, however, the ultimate reason for the end of imperial rule was the structural instability of the Roman monarchy; cf. Börm (2018a) 156–65.

29 Amm. Marc. 25.7.1–8.7. Cf. Mosig-Walburg (2009) 305–20.

30 Cf. Mosig-Walburg (2009) 322.

31 Cf. Blockley (1987); Greatrex (2000).

32 Cf. Lerza (1982). See also Canepa (2009) 7–33.

33 Cf. Börm (2007) 299–304; Payne (2015) 296–99.

In recent years scholars have rightly emphasised that the Sasanian Empire was focused more on the East than on the West,[34] even though the royal residence of Ctesiphon was located quite close to the Roman border and Mesopotamia was of great importance to the rulers. All in all, Iran was usually much less focused on Rome than the Romans were on Iran.[35] Moreover, it appears that the Iranian aristocracy never supported aggressive western policies aiming at conquests rather than booty; presumably, the magnates[36] feared that the rulers would become too powerful if they were able to permanently conquer Syria and Egypt.[37] From the nobility's point of view, the kings' primary task was to defend the Iranian highlands against attacks from the North and East,[38] and during the fifth century rulers such as Wahram V, Yazdgird II and Peroz did indeed focus on these regions, the true heartland of the Sasanian Empire. Usually, it is believed that Hunnic groups such as the Hephthalites[39] were a major threat to Iran during this time; and as a matter of fact these peoples proved to be formidable enemies on more than one occasion.[40]

However, the large ramparts that were built near the Caspian Sea during the fifth century[41] are by no means a clear indication that Hunnic invasions were expected: just like the Romans and the Chinese, the Iranians undoubtedly knew that long walls are quite unsuitable to serve as defences against major attacks, since it is impossible to man them with enough soldiers. Thus the Gorgan Wall, not unlike Hadrian's Wall,[42] was probably used to oversee a peaceful exchange of goods and people, serving as a means of demarcating the limits of the Sasanian Empire, as a barrier to smugglers and looters, as a

34 Cf. Howard-Johnston (2016) 602–603; Rezakhani (2017) 185–93.

35 For the Roman Empire, Iran was the only serious rival. In addition, a special fascination for Persia had arisen already in Classical Greece, with the Iranian kings in particular serving as a projection screen for Greek and Roman discourses on monarchy; cf. Börm (2019a).

36 On the relations between the high nobility and the kings see Rubin (2004) 244–46; McDonough (2011a).

37 On this, see Börm (2019b) 102–104.

38 Cf. Howard-Johnston (2016): 'The prime earthly task of Iranian kings (hence the principal constituent of Sasanian grand strategy) was to keep Turan at bay, to protect Iran from pollution' (602).

39 On the Hephthalites, also called the 'White Huns', see Bivar (2004); Kurbanov (2013); Rollinger and Wiesehöfer (2020).

40 Cf. Rezakhani (2017) 125–46.

41 Cf. Rekavandi et al. (2008). See also Payne (2015) 293–96. These fortifications have been extensively researched by archaeologists in recent years; cf. Rekavandi et al. (2013).

42 It is by now almost universally accepted that the Roman *limites* were intended to monitor border traffic and to deter small, unorganised bands of raiders; cf. Schallmayer (2006) 31–33.

customs border, and as a symbol of the Iranian king's might. To be sure, the strong fortresses built by the Iranians in northeastern Eranshahr in addition to the long walls were far more useful from a military perspective. But just as in Mesopotamia, the troops stationed in these locations could not only be used defensively but also offensively. Since contemporary authors such as Procopius portray the Hephthalites as civilised and reliable partners,[43] it is therefore possible that the aggression leading to war with the Huns often – albeit not always – originated from the Sasanian kings who wanted to stage themselves as the conquerors of Turan.[44] On the other hand, however, it seems that pressure on the Sasanian borders in Transcaucasia and the East did in fact increase since the middle of the fifth century, as there is evidence of growing instability during this period in all of Eurasia.[45] Peaceful cooperation between Romans and Iranians would therefore have been more necessary than ever.

2 The Renewal of Conflict: Kavad I and the Romans

Against this background, the Sasanian surprise attack on the Romans in 502 CE demands an explanation. Relations between the two powers had in fact deteriorated after Peroz's violent death in 484,[46] but it is nevertheless evident that the Roman emperor Anastasius (491 to 518), like his predecessors, had no interest in starting a full-scale Persian war. Instead, there is much to suggest that the Sasanian invasion had primarily domestic causes: Kavad I had been overthrown in 496 after a power struggle within the Iranian nobility, a conflict that was in some way connected to the enigmatic Mazdakite movement.[47] Led by the *kanārang* Gušnaspdad, whose family was responsible for Iran's defence against the Hephthalites,[48] a group of nobles had eventually removed the king from power. However, supported by a party that was apparently allied with the Mazdakites, Kavad managed to escape to the Hephthalites, with whose support

43 Procop. *Pers.* 1.3.1–7. Cf. Kaldellis (2004) 73; Börm (2007) 206–208.

44 Cf. Shayegan (2013) 807–809. See also Lawrence and Wilkinson (2017).

45 Cf. Meier (2020) 306–15.

46 Jos. Styl. 11; Łazar 154–156, 213–215; Ṭab. 1.876. Cf. Syvänne (2021) 108–13.

47 The Mazdakites have attracted a great deal of scholarly interest. However, neither the chronology of events, the specific content of their teaching, their attitude towards Rome, nor the role played by Kavad have yet been satisfactorily clarified; cf. Yarshater (1983b); Crone (1991); Rubin (1995) 179–85; Börm (2007) 230–33; Pourshariati (2008) 85–88; Wiesehöfer (2009); Daryaee (2009) 86–91; De Blois (2015); Rezakhani (2015).

48 Cf. Pourshariati (2008) 266–71.

he was able to return in 498 or 499 after a short civil war.[49] The whole situation is difficult for us to understand, but it seems that especially the *artēštārān sālār* Siyavuxš,[50] the new strong man at the royal court,[51] encouraged the king to behave aggressively towards the Romans. Finally, Kavad found a suitable pretext to start the war – Anastasius probably refused to pay the tributes to which the Sasanians believed they were entitled.[52] For the first and only time, Iranians and Hephthalites allied to fight the emperor.[53] The Roman envoy, the *patricius* Rufinus,[54] who was on his way to Ctesiphon, had to turn around when he heard of the surprise attack.[55]

In the beginning, the invaders were very successful; among other things, Kavad succeeded in conquering the important fortress of Amida in January 503.[56] On this and other occasions, the Sasanians proved to be masters of siege craft, using the entire arsenal common to late antique poliorcetics, including war towers, ballistae, and mounds.[57] It is not without reason that the *Strategicon*, written in the later sixth century, states that the Persians were outstanding both as besiegers and as defenders.[58] A major Roman counteroffensive failed a few months later,[59] but afterwards the imperial army gradually managed to stabilise the situation, a Sasanian attack on Edessa resulted in failure,[60] and in 506 a truce was made.[61] Kavad's alliance with the Hephthalites ended and the king seems to have lost interest in the Roman war. Nevertheless, the intense fighting had not been without consequences, because Anastasius

49 Cf. Bonner (2020) 144–47.

50 As far as I can tell, Siyavuxš (Σεόσης) is mentioned only by Procopius (*Pers.* 1.6.18f). Cf. Bonner (2020) 146, who thinks that this person 'may perhaps be identified with Zamihr the younger' mentioned by Ṭab. 1. 883–84.

51 Cf. Börm (2007) 318–25 for a detailed discussion.

52 Cf. Börm (2007) 321–22.

53 Cf. Haarer (2006) 29–72; Meier (2009) 194–206. The best modern analysis of the war is Greatrex (1998).

54 Cf. *PLRE* II, 954–57.

55 Jos. Styl. 50.

56 Cf. Lenski (2007); Greatrex (2010). After the conquest, the city's male population was decimated and the king left behind a strong garrison that was subsequently besieged by the Romans.

57 Procop. *Pers.* 1.7.12–29. On siege warfare in Late Antiquity, see Whitby (2013).

58 *Strateg.* 11.1.9–12.

59 Cf. Greatrex (1998) 96–7. The fateful division of the Roman supreme command between five men – Patricius, Hypatius, Celer, Justin, and Areobindus (Zach. Rhet. *HE* 7.4) – for the massive offensive of 503 is probably a sign of Anastasius' concern that all the glory of a potentially successful victory over the Persians would fall to a single general.

60 Greatrex (1998) 105–106.

61 Cf. Bonner (2020) 157–58.

had been startled by the Sasanian offensive. For a long time, the Romans had clearly assumed that they would not have to expect an unprovoked Iranian attack; therefore no attempt had been made to replace Nisibis with a new strong fortress. But now the emperor was obviously no longer certain that the Sasanians would return to their defensive western policy; he therefore reorganised the Roman border troops (*limitanei*),[62] and a gigantic fortress was built in Dara in Northern Mesopotamia. In doing so, however, the Romans violated agreements that prohibited the construction of new fortifications close to the border, and Dara was to become the bone of contention between the two great powers.[63]

Still, the armistice of 506 lasted longer than expected. Even the change of sides of the king of Lazica who defected to the Romans in around 522 did not lead to renewed fighting.[64] It seems that Kavad I by that time had emancipated himself from the party of Siyavuxš; the monarch therefore tried to conclude a new *foedus* with Rome in around 525.[65] According to the Greek historian Procopius,[66] he intended to designate his third son Khosrow as his successor, and to secure this arrangement he asked the emperor Justin I (518 to 527) to adopt the prince.[67] However, there were powerful men in both Constantinople and Ctesiphon who obviously wanted to prevent this agreement.[68] The *quaestor sacri palatii* Proculus[69] allegedly claimed that the Sasanian proposal was a trap, arguing that if the emperor adopted Khosrow he would in effect make the prince heir to the Roman Empire – an absurd, legally untenable argument,[70] but one that nevertheless left an impression.[71] In the end, high-ranking representatives of both powers met to negotiate peace; the delegations were mixed on both sides, since both supporters of an agreement, such as Rufinus and a certain Mehbod, and hawks such as the *magister militum* Hypatius and Siyavuxš, took part in the peace talks, with the latter managing to sabotage the negotiations. Khosrow, who had already prepared to travel to Constantinople

62 A Greek inscription discovered in Jordan (SEG 32.1554) contains precise regulations.

63 Procop. *Pers.* 1.10.15–16. Cf. Bonner (2020) 163.

64 Cf. Braund (1994) 276–77.

65 For the date see Greatrex (1998) 137–38.

66 On Procopius, our most important source for these events, see Cameron (1985); Kaldellis (2004); Börm (2007); Stewart (2020) 31–67; Börm (2022).

67 Procop. *Pers.* 1.11.7–11. Cf. Pieler (1972).

68 Cf. Lounghis, *Partei*, and Börm (2007) 311–17.

69 Cf. Pazdernik (2014).

70 Cf. Börm (2015) 260–61.

71 Procop. *Pers.* 1.11.13–22. According to the chronicler Theophanes (AM 6013), the emperor subsequently assembled the senate, which is said to have largely followed Proculus' opinion.

to be adopted by Justin I, saw himself humiliated.[72] The relations between the Romans and Sasanians deteriorated so much following this experience that fighting was resumed in the autumn of 526, focusing on the Caucasus region.[73]

The following summer, Justin's nephew Justinian (527 to 565) became Roman emperor.[74] Although the new ruler was apparently interested in peaceful relations with Iran – according to Procopius, he had been an advocate of Khosrow's adoption – the war between the two great powers initially intensified.[75] Justinian sent his friend Belisarius to the East, who was first named *dux Mesopotamiae* and then *magister militum per Orientem*.[76] Initially his troops were defeated in two battles, at Thannuris and Mindouos, but in 530 the young general managed to defeat a large Sasanian army at the battle of Dara. Since Belisarius did not trust the walls of the fortress that were currently being rebuilt, he and his 25,000 soldiers faced the Persian army, which was about twice the size, in a pitched battle after preparing the battlefield with hidden trenches. The Iranian commander Peroz used the elite unit of the 'Immortals',[77] and the fighting was fierce; but eventually the surprise attack of imperial horsemen who had been hidden by Belisarius brought victory to the Romans.[78] At about the same time, the general Sittas,[79] holding the newly created post of *magister militum per Armeniam*,[80] was able to claim victories over Iranian troops as well.[81] Justinian tried to exploit this success domestically; he held triumphal

72 Procop. *Pers.* 1.11.25–27.

73 Cf. Braund (1994) 282–87; Bonner (2020) 166–69.

74 Arguably the best monographs on Justinian are currently Leppin (2011) and Sarris (2023).

75 Greatrex (1998) 168–212; Whitby (2021) 115–72. Cf. Heather (2018) 103–109. However, I do not share Heather's view that Justinian had wanted war with the Sasanians all along.

76 Cf. *PLRE* III, 181–224.

77 On the Sasanian 'Immortals', cf. Charles (2011). It has long been suspected that the Greeks had probably already made a mistake in the time of Herodotus (7.83.1), since the name of the unit probably actually derives from an Old Iranian word for 'follower' instead of 'immortal'; cf. Huyse (2002) 208. However, the question arises whether this must also apply to the Sasanian period as well, as 'immortals' (ἀθάνατοι) are mentioned not only by Procopius (*Pers.* 1.16.31), but also by other late antique authors such as Socrates (*Hist. eccl.* 7.20) or Malalas (14.23). Although it cannot be ruled out that they all merely imitate Classical models such as Herodotus, this seems rather unlikely.

78 Cf. Greatrex (1998) 169–85. Overall, Procopius' information on Iranian operations at Dara is strikingly in line with a fragment of a Sasanian military manual that seems to have its origin in the sixth century; cf. Inostrancev (1926) 27–28.

79 Cf. *PLRE* III, 1160–63.

80 Cf. Preiser-Kapeller (2004).

81 Cf. Greatrex (1998) 185–90.

celebrations in Constantinople,[82] and an equestrian statue, lost today, praised him as conqueror of the 'Medes'.[83]

The next spring, Kavad I ordered another invasion, involving above all the most important Arab allies of the Sasanians, the Lakhmids.[84] The plan was not to linger on the siege of the strong Roman fortresses in Mesopotamia but to circumvent them, aiming at the Syrian provinces instead.[85] This tactic was possible because heavy cavalry consisting of *clibanarii* and mounted archers (ἱπποτοξόται) now determined the course of the war.[86] The armament and the way of fighting of these elite forces was so similar on both sides that defectors could easily be integrated into both armies.[87] Belisarius was able to outmanoeuvre the invaders and force them to retreat, and in April 531 he attacked the Iranians while they tried to cross the Euphrates at Callinicum. That was a mistake. Both armies comprised a good 20,000 soldiers each, with Arab cavalry playing an important role on both sides. The Jafnid allies (σύμμαχοι),[88] whom Belisarius had positioned on his right flank, however, could not withstand a frontal attack by the Iranians and took flight; the imperial front collapsed, and it was only with great difficulty that Belisarius could prevent the Roman defeat from becoming an outright catastrophe.[89] While the eye-witness Procopius was defending the general, other authors blame Belisarius for the failure,[90] and Justinian ordered him to come straight to Constantinople to personally justify himself.[91] However, the Sasanian troops also seem to have suffered heavy losses, so much so that the general in charge was said to have been degraded and humiliated after his return to Ctesiphon.[92]

---

82 Cf. Meier (2003) 151–52.

83 Anth. Palat. 16.63.

84 Cf. Bosworth (1983).

85 According to Procopius (*Pers.* 1.17.37–43), this suggestion came from the Lakhmids.

86 Cf. Syvänne (2004); Koehn (2018) 115–45.

87 Thanks to Ṭabarī (1.964), the equipment of an ideal Iranian *clibanarius* at the time of Khosrow I is known, and despite the anecdotal context in which it is reported – the king himself is patterned – it probably reflects the late Sasanian reality: armour for the horse, chain mail, breastplate and greaves for the rider, who is armed with sword, lance, shield, club, axe, or butt, as well as a quiver with two bows and thirty arrows. It is unclear whether the stirrup, which is first mentioned in the *Strategicon*, was already known in Khosrow's time. On the late Sasanian army, cf. Howard-Johnston (2012).

88 Cf. Whittow (2002); Greatrex (2015); Lewin (2015).

89 On this battle, see Greatrex (1998) 193–99.

90 Procop. *Pers.* 1.17.30–56; Mal. 18.60. Cf. Brodka (2011).

91 Zach. Rhet. *Hist. eccl.* 9.17.

92 Procop. *Pers.* 1.18.56.

3 Khosrow I and the Eternal Peace of 532

A little later the *shāhānshāh* died and Khosrow I Anuširvan (531 to 579) ascended the throne. However, the Iranian aristocracy was still divided, so resistance came within a short space of time. After the Mazdakites had tried to install Khosrow's eldest brother Kawus on the throne, the king seems to have prosecuted them with great harshness,[93] with an attempt to replace him with his nephew Kavad failing as well.[94] This presumably increased the influence of the men around the *sar-naxvēragān* Mehbod, who not only played an important role in Khosrow's accession to the throne but who had advocated peace with the Romans as early as 525. Siyavuxš, meanwhile, was executed now at the latest,[95] so those who wanted to end the war had finally gained the upper hand in Ctesiphon. Around the same time, in January 532, the Nika Riot broke out in Constantinople, with Hypatius trying to become *Augustus* himself before the rebellion was suppressed by Belisarius and other generals.[96] He was therefore put to death, and thus one of the most important protagonists of the 'hawks' at the imperial court had gone. All this cleared the way for a new *foedus* to finally end the war that had started in 502.[97] Rufinus, assisted by the *magister officiorum* Hermogenes, conducted the negotiations on behalf of the Romans,[98] and in the autumn of 532 the so-called 'Eternal Peace' was signed.

We are not aware of all the details of the treaty.[99] It is, however, certain that attempts were made to regulate the course of the border in order to meet the strategic interests of both sides. The Romans were allowed to keep the city of Dara but no longer used it as a base for the *dux Mesopotamiae* because the Sasanians interpreted the presence of the *comitatenses* he commanded as a threat to Nisibis. Several forts that had been captured during the fighting since 502 were to be handed over,[100] and Khosrow gave up Lazica. The question of

93 Mal. 18.30; Theoph. AM 6016. Cf. Christensen (1944) 353. The Μανιχαῖοι mentioned in the Greek sources are probably Mazdakites, although this cannot be said with absolute certainty.

94 Procop. *Pers.* 1.23.1–6.

95 Procop. *Pers.* 1.12.31–38.

96 Cf. Meier (2014).

97 Cf. Börm (2007) 322–23.

98 Chr. Ede. 104. The former *magister militum* Rufinus was considered a trusted friend of the Sasanian dynasty. Procopius even claims that Khosrow, in response to false news that Justinian had Rufinus executed, decided to attack the Romans and only changed his mind when the emissary appeared before the king in person; Procop. *Pers.* 1.22.9.

99 For a brief assessment, see Bonner (2020) 176–80.

100 Rufinus had to return to Justinian in order to acquire permission to concede the forts to Khosrow; Procop. *Pers.* 1.22.1–8.

the annual payments the Sasanians had been demanding from the Romans for decades was particularly sensitive. The legal basis for this is somewhat unclear – the payments were apparently justified in part by the Iranians defending the Caucasian and Caspian Gates[101] on behalf of the Romans[102] – but it is obvious that the gold demanded by the Iranians was primarily a symbol of the king's superiority over the emperor, while the economic aspect of the transaction was of secondary importance.[103] For this reason, the kings always attached great importance to regular payments, usually around 30,000 gold coins (*solidi*), while the emperors wanted to avoid this in order not to give the impression that the Romans had to pay tribute to the king of kings.[104] In 532 Justinian attempted to permanently eliminate the problem by agreeing to pay a lump sum of 11,000 pounds of gold to the Persians.[105] It would soon become apparent that Khosrow did not believe the matter to be settled; but at first both parties agreed to this compromise.

The Greek historian Agathias[106] reports that shortly before the peace agreement, several pagan philosophers had fled the Roman Empire and had come to Ctesiphon,[107] where they had hoped for tolerance.[108] Justinian apparently demanded that these men, whom he presumably believed to be defectors, be extradited; Khosrow agreed, but had the emperor guarantee that nothing would be done to the philosophers after their return.[109] And the contemporary chronicler Malalas claims that the emperor and king not only made peace but entered into a real alliance that obliged both monarchs to support each other with money or soldiers.[110] However, since the handling of the tributes shows that Justinian was clearly trying not to make any commitments to the Sasanians, one can doubt Malalas' claim. This applies even more to his statement that the *foedus* was limited in time (ἄχρι τῆς τῶν ἀμφοτέρων ζωῆς), because both Procopius and Justinian themselves emphasise that the peace

101 Cf. Hansman (1992).

102 Cf. Braund (1994) 270: 'Persia had good reason to defend the Dariel Pass, but Byzantium did not'.

103 For a detailed discussion, cf. Börm (2008).

104 The Romans usually either wanted to downplay the importance of the annual tribute or to find a way to stop payments altogether; cf. Colvin (2018) 206–209.

105 Cf. Greatrex (1998) 216: 'The size of the Roman payment should not be exaggerated. It certainly did not represent an excessive blow to the Roman treasury'.

106 Cf. Zambarbieri (2017).

107 Cf. Walker (2002); Hartmann (2002).

108 Agath. *Hist.* 2.36.

109 Cf. Nechaeva (2017).

110 Mal. 18.76.

was to last indefinitely: *pacem cum Persis in æternam confirmavimus*.[111] Indeed, everything that is known about the treaty of 532 suggests that at least the emperor was seriously interested in finding a permanent solution to restore the peaceful conditions that had existed before 502.[112] However, whether Khosrow I planned to stick to the agreement in the long term, too, is another question. In any case, the effect of the *foedus* was soon so noticeable in the border region that it was said that you could cross from one empire to the other without even realising it.[113]

While the Roman troops fought Vandals and Goths in the following years, changes appear to have taken place yet again at the royal court in Ctesiphon. Mehbod, the man who had played a crucial role in Khosrow's accession to the throne, fell out of favour and was executed, and from 539 at the latest, the Sasanians were preparing for an attack.[114] Apparently the *shāhānshāh* had once again turned to those aristocrats who advocated an aggressive policy towards the Roman Empire; one of them, Zabergan,[115] is said to have been responsible for Mehbod's downfall.[116] The exact motives of the king are unknown. Maybe he wanted to get rid of the suspicion that he was just a puppet of the Romans; perhaps he wanted to take revenge for the humiliation that had befallen him when the adoption failed; perhaps he wanted to secure his rule, which was still not accepted by everyone, by military success. Procopius claims that Khosrow was concerned about the military successes Belisarius had achieved in the West,[117] and reports that the Gothic *rex* Vitigis secretly sent envoys to Ctesiphon, warning the Sasanians that the emperor was in danger of becoming too powerful.[118] In addition, some Armenian aristocrats from the Roman part of the country are said to have asked Khosrow to intervene militarily.[119] Justinian, who had apparently not acted very wisely at first, seems to have finally noticed that a threatening situation had arisen; but since his best soldiers were bound in the battles with the Goths in Italy, there was little he could do.

111 *Cod. Iust.* 1.27.2 praef. Procopius explicitly calls the peace ἀπέραντος (*Pers.* 1.22.17).

112 Cf. Börm (2007), 328–30.

113 Elias *Vit. Joh. Tel.* 71.21–73.11; translated in Greatrex and Lieu (2002) 98–99. See also Procop. *Aed.* 3.3.9–10.

114 According to Procopius (*Pers.* 2.3.54f), the decision to attack the Romans was made by the king and the assembled nobility.

115 Cf. *PLRE* III, 1410.

116 Procop. *Pers.* 1.23.25–29.

117 In 533 Justinian's troops had destroyed the Vandal *regnum* in Africa after just two battles; cf. Leppin (2011) 150–60; Heather (2018) 123–46.

118 Procop. *Pers.* 2.2.1–12.

119 Procop. *Pers.* 2.3.32–53. Cf. Kruse (2013).

At about that time, Khosrow apparently carried out a series of internal reforms, in part already initiated by his father, which were intended to increase the power of the crown.[120] Among other things, the tax system was altered in order to make royal revenues more predictable – whether the system introduced by emperor Diocletian in the Roman Empire served as a model for this is controversial.[121] In addition, the Sasanian army was reorganised; based on sigillographic evidence, it can now be said with certainty that Khosrow set up four regional commands, each subject to a *spāhbed*.[122] Apparently, a 'landed gentry' was created and equipped with lands, and these *dehgānān* were obliged to fight for the ruler.[123] Overall, the number of troops directly subordinate to the king and permanently at his disposal seems to have been increased by these measures; and the government's dependence on military support from the *vuzurgān* had thus decreased – at least temporarily. These reforms were probably not quite completed when Khosrow attacked the Roman Empire; nevertheless, in the following years it became apparent that the fighting strength of the royal army was no doubt great enough to withstand even a prolonged war with the Romans.

4 The Sasanian Invasion of Roman Syria in 540 CE

In the end, it seems to have been disputes between the Arab allies of the two great powers that provided Khosrow I with a pretext for war.[124] Lakhmids and Jafnids, not mentioned in the *foedus* of 532,[125] quarrelled over some territories; the emperor acted clumsily, provoking Al-Mundhir, the Lakhmid sheikh;[126] and Khosrow seized the opportunity to announce that the Romans had broken the treaty. Justinian sent another envoy to Ctesiphon, but it was too late. In April 540, the *shāhānshāh*, at the head of a large army,[127] invaded the Roman Empire.[128] As in 531, the Iranian troops circumvented the imperial fortresses in

120 Cf. Rubin (1995); Gariboldi (2015); Bonner (2020) 180–82.

121 Cf. Howard-Johnston (1995) 215.

122 Cf. Gyselen (2001).

123 Cf. Tafazzoli (1994); Börm (2018b), 37–39.

124 On the role played by Arabian tribes between the Sasanians and Romans, cf. Fisher (2011). Cf. also Millar (2010); Liebeschuetz (2015).

125 Procop. *Pers.* 2.1.5.

126 Cf. Shahîd (1995) 209–18; Bonner (2020) 182–83.

127 Ṭabarī (1.960) claims that as many as 90,000 soldiers accompanied the king, but this number is likely to be greatly exaggerated. Cf. Downey (1953) 343–44.

128 I have tried to analyse the events in more detail elsewhere; cf. Börm (2006). See also Meier (2003) 313–20; Mitchell and Greatrex (2023) 411–14.

Mesopotamia since they clearly had no plans to make permanent conquests anyway: Khosrow was striving for glory and booty, and was not interested in acquiring new territory but in forcing Justinian to agree to make annual payments once again. Nowhere were these goals easier to achieve than in the Roman provinces of Syria. Clearly, the king knew full well that Justinian's best soldiers were fighting in the West; however, after crossing the Euphrates at Circesium, he was probably still surprised at how little resistance his rapidly advancing army encountered. With Belisarius still in Italy, the *magister militum* Buzes was unable to organise effective resistance, so that the Roman cities were largely left to their own devices.

First, Sura was sacked by the Sasanians,[129] then Khosrow talked to Roman envoys and demanded 10,000 pounds of gold in return for his withdrawal back to his own empire; but since the negotiations led to nothing, he continued the invasion. Hierapolis was able to buy itself free, but Beroea was unable to pay the requested amount and was subsequently attacked by the Iranians. However, after mediation by a bishop, Khosrow granted free withdrawal to those men who had entrenched themselves on the acropolis.[130] Next, the Sasanian army reached Antioch on the Orontes, still one of the largest and most important cities of the later Roman Empire. However, the city walls had been damaged by a severe earthquake, and although 6000 *limitanei* had been moved to Antioch shortly before the arrival of the enemy, the site was poorly prepared for an attack.[131] Justinian's nephew Germanus was supposed to organise the defence, but he soon recognised the hopelessness of the situation and withdrew. Despite this, several citizens of the city mocked the king, who had initially showed himself merciful and willing to negotiate, thereby provoking an attack that the defenders could not withstand.[132] Zabergan is said to have urged Khosrow not to show mercy,[133] and the king had the city sacked and set on fire; a considerable portion of the inhabitants were taken prisoner and subsequently settled in a newly built town not far from Ctesiphon. The place, called *Rumagān* or *Weh-Antiok-Khosrow*, was apparently designed according to the model of Roman cities;[134] its inhabitants were subject to the king alone.[135]

129 Procop. *Pers.* 2.5.17–24.

130 Procop. *Pers.* 2.7.1–36.

131 Cf. Trombley (2005) 393. Brands (2019).

132 John Lyd. *De Mag.* 3.54; Mal. 18.87; Jord. Rom. 376; John Eph. *Hist. eccl.* 6.6.

133 Procop. *Pers.* 2.9.30–32.

134 Procop. *Pers.* 2.14.1–4; Theoph. Sim. 5.6.9; Sebeos 9.69; Ṭab. 1.959; Thaʿālibī 613–14. On deportations in the Sasanian period cf. Kettenhofen (1994); Morony (2004); Börm (2007) 171–77.

135 Procop. *Pers.* 2.14.3.

The sack of Antioch marked a turning point. By now at the latest Khosrow must have realised that the Romans were unable to defend themselves; for the time being he and his soldiers were free to move about freely in Syria and do whatever they wanted. Instead of returning to his own empire after this overwhelming victory, the king of kings decided to do his best to humiliate Justinian, presumably in order to demonstrate once and for all the superiority of the Sasanian dynasty.[136] The first action Khosrow took to make his point clear was characterised by a reference to ancient Near Eastern traditions: When the Iranians reached Seleucia Pieria, the *shāhānshāh* bathed himself in the waters of the Mediterranean;[137] subsequently, according to Procopius, he sacrificed 'to Helios and such other divinities as he wished' – probably a description of Zoroastrian rituals in the middle of the Christian Roman Empire.[138] The king then went to the famous pagan sanctuary at Daphne, where he worshiped the nymphs,[139] and moved on to Apamea. Here, Khosrow's mockery of the Roman emperor reached its climax: he staged *ludi circenses* in the city's hippodrome, placing himself in the imperial box,[140] and he is said to even have ordered the *factio* of which Justinian was known to be a supporter to be denied victory through an obvious manipulation.[141] This mocking *imitatio Augusti* was hard to beat in audacity; Khosrow had deliberately ridiculed Justinian. After this spectacle the slow march home began; the Sasanian army moved east, passing Chalcis, which could buy itself out by paying a high ransom, and again crossed the Euphrates, entering Roman Mesopotamia. There, Edessa could save herself with money, while Carrhae was spared by Khosrow without having to pay anything: remarkably, he justified this with the fact that most of the inhabitants of the city were not Christians – clearly another deliberate provocation aimed at the *imperator christianissimus* Justinian.[142] Afterwards the king demanded tribute from Constantia, claiming that the city belonged to the Sasanians anyway because it had been conquered by Kavad some time ago.[143] All in all, Khosrow suffered only one small setback: he had to abort an unsuccessful siege of Dara after a short time before returning to the Sasanian Empire.

---

136 Cf. Canepa (2009) 172–73.

137 On the Near Eastern context of this very ancient ritual, cf. Rollinger (2012).

138 Procop. *Pers.* 2.11.1.

139 Procop. *Pers.* 2.11.6.

140 On the special importance of *ludi circenses* for late Roman emperors, see Canepa (2009) 167–74.

141 Procop. *Pers.* 2.11.31–33.

142 Procop. *Pers.* 2.13.7. Other authors also confirm that the city, which the Christians called 'the city of the pagans' (Ἑλληνόπολις), was known to be a retreat for adherents of the traditional cults; cf. Theod. *Hist. eccl.* 4.18.14.

143 Procop. *Pers.* 2.13.13–20.

Justinian had made an offer of peace to the king early on; but when the emperor learned of the fate of Antioch and of the subsequent behaviour of the *shāhānshāh*, he seems to have decided that after such humiliation he could not simply conclude a new *foedus*. Khosrow, who had achieved all of his goals, was probably interested in a new treaty; but Justinian simply could not afford to lose face. All in all, the Sasanian attack and the ensuing bloody war seem to have led to a fundamental change of mood among the Romans. There are numerous indications that not only a negative, but also a positive image of the Persians had existed before the middle of the sixth century.[144] Since 540, however, those who no longer believed in the possibility of peaceful coexistence between the two great powers seem to have dominated the political discourse.[145] Both sides were preparing to continue the war in 541.[146]

5 The Lazic War

Apparently, Justinian expected the Sasanians to begin conquering Northern Mesopotamia. But Khosrow was once again able to take the Romans by surprise; reacting to a call for help from the king of Lazica, he personally led an invasion force, occupying most of the country.[147] Since the Iranians had thus gained access to the Black Sea for the first time, the Romans feared that from now on Sasanian warships could threaten Constantinople.[148] It is unclear whether these concerns were really justified, but in any case the imperial troops tried with great effort to get Lazica back under their control. The fighting soon concentrated on the fortress of Petra, stubbornly defended by Sasanian troops.[149] Meanwhile, Belisarius failed to conquer Nisibis, and in 542 Khosrow attacked again, this time in the south; but through skilful tactical manoeuvres the Romans managed to force the royal army to retreat.[150] Although no pitched battle had been fought, Belisarius celebrated this success with a large victory

144 Cf. Börm (2007) 276–89.

145 In their portrayal of Khosrow, Roman authors henceforth fluctuate between disgust and admiration; cf. Brodka (1998) and Huyse (2015).

146 Cf. Dignas and Winter (2007) 37–49.

147 Procop. *Pers*. 2.17.1–2.

148 Procop. *Pers*. 2.28.23; Agath. *Hist*. 2.18.7. Cf. Braund (1994) 297. According to the fifth-century historian Priscus there were no harbours in Lazica; Prisc. fr. 33.1 (Blockley).

149 Cf. Bonner (2020) 199–200; Schleicher (2021) 199–202.

150 Procop. *Pers*. 2.21.1–30.

parade in Antioch which was currently being rebuilt.[151] However, Khosrow was still able to conquer Callinicum and had numerous inhabitants deported.[152]

Despite the fact that since 541 the Justinianic Plague was affecting both empires, and both Khosrow and Justinian fell ill, the war was continued in 543. This time the Romans launched an offensive in Persarmenia but suffered a catastrophic defeat: although the new *magister militum per Orientem*, Martinus, had at his disposal a strong army of 30,000 soldiers, he made several serious mistakes, allowing the Sasanian general Nabed, who according to Procopius commanded no more than 4,000 men, to successfully ambush him near the village of Anglon.[153] The following year, Khosrow, having recovered from his illness, took the initiative again. He was accompanied by Zabergan.[154] But the king's attempt to conquer Edessa failed after a long siege,[155] with the Christians attributing the victory to divine assistance.[156] After that, Khosrow seems to have lost interest in the war against the Romans. An armistice that did not include Lazica was concluded in 545 with Justinian, who needed his best soldiers to fight the Goths in Italy, grudgingly agreeing to pay tribute. Apparently the *shāhānshāh* was still convinced that he was entitled to the regular payment of approximately 500 pounds of gold (36,000 *solidi*) per year.[157]

In 551 the truce was renewed. Lazica, whose king Gubaz had switched sides again in 548,[158] was still exempt, which triggered new heavy fighting. Khosrow had apparently tried to bind the country closer to *Eranshahr* by, among other things, promoting Zoroastrianism, and this is said to have upset many Christian aristocrats. All in all, religion now seems to have become an increasingly important policy instrument of both great powers,[159] especially in the Caucasus region.[160] Already in 549, the Roman commander Dagisthaeus had

---

151 Cf. Laniado (2010) 275–78.

152 Procop. *Pers.* 2.21.30.

153 Cf. Trombley (2005) 404–11.

154 According to Procopius (*Pers.* 2.26.16–19), he again appeared as an advocate of tough action against the Romans.

155 The defenders managed to destroy the Sasanian mound by undermining it and then setting fire to it; Procop. *Pers.* 2.27.1–17.

156 Procopius (*Pers.* 2.26.1) claims that the king wanted to conquer Edessa mainly because the Christians said that their god protected the city.

157 Procop. *Pers.* 2.28.7–11; Marc. *Com. ad ann.* 546.

158 Allegedly the king had found out about Sasanian plans to have him killed; cf. Braund (1994) 298–300.

159 On the religious aspects of the conflict, see most recently Briscoe (2014).

160 Cf. Braund (1994) 281–87; Greatrex (2005) 490–500: 'There were missions sent out by imperial initiative; both the Tzani and the Abasgi in the Caucasus were converted by this

won an important victory,[161] and in 551 the fortress of Petra was conquered and destroyed.[162] In 552 the imperial general Narses was finally able to defeat the Goths in Italy; henceforth Justinian could deploy sufficient troops in the East.[163] In 554 the Iranians nevertheless won a major victory against strong Laz-Roman forces, forcing the latter to withdraw to the western parts of the country, and soon afterwards, Roman officers killed Gubaz, causing outrage among his subjects.[164] Afterwards, however, the Romans slowly gained the upper hand in Lazica, and when a large Sasanian offensive failed in 556 because the Iranians were unable to conquer the important fortress of Phasis, for the first time Khosrow appeared ready to give up the region. A little later, the Iranian general Nakhvaragan, having suffered another defeat, was cruelly executed by the king, who was obviously losing patience. Another truce was made in 557, this time including Lazica,[165] and the *shāhānshāh* subsequently turned his attention to the East, seizing the opportunity to destroy the Hephthalite Empire in alliance with the Gökturks.[166] The Turks probably bore the main burden of this war, but the Sasanians nevertheless claimed victory;[167] and after this triumph, Khosrow could afford to make concessions in the West.

Emperor and king had intensified diplomatic contacts again for a while; embassies were exchanged regularly, and the Sasanian envoy Yazdgušnasp repeatedly stayed in Constantinople for an extended period of time,[168] with Justinian treating him very courteously.[169] On the Roman side, the *magister officiorum* and *patricius* Petrus acted as the most important diplomat.[170] In 561 or 562, after difficult negotiations, a new *foedus* between Sasanians and Romans was finally made, the contents of which we are well informed about thanks to the Greek historian Menander Protector.[171] Khosrow ceded Lazica to the emperor; and this time it was explicitly stated that allies on both sides had

method. When, in both cases, the peoples proved recalcitrant, troops were deployed and the people subjugated by force' (491).

161 Procop. *Pers.* 8.8.21–38.

162 Cf. Bonner (2020) 201–203.

163 Cf. Börm (2018a) 152–54.

164 Cf. Bonner (2020) 204–206.

165 Agath. *Hist.* 4.30.7–10.

166 On the Gökturks, cf. Beckwith (2009) 112–39. It is also possible that Khosrow was alarmed by the sudden emergence of this new empire and therefore decided to turn his attention to the East; cf. Howard-Johnston (2016) 603.

167 Cf. Bonner (2020) 214–18.

168 Cf. *PLRE* III, 722–723.; Diebler (1995) 199–200.

169 Procop. *Pers.* 8.25.20–21.

170 Cf. *PLRE* III, 994–98.

171 Men. Prot. fr. 6.1.314–397. Cf. Bonner (2020) 207–209.

to abide by the peace as well. Tolerance was assured for the Christians in the Iranian sphere of influence.[172] The peace was intended to last for fifty years, although Khosrow was originally against a time limit. From his point of view, one aspect of the *foedus* was certainly particularly important: the emperor had to acknowledge that the Sasanians were entitled to annual tributes of 30,000 *solidi*, making the hierarchy between the *shāhānshāh* and the *Augustus* obvious to everyone.[173] However, in order to disguise this humiliation, Justinian managed to have the amounts for the first ten years paid in two instalments. Only from 572 should payments be made regularly every year; Justinian, who was 80 years old in 562, may have hoped that his successor would have to deal with this embarrassing problem. He was right. He died in 565 and his nephew Justin II (r. 565 to 578) became the next Roman emperor.[174]

6 Khosrow's Final War with the Roman Empire

As early as 563, Arab *foederati* complained in Constantinople about alleged provocations on the part of their Lakhmid rivals,[175] but apparently the aged emperor was careful not to endanger the peace with *Eranshahr*. Justin II, however, seems to have seen things differently. His claim to the throne was not without controversy, and therefore the new *Augustus* appears to have pursued an aggressive foreign policy in order to give legitimacy to his rule. When the *shāhānshāh* congratulated him in writing on his accession to the throne, as was customary,[176] Justin II ostensibly reacted unkindly.[177] Everything indicates that the emperor and his entourage were dissatisfied with the *foedus* of 562 and, above all, considered the annual tributes, due from 572, to be unacceptable. So when a Sogdian named Maniach appeared in Constantinople in 568 to offer the Romans an alliance against the Sasanians on behalf of the Turks, he was met with open ears by Justin II.[178] Relations between Khosrow and the Turks had deteriorated dramatically because, above all, the lucrative silk trade could not be agreed upon. In 569 the *magister militum per Orientem* Zemarchus travelled to the *khagan* Istemi to conclude a *foedus* with the Turks. An attempt

172 Men. Prot. fr. 6.1.398–407.
173 Men. Prot. fr. 6.1.145–154.
174 Cf. Rosen (1999).
175 Theoph. AM 6056.
176 Cf. Güterbock (1906) 9.
177 Cf. Bonner (2020) 221–22.
178 Men. Prot. fr. 10.1.86–88.

by the Iranians to intercept Zemarchus on the way back failed.[179] This created a very threatening situation for Khosrow and the Sasanian Empire; however, around the same time the king was able to establish an Iranian presence in today's Yemen, somewhat improving his strategic position.[180]

Much as in 539, when the Iranian ruler did not have to wait long for the smaller peoples between the two great powers to provide him with a pretext for war, Justin II did not have to wait long for a suitable opportunity to begin a *bellum iustum* against Khosrow.[181] Actually, it happened sooner than he expected: in 571 a revolt broke out in Persarmenia; apparently the Sasanian governor (*marzbān*) had behaved very clumsily, humiliating certain Armenian noblemen. In this tense situation, his order to erect a large *pyreion* was interpreted as an attempt to force the country's Christians to accept Mazdaism. A local nobleman murdered the *marzbān* and fled to Constantinople, asking Justin II for help.[182] The emperor did not hesitate long and decided to send Roman forces to Persarmenia.[183]

Surprisingly, Khosrow initially seemed ready to overlook this outrageous provocation.[184] He informed the emperor that if the Romans retreated from Persamenia and the annual tributes were paid from 572 onwards, peace could still prevail. However, Justin II, trusting in his alliance with the Turks,[185] replied that he would never pay tribute to the king, but instead Khosrow should return the 300,000 *solidi* he had already received.[186] Under these circumstances, war was inevitable, and in 572 *Eranshahr* was threatened simultaneously on three fronts: in the northeast by the Turks, in the southwest by the Arab *foederati* of the Romans, and in the northwest by imperial troops.[187] Although Khosrow was by now an elderly man, he personally took command of the Sasanian army. In a way unknown to us, he or the responsible *spāhbed* must have been able to ward off the Turkish attack.[188] Then the king took care of western affairs. At the beginning of 573 at the latest, he was able to convince the leader of the Arab σύμμαχοι, the Jafnid Al-Mundhir, to abandon the emperor.[189]

179 Cf. Bonner (2020) 224–25.
180 Cf. Shahîd (1995) 364–66.
181 Cf. Elton (2018) 287–89.
182 Greg. Tur. *Hist.* 4.40; John Eph. *Hist. eccl.* 2.20.
183 Evagrius, *Hist. eccl.* 5.7.
184 Interestingly, Theophanes (AM 6064) instead blames Arab conflicts over the lucrative trade with India for the outbreak of war.
185 Men. Prot. fr. 16.1.51–57.
186 Mich. Syr. 10.1. Cf. Börm (2008) 327.
187 Cf. Howard-Johnston (2010) 51–57.
188 Perhaps the Sasanians were able to pay off the Turks; Theoph. Sim. 3.6.10–11.
189 Cf. Whitby (1988) 257–58.

The imperial troops, meanwhile, had been unable to capitalise on their initial advantage. The newly appointed *magister militum per Orientem* Marcianus did not arrive in Northern Mesopotamia until August 572 and had to spend precious time preparing his army for attack.[190] Clearly, Justin II had overestimated the effectiveness of the Roman troops. When Marcianus finally advanced in 573, the Sasanians were prepared. Nisibis proved impregnable, and when the *shāhānshāh* himself suddenly appeared with a large army, the surprised Romans fled. Khosrow crossed the border into Roman territory, while his general Adurmahan devastated Syria. Apamea was conquered and sacked, and numerous inhabitants were deported to Iran.[191] Meanwhile, Khosrow used the siege engines the Romans had left at Nisibis to attack Dara. The fortress fell after six months.[192] When the emperor heard of this defeat, he seems to have suffered a nervous breakdown, occasionally making animal sounds while sitting on the throne.[193] It is unclear whether these reports are correct; it is however certain that Justin II was effectively disempowered in 574. He was allowed to remain *Augustus*, but the government was taken over by the *comes excubitorum* Tiberius Constantinus (574 to 582), who was now raised to the rank of *Caesar*.[194] His main task was to stabilise the Roman military position towards the Iranians.

In the meantime, Khosrow, who in the crisis of 572/73 had once again proven himself to be an outstanding general and diplomat, opened peace negotiations. More than ever, the Sasanians attached importance to Roman tributes as a symbol of their superiority. The Romans indeed agreed to pay 30,000 *solidi* each year in return for a three-year truce in 575, which, however, only applied to Mesopotamia.[195] In 576 Khosrow therefore led an army into Armenia, but this time the fortunes of war turned. The Sasanian siege of Theodosiopolis was a failure, and the able *magister militum* Justinian, a relative of Justin II, managed to fend off an attack on Caesarea in Cappadocia. Eventually Khosrow had to retreat, and when his army crossed the Euphrates at Melitene, Justinian attacked it. In the ensuing chaos many Iranian soldiers drowned, and the king himself was in mortal danger. It is no longer possible to say with certainty how serious the Sasanian defeat really was, but in any case, Khosrow had to leave behind considerable treasures and equipment, and the Roman general sent

190 Evagrius, *Hist. eccl.* 5.8.

191 Khosrow reportedly sent female captives to the *khagan*, an indication that the Turks had switched sides; John Eph. *Hist. eccl.* 6.7.

192 John Eph. *Hist. eccl.* 6.5–6; Theoph. Sim. 3.10–11.2.

193 John Eph. *Hist. eccl.* 3.2–5.

194 Cf. Cameron (1976).

195 Cf. Bonner (2020) 230–31.

the booty,[196] along with captured war elephants,[197] to Constantinople, where it was presented in a triumphal celebration.[198] After this defeat, Khosrow allegedly even ordered that kings should no longer be allowed to act as generals in the future.[199] It is uncertain whether this information is correct; however, it is clear that after 576 Khosrow and his successors no longer acted themselves as leaders of invasions, but from then on entrusted generals with this task – just as the Roman *Augusti* had done for a long time.[200]

The imperial troops took advantage of their victory at Melitene to pillage Sasanian territory; after that, peace negotiations started again. Mainly because of the delicate question of Roman tributes the talks dragged on, and fighting started again in 578. The new *magister militum* Maurice in particular was able to inflict several minor defeats on the Sasanians, while both sides plundered and devastated the territory of the other.[201] In 579, when Khosrow was on his deathbed, the peace negotiations were finally coming to an end; the Romans were ready to withdraw entirely from Persarmenia, while the *shāhānshāh* had in return apparently agreed to waive the annual tributes. The final controversial issue was the return of the fortress of Dara to the Romans. But Khosrow's death prevented peace, because his son and successor Hormizd IV (579 to 590) informed the imperial envoys that Dara had been conquered by his father, and since he had to fully preserve the empire he had inherited, the new king declared that it was impossible for him to return the fortress to the Romans.[202] The war continued.

7 The Aftermath

In the years after Khosrow's death, Iranians and Romans continued to fight each other without any side gaining the upper hand. However, when the royal general Wahram Čobin suffered a minor defeat against imperial troops in 589, this led to a rapidly escalating conflict with Hormizd IV. Numerous aristocrats joined the rebellion, and before Wahram reached Ctesiphon, the king was overthrown and replaced by his son Khosrow II (590 to 628).[203] This event could

196 Theoph. Sim. 3.14.10.

197 On war elephants in Late Antiquity, used quite frequently by the Sasanians (Amm. Marc. 19.2.3; Jos. Styl. 62; Procop. *Pers.* 8.13.4; Procop. *Aed.* 2.1.11; Sebeos 8.68), cf. Rance (2003).

198 John Bicl. *ad ann.* 575. Cf. McCormick (1986) 68–69.

199 Evagrius, *Hist. eccl.* 5.15; Theoph. Sim. 3.14.11; John Eph. *Hist. eccl.* 6.9.

200 Cf. Whitby (1994) 228–29.

201 Cf. Bonner (2020) 231–32.

202 Men. Prot. fr. 23.9.

203 Theoph. Sim. 4.3.4–4.7.4.

not appease the rebels, and after a first battle Khosrow II fled to the Roman Empire, now ruled by Maurice (582 to 602), while Wahram, who claimed to be of Arsacid ancestry, had himself proclaimed *shāhānshāh*. After some hesitation, the emperor decided to support Khosrow II against the usurper, and a large Roman army was sent to fight Wahram together with Sasanian loyalists.[204] In 591 the victorious Khosrow II concluded a new *foedus* with Maurice; he once and for all dispensed with Roman tributes as well as most of Persarmenia. Perhaps he even became the adoptive son of the emperor, over sixty years after his grandfather was denied this honour. Thus ended the last war in which Khosrow I had fought the Romans.[205]

Peace, however, was not long-lasting, because even though he outwardly tried to achieve harmony, Khosrow II seems to have considered the *foedus* of 591 to be quite unacceptable. Until 602 he was busy dealing with internal resistance to his rule, but in 603 he used the violent death of Maurice to declare war on his successor Phocas (602 to 610). By 610, the Iranians, exploiting a Roman civil war,[206] had conquered Northern Mesopotamia and Persarmenia, and in 611, Khosrow's troops began to conquer Syria and Egypt, which were subdued until 619.[207] By appointing his own governors the *shāhānshāh* made it clear that this time he wanted to permanently integrate these areas into his empire.[208] The situation of the Romans seemed hopeless, but the new emperor Heraclius (610 to 641)[209] decided to offer desperate resistance, making sure that the war dragged on.[210] In 626 the Sasanians' attempt to conquer Constantinople in alliance with the Avars failed, and in 627 Heraclius renewed the alliance with the Turks, who now attacked *Eranshahr* in the north and east. When Khosrow II stubbornly refused to end the war with the Romans, he was overthrown and killed.[211] His son Kavad II Shiroye died after a few months and civil wars broke

204 Cf. Riedlberger (1998).

205 Cf. Bonner (2020) 271–77.

206 Theoph. AM 6102. Cf. Sarris (2011) 243–45.

207 Cf. Meier (2019) 1027–35, and especially Howard-Johnston (2021).

208 Cf. Altheim-Stiehl (1992); Foss (2003); Weber (2013). The reasons for this amazing decision are controversial. Some scholars believe that the king wanted to eliminate the danger of a two-front war once and for all by eliminating the Roman Empire, cf. Howard-Johnston (2016) 603. But perhaps Khosrow II simply took advantage of the temporary weakness of the aristocratic opposition in Iran whose ability to prevent the king from pursuing an extensive policy of conquest in the West was severely limited after the bloody civil wars between 590 and 602.

209 On Heraclius, see now Viermann (2021).

210 Cf. Sarris (2011) 249–58; Meier (2019) 1027–35; Raum (2021) 136–98, Decker (2022) 186–204.

211 Turkish invasions in eastern Iran must have troubled the Iranian magnates; cf. Howard-Johnston (2010) 65–67; Potts (2018).

out while the Iranian troops retreated from Syria, Asia Minor and Egypt in an orderly fashion. Thus, after huge casualties, the last war between the Sasanians and the Romans ended with a return to the *status quo ante*. Soon afterwards the Arab invasions began, bringing an end to the Sasanian Empire.[212]

On the one hand, Khosrow I had been able to bring the power of the Sasanian kings to a climax; it is not without reason that to this day he is well remembered in Iran. On the other hand, the invasion of 540 and the unnecessarily provocative behaviour displayed by the *shāhānshāh* in the aftermath proved to be fatal. The Sasanian attack of 502 had already disrupted the peaceful coexistence of the Romans and Iranians, but at first there was still enough reason and good will on both sides to make a return to good relations seem possible. The Eternal Peace concluded with Justinian in 532 was the result of these efforts. But when Khosrow decided to attack the Romans only a few years later, he permanently ruined the relationship with the emperor. There is no evidence that he planned to make permanent conquests west of the Euphrates; instead, the *shāhānshāh* was simply concerned with increasing his own glory by humiliating the Romans. However, this turned out to be a short-sighted miscalculation, because otherwise a secure peace in the West would have been possible, since the *Augusti* had no interest in a major Persian war – the Roman attack of 572 was an exception that was due to special circumstances. Thus it was mainly Khosrow's fault that between 540 and 630 there were only a few precious years of peace between *Ērānshahr* and the *Imperium Romanum*.[213] The endless war between the two powers that he started contributed significantly to the fall of the Sasanian Dynasty: had the Romans and Iranians cooperated with each other in the seventh century as they had in the fifth, the Islamic conquests would probably have been impossible.

Bibliography

Alston, R. and Lieu, S.N.C. (eds) (2007) *Aspects of the Roman East*. Turnhout.

Altheim-Stiehl, R. (1992) 'The Sasanians in Egypt'. *Bulletin de la Société d'Archéologie Copte* 31: 87–96.

Andres, H. (2022) *Bruderzwist: Strukturen und Methoden der Diplomatie zwischen Rom und Iran von der Teilung Armeniens bis zum Fünfzigjährigen Frieden*. Stuttgart.

212 Among the many accounts of early Arab expansion, Kennedy (2007) and Berger (2016, 154–84) are, in my opinion, the best. In addition, Hoyland (2015) is worth reading, sometimes representing somewhat idiosyncratic positions.

213 Cf. Isaac (1995) 137: 'It was more important to humiliate the enemy than to disarm him'.

Arnason, J. and Raaflaub, K. (eds) (2011) *The Roman Empire in Context*. Chichester.

Beckwith, C. (2009) *Empires of the Silk Road: A History of Central Eurasia from the Bronze Age to the Present*. Princeton.

Berger, L. (2016) *Die Entstehung des Islam*. Munich.

Bernheimer, T. and Silverstein, A. (eds) (2012) *Late Antiquity: Eastern Perspectives*. Exeter.

Binder, C., Börm, H., and Luther, A. (eds) (2016) *Diwan: Studies in the History and Culture of the Ancient Near East and the Eastern Mediterranean*. Duisburg.

Bivar, A.D.H. (2004) 'Hephthalites'. *EncIr* 12: 198–201.

Blockley, R.C. (1985) *The History of Menander the Guardsman*. Liverpool.

Blockley, R.C. (1987) 'The Division of Armenia between the Romans and the Persians at the End of the Fourth Century A.D.'. *Historia* 36: 222–34.

Blockley, R.C. (1992) *East Roman Foreign Policy*. Leeds.

Bonner, M.J. (2020) *The Last Empire of Iran*. Piscataway.

Börm, H. (2006) 'Der Perserkönig im Imperium Romanum. Chosroes I. und der sasanidische Einfall in das Oströmische Reich 540 n. Chr'. *Chiron* 36: 299–328.

Börm, H. (2007) *Prokop und die Perser*. Stuttgart.

Börm, H. (2008) '"Es war allerdings nicht so, dass sie es im Sinne eines Tributes erhielten, wie viele meinten ..." Anlässe und Funktion der persischen Geldforderungen an die Römer'. *Historia* 57: 327–46.

Börm, H. (2015) 'Born to be Emperor: The Principle of Succession and the Roman Monarchy', in Wienand (2015) 239–64.

Börm, H. (2016) 'A Threat or a Blessing? The Sasanians and the Roman Empire', in Binder, Börm, and Luther (2016) 615–46.

Börm, H. (2017) 'Kontinuität im Wandel. Begründungsmuster und Handlungsspielräume der iranischen Monarchie in arsakidischer und sasanidischer Zeit', in Rebenich (2017) 545–64.

Börm, H. (2018a) *Westrom: Von Honorius bis Justinian*. Stuttgart.

Börm, H. (2018b) 'König und Gefolgschaft im Sasanidenreich. Zum Verhältnis zwischen Monarch und imperialer Elite im spätantiken Persien', in Drews (2018) 23–42.

Börm, H. (2019a) 'Barbaren als Tyrannen: Das Perserbild in der klassizistischen griechischen Historiographie', in Rollinger, Ruffing, and Thomas (2019) 15–34.

Börm, H. (2019b) 'Die Grenzen des Großkönigs? Überlegungen zur arsakidisch-sasanidischen Politik gegenüber Rom', in Schleicher, Hartmann, and Stickler (2019) 99–122.

Börm, H. (2022) 'Procopius and the East', in Meier and Montinaro (2022) 310–36.

Börm, H. (forthcoming) 'Sasanian Iran and the Roman Empire', in Daryaee and Shayegan (forthcoming).

Börm, H. and Wiesehöfer, J. (eds) (2010) *Commutatio et Contentio. Studies in the Late Roman, Sasanian, and Early Islamic Near East*. Düsseldorf.

Bosworth, C.E. (1983) 'Iran and the Arabs before Islam', in Yarshater (1983a) 593–612.

Braund, D. (1994) *Georgia in Antiquity*. Oxford.

Briscoe, T. (2014) 'Rome and Persia: Rhetoric and Religion', in Dzino and Parry (2014) 155–68.

Brodka, D. (1998) 'Das Bild des Perserkönigs Chosroes I. in den *bella* des Prokopios von Kaisareia', in Styka (1998) 115–24.

Brodka, D. (2011) 'Prokopios und Malalas über die Schlacht bei Callinicum'. *Classica Cracoviensia* 14: 71–93.

Cameron, A. (1976) 'An Emperor's Abdication'. *Byzantinoslavica* 37: 161–67.

Cameron, A. (1985) *Procopius and the Sixth Century*. London.

Cameron, A. (ed) (1995) *The Byzantine and Early Islamic Near East. Vol. 3: States, Resources and Armies*. Princeton.

Campbell, B. and Tritle, L. (eds) (2013) *The Oxford Handbook on Warfare in the Classical World*. Oxford.

Canepa, M.P. (2009) *The Two Eyes of the Earth: Art and Ritual of Kingship between Rome and Sasanian Iran*. Berkeley.

Carile, A. (ed) (2004) *La Persia e Bisanzio*. Rome.

Charles, M. (2011) 'The Sasanian "Immortals"'. *IrAnt* 46: 289–311.

Christensen, A. (1944) *L'Iran sous les Sassanides*.[2] Copenhagen.

Christie, N. and Sarantis, A. (eds) (2013) *War and Warfare in Late Antiquity*. Leiden/Boston.

Chrysos, E.K. (1993) 'Räumung und Aufgabe von Reichsterritorien: Der Vertrag von 363'. *BJ* 193: 165–202.

Colvin, I. (2018) 'Comparing Procopius and Malalas', in Lillington-Martin and Turquois (2018) 201–13.

Crone, P. (1991) 'Kavad's Heresy and Mazdak's Revolt'. *Iran* 29: 21–42.

Dąbrowa, E. (ed) (1994) *The Roman and Byzantine Army in the East*. Krakow.

Daryaee, T. (2009) *Sasanian Persia: The Rise and Fall of an Empire*. London.

Daryaee, T. (2022) 'The End of Ērānšahar: The Decline of the Sasanian Empire', in Gehler, Rollinger and Strobl (2022) 249–59.

Daryaee, T. and Shayegan, M.R. (eds) (forthcoming) *The Blackwell Companion to the Sasanian Empire*. Malden, MA.

De Blois, F. (2015) 'Mazdak the Ancient and Mazdak the Last: Further Remarks on the History and Religious Typology of Mazdakism', in Jullien (2015) 141–53.

Decker, M.J. (2022) *The Sasanian Empire at War: Persia, Rome, and the Rise of Islam 224–651*. Yardley.

De Romanis, F. and Sorda, S. (eds) (2006) *Dal denarius al dinar*. Rome.

De Sena, E. (ed) (2013) *The Roman Empire during the Severan Dynasty: Case Studies in History, Art, Architecture, Economy and Literature*. Piscataway.

Di Cosmo, N. and Maas, M. (eds) (2018) *Empires and Exchanges in Eurasian Late Antiquity: Rome, China, Iran, and the Steppe, ca. 250–750*. Cambridge.

Diebler, S. (1995) 'Les hommes du roi. Sur la représentation souveraine dans les relations diplomatiques entre Byzance et les Sassanides d'après les historiens byzantins du sixième siècle'. *StIr* 24: 187–218.

Dignas, B. and Winter, E. (2007) *Rome and Persia in Late Antiquity*. Cambridge.

Downey, G. (1953) 'The Persian Campaign in Syria in A.D. 540'. *Speculum* 28: 340–48.

Drews, W. (ed) (2018) *Die Interaktion von Herrschern und Eliten in imperialen Ordnungen des Mittelalters*, Berlin.

Drijvers, J.W. (2009) 'Rome and the Sasanid Empire: Confrontation and Coexistence', in Rousseau (2009) 441–54.

Dzino, D. and Parry, K. (eds) (2014) *Byzantium, Its Neighbours and Its Cultures*. Brisbane.

Edwell, P. (2013) 'Sasanian Interactions with Rome and Byzantium', in Potts (2013) 840–55.

Edwell, P. (2020) *Rome and Persia at War: Imperial Competition and Contact, 193–363 CE*. London.

Elton, H. (2018) *The Roman Empire in Late Antiquity: A Political and Military History*. Cambridge.

Erdkamp, P. (ed) (2007) *A Companion to the Roman Army*. Oxford.

Feder, F. and Lohwasser, A. (eds) (2013) *Ägypten und sein Umfeld in der Spätantike*. Wiesbaden.

Fisher, G. (2011) *Between Empires: Arabs, Romans, and Sasanians in Late Antiquity*. Oxford.

Foss, C. (2003) 'The Persians in the Roman Near East (602–630 AD)'. *Journal of the Royal Asiatic Society* 13: 149–70.

Gariboldi, A. (2015) 'The Great "Restoration" of Husraw I', in Jullien (2015) 47–84.

Gehler, M., Rollinger, R. and Strobl, P. (eds) (2022) *The End of Empires*. Wiesbaden.

Genequand, D. and Julien Robin, C. (eds) (2015) *Les Jafnides: Des rois arabes au service de Byzance*. Paris.

Gignoux, P., Jullien, C., and Jullien, F. (eds) (2009) *Trésors d'Orient*. Paris.

Gradoni, M. (2013) 'The Parthian Campaigns of Septimius Severus: Causes, and Roles in Dynastic Legitimation', in De Sena (2013) 3–23.

Greatrex, G. and Lieu, S.N.C. (2002) *The Roman Eastern Frontier and the Persian Wars II, AD 363–630*. London.

Greatrex, G. (1993) 'The Two Fifth-century Wars between Rome and Persia'. *Florilegium* 12: 1–14.

Greatrex, G. (1998) *Rome and Persia at War, 502–532*. Leeds.

Greatrex, G. (2000) 'The Background and Aftermath of the Partition of Armenia in A.D. 387'. *AHB* 14: 35–48.

Greatrex, G. (2005) 'Byzantium and the East in the Sixth Century', in Maas (2005) 477–509.

Greatrex, G. (2007) 'Roman Frontiers and Foreign Policy in the East', in Alston and Lieu (2007) 103–73.

Greatrex, G. (2010) 'Procopius and Pseudo-Zachariah on the Siege of Amida and Its Aftermath (502–506)', in Börm and Wiesehöfer (2010) 227–51.

Greatrex, G. (2014) 'L'influence Byzantine sur la Perse Sassanide', in Sakel (2014) 164–74.

Greatrex, G. (2015) 'Les Jafnides et la defence de l'Empire au VIe siècle', in Genequand and Julien Robin (2015) 121–54.

Gruen, E.S. (ed) (2005) *Cultural Borrowings and Ethnic Appropriations in Antiquity*. Stuttgart.

Güterbock, K. (1906) *Byzanz und Persien in ihren diplomatisch-völkerrechtlichen Beziehungen im Zeitalter Justinians. Ein Beitrag zur Geschichte des Völkerrechts*. Berlin.

Gyselen, R. (2001) *The Four Generals of the Sasanian Empire: Some Sigillographic Evidence*. Rome.

Haarer, F.K. (2006) *Anastasius I*. Leeds.

Haldon, J. and Conrad, L. (eds) (2004) *The Byzantine and Early Islamic Near East VI*. Princeton.

Hansman, J. (1992) 'Caspian Gates'. *EncIr* 5: 61f.

Hartmann, U. (2002) 'Geist im Exil: Römische Philosophen am Hof der Sasaniden', in Schuol, Hartmann, and Luther (2002) 123–60.

Hartmann, U. (2007) 'Wege des Wissens. Formen des Gedankenaustauschs und der kulturellen Beeinflussung zwischen dem spätantiken Rom und dem Sāsānidenreich', in Rollinger, Luther, and Wiesehöfer (2007) 50–107.

Heather, P. (2018) *Rome Resurgent*. Oxford.

Hoffmann, L. (ed) (2005) *Zwischen Polis, Provinz und Peripherie*. Wiesbaden.

Hörandner, W., Koder, J., and Stassinopoulou (eds) (2004) *Wiener Byzantinistik und Neogräzistik*. Vienna.

Howard-Johnston, J.D. (1995) 'The Two Great Powers in Late Antiquity: A Comparison', in Cameron (1995) 157–226.

Howard-Johnston, J.D. (2010) 'The Sasanians' Strategic Dilemma', in Börm and Wiesehöfer (2010) 37–70.

Howard-Johnston, J.D. (2012) 'The Late Sasanian Army', in Bernheimer and Silverstein (2012) 87–127.

Howard-Johnston, J.D. (2016) 'The Grand Strategy of the Sasanian Empire', in Binder, Börm, and Luther (2016) 591–613.

Howard-Johnston, J.D. (2021) *The Last Great War of Antiquity*. Oxford.

Hoyland, R. (2015) *In God's Path: The Arab Conquests and the Creation of an Islamic Empire*. Oxford.

Humphrey, J. (ed) (2002) *The Roman and Byzantine Near East II*. Portsmouth.

Huyse, P. (2002) 'Sprachkontakte und Entlehnungen zwischen dem Griechisch/Lateinischen und dem Mitteliranischen', in Schuol, Hartmann, and Luther (2002) 197–234.

Huyse, P. (2015) 'La règne de Husraw Ier aux yeux des historiographes protobyzantines', in Jullien (2015) 195–216.

Inostrancev, C.A. (1926) 'The Sasanian Military Theory'. *Journal of the K.R. Cama Oriental Institute Bombay* 7: 7–52.

Isaac, B. (1995) 'The Army in the Late Roman East: The Persian Wars and the Defence of the Byzantine Provinces', in Cameron (1995) 125–55.

Jullien, C. (ed) (2015) *Husraw Ier: Reconstructions d'un règne*. Paris.

Kaldellis, A. (2004) *Procopius of Caesarea: Tyranny, History, and Philosophy at the End of Antiquity*. Philadelphia.

Kennedy, H. (2007) *The Great Arab Conquests*. Philadelphia.

Kettenhofen, E. (1994) 'Deportations II: In the Parthian and Sasanian Periods'. *EncIr* 7: 297–308.

Koehn, C. (2018) *Justinian und die Armee des frühen Byzanz*. Boston.

Kruse, M. (2013) 'The Speech of the Armenians in Procopius: Justinian's Foreign Policy and the Transition between Books 1 and 2 of the "Wars"'. *CQ* 63: 866–81.

Kühnen, A. (2008) *Die imitatio Alexandri in der römischen Politik*. Münster.

Kurbanov, A. (2013) *The Archaeology and History of the Hephthalites*. Bonn.

Lanfranchi, G., Bonacossi, D., Pappi, C., Ponchia, S. (eds) (2012) *LEGGO! Studies Presented to Prof. Frederick Mario Fales on the Occasion of his 65th Birthday*. Wiesbaden.

Laniado, A. (2010) 'Belisarius in the City of God: Stephanus Antecessor on Three Ceremonies in Constantinople and Antioch', in Börm and Wiesehöfer (2010) 272–92.

Lawrence, D. and Wilkinson, T.J. (2017) 'The Northern and Western Borderlands of the Sasanian Empire', in Sauer (2017) 99–125.

Lee, A.D. (2013) 'Roman Warfare with Sasanian Persia', in Campbell and Tritle (2013) 708–25.

Lenski, N. (2007) 'Two Sieges of Amida (AD 359 and 502–503) and the Experience of Combat in the Late Roman Near East', in Lewin and Pellegrini (2007) 219–36.

Leppin, H. (2011) *Justinian. Das christliche Experiment*. Stuttgart.

Lerza, P. (1982) 'L'immagine "due potenze = due occhi/luci" nelle testimonianze antiche'. *Maia* 34: 219–24.

Lewin, A.S. (2015) 'Did the Roman Empire have a Military Strategy and Were the Jafnids Part of It?', in Genequand and Julien Robin (2015) 155–92.

Lewin, A. and Pellegrini, P. (eds) (2007) *The Late Roman Army in the Near East from Diocletian to the Arab Conquest*. Oxford.

Liebeschuetz, J.H.W.G. (2015) 'Arab Tribesmen and Desert Frontiers in Late Antiquity'. *JLA* 8: 62–96.

Lillington-Martin, C. and Turquois, E. (eds) (2018) *Procopius of Caesarea: Literary and Historical Interpretations*. London.

Lounghis, T. (2005) 'Die kriegerisch gesinnte Partei der senatorischen Opposition in den Jahren 526 bis 529', in Hoffmann (2005) 25–36.

Luther, A. (2014) 'Ein "übersehener" römisch-persischer Krieg um 416/417?' *Gymnasium* 121: 183–93.

Maas, M. (ed) (2005) *The Cambridge Companion to the Age of Justinian*. Cambridge.

Maas, M. (ed) (2015) *The Cambridge Companion to the Age of Attila*. Cambridge.

Maier, F.K. (2019) *Palastrevolution: Der Weg zum hauptstädtischen Kaisertum im Römischen Reich des vierten Jahrhunderts*. Munich.

Maksymiuk, K. (2018) 'The Two Eyes of the Earth: The Problem of Respect in Sasanid-Roman Relations'. *GRBS* 58: 591–606.

Mathisen, R. and Shanzer, D. (eds) (2011b) *Romans, Barbarians, and the Transformation of the Roman World*. Farnham.

McCormick, M., (1986) *Eternal Victory: Triumphal Rulership in Late Antiquity, Byzantium, and the Early Medieval West*. Cambridge.

McDonough, S. (2011a) 'The Legs of the Throne: Kings, Elites, and Subjects in Sasanian Iran', in Arnason and Raaflaub (2011) 290–321.

McDonough, S. (2011b) 'Were the Sasanians Barbarians? Roman Writers on the "Empire of the Persians"', in Mathisen and Shanzer (2011) 55–66.

Meier, M. (2003) *Das andere Zeitalter Justinians*. Göttingen.

Meier, M. (2009) *Anastasios I*. Stuttgart.

Meier, M. (2014) 'Flavius Hypatios: Der Mann, der Kaiser sein wollte', in Patzold and Ubl (2014) 73–96.

Meier, M. (2017) 'Der Monarch auf der Suche nach seinem Platz. Kaiserherrschaft im frühen Byzanz (5 bis 7. Jahrhundert n. Chr.)', in Rebenich (2017) 509–44.

Meier, M. (2019) *Geschichte der Völkerwanderung*. Munich.

Meier, M. (2020) 'Das Ende des weströmischen Kaisertums – ein Ereignis der chinesischen Geschichte? Auswirkungen von Mobilität in eurasischer Perspektive'. *HZ* 311, 275–320.

Meier, M. and Montinaro, F. (eds) (2022) *A Companion to Procopius of Caesarea*. Leiden.

Meißner, B. and Schmitt, O. (eds) (2005) *Krieg, Gesellschaft, Institutionen*. Berlin.

Millar, F. (2010) 'Rome's Arab Allies in Late Antiquity', in Börm and Wiesehöfer (2010) 199–226.

Mitchell, S. and Greatrex, G. (2023) *A History of the Later Roman Empire AD 284–700*. Hoboken.

Morony, M. (2004) 'Population Transfers between Sasanian Iran and the Byzantine Empire', in Carile (2004) 161–80.

Mosig-Walburg, K. (2009) *Römer und Perser: Vom 3. Jahrhundert bis zum Jahr 363 n. Chr.* Gutenberg.

Nechaeva, E. (2017) 'Seven Hellenes and One Christian in the Endless Peace Treaty of 532'. *SLA* 1: 359–80.

Patzold, S. and Ubl, K. (eds) (2014) *Verwandtschaft, Name und soziale Ordnung*. Berlin.

Payne, R. (2015) 'The Reinvention of Iran: The Sasanian Empire and the Huns', in Maas (2015) 282–99.

Pazdernik, C.F. (2015) 'The Quaestor Proclus'. *GRBS* 55, 221–49.

Pfeilschifter, R. (2013) *Der Kaiser und Konstantinopel*. Berlin.

Pieler, P. (1972) 'L'aspect politique et juridique de l'adoption de Chosroes proposée par les Perses à Justin'. *Revue Internationale des Droits de l'Antiquité* 3: 399–433.

Potts, D. (ed) (2013) *The Oxford Handbook of Ancient Iran*. Oxford.

Potts, D. (2018) 'Sasanian Iran and its Northeastern Frontier: Offense, Defence, and Diplomacy', in Di Cosmo and Maas (2018) 287–301.

Pourshariati, P. (2008) *The Decline and Fall of the Sasanian Empire*. London.

Preiser-Kapeller, J. (2004) 'Magister militum per Armeniam. Überlegungen zum armenischen Kommando im 6. und 7. Jahrhundert', in Hörandner, Koder, and Stassinopoulou (2004) 348–65.

Rance, P. (2003) 'Elephants in Warfare in Late Antiquity'. *Acta Antiqua Academiae Scientiarum Hungaricae* 43: 355–84.

Raum, T. (2021) *Szenen eines Überlebenskampfes. Akteure und Handlungsspielräume im Imperium Romanum 610–630*. Stuttgart.

Rebenich, S. (ed) (2017) *Monarchische Herrschaft im Altertum*. Munich.

Rekavandi, H.O., Sauer, E., Wilkinson, T., Abbasi, G.A., Priestman, S., Tamak, E.S., Ainslie, R., Mahmoudi, M., Galiatsatos, N., Roustai, K., Van Rensburg, J., Ershadi, M., MacDonald, E., Fattahi, M., Oatley, C., Shabani, B., Ratcliffe, J., and Usher-Wilson, L. (2008) 'Sasanian Walls, Hinterland Fortresses and Abandoned Ancient Irrigated Landscapes: The 2007 Season on the Great Wall of Gorgan and the wall of Tammishe'. *Iran* 46: 151–78.

Rekavandi, H.O., Sauer, E., Wilkinson, T., and Nokandeh, J. (2013) *Persia's Imperial Power in Late Antiquity: The Great Wall of Gorgan and the Frontier Landscapes of Sasanian Iran*. Oxford.

Rezakhani, K. (2015) 'Mazdakism, Manichaeism and Zoroastrianism: In Search of Orthodoxy and Heterodoxy in Late Antique Iran'. *Iran. Stud.* 48: 55–70.

Rezakhani, K. (2017) *ReOrienting the Sasanians: East Iran in Late Antiquity*. Edinburgh.

Riedlberger, P. (1998) 'Die Restauration von Chosroes II'. *Electrum* 2: 161–75.

Robinson, C. (ed) (2010) *The New Cambridge History of Islam 1*. Cambridge.

Rollinger, R. and Wiesehöfer, J. (2020) 'The "Empire" of the Hephthalites', in Rollinger, Degen, and Gehler (2020) 317–29.

Rollinger, R. (2012) 'From Sargon of Agade and the Assyrian Kings to Khosrau I and Beyond: On the Persistence of Ancient Near Eastern Traditions', in Lanfranchi et al. (2012) 725–43.

Rollinger, R., Degen, J., and Gehler, M. (eds) (2020) *Short-Term Empires in World History*. Wiesbaden.

Rollinger, R., Luther, A., and Wiesehöfer, J. (eds) (2007) *Getrennte Wege? Kommunikation, Raum und Wahrnehmung in der Alten Welt*. Frankfurt.

Rollinger, R., Ruffing, K., and Thomas, L. (eds) (2019) *Das Weltreich der Perser*. Wiesbaden.

Rosen, K. (1999) 'Iustinus II'. *RAC* 19: 778–801.

Rousseau, P. (ed) (2009) *A Companion to Late Antiquity*. Malden, MA.

Rubin, Z. (1995) 'The Reforms of Khusro Anushirwan', in Cameron (1995) 227–98.

Rubin, Z. (2004) 'Nobility, Monarchy and Legitimation under the later Sasanians', in Haldon and Conrad (2004) 235–73.

Rubin, Z. (2008) 'Persia and the Sasanid Monarchy', in Shepard (2008) 130–55.

Sakel, D. (ed) (2014) *Byzantine Culture*. Ankara.

Sarris, P. (2011) *Empires of Faith: The Fall of Rome to the Rise of Islam, 500–700*. Oxford.

Sarris, P. (2023) *Justinian: Emperor, Soldier, Saint*. New York.

Sauer, E.W. (ed) (2017) *Sasanian Persia*. Edinburgh.

Schallmayer, E. (2006) *Der Limes*. Munich.

Schindel, N. (2006) 'Sasanian Gold Coinage: An Overview', in De Romanis and Sorda (2006) 105–29.

Schindel, N. (2013) 'Kawād I'. *EncIr* 16: 136–41.

Schleicher, F., Hartmann, U., and Stickler, T. (eds) (2019) *Iberien zwischen Rom und Iran*. Stuttgart.

Schleicher, F. (2021) *Iberia Caucasia. Ein Kleinkönigreich im Spannungsfeld großer Imperien*. Stuttgart.

Schuol, M., Hartmann, U., and Luther, A. (eds) (2002) *Grenzüberschreitungen*. Stuttgart.

Shahbazi, A.S. (1990) 'Byzantine-Iranian Relations'. *EncIr* 4: 588–99.

Shahîd, I. (1995) *Byzantium and the Arabs in the Sixth Century*. Washington.

Shayegan, M.R. (2013) 'Sasanian Political Ideology', in Potts (2013) 805–13.

Shepard, J. (ed) (2008) *The Cambridge History of the Byzantine Empire*. Cambridge.

Stewart, M.E. (2020) *Masculinity, Identity, and Power Politics in the Age of Justinian: A Study of Procopius*. Amsterdam.

Stickler, T. (2007) 'The Foederati', in Erdkamp (2007) 495–514.

Strootman, R. and Versluys, M.J. (eds) (2017) *Persianism in Antiquity*. Stuttgart.

Styka, J. (ed) (1998) *Studies of Greek and Roman Civilization*. Krakow.

Syvänne, I. (2004) *The Age of Hippotoxotai: Art of War in Roman Military Revival and Disaster (491–636)*. Tampere.

Syvänne, I. (2021) 'The Three Hephthalite Wars of Peroz 474/5–484'. *Historia i Świat* 10: 95–116.

Tafazzoli, A. (1994) 'Dehqān'. *EncIr* 7: 223–26.

Trombley, F. (2005) 'The Late Roman Practice of War on the Syrian Frontier (A.D. 502–641): Leadership, Infrastructure and Operations', in Meißner and Schmitt (2005) 387–416.

Viermann, N. (2021) *Herakleios, der schwitzende Kaiser. Die oströmische Monarchie in der ausgehenden Spätantike*. Berlin.

Walker, J.T. (2002) 'The Limits of Late Antiquity: Philosophy between Rome and Iran'. *Ancient World* 33: 45–69.

Weber, D. (2013) 'Die persische Besetzung Ägyptens 619–629 n. Chr.: Fakten und Spekulationen', in Feder and Lohwasser (2013) 221–46.

Whitby, M. (1988) *The Emperor Maurice and His Historian: Theophylact Simocatta on Persian and Balkan Warfare*. Oxford.

Whitby, M. (1994) 'The Persian King at War', in Dąbrowa (1994) 227–65.

Whitby, M. (2013) 'Siege Warfare and Counter-Siege tactics in Late Antiquity (ca. 250–640)', in Christie and Sarantis (2013) 433–59.

Whitby, M. (2021) *The Wars of Justinian*. Barnsley.

Whittow, M., (2002) 'Rome and the Jafnids: Writing the History of a 6th-century Tribal Dynasty', in Humphrey (2002) 207–24.

Wienand, J. (ed) (2015) *Contested Monarchy*. Oxford.

Wiesehöfer, J. (2005) 'Rūm as Enemy of Iran', in Gruen (2005) 105–20.

Wiesehöfer, J. (2009) 'Kawad, Khusro I and the Mazdakites: A New Proposal', in Gignoux, Jullien, and Jullien (2009) 391–409.

Wiesehöfer, J. (2010) 'The Late Sasanian Near East', in Robinson (2010) 98–152.

Wiesehöfer, J. (2017) 'Ērān ud Anērān: Sasanian Patterns of Worldview', in Strootman and Versluys (2017) 381–92.

Yarshater, E. (ed) (1983a) *The Cambridge History of Iran Volume 3(2): The Seleucid, Parthian, and Sasanid Periods*. Cambridge.

Yarshater, E. (1983b) 'Mazdakism', in Yarshater (1983a) 991–1024.

Zambarbieri, N. (2017) 'La Persia nelle Storie di Agazia: costruzione dell'alterità e polemica interna'. *Adamantius* 23: 263–77.

CHAPTER 19

The First Great War of the Middle Ages: the Sasanian Conquest of the Eastern Mediterranean (602–630 CE) as a Background to the Islamic *Futūḥāt*

Khodadad Rezakhani

The first half of the seventh century CE witnessed one of the greatest changes in the history of the world, and a transformation of the political, social, economic, and most importantly religious contexts of Central and West Asia and North Africa. This was the rise of Islam in Arabia and its astonishing success in conquering and controlling territories beyond its Arabian homeland. By 650 CE, the nascent Islamic empire, less than thirty years old, appears to have conquered territories extending from the central Iranian plateau to Egypt and from Yemen to Armenia. In various studies of these events, several theories have been applied to the Islamic conquests and their consequences in their Western Eurasian/North African settings. Among these is the idea that the rise of Islam and its control of the Mediterranean was the catalyst for the Middle Ages. While this term is perhaps a problematic one in a West Asian/Near Eastern context, it is widely used to describe an Islamic or medieval Middle East.[1] So, as much as such chronological divisions can be criticised, we might still be justified in understanding the rise of Islam as the event that ushered in the Middle Ages in the Mediterranean and beyond.[2]

Contemporaneous to the rise of Islam, a series of wars taking place at the same time that Muhammad was starting his ministry (610 CE) and establishing his political rule in Medina (622 CE) are commonly dubbed 'the Last Great War of Antiquity'.[3] These wars – the conquests of eastern and central Anatolia, Syria,

1 It is perhaps more justifiable to use 'the Middle Ages' for the non-European world than it is to use 'the Ancient World' or 'Antiquity' for the Near East, where chronologies of 'the ancient' far surpasses the documented European antiquity, yet are often lumped together with it.

2 This is perhaps most influentially presented in the classic work of Pirenne (1937) and its many critics and continuators, of which the most applicable to the current discussion is Hodges and Whitehouse (1983).

3 This is made most popular by James Howard-Johnston, e.g., the titles of Howard-Johnston (2006) and Howard-Johnston (2021). Foss (1975) had called it, in a grander tone, 'the last world war of antiquity'.

 | DOI:10.1163/9789004710771_020

Palestine, and Egypt by the Sasanian troops of Khosrow II Aparwez – started in 602 and concluded with the removal and execution of Khosrow II in February 628. From a historiographic point of view, they have been seen as quintessentially different affairs than the Islamic *Futūḥāt*, the conquests that initiated 'the Middle Ages' in the Mediterranean world. This separation, like much of what divides the transition period between pre-Islamic and Islamic periods, seems artificial and ill-justified, beyond simple bias or disciplinary demarcations.[4] This paper, examining the close connections between the two wars, suggests that the Islamic *Futūḥāt* should be considered within the context of the wars that preceded them. Consequently, Khosrow's conquests, instead of being seen as the last great war of antiquity and the conclusion of centuries-old conflicts between Romans and Persians, need to be seen in the new light of changing Mediterranean realities and the start of the Middle Ages.

1 The Sasanian Grand Strategy and the Seventh Century Wars

Among the most prominent historiographical characterisations of the Sasanian-Byzantine wars of the seventh century is a tendency to consider them the latest, indeed the last, episode in a series of long wars between 'Persia' and 'Rome'. This characterisation essentially serves to present the wars in the context of established clichés and patterns governing the relationships between the two empires, including a clash between 'the East' and 'the West'.[5] However, it is not too difficult to see that a new form of competition between the two empires had already emerged, a new state of 'total war'.[6] This new stage, partially

4 Garth Fowden in *Before and After Muhammad*, Ch.3 (esp. pp. 54–55) has emphsised the continuities between the events in the Sasanian sphere and the rise of Islam and advocated for seeing them as essentially interconnected, proposing the first millennium as a parallel to late antiquity.

5 One could argue that the constant search by modern historians to find signs of Byzantines harking back to Alexander or the Sasanians claiming their Achaemenid inheritance is part of a tradition of ethnicising both sides and arguing for the 'antique' background of the conflicts. On Sasanian remembrance of the past, see Canepa (2010). For specific appeals to the past in situations of war, see Dignas and Winter (2007, 53–62), most specifically in their attention to Shapur's ŠKZ inscription (56–59), and Wiesehöfer (2007). I find most of these arguments to be spurious and based essentially on the Roman historiographical tradition, such as that of Herodian (discussed in Dignas and Winter (2007) 54–55), which expects Rome's Sasanian rivals to have the same view of history. In the absence of any primary Sasanian evidence, or any reflection of such claims in later Islamic-period narratives (which could reasonably be expected to have preserved them), I am inclined to consider this a modern historiographical debate betraying an over-reliance on Greco-Roman sources.

6 Isaac (1995) 127.

initiated by a complete overhaul of the Sasanian state following their defeat at the hands of the Hephthalites in the late fifth century,[7] put the two empires in a state where 'coexistence' was no longer a desired option.[8] Instead, a state of war, occasionally interrupted by periods of impromptu or planned peace through fragile occasional treaties, regulated the relations between the two empires in the sixth and seventh centuries.[9] Starting in 502 with the so-called Anastasian Wars, this state of total war largely acted as means of testing the limits of each empire's defences and forming ideologies for each empire, in preparation for a final, cataclysmic confrontation. Seen in the light of what preceded them, the wars of 602–630 were indeed a continuation of a century of war that preceded them, but also the starting point of what followed them, namely the early Islamic conquests that ushered in the medieval period in the Mediterranean basin.

2 Wars of the Sixth Century: Launching the Invasions

The loss of 'East Iran'[10] to the Hephthalites in the late fifth century forced the Sasanian hand to a western-oriented policy, one focused on strengthening their presence in Asurestan (Mesopotamia/al-Iraq) and developing the economic potential of the west, from Asurestan to Khuzistan.[11] The first sign of a Sasanian shifting of focus to the west is the initiation of the war on Byzantium by Kavad I in 502, after over a century of relative peace in the region. Seizing on an opportunity presented by Anastasius' refusal to pay a protection fee for the Caucasus gate, Kavad attacked Theodosiopolis and Amida, launching a new invasion of Rome's eastern territories.[12] This is usually considered as a thinly disguised excuse to replenish some of the funds that were paid to the

7 For the Hephthalite defeat of the Sasanians and its consequences, see Rezakhani (2017) 126–34; Rezakhani (2019) 229–34.

8 Dignas and Winter (2007, 115–18) provide a useful summary that compared with the previous stages of Sasanian-Roman conflicts quite clearly highlights the end of 'coexistence', at least in the seventh century.

9 Dignas and Winter (2007, 138–48) study the extensive peace treaty of 562 between Khosrow I and Justinian. The tone of the peace treaty and its attention to minute details such as trade-points highlights, as well as the details provided by Menander Protector (fr. 6.1–3) on the negotiation process, reveal a new awareness of the gravity of the situation for both empires.

10 For this term, see Rezakhani (2017) 11–14, 14–23.

11 See Rezakhani (2021) for a discussion of Sasanian responses to their losses in the late fifth century.

12 Jos. Styl. 48, 50. See also Börm, this volume.

Hephthalites as reparation.[13] However, the war that started in 502 can be considered the first stage of a state of the war that continued with interruptions throughout the sixth century.[14] The conflict between the two powers involved occasional and persistent campaigning and efforts at territorial gains, particularly in Armenia and upper Mesopotamia, but also on the Arabian desert frontier.[15]

A second phase of the war started in 540 during the reign of the famous Sasanian king of kings, Khosrow I, who had taken over the throne after his father's death in 531.[16] The war that started in 540 began with the quality of an expeditionary campaign, meant to test the Byzantine defences deep into their territory and as far west as the Mediterranean. The initial stages of the war show Khosrow determined to go as far west as he could, ultimately arriving at the gates of Antioch in June 540. Before this, he had attacked several fortresses and smaller cities such as Sura and Hieropolis, but was easily bought off by smaller payments.[17] He seems not have intended to conquer or cause great destruction, often quickly agreeing to the proposals of negotiating bishops and leaving the inhabitants without harming them, and even allowing local soldiers to join him and move into Sasanian territories.[18]

Khosrow's main objective might have indeed been reaching the Mediterranean and invading Antioch, as his conquest of the city and capture of booty appears quite intentional.[19] This plan to capture Antioch is best exemplified by the transfer of its population to a newly built city on the outskirts of Ctesiphon, the Sasanian capital. Named *Weh-Antiok-Khosrow* ('Better than Antioch of

13 Dignas and Winter (2007) 37.

14 For this initial phase of the war, see Greatrex (1998). A detailed selection of the sources and context is provided by Greatrex and Lieu (2005). Dignas and Winter (2007, 100–15) break the wars down to three stages, interrupted by periods of peace enforced by ephemeral treaties.

15 See Greatrex and Lieu (2005, 88–95) for a summary of the events and the sources. Dignas and Winter (2007, 152–72) provide a summary of the place of 'Arabs' between the two empires throughout the four centuries of Sasanian rule. See a discussion of the place of Arab populations and their importance in the wars of Khosrow II further below.

16 Dignas and Winter (2007) 106–109.

17 Greatrex and Lieu (2005) 103.

18 As was the case with soldiers in Beroea: 'Then the Beroeans, after coming into such great danger, left the acropolis free from harm, and departing went each in the direction he wanted. Among the soldiers a few followed them, but most came as willing deserters to Khusro, putting forth as their grievance that the government owed them their pay for a long time; and with him they later went into the land of Persia' (Procop. *Pers.* 2.7.36–7, quoted in Greatrex and Lieu (2005) 104).

19 See Greatrex and Lieu (2005, 104–105) for a succinct account.

Khosrow')[20] and known as Rumiya ('Romanesque') in the later Islamic literature, it was supposedly an exact copy of Antioch, to such a scale that the newly arriving inhabitants from Antioch could recognise the replicas of their houses and reside in them.[21] Khosrow might have actually wanted to attract the Antiochenes and the rest of the population of Syria to his new venture, as he decreed that all Roman captive slaves related to the Antiochene population could claim freedom from their owners and reside in the new city.[22] Even if largely mythologised, this shows a planned effort to control Syria and its population, as John Lydus says with awe: 'It was the whole of Syria that he carried off to Persia'.[23] While waiting for Justinian's response to a withdrawal agreement, Khosrow continued his slow march through Syria. After Antioch, he arrived at the port city of Seleucia, where 'he bathed himself alone in the sea-water, and after sacrificing to the sun and such other divinities as he wished, and calling upon the gods many times, he went back'.[24] By doing this, Khosrow seems to have ritually announced his presence at the Mediterranean, perhaps to mark a moment in which the sea was no longer simply a 'Roman pond'. Following the arrival of an agreement from Justinian, Khosrow slowly marched back toward Sasanian territory via Edessa, on the way making a half-hearted attempt at capturing Dara.[25]

In the meantime, war over the Caucasian territories such as Lazica and Suani continued until 557[26] and did not reach any conclusions until the peace treaty of 562. The Arabian front further south in Syria appears to have been left to its own devices, with the competition between the Ghassanids and Lakhmids developing into its own war.[27] This eventually took the form of competition between the Arabs, outside any effective control of either empire.[28] Skirmishes in the Caucasus also continued, without any major battles or significant gains for either side, leading to a shaky armistice.[29] However, the continuing conflict

20 Procop. *Pers.* 2.14.1 gives it as Ἀντιόχειάν τε τὴν Χοσρόου ('Antioch of Khosrow').

21 Ṭab. 1.898. On deportation and forceful movement of population, see Morony (2004) 173–74.

22 Procop. *Pers.* 2.14.4. He also provided the Antiochenes with special privileges and even tax breaks (2.14.3).

23 John Lyd. *De Mag.* 3.54 (quoted in Greatrex and Lieu (2005) 105).

24 Procop. *Pers.* 2.11.1.

25 Procop. *Pers.* 2.13.16–29.

26 Greatrex and Lieu (2005) 115–22.

27 Procop. *Pers.* 2.28.12–14. Ṭab. 1.858–59 puts this dispute at the root cause of the entire war between Khosrow I and Justinian.

28 See Fisher (2011, 49–70) for an assessment of the Jafnid (Ghassanid) and Nasrid (Lakhmid) world on the frontiers of Rome and Iran.

29 Agath. *Hist.* 4.30.7–10, quoted in Greatrex and Lieu (2005) 130–31.

on the frontier created much uncertainty, highlighting the need for a stable peace in the region.[30]

3 Financing the Wars: Military Reforms and Economic Growth

While the wars of the first half of the sixth century were not quite successful in gaining many territorial gains for Kavad or Khosrow I, we know that the economic resources of the empire provided adequately for the support of these wars.[31] The economic dynamism of the reign of Kavad and Khosrow, coming in the wake of the devastating wars with the Hephthalites and their remuneration, was most impressive. Later Islamic sources give a long list of city-foundations for Kavad, as well as administrative reforms such as redistricting of various regions.[32] He is in fact credited for having established the district division of *Kurah* in the Sasanian administration.[33] Further land and administrative reforms, including subjugation of allied kingdoms and their integration into a more centralised Sasanian system, particularly in military sense, are reported for Khosrow I.[34] This includes an extensive build-up of defensive structures, as well as frontier zone population centres, on both the Caucasus and the northeastern fronts.[35]

The main Sasanian reforms in the sixth century, however, are often credited to Kavad's son, Khosrow I.[36] These include a set of sweeping financial reforms set around a major new taxation system that has often been the focus

30 Greatrex and Lieu (2005) 131.
31 Howard-Johnston (1995) 202–15.
32 *Nihayat ul-Irab* 298; Ṭab. 1.885.
33 Ṭab. 1.888.
34 Ṭab. 1.895.
35 Ṭab. 1.895. On the Sasanian fortifications in the northeast and the Caucasus, see Nokandeh et al. (2006); Rekavandi et al. (2008); Sauer et al. (2013); Lawrence and Wilkinson (2017).
36 The main source is often assumed to be al-Ṭabarī 1.960–963 (e.g., Howard-Johnston (1995) 214). For a detailed treatment of sources, see Rubin (1995) 239–49. Rubin, of course, correctly points out that there are no contemporary evidence for the reforms of Khosrow I. This might say something about how we understand the so-called revolt of Mazdak, for which see Crone, "Kavad's Heresy and Mazdak's Revolt" 1991 and Rezakhani "Mazdakism, Manichaeism, Zoroastrianism" 2015. See Grignaschi (1971) for another treatment, much criticised by Rubin, based on the so-called *Sīrat Anūšīrvān*. The reforms are mentioned by most Islamic historians, although as Rubin points out (1995, 229), among the contemporary accounts, only Procopius makes any mention of them. As we might see further, this might be due to a preference among modern historians, following medieval historians, to attribute the reforms to Khosrow I.

of both medieval and modern historians.[37] However, the reforms seem to have extended to a major administrative reorganisation as well.[38] This included the division of the empire's military administration into four parts, roughly corresponding to the four cardinal directions.[39] This strengthened the fighting abilities of the Sasanian army and helped it better confront enemies in different directions.[40] Improving the army, and supplying it more effectively, appears to have been the main objective of the reforms as a whole.[41]

The administrative reforms were not the sole tool of the Sasanian state in managing its resources and increasing its revenues for military purposes. Scattered, but convincing, evidence of a conscious effort to improve the empire's economic output can be observed in textual and archaeological sources. The intensified occupation of Mesopotamia, noted in archaeological records, can be understood as a conscious effort to engender extensive growth in the region.[42] This demographic increase is perhaps best exemplified by the forced movement of population from Syria during the reign of Khosrow I, as previously discussed. Nor was this growth limited to Mesopotamia, as similar surveys provide us with corresponding population growth, as well as increasing agricultural activities, in neighbouring Khuzistan.[43] The integration of the Persian Gulf region into the economy of the Sasanians in the sixth century is also part of this effort, evident from the rise of ports in the Iranian coast, as well as increased presence of Persian merchants, miners, and soldiers on the Arabian and West Indian coasts of the Indian Ocean.[44]

Perhaps worth mentioning here is the fact that areas outside Mesopotamia/Asurestan and Khuzistan are often left out of such considerations, leading to

37 Rubin (1995) 268–72.

38 The issue of justice and fair extraction of taxes seems to be at the core of the reforms, as set out by Ṭab. 1.960 and others; see Rubin (1995) 260. Other aspects of the administrative reforms are discussed by Rubin (1995) 272–77.

39 Ṭab. 1.894; Gyselen (2001).

40 Ṭab. 1.894–95.

41 Rubin (1995) 283–84; this is also the general argument of Howard-Johnston (1995).

42 The main sources for this are the surveys of Robert McCormick Adams (1965, 71; 1981, 209–11), although certain aspects of his late Sasanian dating might need re-dating to the early Islamic period. As Howard-Johnston (1995, 205) suggests, Mesopotamia, along with Persian Gulf ports, can be considered the Sasanian 'Egypt' (i.e., a bread-basket).

43 Wenke (1975–1976); Neely (1974). A problem with the inconclusive results of the works of Wenke and Neely, and perhaps Adams and Hansen (1968), is the total ignoring of the areas south of the Susiana Plain, the region that in fact gains much attention in the early Islamic period: Moghaddam and Miri (2003); Moghaddam and Miri (2007); Moghaddam (2012).

44 Howard-Johnston (1995) 205; Whitehouse and Williamson (1973); Morony (2002); Morony (2019).

a lopsided view of the Sasanian world.[45] In this view, conclusions drawn from Asurestan are freely applied and generalised to the rest of the empire, while Asurestan's primacy as the seat of the empire is taken for granted.[46] The fact is that the dearth of archaeological surveys inside Iran have prevented us from understanding the extent of Sasanian investments and efforts to encourage economic production.[47] It would be strange to imagine that the reforms and royal investments were purely done for the economic benefit of only one part of the empire. This is something that the archaeological excavations in Persis/Fars, Central Iran/Jibal, Khurasan, and Adhurbaigan make clear. The efforts of the Sasanians at protecting their northeastern frontier, and the military activities in the east after 560 and the destruction of the Hephthalites by the hand of the Western Türk Empire, are further evidence for the importance of Sasanian interests inside the Iranian plateau. The statement that the Sasanians did not rely on an urban setting for their economic and administrative activities is also indefensible. Perhaps the most prominent feature of the narratives of the life of each Sasanian king of kings in sources ranging from al-Ṭabarī to Ferdowsi is the enumeration of the cities founded by each king. A concentrated example is the Middle Persian treatise, *Šahrestanīhā-ī Ērānšahr*, which provides a list of the cities in the Sasanian Empire, attributed to real and legendary founders, but still demonstrating the extent of urban settlement in the empire.[48]

Another aspect of the Sasanian economic expansion that we must consider is the efforts to increase commercial activities. This is the context in which both expansion of the Sasanian interest to the Persian Gulf/Arabian Sea region and the competition with the Sogdian network of Central Asia needs to be considered. On the sea front, a commercial presence of 'Persian' merchants in India was doubled with increasing Sasanian interests in controlling the coastal regions. Textual evidence for the presence of Sasanian administrators in East Arabia is occasionally countered with the lack of archaeological data,

---

45 Howard-Johnston (1995, 206) compares the Roman Empire's access to resources in areas other than Egypt, suggesting a relative advantage to the Sasanian Empire, and that 'There is nothing to indicate that the city had assumed anything like this importance in the life of Iran, the second core component of the Sasanian Empire, despite government efforts to promote it'.

46 Howard-Johnston (1995) 181, where an argument is made for Mesopotamia (Asurestan) as the region which assumed a 'leading role' within the Sasanian Empire from its inception, and particularly the role of Ctesiphon as its centre. This is a problematic assumption that compounds most future misunderstandings, including the issue of the fall of the Sasanians, for example in Crone (2012, 1–2); see Rezakhani (2019) for a partial refutation.

47 Howard-Johnston (1995) 202.

48 Markwart (1931); Daryaee (2002).

or perhaps a need for redating the data relating to this region.[49] However, the population movement to Arabia, as well as the mid-sixth century competition between the Sasanians and the Axumites for the control of South Arabia, are perhaps more than enough to show the interest in controlling the region.[50]

The investment in the economy and resulting economic growth, particularly in the region of Asurestan, was thus a conscious effort of the Sasanian state and an attempt at ensuring the resources needed for its newly conceived expansion plans on its western borders. In this sense, Khosrow's reforms – and even those of his father – were aimed at preparing the Sasanian state for not just simple raids into eastern Roman territories, but a desire to provide the economic and administrative means to control the region too. While in the larger scale of things, this brought the Sasanian Empire into a collision with its Roman rival, in a regional context, it had consequences for Sasanian relations with Syria. The campaigns of Khosrow I in the region, including his conquest of Antioch and deportation of its population, thus present themselves as expeditionary campaigns to take stock of the Roman presence in the region. Sources such as the treaty of 562, preserved in fragments of the History of Menander Protector, show this careful attention paid to the situation to the west of the Sasanian-Byzantine borders.[51] In a sense, the Sasanian Empire seems to have been preparing for a strategy to integrate greater Syria into its own domains, treating it as a natural continuation of Asurestan, 'dil-i Ērānshahr'. In turn, this 'Grand Strategy' appears to have finally been launched by Khosrow II in his seventh-century wars.

4 The Accession of Khosrow II and the Background of the Wars

Providing a context for the Sasanian-Byzantine wars of the early seventh century, most historical accounts start with the accession of Khosrow II to the throne in 590 CE.[52] The young Khosrow was either willingly or unwillingly pulled into a coup against his father Hormizd IV, organised by high members of the court, and was raised as the leading candidate to replace Hormizd. The rebellion, often attributed to Hormizd's mistreatment of the nobility, created a rift in the higher echelons of Sasanian nobility. The conspirators were soon

49 Kennet (2002); Kennet (2007).

50 Morony (2002); Potts (2008); Munt (2017).

51 Dignas and Winter (2007) 138–48 with discussion of part of the treaty.

52 The primary sources include, among others, Sebeos 10.73–75; Ṭab. 1.995–1000; Balʿami 1079–86.

confronted by the forces of Wahram Chobin, the commander in chief of the Sasanian troops in the north, who had recently won a victory against 'the Turks' and their king, Saweh/Sabeh or Shabeh,[53] presumably on the Sasanian northeastern frontier.[54] Sources are divided as to whether Wahram himself was involved or complacent in the initial rebellion or whether he simply entered the scene as an opponent of Khosrow.[55] In either case, he eventually poses as the main opponent to Khosrow and accuses the latter of plotting the coup and removing his father, and of his incapacity to become the next *shāhānshāh*. The tale of Wahram's march on Ctesiphon and the flight of the young Khosrow II thus forms the first major event in Khosrow's reign, detailed famously in Islamic period sources, including in popular romance.[56] The purpose it serves here is that its continuation sets up the relationship between Khosrow II and the Byzantine court, and thus the justification for the later wars.

A truncated version is as follows. After Wahram's march on Iraq/Asurestan, Khosrow was forced to abandon the capital and seek refuge in Byzantine territories in Syria – possibly Hierapolis[57] – from where he sought help from the Byzantine emperor Maurice against Wahram. Both Wahram, now crowning himself as Wahram VI, and Khosrow made offers of territorial concessions to Maurice, that of Khosrow including Martyropolis, northern Asurestan (Mesopotamia) down to Nisibis, and parts of Persarmenia.[58] This essentially reversed all of the Sasanian gains in the sixth century and returned the *status quo* to the time of Kavad and Justinian. Despite some internal opposition at the court, Maurice decided to support Khosrow's claim and sent troops to restore the deposed king.[59] Khosrow's advance started from northern Mesopotamia, the area around Mardin, where he was declared king in February 591.[60] The

53 Ṭab. 1.992 has Shabeh while Balʿami, who claims better knowledge of Central Asian affairs and should be believed since he writes from Bukhara, has Saweh (1077).

54 An earlier encounter with the Turk, presumably the Western Türk Empire under the Khaghan Istämi (Sinor 1990b, 297–301).

55 Ṭab. 1.993 puts the blame on Wahram and his distrust of Hormuz, while Dīnawarī (ʿAmir (1960) 84) simply says that the grandees removed Hormuz from the throne and sent his regalia to Khosrow in Azerbaijan.

56 The *Romance of Wahram Chobin* is reconstructed and discussed by Christensen (1907).

57 Goubert (1949) 85–86. Ṭab. 1.999 has Antioch; Dīn. 91 says that he proceeded to Yarmuk where he was welcomed by Khalid b. Jabala al-Ghassani. Sebeos 75 (Thompson 18) says that he went to Aleppo (*Khałab*), while Theophylact Simocatta, who might have been best informed, says that Khosrow stopped in Circesium and sent a letter to Maurice from there (4.10.4).

58 Dignas and Winter (2007) 43. Theoph. Sim. 4.13.24.

59 Dīn. 91–92; Sebeos 76–77 (Thompson 19).

60 Greatrex and Lieu (2005) 173.

main encounters between the two sides took place in Atropatene, one of the centres of Wahram's power in the northwest. A series of engagements resulted in a final battle at Ganzak in which Wahram's forces were decimated and the general himself was put to flight.[61] Wahram's subsequent presence at the court of the Khaqan of the Western Türk, his involvement in court intrigues, and his final assassination instigated by Khosrow II, are the focus of a medieval east Iranian romance preserved in Balʿami's Persian translation of al-Ṭabarī's history.[62]

Maurice's assistance in the restoration of Khosrow II, and the latter's return of Dara to Byzantium, created a bond between the two rulers – or at least that is what Theophylact Simocatta suggests when he tells us about the use of the term 'son' by both rulers.[63] The immediate outcome was a cessation of war and impromptu period of peace between the two empires. Part of this peace included Khosrow's more lenient behaviour toward the Christian population of Asurestan and the border region, including a much-publicised patronage of the shrine of St. Sergius.[64] Khosrow's pro-Christian policies, at least as reflected in the Syriac and Byzantine sources, also included marriage to two Christian wives, Maria ('Maryam') and the famed Shīrīn.[65] In the later Islamic sources, after Khosrow's restoration, the tale of his dealing with Wahram and then another rebellion by his uncles Wistahm and Windoy dominate the narrative of his rule.[66] The start of the wars in 602 is mentioned through a reference to the Roman rebellion against Maurice and the removal of the emperor.[67] A relatively quick résumé of the events of the war, including Khosrow's support for Maurice's son and his conquest of Jerusalem and removal of the True Cross, is given by some sources.[68] Islamic sources focus instead on Khosrow's removal of the last Lakhmid ruler of al-Hira and the resulting Battle of Dhu Qar and

61 The prominence of Adurbadegan/Azerbaijan and the city of Ganzak in later narratives of Heraclius' counterattacks (for which see further below) is worth noticing here; Sebeos 10.77–80.

62 See above; Balʿami 1086–89.

63 Theoph. Sim. 5.3.11; Greatrex and Lieu (2005) 174 n. 50.

64 Fowden (1999) 58–59, 134–40.

65 Fowden (1999) 140–41; on Shīrīn's involvement in Sasanian imperial policy, particularly as it concerned the imperial attitude toward the Christians, see also Flusin (1992) 106–18, and a deep analysis of how scholarship has usually treated this involvement in Wood (2013) 184–99.

66 Ṭab. 1.1000–1002.

67 Ṭab. 1.1002.

68 Ṭab. 1.1002–1004; Dīn. 105–106. Much less time is spent on both accounts on the entire war, including the restoration under Heraclius (Ṭab. 1.1004–1005; Dīn. 106), than is given to details of the rebellions of Wahram Chobin or Wistahm and Windoy.

seem much more interested in local affairs in southern Mesopotamia. This episode, apart from shedding light on a neglected theatre of the subsequent wars, also illustrates the selective historiographical tradition regarding the wars of the early seventh century.

4.1 *The Arabian Prelude: Khosrow's Frontier Policy and Islamic Historiography*

As we saw, the majority of Islamic period sources, as a rule, are quite subdued about the Sasanian-Byzantine wars of the early seventh century and seem more interested in the affairs on the Arabian frontier. While at first it might sound strange, or simply negligent, to ignore such great events, a consideration of the sources and how they report the wars perhaps helps explain this apparent lack of attention toward the wars. A central aspect of the accounts offered by the later Islamic sources is the geographical span they consider, or the territories toward which they orient themselves, namely southern Iraq and the Syrian desert.[69] This geography is far removed, and indeed almost fully outside consideration, for the geographical expanse considered by the Byzantine, Syriac, and Armenian narratives. The latter, from the *Chronicon Paschale* and the *Chronographia* of Theophanes to *the Chronicle of Khuzistan* and the *History of Armenians* of Pseudo-Sebeos, are normally concerned with events in northern Mesopotamia and Syria. Their gaze, cast from Constantinople, Theodosiopolis, Nisibis, or Dvin, is on the route of Sasanian invasions in northern Mesopotamia and Western Armenia, running along the old roads in northern Syria and eastern Anatolia. As we have seen before, in the sixth century Sasanian campaigns almost universally began in the Nisibis-Dara region and advanced either west with the aim of harassing and capturing Roman fortresses in northern Syria, or north toward Amida and Theodosiopolis. An exception might be the campaigns of Khosrow I after 540, detailed before, which seem to have an exploratory nature, identifying Roman weaknesses in areas such as Callinicum, while at the same time taking the opportunity to exact tribute. In their narratives of the campaigns of Khosrow II in the seventh century, Byzantine, Syrian, and Armenian sources again mainly pay attention to affairs in the north, while their reports of the southern theatre are often confusing and truncated. The further away from northern Syria-Mesopotamia that these sources get, the less detailed their information becomes. This was already true for earlier sources such as Procopius, Joshua the Stylite, Malalas,

69 On the geographical gaze of the Islamic sources, including their view of Mesopotamia and Syria, see Webb (2015, 57–61).

Menander Protector, and Theophylact Simocatta as well, and is a common historiographical marker of Greek, Syriac, and Armenian sources.[70]

In contrast, Islamic sources appear to not concern themselves at all with the affairs of the north, and thus the better-known narratives of the events in Armenia, Anatolia, and northern Syria. Instead, the bulk of the narratives is devoted to the affairs of the eastern part of the Syrian desert and the southern Iraqi zone, essentially the northern Arabian zone adjacent to the Hijaz.[71] In their narrative, the history of this region in the sixth and seventh centuries is mainly a tale of events happening in the Lakhmid, and less so Kindite, zones of influence, as well as that of South Arabia, particularly following the fall of Himyar. Such narratives, interwoven with the history of the Sasanian Empire, continue seamlessly to the seventh century and almost completely ignore the wars, instead focusing on local affairs, in particular the events leading to the fall of the Lakhmid house.[72] Eventually, the gaze of the Islamic historians diverts away from the Lakhmids, instead shifting from the Iraqi zone down to Mecca and Medina, viewing the events in light of the rise of Islam and Muhammad's Medinan state.[73]

Considering this, modern scholarship largely appears to follow the lead set by each set of sources. Those relying on Greek, Syriac, and Armenian narratives tend to concentrate on Sasanian-Byzantine affairs in in Armenia, Anatolia, and Syria, and largely generalise these interactions as covering the full gamut of relationships between the two powers. This set of histories essentially end

70 I am here excluding the lesser studied Georgian sources, mainly due to the fact that they still remain beguilingly out of reach, and second because some, such as Strategus' account of the Sasanian conquest of Jerusalem, appear to have preserved local accounts of the events in the 'south'.

71 For the use of geography, particularly the way al-Sham/Syria was imagined, perhaps surprisingly as an outside space in the rise of Islam, see Webb (2015) 137–54.

72 Webb (2015, 144–45) demonstrates how the Islamic sources are largely uninterested in the affairs of Syria, behind its extreme southeastern border city of Bostra. It also shows that in stark contrast to the modern historical fascination with the Ghassanids as clients of Rome and the quintessential Christian Arab power in Syria (see Shahid (1995); Fisher (2011)), the Islamic sources are in fact quite disinterested in them and seem not to count them as significant in the formation of the Arabs and Arab identity.

73 In al-Ṭabarī, Dīnawarī, and other Islamic sources, the stories of Sasanian kings are mixed closely with stories of Lakhmid or Himyarite history (for examples of which, see the entire volume of Ṭabarī's history translated in Bosworth (1999), extensively referenced throughout this paper). The narrative, except on the episode of the birth of Muhammad (Ṭab. 1.967–81) keeps its focus on the Sasanian world, largely telling the story of the Lakhmids and even Yemen from a Sasanian viewpoint. However, the fall of the Sasanians in fact causes a chronological backtracking to the rise of Islam and a continuation of events from the viewpoint of Medina (and later Kufa/al-Hirah).

their narratives – except when concerned with the Byzantine state itself after Heraclius – with the Islamic conquests.[74] In turn, scholars working with Arabic sources, commonly historians of early Islam, view the events from the southern angle, focusing on the rise of Islam while generally ignoring the wars and considering the *Futūḥāt* as *ex nihilo* events unrelated to the previous events.[75] The 'Byzantine gaze' thus misses the events of the southern zone, as well as seemingly overlooking the consequences of the wars of the sixth and seventh centuries for the coming Islamic conquests, leaving the latter almost without context.[76] The 'Islamic gaze', seemingly unaware of what was happening in Armenia, Anatolia, and northern Syria, similarly treats the Islamic conquests as affairs entirely planned and carried out in Hijaz, and ignores their background in the sixth- and seventh-century wars. In this way, such works contribute to isolating the 602–628 wars within the context of antiquity and create a division between it and the events leading to the rise of Islam, and thus the rise of the medieval ages.

It is true, of course, that recent scholarship has afforded the southern zone more consideration, shedding some light on the history of the region and aiming to put it in communication with the events in the imperial centres in Constantinople and Ctesiphon. The work of the previous generation of scholars such as Irfan Shahid allowed us to better consider the region by making a wealth of primary sources about the area available.[77] However, their attention was focused mainly on the relationship between the local Arabs and the Byzantines. A preoccupation with Arab-Roman relations is in fact a major feature of the studies of the region.[78] Studies addressing the situation of the Arabs in the region also commonly spend much time on Arab conversions to Christianity and their situation vis-à-vis Byzantine Christianity.[79] Isabell

74 For examples of this in the historiography of the Iranian world, see the arrangement of chapters in Yarshater (1983) or Daryaee (2009).

75 See Hoyland (2014, 12), where the 602–630 wars are essentially presented in about three sentences. Even when the affairs of the Syrian desert are considered in the historiography of late antiquity, they are often concerned with the issues of religion, such as the Arab position vis-à-vis the Christian church struggles in the Byzantine Empire. On this, see Fowden (1999); Fisher (2011) 49–71.

76 Howard-Johnston (2021, 361–87) spends a bit of time on this (including a very acute observation about the narratives of decline and exhaustion at 372). See also Kaegi (1992) 43–46.

77 Most related to the present paper is Shahid (1995).

78 Fisher (2011) 173–93. Similar to many other works, its main focus is also on the Jafnids/Ghassanids, the Arab clients of Rome, in the Palestine and Syrian region, again ignoring much of the Nasrid/Lakhmid side, as well as the Kindis who in fact occupy a primary place in the Islamic narratives of the pre-Islamic events.

79 Fowden (1999); Fisher (2011); Toral-Niehoff (2013a) 151–211.

Toral-Niehoff's study of the city of al-Hira and its placement on the desert frontier brings together much information about the area and its local social and political setting, shedding light on local history. It is, of course, a positive advancement shedding light on the neglected history of the Lakhmids of al-Hīra, although it says little about their role at the centre of Sasanian policies on the desert frontier.[80]

A few sources highlight the close relationship between the Sasanian court and their Arab clientele, a relationship that appears closer than the one enjoyed by the Byzantines with their clients in southern Syria and Palestine. Procopius mentions an episode when the Lakhmid king Nuʿman advises Khosrow I on how to undertake a campaign through the desert against northern Syria.[81] Islamic sources give much detail about the interactions between Kavad and the Kindi leader al-Harith, as well as close involvement of Nasrid rulers in the wars of the sixth century. The intensification of contact between the Lakhmid rulers of al-Hira and the Sasanian court is perhaps best exemplified by Qābūs b. al-Mundhir III, who ruled al-Hira between 569–573. The Persian name of Qābūs (from MP Kāwūs, also the name of an older brother of Khosrow I) shows the influence of the Sasanians on the Lakhmids as well as the close connection between the two courts.[82]

In the context of the Battle of Callinicum, the successful campaigns of al-Mundhir on the southern front, in southern Syria and Palestine, demonstrate the Arabian frontier policy of the Sasanians. Procopius' comments, put in the mouth of al-Mundhir, on the emptiness of the southern frontier and its lack of defence, can be interpreted as Procopius taking a jab at Justinian's neglect of this *limes*.[83] However, it can also be seen as a reflection of the reality of Sasanian campaigns, seeking various ways to control Syria and invade Antioch, something that was finally achieved by Khosrow I. But the increasing importance of the Arab frontier and the portrayal of al-Mundhir as a great enemy of Rome in his own right[84] also emphasises that this frontier region needed to somehow be controlled. It is significant that Mundhir, in justifying

80 Toral-Niehoff (2013a) 59–68.

81 Procop. *Pers.* 1.17.29–48.

82 Ṭab. 1.1038, with Bosworth (1999) 370 n. 906. Despite what Bosworth, and Nöldeke ((1879) 346 n. 1), suggest, there is no reason to assume that the next ruler mentioned by al-Ṭabarī, al-Suhrab, was not also a member of the Nasrid family and was instead a 'Persian' official sent by the Sasanians (Bosworth (1999) 371 n. 908). The conjecture that the name is otherwise unattested amongst the Nasrids can be dismissed, as indeed Qābūs itself is otherwise unattested, and yet not doubted.

83 Procop. *Pers.* 1.17–34–35. See Greatrex (2022) 238–39.

84 Procop. *Pers.* 1.17–40–43.

his incursions into Byzantine territory, mentions that he was not violating the terms of the 532 treaty between the Sasanians and the Byzantines, because 'neither one of them had included him in it'.[85] As Procopius too concurs, this was indeed true, 'for no mention of Saracens was ever made in treaties, on the ground that they were included under the names of Persians and Romans'.[86] However, this shows that the inhabitants of the Byzantine-Sasanian borderlands alongside lower Asurestan/Iraq facing the Syrian desert felt less and less bounded to the imperial powers that controlled them.

For the rest of the sixth century, the Sasanian policy toward the Arab frontier seems to have concentrated on both containment and exploitation. Alliance and cooperation by strong and active leaders such as al-Mundhir III and his son ʿAmr III allowed the Sasanians to take advantage of the southern frontier and make excursions against the less protected cities in the area. This did not, however, encourage the Sasanians to invade the province of Palestine or to advance toward Egypt. Eventually, the turning of the tides during the reign of Qabus, son of al-Mundhir III, and his death at the hands of his Ghassanid rival, might have rendered the Lakhmids more of a burden. This is evidenced by the sidelining of the Nasrid family from the throne of al-Hira around 602 and the transfer of power to governors appointed by the Sasanian court. The result, outside the concerns of the primary sources, highlights the importance of al-Hira in Sasanian strategy regarding the western frontiers.

The removal of the Lakhmid house from the throne of al-Hira, perhaps a shrewd policy of directly controlling the Arab fighting forces from the Sasanian point of view, left a deep impression in the local narratives. Sources such as *the Chronicle of Khuzistan* tell a tale of personal enmity between Khosrow and the Nasrid ruler al-Nuʿman III b. al-Mundhir IV, leading to the latter's removal from the throne of al-Hira and his replacement by a Sasanian governor. The removal might have been related to al-Nuʿman's conversion to Nestorian Christianity, although this is not the sense one gets from the Islamic sources.[87] The events that follow al-Nuʿman's removal and the subsequent Battle of Dhu Qar essentially dominate the history of the wars in sources such as al-Ṭabarī, creating a bridge to connect the fall of the Lakhmids to the rise of Islam.

Al-Ṭabarī provides much detail for the episode, naming a nemesis of al-Nuʿman, Iyās b. Qābiṣah al-Ṭāʾī, as his successor, alongside a Sasanian commander named al-Nakhīrajān, whose name might in fact be a title related to his

85 Procop. *Pers.* 2.1.4.
86 Procop. *Pers.* 2. 1.5.
87 Bosworth (1999) 371 n. 910; compare with *Khuz. Chron.* 9 (Rezakhani/Amiri Bavandpour (2016) 64–66).

position as the Master of the Hunt.[88] Iyās's nine year reign came to an end around 611,[89] thus at the beginning of Khosrow's second phase of campaigns against Byzantium.[90] The next ruler of al-Hira was a Sasanian commander, presumably a nobleman, called Azadbih son of Adhur-Mahan,[91] son of Mihrbundādh, who ruled for seventeen years, based on al-Ṭabarī's chronology, until a month into the rule of Būrān-duxt/Bōrān.[92]

The removal of al-Nuʿman was only the latest event in the Sasanian policy of controlling the southern frontier. It is usually presented in the context of Khosrow's dissatisfaction with al-Nuʿman when the former fled Asurestan before the advance of his rival Wahram Chobin in 591. This is in fact connected closely to how Muslim chroniclers saw the position of the court of al-Hira in relation to the Sasanian imperial court. Presented in an anecdotal manner, involving an almost legendary story that involves Khosrow's desire for a Nasrid princess, the narrative takes the conflict out of the context of frontier policies and larger imperial concerns and renders it as personal vendetta. This emphasises the commanding role of the court of al-Hira in imperial affairs, perhaps reflective of the attitude of Baghdadi historians of the tenth century and efforts at extending Arab superiority to the pre-Islamic period. But the flow of the story and its immediate continuation to the Battle of Dhu Qar is indicative of how the power struggle among the Arab tribes at the periphery of the Sasanian Empire and its relation to the Sasanian court was imagined by the later historians. Thus, the dignity of the Nasrids and their allies, in this case the Banu Shayban, is contrasted with that of the pro-Sasanians, such as Iyas b. Qubisah and the Taʿi.

In the context of the wars of the seventh century, the removal of the Nasrids from the throne of al-Hira is seen, perhaps teleologically, as a weakness in Sasanian policy, removing a 'buffer zone' through which the Islamic conquests penetrated the empire. Seen from the viewpoint of Sasanian strategy, however,

88 Ṭab. 1.1038 (with Bosworth (1999) 372 n. 911). Al-Nakhīrajān النخيرجان could be a misreading of Nakhjirajān (Persian Nakhjīrgān) النخجيرجان, or 'keeper of *Nakhjīr*', from Middle Persian Naxčīr 'game, chase' (McKenzie (1971) 58).

89 Bosworth (1999) 372 n. 911.

90 See also Ṭab. 1.1038 for a suggestion that Muhammad's mission started 'one year and eight months' into the start of Iyās' rule, as well as Bosworth's suggestion (1999, 372 n. 912) that this is 'too early' as the Mab'ath should be placed in 610.

91 See Bosworth (1999, 372 n. 913) for problems in reading of this name.

92 The dating is problematic, as is that of Iyās before. Bosworth (1999, 372 n. 913) suggests that this should be nineteen years, an understandable suggestion as Bōrān is supposed to have ascended the throne after the murder of the usurper Shahrwaraz, in spring 630 (Chaumont (1989)).

this was a decisive action to use the desert frontier as an offensive front, exactly the way al-Mundhir had advised Kavad seventy years earlier. In what follows, we will discuss the wars of the seventh century and their continuation in the form of the Arab/Muslim Conquests from the vantage point of the Sasanians. The purpose will be to argue that for the Sasanians, and Khosrow II particularly, these wars were the conclusion of a century of planning.

4.2 *Dhu Qar, the Sasanian Conquests, and the* Futūḥāt

In Islamic sources, the removal of al-Nuʿman is seen as the main impetus behind the Battle of Dhu Qār, a local affair which matters more for the preservation of its memory than for its actual military achievements.[93] Essentially, Dhu Qar is presented as the prelude to the Islamic conquests and a demonstration of essential Sasanian weakness, as well as an example of Arab unity on the desert frontier. In al-Ṭabarī's telling, however, the reason for the conflict is presented as personal vendetta involving al-Nuʿman III, whose elevation to the throne of al-Hīra was the result of Khosrow's direct interference in the succession of the Nasrids, through his advisor, ʿAdī b. Zayd.[94] The eventual removal of al-Nuʿman is also presented as a personal strife between two rival families within al-Hīra, further helping to set the episode outside the context of the Sasanian conquests.[95] While the story contains many romantic details, the one of interest to us is that upon receiving the news of his removal, Al-Nuʿman left his family in the care of a Shaybanid leader, Hanī b. Masʿūd, before proceeding to Ctesiphon. There, he was put in fetters and sent to prison in Khaniqin, where he died from a plague shortly before the advent of Islam.[96] Initially, the response of the Sasanian court seems to have been mild, allowing al-Nuʿman's Shaybanid protectors to roam freely.[97] However, the refusal of Hanī to send the family of al-Nuʿman to Khosrow finally led to a battle, involving Sasanian troops, near the site of Dhu Qar, resulting in a resounding Sasanian defeat. Despite the description of al-Ṭabarī, the battle must have been a small and

93 As Bosworth mentions, the encounter at Dhu Qār is more important as part of Islamic historiography, particularly the historiography of the conquests, than for its actual military importance, which was probably nothing more than a border skirmish (Ṭab. 1.1016; Bosworth (1999) 338 n. 794). It might also show the alignment of the Arab tribes at the border region such as the Shaybanis and the Bakrīs, although these alliances and infightings necessarily shifted in the subsequent period (Donner (1981) 189–90).

94 Ṭab. 1017–18. Bosworth (1999, 342 n. 812) suggests that the person involved in the elevation of Al-Nuʿman to the throne of al-Hira was Khosrow's father, Hormizd IV.

95 Ṭab. 1.1024–28.

96 Ṭab. 1.1028–29.

97 Ṭab. 1.1030.

limited affair.[98] This of course explains the romanticised tale of the actual battle and heroic acts of the members of the Shaybani and Bakr ibn Wa'il tribes, which takes up several pages of al-Ṭabarī, accompanied by appropriate poetry.[99]

What the Battle of Dhu Qar demonstrates, despite its limited actual effects, is both the relative freedom of action and the sometimes precarious situation of the Arab actors living in the southwestern Sasanian borderlands. The need of the Sasanian system to control and tap into the fighting resource that this population provided is also evident in the lead up to the battle, as the Sasanian king attempted to suppress its political autonomy. It is here that we can glimpse the involvement of these Arab actors in the Sasanian-Byzantine wars and the importance of the Iraqi-Syrian desert region for the campaigns on other frontiers. Ignoring the events of the southern theatre of war thus results in a myopic view of the events of the early seventh century, and an unnecessary division between the Sasanian conquests and the Islamic *futūḥāt*. The rest of this paper presents the wars of 602–630 with awareness of the events of the desert frontier and the rising Islamic state, and the subsequent *futūḥāt* that let the nascent Islamic empire establish itself in the same territories previously conquered by the Sasanians.

5 The Conquests of Khosrow II

5.1 *The War of Khosrow and Phocas*

The sources are unanimous in pointing to a bond between Khosrow and Maurice that provided the impetus for the start of a new war between the two powers.[100] In 602, a rebellion against Maurice started in Thrace, with the rebels soon choosing a mid-level officer, Phocas, as their leader. Phocas and the army marched on Constantinople and captured and executed Maurice and his family, raising Phocas to the throne.[101] Following this event, some sources, mostly eastern in nature and including Syriac and Armenian accounts, mention the escape of a son of Maurice by the name of Theodosius and his arrival at the Sasanian court. Khosrow II, thus poised as the avenger of his former benefactor Maurice, rose in action and sent an army in support of Theodosius toward northern Syria. The similarity of the situation to the events of 590–591 is

98 Ṭab. 1.1031–32.

99 Ṭab. 1.1032–38.

100 Howard-Johnston (2021) 17–18; Theoph. Sim. 8.15.7; Sebeos 31; Ṭab. 1.1000.

101 See Howard-Johnston (2021, 8–18) for a detailed retelling of the rebellion that brought Phocas to power.

remarkable and can scarcely be coincidental, although most modern Western commentators have considered this Theodosius to be an imposter and a puppet of Khosrow.

The mobilisation against Phocas was strengthened by the rebellion of Narses, the Roman general who commanded the Byzantine army in the restoration of Khosrow, perhaps another sign that the preceding events were not pure coincidences. Acting at the time as the *Magister Militum per Orientem*, Narses left Antioch, his base, and occupied Edessa, which appears to have supported and favoured him. Khosrow's initial campaign, conducted from Nisibis, was successful in conquering the important strategic border town of Dara in 604,[102] leading to the death of the Roman general Germanus,[103] and allowing Khosrow to meet Narses in Edessa. The rebellious general welcomed Khosrow and initiated the coronation of Theodosius as the Roman emperor, under Khosrow's protection. In this initial phase of the war, Khosrow himself was in charge of the operations and supervised both the siege and fall of Dara, as well as the procession for Theodosius' coronation.

Following this, Khosrow appears to have returned to Nisibis and eventually his favourite residence in Dastgird, to the north of Ctesiphon, leaving the war in the hands of his generals.[104] The campaigns up to 610, the year of Phocas' removal from the throne in Constantinople, were in fact limited in geographical scope, mostly focused on Armenia and northern Asurestan/Iraq.[105] The earliest campaigns in the east of Armenia and the area of Dvin were led by a general called Juan Veh, aiming to secure southwestern Caucasus and block any Byzantine incursion from that direction.[106] The Armenian campaigns were undertaken at the same time as the campaign against Dara and constituted parts of a two-pronged campaign against Byzantine presence east of the Euphrates. They were in fact attempts to reverse their territorial losses to the Byzantines following the restoration of Khosrow in 591 and the concessions made after that in payment for Maurice's support.

The strategy of two-pronged attacks was solidified in a new season of campaigns starting in 607. A commander, called Khorream in Armenian sources and presumably same as the one later known as Shahrwaraz, was put in charge of the army stationed in northern Asurestan/Iran.[107] Khorream's campaigns,

102 Greatrex and Lieu (2005) 184–85; Theoph. 292.6–25.

103 Howard-Johnston (2021) 23.

104 Howard-Johnston (2021) 24.

105 Sebeos 31–34.

106 Sebeos 31.

107 Sebeos 33.

starting in the summer of 607, were first directed toward the area to the north of Nisibis and Dara, the so-called Tur Abdin/Mount Izla region. Initial success for Khorream came with the fall of Mardin in 608, bringing the whole of the region of Mt. Izla under Sasanian control.

Further north, Theodosiopolis capitulated to another general, Shahin, in 607, and his Sasanian army penetrated as far west as Nicopolis, to the southwest of Trebizond.[108] Shahin was put in charge of the Armenian troops following the death of Juan Veh and was to become one of the most famous of Sasanian generals.[109] The fall of Mardin and control of Nicopolis provided Shahin with a prominent foothold in the Taurus region and blocked Byzantine access to the northern route toward Armenia. This strategy would eventually become quite central in Sasanian successes, as well as future Byzantine counterattacks.[110] By 608, all of Roman Armenia was in Sasanian hands, with the route to invade the west lying open. At the same time, Edessa also fell in 609, which hastened the capitulation of Amida, Rasaina, and Constantia.

At the same time, another rebellion had started in North Africa, initiated by Heraclius, the exarch of Africa, in the name of his son Heraclius the Younger. Heraclius' brother Gregory and his son Nicetas also rose in rebellion.[111] While the Sasanian armies were capturing city after city in northern Iraq and Eastern Anatolia, the African rebels moved toward Egypt, which they managed to conquer by the summer of 609. Heraclius the Younger was recognised as the main leader of the rebels, while his cousin Nicetas acted as the rival and also the general in charge of subduing Egypt and bringing it to the rebels' cause. A series of local disturbances and rebellions in Antioch and other cities on the Levantine seaboard allowed for the quick spread of the rebellion. Nicetas, the son of Gregory, set out toward Constantinople via land, while Heraclius the Younger proceeded via the sea, arriving in the vicinity of the capital in the autumn of 610. A series of land and sea attacks brought the entire area around Constantinople under Heraclius' control, and on 5 October 610, saw his

108 Sebeos 33; for detailed discussion, see Howard-Johnston (2021) 21–28.

109 Both Shahrwaraz and Shahin are associated with regional military rule as *spahbed*s in the North (Adurbadegan) and the West (Khwarewaran) respectively (Howard-Johnston (2021) 177–79). On the regional commands, see Gyselen (2001). In the episode of the execution of Khosrow, the involvement of a son of a previous ruler of the region of the West, a *pādgūspān* or civil governor named Mardānšāh is notable: Ṭab. 1.1058–59. Since the *kust* of the West included Ctesiphon itself, its centrality in the wars was perhaps precluded.

110 The initial counteroffensive of Heraclius in 625–626 depended on access to this northern route, something that has not been given proper attention up to this point. See below on this.

111 See Howard-Johnston (2021) 37–60 for details.

entry into the city.[112] Phocas was summarily arrested, mutilated, and executed, henceforth inscribed in Heraclian-sponsored historiography as a vile 'usurper'.

Heraclius' success and the resulting narratives, aiming to establish Heraclius' legitimacy, also tend to consider the initial stage of the Sasanian conquests as a pure war of revenge for the murder of Maurice. Consequently, they suggest that Khosrow should have abandoned his campaigns of conquest and ceased further action following Heraclius' success in removing Phocas.[113] This narrative considers Heraclius to be the legitimate avenger of Maurice, as the rightful Roman emperor, and thus superior in moral claim to Khosrow, a foreign ruler. This means that the official Byzantine historiography treats Khosrow's subsequent campaigns as both illogical and illegal punitive actions against the legitimate Byzantine ruler, Heraclius, and not a legitimate continuation of wars aiming to restore the house of Maurice to the Roman throne. However, this view ignores the existence of Theodosius as a claimant to the throne and the dynasty of Maurice in Khosrow's framing of the war. It also discounts what was clearly a long and careful preparation for war, since at least the reign of Khosrow I, and the main objective of the Sasanians and Khosrow II himself, that is the conquest of the entire eastern Mediterranean. Such a take, largely adopted by modern historians steeped in Byzantine historiography and Heraclian propaganda, treats the Sasanian campaigns as blind actions with short-term objectives in mind. However, as demonstrated before, the Sasanian 'grand strategy' at least since the sixth century was aimed at providing the economic, financial, and human resources needed to annihilate the Byzantine Empire or to reduce it to vassal-state status.[114] Khosrow's subsequent actions leave little doubt about this, and his tactics, read through later sources and often observed through an early Islamic gaze, lay this out clearly as well.

5.2 *Total War in the Eastern Mediterranean: Khosrow and Heraclius*

The rebellion of Heraclius and his toppling of Phocas, as well as the disturbances in the Levant and the Mediterranean seaboard, left the areas to the west of the Euphrates openly exposed to the Sasanian troops. Khosrow's troops in Armenia and northern Mesopotamia were well placed to take advantage.

112 Howard-Johnston (2021) 61–71.

113 This is the general feeling one might read behind the famous letter by the Senate to Khosrow in 614, for which see the discussion in Howard-Johnston (2021) of 603–612 and what Howard-Johnston calls Khosrow's 'Fateful decision'.

114 Or as Howard-Johnston calls it, a 'rump-state', which in fact happens in the subsequent period of Islamic conquests; Howard-Johnston (2021) 113, 367. This is another reason for the present paper's insistence that the two events, the Sasanian conquests and the Islamic *futūḥāt*, are closely related and need to be considered in relation to each other.

By 610, the Sasanian troops had conquered all the cities to the east of the Euphrates and were ready to advance westward across the Euphrates into Syria and Anatolia. In fact, by the summer of 610, Zenobia to the west of the Euphrates had already fallen into Sasanian hands.[115] Here, it becomes quite evident that the Sasanian objective was indeed conquest, and not simply a war of revenge. The double (or three-)pronged attack shows that the Sasanian strategy was set from the beginning of the campaigns of Khosrow II on advancing toward the Mediterranean and possibly liquidating the Byzantine state.

This state of total war was in fact a logical continuation of the exploratory campaigns of Khosrow I half a century earlier. In this view, the actions of Kavad I and Khosrow I in the sixth century, from military campaigns to economic, military, and social reforms, can be seen as conscious preparations for the devastating campaigns of Khosrow II. It is significant that at no points in the campaigns is there a mention of a shortage of funds or fighting troops for the Sasanian military actions. In fact, in the record of the trial of Khosrow II following his removal in 628, a striking complaint of the court grandees is the fact that the royal treasury is kept full, even at that late stage in the wars, and that Khosrow still refuses to share the wealth despite being successful in his campaigns.[116]

The unwillingness of Khosrow II, slowly coming to earn his nickname of *Aparwez* ('Invincible'), to recognise the new Byzantine regime and Heraclius as its legitimate head, was revealed by the advance of the Sasanian troops against Byzantine positions west of the Euphrates. Such a refusal was evidently well understood by Heraclius himself, as well as his court, as evident from the testimony of the *Chronicon Paschale*.[117] Here, Heraclius seems to admit and apologise for, a few years later, his assumption, acknowledging Khosrow's right to withdraw such recognition and uncharacteristically asking for his

115 Greatrex and Lieu (2005) 187–89.

116 See Ṭab. 1.1047, items 5 and 6 regarding extraction of taxes and amassing of a great wealth, and Khosrow's answer, Ṭab. 1.1055–57, where he explains his financial policies, summarised in this statement: 'From the thirtieth year of our reign to the thirty-eighth, which is the present year, our stores of wealth have not ceased growing in extent and richness, our lands in florescence, our subjects in security and tranquility, and our frontiers and peripheral regions in impregnability and strength of defenses'.

117 See Howard-Johnston (2021, 105–12) for a detailed discussion of this episode. However, I am sceptical about the accuracy of this letter and how much weight we should give to its authenticity. Such emphasis perhaps serves more to advance arguments that Khosrow's pride or blind ignorance of facts of war led to a fateful reversal of fortunes (twelve years later?) and were thus the consequences of his own actions. It serves well to notice that such pride was presented before the fall of Jerusalem or the conquest of Egypt, and perhaps was well justified at this particular juncture.

forgiveness.[118] This political confrontation of course led to an intensification of the Sasanian war efforts and a new vigour in continuing their conquests.

5.3 *Campaigns in Anatolia*

The accounts of the war are often uneven in their treatment of the routes of conquest and the order of the events. The multiplicity of names also creates some problems in reconstruction. Shahin, the main general in charge of the Armenian and Anatolian campaigns, appears to have successfully advanced from Caesarea in Cappadocia, despite some early setbacks.[119] Despite great praise given to Heraclius and his commanders Philippicus (in Armenia), his brother Theodore, and Nicetas, and a 'bold' counterattack in Syria which led to a resounding victory for Shahin,[120] the Sasanian commander seems to have taken over Cilicia by 613. By 614, Shahin had conquered Ephesus and reached Chalcedon, across the Bosphorus from Constantinople, putting direct pressure on the Byzantine capital.[121]

Heraclius is said to have personally boarded a ship – despite his reputed fear of sea travel – and have come across the bay to meet with Shahin, at a distance.[122] Heraclius' obeisance and even an offer to step aside in favour of a candidate chosen by Khosrow, appears to demonstrate the lowest point in Byzantine political and military power. Shahin's agreement to cease hostilities and refer the case to Khosrow himself perhaps demonstrates the shortcomings of the Sasanian troops, despite their superiority at this moment, particularly in their ability to conquer the well-defended Byzantine capital on their own and without a major navy, a fact that was proven definitively a few years later.

118 If Sebeos' report (34) of such an embassy is to be believed, then the letter sent by Heraclius to Khosrow four years later during the siege of Constantinople, recorded in *Chron. Pasch.* 707.1–709.23, would not make sense. It is somewhat more logical that the *Chron. Pasch.*'s testimony about Heraclius' neglect in informing Khosrow of his accession, based on an established custom, is genuine. In that case, the testimony of Sebeos about Khosrow's refusal to accept Heraclius as a legitimate ruler and ignoring his request for peace needs to be considered inaccurate. We will discuss this letter further in its place further below.

119 See Greatrex and Lieu (2005, 188) with references to sources, some of which are as late as the thirteenth century (*Chr. 1234*). Howard-Johnston (2021, 82–85) uses these sources creatively to provide a reconstruction of the events that led to the presentation of Heraclius himself in the area of Caesarea and his leadership of the campaign.

120 Greatrex and Lieu (2005, 189) choose a strange turn of phrase to summarise Heraclius' defeat; see Howard-Johnston (2021, 85–86) for a narrative treatment of the sources.

121 Foss (1975) 743; Greatrex and Lieu (2005) 193–94; *Chron. Pasch.* 706.

122 Greatrex and Lieu (2005) 194; see also Viermann (2021) 178–79 for a mention of these events and their place within Heraclius' reign.

After Shahin's withdrawal from Chalcedon, the Sasanians appear to have concentrated their focus on consolidating their power around Anatolia, including in Cilicia and Isauria. They might have even developed a fleet to attack the Byzantines on the sea and attempt the conquest of various islands.[123] This attempt at slowing Sasanian conquests via diplomacy was obviously not in the purview of Khosrow or his policy to conquer and control the Byzantine territories.

5.4 *Campaigns West of the Euphrates*

On the northern Syrian front, the armies of Khorream, henceforth more commonly referred to as Shahrwaraz, managed to capture Antioch, the most important Byzantine city in Syria, in a swift campaign directly proceeding from Edessa. Apamea and Emesa were also conquered in 610, bringing the Sasanian army to the coast of the eastern Mediterranean. Thus far, the Sasanian troops had repeated what was essentially the strategy of Khosrow I half a century earlier, controlling northern Syria and establishing themselves on the eastern coast of the Mediterranean to challenge the Roman hegemony in the *Mare Nostrum*. This initial success forced Heraclius to try and take some action to limit the damage, starting a short, unsuccessful attempt at turning the tide. Some sources suggest that Heraclius had already asked Khosrow to cease his invasion and recognise his removal of Phocas as an official end of the war of revenge. But the Sasanian *shāhānshāh* had refused this request and evidently killed Heraclius' ambassador.[124] In their initial campaign, the forces of Heraclius faced the Sasanian armies in the field, but were defeated near Antioch in 613 and withdrew to Constantinople.[125] Damascus, considerably south of Antioch and in fact on the edge of the desert zone, fell to the Sasanians in the same year, opening the road to Palestine and the conquest of Jerusalem. The next five years were a period of unprecedented success for the Sasanians. The pace of the conquests did not slow down, with advances toward Jerusalem and eventually Egypt.

The further expansion of the campaigns to southern Syria initially resulted in the conquest of Damascus in 613, for the first time putting the Sasanians in the region south of Emesa. A rebellion in Jerusalem in 614 – variably attributed to the competition between the Blue and Green circus factions or the local

123 Greatrex and Lieu (2005) 195. See Howard-Johnston (2021, 124–25) for a no less confusing narrative based on sources which, however, seem to be a repetition of campaigns that had taken place earlier, before Shahin's arrival in Chalcedon.

124 Greatrex and Lieu (2005) 188.

125 Foss (1975) 743.

Jewish population – provided the forces under Shahrwaraz with the excuse to invade the city, which fell to them in May 614. Syriac Christian sources, in a customary antisemitic tone, accuse the Jewish population of the city of having collaborated with the Sasanian troops and welcoming them as liberators. Gross and obviously malignant reports of murder of Christians by Jews, with the agreement of the Sasanian occupiers, demonstrate the hyperbolic and symbolically hostile tone of these sources.[126] These reports are then to be treated with much care, as are the reports of the violent destruction of the local Christian churches and structures.[127]

It is also the local Jews who are said to have shown the Sasanian commanders the location of the True Cross, hidden by the local priests in a vegetable garden.[128] Playing an important role in the religious propaganda of Heraclius in his subsequent counterattacks, the True Cross came to signify the captivity of the whole of Syria for the Byzantines and in turn, mark the success of the Sasanians in their campaigns.[129] The True Cross was sent to Ctesiphon with honour, where it was initially handed over to Yazdin, Khosrow's trusted Christian treasurer, for safe keeping. We are told that Yazdin placed small pieces of the True Cross inside bejewelled crucifixes and sent them to important Christian leaders of the Sasanian Empire, thus demonstrating the victory of the *shāhānshāh* and displaying his patronage of Christianity.[130]

Between 615 and 618, the Sasanian troops secured their control over the countryside of Syria. The fact that Shahin and Shahrwaraz appear to have conducted joint expeditions and raids around Anatolia[131] is probably indicative of the fact that their control over Mesopotamia, Syria, and Palestine was quite secure and unchallenged. This is a testament to the success of the policies of Khosrow II in the area and his attempts at attracting the support of the local population, many of whom were Christians, by patronising local shrines such as that of St. Sergius.[132] Commanders such as Shahrwaraz also extended patronage to local shrines and churches and participated in restoration of sanctuaries.[133] The reports on the welcoming of the Sasanians by the local Jews, apart from their obvious antisemitic undertones, can also be interpreted

126 See the account by Strategius of Mar Saba (Phenix and Horn (2022)).

127 On archaeological evidence, or the lack thereof, from the Sasanian conquest of Jerusalem, see Stoyanov (2011) 11–23.

128 *Khuz. Chron.* 79; Ṭab. 1.1002.

129 On the religious propaganda, see Stoyanov (2011) 45–75.

130 *Khuz. Chron.* 25.

131 Howard-Johnston (2021) 125–27.

132 Fowden (1999) 58–59.

133 *Khuz. Chron.* 28.

as a reflection of the reality, in which the Jewish population found life under Sasanian rule more agreeable, much as their Babylonian brethren had done in the centuries before.

The next stage of the conquests took the Sasanians, under the command of Shahrwaraz, to Egypt. We do not know the exact route taken, but the goal of the expedition appears to have been Alexandria. A difficult and prolonged siege provided enough time for some communication regarding the defence of the city to be exchanged between Constantinople and Alexandria. In a manner typical of many contemporary chronicles, the author of the *Chronicle of Khuzistan* blames the fall of Alexandria on the treachery of a foreign student showing a weak point in the defences to the Sasanians. A story from later Arabic and Persian literature tells the tale of an unfortunate wind diverting a number of boats carrying the wealth of the city to Constantinople, toward the Sasanian position – presumably in the eastern delta. The diverted boats and their treasures then fell into the lap of the waiting Sasanian troops who shipped them back to Ctesiphon, where they were deposited in the treasury, earning the booty the title of 'Wind Brought Treasure' and becoming a famous saying in New Persian literature.[134]

In general, the booty gained from the conquests appears prominently in the contemporary and later sources. Treasures from Syria, Palestine, and Egypt were sent back to Ctesiphon. There, they were entrusted to Yazdin, the Christian treasurer of Khosrow II, who appears to have amassed them in the treasury.[135] Of course, this can scarcely have had a real economic effect on the wars, but was certainly important enough to have affected the general perception of the wars as quite successful.[136]

However, the Sasanian success in establishing control of Syria and Palestine should be seen, more than anything else, in light of their successful alliance and control of the population of the desert zone in southern Iraq and Syria. Whether we want to recognise them as 'Arabs' at this point or not, these are the people who in later Islamic sources are recognised as belonging to various Arab tribal confederations.[137] These confederations lived in various levels of autonomy in relation to the two great powers, and as discussed before, presented a serious challenge to both empires. Consequently, we are well justified here to look deeper into their position within the context of the Sasanian conquests.

---

134 Ibn Balkhi *Farsnama* 253 (LeStrange et al. (1995)).

135 *Khuz. Chron.* 21.

136 Ṭab. 1.1055–57.

137 See Webb (2016), particularly the introduction, for a discussion of the ethnogenesis of the Arabs that goes beyond the concern of the present paper.

We must consider the geopolitics of the campaigns in the Iraqi-Syrian desert and the place of the local populace within the political and military settings of the region. The suggestion here is that unless the Sasanian troops were secure in their southeastern flank, they would not have risked attacking the regions south of Emesa and to the west of the lower Euphrates, such as the region of Damascus and further west. This would then beg the question of the placement and possible involvement of the 'Arab' tribes of the Iraqi-Syrian desert in the campaigns of Khosrow.

5.5 *The Desert Frontier in the Seventh Century*

Based on mainly Byzantine and Armenian sources, the modern scholarship discussing the conquest of Damascus and Jerusalem suggests that these campaigns were undertaken by troops coming southwards from Antioch toward Emesa and beyond. Of course, as we have seen, the actual information provided in sources originating from Armenia or Constantinople is quite vague about the campaigns the further they get from northern Syria. The information here is sometimes quite thin and spends much less time on details of sieges and conquests, unlike that provided for the campaigns in Armenia or Tur Abdin. It is thus worth noticing that some sources, such as the *Chronicle of Khuzistan*, suggest that Shahrwaraz came to conquer Jerusalem 'directly'. This presumably means that Shahrwaraz proceeded directly from the Sasanian territories and not as part of a long campaign. As mentioned before, the relative ease of the conquests of Damascus and Jerusalem also suggests that, whichever route that they might have taken, the Sasanian troops were feeling assured of their safety to the east, the desert hinterland of Damascus, including a reliable supply of provisions. This would thus highlight the importance of the Iraqi-Syrian desert and its role in the wars.[138]

Physically, the Iraqi-Syrian desert, past the Euphrates, presents a desolate space with little urbanisation. Apart from a few outposts such as Bostra or Anbar, neither empire seems to have invested much in developing the region, populating it, or properly controlling it. Some recent archaeological surveys in the area west of the Euphrates to the south of Perōz-Šāpūr (Anbar, formerly Misikhe) have revealed the presence of structures that appear to mark a defensive line in the desert zone. These structures, a complex of irrigation canals, walls, and defensive fortresses, have been interpreted as lines of imperial demarcation.[139] They are also geographically close to al-Hira, the capital of the Lakhmid client kingdom, as discussed above. Arabic sources provide much

138 Morley (2017).

139 Lawrence and Wilkinson (2017).

romanticised information about desert palaces built by rulers of al-Hira – one supposedly as a gift for Wahram v who was raised by a 'Lakhmid' foster parent – showing both the glory of the rulers of al-Hira and their participation in the Sasanian imperial project.[140]

By the time of Khosrow's campaigns of conquest after 602, the independent rulers of al-Hira were long gone, vanished in the aftermath of the Battle of Dhu Qar, and with their capital being controlled by Sasanian marzbāns ('margraves').[141] But these 'desert palaces' and their archaeological remains clearly hint at the presence of a desert-frontier policy at the Sasanian court in this period. The autonomous kings of al-Hira, appearing as advisors and military commanders for Khosrow i in the sixth century, had disappeared under Khosrow ii, replaced by Sasanian imperial administrators. This is often seen as a mistake, a woeful ignorance of the importance of the Arabs in Hijaz and other places in the Arabian peninsula, and removing a buffer zone – the Lakhmid state – which, by disturbing a regional balance of power, hastened the Islamic conquest of the Sasanian domains later.[142]

However, assuming that Khosrow's removal of the Lakhmids was not purely out of spite – as presented by the Syriac or Islamic sources[143] – or pure shortsightedness, it might be opportune for us to re-examine the situation in al-Hira and the desert to the west of it in the seventh century. The seemingly quick and definitive success of the Sasanian campaigns in Damascus and Jerusalem might indicate a calm situation in the desert that lies between al-Hira and Damascus. We have relatively little information about the actual presence of the Arab tribesmen in the wars, since as we have seen before, Islamic sources – our main source of information about the situation in the desert – are mostly silent about the wars.[144] So, we are forced to dig into other sources in order to recover some information about the area and its population.

Perhaps the most intriguing information about the presence of 'Arab' fighters in the Sasanian-Byzantine wars is the mention of 'Bedouin' attacks on Palestine, particularly Jerusalem, immediately prior to the conquest of the city by Shahrwaraz. Sources appear to suggest that these attacks were random acts, without co-ordination with the Sasanian army, and are vague about their effect

140 Ṭab. 1.855–58.

141 Toral-Niehoff (2013b) 115–26.

142 Dignas and Winter (2007) 172.

143 *Khuz. Chron.* 19–20.

144 Various notices about the presence of Arab contingents in both Sasanian and Byzantine armies are available, for example in the army of Heraclius when he is in Armenia (Howard-Johnston (2021), 226) and in the army of Shahrwaraz in 629 (Howard-Johnston (2021) 371; Movses Dasxuranci 166.14–168.17).

on the conquest itself. It is probably reasonable for us to assume, however, that at least the weakness of the (Byzantine) defences of Jerusalem was known to these 'Bedouins' who were conducting such attacks.[145] The awareness of the desert-dwelling tribes of the conditions in Syria can perhaps be conjecturally considered by referencing the quotation of Procopius mentioned above. Even though in the 610s, there was no more Nuʿman to lead Khosrow and his troops across the desert for an assault on Syria, there is little reason to assume that the tribes previously under Lakhmid suzerainty were unaware of the same geopolitical setting and did not take advantage of opportunities.[146] A strategic alliance of these Bedouin tribes with the Sasanians is hard to dismiss in the account of the siege of Jerusalem in 614, when the Bedouin attacks essentially emptied Jerusalem's hinterland of all population and allowed the Sasanians to cut off supplies.[147]

Further information is provided by Ibn Aʿtham al-Kufi in his *Kitab al-Futuh*, one of the earliest accounts of the Islamic conquests.[148] In his description of the capitulation of al-Hira itself, Ibn Aʿtham tells us that the Muslim troops were greeted by a grandee of al-Hira, a certain Abdul-Masih al-Ghassani.[149] This Abdul-Masih ('Servant of Messiah') stood at the gates of al-Hira, greeting the Muslim troops, praising the achievements of his Ghassanid ancestors, and offered the submission of al-Hira in exchange for favourable terms. One might then wonder about the presence of a Ghassanid scion, usually expected in Palestine/Syria, in al-Hira and acting as a representative of the city. Considering the removal of the Ghassanids from power by the Byzantines in the 580s, could we perhaps assume the patronage of the Ghassanids by the Sasanians? We might suggest that the subsequent removal of the Lakhmids from the throne of al-Hira by Khosrow II, following his return from Syria in 591 to recover his crown, might be an indication of favour shown toward remnants of the Ghassanids in Syria. Was Abdul-Masih then a representative of this Sasanian-Ghassanid alliance, replacing the Lakhmids in al-Hira? This would then suggest a new regime of controlling the tribes across the desert, and more

145 Howard-Johnston (2021) 88, 91, 94, 97.

146 Howard-Johnston (2021, 98) seems to suggest that the Bedouins were as much of an undesirable headache for Shahrwaraz and his troops as they were for the Byzantines, and a 'problem' to be dealt with. Here, I suggest otherwise, that the Bedouins were already allies of Shahrwaraz and the Sasanians.

147 Howard-Johnston (2021) 94.

148 Shiri (1990).

149 Ibn Aʿtham 76.

than anything, a tribal unity across the desert, acting as allies of the Sasanians in their conquest of southern Syria, Palestine, and ultimately even Egypt.[150]

Another set of evidence is provided in some accounts of the rise of Islam in Mecca. In his description of the early life and ministry of Muhammad, al-Ṭabarī tells us that most of the people of Mecca – particularly the Quraysh tribal members – were supporters of the Persians and were often happy about the news of Sasanian conquests in Syria and Egypt. In al-Ṭabarī's account, Muhammad and his early believers were in turn supporters of Byzantium, as the latter were Christians and thus believed in one god, as opposed to polytheistic Zoroastrians whom the prophet disliked.[151] This is the reason al-Ṭabarī gives for the famous line in the Surat al-Rūm in the Qurʿan, where a final victory of the Romans is promised by Allah.[152] However, we need to balance this evidence by the evidence of Muhammad's last war, the Battle of Tabouk, which although militarily a non-event, has a fascinating justification.[153] The battle takes place in the year 630 (or 9 Hijri), thus in the final stages of the Sasanian-Byzantine wars. Muhammad and his followers heard of a possible approach of Heraclius toward Aqaba. Now based in Medina and ruling the nascent Islamic state, Muhammad became worried about the Byzantine threat and rushed up to face Heraclius and possibly stop him from reaching Aqaba and further west. The expedition apparently stayed around in Tabouk for twenty days, but in fact saw no battle and returned to Medina.[154] This appears to suggest that Muhammad was not too enthusiastic about the possibility of a full restoration of Byzantine control of Syria and Palestine and tried to stop it from occurring, while we have no similar episodes in response to the Sasanian control of the region. Consequently, the involvement of the Arabs, whether in distant Mecca and Medina or in al-Hira, and their participation in the wars is to be expected. However, due to concentration on the northern theatre of war and the Nisibis-Antioch axis as the main route of invasion, we appear to ignore this particular frontier and its presence in the wars. Perhaps this is the shortcoming that also creates much confusion when we come to discuss the rise of Islam and its early conquests, which indeed took place mostly in this exact same area, namely the southern Iraqi-Syrian desert frontier. My suggestion would be that the next stage of the conquest beyond Palestine, namely

150 This is already suggested by Howard-Johnston (2021, 98), who suggests that this revival of 'Jafnid-led Ghassan' was after their initial intervention in Jerusalem around 612–613.

151 Ṭab. 1.1005–1006.

152 Ṭab. 1.1006; see El-Cheikh (2004, 26–29) for a discussion of various readings of this line.

153 Hoyland (2014) 39; Ṭab. 4.1292–94.

154 Al-Waqidi, *Kitab al-Maghazi* (= Jones (1966)), vol. 3, 1014–15.

that of Egypt, was particularly successful because it was undertaken from this direction and was much swifter than expected.[155]

5.6 *The Establishment of Sasanian Administration*

Between the fall of Alexandria and start of Byzantine counterattacks in 626, little action is reported from either side of the conflict. This appears to be the period when the Sasanian administration, already hard at work at least since 610, made attempts to control the conquered regions. Little evidence of this effort remains, however, as the occupying administration evidently did not attempt to issue coins in the new territories. The presence of coin hoards containing Sasanian silver issues might indicate the circulation of Sasanian silver,[156] but this is too early to talk about the integration of the Roman gold and Sasanian silver drachm zones.[157] However, some of the coins issued during this period, showing obverses copied from the coins of Heraclius with reverses showing Sasanian iconography and surrounded by Pahlavi inscriptions, might be a sign of Sasanian local administration or the power of local war-chiefs.[158]

Our evidence is thus limited to some administrative papyri, as well as few textile and parchment pieces, in the form of correspondence and rulings, occasionally involving Shahr-Alanyozan (presumably Shahrwaraz) and presumably dated from 619–629 CE.[159] The Sasanian commander in the southern front also acted as the military governor of Egypt – and possibly the rest of the conquered territories in the south. These letters show us that the actual zone of Sasanian control went beyond Alexandria and extended to the Fayyum in Middle Egypt, where most of the documents were found.[160] The occurrence of Zoroastrian names among those mentioned in the documents, as well as the use of late Sasanian Pahlavi cursive, shows the depth of Sasanian administrative reach.[161] Also interesting for our current purposes is the presence of references to provisions for the Sasanian troops (MP *gūnd*) in some of the documents.[162] At the

155 This again harks back to the episode of al-Nuʿman offering to lead Khosrow I through a desert expedition, mentioned above: Procop. *Pers.* 1.17.29–48.

156 Howard-Johnston (2021) 157.

157 Heidemann (1998).

158 Rezakhani (2020) 140–41.

159 Weber (2003); Weber (2005); Weber (2007); Weber (2011). Few documents actually bear dates. The collection is assumed to date to 619–629 because those are the dates of the Sasanian occupation of Egypt (Weber (2007) 725), but could potentially be dated to the later periods as well (for an example, see Weber (2005) 225).

160 Weber (2007).

161 Weber (2007) 730.

162 Weber (2007) 736–37.

same time, the suspension of *Annona* in Constantinople shows the complete breaking of Egyptian relations with the Byzantine centre and the redirection of its resources toward the Sasanian world.[163]

Such information is scarce for the area of Syria and the Sasanian control there and can only be observed through conjecture. A lack of major conflicts in the region, evidence of Sasanian rebuilding of destroyed buildings,[164] and perhaps the presence of people with semitic names in the Egyptian Pahlavi papyri,[165] presumably merchants from Palestine, might be taken as signs of such administration in Syria.[166] Sasanian troops were garrisoned at newly conquered cities such as Caesarea, Alexandria, and Thebes, and the presence of Sasanian nobles and military administrators is evidenced in the Pahlavi papyri.[167] In general, the Sasanians seem not to have disturbed the administration in Syria, leaving local administrators in charge in large cities such as Edessa and Caesarea and attempting to win over the local administrative nobility.[168] Whether this included the Ghassanids and an expansion of the latter's role in the region is unclear. Sasanian involvement in administration and changes, at least at the top level, may be seen in anecdotal stories such as the deportation of the family of nobles from Constantia who had previously insulted Khosrow when he was taking refuge there following his removal by Wahram Chobin in 590. It is uncertain much weight should be given to Khosrow's effort at cultivating the relationship with Monophysite bishops[169] or his patronage of the shrine of St. Sergius – presumably important to the Arabs in Syria.[170] Sasanians also might have dealt with the local Arab tribes and their administration in the same way as that of the Romans, including continuing the system of the phylarchate for controlling the Arab nobility of the region.[171]

5.7 *Advances in Asia Minor and the Siege of Constantinople*

While the advances in Palestine and Egypt were taking place, the Sasanian troops in Asia Minor were making their way toward Constantinople. Ancyra, in central Anatolia, fell to the Sasanian troops in 620 (or 622)[172] allowing them

163 *Chron. Pasch.* 711.
164 Howard-Johnston (2021) 161.
165 Weber (2007) 730.
166 See the extensive discussion in Howard-Johnston (2021) 153–73.
167 Howard-Johnston (2021) 154.
168 Howard-Johnston (2021) 153.
169 Howard-Johnston (2021) 161–62.
170 Greatrex and Lieu (2005) 176–78. For further discussion of St. Sergius, see Fowden (1999) 133–49; Fisher (2011) 199–203.
171 Howard-Johnston (2021) 171–73.
172 Foss (1975) 725.

direct control over the main road going from Constantinople to the east, and thus limiting Byzantine movement toward Armenia.[173] The Sasanians appear to have even formed a navy, as they managed to conquer the economically important island of Rhodes, off the southwestern coast of Anatolia, in 622–623, as well as Samos.[174] Cyprus, of course, was conquered earlier in 617 and during a previous phase of the war.[175] Sasanian naval activities were probably aimed at both controlling sea access to the Ionian coast and the eastern Aegean, and taking advantage of the resources offered both by the Anatolian coast and the islands.[176] The Byzantine losses can be seen in the minting of coins, which slowly came to a halt in Anatolia after 618, with Nicomedia, the last working mint, ceasing production in 619.[177] However, the Sasanians seem to have been less interested in controlling Anatolia than in Syria and Egypt, aiming to only advance against Constantinople.[178]

These conquests are credited in later sources, in fact the only sources available, to Shahrwaraz, whom we expect to find in Palestine and Egypt. His involvement is in fact reported by the thirteenth-century *Chronicle to 1234*[179] and is understood by modern scholars to be a response to a counterattack by Heraclius in 622. The sole narrative of this campaign is a poem by Heraclius' panegyrist, George of Pisidia, called *Expeditio Persica*.[180] Without getting into the validity of George's poem as a reliable account of an actual expedition, which despite its claimed success in fact came to naught and had little results, one might point to its extreme romantic tone to cast a serious doubt on the entire event.[181]

173 Foss (1975) 725.
174 Greatrex and Lieu (2005) 197.
175 Foss (1975) 724.
176 Foss (1975) 725.
177 Foss (1975) 729, 729–35 for an extensive discussion of the coin hoards.
178 Greatrex and Lieu (2005) 197; for a particularly negative take, see Foss (1975) 728, 744–45.
179 *Chron.1234*, 230.7–9 (trans. Hoyland, 66), quoted in Howard-Johnston (2021) 210–11.
180 Howard-Johnston (2021) 192–200 makes a copious and very erudite use of the poem to re-create the campaign. However, the poem itself, and the lack of any corroborating evidence for the campaign, is quite suspect, possibly serving only as an 'excuse' to justify Heraclius' absence during the subsequent Avar attack on Illyricum (Howard-Johnston (2021) 202–207). Militarily, it makes little sense for Heraclius to rush so far east, pass the troops of Shahin and Shahrwaraz – assuming the latter was really there – and run a punitive attack in Armenia while his route back to Constantinople could have easily been cut off. For a military genius, which most ancient and modern historians insist Heraclius was, this would be extremely careless and irresponsible!
181 See Howard-Johnston (2021, 197–99), where George's purpose in presenting Heraclius as a general of unparalleled sagacity is discussed in detail. As mentioned by Greatrex and Lieu (2005, 198–99), the account of the campaign perhaps did more to boost the morale of the Byzantine troops.

At this point, Heraclius set out to the west to counter an Avar attack on Thrace, and was set to meet the Khaghan of the Avars for peace negotiations. The event, however, turned around quite surprisingly, with a trap set for the person of Heraclius, which he managed to escape. A subsequent attack on Constantinople itself put extreme pressure on Heraclius and his forces in Anatolia, which now had to be moved to the west of the city in order to meet the Avar onslaught.[182] This western distraction obviously further helped the Sasanians in advancing into Anatolia and getting closer to the walls of Constantinople.[183]

6 The Byzantine Counteroffensives

6.1 *The 624–625 Campaigns: an Exercise in Futility?*

Sources about the Byzantine counteroffensive provide two different stages of for the campaigns of Heraclius. The first of these has been dated by modern historians to the period 624–625 and includes a series of swift movements by the Byzantine emperor to the area of the Caucasus. The majority of the narrative of the 624–625 campaign is based on the much later compilation, *Chronographia*, put together by Theophanes Confessor.[184] This is essentially the narrative of a swift campaign by Heraclius eastwards, penetrating into Persarmenia and plundering the region close to Adurbadegan and getting close to Albania.[185]

The initial motive for Heraclius' counteroffensive, and evidently the reason many modern historians take the entire episode to be genuine, is a supposed provocation by Khosrow II.[186] The provocation is then adopted for an all-out war, which includes Heraclius' entire family moving over the Bosphorus to Anatolia in March 624 to see their heroic general-emperor off to war.[187]

182 Howard-Johnston (2021) 207–211; Greatrex and Lieu (2005) 199.

183 Greatrex and Lieu (2005) 199–200. Howard-Johnston (2021) 211–12.

184 See the edition of Mango and Scott (1997); Proudfoot (1974) provides a source-critical study.

185 The basic narrative, based essentially on Theophanes and supplemented by undated excerpts from Movses Dasxurantsi's *History of Aghuank'*, is provided in Greatrex and Lieu (2005) 200–205.

186 See Howard-Johnston (2021) 215–16 for a detailed analysis and a correct dismissal of the letter as a 'spurious' creation. However, I do not agree with his justification, that Ps.-Sebeos was 'an admirably modest historian who sought simply to preserve for posterity material which he judged to be of historical value with a minimum of clearly demarcated editorial commentry' (216), for dismissing the possibility that he had simply conjured the letter and its overly Christian content himself.

187 Howard-Johnston (2021) 217.

Modern historians then take on the cause of the general-emperor at this point, spending much time digging into the few sources we have and painstakingly reconstructing a campaign season that supposedly lasted over a year.[188] Details of these reconstructions and disputing each one would require an entirely different paper on the historiography of the period. Suffice it to say that most historians would admit that it would take much creative work to draw details from the events as described by Theophanes and others.[189]

The reconstruction of the events, particularly the campaigns of Heraclius at this point, often takes the form of a panegyric to the general-emperor, who has suddenly transformed into a master propagandist,[190] genius tactician, and unique strategist despite very little evidence for it up to this point.[191] In any case, the course of the campaign is described as a an exercise for the campaigns of 626–627, which did in fact result in the collapse of the Sasanian war-machine and the removal of Khosrow II, while here, it seems to have had little actual territorial gains. Essentially, the 624 campaign is described as a swift and bold move by Heraclius directly against Sasanian troops located to the south-east of Lake Urmia in southern Adurbadegan, about 1,500 km from Caesarea in Cappadocia, the proposed gathering place for Heraclius' troops.[192] Heraclius' route has been reconstructed as going to Theodosiopolis, 600 km to the east, then another 500 km to Dvin, then 250 km south to Nakhchiwan.[193]

188 Howard-Johnston (2021) 217–22.

189 Howard-Johnston (2021) 216–21. The number of times that Howard-Johnston has to correct Theophanes, who for example seems to have forgotten the Avar attack of 623 otherwise so prominent in other sources, based on a problematic notice in *Chronicon Paschale*, corroborating a note transposed from another event in Ps.-Sebeos, even makes the author himself admit that the reconstruction is often based on conjecture: 'By this stage, the reader may be losing patience. Conjecture is being stretched, perhaps beyond breaking point, as the search moves on first from extant sources to their sources, then to the sources of those sources' (268).

190 The generous use of the phrase 'black propaganda', as an underhanded way of both crediting and gleefully admonishing Heraclius's supposed spread of misinformation, is quite noticeable and surprising (Howard-Johnston (2021) 214, 215, 224 for just this one episode!). Along with 'strength of diplomacy', it is generally used to dismiss the intelligence of the enemy and suggest the ability of the Byzantine emperor to run circles around them, so to speak!

191 See Howard-Johnston (2021) 214–15 for a suggestion that Heraclius and his generals essentially went back to classical manuals of war to prepare themselves for the upcoming war. The simplicity of such an act is striking, if nothing else for the fact that it seems to have left little effect on the actual outcome, and also for its lack of consequence. Surely such academic study of military tactics resulting in ingenious outcomes should have somehow resulted in foundation of a military academy of sorts?

192 Howard-Johnston (2021) 224.

193 Greatrex and Lieu (2005) 200.

Our almost unique source, Theophanes, has the Sasanian general Shahrwaraz campaigning in Anatolia while Heraclius is covering well over 1,000 km in 'swift marches', exposing his western flank and allowing himself to be easily cut off from supplies in a hostile territory.[194]

The next stage, based on Theophanes' testimony, is Heraclius becoming aware of the presence of Khosrow in Gazakos (Ganzak) with 40,000 men, a number that is prudently dismissed as exaggerated.[195] Hearing of this, Heraclius evidently rushed down the 500 km distance and surprised the hapless Sasanian *shāhānshāh* and his troops, chasing them to the fire temple of 'Thebarmais' which allegedly contained the treasures of Croesus, king of Lydia 1,200 years earlier.[196] It is fascinating how Theophanes' description of Heraclius' surefooted march over a plain, described as difficult for Khosrow, is trusted here to suggest the Byzantines striking a great success at essentially the heart of the province of Adurbadegan, one of the four main regions of the Sasanian Empire, 700 km northeast of Baghdad and 900 southeast of the traditional Byzantine-Sasanian border at Theodosiopolis.[197] The details of the campaign against Khosrow II in eastern Adurbadegan, so deep into Sasanian territory, are indeed astonishing enough. Putting the Sasanian emperor to flight in his own territories, carrying the treasures of his most important fire temple across high mountains to his capital in the southwest, sounds almost impossible to believe.[198] Relying on a single source, written a few centuries later based on lost material, might also not be the most prudent way of doing history. However, it is not in the capacity of a single paper to dispute a large tradition of scholarship by some of the most important historians of the late Roman/Early Byzantine period. Instead, perhaps it is more sensible to modestly suggest that the entire narrative serves a historiographical purpose, possibly one of explaining, or justifying, Heraclius' absence from Constantinople during the subsequent joint Sasanian-Avar siege of Constantinople.

In any case, the narrative of Heraclius' campaign continues with the latter's destruction of the fire temple in Thebarmais, identified as the modern

194 Theoph. 306.21–23, 306.27–307.1, referenced in Greatrex and Lieu (2005, 200).

195 Theoph. 308.6; Howard-Johnston (2021) 222.

196 Theoph. 308.7. It is probably not too modest to suggest that Theophanes is not quite a reliable source here, since what he seems to know about the Sasanian Empire is based on his readings in the history of the Achaemenid Empire written over 1,200 years earlier by Herodotus!

197 Theoph. 308.8–9. The entire first phase of the campaign, peppered with phrases from George of Pisidia's panegyric (Heraclius speeding past fortified cities full of hostile troops '*like a fast running breeze*!') is described in more detail by Howard-Johnston (2021) 225.

198 Howard-Johnston (2021) 225–26.

archaeological site of Takht-e Soleyman, southeast of Takab (Ganzak) in the Iranian province of East Azerbaijan.[199] He subsequently withdrew to Caucasian Albania, where he supposedly wintered after his campaign of 'plundering, ruining, and enslaving the whole land' further south.[200] The details provided in modern reconstructions for the presence of Heraclius and his troops in Aghuankʻ/Caucasian Albania again demonstrates our dependence on the availability of sources and their particular testimony. The account of wintering is based on Movses Dasxurantsi's history of Caucasian Albania, and thus specific references to actions in the region, including the author's native village of Kałankatukʻ.[201]

The next phase of the campaign, Heraclius' return, is reconstructed based on a series of evidence buried in Theophanes – supposedly based on a lost 'Official History' by George of Pisidia – supplemented by Pseudo-Sebeos.[202] This consists of an ingenious quick march to the west by Heraclius, avoiding three major Sasanian armies led by Shahrwaraz in northern Mesopotamia, Shahin, as the supreme commander, in western Armenia, and a new commander, Sharpalakan,[203] in Siwnikʻ and Sisagan.[204] Instead of returning home to Constantinople, the heroic Byzantine general-emperor marched north to eastern Transcaucasus, supposedly for a rendezvous with the leader of the Huns/Turks to extract a promise of co-operation against the Sasanian troops.[205] In the process, he managed to defeat the armies of both Shahin and Shahrwaraz, as well as to defeat and kill Sharpalakan himself.[206]

Following this, in an attempt to escape south and west through Armenia, Heraclius came face to face with a Sasanian army made up of the combined troops of Shahrwaraz and Shahin. A moving speech by Heraclius, perhaps borrowing from Armenian notions of martyrdom, raised the morale of the Roman troops, and evidently was effective even on the Sasanian troops who were, based in the evidence of George of Pisidia, intimidated enough not to engage

199 Hozhabri (2013) 11.

200 *History to 682*, quoted in Howard-Johnston (2021, 228).

201 See Howard-Johnston (2021) 229–31, carefully mining Movses for details that can be discerned here.

202 Howard-Johnston (2021) 233–36.

203 Supposedly Middle Persian *šahr-palangān* ('the Panther of the Realm'), created as a parallel to *šahrwarāz* ('the Boar of the Realm').

204 The creativity needed for following, and making sense of, Heraclius' successful march back is acknowledged by Howard-Johnston (2021, 236).

205 Howard-Johnston (2021) 238–39.

206 Howard-Johnston (2021) 239.

in battle.[207] This supposedly allowed Heraclius and his troops to escape southwards and put a day's worth of marching between themselves and the Sasanian troops. A surprise attack by Heraclius against Shahrwaraz's headquarters also left the Sasanian general paralysed, but allowed the Byzantine emperor to leave Transcaucasia and move back into Anatolia.[208] The territorial gains against the Sasanians seem to have been zero, as the Sasanian troops indeed started their campaign against Constantinople from central Anatolia, and any attempt at destroying Sasanian forces also appears to have been unsuccessful, given their readiness to lay siege to the Byzantine capital. It is surprising that considering their great difficulty in facing the heroic general-emperor, almost slapstick withdrawals, chases, and their inability to engage, and crushing defeats dealt by Heraclius, the Sasanian generals managed to get enough initiative together to gather a large army for a siege of Constantinople in 626, with the cooperation of the Avars.

6.2 *The Siege of Constantinople in 626*

The entire narrative of the 624–625 campaigns by Heraclius seems to serve one single purpose, that of creating a connection to the Turkic troops beyond the Caucasus.[209] That this meeting should have taken place in the eastern Caucasus is surprising, and the fact that one of the few sources about it, the *History to 682*, seems to report the meeting as taking place in Constantinople, is interpreted as 'something going wrong' with the account of this source.[210] But I wish to turn here to the Sasanians, whose agency in the entire episode seems to have been reduced to incompetent amateurs following, and missing, the initiatives of Heraclius.

The next episode of the war, the great siege of Constantinople by a joint Avar-Sasanian force, seems to be in fact the result of a parallel Sasanian policy. If the Byzantine plan had been to mobilise the Turkic power in the Transcaucasus against the Byzantines, the Sasanians appear to have succeeded sooner in moving the Avars of the Balkans against Constantinople. The unsuccessful Avar attack in the Balkans in 620–622 and the surprise attack on Constantinople in 623, only diverted by Heraclius' paying a high tribute and sending high-level hostages to the Avar court, indeed might have been an exercise for the siege of 626.[211]

207 Howard-Johnston (2021) 240–42.
208 Howard-Johnston (2021) 242–45.
209 Howard-Johnston (2021) 238–39.
210 Howard-Johnston (2021) 239.
211 Howard-Johnston (2021) 248 speculates that the Sasanian-Avar coordination dates to 625.

The command of the Sasanian troops who advanced to Chalcedon was assigned to Shahrwaraz, the famed commander who was supposedly defeated by Heraclius in eastern Anatolia, but now had managed to escape further west and reach the Bosphorus. Heraclius himself was staying in the plain of Taron, presumably preparing for a new season of campaigning in Armenia.[212] The suggestion of the ceremonial role of the Sasanians in the siege and their concentration on driving Heraclius further west seems strange.[213] Shahrwaraz is supposed to have chased Heraclius into central Anatolia, where the general-emperor struck back against his pursuer and defeated him, having enough troops to split his army into two and send reinforcements to defend Constantinople.[214] The account of all these extraordinary events is so confused and unbelievable, and perhaps so much reworked and edited in order to put Heraclius and his actions at the centre of all events, that at least one modern historian decided to take complete leave of him, explaining this decision in eleven pages, while having used it as almost the sole source for reconstruction of the events in the previous fifty-five pages![215]

The actual events, sometimes told through the eyewitness account of Theodore Syncellus, refocus our attention from the heroic general-emperor to Constantinople and to the actions of the local defences there, including the abilities of his son, Heraclius Constantine. Relying on the impressive walls of the city, local defence was organised in the form of a fleet to prevent any Avar, or possibly Sasanian attack through the sea. Hoarding of supplies allowed the city to withstand a siege, giving time to both military and diplomatic solutions to divert the danger. A joint attack by the Sasanians and Avars, starting on 31 July, 626, lasted several days. A series of mistakes, including unsuccessful naval actions by the Avars as well as failed attacks on the walls of the city by both the Avars and the Sasanians, resulted in the loss of power for both the Sasanians and the Avars.[216] In the meantime, a diversion was created among the Sasanian troops when a letter from Khosrow II – possibly a false letter by the Byzantines – asked Shahrwaraz's second-in-command to kill the Sasanian general. Heraclius Constantine is supposed to have lured Shahrwaraz into Constantinople, showing him the impregnable defences of the city, and thus dissuading him from continuing the siege.[217] After failing to defeat the

212 Howard-Johnston (2021) 248–49.
213 Howard-Johnston (2021) 250.
214 Howard-Johnston (2021) 256.
215 Howard-Johnston (2021) 257–68.
216 Howard-Johnston (2021) 279–81.
217 Greatrex and Lieu (2005) 205, with much greater detail in Howard-Johnston (2021) 269–84.

Byzantines, and supposedly egged on by Byzantine misinformation ('black propaganda'), the Avars withdrew from the city on 8 August 626 and returned to the Balkans.[218]

6.3 *War and Intrigue: Heraclius' 626–628 Campaign and the Fall of Khosrow's Regime*

While many modern studies date the turning point to 624 and the 'initial' campaign of Heraclius, as discussed above, the actual turning of the tide appears to have started following the failure of the siege of Constantinople in 626. The withdrawal of Shahrwaraz from the siege probably had great consequences for the Sasanian armies, whose other great commander, Shahin, had died earlier that year.[219] An initial attack by the Yabghu (Jabu) of the Khazars (or possibly Turks)[220] was diverted by a strongly worded letter from Khosrow II.[221]

The actual counteroffensive by Heraclius in fact started late in 626, facing the coming winter. The Turkish army, supposedly commanded by a nephew of the Khaghan known as Shad, withdrew north of the Caucasus.[222] Heraclius had returned to Constantinople, celebrating the repulse of the Avar siege of the city in September. Reconstruction of the earlier parts of the campaign during the winter of 626–627 is quite difficult, but seems to be built on the assumption that Heraclius' main objective was to meet the Khaghan of the Turks near Tiflis/Tbilisi in eastern Georgia.[223] Much weight is given here to the role of the Turks in assuring Byzantine victory, and furthermore, to Heraclius' leading role in devising the strategy.[224] However, the question remains that if we assume that the Western Turkic Khaghanate was in fact involved in the events in the Caucasus, why would it not have simply put a lot of pressure on the Sasanians much further east, in Transoxiana and Central Asia? This was in fact the place where the Turks had managed to confine Sasanian control to the south of the Atrak river, half a century earlier, and enjoyed much strategic advantage.

---

218 Howard-Johnston (2021) 282.

219 Theoph. 315 suggests Shahin was defeated by Heraclius' brother and died of grief. It is unclear who was in fact in charge of the situation and where the engagement happened, see Howard-Johnston (2021) 285.

220 Howard-Johnston (2016b) 9–28.

221 Howard-Johnston (2021) 289–90. Unlike the narrative being advanced by Howard-Johnston, the Khazar/Turkic campaigns seem to have been quite careful and never interested in getting out of the Caucasus.

222 Howard-Johnston (2021) 294.

223 Howard-Johnston (2021) 294–96 on the lack of sources and their unreliability in creating a narrative.

224 Howard-Johnston (2021, 297–304) mostly takes leave of Theophanes' narrative, which is utilised by others (e.g., Greatrex and Lieu (2005) 205–209), and works with the *History to 682*, read through Movses Dasxurantsi's recycling of the material from that source (296).

The important phase of Heraclius' counteroffensive, however, started in autumn 627, when Heraclius started to march down from Armenia, keeping to the west of Lake Urmia, and heading for northern Mesopotamia to the area of Nineveh.[225] He was constantly skirted by a Sasanian general, Rahzād, who did not risk engagement, but probably just took care to prevent Heraclius' movement to the east, to attack the eastern side of Lake Urmia. To this end, Rahzād moved down the east side of Lake Urmia and stationed himself in Ganzak, thus not allowing Heraclius to take over the important city.[226] Closely chasing Heraclius down the course of the Greater Zab River, Rahzād – reinforced by fresh forces sent by Khosrow – finally engaged Heraclius on 12 December. The result was a disaster for the Sasanians, with Rahzād and many of his troops being killed in the battle, and their provisions and even armour being taken by the Byzantine army.[227]

The battle laid open the road toward Ctesiphon for Heraclius. No major Sasanian army was in the region, with Shahrwaraz, the only commander with a major army in the vicinity, being hundreds of kilometres to the west, presumably in Syria. Marching down the Tigris, the Byzantines engaged a smaller Sasanian army at the junction of the Lesser Zab, and defeating them on 22 December 627, captured the bridge over the river.[228] This is when sources start reporting Khosrow's flight from his favourite residence, Dastgird-e Khosrow, and his withdrawal to Ctesiphon. Heraclius then decided to slowly march on Ctesiphon, destroying and plundering on his way, including a pleasure palace with exotic animals, which were duly killed and consumed as food by Heraclius' troops.[229]

Facing the formidable defences of Ctesiphon, including the complicated network of canals and rivers on the hinterland of the city, Heraclius decided to halt his attack. He then started a march backward to the north, in the process plundering more places, including the palace of Dastgird and its treasures in early January 628.[230] The effect of these actions was as Heraclius perhaps intended. This was to strike fear at the heart of the Sasanian Empire, and to act as a catalyst in political discord that must have been forming at the Sasanian court.

---

225 Compared with the description of the campaign of 624–625, the actions of Heraclius now make sense, as the path from the west of Lake Urmia leads directly to northern Mesopotamia, instead of the unbelievable campaign of 624 when the general-emperor had to cross the mountains east to attack Ganzak.

226 Howard-Johnston (2021) 304–305.

227 Howard-Johnston (2021) 308.

228 Howard-Johnston (2021) 310.

229 Howard-Johnston (2021) 313.

230 Howard-Johnston (2021) 313–14.

We get some wind of this political discord in Islamic records of the fall of Khosrow. On 9 February 628, about a month after Heraclius started his march north from Dastgird, a palace *coup d'état* brought Khosrow II down from his throne. Officially headed by his son Shiroye, quickly crowned as Kavad II, the coup was most likely organised by Shamta, the son of Yazdin, Khosrow's old treasurer. That somehow Shahrwaraz's agreement was considered in the planning is implied by the latter's tacit agreement and inaction. The Sasanian emperor was caught by surprise, captured in the garden of his palace, and kept under arrest while the leaders of the coup figured out what to do. A while later, perhaps even as long as a month later, the *shāhānshāh* was confronted with a list of grievances officially drafted by his son, Shiroye. While some of the charges are personal (preventing his sons from marrying), a few relate to the war.[231] Perhaps most important of these are:

1. Length of the war: the coup leaders complain about the fact that the war has lasted too long and that, beyond initial conquests, not much has happened, resulting in the unhappiness of soldiers who have had to live away from their families for many long years.
2. The captivity of the True Cross: while a strange charge to come from the Sasanian nobility, this charge makes sense, as it blames Khosrow for giving an excuse to Heraclius to mobilise his troops behind a religious cause.
3. The imprisonment of a large number of soldiers: this charge seems to be related to imprisonment of soldiers who had either refused to fight or had neglected their duties. While not specified, it is possible to imagine that these might have been the soldiers involved in Shahin's last campaign, which appears to have been a disaster that had angered Khosrow.
4. Hoarding of treasures: this charge, by itself an important issue in the narratives of the fall of the Sasanians, again appears as strange. One would expect the nobility to blame the emperor for emptying the treasury to pay for the wars. Instead, the complaint is that he has hoarded much money in his coffers and refuses to distribute them!

The replies, and defences, of Khosrow makes the entire exchange appear like a trial. Khosrow's answers are harsh, mostly using choice words for Shiroye and his naïveté, but also measured and reasonable. He defends his imprisonment of the soldiers and calls it a just punishment, and credits the wars for having brought much glory to the empire and expanded the realm. His most interesting defence, however, is about the True Cross, which he warns Shiroye not to surrender to Heraclius. He says that by removing the True Cross, he

231 Ṭab. 1.1046–58.

has removed an important claim to legitimacy from Heraclius and has in fact given his own empire a great claim. He also warns Shiroye not to distribute the wealth amongst greedy nobles, so as not to bankrupt the empire and reverse all of his achievements.

Of course, none of these defences seem to have worked. Shiroye still found Khosrow guilty, and under pressure by the nobles who threatened to remove him and reinstall Khosrow, had to relent to the execution of his father. Shamta, the leader of the coup, evidently took the task into his own hands, although the son of another official actually struck and killed the *shāhānshāh*.[232]

In the meantime, Heraclius had marched up toward Adurbadegan and was wintering with his troops in Ganzak.[233] He got the news of the coup a few days after it happened, and waited there for envoys from the new *shāhānshāh* to arrive. What we know about the rest of the events is partly based on an incomplete account of a peace treaty mentioned in the *Chronicon Paschale*.[234] It is perhaps not surprising that Heraclius made no attempt to go west, through northern Syria, to recover the cities in Anatolia and Syria. Instead, in March 628, he started a march toward Armenia, then southward toward the northern Taurus mountains toward Pontus. It was during this march that he sent a message containing a tale of his campaign that was read aloud in Constantinople, as preserved in the *Chronicon Paschale*.[235]

The fighting had ended, with the Sasanian emperor removed and the Byzantine emperor surviving. The extent of Byzantine recovery of lost territories after this point seems to have been limited to parts of Anatolia, while even cities such as Edessa refused to allow Byzantine troops back in. What we know about the recovery is mostly based on narratives of the peace treaty between Shiroye and Heraclius, and little is known about actual military campaigns. Despite a meeting between Heraclius and Shahrwaraz at Arabissus to negotiate the return of the Syrian territories,[236] the Sasanian commander appears to have stayed in control of the region throughout the reign of Shiroye and that of his son, Ardashir III. Indeed, when in 630 the Sasanian general moved east from Syria to invade Ctesiphon to remove Ardashir III and claim the throne for only forty days, he seems to have done this from northern Syria. The True Cross too, the bone of contention and symbol of the war, was not returned until the reign of Boran, Khosrow's daughter and queen after the murder of Shahrwaraz,

232 Ṭab. 1.1057; *Khuz. Chron.* 28–29.

233 Howard-Johnston (2021) 318.

234 See Howard-Johnston (2021) 321–59 for a complete recap of the negotiations.

235 *Chron. Pasch.* 727–736 (Whitby and Whitby (1989) 182–88).

236 Kaegi and Cobb (2008).

in 630.[237] The recovery seems to have been quite slow, and in most of southern Syria, Palestine, and even Egypt, in fact quite scarce. Of course, whatever this recovery might have meant, it was soon reversed by a new wave of invasions.

7 Excursus: Islamic *Futūhāt* and Sasanian Conquests

Less than a decade after the removal and death of Khosrow II, most of the same territories that he had conquered were controlled by a new power, that of the Islamic caliphate.[238] This is at least how later histories, particularly Islamic histories, define the process of events that lead to the control of Iraq, Syria, Palestine, and Egypt by the power based in Medina.[239] The success of the nascent Islamic empire and the astonishing speed with which this success was achieved has occupied the minds of historians almost since the events themselves took place. Various reasons, most importantly the exhaustion of Sasanian and Byzantine military, a result of over two decades of war, have been presented for this success and speed.[240]

However, as we have seen above, this is a reason that barely holds up against the evidence. Despite a long war, neither the Sasanian nor the Byzantine armies, however we try to define them, were exhausted enough not to withstand an onslaught by a force originating from Arabia. Sasanian armies had shown themselves quite competent in conquering and successfully controlling territories stretching from northern Iraq to Antioch and from Damascus to Alexandria. Except in particular settings, such as Heraclius' surgical strikes at Nineveh or occasional battles, the Sasanian army was not really defeated in Syria, Palestine, or Egypt, provinces which seems to have proceeded directly to the Arab conquests without a significant restoration of Byzantine administration. The wars, as we have seen, ended not as a result of pitched battles and decimation of an army by another, but rather as the result of a *coup d'état* and a series of peace treaties.

Byzantine troops too could have barely been exhausted enough not to be able to withstand the attacks from Medina. Apart from winning victories in Nineveh and other places, as well as successfully repelling the threat of the

237 Daryaee (2009) 35–36.

238 For a different take on the dating of the events, see Pourshariati (2008) 190–219.

239 See Hoyland (2014, 31–65) and Donner (1981, 51–220) for the details of the conquests of Syria and Iraq, although surprisingly detached from what had happened before.

240 Once again, credit is due to Howard-Johnston (2021, 372–73) for bringing doubt to this convenient but very problematic assumption; see Zarrinkoub (1975, 1–56) for a standard narrative.

Avars in 626, the Byzantine armies had also had the opportunity of avoiding war since at least the loss of Egypt in 619. Heraclius' effective counteroffensive in 627, based on the use of effective strategies and old tactics, had clearly demonstrated that Byzantine troops were capable of innovation and effective fighting.

Consequently, we would circle back to the original question of how to explain the astonishing success of the Islamic conquests in its initial phase, particularly the campaigns between 636 and 639. This was the period in which the Muslims, in campaigns supposedly originating from Medina, managed to defeat Sasanian and Byzantine troops and take control of Iraq, most of Syria, Palestine, and Egypt. Again, it would not escape the historian's eyes that these are exactly the territories either originally controlled by Khosrow II or conquered during his campaigns between 602–619 and controlled by the Sasanians until 630.

As this paper has previously argued, two historiographical details need to be noticed in this discussion. The first is the tendency of the historians concentrating on the Sasanian-Byzantine wars of 602–630 to view the wars from a northern, largely Armenian and Syrian, as well as Anatolian, viewpoint. At the same time, Islamic histories, which tend to view the early *futūḥāt* from a 'southern' gaze, focusing on conquests of Sawad (lower Iraq), the Syrian desert, and Palestine, generally ignore the Sasanian-Byzantine wars. This artificial division, and mutual ignorance of the other historiography, in fact has important historical consequences.

If we agree with the earlier points of this paper, that a great part of the Sasanian campaigns in the 'southern' theatre of war was in fact undertaken by the 'Arab' residents of the Iraqi-Syrian desert, acting as *foederati* of the Sasanians, a possible solution can be offered. The suggestion here is that these 'Arab' troops, fighting on behalf of the Sasanians, became autocephalous agents, headed by warlords, who attempted at consolidating their control over conquered territories following the disappearance of the Sasanian authority. As we have seen before, many of these tribes and their heads, including the kings of the Lakhmids, already thought of themselves as autonomous rulers, perhaps on a par with the Sasanian *shāhānshāh* and the Byzantine *basileus*, in the treaties concluded in the sixth century. It was thus natural for them to simply attempt to gain territories in the absence of imperial control, when the Sasanian emperor was executed and the Byzantine one was over a thousand miles away in Constantinople.

I would, then, like to suggest that what is seen as the astonishing speed of the Islamic *futūḥāt* needs to be seen as a period of consolidation of the already existing conquests by autonomous local actors. In a forthcoming paper, I argue that the Medinan State needs to be seen as one of these autonomous actors, one that managed to gain supremacy over others either through coercion or through extension of patronage and political legitimacy. It was this process, consolidation of local power and the growing legitimacy of the Medinan State, that was later put in the rubric of military conquests by Islamic historians, who then placed the initiative for all campaigns in the hands of the caliphs in Medina, particularly Omar. In fact, as we can see from cases such as Muthanna al-Haritha al-Shaibani, a local warlord in southern Iraq who appealed to Caliph Abu Bakr for support, it was local actors who initiated the *futūḥāt*, taking advantage of the political vacuum and consolidating territories under their control.[241]

8 Conclusion

The Sasanian-Byzantine wars of the early seventh century are usually treated as simply one episode in a long history of wars between the two empires, and even the longer wars between the powers of central and west Asia and those in the Mediterranean. However, historians have long realised the special character of these wars, both because of their style of fighting and their chronological position immediately prior to the Islamic *futūḥāt* and the rise of the Caliphal empire. The prominence of religious propaganda in the wars, particularly that of the Byzantines, has been an important aspect of the wars distinguishing them from the events that preceded them. In particular, the use of apocalyptic language and visions of martyrdom sets the wars apart. On the other hand, the rise of Islam in the aftermath of these wars also bears a heavy religious imagery, as well as a sense of inevitability for the outcome of these wars.

In both of these visions, the role of the Sasanians as initiators of the wars who dominated events for the majority of the period is often discounted. It is assumed that the Sasanians were undertaking a series of punitive attacks without much advance planning and simply aiming to conquer territories in the same vein as they did in the centuries before, perhaps hoping to capture booty and temporary control of the territories. In turn, the Byzantine side is seen as a purposeful, organised machine that, despite earlier setbacks and over a decade of losing ground to the Sasanians, finally managed to regain the initiative, turn the tide of war, and come out as ultimately victorious, even if ephemerally.

This paper, positing that the early Islamic conquests essentially emulated and replicated the Sasanian conquests of 602–628, sets out to identify the

241 Donner (1981) 175–76. The presence of such actors, pivoting between the Sasanian conquests and the Islamic *futūḥāt*, is the best demonstration of the presence of autocephalous warlords that spanned the period in between the two events.

roots of Sasanian war efforts and the reasons for their success. In the process, it suggests that a long period of planning and careful testing of strategies had prepared the Sasanians to undertake the conquests in the early seventh century. These preparations, part of what I have otherwise called 'the Sasanian Grand Strategy', included a series of economic, social, and military reforms that provided the empire with the financial, material, and human resources that allowed it to attempt such larger scale invasions. Furthermore, by highlighting the relationship between the Sasanian imperial administration and the inhabitants of the Iraqi-Syrian Desert zone, the paper calls attention to the quintessential connection between the 'Arabs' and the Sasanians, and the possibility of close cooperation between the two. Positing that the conquests of Damascus and Palestine, and subsequently that of Egypt, took place through a southern, desert route, the paper argues for the involvement of the population of the desert zone in the wars themselves. It thus follows logically that with the disappearance of Sasanian imperial control following the removal and execution of Khosrow II in early 628, the fighting forces, including the Arab *foederati* of the Sasanians, might have acted in their own self-interest. Consequently, the leaders of these groups, acting as warlords, undertook efforts to consolidate power and take control of the regions which were slipping from the waning Sasanian control. In a forthcoming article, I will suggest that what are generally understood as the early Islamic conquests, the speed of which often leaves scholars flabbergasted, have to be seen as part of the same Sasanian-Byzantine wars that preceded them. In fact, by creating an artificial difference between the Sasanian conquests and Islamic *futūḥāt*, later Islamic historiography simply attempted to do what is familiar to us from other contexts, namely to drive a wedge between the *Jahiliyya* and Islamic periods. Creating a sense of renewal and uniqueness of the revolution of Islam and Muhammad was indeed a well-established tradition in Islamic historiography. Thus, distinguishing between the conquests of a pre-Islamic empire, acting out of greed, and an Islamic caliphate, acting to spread the word of Islam, was the natural tendency of Islamic sources.

Bibliography

Adams, R.M. (1965) *Land Behind Baghdad: A History of Settlement on the Diyala Plains*. Chicago.

Adams, R.M. (1981) *Heartland of Cities. Surveys of Ancient Settlement and Land Use on the Central Floodplain of the Euphrates*. Chicago.

Adams, R.M. and Hansen, D.P. (1968) 'Archaeological Reconnaissance and Soundings in Jundi Shahpur'. *Ars Orientalis* 7: 53–70.

ʾAmir, A. (1960) *Dinawari. Al-Akhbār al-ṭiwāl*. Cairo.

Bahar, M.-T. and Gonabadi, M.P. (1934) *Tarikh-e Bal'ami* (*History of Bal'ami*). Tehran.

Balatti, S., Klinkott, H., and Wiesehöfer, J. (eds) (2021) *Paleopersepolis. Environment, Landscape and Society in Ancient Fars*. Stuttgart.

Binder, C., Börm, H., and Luther, A. (eds) (2016) *Diwan: Studies in the History and Culture of the Ancient Near East and the Eastern Mediterranean. Festschrift für Josef Wiesehöfer zum 65. Geburtstag*. Duisburg.

Blockley, R.C. (1985) *The History of Menander the Guardsman*. Liverpool.

Bosworth, C.E. (1999) *The History of Ṭabarī: Vol. 5: The Sāsānids, the Byzantines, the Lakhmids, and Yemen*. Albany, NY.

Cameron, A. (1985) *Procopius and the Sixth Century*. Berkeley.

Cameron, A. (ed) (1995) *The Byzantine and Early Islamic Near East. Vol. 3: States, Resources and Armies*. Princeton.

Canepa, M.P. (2010) 'Technologies of Memory in Early Sasanian Iran: Achaemenid Sites and Sasanian Identity'. *AJA* 114: 563–96.

Carile, A. (ed) (2004) *La Persia e Bisanzio: Convegno Internazionale, Roma 14–18 Ottobre 2002*. Rome.

Chaumont, M. (1989). 'Bōrān'. *Enc Ir* IV.4: 366.

Christensen, A. (2007) *Romanen om Bahrâm Tschôbîn: et Rekonstruktions-Forsøg*. Copenhagen.

Crone, Patricia. 1991. 'Kavād's Heresy and Mazdak's Revolt'. *Iran* 29: 21–42.

Crone, P. (2012) *The Nativist Prophets of Early Islamic Iran: Rural Revolt and Local Zoroastrianism*. Cambridge.

Daryaee, T. (2002) *Šahrestānīhā Ī Ērānšahr: A Middle Persian Text on Late Antique Geography*. Costa Mesa, CA.

Daryaee, T. (2009) *Sasanian Persia: The Rise and Fall of an Empire*. London.

Daryaee, T. and Compareti, M. (eds) (2019). *Studi Sulla Persia Sasanide e Suoi Rapporti Con La Civilta Attigue*. (Studia Persica 1.) Bologna.

Dignas, B. and Winter, E. (2007) *Rome and Persia in Late Antiquity: Neighbours and Rivals*. Cambridge.

Donner, F.M. (1981) *The Early Islamic Conquests*. Princeton.

Downing, T. and Gibson, M. (eds) (1974) *Irrigation's Impact on Society*. Tucson.

El Cheikh, N. (1998) 'Sūrat Al-Rūm: A Study of the Exegetical Literature'. *JAOS* 118: 356–64.

El-Cheikh, N. (2004) *Byzantium Viewed by the Arabs*. Cambridge, MA.

Fisher, G. (2011) *Between Empires: Arabs, Romans, and Sasanians in Late Antiquity*. Oxford.

Flusin, B. (1992) *Saint Anastase Le Perse et l'histoire de La Palestine Au Début Du VIIe Siècle*. Paris.

Foss, C. (1975) 'The Persians in Asia Minor and the End of Antiquity'. *English Historical Review* 90 (357): 721–47.

Fowden, E.K. (1999) *The Barbarian Plain: Saint Sergius between Rome and Iran*. Berkeley.

Fowden, G. (2013) *Before and After Muhammad: The First Millennium Refocused*. Princeton.

Frye, R.N. (ed) (1975) *The Cambridge History of Iran. Vol. 4: The Period from the Arab Invasion to the Saljuqs*. Cambridge.

Goubert, P. (1949) 'Les rapports de Khosrau II, roi des rois sassanide, avec l'empereur Maurice'. *Byzantion* 19: 81–98.

Greatrex, G. (1998) *Rome and Persia at War, 502–532*. Leeds.

Greatrex, G. (2007) 'Dukes of the Eastern Frontier'. *BICS* 50: 87–98.

Greatrex, G. (2022) *Procopius of Caesarea: The Persian Wars. A Historical Commentary*. Cambridge.

Greatrex, G. and Lieu, S.N.C. (2005) *The Roman Eastern Frontier and the Persian Wars AD 363–628*. London.

Grignaschi, M. (1971) 'La riforma tributaria di Hosro I e il feudalesimo sasanide'. *Accademia Nazionale dei Lincei Quaderno* 160: 87–147.

Gyselen, R. (2001) *The Four Generals of the Sasanian Empire: Some Sigillographic Evidence*. Rome.

Gyselen, R. (ed) (2016) *Sasanian Persia and the Tabarestān Archive*. (*Res Orientales* XXVII.) Bures-sur-Yvette.

Heidemann, S. (1998) 'The Merger of Two Currency Zones in Early Islam: The Byzantine and Sasanian Impact on the Circulation in Former Byzantine Syria and Mesopotamia'. *Iran* 36: 95–112.

Hodges, R. and Whitehouse, D. (1983) *Mohammed, Charlemagne & the Origins of Europe: Archaeology and the Pirenne Thesis*. Ithaca.

Howard-Johnston, J. (1995) 'The Two Great Powers in Late Antiquity: Comparison', in Cameron (1995) 157–226.

Howard-Johnston, J. (1999) 'Heraclius' Persian Campaigns and the Revival of the East Roman Empire, 622–630'. *War in History* 6: 1–44.

Howard-Johnston, J. (2004) 'Pride and Fall: Khusro II and His Regime, 626–628', in Carile (2004) 93–113.

Howard-Johnston, J. (2006a) 'Al-Tabari on the Last Great War of Antiquity', in Howard-Johnston (2006b) 1–22.

Howard-Johnston, J. (2006b) *East Rome, Sasanian Persia, and the End of Antiquity*. Aldershot.

Howard-Johnston, J. (2010) *Witnesses to a World Crisis: Historians and Histories of the Middle East in the Seventh Century*. Oxford.

Howard-Johnston, J. (2016a) 'The Grand Strategy of the Sasanian Empire', in Binder, Börm, and Luther (2016) 591–613.

Howard-Johnston, J. (2016b) 'World War in Eurasia at the End of Antiquity', in Gyselen (2016) 9–28.

Howard-Johnston, J. (2021) *The Last Great War of Antiquity*. Oxford.

Howard-Johnston, J., Mango, C., and Dagron, G. (1995) 'The Siege of Constantinople in 626', in Mango, Dagron, and Greatrex (1995) 131–42.

Hoyland, R.G. (2011) *Theophilus of Edessa's Chronicle and the Circulation of Historical Knowledge in Late Antiquity and Early Islam*. Liverpool.

Hoyland, R.G. (2014) *In God's Path: The Arab Conquests and the Creation of an Islamic Empire*. Oxford.

Hozhabri, A. (2013) 'The Evolution of Religious Architecture in the Sasanian Period'. *Sasanika Online, Archaeology* 18: 1–40.

Hunter, E.C.D. (2008) 'The Christian Matrix of Al-Hira', in Jullien (2008) 41–56.

Isaac, B. (1995) 'The Army in the Late Roman East: The Persian Wars and the Defence of the Byzantine Provinces', in Cameron (1995) 125–56.

Jones, M. (ed) (1966) *Kitab Al-Maghazi of al-Waqidi*. London.

Jullien, C. (ed) (2008) *Les Controverses Des Chretiens Dans l'Iran Sassanide*. Paris.

Kaegi, W.E. (1992) *Byzantium and the Early Islamic Conquests*. Cambridge.

Kaegi, W.E. (2003) *Heraclius, Emperor of Byzantium*. Cambridge.

Kaegi, W.E. and Cobb, P.M. (2008) 'Heraclius, Shahrbaraz, and al-Tabari', in Kennedy (2008) 95–112.

Kennedy, H. (ed) (2008) *Al-Tabari: A Medieval Muslim Historian and His Work*. Studies in Late Antiquity and Early Islam 15. Princeton.

Kennet, D. (2002) 'Sasanian Pottery in Southern Iran and Eastern Arabia'. *Iran* 40: 153–62.

Kennet, D. (2007) 'The Decline of Eastern Arabia in the Sasanian Period'. *Arabian Archaeology and Epigraphy* 18: 86–122.

Lawrence, D. and Wilkinson, T.J. (2017) 'The Northern and Western Borderlands of the Sasanian Empire: Contextualising the Roman/Byzantine and Sasanian Frontier', in Sauer (2017) 99–125.

Le Strange, G., Nicholson, R., and Fasaei, M.R. (eds) (1995) *Farsnama*. Shiraz.

Mango, C. and Scott, R. (1997) *The Chronicle of Theophanes Confessor*. Oxford.

Mango, C., Dagron, G., and Greatrex, G. (eds) (1995) *Constantinople and Its Hinterland*. Aldershot.

Markwart, J. (1931) *A Catalogue of the Provincial Capitals of Eranshahr*. Rome.

McKenzie, D.N. (1971) *A Concise Pahlavi Dictionary*. Oxford.

Moghaddam, A. (2012) 'Southwestern Iran', in Potts (2012) 512–30.

Moghaddam, A. and Miri, N. (2003) 'Archaeological Research in the Mianab Plain of Lowland Susiana, South-Western Iran'. *Iran* 41: 99–137.

Moghaddam, A. and Miri, N. (2007) 'Archaeological Surveys in the "Eastern Corridor", Southwestern Iran'. *Iran* 45: 23–55.

Morley, C. (2017) 'The Arabian Frontier: A Keystone of the Sasanian Empire', in Sauer (2017) 268–83.

Morony, M.G. (2002) 'The Late Sasanian Economic Impact on the Arabian Peninsula'. *Nāme-ye Irān-e Bāstān* 2: 25–37.

Morony, M.G. (2004) 'Population Transfers between Sasanian Iran and the Byzantine Empire', in Carile (2004) 161–79.

Morony, M.G. (2019) 'The Early Islamic Mining Boom'. *JESHO* 62: 166–221.

Munt, H. (2017) 'Oman and Late Sasanian Imperialism'. *Arabian Archaeology and Epigraphy* 28: 264–84.

Neely, J. (1974) 'Sassanian and Early Islamic Water-control and Irrigation Systems on the Deh Luran Plain, Iran', in Downing and Gibson (1974) 21–42.

Nokandeh, J., Sauer, E.W., Wilkinson, T., Abbasi, G.A., Schwenninger, J.-L., Mahmoudi, M., Parker, D., Fattahi, M., Usher-Wilson, L., Ershadi, M., Ratcliffe, J., and Gale, R. (2006) 'Linear Barriers of Northern Iran: The Great Wall of Gorgan and the Wall of Tammishe'. *Iran* 44: 121–73.

Nöldeke, T. 1879. *Geschichte der Perser und Araber zur Zeit der Sasaniden: aus der arabischen Chronik des Tabari übersetzt und mit ausführlichen Erläuterungen und Ergänzungen versehen*. Vol. 1. Leiden.

Palme, B. (ed) (2007) *Akten des 23. Internationalen Papyrologenkongress (Wien, 22.–28. Juli 2001)*. Vienna.

Palmer, A. (1993) *The Seventh Century in the West-Syrian Chronicles*. Liverpool.

Phenix, R. and Horn, C. (eds) (2022) *Strategius of Mar Saba. On the Captivity of Jerusalem by the Persians in 614 CE*. Chesterfield, MO.

Pirenne, H. (1937) *Mahomet et Charlemagne*. Paris.

Pirenne, H. (2013) *Mohammed and Charlemagne*. New York.

Potts, D.T. (2008) 'The Sasanian Relationship with South Arabia: Literary, Epigraphic and Oral Historical Perspectives'. *StIr* 37: 197–213.

Potts, D.T. (ed) (2012) *A Companion to the Archaeology of the Ancient Near East*. Malden, MA.

Pourshariati, P. (2008) *Decline and Fall of the Sasanian Empire: The Sasanian-Parthian Confederacy and the Arab Conquest of Iran*. London.

Proudfoot, A. (1974) 'The Sources of Theophanes for the Heraclian Dynasty'. *Byzantion* 44: 367–439.

Raaflaub, K.A. (ed) (2007) *War and Peace in the Ancient World*. Oxford.

Rekavandi, H.O., Sauer, E.W., Wilkinson, T., and Abbasi, G.A. (2008) 'Sasanian Walls, Hinterland Fortresses, and Abandoned Ancient Irrigated Landscapes: The 2007 Season on the Great Wall of Gorgan and the Wall of Tammishe'. *Iran* 46: 151–78.

Rezakhani, Khodadad (2015). 'Mazdakism, Manichaeism and Zoroastrianism: In Search of Orthodoxy and Heterodoxy in Late Antique Iran'. *Iranian Studies* 48 (1): 55–70.

Rezakhani, K. (2017a) *ReOrienting the Sasanians: East Iran in Late Antiquity*. Edinburgh.

Rezakhani, K. (2017b) 'Arab Conquest of Sasanian Iran'. *History Today* 67.4: 28–36.

Rezakhani, K. (2019) 'The End of Sasanian Rule: The Center and Periphery of Ērānšahr in the Seventh Century', in Daryaee and Compareti (2019) 229–55.

Rezakhani, K. (2020) 'History from the Edge: Teaching Early Islamic History from Sub-Sasanian Coins'. *J. Numis. Assoc. Aust.* 30: 127–47.

Rezakhani, K. (2021) 'Nobles and Land: The Formation of the South (Nēmrōz) and the Politics of the Elite in the Sasanian Empire', in Balatti, Klinkott, and Wiesehöfer (2021) 241–66.

Rezakhani, K. and Amiri Bavandpour, S. (2016) *Rūdādnāme-ye Soryāni Mosūm be Rūydādnāme-ye Khūzestān* (*The Syriac Chronicle Known as the Khuzistan Chronicle*). Tehran.

Rubin, Z. (1995) 'The Reforms of Khusro Anushirwan', in Cameron (1995) 227–98.

Sauer, E.W., Rekavandi, H.O., Wilkinson, T.J., and Nokandeh, J. (2013) *Persia's Imperial Power in Late Antiquity: The Great Wall of Gorgān and Frontier Landscapes of Sasanian Iran*. Oxford.

Sauer, E.W. (ed) (2017) *Sasanian Persia: Between Rome and the Steppes of Eurasia*. Edinburgh.

Shahid, I. (1995) *Byzantium and the Arabs in the Sixth Century*. Washington, DC.

Shiri, A. (1990) *Kitab ul-Futuh*. Beirut.

Sinor, D. (ed) (1990a) *The Cambridge History of Early Inner Asia*. Cambridge.

Sinor, D. (1990b) 'The Establishment and Dissolution of the Türk Empire', in Sinor (1990a) 285–316.

Stoyanov, Y. (2011) *Defenders and Enemies of the True Cross*. Vienna.

Thomson, R.W. (1999) *The Armenian History Attributed to Sebeos*. Liverpool.

Toral-Niehoff, I. (2013a) *Al-Ḥīra: Eine Arabische Kulturmetropole Im Spätantiken Kontext*. Leiden.

Toral-Niehoff, I. (2013b) 'Late Antique Iran and the Arabs: The Case of al-Hira'. *JPS* 6: 115–26.

Viermann, N. (2021) *Herakleios, der schwitzende Kaiser. Die oströmische Monarchie in der ausgehenden Spätantike*. Berlin.

Webb, P. (2015) 'Pre-Islamic "al-Shām" in Classical Arabic Literature: Spatial Narratives and History-Telling'. *Studia Islamica* 110: 135–64.

Webb, P. (2016) *Imagining the Arabs: Arab Identity and the Rise of Islam*. Edinburgh.

Weber, D. (2003) *Berliner Papyri, Pergamente Und Leinenfragmente in Mittelpersischer Sprache*. Vol. 3. London.

Weber, D. (2005) 'A Pahlavi Papyrus from Islamic Times'. *BAI* 19: 225–31.

Weber, D. (2007) 'The Vienna Collection of Pahlavi Papyri', in Palme (2007) 725–38.

Weber, D. (2011) 'An Interesting Pahlavi Letter from Early Islamic Times (Berk. 245)'. *Zeitschrift Der Deutschen Morgenländischen Gesellschaft* 161: 91–98.

Wenke, R.J. (1975–1976) 'Imperial Investments and Agricultural Developments in Parthian and Sasanian Khuzestan. 1509 B.C. to A.D. 640'. *Mesopotamia* 10–11: 31–221.

Whitehouse, D. and Williamson, A. (1973) 'Sasanian Maritime Trade'. *Iran* 11: 29–49.

Wiesehöfer, J. (2007) 'From Achaemenid Imperial Order to Sasanian Diplomacy: War, Peace, and Reconciliation in Pre-Islamic Iran', in Raaflaub (2007) 121–40.

Winter, E. and Dignas, B. (2001) *Rom und das Perserreich: zwei Weltmächte zwischen Konfrontation und Koexistenz*. Berlin.

Wood, P. (2013) *The Chronicle of Seert: Christian Historical Imagination in Late Antique Iraq*. Oxford.

Yarshater, E. (ed) (1983) *The Cambridge History of Iran. Vol. 3(2): The Seleucid, Parthian, and Sasanid Periods*. Cambridge.

Zarrinkoub, A. (1975) 'The Arab Conquest of Iran and Its Aftermath', in Frye (1975) 1–56.

CHAPTER 20

Epilogue: the Evolution of the Ancient Iranian Army

Touraj Daryaee

It is rather difficult to outline the development of Iranian fighting forces from the earliest literary and archaeological evidence to the coming of the Muslims on the Iranian plateau. However, we can attempt to contextualise specific periods where developments and changes took place, from the first millennium BCE to the seventh century CE. One can point to several specific phases in the development of the Iranian army in ancient times, with new tactics and new forces.

The first phase of what can be termed a distinctive Iranian military force, and its specific technology and weapons, can be identified with the Indo-European Iranians.[1] The Sintashta Culture is usually associated with the first signs of Iranian presence on the Russian steppes. What is striking is the importance of two-wheeled chariots, which were most probably used for movement and combat in this Indo-Iranian culture.[2] There are large numbers of such chariots which were buried in this region, providing some of the earliest clues on the use of chariots in the third millennium BCE.[3] Indo-Iranian poetics also provide us with some of the most complete descriptions of chariots, their parts, and their use in terms of racecourse competition and winning,[4] as metaphorically used in the oldest stratum of Iranian poetic tradition in *Yasna* 30.10,[5] where men achieve glory and imperishable fame.[6] The chariot also appears metaphorically when the movement of the sun is likened to the ritual chariot race.[7] In the Avestan hymn to Mithra (X.76), the chariot (*raϑa-*) and charioteer (*raϑi-*) are well attested, as in the reference to 'good horses and a good chariot-driver'.[8] The term for 'warrior' in Indo-Iranian is derived from *raϑaēštā*, which means 'one who stands on the chariot'.

1 Pourdavoud (1968).
2 Pyankov (2002) 46–47.
3 Anthony and Vinogradov (1995) 38.
4 Spaareboom (1985).
5 Schwartz (2000) 2; Louchakova-Schwartz (2018) 480, 488.
6 Watkins (1995) 13, 415.
7 Skjærvø (2015) 62.
8 Gershevitch (1967) 109; Lubotsky (2023) 259.

 | DOI:10.1163/9789004710771_021

The weapon of choice in the hymn to Mithra appears to have been the spear (*aršti-*) (10.20: 'back flies the spear').[9] Other important weapons wielded by Mithra in battle are the 'mace' (*vazra-*) with knobs and blades (10.132),[10] the 'battle axe' (*čakuš-*) (10.130),[11] and the 'club' (*gadā-*),[12] which is part of the name for the city of Cyrus II and the Teispids, that is, Pasargadae (**Pāthra-gadā-*), meaning 'those who wield heavy clubs' or 'protective mace/club'.[13] Thus, we are speaking of early Iranian or Indo-Iranian fighting forces who used specific weapons, such as the stone mace or axe, which are found in kurgan burials in Southern Russia and were brought to the Iranian plateau.

As these Iranian speakers entered the plateau with their armies, they encountered autochthonous polities, but the exception was the kingdom of Elam, which was organised on an imperial scale.[14] The military might of Elam is clear in its regional warfare and the alliance with the Babylonians which stood against the Assyrian Empire in the eighth century BCE.[15] While there were more defeats at the hands of the Assyrians, the Elamites recruited forces from the neighbouring regions to create an imperial army, similar to the Teispid-Achaemenids. This is apparent with Humban-menana III in the seventh century BCE, who had an Elamite army composed of Elamites, Ellipi, Parsumaš, and Anshanites.[16] Thus, a hybrid army on the Iranian plateau began with the Elamites and then the Teispids, and finally the Achaemenids.

The appearance of the Medes, whose forces were often dreaded in the Assyrian annals as 'barbarians' (*Ummān-Mānda*),[17] or what could be called a *Männerbund* force, is amply attested in the Biblical tradition. Not only did the Medes enter into the military service of the Assyrians, but they were also important cavalrymen, who along with the Babylonians were responsible for bringing down the Assyrian Empire,[18] and made themselves masters of the northern Iranian plateau. Their power as an elite fighting force in the Teispid-Achaemenid Empire is clear from the Old Persian conjunction of the '*kāra-* of the Persians and Medes' (*kārum Pārsam utā Mādam*).[19] Darius I's inscriptions provide

9 Malandra (1973) 270; Gershevitch (1967) 83.
10 Malandra (1973) 282–83.
11 Malandra (1973) 273.
12 Malandra (1973) 273.
13 Tavernier (2007) 392.
14 See Álvarez-Mon and Wicks, this volume.
15 Brinkman (1965) 161.
16 Vallat (2011).
17 Adali (2011).
18 Lanfranchi (2021) 144.
19 Klein (1988) 388.

a window into the transformation of the Teispid-Achaemenid army from a 'militia-type' *kāra-* ('the people in arms'),[20] which goes back to Indo-European **koryos* ('army, war-band'),[21] to a professional army (*spādā-*).[22] There are clear signs that the Persians attempted to place their own imprint on the organization of the new imperial army, but the Medes played a significant part in the new army of the empire.[23] Still, from Darius' inscription at Persepolis, it is clear that the Persian *kāra-* was the nucleus of support for the Great King, and one may suggest that many of the Immortals were conscripted from the Medes and the Persians.[24] Even the Great King Darius I calls himself a good 'warrior/fighter' (*hamaranakara*) (DNb 34).[25]

The Teispid-Achaemenid army was an imperial army on a scale that was not known before, whose contingents and men were from the various satrapies and included many mercenaries. In the Greco-Persian Wars, there were as many, if not more Greeks fighting on the side of the Persians, including the empire's navy, than for the Greek city-states.[26] The warped narrative of the Persian army which appears in Greek sources, assuming the superiority of Greek military forces, has now been reevaluated.[27] The way of war in the Teispid-Achaemenid Empire has already been studied brilliantly by P. Briant, who has shown the quality of the warriors beyond the Greek stereotypes, including Persian leadership in battle, their training and education, and the different types of forces available to the Great King.[28] We are also fortunate that the Achaemenid military has now also been dealt with in terms of clothing and weapons in much more detail.[29] By no means was the Teispid-Achaemenid army static in its two centuries, and as an example we know that for the Battle of Gaugamela against Alexander of Macedon, Darius III attempted to draw on his eastern satrapies, including Scythians as well as Indian forces and elephants. The detailed study of Manning on the types of troops and technology has given a fresh view, so that the influence of Greek warfare, scythed chariots, and weapons is now better understood in the context of Teispid-Achaemenid military history.[30]

---

20 Benveniste (2016) 82.
21 Mallory (2007) 91.
22 Frye (1972) 86.
23 Harmatta (1971) 13–14.
24 Daryaee (2021) 73–74; see also Hyland, this volume.
25 Schmitt (2000) 40.
26 R. Rollinger, personal note.
27 Manning (2021).
28 Briant (2018a).
29 Tuplin and Jacobs (2021) 1161–81.
30 Manning (2021) 110–11.

The next phase came with the Seleucid army, which took over the spoils of Alexander of Macedon's Near Eastern territory, created a number of cities, and settled its forces while fostering alliances with the local nobility and satraps.[31] This was first and foremost a military force, which through marriage with the Iranian nobility created a web of connections and control over the Iranian plateau.[32] It was once thought that the infantry phalanx was the key to Macedonian superiority over the light-armoured Persian troops, and the Greco-Macedonian cavalry was fast and well-organised.[33] However, in a reassessment of the sources, it appears that after his battlefield victories over Darius III, Alexander had to reorganise his army for war on the Iranian plateau, where javelinists, archers, and light infantry took precedence.[34] Many of the Iranian forces joined the Seleucid army to keep their power and privilege in the new Hellenistic Empire. Olbrycht believes that it was the Iranian cavalry which was the mainstay of the Seleucid army and the reason for its success, providing victory over others.[35] The Seleucids, who had continued the Iranian way of warfare in the first half of their history, had to change their tactics in the second half of Seleucid rule, because of their defeat at hands of the Romans in the second century BCE.[36]

The Seleucids met their end on the Iranian plateau in encounters with new rapid armies from Eastern Iran. The two types of Arsacid cavalry, mounted archers (*hippotoxotai*) and the heavy cavalry (*kataphraktoi*), gave them an advantage over the Seleucids and others throughout their history for some four centuries. Thus, the new Arsacid way of rapid and mobile warfare became superior to the Seleucid army and brought about its defeat.[37] The nature of the Arsacid army has been subject to debate, in its structure and composition for the new empire. It appears to be an accurate assessment that the army was composed of only a relatively small number of nobles (*liberi*), with a large number of cavalry forces, potentially up to 50,000.[38] Hauser has suggested that the Arsacid army, while relying on local kings and nobility, was firmly controlled by imperial commanders.[39]

31 See Coloru, this volume.
32 Strootman (2023) 11.
33 Shahbazi (2011).
34 Olbrycht (2018) 81.
35 Olbrycht (2001) 109.
36 Sekunda (1994).
37 Olbrycht (2001) 108–109.
38 Hauser (2006) 304.
39 Hauser (2006) 312.

Of course, the capability of the Arsacid army was shown in its battle against the Romans in 53 BCE, at the Battle of Carrhae.[40] However, this relative parity with Rome throughout the second half of its history suggests a much more structural strength in its military might than usually assumed.[41] While the Romans were never able to strike a final blow against the Arsacids, it was rather the army of Ardashir I and the Sasanian forces in the third century which brought down the Arsacid Empire. Beside the well-organised and well-funded Persian army of the Sasanians, ideology may also have had a hand in bringing down the house of Arsaces. In this line, the idea of 'militant piety' may be a matter which allowed the Mazda worshippers of the Iranian plateau to band together to bring a new vision of the Iranian world, that is, the Ērānšahr or 'Empire of the Iranians'. Thus, one may suggest that Ardashir and his battle against Ardawan IV took on a semblance of 'Holy War' (*ardīg*), which is reflected pictorially on his rock relief at Naqš-e Rostam.[42] If one was to lose his life for 'religion and empire', it would have been sanctioned by the spiritual commitment to give one's life selflessly, or what is known in Middle Persian as 'self-sacrificing' (*gyān-abespār*).[43]

This religious zeal coupled with military organisation became amply clear in the destruction of the Kushan Empire to the east, and the bringing down of the caravan towns and checking the Roman forces in the west.[44] By the fourth century CE, both the Sasanian heavy cavalry and its infantry were feared, a change from the Arsacid period, if we are to believe the Roman sources.[45] The expansion of the Sasanians in all directions in the Near East, and at times into Anatolia, suggests also a move towards a more centralised army, but only gradually. This Sasanian military, while it held its own in its wars with the Romans,[46] encountered serious problems with the new nomadic forces from the east, whereby after the death of King Peroz in the fifth century BCE, it became a tributary state to the Hephthalites for a short time.[47] While we are not sure if the use of stirrups gave the decisive advantage to the Hephthalites,[48] the reforms of Kavad I and his son Khosrow I in the sixth century CE reorganised

40 Traina (2011); Overtoom, this volume.
41 See Gregoratti, this volume.
42 Choksy (2012) 170; Choksy and Nematollahi (2017) 883.
43 Azarnouche and Petitjean (2022) 29.
44 Daryaee (2022) 153.
45 See Maksymiuk, this volume.
46 Maksymiuk (2015) 16; Maksymiuk (2018).
47 Rezakhani (2017) 128–29.
48 Bivar (1955) 61–68.

and reinvigorated the Sasanian army.[49] This attention to the army is found in a variety of sources, from the *Shāhnāmeh*, a medieval Persian rendition of the lost Sasanian *Xwadāy-nāmag* (Book of Lords), to Byzantine observations in their military manuals on Persian fighting forces and tactics,[50] to sigillographic evidence[51] and the epitome of a Sasanian military manual, the 'Book of Warriors' (*Artēštārestān*).[52]

It is no small feat for the Sasanians to have matched the Eastern Roman Empire in strength, although Sasanian resources seems to have been less.[53] The Sasanian military was headed by 'generals' (*spāhbeds*), and protected at its boundaries by 'march wardens' (*marzbāns*), 'cavalry commanders' (*aspbeds*), and many provincial officers who could defend the empire; their forces were drawn from a regular army, guard regiments in strategic locations, and foreign auxiliaries.[54] The army's wealth and capabilities are also discernible from a list of its weapons and arms;[55] Ṭabarī also provides us with a list of weapons for cavalrymen after the reforms of the sixth century CE, which demonstrates that the amount of expenditure for the army must have been huge.[56]

While most of the reorganisation of the Sasanian army allowed it to thrive and nearly conquer the Eastern Roman Empire, it also exhibited certain weaknesses. It has been suggested that the new army was more defensive in its structure than offensive and gave too much power to the generals.[57] Most important was the beginning of what has been called the 'Barbarization' of the army,[58] where new conscription from nomadic groups and forces from mountainous areas (Daylamites), along with charismatic generals, brought new loyalties.[59] The army of the Sasanians was large in the sixth and early seventh centuries CE, but the coup and death of Khosrow II in 628 certainly damaged the empire. The fractured army with its divided loyalties, and some cavalry (*aswār*/*aswāriyya*) joining the enemy,[60] was a main reason for the total collapse of the Sasanian Empire in the face of the Arab Muslim conquest.

49 See Börm, this volume; Rezakhani, this volume.
50 Dennis (2001).
51 Gyselen (2001).
52 Azarnouche and Petitjean (2023).
53 Howard-Johnston (2012) 91.
54 Howard Johnston (2012) 123.
55 Tafazoli (1993–1994).
56 Bosworth (1999) 262–63.
57 Farrokh, Karamian, and Maksymiuk (2018) 25–26.
58 Rubin (1995) 285.
59 Daryaee (2010).
60 Zakeri (1995).

As J. Howard-Johnston has stated, the Arab Muslim armies did not really conquer the Ērānšahr, but rather after several battles, 'inserted' themselves into it.[61]

Thus, to understand the ancient Iranian military forces, one needs to consider elements of continuity, but more importantly the many changes brought to it over time by various contexts and by new outside forces. Adoption and adaptation appear to have been important reasons for the development and strength of the army in pre-Islamic Iran. Some of these experiences came at the cost of defeats at the hands of their settled enemies (Greeks, Romans), while others involved encounters with new and unknown nomadic powers from the east and the south (Huns, Turks, Arabs).

What I think is still worth pursuing is the idea of the ritual of war in the Iranian tradition.

I would like to provide two brief case studies for the importance of ritual to give meaning to the ancient Iranian idea of war. The first is from the Avestan hymn to the god of 'offensive victory', Wahram (*Vərəϑraγna*). Wahram was an important deity from remote antiquity through the historical dynasties of the Achaemenids, Arsacids, and the Sasanians. In his Hellenic garb, he took on the form of Hercules, and was mentioned as such in bilingual Greco-Iranian inscriptions 'Ηρακλέους θεου/WRTRGN 'LḤ'.[62] On the coins of Khosrow II, the wings of the bird symbolise Wahram, and the wearing of such a crown would invoke the presence of the deity in battle and victory against enemies.[63] In the Avesta we find the full idea of the importance of Wahram at the time of battle, where specific invocations are called for (14.62):

> We worship Ahura-created *Wərəϑ*raγna (Wahram), who destroys the battle lines, who cuts the battle lines, who tramples the battle lines, who throws the battle lines into confusion.[64]

It is clear that the invocation of such a deity in battle and the associated rituals were important, which often is missed in the historical documentation. This ritualistic importance of war can also be gleaned from the Sasanian military manual, the *Artēštārestān*, in which such ceremonies must take place before the war (*Dēnkard* VIII.37.29):

> On the ceremony of the day of the fight (and) the confrontation against the evildoers, on the barsom plant (used in) this ceremony, and (on) the

61 Howard-Johnston (2022) 109.
62 Moreno (1990) 231; Potter (1991) 278–79.
63 Daryaee (1997) 49–51.
64 Malandra (1983) 87.

> Avestan (prayer recited) while fighting (and) shooting the first arrow at a target, (and) the ceremony (performed) by carrying the offering to the water that is nearest to the battlefield.[65]

Obviously, the Greco-Roman sources do not mention such ceremonies and rituals, which are part of the ancient Iranian mores and values for war. By not keeping these matters in mind, one may miss the entire constructed worldview of the Iranian warriors whose way of war we are trying to understand. There is much more that needs to be done in this line of research, which obviously makes the notions of warfare better understood. For the ancient Iranians, war was a tool to fight the injustice which had been unleashed into their world through the agents of chaos and evil (*druj-*), and justice and order, under the watchful eyes of the king and his supreme deity, Ahuramazda/Ohrmazd, were to be restored by any means.[66] What is also noteworthy is that in pre-Islamic Iranian tradition, the idea of 'war and treaty' is the main concern, rather than 'war and peace', which one normally expects.[67]

All societies engage in battle and there are a number of terms for it in Iranian, such as *razm*, *kārezār*, *stēzak*, *pahīkār*, and *jang*,[68] but the ways, rules, rituals and manner of war differ from one another. If one was to try to understand the rituals of war for the ancient Iranian world, one would have to take into consideration the *Artēštārestān* and the *Avesta*, the *Ayādgār ī Zārērān* and the *Shāhnāmeh*, as the theatres where they play out. This would not be too dissimilar from the Indian tradition of the *Dhanurvedic* manual of war,[69] along with that of the *Mahbharata*, where the Battle of Kurukeshtra is one of the main components of the text.

Bibliography

Adali, S. (2011) *The Scourge of God: The Umman-manda and Its Significance in the First Millennium BC*. Winona Lake.

Agut-Labordère, D., Boucharlat, R., Joannès, F., Kuhrt, A., and Stolper, M.W. (eds) (2021) *Achemenet. Vingt-ans après: Études offertes à Pierre Briant à l'occasion des vingt ans du Programme Achemenet*. Leuven.

65 Azarnouche and Petitjean (2023) 42.
66 See Rung, this volume.
67 Utas (1986) 295.
68 Utas (1986) 292.
69 Arya (2015).

Anthony, D.W. and Vinogradov, N.B. (1995) 'Birth of the Chariot'. *Archaeology* 48/2: 36–41.

Arya, R.P. (2015). *Dhanurveda: The Vedic Military Science.* Indian Foundation for Vedic Science.

Azarnouche, S. and Petitjean, M. (2022) 'Sasanian Warriors in Context: Historical and Religious Commentary on a Middle Persian Chapter on *Artēštārān* (*Dēnkard* VIII.26)'. *Revue Internationale d'Histoire Militaire Ancienne* 11: 331–84.

Benveniste, E. (2016) *Dictionary of Indo-European Concepts.* Chicago.

Bernheimer, T. and Silverstein, A. (eds) (2012) *Late Antiquity: Eastern Perspectives.* Exeter.

Bosworth, C.E. (1999) *The History of al-Tabarī: The Sāsānids, the Byzantines, the Lakmids, and Yemen* (Vol. V). Albany, NY.

Briant, P. (2018a) 'The Achaemenid Empire', in Briant (2018b) 181–204.

Briant, P. (2018b) *From Cyrus to Seleukos: Studies in Achaemenid and Hellenistic History.* Leiden.

Brinkman, J.A. (1965) 'Elamite Military Aid to Merodach-Baladan'. *JNES* 24: 161–66.

Cameron, A. (ed) (1995) *The Byzantine and Early Islamic Near East. Vol. III: States, Resources and Armies.* Princeton.

Choksy, J.K. (2012) 'Justifiable Force and Holy War in Zoroastrianism', in Renard (2012) 158–76.

Choksy, J.K. and Nematollahi, N. (2017) 'Zoroastrianism and War', in Shaw and Demy (2017) 882–84.

Daryaee, T. (1997) 'The Use of Religio-Political Propaganda on the Coinage of Xusrō II'. *American Numismatic Society* 9: 41–53.

Daryaee, T. (2010) 'Barbarized Armies and Barracks Kings of Late Antique Iran', in Macuch, Weber, and Durkin-Meisterernst (2010) 43–52.

Daryaee, T. (2021) '*Männerbund* Aspects of Old Persian *anušiya*', in Agut-Labordère, Boucharlat, Joannès, Kuhrt, and Stolper (2021) 73–78.

Daryaee, T. (2022) 'Ardashīr the Argbed: Sasanian Economic Resources and Finances in the Third Century CE'. *Antiquite Tradive* 30: 151–59.

Daryaee, T. and Rollinger, R. (eds) (2021) *Iran and Its Histories: From the Beginnings through the Achaemenid Empire.* Wiesbaden.

Daryaee, T., Rollinger, R., and Canepa, M.P. (eds) (2023) *Iran and the Transformation of Ancient Near Eastern History: The Seleucids (ca. 312–150 BCE).* Wiesbaden.

Dennis, G.T. (2001) *Maurice's Strategikon: Handbook of Byzantine Military Strategy.* Philadelphia.

Farrokh, K., Karamian, G., Maksymiuk, K. (2018) *A Synopsis of Sasanian Military Organization and Combat Units.* Siedlce.

Frye, R.N. (1972) 'The Institutions of the Achaemenids', in Walser (1972) 83–93.

Harmatta, J. (1971) 'The Rise of the Old Persian Empire: Cyrus the Great'. *Acta Antiqua Academiae Scientiarum Hungaricae* 19: 1–15.

Howard-Johnston, J. (2012) 'The Late Sasanian Army', in Bernheimer and Silverstein (2012) 87–127.

Howard-Johnston, J. (2022) 'The Coming of Arabs to Iran'. *Sasanian Studies* 1: 99–126.

Jacobs, B. and Rollinger, R. (eds) (2021) *A Companion to the Achaemenid Persian Empire*. Hoboken, NJ.

Jones-Bley, K. and Zdanovich, D.G. (eds) (2002) *Complex Societies of Central Eurasia from the 3rd to the 1st Millennium BC*. Washington, DC.

Kahrs, E. (ed) (1986) *Kalyanamitraraganam: Essays in Honour of Nils Simonsson*. Oslo.

Klein, J.S. (1988) 'Coordinate Conjunction in Old Persian'. *JAOS* 108: 387–417.

Lanfranchi, G.B. (2021) 'Training for Empire: The Assyrian Pressure on Western Iran (9th–7th BCE)', in Daryaee and Rollinger (2021) 133–48.

Louchakova-Schwartz, O. (2018) 'Intersubjectivity and Multiple Realities in the Zarathushtra's Gathas'. *Open Theology* 4: 471–88.

Macuch, M., Weber, D., and Durkin-Meisterernst, D. (eds) (2010) *Ancient and Middle Iranian Studies: Proceedings of the 6th European Conference of Iranian Studies*. Wiesbaden.

Maksymiuk, K. (2015) *Geography of Roman-Iranian Wars: Military Operations of Rome and Sasanian Iran*. Siedlce.

Malandra, W.W. (1973) 'A Glossary of Terms for Weapons and Armor in Old Iranian'. *Indo-Iranian Journal* 15: 264–89.

Malandra, W.W. (1983) *An Introduction to Ancient Iranian Religions: Readings from the Avesta and the Achaemenid Inscriptions*. Duluth.

Mallory, J.P. (2006) 'Indo-European Warfare'. *Journal of Conflict Archaeology* 2: 77–98.

Manning, S. (2021) *Armed Force in the Teispid-Achaemenid Empire: Past Approaches, Future Prospects*. Stuttgart.

Masson, V. (ed) (2001) *The Role of Ahalteke Horse in the Formation of World Horse-Breeding*. Ashgabat.

Olbrycht, M.J. (2001) 'The Origins of the Arsacid Parthian Cavalry: Some Remarks', in Masson (2001) 108–11.

Olbrycht, M.J. (2018) 'Alexander the Great in Sittakene and the Reorganization of His Army (331 B.C.)'. *Anabasis* 9: 80–92.

Potter, D. (1991) 'The Inscription on the Bronze Herakles from Mesene: Vologeses IV's War with Rome and the Date of Tacitus' Annales'. *ZPE* 88: 277–90.

Pyankov, I.V. (2002) 'Arkaim and the Indo-Iranian Var', in Jones-Bley and Zdanovich (2002) 42–52.

Pourdavoud, E. (1968) *Zēn-abzār*. Tehran.

Renard, J. (ed) (2012) *Fighting Words: Religion, Violence, and the Interpretation of Sacred Texts*. Berkeley.

Rezakhani, K. (2017) *ReOrienting the Sasanians: East Iran in Late Antiquity*. Edinburgh.

Rubin, Z. (1995) 'The Reforms of Khusro Anūshirwān', in in Cameron (1995) 227–98.

Schwartz, M. (2000) 'Revelations, Theology, and Poetics in the Gathas'. *Bulletin of the Asia Institute* 14: 1–18.

Sekunda, N. (1994) *Seleucid and Ptolemaic Reformed Armies 168–145 BC. Vol. 1: The Seleucid Army under Antiochus IV Epiphanes*. London.

Shahbazi, A.S. (2011) 'Army i: Pre-Islamic Iran'. *Encyclopaedia Iranica*. https://iranicaonline.org/articles/army-i.

Shaw, J. and Demy, T. (2017) *War and Religion: An Encyclopedia of Faith and Conflict*. Vol. 3. Santa Barbara.

Skjærvø, P.O. 'The Gāthās as Myth and Ritual', in Stausberg and Vevaina (2015) 59–68.

Stausberg, M. and Vevaina, Y.S.-D. (2015) *The Wiley Blackwell Companion to Zoroastrianism*. Malden.

Sparreboom, M. (1985) *Chariots in the Veda*. Leiden.

Strootman, R. (2023) 'How Iranian was the Seleucid Empire?', in Daryaee, Rollinger, and Canepa (2023) 11–36.

Syvanne, I and Maksymiuk, K. (2018) *The Military History of the Third Century Iran*. Siedlce.

Tafazzoli, A. (1993–1994) 'A List of Terms for Weapons and Armour in Western Middle Iranian'. *Silk Road Art and Archaeology* 3: 187–98.

Tavernier, J. (2007) *Iranica in the Achaemenid Period (ca. 550–330 B.C.): Lexicon of Old Iranian Proper Names and Loanwords, Attested in Non-Iranian Texts*. Peeters.

Tuplin, C. and Jacobs, B. (2021) 'Military Organisation and Equipment', in Jacobs and Rollinger (2021) 1161–81.

Utas, A. (1986) '*Jang u ašti*: War and peace in Iran', in Kahrs (1986) 287–302.

Vallat, F. (2011) 'Elam i: The History of Elam'. *Encyclopaedia Iranica*. https://www.iranicaonline.org/articles/elam-i.

Walser, G. (ed) (1972) *Beiträge zur Achämenidengeschichte*. Historia Einzelschriften 18. Wiesbaden.

Zakeri, M. (1995) *Sāsānid Soldiers in Early Muslim Society: The Origins of 'Ayyārān and Futuwwa*. Wiesbaden.

Index

www.ingramcontent.com/pod-product-compliance
Lightning Source LLC
LaVergne TN
LVHW010657300926
849161LV00005B/1156
* 9 7 8 9 0 0 4 7 0 8 2 1 1 *